MONOIDAL TOPOLOGY

Monoidal Topology describes an active research area that, after various past proposals on how to axiomatize "spaces" in terms of convergence, began to emerge at the beginning of the millennium. It combines Barr's relational presentation of topological spaces in terms of ultrafilter convergence with Lawvere's interpretation of metric spaces as small categories enriched over the extended real half-line. Hence, equipped with a quantale $\mathcal{V}$ (replacing the reals) and a monad $\mathbb{T}$ (replacing the ultrafilter monad) laxly extended from set maps to $\mathcal{V}$-valued relations, the book develops a categorical theory of $(\mathbb{T}, \mathcal{V})$-algebras that is inspired simultaneously by its metric and topological roots. The book highlights in particular the distinguished role of equationally defined structures within the given lax-algebraic context and presents numerous new results ranging from topology and approach theory to domain theory. All the necessary pre-requisites in order and category theory are presented in the book.

Encyclopedia of Mathematics and Its Applications

This series is devoted to significant topics or themes that have wide application in mathematics or mathematical science and for which a detailed development of the abstract theory is less important than a thorough and concrete exploration of the implications and applications.

Books in the **Encyclopedia of Mathematics and Its Applications** cover their subjects comprehensively. Less important results may be summarized as exercises at the ends of chapters. For technicalities, readers can be referred to the bibliography, which is expected to be comprehensive. As a result, volumes are encyclopedic references or manageable guides to major subjects.

ENCYCLOPEDIA OF MATHEMATICS AND ITS APPLICATIONS

All the titles listed below can be obtained from good booksellers or from Cambridge University Press. For a complete series listing visit www.cambridge.org/mathematics.

106 A. Markoe *Analytic Tomography*
107 P. A. Martin *Multiple Scattering*
108 R. A. Brualdi *Combinatorial Matrix Classes*
109 J. M. Borwein and J. D. Vanderwerff *Convex Functions*
110 M.-J. Lai and L. L. Schumaker *Spline Functions on Triangulations*
111 R. T. Curtis *Symmetric Generation of Groups*
112 H. Salzmann et al. *The Classical Fields*
113 S. Peszat and J. Zabczyk *Stochastic Partial Differential Equations with Lévy Noise*
114 J. Beck *Combinatorial Games*
115 L. Barreira and Y. Pesin *Nonuniform Hyperbolicity*
116 D. Z. Arov and H. Dym *J-Contractive Matrix Valued Functions and Related Topics*
117 R. Glowinski, J.-L. Lions and J. He *Exact and Approximate Controllability for Distributed Parameter Systems*
118 A. A. Borovkov and K. A. Borovkov *Asymptotic Analysis of Random Walks*
119 M. Deza and M. Dutour Sikirić *Geometry of Chemical Graphs*
120 T. Nishiura *Absolute Measurable Spaces*
121 M. Prest *Purity, Spectra and Localisation*
122 S. Khrushchev *Orthogonal Polynomials and Continued Fractions*
123 H. Nagamochi and T. Ibaraki *Algorithmic Aspects of Graph Connectivity*
124 F. W. King *Hilbert Transforms I*
125 F. W. King *Hilbert Transforms II*
126 O. Calin and D.-C. Chang *Sub-Riemannian Geometry*
127 M. Grabisch et al. *Aggregation Functions*
128 L. W. Beineke and R. J. Wilson (eds.) with J. L. Gross and T. W. Tucker *Topics in Topological Graph Theory*
129 J. Berstel, D. Perrin and C. Reutenauer *Codes and Automata*
130 T. G. Faticoni *Modules over Endomorphism Rings*
131 H. Morimoto *Stochastic Control and Mathematical Modeling*
132 G. Schmidt *Relational Mathematics*
133 P. Kornerup and D. W. Matula *Finite Precision Number Systems and Arithmetic*
134 Y. Crama and P. L. Hammer (eds.) *Boolean Models and Methods in Mathematics, Computer Science, and Engineering*
135 V. Berthé and M. Rigo (eds.) *Combinatorics, Automata and Number Theory*
136 A. Kristály, V. D. Rădulescu and C. Varga *Variational Principles in Mathematical Physics, Geometry, and Economics*
137 J. Berstel and C. Reutenauer *Noncommutative Rational Series with Applications*
138 B. Courcelle and J. Engelfriet *Graph Structure and Monadic Second-Order Logic*
139 M. Fiedler *Matrices and Graphs in Geometry*
140 N. Vakil *Real Analysis through Modern Infinitesimals*
141 R. B. Paris *Hadamard Expansions and Hyperasymptotic Evaluation*
142 Y. Crama and P. L. Hammer *Boolean Functions*
143 A. Arapostathis, V. S. Borkar and M. K. Ghosh *Ergodic Control of Diffusion Processes*
144 N. Caspard, B. Leclerc and B. Monjardet *Finite Ordered Sets*
145 D. Z. Arov and H. Dym *Bitangential Direct and Inverse Problems for Systems of Integral and Differential Equations*
146 G. Dassios *Ellipsoidal Harmonics*
147 L. W. Beineke and R. J. Wilson (eds.) with O. R. Oellermann *Topics in Structural Graph Theory*
148 L. Berlyand, A. G. Kolpakov and A. Novikov *Introduction to the Network Approximation Method for Materials Modeling*
149 M. Baake and U. Grimm *Aperiodic Order I: A Mathematical Invitation*
150 J. Borwein et al. *Lattice Sums Then and Now*
151 R. Schneider *Convex Bodies: The Brunn–Minkowski Theory (Second Edition)*
152 G. Da Prato and J. Zabczyk *Stochastic Equations in Infinite Dimensions (Second Edition)*
153 D. Hofmann, G. J. Seal and W. Tholen (eds.) *Monoidal Topology*
154 M. Cabrera-García and Á. Rodríguez Palacios *Non-Associative Normed Algebras I: The Vidav–Palmer and Gelfand–Naimark Theorems*
155 C. F. Dunkl and Y. Xu *Orthogonal Polynomials of Several Variables (Second Edition)*

ENCYCLOPEDIA OF MATHEMATICS AND ITS APPLICATIONS

Monoidal Topology

A Categorical Approach to Order, Metric, and Topology

Edited by

DIRK HOFMANN
Universidade de Aveiro, Portugal

GAVIN J. SEAL
Swiss Federal Institute of Technology

WALTER THOLEN
York University, Toronto

CAMBRIDGE
UNIVERSITY PRESS

CAMBRIDGE
UNIVERSITY PRESS

University Printing House, Cambridge CB2 8BS, United Kingdom

One Liberty Plaza, 20th Floor, New York, NY 10006, USA

477 Williamstown Road, Port Melbourne, VIC 3207, Australia

314-321, 3rd Floor, Plot 3, Splendor Forum, Jasola District Centre, New Delhi - 110025, India

103 Penang Road, #05-06/07, Visioncrest Commercial, Singapore 238467

Cambridge University Press is part of the University of Cambridge.

It furthers the University's mission by disseminating knowledge in the pursuit of education, learning and research at the highest international levels of excellence.

www.cambridge.org
Information on this title: www.cambridge.org/9781107063945

First published 2014

A catalogue record for this publication is available from the British Library

Library of Congress Cataloging in Publication data
Monoidal topology : a categorical approach to order, metric, and topology / edited by Dirk Hofmann, Universidade de Aveiro, Gavin J. Seal, Swiss Federal Institute of Technology, Walter Tholen, York University, Toronto.
pages cm. – (Encyclopedia of mathematics and its applications)
ISBN 978-1-107-06394-5 (hardback)
1. Topological semigroups. 2. Group theory. I. Hofmann, Dirk, 1970–
II. Seal, Gavin J. III. Tholen, W. (Walter), 1947–
QA387.M65 2014
514′.32–dc23
2013046221

ISBN 978-1-107-06394-5 Hardback

To Horst Herrlich

Summary of contents

Preface *page* xv

I Introduction **1**
Robert Lowen and Walter Tholen

II Monoidal structures **18**
Gavin J. Seal and Walter Tholen

III Lax algebras **145**
Dirk Hofmann, Gavin J. Seal, and Walter Tholen

IV Kleisli monoids **284**
Dirk Hofmann, Robert Lowen, Rory Lucyshyn-Wright, and Gavin J. Seal

V Lax algebras as spaces **375**
Maria Manuel Clementino, Eva Colebunders, and Walter Tholen

Bibliography 467
Selected categories 480
Selected functors 484
Selected symbols 487
Index 491

Contents

Preface *page* xv

I Introduction **1**
I.1 The ubiquity of monoids and their actions 1
I.1.1 Monoids and their actions in algebra 2
I.1.2 Orders and metrics as monoids and lax algebras 3
I.1.3 Topological and approach spaces as monoids and lax algebras 5
I.1.4 The case for convergence 7
I.1.5 Filter convergence and Kleisli monoids 9
I.2 Spaces as categories, and categories of spaces 10
I.2.1 Ordinary small categories 10
I.2.2 Considering a space as a category 11
I.2.3 Moving to the large category of all spaces 13
I.3 Chapter highlights and dependencies 14

II Monoidal structures **18**
II.1 Ordered sets 18
II.1.1 The Cartesian structure of sets and its monoids 18
II.1.2 The compositional structure of relations 19
II.1.3 Orders 21
II.1.4 Modules 22
II.1.5 Adjunctions 22
II.1.6 Closure operations and closure spaces 24
II.1.7 Completeness 25
II.1.8 Adjointness criteria 27
II.1.9 Semilattices, lattices, frames, and topological spaces 28
II.1.10 Quantales 30

	II.1.11	Complete distributivity	32
	II.1.12	Directed sets, filters, and ideals	34
	II.1.13	Ultrafilters	35
	II.1.14	Natural and ordinal numbers	37
	Exercises		38
II.2	Categories and adjunctions		42
	II.2.1	Categories	42
	II.2.2	Functors	44
	II.2.3	Natural transformations	46
	II.2.4	The Yoneda embedding	48
	II.2.5	Adjunctions	49
	II.2.6	Reflective subcategories, equivalence of categories	53
	II.2.7	Initial and terminal objects, comma categories	54
	II.2.8	Limits	56
	II.2.9	Colimits	60
	II.2.10	Construction of limits and colimits	61
	II.2.11	Preservation and reflection of limits and colimits	63
	II.2.12	Adjoint Functor Theorem	65
	II.2.13	Kan extensions	67
	II.2.14	Dense functors	69
	Exercises		70
II.3	Monads		74
	II.3.1	Monads and adjunctions	74
	II.3.2	The Eilenberg–Moore category	76
	II.3.3	Limits in the Eilenberg–Moore category	78
	II.3.4	Beck's monadicity criterion	79
	II.3.5	Duskin's monadicity criterion	82
	II.3.6	The Kleisli category	84
	II.3.7	Kleisli triples	85
	II.3.8	Distributive laws, liftings, and composite monads	86
	II.3.9	Distributive laws and extensions	90
	Exercises		92
II.4	Monoidal and ordered categories		96
	II.4.1	Monoidal categories	96
	II.4.2	Monoids	98
	II.4.3	Actions	99
	II.4.4	Monoidal closed categories	101
	II.4.5	Ordered categories	104
	II.4.6	Lax functors, pseudo-functors, 2-functors, and their transformations	105
	II.4.7	Maps	106
	II.4.8	Quantaloids	107

II.4.9 Kock–Zöberlein monads 109
II.4.10 Enriched categories 111
Exercises 112
II.5 Factorizations, fibrations, and topological functors 114
II.5.1 Factorization systems for morphisms 114
II.5.2 Subobjects, images, and inverse images 117
II.5.3 Factorization systems for sinks and sources 119
II.5.4 Closure operators 122
II.5.5 Generators and cogenerators 125
II.5.6 U-initial morphisms and sources 127
II.5.7 Fibrations and cofibrations 129
II.5.8 Topological functors 130
II.5.9 Self-dual characterization of topological functors 132
II.5.10 Epireflective subcategories 134
II.5.11 Taut Lift Theorem 136
Exercises 137
Notes on Chapter II 143

III Lax algebras **145**
III.1 Basic concepts 145
III.1.1 $\mathcal{V}$-relations 145
III.1.2 Maps in $\mathcal{V}$-Rel 147
III.1.3 $\mathcal{V}$-categories, $\mathcal{V}$-functors, and $\mathcal{V}$-modules 150
III.1.4 Lax extensions of functors 154
III.1.5 Lax extensions of monads 157
III.1.6 $(\mathbb{T}, \mathcal{V})$-categories and $(\mathbb{T}, \mathcal{V})$-functors 158
III.1.7 Kleisli convolution 161
III.1.8 Unitary $(\mathbb{T}, \mathcal{V})$-relations 163
III.1.9 Associativity of unitary $(\mathbb{T}, \mathcal{V})$-relations 165
III.1.10 The Barr extension 169
III.1.11 The Beck–Chevalley condition 172
III.1.12 The Barr extension of a monad 175
III.1.13 A double-categorical presentation of lax extensions 178
Exercises 180
III.2 Fundamental examples 185
III.2.1 Ordered sets, metric spaces, and probabilistic metric spaces 185
III.2.2 Topological spaces 186
III.2.3 Compact Hausdorff spaces 190
III.2.4 Approach spaces 192
III.2.5 Closure spaces 198
Exercises 199

III.3 Categories of lax algebras 201
III.3.1 Initial structures 201
III.3.2 Discrete and indiscrete lax algebras 203
III.3.3 Induced orders 203
III.3.4 Algebraic functors 207
III.3.5 Change-of-base functors 209
III.3.6 Fundamental adjunctions 212
Exercises 213
III.4 Embedding lax algebras into a quasitopos 216
III.4.1 $(\mathbb{T}, \mathcal{V})$-graphs 216
III.4.2 Reflecting $(\mathbb{T}, \mathcal{V})$-RGph into $(\mathbb{T}, \mathcal{V})$-Cat 221
III.4.3 Coproducts of $(\mathbb{T}, \mathcal{V})$-categories 223
III.4.4 Interlude on partial products and local Cartesian closedness 227
III.4.5 Local Cartesian closedness of $(\mathbb{T}, \mathcal{V})$-Gph 230
III.4.6 Local Cartesian closedness of subcategories of $(\mathbb{T}, \mathcal{V})$-Gph 234
III.4.7 Interlude on subobject classifiers and partial-map classifiers 237
III.4.8 The quasitopos $(\mathbb{T}, \mathcal{V})$-Gph 245
III.4.9 Final density of $(\mathbb{T}, \mathcal{V})$-Cat in $(\mathbb{T}, \mathcal{V})$-Gph 246
Exercises 249
III.5 Representable lax algebras 252
III.5.1 The monad $\mathbb{T}$ on $\mathcal{V}$-Cat 252
III.5.2 $\mathbb{T}$-algebras in $\mathcal{V}$-Cat 254
III.5.3 Comparison with lax algebras 257
III.5.4 The monad $\mathbb{T}$ on $(\mathbb{T}, \mathcal{V})$-Cat 261
III.5.5 Dualizing $(\mathbb{T}, \mathcal{V})$-categories 264
III.5.6 The ultrafilter monad on Top 266
III.5.7 Representable topological spaces 270
III.5.8 Exponentiable topological spaces 273
III.5.9 Representable approach spaces 276
Exercises 278
Notes on Chapter III 280

IV Kleisli monoids 284
IV.1 Kleisli monoids and lax algebras 284
IV.1.1 Topological spaces via neighborhood filters 285
IV.1.2 Power-enriched monads 287
IV.1.3 $\mathbb{T}$-monoids 289
IV.1.4 The Kleisli extension 291
IV.1.5 Topological spaces via filter convergence 294
Exercises 298

IV.2 Lax extensions of monads 300
IV.2.1 Initial extensions 301
IV.2.2 Sup-dense and interpolating monad morphisms 304
IV.2.3 $(\mathbb{S}, \mathsf{2})$-categories as Kleisli monoids 306
IV.2.4 Strata extensions 310
IV.2.5 $(\mathbb{S}, \mathcal{V})$-categories as Kleisli towers 313
Exercises 317
IV.3 Lax algebras as Kleisli monoids 320
IV.3.1 The ordered category $(\mathbb{T}, \mathcal{V})$-$\mathsf{URel}$ 320
IV.3.2 The discrete presheaf monad 323
IV.3.3 Approach spaces 328
IV.3.4 Revisiting change of base 332
Exercises 334
IV.4 Injective lax algebras as Eilenberg–Moore algebras 336
IV.4.1 Eilenberg–Moore algebras as Kleisli monoids 336
IV.4.2 Monads on categories of Kleisli monoids 337
IV.4.3 Eilenberg–Moore algebras over $\mathbb{S}$-Mon 341
IV.4.4 Continuous lattices 344
IV.4.5 Kock–Zöberlein monads on $\mathbb{T}$-Mon 347
IV.4.6 Eilenberg–Moore algebras and injective Kleisli monoids 349
Exercises 353
IV.5 Domains as lax algebras and Kleisli monoids 355
IV.5.1 Modules and adjunctions 355
IV.5.2 Cocontinuous ordered sets 356
IV.5.3 Observable realization spaces 358
IV.5.4 Observable realization spaces as lax algebras 361
IV.5.5 Observable specialization systems 362
IV.5.6 Ordered abstract bases and round filters 364
IV.5.7 Domains as Kleisli monoids of the ordered-filter monad 364
IV.5.8 Continuous dcpos as sober domains 366
IV.5.9 Cocontinuous lattices among lax algebras 370
Exercises 372
Notes on Chapter IV 372

V Lax algebras as spaces 375
V.1 Hausdorff separation and compactness 375
V.1.1 Basic definitions and properties 376
V.1.2 Tychonoff Theorem, Čech–Stone compactification 379
V.1.3 Compactness for Kleisli-extended monads 381
V.1.4 Examples involving monoids 385
Exercises 388

V.2 Low separation, regularity, and normality 389
V.2.1 Order separation 389
V.2.2 Between order separation and Hausdorff separation 391
V.2.3 Regular spaces 393
V.2.4 Normal and extremally disconnected spaces 397
V.2.5 Normal approach spaces 402
Exercises 404
V.3 Proper and open maps 405
V.3.1 Finitary stability properties 406
V.3.2 First characterization theorems 409
V.3.3 Notions of closure 415
V.3.4 Kuratowski–Mrówka Theorem 419
V.3.5 Products of proper maps 423
V.3.6 Coproducts of open maps 426
V.3.7 Preservation of space properties 429
Exercises 430
V.4 Topologies on a category 432
V.4.1 Topology, fiberwise topology, derived topology 433
V.4.2 $\mathcal{P}$-compactness, $\mathcal{P}$-Hausdorffness 437
V.4.3 A categorical characterization theorem 439
V.4.4 $\mathcal{P}$-dense maps, $\mathcal{P}$-open maps 442
V.4.5 $\mathcal{P}$-Tychonoff and locally $\mathcal{P}$-compact Hausdorff objects 447
Exercises 451
V.5 Connectedness 453
V.5.1 Extensive categories 453
V.5.2 Connected objects 457
V.5.3 Topological connectedness governs 459
V.5.4 Products of connected spaces 461
Exercises 463
Notes on Chapter V 464

Bibliography 467
Selected categories 480
Selected functors 484
Selected symbols 487
Index 491

Preface

Monoidal topology describes an active research area that, after many proposals throughout the past century on how to axiomatize "spaces" in terms of convergence, started to emerge at the beginning of the millennium. It provides a powerful unifying framework and theory for fundamental ordered, metric, and topological structures. Inspired by the topological concept of filter convergence, its methods are lax-algebraic and categorical, with generalized notions of monoid recurring frequently as the fundamental building blocks of its key notions. Since the main components of this new area have to date been available only in a scattered array of research articles, the authors of this book hope that a self-contained and consistent introduction to the theory will serve a broad range of mathematicians, scientists, and their graduate students with an interest in a modern treatment of the mathematical structures in question. With all essential elements from order and category theory provided in the book, it is assumed that the reader will appreciate a framework which highlights the power of equationally defined algebraic structures as particularly important elements of the broader lax-algebraic context which, roughly speaking, replaces equalities by inequalities.

There are two principal roots to the theory presented in this book: Barr's 1970 relational presentation of topological spaces which naturally extends Manes' 1969 equational presentation of compact Hausdorff spaces as the Eilenberg–Moore algebras of the ultrafilter monad, and Lawvere's 1973 description of metric spaces as (small individual) categories enriched over the extended non-negative real half-line. In hindsight, it seems surprising that it took some thirty years until the two general parameters at play here were combined in a compatible fashion, given by a monad $\mathbb{T}$ replacing the ultrafilter monad and a quantale (or, more generally, a monoidal closed category) $\mathcal{V}$ replacing the half-line. Of course, when considered separately, these two pivotal papers triggered numerous important developments. Lawvere's surprising discovery quickly became a cornerstone of enriched category theory, with his characterization of Cauchy completeness in purely enriched-categorical terms enjoying most of the attention, and Barr's paper was followed by at least two major but quite distinct attempts to develop a general topologically inspired theory using a lax-algebraic monad approach, by Manes [1974] and Burroni [1971]. However, the uptake of these articles in terms of follow-up work remained sporadic, perhaps because not many strikingly new applications beyond Barr's work came to the fore, with one prominent exception: the inclusion of Lambek's 1969 multicategories in addition to Barr's topological spaces provides a powerful motivation for Burroni's elegant setting.

In 2000, Bill Lawvere was the first to suggest (in a private communication to Walter Tholen) that, in the same way as topological spaces generalize ordered sets, Lowen's 1989 approach spaces should be describable as generalized metric spaces "using $\mathcal{V}$-multicategories in a good way" instead of just $\mathcal{V}$-categories, thus implicitly envisioning a merger of the parameters $\mathbb{T}$ and $\mathcal{V}$. At about the same time, following a suggestion by George Janelidze, Clementino and Hofmann [2003] gave a lax-algebraic description of approach spaces using a "numerical extension" of the ultrafilter monad. Both suggestions set the stage for Clementino and Tholen [2003] to develop a setting that combines the two parameters efficiently, especially when the monoidal-closed category $\mathcal{V}$ is just a quantale. As emphasized in [Clementino, Hofmann, and Tholen, 2004b], this setting suffices to capture ordered, metric, and topological structures. In a slightly relaxed form, as presented in [Seal, 2005], it also permits to replace ultrafilter convergence by filter convergence (and its "approach generalization") for its key applications, and it is this setting that has been adopted in this book.

When, following a meeting in Barisciano (Italy) in 2006, the authors of this book began to embark decisively on a project to give a self-contained presentation of the emerging theory, the heterogeneous make-up of the group itself made it necessary to document clearly all needed ingredients in a coherent fashion. Hence, this book contains:

- a "crash course" on order and category theory that highlights many aspects not readily available in existing texts and of interest beyond its use for order, metric, and topology;
- an in-depth presentation of the syntactical framework involving the monad $\mathbb{T}$ and the quantale $\mathcal{V}$ needed for a unified treatment of the principal target categories;
- some novel applications leading to new insights, even in the context of ordinary topological spaces, with ample directions to additional or subsequent work that could not be included in this book.

In acknowledging the valuable advice and contributions received from many colleagues, we should highlight first some theses written on subjects pertaining to this book and to various degrees influencing its development, including the Ph.D. theses of Van Olmen [2005], Schubert [2006], Cruttwell [2008], and

Reis [2013], and the Master's theses of Akhvlediani [2008] and Lucyshyn-Wright [2009]. We are grateful especially to Christoph Schubert and Andrei Akhvlediani, who respectively helped to transform Walter Tholen's lecture notes for courses given at the University of Bremen (Germany) in 2003 and at a workshop organized by Francis Borceux at Haute Bodeux (Belgium) in 2007 into something legible and digestible. Christoph was also an active contributor to the various meetings that the group of authors held at the University of Antwerp until 2009, generously organized by Eva Colebunders and Robert Lowen.

The long but surely incomplete list of names of colleagues who offered helpful comments at various stages includes those of Bernhard Banaschewski, Francis Borceux, Franck van Breugel, Marcel Erné, Cosimo Guido, Eraldo Giuli, Horst Herrlich, Kathryn Hess, George Janelidze, Bill Lawvere, Frédéric Mynard, Robert Paré, Hans Porst, Sergejs Solovjovs, Isar Stubbe, Paweł Waszkiewicz, and Richard Wood; we thank them all. We also appreciate the help in proofreading provided by Luca Hunkeler, Valentin Mercier, and Eiichi Piguet.

The authors of this book gratefully acknowledge the financial assistance received from various sources while working on this book project, including the Research Foundation Flanders (FWO) research project G.0244.05, the Centre for Mathematics of the University of Coimbra, the European Regional Development Fund through the program COMPETE, the Portuguese Government through the Fundação para a Ciência e a Tecnologia, the Center for Research and Development in Mathematics and Applications of the University of Aveiro, the European Union through a Marie-Curie International Reintegration Grant, the Swiss National Science Foundation through an Advanced Researcher Fellowship, and the Natural Sciences and Engineering Council of Canada through Discovery Grants. We also thank Roger Astley and his colleagues at Cambridge University Press for having been receptive and open to our wishes regarding the publication of this book.

We dedicate this book to Horst Herrlich, whose work and dedication to mathematics have had formative influence on all authors of this book.

I

Introduction

Robert Lowen and Walter Tholen

In this introductory chapter we explain, in largely non-technical terms, not only how monoids and their actions occur everywhere in algebra, but also how they provide a common framework for the ordered, metric, topological, or similar structures targeted in this book. This framework is categorical, both at a micro level, since individual spaces may be viewed as generalized small categories, and at a macro level, as we are providing a common setting and theory for the categories of all ordered sets, all metric spaces, and all topological spaces – and many other categories.

Whilst this Introduction uses some basic categorical terms, we actually provide all required categorical language and theory in Chapter II, along with the basic terms about order, metric, and topology, before we embark on presenting the common setting for our target categories. *Many readers may therefore want to jump directly to Chapter III, using the Introduction just for motivation and Chapter II as a reference for terminology and notation.*

I.1 The ubiquity of monoids and their actions

Nothing seems to be more benign in algebra than the notion of *monoid*, i.e. of a set M that comes with an associative binary operation $m : M \times M \longrightarrow M$ and a neutral element, written as a nullary operation $e : 1 \longrightarrow M$. If mentioned at all, normally the notion finds its way into an algebra course only as a brief precursor to the segment on group theory. However, with the advent of monoidal categories, as first studied by Bénabou [1963], Eilenberg and Kelly [1966], Mac Lane [1963], and others, came the realization that monoids and their actions occur everywhere

Monoidal Topology: A Categorical Approach to Order, Metric, and Topology, eds. Dirk Hofmann, Gavin J. Seal, and Walter Tholen. Published by Cambridge University Press.

in algebra, as the fundamental building blocks of more sophisticated structures. This book is about the extension of this realization from algebra to topology.

I.1.1 Monoids and their actions in algebra

Every algebraist of the past hundred years would subscribe to the claim that free algebras amongst all algebras of a prescribed type contain all the information needed to study these algebras in general. However, what "contain" means was made precise only during the second half of this period. First, there was the observation of the late 1950s [Godement, 1958; Huber, 1961] that the endofunctor $T = GF$ induced by a pair $F \dashv G : \mathsf{A} \to \mathsf{X}$ of adjoint functors comes equipped with natural transformations

$$m : TT \to T \qquad \text{and} \qquad e : 1_{\mathsf{X}} \to T\ ,$$

which, when we trade the Cartesian product of sets and the singleton set 1 for functor composition and the identity functor on X, respectively, are associative and neutral in an easily described diagrammatic sense. Hence, they make T a monoid in the monoidal category of all endofunctors on X, i.e. a *monad* on X [Mac Lane, 1971]. If G is the underlying-set functor of an algebraic category, like the variety of groups, rings, or a particular type of algebras, the free structure TX on X-many generators is just a component of that monad.

On the question of how to recoup the other objects of the algebraic category from the monad they have induced, let us look at the easy example of *actions* of a fixed monoid M in Set. Hence, our algebraic objects are simply sets X equipped with an action $a : M \times X \to X$ making the diagrams

$$\begin{array}{ccc} M\times M\times X & \xrightarrow{1_M\times a} & M\times X \\ {\scriptstyle m\times 1_X}\downarrow & & \downarrow{\scriptstyle a} \\ M\times X & \xrightarrow{a} & X \end{array} \qquad \begin{array}{ccc} X & \xrightarrow{\langle e,1_X\rangle} & M\times X \\ & {\scriptstyle 1_X}\searrow & \downarrow{\scriptstyle a} \\ & & X \end{array}$$

commutative. Realizing that $TX = M \times X$ is in fact the carrier of the free structure over X, we may now rewrite these diagrams as

$$\begin{array}{ccc} TTX & \xrightarrow{Ta} & TX \\ {\scriptstyle m_X}\downarrow & & \downarrow{\scriptstyle a} \\ TX & \xrightarrow{a} & X \end{array} \qquad \begin{array}{ccc} X & \xrightarrow{e_X} & TX \\ & {\scriptstyle 1_X}\searrow & \downarrow{\scriptstyle a} \\ & & X\ . \end{array} \tag{I.1.1.i}$$

Using a similar presentation of the relevant morphisms, i.e. of the action-preserving or equivariant maps, Eilenberg and Moore [1965] realized that with every monad $\mathbb{T} = (T, m, e)$ on a category X (in lieu of Set) one may associate the category $\mathsf{X}^{\mathbb{T}}$ whose objects are X-objects X equipped with a morphism $a : TX \to X$ making the two diagrams (I.1.1.i) commutative. Furthermore, there is an adjunction $F^{\mathbb{T}} \dashv G^{\mathbb{T}} : \mathsf{X}^{\mathbb{T}} \to \mathsf{X}$ inducing $\mathbb{T}$, such that, when $\mathbb{T}$ is induced by any adjunction $F \dashv G : \mathsf{A} \to \mathsf{X}$, there is a "comparison functor" $K : \mathsf{A} \to \mathsf{X}^{\mathbb{T}}$ which, at least for $\mathsf{X} = \mathsf{Set}$, measures the "degree of algebraicity" of A over X. In fact, for any variety of general algebras (with "arities" of operations allowed to be arbitrarily large, as long as the existence of free algebras is guaranteed), K is an equivalence of categories and therefore faithfully recoups the algebras from their monad. By contrast, an application of this procedure to the underlying-set functors of categories of ordered sets or topological spaces in lieu of general algebras would just render the identity monad on Set whose Eilenberg–Moore category is Set itself, i.e. all structural information would be lost.

Whilst all categories of general algebras allowing for free structures may be seen as categories of generalized monoid actions as just described, this fact by no means describes the full extent of the ubiquity of monoids and their actions in algebra. For example, a unital ring R is nothing but an Abelian group R equipped with homomorphisms

$$m : R \otimes R \to R \qquad \text{and} \qquad e : \mathbb{Z} \to R\,,$$

which are associative and neutral in a quite obvious diagrammatic sense. Hence, when one trades the Cartesian category $(\mathsf{Set}, \times, 1)$ for the monoidal category $(\mathsf{AbGrp}, \otimes, \mathbb{Z})$, monoids R are simply rings, and their actions are precisely the left R-modules. This example, however, is just the tip of an iceberg which places the systematic use of monoidal structures, monoids, and their actions at the core of post-modern algebra.

I.1.2 Orders and metrics as monoids and lax algebras

Although trying to describe ordered sets via the monad induced by the forgetful functor to Set is hopeless, since it induces just the identity monad on Set, a "monoidal perspective" on structures is nevertheless beneficial. First, departing from the notion of a monad, but trading endofunctors T on a category X for relations a on a set X, one can express transitivity and reflexivity of a by

$$a \cdot a \leq a \qquad \text{and} \qquad 1_X \leq a\,, \tag{I.1.2.i}$$

with $\leq$ to be read as set-theoretical inclusion if a is presented as $a \subseteq X \times X$. Hence, with the morphisms $m : a \cdot a \to a$ and $e : 1_X \to a$ simply given by $\leq$, what we regard as the two indispensable requirements of an *order* a on X, transitivity and reflexivity, are expressed by a carrying the structure of a monoid in the monoidal category of endorelations of X.[1] (The fact that such a relation actually satisfies the equation $a \cdot a = a$ is of no particular concern at this point.)

[1] In this book, in order to avoid the proliferation of meaningless prefixes, we refer to what is usually called a preorder as an order, considering the much less used antisymmetry axiom as an add-on separation condition whenever needed. In fact, with respect to the induced order topology, antisymmetry amounts to the T0-separation requirement.

But it is also possible to consider an order a on X in its role as a structure on X in the spirit of Section I.1.1 as follows. Replacing Set by the category Rel of sets with relations as morphisms and choosing for $\mathbb{T}$ the identity monad on Rel, we see that the inequalities (I.1.2.i) are instances of lax versions of the Eilenberg–Moore requirements (I.1.1.i). Indeed, when formally replacing strict ("=") by lax ("≤") commutativity in (I.1.1.i), we obtain

$$\begin{array}{ccc} TTX & \xrightarrow{Ta} & TX \\ {\scriptstyle m_X}\downarrow & \geq & \downarrow{\scriptstyle a} \\ TX & \xrightarrow{a} & X \end{array} \qquad \begin{array}{ccc} X & \xrightarrow{e_X} & TX \\ & {\scriptstyle 1_X}\searrow \;\; \leq & \downarrow{\scriptstyle a} \\ & & X\,. \end{array} \qquad \text{(I.1.2.ii)}$$

In doing so, we suppose that the ambient category X (which is Rel in the case at hand) is ordered, so that its hom-sets are ordered, compatibly with composition. Briefly: ordered sets are precisely the lax Eilenberg–Moore algebras of the identity monad on the ordered category Rel.

Next, presenting relations a on X as functions $a : X \times X \to \mathsf{2} = \{\bot < \top\}$ with at most two truth values, let us rewrite the transitivity and reflexivity requirements as

$$a(x, y) \wedge a(y, z) \leq a(x, z) \qquad \text{and} \qquad \top \leq a(x, x)$$

for all $x, y, z \in X$. In this way, there appears a striking formal similarity with what we regard as the two principal requirements of a *metric* $a : X \times X \to [0, \infty]$ on X, the triangle inequality and the 0-distance requirement for a point to itself:[2]

$$a(x, y) + a(y, z) \geq a(x, z) \qquad \text{and} \qquad 0 \geq a(x, x)\,.$$

Hence, the set $\mathsf{2}$ with its natural order $\leq$ and its inherent structure $\wedge$ and $\top$ has been formally replaced by the extended real half-line $[0, \infty]$, ordered by the natural $\geq$ (!), and structured by $+$ and 0. Just as for orders, one can now interpret metrics as both monoids and lax Eilenberg–Moore algebras with respect to the identity monad, after extending the relational composition

$$(b \cdot a)(x, z) = \bigvee\nolimits_{y \in Y} (a(x, y) \wedge b(y, z))$$

for $a : X \times Y \to \mathsf{2}$, $b : Y \times Z \to \mathsf{2}$ and all $x \in X$, $y \in Y$, by[3]

$$(b \cdot a)(x, z) = \inf\nolimits_{y \in Y} (a(x, y) + b(y, z))$$

for $a : X \times Y \to [0, \infty]$, $b : Y \times Z \to [0, \infty]$ and all $x \in X$, $y \in Y$.

[2] Similarly to the use of the term ordered set, in this book we refer to a distance function a satisfying these two basic axioms as a metric, using additional attributes for the other commonly used requirements when needed, like finiteness, symmetry, and separation.

[3] Although we use $\bigwedge$, $\bigvee$ to refer to infima and suprema in general, in order to avoid ambiguity arising from the "inversion of order" in $[0, \infty]$, we use sup and inf when denoting suprema and infima with respect to the natural order.

The generalized framework encompassing both structures that we will use in this book is provided by a unital *quantale* $\mathcal{V}$ in lieu of $\mathsf{2}$ or $[0, \infty]$; i.e. of a complete lattice equipped with a binary operation $\otimes$ (in lieu of $\wedge$ or $+$) respecting arbitrary joins in each variable, and a $\otimes$-neutral element k (in lieu of $\top$ or 0). The role of the monad $\mathbb{T}$ that appears to be rather artificial in the presentation of ordered sets and metric spaces will become much more pronounced in the presentation of the structures discussed next.

I.1.3 Topological and approach spaces as monoids and lax algebras

In Section I.1.2 we described ordered sets and metric spaces as lax algebras with respect to the identity monad on the category of relations and "numerical" relations, respectively. Taking a historical perspective, we can now indicate how topological spaces fit into this setting once we allow the identity monad to be traded for an arbitrary "lax monad," and how the less-known approach spaces [Lowen, 1997] emerge as the natural hybrid of metric and topology in this context.

Although the axiomatization of topologies in terms of convergence, via filters or nets, has been pursued early on in the development of these structures since Hausdorff [1914], notably by Fréchet [1921] and others, the geometric intuition provided by the open-set and neighborhood perspective clearly dominates the way in which mathematicians perceive topological spaces. Nevertheless, the proof by Manes [1969] that compact Hausdorff spaces are precisely the Eilenberg–Moore algebras of the ultrafilter monad $\mathbb{\beta} = (\beta, m, e)$ on Set could not be ignored, as it gives the ultimate explanation for why the category $\mathsf{CompHaus}$ behaves in many ways just like algebraic categories do. (For example, just as in algebra, but unlike in the case of arbitrary topological spaces, the set-theoretic inverse of a bijective morphism in $\mathsf{CompHaus}$ is automatically a morphism again.) In this description, a compact Hausdorff space is a set X equipped with a map $a : \beta X \to X$ assigning to every ultrafilter $\mathfrak{x}$ on X (what turns out to be) its point of convergence in X, requiring the two basic axioms of an Eilenberg–Moore algebra:

$$a(\beta a(\mathfrak{X})) = a(m_X(\mathfrak{X})) \qquad \text{and} \qquad a(e_X(x)) = x \tag{I.1.3.i}$$

for all $\mathfrak{X} \in \beta\beta X$ and $x \in X$; here the following ultrafilters on X are used:

$$e_X(x) = \dot{x} = \{A \subseteq X \mid x \in A\}$$

is the principal filter on x;

$$m_X(\mathfrak{X}) = \sum \mathfrak{X} = \left\{A \subseteq X \mid \{\mathfrak{x} \in \beta X \mid A \in \mathfrak{x}\} \in \mathfrak{X}\right\}$$

is the *Kowalsky sum* of $\mathfrak{X}$; and

$$\beta a(\mathfrak{X}) = a[\mathfrak{X}] = \left\{A \subseteq X \mid \{\mathfrak{x} \in \beta X \mid a(\mathfrak{x}) \in A\} \in \mathfrak{X}\right\}$$

is simply the image filter of $\mathfrak{X}$ under the map a. Writing $\mathfrak{x} \longrightarrow y$ instead of $a(\mathfrak{x}) = y$ and $\mathfrak{X} \longrightarrow \mathfrak{y}$ instead of $a[\mathfrak{X}] = \mathfrak{y}$, the conditions (I.1.3.i) take the more intuitive form

$$\exists \mathfrak{y} \in \beta X\ (\mathfrak{X} \longrightarrow \mathfrak{y}\ \&\ \mathfrak{y} \longrightarrow z) \iff \sum \mathfrak{X} \longrightarrow z \qquad \text{and} \qquad \dot{x} \longrightarrow x$$

for all $\mathfrak{X} \in \beta\beta X$ and $x \in X$. In fact, in the presence of the implication " $\Longrightarrow$ " in the displayed equivalence, the implication " $\Longleftarrow$ " comes for free (as $\mathfrak{y} = a[\mathfrak{X}]$ necessarily satisfies $\mathfrak{y} \longrightarrow z$ when $\sum \mathfrak{X} \longrightarrow z$), and conditions (I.1.3.i) take the form

$$\mathfrak{X} \longrightarrow \mathfrak{y}\ \&\ \mathfrak{y} \longrightarrow z \implies \sum \mathfrak{X} \longrightarrow z \qquad \text{and} \qquad \dot{x} \longrightarrow x \qquad \text{(I.1.3.ii)}$$

for all $\mathfrak{X} \in \beta\beta X$, $\mathfrak{y} \in \beta X$, $x, z \in X$.

As Barr [1970] observed, if one allows a to be an arbitrary relation between ultrafilters on X and points of X, rather than a map, so that we are no longer assured that every ultrafilter has a point of convergence (compactness) and that there is at most one such point (Hausdorffness), then the relations $\longrightarrow$ satisfying (I.1.3.ii) describe arbitrary topologies on X, with continuous maps characterized as convergence-preserving maps. Furthermore, given the striking similarity of (I.1.3.ii) with the transitivity and reflexivity conditions of an ordered set, it is not surprising that (I.1.3.ii) gives rise to the presentation of topological spaces as both monoids and lax algebras of the ultrafilter monad.

In this statement, however, we glossed over an important point: having the Set-functor β, one knows what βa is when a is a map, but not necessarily when a is just a relation. Whilst there is a fairly straightforward answer in the case at hand, in general we are confronted with the problem of having to extend a monad $\mathbb{T} = (T, m, e)$ on Set to Rel or, even more generally, to $\mathcal{V}$-Rel, the category of sets and $\mathcal{V}$-relations $r : X \nrightarrow Y$, given by functions $r : X \times Y \rightarrow \mathcal{V}$. Although for our purposes it suffices that this extension be lax, i.e. quite far from being a genuine monad on $\mathcal{V}$-Rel, the study of the various needed methods of just laxly extending monads on Set to $\mathcal{V}$-Rel can be cumbersome and takes up significant space in this book.

The general framework that emerges as a common setting is therefore given by a unital (but not necessarily commutative) quantale $(\mathcal{V}, \otimes, k)$ and a monad $\mathbb{T} = (T, m, e)$ on Set laxly extended to $\mathcal{V}$-Rel, with the lax extension usually denoted by $\hat{T} : \mathcal{V}\text{-Rel} \rightarrow \mathcal{V}\text{-Rel}$ (although a given $\mathbb{T}$ may have several lax extensions). The lax algebras considered are sets X equipped with a $\mathcal{V}$-relation $a : TX \nrightarrow X$ satisfying the two basic axioms

$$\hat{T}a(\mathfrak{X}, \mathfrak{y}) \otimes a(\mathfrak{y}, z) \leq a(m_X(\mathfrak{X}), z) \qquad \text{and} \qquad k \leq a(e_X(x), x) \qquad \text{(I.1.3.iii)}$$

for all $\mathfrak{X} \in TTX$, $\mathfrak{y} \in TX$, $z \in Z$. The lax algebras are to be considered as generalized categories enriched in $\mathcal{V}$, with the domain $\mathfrak{x}$ of the hom-object $a(\mathfrak{x}, y)$

not lying in X but in TX. Furthermore, relational composition can be generalized to *Kleisli convolution* for $\mathcal{V}$-relations $r : TX \times Y \to \mathcal{V}$, $s : TY \times Z \to \mathcal{V}$ via

$$(s \circ r)(\mathfrak{x}, z) = \bigvee_{\substack{\mathfrak{X} \in TTX \\ m_X(\mathfrak{X}) = \mathfrak{x}}} \bigvee_{\mathfrak{y} \in TY} \hat{T}r(\mathfrak{X}, \mathfrak{y}) \otimes s(\mathfrak{y}, z)$$

for all $\mathfrak{x} \in TX$, $z \in Z$. The lax algebra axioms for (X, a) are then represented via the monoidal structures

$$a \circ a \leq a \qquad \text{and} \qquad 1_X^\sharp \leq a\,,$$

where $1_X^\sharp$ is neutral with respect to the Kleisli convolution.

In this general framework we have so far encountered the objects in the following table, displayed with the corresponding monad $\mathbb{T}$ and quantale $\mathcal{V}$ (here, $\mathsf{P}_+ = (([0, \infty], \geq), +, 0)$ is the extended non-negative real half-line):

$\mathbb{T}$ \ $\mathcal{V}$	$\mathsf{2}$	P_+
Identity monad	ordered sets	metric spaces
Ultrafilter monad	topological spaces	?

Fortunately, the field left blank is filled with a well-studied, but much less familiar, structure, called *approach space*. It is perhaps easiest described in metric terms: an approach structure on a set X can be given by a point-set distance function $\delta : X \times PX \to [0, \infty]$ satisfying suitable conditions. A metric space (X, d) becomes an approach space via

$$\delta(x, B) = \inf_{y \in B} d(y, x)$$

for all $x \in X$, $B \subseteq X$. When an approach space is presented as a lax algebra (X, a) with $a : \beta X \times X \to [0, \infty]$, one can think of the value $a(\mathfrak{x}, y)$ as the distance that the point y is away from being a limit point of $\mathfrak{x}$. Indeed, a topological space X has its approach structure given by

$$a(\mathfrak{x}, y) = \begin{cases} 0 & \text{if } \mathfrak{x} \longrightarrow y, \\ \infty & \text{otherwise.} \end{cases}$$

As for topological spaces, the more categorical view of approach spaces in terms of convergence proves useful.

I.1.4 The case for convergence

A topology (of open sets) on a set X is most elegantly introduced as a subframe of the powerset X, i.e. a collection of subsets of X closed under finite intersection and arbitrary union. Via complementation, a topology (of closed sets) is equivalently described as a collection closed under finite union and arbitrary intersection, and this simple tool of Boolean duality (switching between open and closed sets) proves to be very useful. There is, however, an unfortunate breakdown of this duality when it comes to morphisms. Although continuous maps are equivalently described by their inverse-image function preserving openness or closedness of subsets, the seemingly most important and natural subclasses of morphisms, namely those continuous maps whose image functions preserve openness or closedness (open or closed continuous maps) behave very differently: whilst open maps are stable under pullback, closed maps are not; not even the subspace restriction $f^{-1}B \to B$ of a closed map $f : X \to Y$ with $B \subseteq Y$ will generally remain closed. Hence, as recognized by Bourbaki [1989], more important than the closed maps are the proper maps, i.e. the morphisms f that are stably closed, so that every pullback of the map f is closed again, also characterized as the closed maps f with compact fibers.

Although under the open- or closed-set perspective no immediate "symmetry" between open and proper maps becomes visible, their characterization in terms of ultrafilter convergence reveals a remarkable duality: a continuous map $f : X \to Y$ is

- *open* if $\mathfrak{y} \longrightarrow f(x)$ (with $x \in X$ and $\mathfrak{y} \in \beta Y$) implies $\mathfrak{y} = f[\mathfrak{x}]$ with $\mathfrak{x} \longrightarrow x$ for some $\mathfrak{x} \in \beta X$,

$$\begin{array}{ccc} \mathfrak{x} & \dashrightarrow & x \\ \vdots & & | \\ \mathfrak{y} & \longrightarrow & f(x) \end{array}$$

- *proper* if $f[\mathfrak{x}] \longrightarrow y$ (with $\mathfrak{x} \in \beta X$ and $y \in Y$) implies $y = f(x)$ with $\mathfrak{x} \longrightarrow x$ for some $x \in X$.

$$\begin{array}{ccc} \mathfrak{x} & \dashrightarrow & x \\ | & & \vdots \\ f[\mathfrak{x}] & \longrightarrow & y \end{array}$$

In fact, once presented as lax homomorphisms between lax Eilenberg–Moore algebras with respect to the ultrafilter monad (laxly extended from Set to Rel), these two types of special morphisms occur most naturally as the ones for which an inequality characterizing their continuity may be replaced by equality, i.e. by a strict homomorphism condition.

Another indicator why convergence provides a most useful complementary view of topological spaces is the following. For a set X and maps $f_i : X \to Y_i$ into topological spaces Y_i, $i \in I$, there is a "best" topology on X making all f_i continuous, often called "weak," but "initial" in this book. Its description in terms of open sets is a bit cumbersome, as it is *generated* by the sets $f^{-1}(B)$, $B \subseteq Y_i$ open, $i \in I$, whereas the characterization in terms of ultrafilter convergence is *immediate*: $\mathfrak{x} \longrightarrow x$ in X precisely when $f_i[\mathfrak{x}] \longrightarrow f(x)$ for all $i \in I$. For example, when $X = \prod_{i \in I} Y_i$ with projections f_i, so that the topology on X just described is the product topology, a proof of the Tychonoff Theorem (on the stability of compactness under products) becomes almost by necessity cumbersome when performed in the open-set environment, but is in fact a triviality in the convergence setting.

We stress, however, the fact that the roles of open sets versus convergence relations are reversed in the dual situation, when one wants to describe the "best" (or "final") topology on a set Y with respect to given maps $f_i : X_i \to Y$ originating from topological spaces $X_i, i \in I$. Its description in terms of open sets is immediate, as $B \subseteq Y$ is declared open whenever all $f_i^{-1}(B)$ are open, whereas a characterization in terms of convergence involves a cumbersome generation process.

In conclusion, we regard the two perspectives not at all as mutually exclusive but rather as complementary to each other. Consequently, this book provides a number of results on topological and approach spaces which arise naturally from the general convergence perspective, but which are far from being obvious when expressed in the more classical open-set or point-set distance language.

I.1.5 Filter convergence and Kleisli monoids

To what extent is it possible to trade ultrafilter convergence for filter convergence when presenting topological spaces as in Section I.1.3 or characterizing open and proper maps as in Section I.1.4? In order to answer this question, it is useful to axiomatize topologies on a set X in terms of maps $\nu : X \to FX$ into the set FX of filters on X, to be thought of as assigning to each point its neighborhood filter. Ordering such maps pointwise by reverse inclusion and using the same notation as in Section I.1.3, except that now $\circ$ denotes the *Kleisli composition* rather than the Kleisli convolution, one obtains another (and, in fact, more elementary) monoidal characterization of topologies on a set X:

$$\nu \circ \nu \leq \nu \qquad \text{and} \qquad e_X \leq \nu\,;$$

in pointwise terms, this reads as

$$\sum \nu[\nu(x)] \supseteq \nu(x) \qquad \text{and} \qquad \dot{x} \supseteq \nu(x)$$

for all $x \in X$. We say that topological spaces are represented as *Kleisli monoids* (X, ν), or simply as $\mathbb{F}$*-monoids*, since the filter monad $\mathbb{F} = (F, m, e)$ may be traded for any monad $\mathbb{T}$ on Set such that the sets TX carry a complete-lattice order, suitably compatible with the monad operations. As such a monad $\mathbb{T}$ may be characterized via a monad morphism $\tau : \mathbb{P} \to \mathbb{F}$, with $\mathbb{P}$ the powerset monad, we call $\mathbb{T}$ *power-enriched*. The basic correspondence between filter convergence and neighborhood systems, given by

$$f \longrightarrow x \iff f \supseteq \nu(x)\,,$$

may now be established at the level of a power-enriched monad $\mathbb{T}$. With a suitable lax extension of $\mathbb{T}$ to Rel, it yields a presentation of $\mathbb{T}$-monoids as lax algebras. For $\mathbb{T} = \mathbb{F}$ it tells us that, remarkably, the characterization (I.1.3.ii) of topological spaces remains valid if we trade ultrafilters for filters. This fact, although established by Pisani [1999] in slightly weaker form, remained unobserved until proved by Seal [2005]. All previous axiomatizations of the notions of topology in terms of filter convergence entailed redundancies.

The answer to our initial question is therefore affirmative with respect to the convergence presentation of topological space. Also, the characterization of open maps given in Section I.1.4 survives the filters-for-ultrafilters exchange, but that of proper maps does not. Hence, we must be cognizant of the fact that the notions introduced for lax algebras will in general depend on the parameters $\mathbb{T}$ and $\mathcal{V}$, not just on the category of lax algebras described by them, such as the category of topological spaces considered here.

I.2 Spaces as categories, and categories of spaces

It has been commonplace since the very beginning of category theory to regard individual ordered sets as categories: they are precisely the categories whose hom-sets have at most one element. By contrast, it was a very bold step for Lawvere [1973] to interpret the distance $a(x, y)$ in a metric space as $\hom(x, y)$. To understand this interpretation, we first recall how ordinary categories fare in the context of orders and metrics as described in Section I.1.2. We then indicate how the consideration of individual ordered sets, metric spaces, topological spaces, and similar objects as small generalized categories leads to new insights and cross fertilization between different areas, as does the investigation of the properties of the category of all such small categories of a particular type.

I.2.1 Ordinary small categories

Replacing "truth values" (2-valued or $[0, \infty]$-valued) by arbitrary sets, for a given set X of "objects" let us consider functions

$$a : X \times X \longrightarrow \mathsf{Set} .$$

X is then the set of objects of a category with hom-sets $a(x, y)$ if there are families of maps

$$m_{X,Y,Z} : a(x, y) \times a(y, z) \longrightarrow a(x, z) \qquad \text{and} \qquad e_X : 1 \longrightarrow a(x, x)$$

satisfying the obvious associativity and neutrality conditions, expressible in terms of commutative diagrams. Hence, the notion of small category fits into the same structural pattern already observed for orders and metrics, where now the composition of functions $a : X \times Y \longrightarrow \mathsf{Set}$, $b : Y \times Z \longrightarrow \mathsf{Set}$ is given by

$$(b \cdot a)(x, z) = \textstyle\coprod_{y \in Y} (a(x, y) \times b(y, z))$$

for all $x \in X$, $z \in Z$.

Briefly, if one allows the above-mentioned setting of a unital quantale $(\mathcal{V}, \otimes, k)$ to be extended to that of a monoidal closed category, ordinary small categories occur as monoids or lax Eilenberg–Moore algebras of an identity monad when V is taken to be $(\mathsf{Set}, \times, 1)$.

We note in passing that the presentation of ordinary small categories just given becomes perhaps more familiar when one exhibits functions $a : X \times X \to \mathsf{Set}$ equivalently as *directed graphs*

$$E \underset{\text{codomain}}{\overset{\text{domain}}{\rightrightarrows}} X$$

(with a fixed set X of vertices), where a and E determine each other via

$$\begin{aligned} &E = \textstyle\coprod_{x,y\in X} a(x, y) \\ \text{and}\quad &a(x, y) = \{f \in E \mid \text{domain}(f) = x, \text{codomain}(f) = y\}\,. \end{aligned}$$

Hence, ordinary categories are monoids in the monoidal category of directed graphs, the tensor product (i.e. composition) of which corresponds to the above composition of Set-valued functions.

We have made clear now that the setting of a monoidal closed category $(\mathsf{V}, \otimes, k)$ and the theory of categories enriched over V (so that their "hom-sets" and structural components live in V rather than Set; see [Kelly, 1982]) provide the right environment for studying not only orders and metrics, but also categories themselves, ordinary or additive (with $\mathsf{V} = \mathsf{AbGrp}$), and much more; in this book, we restrict ourselves to considering the highly simplified case of a quantale $(\mathcal{V}, \otimes, k)$. This is sufficient for reaching the intended target categories, and it makes the theory "technically" simpler since the triviality of 2-cells (given by order in this case) makes all coherence issues disappear as all diagrams in $\mathcal{V}$ commute. Nevertheless, the categorical perspective of interpreting the entity $a(x, y)$ as $\hom(x, y)$ turns out to be very useful even in this simplified situation, as we indicate next.

I.2.2 Considering a space as a category

A key tool of category theory is the *Yoneda embedding*

$$\mathbf{y} : \mathsf{X} \to \mathsf{Set}^{\mathsf{X}^{\mathrm{op}}}\,, \qquad y \mapsto \mathsf{X}(-, y)\,,$$

which assigns to every object y of a category X its contravariant hom-functor $\mathsf{X}(-, y) = \hom_{\mathsf{X}}(-, y) : \mathsf{X}^{\mathrm{op}} \to \mathsf{Set}$. It fully embeds X into a category with all (small-indexed) colimits; moreover, it is dense, in the sense that every object of its codomain is, in a natural way, a colimit of representable functors, i.e. of objects in the image of $\mathbf{y}$. In other words, $\mathsf{Set}^{\mathsf{X}^{\mathrm{op}}}$ serves as a *cocompletion* of X. Other types of cocompletions of X may be found inside $\mathsf{Set}^{\mathsf{X}^{\mathrm{op}}}$ through suitable closure processes, and this statement remains valid even when one moves from ordinary to enriched category theory, trading Set for a monoidal closed category V.

For ordered sets, such that $\mathcal{V} = \mathsf{2} = \{0 < 1\}$ is the two-element chain, monotone maps $\mathsf{X}^{\mathrm{op}} \longrightarrow \mathsf{2}$ correspond to down-closed subsets of the ordered set X, so that

$$\mathbf{y} : X \to \mathrm{Dn}\,X\,, \quad y \mapsto \downarrow y = \{x \in X \mid x \leq y\}$$

is the embedding of X into the *sup-completion* of X, i.e. the lattice with all suprema freely generated by X.

For metric spaces, in the generality adopted in Section I.1.2, it is natural to endow $\mathcal{V} = [0, \infty]$ with its "internal hom" given by the non-symmetric distance function

$$\mu(v, w) = \begin{cases} w - v & \text{if } v \leq w < \infty, \\ 0 & \text{if } w \leq v, \\ \infty & \text{if } v < w = \infty, \end{cases}$$

the symmetrization of which gives the Euclidian metric suitably extended to ∞. Dualization of a metric space $X = (X, a)$ is as trivial as for ordered sets: $X^{\mathrm{op}} = (X, a^\circ)$ with $a^\circ(x, y) = a(y, x)$. Now $\mathbf{y}$ provides an isometric embedding into the space of all non-expansive maps $X^{\mathrm{op}} \longrightarrow [0, \infty]$, provided with the sup-metric. This space inherits various completeness properties from $[0, \infty]$, and inside of it one finds the *Cauchy completion* of X: for every Cauchy sequence $(x_n)_{n \in \mathbb{N}}$ in X, one considers the non-expanding map

$$\psi : X^{\mathrm{op}} \longrightarrow [0, \infty]\,, \quad x \mapsto \lim_{n \to \infty} a(x, x_n)\,,$$

and the subspace formed by all such maps is the Cauchy completion of X. Amazingly, as first observed by Lawvere [1973], this construction may be performed for arbitrary $\mathcal{V}$-categories, since the Cauchy property of $(x_n)_{n \in \mathbb{N}}$ is fully characterized by an adjointness property of ψ when viewed as a *module*, i.e. as a generalized compatible relation.

It is now natural to ask whether such constructions may be performed for topological spaces, presented as lax algebras as in Section I.1.3. The additional parameter given by the monad $\mathbb{T} = (T, m, e)$ does in fact introduce a serious obstacle, which starts with trying to determine what X^{op} should be: a simple switch of arguments of the structure a of X is no longer possible! It turns out that by changing carrier sets from X to TX when forming the dual, it is possible to develop a comprehensive completion theory in the general $(\mathbb{T}, \mathcal{V})$-context, with the Yoneda embedding providing the central tool also at this level of generality. For a topological space, among other constructions, the Yoneda embedding leads to its *sobrification*. Whilst the core of this general completion theory, along with other advanced topics, is the subject of work in progress, many of the needed tools are presented in this book.

I.2.3 Moving to the large category of all spaces

The internally defined property of completeness may be externally characterized within the category of all spaces of a particular type: Banaschewski and Bruns [1967] and Isbell [1964] respectively characterized completeness of ordered sets and metric spaces by injectivity. Remarkably this categorical characterization can be established in the general $(\mathbb{T}, \mathcal{V})$-context for a whole scheme of completeness notions. Whereas this characterization may be seen as depending only on the category of lax algebras of a particular type, hence as independent of the parameters $\mathbb{T}$ and $\mathcal{V}$ presenting them, there is also an equational characterization of completeness, which uses these parameters in a substantial way. Indeed, within the context of (suitably defined) separated lax algebras, the cocomplete objects are precisely the Eilenberg–Moore algebras of a certain monad on the category of all lax algebras.

The characterization of equationally defined objects as the injectives in a category of lax algebras or monoids is in fact a recurring theme in the book. For example, in Chapter IV we present general theorems that entail the identification of *continuous lattices* as the regular-injective objects in the category of T0-spaces. Furthermore, the general context of lax algebras allows us to make precise the connection of the equationally defined versus injective paradigm with the fundamental categorical notion of *exponentiability*. In the context of topological spaces, it facilitates the formation of function spaces, and Day and Kelly [1970] identified the exponentiable objects as the core-compact spaces. When topological spaces are described as lax algebras (X, a) by the inequalities (I.1.3.iii), which, with $\mathcal{V}$-relational composition, may be transcribed as

$$a \cdot \hat{T}a \leq a \cdot m_X \qquad \text{and} \qquad 1_X \leq a \cdot e_X \,, \tag{I.2.3.i}$$

the core-compact spaces are precisely those that make the first of these two inequalities an equality. If in (I.2.3.i) one lets T be the filter monad (with its Kleisli extension), rather than the ultrafilter monad, those spaces which satisfy the multiplicative law (I.2.3.i) up to equality form again an important subclass of spaces, called *observable realization spaces* in this book, for which we give alternative characterizations in Chapter IV.

With these facts in mind, it is not surprising that the category of *reflexive graphs*, given by all pairs (X, a) required to satisfy only the second of the two inequalities (I.2.3.i), form a *quasitopos* which contains the category of lax algebras as a reflexive subcategory, under mild conditions on the parameters $\mathbb{T}$ and $\mathcal{V}$. In fact, for $\mathbb{T} = \beta$ and $\mathcal{V} = \mathsf{2}$, this extension is minimal, producing the category of pseudotopological spaces. In the general $(\mathbb{T}, \mathcal{V})$-context, however, the quasitopos hull of the category of lax algebras will form a proper subcategory of that of all reflexive graphs, which leads to the consideration of important intermediate categories (Chapter III).

I.3 Chapter highlights and dependencies

Chapter II provides a rapid introduction into ordered sets and category theory, to the extent needed in this book. It provides not only the notation, terminology, and theory used in the main body of the book (starting with Chapter III), but also emphasizes areas of importance in the sequel that may play a less prominent role in other introductory texts, such as monadic and topological functors. For enriched and higher-order category theory, we get by with a brief exposition of monoidal and ordered categories.

Whilst the presentation of topics is self-contained, the arguments provided are often quite compact and pitched at a level that requires a degree of mathematical maturity that may at times be challenging for a beginning graduate student. Some of the exercises at the end of each section should help to overcome these challenges. Others are complementary to the main body of the text and may be used later on.

Chapter III, Sections III.1–III.3, provide the first key notions, properties, and examples of the theory and applications of lax algebras. Introduced under the name $(\mathbb{T},\mathcal{V})$-*category* in Section III.1.6, in order to stress their status as individual small generalized categories, they are alternatively called $(\mathbb{T},\mathcal{V})$-*algebras* or $(\mathbb{T},\mathcal{V})$-*spaces*, depending on whether we want to emphasize their algebraic or geometric-topological roles. It is important that the reader does not skip the preceding Subsections III.1.1–III.1.5 in which many of the syntactical tools pertaining to lax extensions of the monad $\mathbb{T}$ are developed.

Topologicity of the resulting large category $(\mathbb{T},\mathcal{V})$-Cat over Set is shown at the beginning of Section III.3, followed by a discussion of the impact of change in the parameters, arising from morphisms $\mathbb{S}\to\mathbb{T}$ and $\mathcal{V}\to\mathcal{W}$.

As $(\mathbb{T},\mathcal{V})$-Cat fails to be Cartesian closed, a presentation of quasitopoi containing $(\mathbb{T},\mathcal{V})$-Cat follows in Section III.4, along with an introduction to the categorical tools on exponentiability of morphisms. For the role model $\mathsf{Top}\cong(\beta,2)$-$\mathsf{Cat}$, the quasitopos extension $(\mathbb{T},\mathcal{V})$-Gph ("$(\mathbb{T},\mathcal{V})$-graphs") of $(\mathbb{T},\mathcal{V})$-Cat leads to the category of pseudotopological spaces. Section III.5 gives a first demonstration of how the general theory feeds into applications and provides new insights. There is a key adjunction

$$(\mathbb{T},\mathcal{V})\text{-}\mathsf{Cat}\ \underset{\longleftarrow}{\overset{\longrightarrow}{\bot}}\ (\mathcal{V}\text{-}\mathsf{Cat})^{\mathbb{T}}$$

which compares $(\mathbb{T},\mathcal{V})$-Cat with the Eilenberg–Moore category of $\mathbb{T}$ extended from Set to $\mathcal{V}$-Cat. In our role model, it relates Top with the category of *ordered compact Hausdorff spaces* and emphasizes the importance of the order

$$\begin{aligned} \mathfrak{x}\le\mathfrak{y} &\iff \forall A\subseteq X \text{ closed } (A\in\mathfrak{x}\implies A\in\mathfrak{y}) \\ &\iff \forall B\subseteq X \text{ open } (B\in\mathfrak{y}\implies B\in\mathfrak{x}) \end{aligned}$$

on the set βX of ultrafilters on the set X, for every topological space X. This approach leads to the powerful notion of *representable* $(\mathbb{T}, \mathcal{V})$*-category*, which, in the role model, entails core-compactness, i.e. exponentiability in the category Top. Whereas Sections III.1–III.3 of Chapter III are a necessary prerequisite for Chapters IV and V, the slightly more demanding Sections III.4 and III.5 will be used only sporadically.

Chapter IV provides powerful alternative descriptions of the category $(\mathbb{T}, \mathcal{V})\text{-}\mathsf{Cat}$, the most striking of which arises from the fact that the quantale $\mathcal{V}$ and the monad $\mathbb{T}$ on Set laxly extended to $\mathcal{V}\text{-}\mathsf{Rel}$ allow for the construction of a new monad $\mathbb{\Pi} = \mathbb{\Pi}(\mathbb{T}, \mathcal{V})$ (read "Pi") on Set laxly extended to $\mathsf{Rel} = 2\text{-}\mathsf{Rel}$ such that

$$(\mathbb{T}, \mathcal{V})\text{-}\mathsf{Cat} = (\mathbb{\Pi}, 2)\text{-}\mathsf{Cat}\,,$$

associativity of the Kleisli convolution granted. Consequently, $(\mathbb{T}, \mathcal{V})$-categories may be presented equivalently as *relational* lax algebras, with respect to a *power-enriched* (see Section I.1.5) monad $\mathbb{\Pi}$. The fact that all relevant information provided by $\mathbb{T}$ and $\mathcal{V}$ can be encoded by a new monad $\mathbb{\Pi}$ gives the parameter $\mathbb{T}$ some prominence over $\mathcal{V}$. This result, presented in Section IV.3 along with applications, including the relational description of approach spaces which initiated this research, needs some preparation from Sections IV.1 and IV.2 that are of independent interest.

Guided by the role model of the filter monad, for a power-enriched monad $\mathbb{T}$ we describe in Section IV.1 the isomorphism

$$(\mathbb{T}, 2)\text{-}\mathsf{Cat} \cong \mathbb{T}\text{-}\mathsf{Mon}\,,$$

presenting relational algebras as Kleisli monoids. In Section IV.2, taking the inclusion $\mathbb{\beta} \longrightarrow \mathbb{F}$ of ultrafilters into filters as the role model, for a suitable morphism $\mathbb{S} \longrightarrow \mathbb{T}$ of monads we present an isomorphism

$$\mathbb{T}\text{-}\mathsf{Mon} \cong (\mathbb{S}, 2)\text{-}\mathsf{Cat}\,,$$

which provides the general framework for the identical description of topological spaces in terms of either filter or ultrafilter convergence. A $\mathcal{V}$-level generalization (in lieu of $\mathcal{V} = 2$) of this last isomorphism is also provided.

In the context of a morphism $\mathbb{S} \longrightarrow \mathbb{T}$ of power-enriched monads one can construct a monad $\mathbb{T}'$ on $\mathbb{S}\text{-}\mathsf{Mon}$ which has the same Eilenberg–Moore category as $\mathbb{T}$:

$$\mathsf{Set}^{\mathbb{T}} \cong (\mathbb{S}\text{-}\mathsf{Mon})^{\mathbb{T}'}\,.$$

For $\mathbb{S} = \mathbb{T}$, the right-hand-side category becomes isomorphic to a category of injective $\mathbb{T}$-monoids, as we show in Section IV.4. For $\mathbb{T}$ the filter monad, so that $\mathbb{T}\text{-}\mathsf{Mon} \cong \mathsf{Top}$, one obtains in particular the simultaneous description of continuous lattices as Eilenberg–Moore algebras and as injectives in Top (with respect to the class of initial morphisms).

Section IV.5 is devoted exclusively to the study of those topological spaces that, when presented as lax algebras with the filter monad, satisfy the multiplicative law (I.2.3.i) up to equality, for which alternative descriptions are given, and the continuous lattices among them are fully characterized.

Chapter V looks at $(\mathbb{T}, \mathcal{V})$-categories as *spaces* and explores topological properties, such as separation, regularity, normality, and compactness in $(\mathbb{T}, \mathcal{V})$-Cat (Sections V.1–V.4). Emphasis is given to those properties that arise "naturally" in the $(\mathbb{T}, \mathcal{V})$-setting, such as the symmetric descriptions of Hausdorff separation and compactness, or the symmetrically described properties of properness and openness for morphisms, as already alluded to in Section I.1.4. There is also a much more hidden symmetry between normality and extremal disconnectedness.

Closure of the relevant properties under direct products (for compact objects or proper morphisms) is a prominent theme (including the Tychonoff Theorem), and so is the generalization of the Kuratowski–Mrówka Theorem characterizing compact spaces in the general $(\mathbb{T}, \mathcal{V})$-context.

Section V.4 gives an axiomatic categorical framework for treating some of these key properties in a most economical fashion, and Section V.5 explores the notion of connectedness in extensive categories in general and in the $(\mathbb{T}, \mathcal{V})$-context in particular.

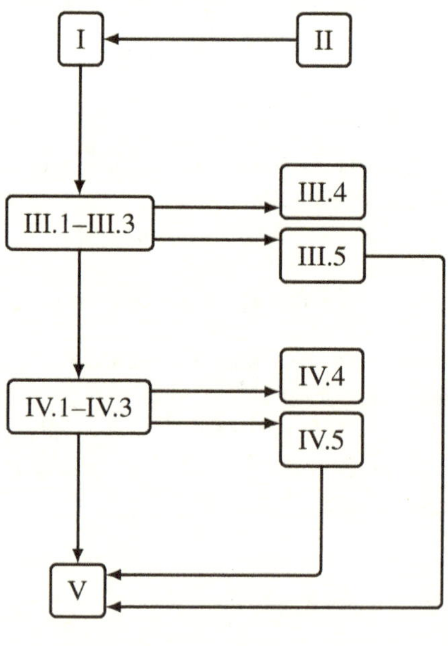

The range of example categories is expanded beyond the realm of metric and topology. It includes *multi-ordered sets*, to be thought of as a "thin" version of

Lambek's *multicategories* [Lambek, 1969] (also known as *colored operads*) that are gaining considerable attention in algebraic topology.

Summary of chapter dependencies

The left column in the diagram indicates the principal stream of suggested reading, and the right column lists the order-theoretic and categorical prerequisites, as well as special topics that may be omitted initially.

A word about sets, classes, and choice

Without reference to any particular kind of set-theoretic foundations, in this book we distinguish between *sets*, *classes*, and *conglomerates*, to be able to form the class of all sets and the conglomerate of all classes, leading us in particular to the category Set and the metacategory SET, respectively (as in Sections II.2.2 and II.2.2). Classes whose elements may be labeled by a set are also called *small*; others are *large* or *proper classes*. We refer to the Notes on Chapter II for suggested further reading on this topic.

We frequently use the *Axiom of Choice* (guaranteeing that surjective maps of sets are retractions). In fact, key results (such as the equivalence of the open-set and the ultrafilter-convergence presentations of topologies) rely on it. We alert the reader to each new use of the Axiom of Choice by putting the symbol

©

in the margin. The symbol merely indicates our use of Choice at the instance in question, without any affirmation that the use is actually essential.

II

Monoidal structures

Gavin J. Seal and Walter Tholen

This chapter provides a compactly written introduction to the order- and category-theoretic tools most commonly used throughout the remainder of the book. A newcomer to the subject may at times need to consult a standard reference on category theory, and the chapter may be skipped by the more advanced reader who might use it just as a reference point for notation and terminology.

It is hoped that all readers will appreciate the ubiquity of "monoidal structures" appearing in the text. We allude to them quite explicitly only in Sections II.1, II.3, and II.4 but note that, after all, categories are generalized monoids.

II.1 Ordered sets

II.1.1 The Cartesian structure of sets and its monoids

Any two sets A, B may be "multiplied" in terms of their *Cartesian product*

$$A \times B = \{(x, y) \mid x \in A, y \in B\} ,$$

and this multiplication extends to maps $f : A \longrightarrow A'$, $g : B \longrightarrow B'$ via

$$f \times g : A \times B \longrightarrow A' \times B' , \quad (x, y) \longmapsto (f(x), g(y)) .$$

The Cartesian product respects identity maps, since

$$1_A \times 1_B = 1_{A\times B} ,$$

as well as composition of maps, since for $f' : A' \longrightarrow A''$, $g' : B' \longrightarrow B''$ one has the *middle-interchange law*

Monoidal Topology: A Categorical Approach to Order, Metric, and Topology, eds. Dirk Hofmann, Gavin J. Seal, and Walter Tholen. Published by Cambridge University Press.

$$(f' \times g') \cdot (f \times g) = (f' \cdot f) \times (g' \cdot g) \,.$$

The Cartesian product is associative "up to isomorphism," and any one-element set $E = \{\star\}$ acts as a neutral element. More precisely, there are obvious natural bijections

$$\alpha_{A,B,C} : A \times (B \times C) \to (A \times B) \times C \,, \quad \lambda_A : E \times A \to A \,, \quad \rho_A : A \times E \to A$$

satisfying the so-called *coherence conditions*

$$\lambda_E = \rho_E \,, \quad (\rho_A \times 1_B) \cdot \alpha_{A,E,B} = 1_A \times \lambda_B \,,$$
$$(\alpha_{A,B,C} \times 1_D) \cdot \alpha_{A,B\times C,D} \cdot (1_A \times \alpha_{B,C,D}) = \alpha_{A\times B,C,D} \cdot \alpha_{A,B,C\times D} \,,$$

for all sets A, B, C, D. Moreover, the Cartesian structure is *symmetric*, since there is a natural bijection

$$\sigma_{A,B} : A \times B \to B \times A$$

with

$$\sigma_{B,A} \cdot \sigma_{A,B} = 1_{A\times B} \,, \quad \rho_A = \lambda_A \cdot \sigma_{A,E} \,,$$
$$\alpha_{C,A,B} \cdot \sigma_{A\times B,C} \cdot \alpha_{A,B,C} = (\sigma_{A,C} \times 1_B) \cdot \alpha_{A,C,B} \cdot (1_A \times \sigma_{B,C}) \,.$$

A *monoid* M (with respect to the Cartesian structure of sets) is a set M that comes with a binary and a nullary operation

$$m : M \times M \to M \,, \quad e : E \to M$$

that are associative and make $e = e(\star)$ a neutral element of M; equivalently, the diagrams

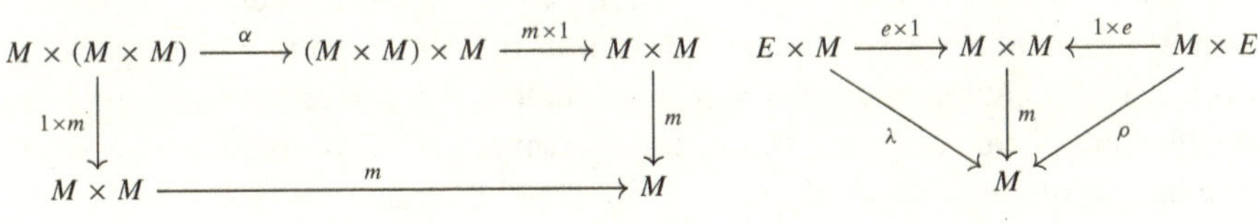

commute. The monoid is *commutative* if $m \cdot \sigma = m$. A *homomorphism* $f : M \to N$ of monoids preserves both operations: $f \cdot m_M = m_N \cdot (f \times f)$ and $f \cdot e_M = e_N$.

II.1.2 The compositional structure of relations

A *relation* r *from* a set X *to* a set Y distinguishes those elements $x \in X$ and $y \in Y$ that are r-related; we write $x \; r \; y$ if x is r-related to y. Hence, depending on whether we display r as a subset, a two-valued function, or a multi-valued function via

$$r \subseteq X \times Y\,, \quad r : X \times Y \to \{\text{true}, \text{false}\}\,, \quad r : X \to PY\,,$$

respectively, $x\ r\ y$ may be equivalently written as

$$(x, y) \in r\,, \quad r(x, y) = \text{true}\,, \quad y \in r(x)\,,$$

where PY denotes the powerset of Y. Writing $r : X \nrightarrow Y$ when r is a relation from X to Y, we can "multiply" r with $s : Y \nrightarrow Z$ via ordinary relational composition:

$$x\ (s \cdot r)\ z \iff \exists y \in Y\ (x\ r\ y\ \ \&\ \ y\ s\ z)\,.$$

Writing $r \leq r'$ (with $r' : X \nrightarrow Y$) when, equivalently,

$$r \subseteq r'\,, \quad \forall x \in X\, \forall y \in Y\ (r(x, y) \models r'(x, y))\,, \quad \forall x \in X\ (r(x) \subseteq r'(x))\,,$$

we see that the multiplication respects $\leq$, since

$$r \leq r', s \leq s' \implies s \cdot r \leq s' \cdot r'\,. \tag{II.1.2.i}$$

Moreover, relational composition is associative, so that

$$t \cdot (s \cdot r) = (t \cdot s) \cdot r$$

when $t : Z \nrightarrow W$, and for the identity relation 1_X (with $x\ 1_X\ x' \iff x = x'$) one has

$$r \cdot 1_X = r = 1_Y \cdot r\,.$$

Hence, comparing with Section II.1.1, we observe that sets $A, B, \ldots$ have been replaced by relations $r, s, \ldots$ and the Cartesian product by composition. While in Section II.1.1 there is room for maps f, g, here we have only $\leq$ between relations, so that the middle-interchange law of Section II.1.1 reduces to a mere property (II.1.2.i). The natural bijections α, λ, ρ of Section II.1.1 have become identities, but the multiplicative structure is no longer "symmetric." However, for $r : X \nrightarrow Y$ one has the *opposite* (or *dual*) relation $r^\circ : Y \nrightarrow X$ with

$$y\ r^\circ\ x \iff x\ r\ y$$

for all $x \in X$, $y \in Y$, which satisfies

$$(s \cdot r)^\circ = r^\circ \cdot s^\circ\,, \quad (1_X)^\circ = 1_X\,, \quad (r^\circ)^\circ = r\,, \quad r \leq r' \implies r^\circ \leq (r')^\circ\,.$$

Note that when r is the *graph* of a map $f : X \to Y$ (so that $x\ r\ y \iff f(x) = y$), then $r^\circ(y) = f^{-1}(y)$ is simply the *fiber* of f over $y \in Y$. In what follows *we make no notational distinction between a map and its graph.*

II.1.3 Orders

An *order* on a set X is a relation $a : X \nrightarrow X$ that carries a monoid structure with respect to the compositional structure of relations; i.e. a satisfies

$$a \cdot a \leq a \,, \quad 1_X \leq a \,.$$

Hence, a is simply a *transitive* and *reflexive* relation on X:

$$(x \leq y \;\&\; y \leq z \implies x \leq z) \,, \quad x \leq x$$

for all $x, y, z \in X$, when we write $x \leq y$ for $x \; a \; y$. The order is

(1) *separated* if $a \cap a^\circ = 1_X$ (so that $x \leq y \;\&\; y \leq x \implies x = y$);
(2) *total* if $a \cup a^\circ = X \times X$ (so that $x \leq y$ or $y \leq x$, for all $x, y \in X$).

In the literature, orders on X are usually called *preorders*, and separated (i.e. *antisymmetric*) orders are often called *partial orders* on X. In this book, an *ordered set* X is simply a set X equipped with an order, and X is *separated* if the order is separated. If a is an order on X (respectively, a separated or total order), then so is a°. A *chain* is a set with a separated total order. A map $f : X \to Y$ of ordered sets is *monotone* (or *order preserving*) if

$$f \cdot a \leq b \cdot f \,,$$

where a, b denote the orders on X, Y, respectively, and f is identified with its graph; hence, if we write $\leq$ for both a and b,

$$x < y \implies f(x) \leq f(y)$$

for all $x, y \in X$. If the implication "$\impliedby$" also holds, so that $a = f^\circ \cdot b \cdot f$, then f is *fully faithful*. For an ordered set X, we write X^{op} for the same set equipped with the opposite order; thus, when $f : X \to Y$ is monotone, so is $f^{\mathrm{op}} : X^{\mathrm{op}} \to Y^{\mathrm{op}}$ (with $f^{\mathrm{op}}(x) = f(x)$ for all $x \in X$).

Every relation $r : X \nrightarrow X$ has an *ordered hull* $\overline{r}$, which may be described as

$$\overline{r} = \bigcup\nolimits_{n \geq 0} r^n$$

(where $r^0 := 1_X$, $r^{n+1} = r \cdot r^n$), and, if r is separated, so is $\overline{r}$. For any order a on X, $a \cap a^\circ$ is an equivalence relation on X, which, when a is written as $\leq$, is denoted by $\simeq$, so that

$$x \simeq y \iff x \leq y \;\&\; y \leq x \,.$$

There is a least order b on the quotient set $X/{\simeq}$ which makes the projection $p : X \to X/{\simeq}$ monotone, namely $b = p \cdot a \cdot p^\circ$, also described by

$$p(x) \leq p(y) \iff x \leq y \,.$$

The point of this construction is that b is separated; we call $X/{\simeq}$ the *separated reflection* of X.

II.1.4 Modules

A relation $r : X \nrightarrow Y$ between ordered sets is a *module* if $(\leq_Y) \cdot r \cdot (\leq_X) \leq r$, i.e. if

$$x' \leq x \;\&\; x \, r \, y \;\&\; y \leq y' \implies x' \, r \, y'$$

for all $x, x' \in X$, $y, y' \in Y$. Hence, the relation r is a module if and only if the map $r : X^{\mathrm{op}} \times Y \to \{\text{true}, \text{false}\}$ is monotone (where $X^{\mathrm{op}} \times Y$ is ordered componentwise). Graphically, we indicate modularity of a relation $r : X \nrightarrow Y$ by

$$r : X \overset{\circ}{\nrightarrow} Y \, .$$

Every monotone map $f : X \to Y$ gives rise to the modules

$$f_* = (\leq_Y) \cdot f : X \overset{\circ}{\nrightarrow} Y \qquad \text{and} \qquad f^* = f^\circ \cdot (\leq_Y) : Y \overset{\circ}{\nrightarrow} X \, ,$$

i.e.

$$x \, f_* \, y \iff f(x) \leq y \qquad \text{and} \qquad y \, f^* \, x \iff y \leq f(x)$$

for all $x \in X$, $y \in Y$. The following rules may be easily verified when $g : Y \to Z$ is monotone:

(1) $1_X^* = (1_X)_* = (\leq_X)$;
(2) $(g \cdot f)_* = g_* \cdot f_*$ and $(g \cdot f)^* = f^* \cdot g^*$;
(3) $1_X^* \leq f^* \cdot f_*$ and $f_* \cdot f^* \leq 1_Y^*$.

Modularity is also closed under relational composition. Indeed, for modules $r : X \overset{\circ}{\nrightarrow} Y$ and $s : Y \overset{\circ}{\nrightarrow} Z$, one has

$$(\leq_Z) \cdot (s \cdot r) \cdot (\leq_X) \leq (\leq_Z) \cdot s \cdot (\leq_Y) \cdot (\leq_Y) \cdot r \cdot (\leq_X) \leq s \cdot r \, ,$$

so that $s \cdot r : X \overset{\circ}{\nrightarrow} Z$ is again a module.

II.1.5 Adjunctions

For ordered sets X, Y, the set

$$\mathsf{Ord}(X, Y) = \{f \mid f : X \to Y \text{ monotone}\}$$

is itself ordered *pointwise* by

$$f \leq f' \iff \forall x \in X \, (f(x) \leq f'(x)) \, .$$

This order is preserved by composition on either side: whenever $h : W \to X$ and $k : Y \to Z$ are monotone, then

$$f \leq f' \implies k \cdot f \cdot h \leq k \cdot f' \cdot h \, .$$

A monotone map $g : Y \to X$ is called

(1) *right adjoint* if there is a monotone map $f : X \to Y$ with $1_X \leq g \cdot f$, $f \cdot g \leq 1_Y$;
(2) an *equivalence* if there is a monotone map $f : X \to Y$ with $1_X \simeq g \cdot f$, $f \cdot g \simeq 1_Y$;
(3) an *isomorphism* if there is a monotone map $f : X \to Y$ with $1_X = g \cdot f$, $f \cdot g = 1_Y$.

By definition, one has the implications

$$\text{isomorphism} \implies \text{equivalence} \implies \text{right adjoint} .$$

The map f occurring in the definition of right adjointness is, up to "$\simeq$," uniquely determined by g: if $1_X \leq g \cdot f'$ and $f' \cdot g \leq 1_Y$, then

$$f' = f' \cdot 1_X \leq f' \cdot g \cdot f \leq 1_Y \cdot f = f ,$$

and dually $f \leq f'$. If g is right adjoint, the corresponding f is called *left adjoint* to g, and one writes

$$f \dashv g .$$

This terminology becomes more plausible when we consider the following fact.

II.1.5.1 Proposition *A map $g : Y \to X$ (not assumed to be monotone a priori) is right adjoint if and only if there is a map $f : X \to Y$ such that*

$$f(x) \leq y \iff x \leq g(y)$$

for all $x \in X$, $y \in Y$.

Proof The necessity of the condition is obvious since $x \leq g(y)$ implies $f(x) \leq f \cdot g(y) \leq y$, and dually for "$\implies$." For its sufficiency, observe that $f(x) \leq f(x)$ implies $x \leq g \cdot f(x)$, and dually $f \cdot g(y) \leq y$. The monotonicity of f follows, since $x \leq x' \leq g \cdot f(x')$ yields $f(x) \leq f(x')$, and likewise for g. □

Calling a pair $(f : X \to Y, g : Y \to X)$ of monotone maps an *adjunction* if f is left adjoint to g, we see that $(g^{\mathrm{op}} : Y^{\mathrm{op}} \to X^{\mathrm{op}}, f^{\mathrm{op}} : X^{\mathrm{op}} \to Y^{\mathrm{op}})$ is also an adjunction. In other words, f is left adjoint (to g) if and only if f^{op} is right adjoint (with left adjoint g^{op}):

$$f \dashv g : Y \to X \iff g^{\mathrm{op}} \dashv f^{\mathrm{op}} : X^{\mathrm{op}} \to Y^{\mathrm{op}} .$$

An adjunction $(f : X^{\mathrm{op}} \to Y, g : Y \to X^{\mathrm{op}})$ is often called a *Galois correspondence* between X and Y.

One sees easily that the following statement holds.

II.1.5.2 Corollary *A right adjoint map g (with left adjoint f) is fully faithful if and only if $f \cdot g \simeq 1_Y$.*

The conjunction of this statement with its dual yields the following corollary.

II.1.5.3 Corollary *Equivalences are given by those adjunctions for which both maps are fully faithful. In this case, each map serves as both a left and a right adjoint.*

All properties for maps between ordered sets discussed so far (monotone, fully faithful, right adjoint, left adjoint, equivalence, isomorphism) are closed under composition.

II.1.6 Closure operations and closure spaces

For any adjunction $f \dashv g : Y \to X$ one has

$$f \cdot g \cdot f \simeq f \qquad \text{and} \qquad g \cdot f \cdot g \simeq g\,,$$

since $1_X \leq g \cdot f$ and $f \cdot g \leq 1_Y$ imply $f \leq f \cdot g \cdot f = (f \cdot g) \cdot f \leq f$; hence the first equivalence holds, and the second follows by duality. Therefore, setting

$$c := g \cdot f \qquad \text{and} \qquad d := f \cdot g\,,$$

one obtains monotone maps $c : X \to X$, $d : Y \to Y$ with

$$c \cdot c \simeq c\,, \quad 1_X \leq c \qquad \text{and} \qquad d \simeq d \cdot d\,, \quad d \leq 1_Y\,.$$

Any such map c is called a *closure operation* on X, and any such d is called an *interior operation* on X. Since $1_X \leq c$ and the monotonicity of c imply $c \leq c \cdot c$, it suffices to ask c to satisfy

$$c \cdot c \leq c\,, \quad 1_X \leq c\,;$$

i.e. a closure operation on X is nothing but an element of $\mathsf{Ord}(X, X)$ that carries a monoid structure with respect to the compositional structure.

With c and d induced by $f \dashv g$ as discussed, one easily sees that f and g can be restricted to an equivalence between the subsets

$$\mathrm{Fix}(c) := \{x \in X \mid c(x) \simeq x\} \qquad \text{and} \qquad \mathrm{Fix}(d) := \{y \in Y \mid d(y) \simeq y\}\,,$$

of c-*closed* and d-*open* elements (or simply *closed* and *open* elements), also referred to as *fixpoints* of c and d, respectively. The following diagram summarizes this situation:

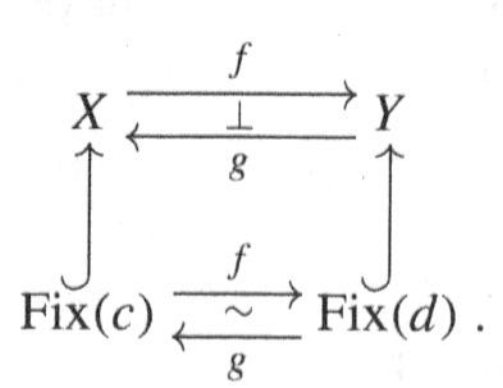

Any closure operation c on X is induced by an adjunction whose left adjoint is defined on X:

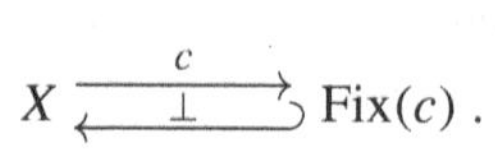

A *closure space* is a set X which comes with a closure operation on the powerset PX, ordered by inclusion. A map $f : X \longrightarrow Y$ is *continuous* if

$$f(c_X(A)) \subseteq c_Y(f(A))$$

for all $A \subseteq X$. Since for any map $f : X \longrightarrow Y$ one has an adjunction

$$PX \underset{f^{-1}(-)}{\overset{f}{\underset{\longleftarrow}{\overset{\longrightarrow}{\bot}}}} PY$$

given by image and preimage along f, the continuity condition is equivalently written as

$$c_X(f^{-1}(B)) \subseteq f^{-1}(c_Y(B))$$

for all $B \subseteq Y$. Via the order isomorphism $(-)^{\complement} : PX^{\mathrm{op}} \longrightarrow PX$ (which maps $A \in PX$ to its *complement* $A^{\complement} := X \setminus A$ in X), any closure operation c on PX corresponds to an interior operation d on PX, and vice versa:

$$c(A)^{\complement} = d(A^{\complement}) ,$$

for all $A \subseteq X$. Therefore, there is a concept of interior space, equivalent to that of a closure space, and in this context a map $f : X \longrightarrow Y$ between interior spaces (X, d_X) and (Y, d_Y) is *continuous* if

$$f^{-1}(d_Y(B)) \subseteq d_X(f^{-1}(B))$$

for all $B \subseteq Y$.

II.1.7 Completeness

For an element x in an ordered set X, let

$$\downarrow_X x = \downarrow x = \{y \in X \mid y \leq x\}$$

be the *down-set* of x in X. The *down-closure* of $A \subseteq X$ is

$$\downarrow_X A = \downarrow A = \textstyle\bigcup_{x \in A} \downarrow x ,$$

and A is *down-closed* (or a *down-set*) if $\downarrow A = A$. There is a fully faithful map

$$\downarrow : X \longrightarrow \mathrm{Dn}\, X = \mathrm{Fix}(\downarrow_X) = \{A \subseteq X \mid \downarrow A = A\} ,$$

where the set of down-sets in X is ordered by inclusion. The ordered set X is *complete* if and only if this map is right adjoint; equivalently, if there is a map $\bigvee_X = \bigvee : \mathrm{Dn}\, X \longrightarrow X$ which for every $A \in \mathrm{Dn}\, X$ satisfies

$$\forall x \in X \; (\textstyle\bigvee A \leq x \iff A \subseteq \downarrow x) . \qquad \text{(II.1.7.i)}$$

Calling x an *upper bound* of A in X whenever $A \subseteq {\downarrow} x$, we may rephrase the characteristic property of the *join* (or *supremum*, or *least upper bound*) $\bigvee A$ of A more familiarly by

(1) $\bigvee A$ is an upper bound of A in X (" $\Longrightarrow$ " of (II.1.7.i)), and
(2) if x is an upper bound of A in X, then $\bigvee A \leq x$ (" $\Longleftarrow$ " of (II.1.7.i)).

Of course, in general $\bigvee A$ is uniquely determined by A only up to "$\simeq$." Note also that our notion of completeness does not only give mere existence of $\bigvee A$, but also comes with a given choice of $\bigvee A$ for every $A \in \mathrm{Dn}\,X$. Finally, the existence of suprema for arbitrary subsets $B \subseteq X$ (not necessarily down-closed) follows from that of down-closed subsets: since B and ${\downarrow} B$ have the same upper bounds, one can put

$$\bigvee B = \bigvee {\downarrow} B\ ;$$

in other words, the adjunctions

$$X \underset{\bigvee}{\overset{\downarrow}{\underset{\top}{\rightleftarrows}}} \mathrm{Dn}\,X \underset{\downarrow}{\overset{\hookrightarrow}{\underset{\top}{\rightleftarrows}}} PX$$

compose!

Exploiting the adjunction $\bigvee_X \dashv {\downarrow}_X$ for X^{op} in lieu of X, we obtain

$$X^{\mathrm{op}} \underset{\bigvee_{X^{\mathrm{op}}}}{\overset{\downarrow_{X^{\mathrm{op}}}}{\underset{\top}{\rightleftarrows}}} \mathrm{Dn}\,(X^{\mathrm{op}})\ ,$$

and dualization of this adjunction yields, with

$$\mathrm{Up}\,X := (\mathrm{Dn}\,(X^{\mathrm{op}}))^{\mathrm{op}}\ , \quad {\uparrow}_X := ({\downarrow}_{X^{\mathrm{op}}})^{\mathrm{op}}\ , \quad \textstyle\bigwedge_X := (\bigvee_{X^{\mathrm{op}}})^{\mathrm{op}}\ ,$$

the adjunction

$$X \underset{\bigwedge_X}{\overset{\uparrow_X}{\underset{\bot}{\rightleftarrows}}} \mathrm{Up}\,X\ .$$

Note that, for ${\uparrow}_X$ to be monotone, $\mathrm{Up}\,X$ is (unlike $\mathrm{Dn}\,X$) ordered by *reverse* inclusion. The dual notions (like *up-set*, *up-closure*, *up-closed*, *lower bound*, *meet*, *infimum*, *greatest lower bound*) are all naturally describable in terms of this adjunction. For example, for $A \in \mathrm{Up}\,X$, $\bigwedge A$ is characterized by

$$\forall x \in X\ (x \leq \textstyle\bigwedge A \iff A \subseteq {\uparrow} x)\ .$$

Moreover, this adjunction exists (equivalently, X^{op} is complete) precisely when X is complete, since the meet of an (up-closed) set can be realized as the join of the set of its lower bounds; a more elegant argument is given in Corollary II.1.8.4.

II.1.8 Adjointness criteria

A monotone map $f : X \to Y$ of ordered sets *preserves the supremum* $\bigvee A$ of $A \subseteq X$ if $f(\bigvee A)$ is a supremum of $f(A) = \{f(x) \mid x \in A\}$ in Y. Moreover, f is a *sup-map* if it preserves every existing supremum in X:

$$f(\bigvee A) \simeq \bigvee f(A)$$

whenever $\bigvee A$ exists. The dual notions are: *preserves an infimum, inf-map*. Sup-preserving maps are useful for detecting left adjoints.

II.1.8.1 Proposition *Every left adjoint map f is a sup-map.*

Proof Indeed, if $\bigvee A$ exists

$$\begin{aligned} f(A) \subseteq \downarrow y &\iff \forall x \in A\ (f(x) \leq y) \\ &\iff \forall x \in A\ (x \leq g(y)) \\ &\iff \bigvee A \leq g(y) \iff f(\bigvee A) \leq y \end{aligned}$$

for all $y \in Y$. □

Dually, a right adjoint map is an inf-map. Being a sup-map (respectively an inf-map) is not only a necessary condition for being left adjoint (respectively right adjoint), but also sufficient, provided that the domain of the map is complete. More precisely:

II.1.8.2 Proposition *A monotone map $f : X \to Y$ is left adjoint if and only if there is a map $g : Y \to X$ such that for all $y \in Y$*

$$g(y) \simeq \bigvee \{x \in X \mid f(x) \leq y\}, \tag{II.1.8.i}$$

and f preserves those suprema. Hence, when X is complete, the map g can be given as the composite map

$$Y \xrightarrow{\ \downarrow\ } \mathrm{Dn}\,Y \xrightarrow{f^{-1}(-)} \mathrm{Dn}\,X \xrightarrow{\ \bigvee\ } X\,.$$

Proof The condition (II.1.8.i) is clearly necessary since $f \dashv g$ yields $f^{-1}(\downarrow y) = \downarrow g(y)$, and $\bigvee \downarrow g(y) \simeq g(y)$ for all $y \in Y$. Conversely, existence of the join (II.1.8.i) gives $x \leq g(y)$ whenever $f(x) \leq y$, and its preservation by f yields $f \cdot g(y) \leq y$, so that $f(x) \leq y$ whenever $x \leq g(y)$. □

II.1.8.3 Corollary *When X is a complete ordered set, a map $f : X \to Y$ is left adjoint if and only if f is a sup-map.*

As an application, let us prove the following result (see Section II.1.7).

II.1.8.4 Corollary

(1) *X^{op} is complete when X is complete.*
(2) *When Y is complete, a map $g : Y \to X$ is right adjoint if and only if g is an inf-map.*

Proof (1): It suffices to show that $\uparrow_X : X \to \mathrm{Up}\, X$ is a sup-map. But

$$\uparrow \bigvee A = \bigcap_{a \in A} \uparrow a$$

for all $A \subseteq X$ is just the defining property for suprema:

$$\bigvee A \leq x \iff \forall a \in A\ (a \leq x) \iff A \subseteq \downarrow x\,.$$

(2): By (1) one may apply Corollary II.1.8.3 with g^{op} in lieu of f. □

II.1.9 Semilattices, lattices, frames, and topological spaces

For a separated ordered set X, the map

$$\downarrow_X : X \to \mathrm{Dn}\, X$$

is an *order-embedding*, i.e. the map is injective and fully faithful (separatedness is not essential, but is assumed for convenience). The set $\mathrm{Dn}\, X$ is complete, with infima given by intersection. In particular, $\mathrm{Dn}\, X$ with the binary operation $\cap$ and the nullary operation X (largest element in $\mathrm{Dn}\, X$) is a commutative monoid. When do these operations restrict to X along $\downarrow_X$? That is, when do we have dotted maps making the diagrams

$$\begin{array}{ccc} X \times X & \xrightarrow{\downarrow \times \downarrow} & \mathrm{Dn}\, X \times \mathrm{Dn}\, X \\ {\scriptstyle m}\,\vdots & & \big\downarrow{\scriptstyle \cap} \\ X & \xrightarrow{\quad\downarrow\quad} & \mathrm{Dn}\, X \end{array} \qquad\qquad \begin{array}{ccc} & E & \\ {\scriptstyle e}\swarrow & & \searrow{\scriptstyle X} \\ X & \xrightarrow{\ \downarrow\ } & \mathrm{Dn}\, X \end{array}$$

commute? Precisely when all finite infima exist in X, and then we must have

$$m(x, y) = x \wedge y = \bigwedge\{x, y\}$$

for all $x, y \in X$, and e must be the largest element of X: $e = \bigwedge \varnothing$.

A *meet-semilattice* X is a separated ordered set with finite infima. A *homomorphism* of meet-semilattices preserves finite infima, i.e. it preserves the binary $\wedge$ and the largest element. Trading infima for suprema (hence $\wedge$ for $\vee$ and largest for smallest), one obtains the notions of *join-semilattice* and *homomorphism* thereof.

Both meet- and join-semilattices have a common algebraic description: $(X, \wedge, \top)$ and $(X, \vee, \bot)$, with $\top := \bigwedge \varnothing$ and $\bot := \bigvee \varnothing$ the *top* and *bottom* elements (or *maximum* and *minimum*) of X, respectively, are simply commutative monoids in which, under mutliplicative notation, every element is *idempotent*, so that $x \cdot x = x$ for all $x \in X$. One calls such monoids $(X, \cdot, e)$ *semilattices*, since they may equivalently be considered as either a meet- or a join-semilattice, depending on whether one puts

$$(x \leq y \iff x \cdot y = x) \qquad \text{or} \qquad (x \leq y \iff x \cdot y = y)\,,$$

in which case one obtains $x \cdot y = x \wedge y$ or $x \cdot y = x \vee y$ (see Exercise II.1.L). *Homomorphisms* of such monoids are equivalently described as homomorphisms of meet- or join-semilattices.

A *lattice* is a separated ordered set X with finite infima and finite suprema. It may be equivalently described as a set X with binary operations $\wedge$, $\vee$, and nullary operations $\top$, $\bot$, such that both $(X, \wedge, \top)$, $(X, \vee, \bot)$ are commutative monoids such that

$$x \wedge x = x = x \vee x\,, \quad x \wedge (x \vee y) = x = x \vee (x \wedge y)\,.$$

A *homomorphism of lattices* is a map that preserves the operations $\wedge, \vee, \top, \bot$.

A *frame* is a complete meet-semilattice X such that, for all $a \in X$,

$$a \wedge (-) : X \to X$$

is a sup-map. Hence, a frame is simply a *complete lattice* (i.e. a complete separated ordered set) which satisfies the infinite distributive law

$$a \wedge \bigvee_{i \in I} b_i = \bigvee_{i \in I} a \wedge b_i\,.$$

In particular, each complete chain is a frame (see Exercise II.1.E). A *homomorphism of frames* must be both a homomorphism of meet-semilattices and a sup-map, i.e. it must preserve finite infima and arbitrary suprema.

For example, if X is an ordered set, then $\mathrm{Dn}\,X$ is a frame, and every monotone map $f : X \to Y$ induces a homomorphism of frames $f^{-1}(-) : \mathrm{Dn}\,Y \to \mathrm{Dn}\,X$ (see Exercise II.1.K). This applies in particular when X is *discrete*, i.e. when its order is given by equality, so that $\mathrm{Dn}\,X = PX$.

A *topology* (of open sets) on X is simply a *subframe* of PX, i.e. a subset of PX that is closed under finite infima and arbitrary suprema, making X a *topological space*; the topology of X is usually denoted by $\mathcal{O}X$. A map $f : X \to Y$ of topological spaces is *continuous* if $f^{-1}(-) : PY \to PX$ restricts to a map $f^{-1}(-) : \mathcal{O}Y \to \mathcal{O}X$. A *base* for $\mathcal{O}X$ is a subset $\mathcal{B} \subseteq PX$ with $\mathcal{O}X = \{\bigcup \mathcal{C} \mid \mathcal{C} \subseteq \mathcal{B}\}$. Every topological space is a closure space, and the notions of continuity given here and in Section II.1.6 are equivalent (Exercise II.1.F). Every topological space can be endowed with its *underlying* (or *induced*) order

$$x \leq y \iff \forall U \in \mathcal{O}X\ (y \in U \implies x \in U)\,.$$

This order is separated precisely when X is a *T0-space*. Every continuous map is monotone with respect to the underlying orders. The dual of this order is called the *specialization order* of a topological space.

II.1.10 Quantales

A *quantale* $\mathcal{V}$ (more precisely, a unital quantale) is a complete lattice which carries a monoid structure with neutral element k (as in Section II.1.1) such that, when the binary operation is denoted as a *tensor* $\otimes$,

$$a \otimes (-) : \mathcal{V} \to \mathcal{V}\,, \quad (-) \otimes b : \mathcal{V} \to \mathcal{V}$$

are sup-maps for all $a, b \in \mathcal{V}$; hence the tensor distributes over suprema:

$$a \otimes \bigvee_{i \in I} b_i = \bigvee_{i \in I} (a \otimes b_i)\,, \quad \bigvee_{i \in I} a_i \otimes b = \bigvee_{i \in I} (a_i \otimes b)\,.$$

A *lax homomorphism of quantales* $f : \mathcal{V} \to \mathcal{W}$ is a monotone map satisfying

$$f(a) \otimes f(b) \leq f(a \otimes b)\,, \quad l \leq f(k)$$

for all $a, b \in \mathcal{V}$ and l the neutral element of $\mathcal{W}$; monotonicity of f means equivalently lax preservation of joins, i.e. $\bigvee f(A) \leq f(\bigvee A)$ for all $A \subseteq V$. For f to be a *homomorphism*, these three inequalities must be identities. A quantale is *commutative* if it is commutative as a monoid. *Every frame becomes a commutative quantale* when we put $\otimes = \wedge$ and let k be the top element. In fact, frames are those commutative quantales $\mathcal{V}$ for which $a \otimes a = a$ for all $a \in \mathcal{V}$ and k is the top element (see Exercise II.1.L). If $\mathcal{V}$ and $\mathcal{W}$ are frames (considered as quantales), a lax homomorphism of quantales $f : \mathcal{V} \to \mathcal{W}$ is a homomorphism precisely when it is a sup-map. In a quantale $\mathcal{V}$, for every $a \in \mathcal{V}$, the sup-map $a \otimes (-)$ is left adjoint to a map $a \multimap (-) : \mathcal{V} \to \mathcal{V}$ which is uniquely determined by

$$a \otimes v \leq b \iff v \leq a \multimap b$$

for all $v, b \in \mathcal{V}$; hence

$$a \multimap b = \bigvee \{v \in \mathcal{V} \mid a \otimes v \leq b\}\,.$$

Likewise, for all $a \in \mathcal{V}$, the sup-map $(-) \otimes a$ is left adjoint to a map $(-) \multimapinv a : \mathcal{V} \to \mathcal{V}$. In the case where $\mathcal{V}$ is commutative, $a \multimap (-)$ and $(-) \multimapinv a$ coincide, and either of the two notations may be used.

The following examples of commutative quantales are used frequently in this book. For examples of not necessarily commutative quantales, see Exercise II.1.M.

II.1.10.1 Examples

(1) The *two-chain* $\mathsf{2} = \{\text{false} \models \text{true}\} = \{\bot, \top\}$ with $\otimes = \wedge$, $k = \top$. Here, $a \multimap b$ is the Boolean truth value of the implication $a \to b$. More generally, we use this arrow notation for any frame considered as a quantale.

(2) The *three-chain* $\mathsf{3} = \{\bot, k, \top\}$ is a complete chain, and therefore a frame (see Exercise II.1.E). However, the quantale structure we will consider is given by choosing k to be the unit for the multiplication. In fact, the

multiplication is now uniquely determined, and it is represented together with the right adjoints $(-) \multimapdotinv a$, for $a \in \mathsf{3}$, in the following tables:

$\otimes$	$\bot$	k	$\top$
$\bot$	$\bot$	$\bot$	$\bot$
k	$\bot$	k	$\top$
$\top$	$\bot$	$\top$	$\top$

$\multimapdotinv$	$\bot$	k	$\top$
$\bot$	$\top$	$\bot$	$\bot$
k	$\top$	k	$\bot$
$\top$	$\top$	$\top$	$\top$

(3) Allowing for an interval of truth values, we consider the extended real half-line $[0, \infty]$ which is a complete lattice with respect to its natural order $\leq$. But *we reverse its order*, so that $0 = \top$ is the top and $\infty = \bot$ is the bottom element, and we consider it a quantale with $\otimes$ given by addition extended via

$$a + \infty = \infty + a = \infty$$

for all $a \in [0, \infty]$, and necessarily $k = 0 = \top$. Briefly, we write

$$\mathsf{P}_+ = ([0, \infty]^{\mathrm{op}}, +, 0) .$$

When working with this quantale, the relation $\leq$ *always refers to the natural order of* $[0, \infty]$*, and we use the symbols* inf *and* sup *when forming infima and suprema in* $[0, \infty]$*, and we use joins* $\bigvee$ *and meets* $\bigwedge$ *when forming these in* P_+. Hence,

$$b \multimapdotinv a = b \ominus a := \inf\{v \in [0, \infty] \mid b \leq a + v\} ,$$

so that $b \ominus a = b - a$ if $a \leq b < \infty$, and $b \ominus a = 0$ if $b \leq a$, while $b \ominus a = \infty$ if $a < b = \infty$.

(4) In (3), addition may be replaced by multiplication extended via

$$a \cdot \infty = \infty \cdot a = \infty$$

for all $a \in [0, \infty]$ (this definition is necessary because the tensor must preserve the empty join, i.e. the bottom element ∞). Hence we obtain the quantale

$$\mathsf{P}_\times = ([0, \infty]^{\mathrm{op}}, \cdot, 1) .$$

Here,

$$b \multimapdotinv a = b \oslash a := \inf\{v \in [0, \infty] \mid b \leq a \cdot v\} ,$$

so that $b \oslash a = b/a$ if $0 < a, b < \infty$, and $b \oslash 0 = \infty = \infty \oslash a$ if $0 < a, b$, while $0 \oslash a = 0 = b \oslash \infty$ for all $a, b \in [0, \infty]$.

(5) Since $[0, \infty]^{\mathrm{op}}$ is (like $[0, \infty]$) a chain, it is a frame, and we may consider it a quantale with its meet operation (which, according to our conventions, is the max with respect to the natural order of $[0, \infty]$):

$$\mathsf{P}_{\max} = ([0, \infty]^{\mathrm{op}}, \max, 0) .$$

Here,

$$a \multimap b = a \to b := \inf\{v \in [0,\infty] \mid b \le \max\{a, v\}\}\ ,$$

which turns out to be $a \to b = 0$ if $b \le a$, and $a \to b = b$ if $a < b$.

There is a sup-map $\iota : \mathsf{2} \to [0,\infty]^{\mathrm{op}}$, sending $\top$ to 0, and $\bot$ to ∞, which gives homomorphisms of quantales

$$\iota : \mathsf{2} \to \mathsf{P}_+ \qquad \text{and} \qquad \iota : \mathsf{2} \to \mathsf{P}_{\max}\ .$$

However,

$$\iota : \mathsf{2} \to \mathsf{P}_\times$$

is only a lax homomorphism since ι does not preserve the neutral elements of the respective monoid structures. This can be corrected if one replaces $\mathsf{2}$ by $\mathsf{3}$; then the sup-map

$$\kappa : \mathsf{3} \to \mathsf{P}_\times$$

sending $\top$ to 0, k to 1, and $\bot$ to ∞ is a homomorphism of quantales.

II.1.11 Complete distributivity

As we saw in Section II.1.7, completeness of the ordered set X is characterized by the existence of an adjunction

$$\bigvee \dashv \downarrow : X \to \mathrm{Dn}\, X\ .$$

We call a complete lattice *completely distributive* if the left adjoint $\bigvee$ has itself a left adjoint, i.e. if there is a map

$$\Downarrow : X \to \mathrm{Dn}\, X$$

with

$$\Downarrow a \subseteq S \iff a \le \bigvee S$$

for all $a \in X$, $S \in \mathrm{Dn}\, X$. Necessarily,

$$\Downarrow a = \bigcap \left\{S \in \mathrm{Dn}\, X \mid a \le \bigvee S\right\}\ ,$$

so that when we write $x \ll a$ (read as: x is *totally below* a) instead of $x \in \Downarrow a$, this relation is given by

$$x \ll a \iff \forall S \in \mathrm{Dn}\, X\ (a \le \bigvee S \implies x \in S)\ .$$

Writing $S = \downarrow A$ with $A \subseteq X$, an equivalent characterization is given by

$$x \ll a \iff \forall A \subseteq X\ (a \le \bigvee A \implies \exists y \in A\ (x \le y))\ .$$

Since $\Downarrow$ is a monotone map with $1_X \le \bigvee \cdot \Downarrow$, one has for all $a, b, x \in X$:

(1) if $x \ll a \le b$, then $x \ll b$;
(2) $a \le \bigvee\{x \in X \mid x \ll a\}$.

In (2), one actually has equality (so that $a = \bigvee \Downarrow a$): from $a \le \bigvee \downarrow a$ follows $\Downarrow a \subseteq \downarrow a$ by adjunction, so that $\bigvee \Downarrow a \le a$. We also note that, by the very definition of $\ll$, every element in X is *$\ll$-atomic* in the sense that

$$x \ll \bigvee S \implies x \in S$$

for all $S \in \mathrm{Dn}\,X$, so that in (2) the join is taken "only" over $\ll$-atomic elements. Keeping this in mind, we see that the existence of the relation $\ll$ with properties (1) and (2) is characteristic of complete distributivity.

II.1.11.1 Proposition *If the complete lattice X allows for some relation $\prec$ satisfying*

(1′) *if $x \prec a \le b$, then $x \prec b$, and*
(2′) $a \le \bigvee\{x \in X \mid x$ *is $\prec$-atomic and* $x \prec a\}$

for all $a, b, x \in X$, then X is completely distributive.

Proof Indeed, assuming these conditions, and setting

$$S_a := \downarrow\{x \in X \mid x \text{ is } \prec\text{-atomic and } x \prec a\}\,,$$

one has $a \le \bigvee S_a$ by (2′), and, whenever $a \le \bigvee S$ for some $S \in \mathrm{Dn}\,X$, every $x \in S_a$ satisfies $x \prec \bigvee S$ by (1′), and therefore lies in S. Consequently,

$$S_a = \bigcap\left\{S \in \mathrm{Dn}\,X \mid a \le \bigvee S\right\} = \Downarrow a\,,$$

as desired. □

The complete lattices $\mathbf{2}$ and $[0, \infty]$ considered in Examples II.1.10.1 are completely distributive, with the "totally below" relation given by ($x \ll a \iff a =$ true) and ($x \ll a \iff x < a$), respectively. For every set X, *the ordered set $\mathsf{P}X$ is completely distributive*; more generally, for every ordered set X, *the ordered set $\mathrm{Dn}\,X$ is completely distributive* (see Exercise II.1.O). We note that $[0, \infty]^{\mathrm{op}}$ is also completely distributive (with "totally below" meaning ">"); in fact, one can prove that every chain is completely distributive and that *X^{op} is completely distributive whenever X has that property* (see [Wood, 2004]).

The term "completely distributive" still deserves some justification. By Corollary II.1.8.3, the map $\bigvee : \mathrm{Dn}\,X \longrightarrow X$ is right adjoint if it is an inf-map:

$$\bigvee\left(\bigcap\nolimits_{i\in I} S_i\right) = \bigwedge\nolimits_{i\in I} \bigvee S_i$$

for all families of down-sets S_i in X, $i \in I$; equivalently, if

$$\bigvee\left(\bigcap\nolimits_{i\in I} \downarrow A_i\right) = \bigwedge\nolimits_{i\in I} \bigvee A_i \qquad \text{(II.1.11.i)}$$

for all families of subsets A_i in X, $i \in I$. Now, assuming the Axiom of Choice, ©
the latter identity may be written equivalently as

$$\bigvee\nolimits_{(a_i)\in\prod_I A_i} \bigwedge\nolimits_{i\in I} a_i = \bigwedge\nolimits_{i\in I} \bigvee A_i\,. \qquad \text{(II.1.11.ii)}$$

Indeed, denoting by s, t the left sides of (II.1.11.i), (II.1.11.ii), respectively, $t \le s$ follows by noting that $(a_i)_{i\in I} \in \prod_{i\in I} A_i$ yields $\bigwedge_{i\in I} a_i \in \bigcap_{i\in I} \downarrow A_i$. Conversely, given $x \in \bigcap_{i\in I} \downarrow A_i$, for every $i \in I$, one has $a_i \in A_i$ with $x \le a_i$; by the Axiom of Choice, this defines an element $(a_i)_{i\in I} \in \prod_{i\in I} A_i$ with $x \le \bigwedge_{i\in I} a_i$, and $s \le t$ follows.

Most authors define complete distributivity via (II.1.11.ii) and reserve the name *constructively completely distributive* for the Choice-free notion given by (II.1.11.i), i.e. by the right adjointness of $\bigvee$.

II.1.12 Directed sets, filters, and ideals

A subset $A \subseteq Z$ of a separated ordered set Z is *down-directed* if every finite subset of A has a lower bound in A; this just means that for all $x, y \in A$ there is a $z \in A$ with $z \le x, z \le y$, and that $A \ne \varnothing$. A subset $A \subseteq Z$ is a *filter in* Z if A is down-directed and up-closed in Z; a filter A is *proper* if $A \ne Z$. When Z has finite infima, filters A in Z are up-closed meet-semilattices of Z; i.e. the filters are those A which satisfy

(1) $x, y \in A \implies x \wedge y \in A$,

(2) $\top \in A$, and

(3) $x \in A,\ x \le y \implies y \in A$

for all $x, y \in Z$. If Z has a bottom element $\bot$, properness then means

(4) $\bot \notin A$.

Every down-directed set A in Z generates the filter $\uparrow A$ in Z, in which case A is also called a *filter base* for $\uparrow A$. In particular, for every element $a \in Z$, one has the *principal filter* $\uparrow a$ in Z. The dual notions are those of *up-directed* set, *ideal, proper ideal, ideal base*, and *principal ideal*.

We use these notions predominantly for $Z = PX$ (for some set X), ordered by inclusion. Hence, $\mathfrak{a}$ is a *filter on* the set X if it is a filter *in* PX, i.e. $\mathfrak{a}$ must satisfy

(1′) $A, B \in \mathfrak{a} \implies A \cap B \in \mathfrak{a}$,

(2′) $X \in \mathfrak{a}$,

(3′) $A \in \mathfrak{a}, A \subseteq B \implies B \in \mathfrak{a}$,

and $\mathfrak{a}$ is proper when

(4′) $\varnothing \notin \mathfrak{a}$.

Here are some frequently used filter-generation procedures.

(a) For every map $f : X \to Y$ and every filter $\mathfrak{a}$ on X, one defines the *image filter* $f[\mathfrak{a}]$ on Y by

$$f[\mathfrak{a}] = \uparrow\{f(A) \mid A \in \mathfrak{a}\} = \{B \subseteq Y \mid f^{-1}(B) \in \mathfrak{a}\} .$$

For a filter $\mathfrak{b}$ on Y, one defines the *inverse image* $f^{-1}[\mathfrak{b}]$ by

$$f^{-1}[\mathfrak{b}] = \uparrow\{f^{-1}(B) \mid B \in \mathfrak{b}\} = \{A \subseteq X \mid \exists B \in \mathfrak{b}\ (f^{-1}(B) \subseteq A)\} ,$$

but this filter on X is proper only when $f^{-1}(B) \neq \varnothing$ for all $B \in \mathcal{b}$. If f is an inclusion map $X \hookrightarrow Y$, then

$$f^{-1}[\mathcal{b}] = \{X \cap B \mid B \in \mathcal{b}\} \iff X \in \mathcal{b} \,;$$

when $X \in \mathcal{b}$, one calls $\mathcal{b}|_X := f^{-1}[\mathcal{b}]$ the *restriction* of $\mathcal{b}$ to X, and says that $\mathcal{b}$ *is a filter on* X. In fact, $\mathcal{b}$ is then determined by $\mathcal{b}|_X$, since $\uparrow f[\mathcal{b}|_X] = \mathcal{b}$.

(b) For every $A \subseteq X$, one has the following *principal filter* on X:

$$\dot{A} = \uparrow A = \{B \subseteq X \mid A \subseteq B\} \,.$$

Obviously, for $f : X \to Y$, $f[\uparrow A] = \uparrow f(A)$.

(c) For a set X, consider a filter $\mathcal{A}$ on the set

$$FX = \{a \mid a \text{ is a filter on } X\} \,.$$

Then the *filtered sum* (or *Kowalsky sum*) $\sum \mathcal{A}$, defined by

$$A \in \sum \mathcal{A} \iff A^{\mathbb{F}} \in \mathcal{A} \,,$$

for all $A \subseteq X$, where $A^{\mathbb{F}} = \{a \in FX \mid A \in a\}$, gives a filter on X. Hence, a set $A \subseteq X$ lies in $\sum \mathcal{A}$ precisely when the set of those filters on X that are actually filters on A lies in $\mathcal{A}$.

II.1.13 Ultrafilters

An *ultrafilter* x on a set X is a *maximal* element within the set of proper filters on X, ordered by inclusion; i.e. x is a proper filter on X such that, if a is a proper filter on X with $x \subseteq a$, then $x = a$. A handier characterization is the following:

II.1.13.1 Lemma *For a proper filter x on X, the following statements are equivalent:*

(i) *x is an ultrafilter on X;*
(ii) *for all $A, B \subseteq X$, if $A \cup B \in x$ then $A \in x$ or $B \in x$;*
(iii) *for every subset $A \subseteq X$, one has $A \in x$ or $A^{\complement} \in x$ (where $A^{\complement} = X \setminus A$ denotes the complement of A in X).*

Proof (i) $\Longrightarrow$ (ii): If $A \cup B \in x$ but $A \notin x$ (for some $A, B \subseteq X$), then $x \subsetneq \uparrow\{A \cap C \mid C \in x\}$, so, by maximality of x, the right-hand side filter cannot be proper; hence, $A \cap C = \varnothing$ for some $C \in x$, and $(A \cup B) \cap C = B \cap C \in x$, so $B \cap C \subseteq B$ tells us that $B \in x$.

(ii) $\Longrightarrow$ (iii): Immediate from $X = A \cup A^{\complement} \in x$.

(iii) $\Longrightarrow$ (i): $x \subsetneq a$ implies that there is an $A \in a$ with $A \notin x$; thus, $A^{\complement} \in x \subsetneq a$ and $A \cap A^{\complement} = \varnothing \in a$, which therefore is not proper. □

The filter generation processes described in Section II.1.12 may be specialized to ultrafilters; more precisely:

(a) For a map $f : X \to Y$ and an ultrafilter $\mathfrak{x}$ on X, the image $f[\mathfrak{x}]$ is also an ultrafilter on Y. When $f : X \hookrightarrow Y$ is an inclusion map, and $\mathfrak{y}$ is an ultrafilter on Y with $X \in \mathfrak{y}$, then $\mathfrak{y}|_X$ is an ultrafilter on X.
(b) For every $x \in X$, the principal filter $\dot{x} = \uparrow\{x\}$ is an ultrafilter on X.
(c) If $\mathcal{X}$ is an ultrafilter on the set

$$\beta X = \{\mathfrak{x} \mid \mathfrak{x} \text{ is an ultrafilter on } X\} ,$$

then $\sum \mathcal{X}$ is an ultrafilter on X.

For the creation of other ultrafilters, one resorts to the Axiom of Choice:

© **II.1.13.2 Proposition** *Every proper filter $\mathfrak{a}$ on X is contained in an ultrafilter $\mathfrak{x}$ on X.*

Proof This is guaranteed by an easy application of Zorn's Lemma; see Exercise II.1.P. □

In fact, this statement can be used to formulate a formally finer assertion as follows.

© **II.1.13.3 Corollary** *For a filter $\mathfrak{b}$ and a proper filter $\mathfrak{a}$ on X such that $\mathfrak{a} \subsetneq \mathfrak{b}$, there is an ultrafilter $\mathfrak{x}$ on X with $\mathfrak{a} \subseteq \mathfrak{x}$ but $\mathfrak{b} \not\subseteq \mathfrak{x}$.*

Proof Indeed, for some $B \in \mathfrak{b}$ with $B \notin \mathfrak{a}$, one considers the filter

$$\mathfrak{a}' = \uparrow\{B^{\complement} \cap A \mid A \in \mathfrak{a}\} ,$$

which is proper since $B^{\complement} \cap A = \varnothing$ would imply $A \subseteq B \in \mathfrak{a}$. So there is an ultrafilter $\mathfrak{x}$ containing $\mathfrak{a}'$, and therefore also $\mathfrak{a}$; as $B^{\complement} \in \mathfrak{x}$, we must have $B \notin \mathfrak{x}$. □

As an important consequence, we obtain:

© **II.1.13.4 Corollary** *Every filter $\mathfrak{a}$ on X is the intersection of all ultrafilters on X containing $\mathfrak{a}$.*

Proof We must show that every filter $\mathfrak{a} \in FX$ may be obtained as

$$\mathfrak{a} = \bigcap \{\mathfrak{x} \in \beta X \mid \mathfrak{a} \subseteq \mathfrak{x}\} .$$

This equality holds trivially when $\varnothing \in \mathfrak{a}$, and the inclusion "$\subseteq$" is also immediate. Moreover, when $\mathfrak{a}$ is proper, so is the filter $\mathfrak{b}$ obtained on the right-hand side; therefore, if $\mathfrak{a} \subsetneq \mathfrak{b}$, there exists an ultrafilter $\mathfrak{x}$ with $\mathfrak{a} \subseteq \mathfrak{x}$ but $\mathfrak{b} \subsetneq \mathfrak{x}$, contradicting the definition of $\mathfrak{b}$. □

One obtains alternatively the following result.

© **II.1.13.5 Corollary** *For a proper filter $\mathfrak{a}$ and an ideal $\mathfrak{j}$ on X such that $\mathfrak{a} \cap \mathfrak{j} = \varnothing$, there is an ultrafilter $\mathfrak{x}$ with $\mathfrak{a} \subseteq \mathfrak{x}$ and $\mathfrak{x} \cap \mathfrak{j} = \varnothing$.*

Proof Since $\mathfrak{a}$ is an up-set, the fact that $\mathfrak{a}$ is disjoint from $\mathfrak{j}$ translates as $A \nsubseteq J$ for all $A \in \mathfrak{a}$ and $J \in \mathfrak{j}$, or equivalently as $A \cap J^{\complement} \neq \varnothing$ for all $A \in \mathfrak{a}, J \in \mathfrak{j}$. Thus, $\mathfrak{b} := \{A \cap J^{\complement} \mid A \in \mathfrak{a}, J \in \mathfrak{j}\}$ is a proper filter containing $\mathfrak{a}$, and Proposition II.1.13.2 yields the existence of an ultrafilter $\mathfrak{x}$ with $\mathfrak{b} \subseteq \mathfrak{x}$, and consequently $\mathfrak{a} \subseteq \mathfrak{x}$. If there were $J \in \mathfrak{j} \cap \mathfrak{x}$, one would conclude $J^{\complement} \notin \mathfrak{x}$, a contradiction. □

II.1.14 Natural and ordinal numbers

We end this introductory section with some foundational remarks. A *natural numbers object for sets* is a set N with a distinguished element 0 and a map $s : N \to N$ such that, for any set X equipped with a map $t : X \to X$ and an element $a \in X$, there is a unique map f making the diagram

$$\begin{array}{ccccc} 0 & & N & \xrightarrow{\;s\;} & N \\ \big\downarrow & & {\scriptstyle f}\,\big\downarrow & & \big\downarrow\,{\scriptstyle f} \\ a & & X & \xrightarrow{\;t\;} & X \end{array} \tag{II.1.14.i}$$

commutative. Such a set N must necessarily have the form

$$N = \{0\} \cup s(N) ,$$

and $\mathbb{N} = (N, s, 0)$ is uniquely determined up to a unique compatible bijection. Briefly, $\mathbb{N}$ is characterized by the requirement to allow for *inductive definitions*, via

$$f(0) = a , \quad f(s(n)) = t(f(n)) \quad (n \in \mathbb{N}) .$$

Alternatively, in categorical language (as introduced in Section II.2.7) $\mathbb{N}$ is *initial amongst all general algebras with one nullary and one unary operation* and no other requirements.

Defining the sets $\mathbb{N}_n$ recursively by

$$\mathbb{N}_0 = \{0\} , \quad \mathbb{N}_{s(n)} = \mathbb{N}_n \cup \{s(n)\} \quad (n \in \mathbb{N}) ,$$

one may define the natural order on $\mathbb{N}$ by

$$n \leq m \iff \mathbb{N}_n \subseteq \mathbb{N}_m .$$

It is the only order that makes $\mathbb{N}$ a chain with $n \leq s(n)$ for all $n \in \mathbb{N}$. Let us now assume that X in (II.1.14.i) is ordered and t is *pointed*, in the sense that $x \leq t(x)$ for all $x \in X$. Then the recursively defined function f is monotone. Consequently, $\mathbb{N}$ *is also initial amongst ordered general algebras with a nullary and unary operation that is pointed.* No such object exists within the realm of *finite* sets or algebras. In other words, the inductive condition may also be seen as an infinity axiom.

Missing a top element, $\mathbb{N}$ fails to be complete. It is therefore natural to ask: is there a separated complete ordered set $\mathcal{O}$ with a distinguished element $0 \in \mathcal{O}$ and a pointed operation $s : \mathcal{O} \to \mathcal{O}$ such that, for any separated complete ordered set X with a distinguished element $a \in X$ and pointed operation $t : \mathcal{O} \to \mathcal{O}$, there is a unique sup-map $f : \mathcal{O} \to X$ with $f(0) = a$ and $t(f(\alpha)) = f(s(\alpha))$ for all $\alpha \in \mathcal{O}$? Naively, such a set should have the form

$$\mathcal{O} = \{0, s(0), s(s(0)), \ldots, \omega = \sup_{n<\omega} n, s(\omega), s(s(\omega)), \ldots\} ,$$

with a considerably increased degree of uncertainty of how to "continue" the description of its elements, in comparison to the description of those of $\mathbb{N}$. In fact, just like the fact that a natural numbers object may not be found within the realm of finite sets, the desired object $\mathcal{O}$ would be "too big" to be a set. But, regardless of which foundational requirements one may adopt to govern the use of "sets," it is reasonable to require the existence of a *class* $\mathcal{O}$ with a separated complete order (so that it has all set-indexed suprema), a distinguished element 0, and a pointed unary operation that is initial amongst all *classes* X structured by a and t in the same way; in other words, such that maps $f : \mathcal{O} \to X$ may be uniquely defined by *ordinal recursion*, via

$$f(0) = a\ , \quad f(s(\alpha)) = t(f(\alpha))\ , \quad f(\sup_{i\in I} \alpha_i) = \sup_{i\in I} f(\alpha_i)$$

for all $\alpha, \alpha_i \in \mathcal{O}$, $i \in I$, and set I.

We will apply this ordinal recursion principle only twice in this book, for the proof of the main result of III.4.2 and for Proposition V.4.4.9, with rather limited impact on other results.

Exercises

II.1.A *Universal property of the separated reflection.* Every monotone map $f : X \to Y$ of ordered sets, with Y separated, factors through the separated reflection $p : X \to X/{\simeq}$ by a uniquely determined monotone map $g : X/{\simeq} \to Y$, so that $f = g \cdot p$.

II.1.B *Adjunctions and fully faithful maps.* For an adjunction $f \dashv g : Y \to X$, the map g is fully faithful if and only if $f \cdot g \simeq 1_Y$. If $h \dashv f \dashv g$, then g is fully faithful if and only if h is fully faithful.

II.1.C *Adjunctions for free.* For a module $r : X \nrightarrow Y$ and an ordered set Z, denote by r_Z the map

$$\mathsf{Mod}(Z, X) \to \mathsf{Mod}(Z, Y)\ , \quad t \mapsto r \cdot t\ ,$$

where $\mathsf{Mod}(Z, X)$ denotes the ordered set of modules $Z \nrightarrow X$. With a module $s : Y \nrightarrow X$, show

$$\forall Z\ (r_Z \dashv s_Z) \iff 1_X^* \le s \cdot r \ \&\ r \cdot s \le 1_Y^*\ .$$

Conclude that for every monotone map $f : X \longrightarrow Y$ and every ordered set Z, one has an adjunction $(f_*)_Z \dashv (f^*)_Z$.

II.1.D *Closure operations on complete ordered sets.* For every closure operation c on an ordered set X, there is an adjunction $e \dashv j : \mathrm{Fix}(c) \hookrightarrow X$. If X is complete, so is $\mathrm{Fix}(c)$, with infima in $\mathrm{Fix}(c)$ formed as in X, while the supremum of $A \subseteq \mathrm{Fix}(c)$ in $\mathrm{Fix}(c)$ is given by $c(\bigvee A)$.

II.1.E *Complete chains are frames.* For elements a, b in a chain X, put

$$a \longrightarrow b := \begin{cases} \top & \text{if } a \leq b, \\ b & \text{otherwise.} \end{cases}$$

Then

$$x \wedge a \leq b \iff x \leq (a \longrightarrow b)\ .$$

Hence, every complete chain is a frame.

II.1.F *Topological spaces as closure spaces.* Every topological space X becomes a closure space via its *Kuratowski closure* operation:

$$\begin{aligned}\overline{A} &= \bigcap \{B \subseteq X \mid A \subseteq B,\, B^{\complement} \in \mathcal{O}X\} \\ &= \{x \in X \mid \forall U \in \mathcal{O}X\ (x \in U \implies A \cap U \neq \varnothing)\}\end{aligned}$$

for all $A \subseteq X$. The map $\overline{(-)} : PX \longrightarrow PX$ is not only a closure operation, but is also *finitely additive*:

$$\overline{A \cup B} = \overline{A} \cup \overline{B}\,, \quad \overline{\varnothing} = \varnothing$$

for all $A, B \subseteq X$. Hence, the map $\overline{(-)}$ is a monoid homomorphism of the join-semilattice PX. A map $f : X \longrightarrow Y$ between topological spaces is continuous if and only if it is continuous as a map of closure spaces. The underlying order may then be written as

$$x \leq y \iff y \in \overline{\{x\}}\ .$$

Of course, since closure spaces and interior spaces are equivalent concepts, every topological space also becomes an interior space (whose interior operation preserves finite intersections).

II.1.G *Closure spaces as topological spaces.* For a set X, any finitely additive closure operation c on PX defines the topology

$$\{A \subseteq X \mid c(A^{\complement}) = A^{\complement}\}$$

on X, and this establishes a bijective correspondence between topologies on X and finitely additive closure operations on PX. The bijective correspondence between topologies on X and interior operations on PX that preserve finite intersections is even more immediate, as it avoids the use of $(-)^{\complement}$.

II.1.H *A simple non-linear quantale.* The *diamond lattice* $\mathsf{2}^2 = \{\bot, u, v, \top\}$, with u and v two incomparable elements, is a frame, and as such is obviously isomorphic to the powerset of $\mathsf{2}$. Compute $a \multimap b$ for all $a, b \in \mathsf{2}^2$.

II.1.I *Lax homomorphisms from the extended real half-line.* There are uniquely determined monotone maps $o, p : [0, \infty]^{\mathrm{op}} \longrightarrow \mathsf{2}$ with

$$o \dashv \iota \dashv p\,,$$

where $\iota(\bot) = \infty$ and $\iota(\top) = 0$. For the "optimist's map," one has $o(v) = \top$ for all $v < \infty$, while for the "pessimist's map," one has $p(v) = \top$ only when $v = 0$. When $[0, \infty]^{\mathrm{op}}$ is considered as a quantale P_{+}, $\mathsf{P}_{\times}$, or $\mathsf{P}_{\max}$, which of these maps turns into lax homomorphisms, and, among these, which are in fact homomorphisms?

II.1.J *Right and left adjoints to* $\mathrm{Dn}\,X \hookrightarrow PX$. For an ordered set X, the inclusion $\mathrm{Dn}\,X \hookrightarrow PX$ has a left adjoint given by the down-closure $\downarrow$, and a right adjoint given by the *down-interior* $\mathring{\downarrow}$, where

$$\mathring{\downarrow} A = \{x \in X \mid \downarrow x \subseteq A\}$$

for $A \subseteq X$. In particular, $\mathrm{Dn}\,X$ is a complete lattice with meets and joins formed as in PX.

II.1.K *Right and left adjoints to* $f^{-1}(-) : \mathrm{Dn}\,Y \longrightarrow \mathrm{Dn}\,X$. For a monotone map $f : X \longrightarrow Y$ of ordered sets, the map $f^{-1}(-) : \mathrm{Dn}\,Y \longrightarrow \mathrm{Dn}\,X$ has a left adjoint $f_!$ given by $f_!(A) = \downarrow f(A)$, and a right adjoint f_* given by $f_*(A) = \mathring{\downarrow} \bigvee\{B \subseteq Y \mid f^{-1}(B) \subseteq A\}$ for all $A \subseteq X$. In particular, $f^{-1}(-)$ is a homomorphism of frames.

II.1.L *Semilattices as monoids; frames as quantales.* Let X be a separated ordered set which is also a (multiplicatively written) commutative monoid in which the multiplication is monotone in each variable. If every element in the monoid X is idempotent, and the neutral element is the top element, the ordered set X is a meet-semilattice: the order and the monoid structure determine each other via

$$(x \le y \iff x \cdot y = x) \qquad \text{and} \qquad x \cdot y = x \wedge y\,.$$

In particular, for a commutative quantale $(\mathcal{V}, \otimes, k)$ with $k = \top$ and $v \otimes v = v$ for all $v \in \mathcal{V}$, one has $\otimes = \wedge$.

II.1.M *Quantales arising from monoids.* For a monoid $M = (M, m, e)$, the powerset PM becomes a monoid with

$$B \cdot A = \{y \cdot x \mid x \in A, y \in B\} \qquad (A, B \subseteq M)$$

and neutral element $\{e\}$. Ordered by inclusion, PM is a quantale which is commutative precisely when M is commutative. For a homomorphism $f : M \to N$ of monoids, $f(-) : PM \to PN$ is a homomorphism of quantales.

II.1.N *Complete lattices as frames.* A complete lattice X is a frame if and only if $\bigvee : \mathrm{Dn}X \to X$ is a meet-semilattice homomorphism. Every completely distributive lattice is a frame.

II.1.O *Complete distributivity of* $\mathrm{Dn}X$. For every ordered set X, $\mathrm{Dn}X$ is completely distributive. In particular, the powerset PX of a set X is completely distributive.
Hint. The join map of $\mathrm{Dn}X$ coincides with $(\downarrow_X)^{-1}(-) : \mathrm{Dn}\mathrm{Dn}X \to \mathrm{Dn}X$ (where $\downarrow_X : X \to \mathrm{Dn}X$), and by Exercise II.1.K it has the left adjoint $(\downarrow_X)_!$.

II.1.P *Zorn's Lemma and ultrafilters.* *Zorn's Lemma*, which is in fact *equivalent* to the Axiom of Choice (for an elementary account, see, for example, [Davey and Priestley, 1990]), states: ©

> If X is a separated ordered set and every chain $A \subseteq X$ has an upper bound in X, then X has a *maximal* element, that is, an element $x \in X$ such that if $y \in X$ satisfies $x \leq y$, then $x = y$.

As a consequence, every proper filter $\mathfrak{a}$ on a set X is contained in an ultrafilter $\mathfrak{x}$ on X.

II.1.Q *Convergence of sequences.* Given a *sequence* $s : \mathbb{N} \to X$ in X (also denoted by $s = (x_n)_{n\in\mathbb{N}}$), one can consider the sets $S_n := \{x_m \in X \mid n \leq m\}$, and the *filter associated* to s

$$\langle s \rangle := \uparrow_{PX}\{S_n \mid n \in \mathbb{N}\}\,.$$

If X is a topological space (with topology $\mathcal{O}X$), then the *neighborhood filter* at $x \in X$ is the filter

$$\nu(x) := \uparrow_{PX}\{U \in \mathcal{O}X \mid x \in U\}\,.$$

A *neighborhood* of $x \in X$ is an element of $\nu(x)$. Show that the open sets of X are exactly the subsets V that are neighborhoods of each of their points, and that a sequence s *converges* to a point x, i.e.

$$\forall V \in \nu(x)\, \exists n \in \mathbb{N}\, \forall m \geq n\ (x_m \in V)$$

if and only if $\langle s \rangle \supseteq \nu(x)$. Similarly, if $s' = (x_{n_i})_{i\in\mathbb{N}}$ is a *subsequence* of s (i.e. $\langle s' \rangle \supseteq \langle s \rangle$), and s converges to x, then s' also converges to x. Finally, the *constant sequence* $s_x : \mathbb{N} \to X$ at x (so that $s_x(\mathbb{N}) = \{x\}$) converges to x for any topology on X.

II.1.R *Continuity and sequences.* The image of a sequence $s : \mathbb{N} \to X$ under a map $f : X \to Y$ is the sequence $f \cdot s : \mathbb{N} \to Y$, or equivalently the sequence

$(f(x_n))_{n\in\mathbb{N}}$ (where $s = (x_n)_{n\in\mathbb{N}}$). The filter associated to $f \cdot s$ is then the image filter of $\langle s \rangle$:

$$f[\langle s \rangle] = \langle f \cdot s \rangle \ .$$

If $f : X \to Y$ is a continuous map between topological spaces X and Y, and a sequence s converges to $x \in X$ (in symbols $\langle s \rangle \to x$), then $f \cdot s$ converges to $f(x)$:

$$\langle s \rangle \to x \implies f[\langle s \rangle] \to f(x) \ ;$$

in other words, *continuous maps preserve convergence of sequences.*

II.1.S *Sequences do not suffice.* In general, topological concepts are not faithfully represented by convergence of sequences. For example, consider the real line $\mathbb{R}$ equipped with the topology given by complements of countable subsets:

$$\mathcal{O}\mathbb{R} := \{U \subseteq \mathbb{R} \mid U^{\complement} \text{ is countable}\} \ .$$

In this topology, a sequence $s = (x_n)_{n\in\mathbb{N}}$ converges to a point x if and only if s is *eventually constant*:

$$\langle s \rangle \to x \iff \exists n \in \mathbb{N}\, \forall m \geq n \ (x_m = x) \ .$$

The identity map $1_{\mathbb{R}} : (\mathbb{R}, \mathcal{O}\mathbb{R}) \to (\mathbb{R}, P\mathbb{R})$ preserves convergence of sequences, but is not continuous (the converse statement is true, however; see Exercise II.1.R).

II.2 Categories and adjunctions

II.2.1 Categories

A *small category* C is given by a set $\mathrm{ob}\,\mathsf{C}$ of *objects* of C, a map $\hom_{\mathsf{C}}$ which assigns to each pair of objects (A, B) a set $\hom_{\mathsf{C}}(A, B)$, called their *hom-set* and more briefly written as $\mathsf{C}(A, B)$, as well as *composition* and *identity* operations:

$$\begin{array}{rcl} \mathsf{C}(A,B) \times \mathsf{C}(B,C) \to \mathsf{C}(A,C) \\ (f,g) \mapsto g \cdot f \end{array} \ , \quad \begin{array}{rcl} \{\star\} \to \mathsf{C}(A,A) \\ \star \mapsto 1_A \end{array}$$

subject to the *associativity* and *right* and *left identity laws*

$$h \cdot (g \cdot f) = (h \cdot g) \cdot f \ , \quad f \cdot 1_A = f = 1_B \cdot f$$

for all $f \in \mathsf{C}(A, B)$, $g \in \mathsf{C}(B, C)$, $h \in \mathsf{C}(C, D)$, $A, B, C, D \in \mathrm{ob}\,\mathsf{C}$. Given objects A, B, one writes

$$f : A \to B$$

instead of $f \in \mathsf{C}(A, B)$, and calls $A = \mathrm{dom}\, f$ the *domain* and $B = \mathrm{cod}\, f$ the *codomain* of the *morphism* (or *arrow*) f in C, with the understanding that *two morphisms f, g can coincide only if*

$$\mathrm{dom}\, f = \mathrm{dom}\, g \qquad \text{and} \qquad \mathrm{cod}\, f = \mathrm{cod}\, g \ .$$

The morphism $f : A \longrightarrow B$ in C is an *isomorphism* of C if $f' \cdot f = 1_A$ and $f \cdot f' = 1_B$ for a (necessarily uniquely determined) morphism $f' : B \longrightarrow A$ in C. One has the self-explanatory rules

$$(g \cdot f)^{-1} = f^{-1} \cdot g^{-1}, \quad 1_A^{-1} = 1_A, \quad (f^{-1})^{-1} = f.$$

An equivalence relation on $\mathrm{ob}\,\mathsf{C}$ is defined by writing $A \cong B$ if there is an isomorphism $f : A \longrightarrow B$ in C. An order on $\mathrm{ob}\,\mathsf{C}$ is obtained by setting $A \leq B$ if there exists a morphism $f : A \longrightarrow B$. In this case, $A \cong B$ implies $A \simeq B$ (meaning $A \leq B$ and $B \leq A$, as in II.1.3), but the converse is not true in general.

In Section II.1, we encountered two important types of small categories. First, every monoid M becomes a one-object category C when one puts $\mathsf{C}(\star, \star) = M$ (with $\star$ denoting the only object of C) and interprets monoid operation as composition. In fact, *monoids can be thought of as precisely the one-object small categories*, and small categories are simply "multi-object monoids." Second, every ordered set X becomes a small category C with $\mathrm{ob}\,\mathsf{C} = X$ and $\mathsf{C}(x, y) = \{\star\}$ if $x \leq y$ in X, and $\mathsf{C}(x, y) = \varnothing$ otherwise; the composition and identity operations appear as the transitivity and reflexivity properties in X. Note that in this case the order on C defined above coincides with that of X. This shows that ordered sets can be thought of as precisely those small categories for which the hom map is $\{\varnothing, \{\star\}\}$-valued, and small categories are simply sets with "multi-valued structured orders."

Arbitrary *categories* are defined just like small categories, but one allows $\mathrm{ob}\,\mathsf{C}$ and $\mathsf{C}(A, B)$ to be *classes* rather than just sets (for all $A, B \in \mathrm{ob}\,\mathsf{C}$). The understanding here is that a set is a particular class, and that we are allowed to form the class of all sets (see I.3). For every category C, the *opposite* or *dual* category C^{op} of C is given by

$$\mathrm{ob}\,\mathsf{C}^{\mathrm{op}} = \mathrm{ob}\,\mathsf{C}, \quad \mathsf{C}^{\mathrm{op}}(A, B) = \mathsf{C}(B, A), \quad \text{and} \quad g \cdot_{\mathsf{C}^{\mathrm{op}}} f = f \cdot_{\mathsf{C}} g.$$

Here are some examples of categories that we encountered (implicitly) in Section II.1:

- Set the category whose objects are *sets*, and whose morphisms are *maps* with ordinary composition of maps
- Rel objects are *sets*, but morphisms are *relations* with ordinary relational composition
- Ord *ordered sets*, with *monotone maps* and ordinary map composition
- Mod *ordered sets*, with *modules* and ordinary relational composition
- $\mathsf{Ord}_{\mathrm{sep}}$ *separated* ordered sets, with *monotone maps* and, as in all the following examples, ordinary map composition; a more common name for this category is PoSet, which stands for "partially ordered sets"
- Sup *complete lattices*, with *sup-maps*
- Inf *complete lattices*, with *inf-maps*

Mon *monoids*, with *monoid homomorphisms*
SLat *semilattices*, with *semilattice homomorphisms*
Lat *lattices*, with *lattice homomorphisms*
Frm *frames*, with *frame homomorphisms*
Qnt *quantales*, with *quantale homomorphisms*
Top *topological spaces*, with *continuous maps*
Cls *closure spaces*, with *continuous maps*
Int *interior spaces*, with *continuous maps*

II.2.2 Functors

A *functor* $F : \mathsf{C} \to \mathsf{D}$ from a category C to a category D is given by functions

$$F : \mathrm{ob}\,\mathsf{C} \to \mathrm{ob}\,\mathsf{D} \qquad \text{and} \qquad F_{A,B} : \mathsf{C}(A, B) \to \mathsf{D}(FA, FB)$$

for all $A, B \in \mathrm{ob}\,\mathsf{C}$, with (when writing $Ff = F_{A,B}(f)$)

$$F(g \cdot f) = Fg \cdot Ff \,, \quad F1_A = 1_{FA}$$

for all $f : A \to B$, $g : B \to C$ in C. There is an obvious *identity functor* $1_{\mathsf{C}} : \mathsf{C} \to \mathsf{C}$, and the *composition* GF of two functors $F : \mathsf{C} \to \mathsf{D}$ and $G : \mathsf{D} \to \mathsf{E}$ is defined by using composition of functions. The functor F is *faithful* if all functions $F_{A,B}$ are injective, and it is *full* if these functions are surjective. *Fully faithful* functors (functors that are both full and faithful) typically arise when forming a *full subcategory* D of a category C, so that $\mathrm{ob}\,\mathsf{D} \subseteq \mathrm{ob}\,\mathsf{C}$, $\mathsf{D}(A, B) = \mathsf{C}(A, B)$ for all $A, B \in \mathrm{ob}\,\mathsf{D}$, and composition in D is as in C; then the *inclusion functor* $F : \mathsf{D} \hookrightarrow \mathsf{C}$ is full and faithful. Here, the functor F has the additional property that its object function is injective, which makes F a *full embedding*. Small categories and functors form a category Cat. With suitable foundational provisions, we may also form the *metacategory* CAT of all categories and functors.

Functors between monoids (considered as categories) are precisely monoid homomorphisms, and functors between ordered sets are precisely monotone maps; in this last case, the "fully faithful" terminology coincides with that introduced in Section II.1.3 for ordered sets. Many functors also arise as *forgetful functors*, forgetting properties or structures of objects and morphisms, such as in the sequence

$$\mathsf{Qnt} \to \mathsf{Sup} \to \mathsf{Ord} \to \mathsf{Set}\,.$$

Some other functors that we implicitly encountered in Section II.1 include the following.

- The *covariant powerset functor* $P : \mathsf{Set} \to \mathsf{Set}$, with $Pf(A) = f(A)$ for $f : X \to Y$, $A \subseteq X$.

- The *contravariant powerset functor* $P^\bullet : \mathsf{Set}^{\mathrm{op}} \to \mathsf{Set}$, with $P^\bullet f(B) = f^{-1}(B)$ for $f : X \to Y$, $B \subseteq Y$.
- The *down-set functor* $\mathrm{Dn} : \mathsf{Ord} \to \mathsf{Ord}$, with $\mathrm{Dn}\, f(A) = \downarrow f(A)$ for $f : X \to Y$, $A \subseteq X$ (see Section II.1.7).
- The *open-set functor* $\mathcal{O} : \mathsf{Top}^{\mathrm{op}} \to \mathsf{Frm}$ (see Section II.1.9).
- The *underlying-order functor* $\mathsf{Top} \to \mathsf{Ord}$ (see Section II.1.9).
- The *Kuratowski-closure functor* $\mathsf{Top} \to \mathsf{Cls}$ (see Exercise II.1.F).
- The functor $\mathsf{Cls} \to \mathsf{Int}$ arising from two-fold complementation (see Section II.1.6), which is an isomorphism in CAT.
- The *dualization functor* $(-)^{\mathrm{op}} : \mathsf{Ord} \to \mathsf{Ord}$ (see Section II.1.3). Since $(X^{\mathrm{op}})^{\mathrm{op}} = X$, this functor is an isomorphism (in CAT), and extends to an isomorphism $(-)^{\mathrm{op}} : \mathsf{CAT} \to \mathsf{CAT}$ (see Section II.2.1). It maps a functor $F : \mathsf{C} \to \mathsf{D}$ to $F^{\mathrm{op}} : \mathsf{C}^{\mathrm{op}} \to \mathsf{D}^{\mathrm{op}}$, where $F^{\mathrm{op}} f = Ff$. A restriction of the dualization functor yields an isomorphism $\mathsf{Inf} \to \mathsf{Sup}$ in CAT.
- The *adjunction functor* $\mathsf{Sup}^{\mathrm{op}} \to \mathsf{Inf}$, which maps objects identically and assigns to a sup-map its right adjoint (see Corollary II.1.8.3); again this is an isomorphism in CAT. Composition with the isomorphism $\mathsf{Inf} \to \mathsf{Sup}$ shows that Sup is *self-dual*: $\mathsf{Sup}^{\mathrm{op}} \cong \mathsf{Sup}$; likewise for Inf.
- The *module functors* $(-)_* : \mathsf{Ord} \to \mathsf{Mod}$ and $(-)^* : \mathsf{Ord}^{\mathrm{op}} \to \mathsf{Mod}$, which map objects identically and assign to a monotone map f the induced modules f_* and f^*, respectively.

For every object A in a category C, one has the *covariant hom-functor* of A

$$\mathsf{C}(A, -) : \mathsf{C} \to \mathsf{SET}$$

(where the objects of the metacategory SET are classes), with

$$\begin{aligned} \mathsf{C}(A,-)(B) &:= \mathsf{C}(A,B)\,, \\ \mathsf{C}(A,-)(g) &:= \mathsf{C}(A,g) : \mathsf{C}(A,B) \to \mathsf{C}(A,C)\,, \quad f \mapsto g \cdot f \end{aligned}$$

for all $f : A \to B$, $g : B \to C$ in C. Similarly, one has the *contravariant hom-functor* of A

$$\mathsf{C}(-, A) = \mathsf{C}^{\mathrm{op}}(A, -) : \mathsf{C}^{\mathrm{op}} \to \mathsf{SET}\,.$$

These functors take value in Set if C is *locally small*, i.e. if all $\mathsf{C}(A, B)$ are sets. Note that when C is a monoid M seen as a one-object category, the hom-functor $M(\star, -) : M \to \mathsf{Set}$ has as values those maps $M \to M$ which are the "left translations by a" ($a \in M$); with a suitable codomain restriction, $M(\star, -)$ is in most algebra books referred to as the *Cayley representation* of M, although M is normally assumed to be a *group* (i.e. a monoid in which every element is invertible). When C is an ordered set X and $x \in X$, then $X(x, -)$ is two-valued and therefore the characteristic function of a subset of X, namely $\uparrow_X x$.

II.2.3 Natural transformations

Let $F, G : \mathsf{C} \to \mathsf{D}$ be functors. A *natural transformation* $\alpha : F \to G$, also depicted by

$$\mathsf{C} \underset{G}{\overset{F}{\underset{\Downarrow\alpha}{\rightrightarrows}}} \mathsf{D}\,,$$

is given by a family of morphisms $\alpha_A : FA \to GA$ in D (with A running through $\mathrm{ob}\,\mathsf{C}$), making the *naturality diagrams*

$$\begin{array}{ccc} FA & \xrightarrow{Ff} & FB \\ {\scriptstyle \alpha_A}\downarrow & & \downarrow{\scriptstyle \alpha_B} \\ GA & \xrightarrow{Gf} & GB \end{array}$$

commute for all $f : A \to B$ in C; hence, writing $\alpha_f : FA \to GB$ for the *diagonal morphism*, we have

$$\alpha_f = Gf \cdot \alpha_A = \alpha_B \cdot Ff\ .$$

For example, forming singleton sets is a natural transformation $1_{\mathsf{Set}} \to P$ (where P is the powerset functor), and forming unions is a natural transformation $PP \to P$. The down-set function defines a natural transformation $\downarrow\, : 1_{\mathsf{Ord}} \to \mathrm{Dn}$, and when we restrict Dn to a functor $\mathrm{Dn} : \mathsf{Sup} \to \mathsf{Sup}$, there is a natural transformation $\bigvee : \mathrm{Dn} \to 1_{\mathsf{Sup}}$.

Vertical pasting of naturality diagrams defines the *vertical composition*

$$\beta \cdot \alpha : F \to H \quad \text{with} \quad (\beta \cdot \alpha)_A = \beta_A \cdot \alpha_A \quad (A \in \mathrm{ob}\,\mathsf{C})\,,$$

for $\alpha : F \to G$, $\beta : G \to H$, and $F, G, H : \mathsf{C} \to \mathsf{D}$; one also has the identity transformation

$$1_F : F \to F \quad \text{with} \quad (1_F)_A = 1_{FA} \quad (A \in \mathrm{ob}\,\mathsf{C})\,.$$

The vertical composition is trivially associative, and makes the functors from C to D the objects of the *functor category* D^{C} whose morphisms are natural transformations. (We note that unless C is small, D^{C} is actually a metacategory.) For functors $S : \mathsf{B} \to \mathsf{C}$ and $T : \mathsf{D} \to \mathsf{E}$, one has the *whiskering functors*

$$\begin{array}{c} (-)S : \mathsf{D}^{\mathsf{C}} \to \mathsf{D}^{\mathsf{B}} \\ (\alpha : F \to G) \mapsto (\alpha S : FS \to GS) \end{array} \quad \text{and} \quad \begin{array}{c} T(-) : \mathsf{D}^{\mathsf{C}} \to \mathsf{E}^{\mathsf{C}} \\ (\alpha : F \to G) \mapsto (T\alpha : TF \to TG) \end{array}$$

which map natural transformations via

$$(\alpha S)_B = \alpha_{SB}\,, \quad (T\alpha)_A = T\alpha_A \quad (B \in \mathrm{ob}\,\mathsf{B},\ A \in \mathrm{ob}\,\mathsf{C})\,,$$

as suggested by

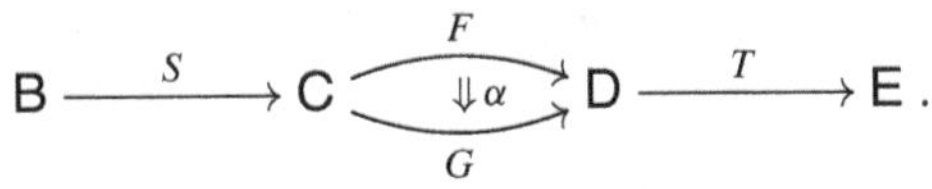

These whiskering functors are of course induced by either $\mathsf{CAT}(-, \mathsf{D})$ or $\mathsf{CAT}(\mathsf{C}, -)$ applied to $S : \mathsf{B} \to \mathsf{C}$ or $T : \mathsf{D} \to \mathsf{E}$, respectively. In fact, whiskering by a functor (from the left or the right) is just an instance of the *horizontal composition* for natural transformations

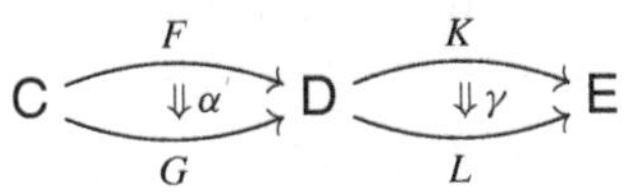

defined by

$$\gamma \circ \alpha : KF \to LG\,, \quad (\gamma \circ \alpha)_A = \gamma_{\alpha_A} \quad (A \in \mathrm{ob}\,\mathsf{C})\,.$$

For a morphism f in C, one obtains

$$(\gamma \circ \alpha)_f = \gamma_{\alpha_f}\,.$$

Moreover, since

$$\begin{aligned}(\gamma \circ \alpha)_A &= L\alpha_A \cdot \gamma_{FA} = \gamma_{GA} \cdot K\alpha_A \\ &= (L\alpha)_A \cdot (\gamma F)_A = (\gamma G)_A \cdot (K\alpha)_A\,,\end{aligned}$$

the natural transformation $\gamma \circ \alpha$ is completely determined by the values of the whiskering functors on α and γ:

$$\gamma \circ \alpha = L\alpha \cdot \gamma F = \gamma G \cdot K\alpha\,.$$

Equivalently, since

$$\gamma \circ \alpha = (1_L \circ \alpha) \cdot (\gamma \circ 1_F) = (\gamma \circ 1_G) \cdot (1_K \circ \alpha)\,, \tag{II.2.3.i}$$

the horizontal composition is determined by the special instances of composition with identity transformations. The equality (II.2.3.i) is itself a special instance of the *middle-interchange law*; i.e. given

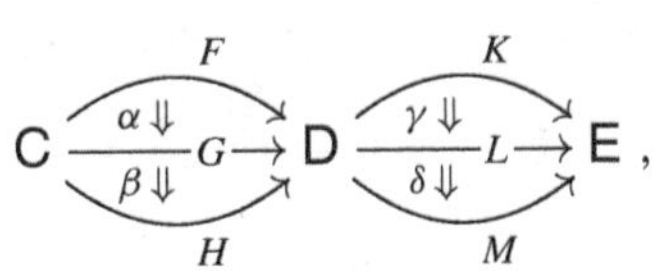

one has

$$(\delta \cdot \gamma) \circ (\beta \cdot \alpha) = (\delta \circ \beta) \cdot (\gamma \circ \alpha)\,.$$

The horizontal composition is associative, and $\alpha \circ 1_{1_\mathsf{C}} = \alpha = 1_{1_\mathsf{D}} \circ \alpha$. With $\alpha : F \to G$, we also have a natural transformation $\alpha^{\mathrm{op}} : G^{\mathrm{op}} \to F^{\mathrm{op}}$, where $(\alpha^{\mathrm{op}})_A = \alpha_A$ is an arrow in D^{op}, for all $A \in \mathrm{ob}\,\mathsf{C}$.

Finally, α is a *natural isomorphism* (i.e. an isomorphism in D^{C}) if and only if every morphism α_A is an isomorphism in D, and then one has $(\alpha^{-1})_A = (\alpha_A)^{-1}$, briefly written as α_A^{-1}.

II.2.4 The Yoneda embedding

For every morphism $g : B \longrightarrow C$ in a category C, one has a natural transformation

$$\mathsf{C}(-, g) : \mathsf{C}(-, B) \to \mathsf{C}(-, C)$$

of contravariant hom-functors. When C is locally small, there is therefore a functor

$$\mathbf{y} : \mathsf{C} \longrightarrow \mathsf{Set}^{\mathsf{C}^{\mathrm{op}}}, \quad g \mapsto \mathbf{y}(g) = \mathsf{C}(-, g) .$$

This functor is faithful, since $\mathsf{C}(-, g) = \mathsf{C}(-, h)$ (for $g, h : B \longrightarrow C$) yields $g = \mathsf{C}(B, g)(1_B) = \mathsf{C}(B, h)(1_B) = h$. Moreover, on objects $\mathbf{y}$ is one-to-one: if $\mathbf{y}B = \mathbf{y}C$, then $1_B \in \mathsf{C}(B, B) = \mathsf{C}(B, C)$ and $B = \mathrm{cod}\, 1_B = C$. It is less obvious that $\mathbf{y}$ is also full, so that

$$\mathbf{y} : \mathsf{C} \to \widehat{\mathsf{C}} = \mathsf{Set}^{\mathsf{C}^{\mathrm{op}}}$$

is a full embedding, called the *Yoneda embedding*. As a preparation for that, we first prove the *Yoneda Lemma.*

II.2.4.1 Lemma *For every object A of a locally small category C, and every functor $F : \mathsf{C}^{\mathrm{op}} \longrightarrow \mathsf{Set}$, the map*

$$\eta : \widehat{\mathsf{C}}(\mathbf{y}A, F) \longrightarrow FA , \quad \alpha \mapsto \alpha_A(1_A)$$

is bijective.

Proof Routine verification shows that the map

$$\delta : FA \longrightarrow \widehat{\mathsf{C}}(\mathbf{y}A, F) ,$$

with $\delta(a)_B : \mathsf{C}(B, A) \longrightarrow FB$ sending f to $Ff(a)$, for all $a \in FA$, $B \in \mathrm{ob}\,\mathsf{C}$, is well defined and inverse to η. □

In the special case where $F = \mathbf{y}B$ for $B \in \mathrm{ob}\,\mathsf{C}$, the map

$$\delta : \mathsf{C}(A, B) \longrightarrow \widehat{\mathsf{C}}(\mathbf{y}A, \mathbf{y}B)$$

is precisely the hom map $\mathbf{y}_{A,B}$ of the functor $\mathbf{y}$. Hence:

II.2.4.2 Corollary *The functor $\mathbf{y} : \mathsf{C} \to \widehat{\mathsf{C}}$ is a full embedding.*

Consequently, every locally small category may be considered as a full subcategory of a functor category over Set.

As we saw at the end of Section II.2.2 in the dual situation, when C is an ordered set X and $x \in X$, the functor $\mathbf{y}(x) = X(-, x)$ may be identified with the down-set $\downarrow_X x$. But the map $\downarrow_X : X \to \mathrm{Dn}X$ is injective only when X is separated, and in that case $\downarrow_X$ is just a codomain restriction of $\mathbf{y}$.

II.2.5 Adjunctions

In generalization of the notion introduced in Section II.1.5, one calls a functor $G : \mathsf{A} \to \mathsf{X}$ *right adjoint* if there is a functor $F : \mathsf{X} \to \mathsf{A}$ and natural transformations $\eta : 1_{\mathsf{X}} \to GF$ and $\varepsilon : FG \to 1_{\mathsf{A}}$ satisfying the *triangular identities*

$$G\varepsilon \cdot \eta G = 1_G \qquad \text{and} \qquad \varepsilon F \cdot F\eta = 1_F\ .$$

These data are, up to isomorphism, uniquely determined by G. Indeed, if F', η', ε' satisfy the corresponding conditions, then

$$\alpha = (F \xrightarrow{F\eta'} FGF' \xrightarrow{\varepsilon F'} F')\ , \qquad \beta = (F' \xrightarrow{F'\eta} F'GF \xrightarrow{\varepsilon' F} F)$$

are inverse to each other, as one shows by repeated applications of the middle-interchange law; for example:

$$\begin{aligned}
\beta\cdot\alpha &= \varepsilon' F\cdot F'\eta\cdot \varepsilon F'\cdot F\eta' & \\
&= \varepsilon' F\cdot \varepsilon F' GF\cdot FGF'\eta\cdot F\eta' & \text{(middle-interchange)}\\
&= \varepsilon F\cdot FG\varepsilon' F\cdot F\eta' GF\cdot F\eta & \text{(middle-interchange, twice)}\\
&= \varepsilon F\cdot 1_{FGF}\cdot F\eta & \text{(triangular identity)}\\
&= 1_F & \text{(triangular identity).}
\end{aligned}$$

Furthermore, $\eta' = G\alpha\cdot\eta$ and $\varepsilon' = \varepsilon\cdot\beta G$. One calls F *left adjoint* to G, and η is the *unit*, while ε is the *counit* of the *adjunction*, denoted by

$$F \underset{\varepsilon}{\overset{\eta}{\dashv}} G : \mathsf{A} \longrightarrow \mathsf{X}\,.$$

A functor F is left adjoint to $G : \mathsf{A} \to \mathsf{X}$ if and only if G^{op} is left adjoint to $F^{\mathrm{op}} : \mathsf{X}^{\mathrm{op}} \to \mathsf{A}^{\mathrm{op}}$; more precisely,

$$F \underset{\varepsilon}{\overset{\eta}{\dashv}} G \iff G^{\mathrm{op}} \underset{\eta^{\mathrm{op}}}{\overset{\varepsilon^{\mathrm{op}}}{\dashv}} F^{\mathrm{op}}\ .$$

II.2.5.1 Examples

(1) Adjunctions between ordered sets (considered as categories) are precisely those described in Section II.1.5.

(2) The forgetful functor $G : \mathsf{Mon} \to \mathsf{Set}$ is right adjoint. Its left adjoint F is the *free-monoid functor*, which for a set X may be constructed as the set of "words over the alphabet" X

$$FX = X^* = \textstyle\bigcup_{n\geq 0} X^n\ ,$$

with concatenation as multiplication; the map $\eta_X : X \to X^*$ (for a set X) considers an element of X as a one-letter word, and the homomorphism $\varepsilon_A : A^* \to A$ (for a monoid A) sends words over A to the actual product of its letters in A.

(3) The forgetful functor $\mathsf{Ord} \to \mathsf{Set}$ has a left adjoint that provides each set with the *discrete order* (given by the graph of 1_X); similarly for $\mathsf{Top} \to \mathsf{Set}$, whose left adjoint provides a set X with the *discrete topology*. Both functors also have right adjoints, which provide a set X with the respective *indiscrete* (or *chaotic*) structure.

(4) For a fixed set A, the hom-functor $\mathsf{Set}(A, -) : \mathsf{Set} \to \mathsf{Set}$ has a left adjoint, namely

$$(-) \times A : \mathsf{Set} \to \mathsf{Set} .$$

Writing $B^A = \mathsf{Set}(A, B)$, the counit ε is given by the *evaluation maps*

$$\varepsilon_B : B^A \times A \to B , \quad (f, x) \mapsto f(x) ,$$

while the unit sends every element x of a set X to the section

$$\eta_X(x) : A \to X \times A , \quad a \mapsto (x, a) .$$

(5) Considering the powerset PX of a set X as a complete lattice (ordered by inclusion) makes $P : \mathsf{Set} \to \mathsf{Sup}$ left adjoint to the forgetful functor $U : \mathsf{Sup} \to \mathsf{Set}$. The unit is described by the singleton maps $\eta_X : X \to PX$ that send x to $\{x\}$, and the components of the counit $\varepsilon_A : PA \to A$ (for a complete lattice A) are simply given by $\bigvee$.

(6) The *contravariant powerset functor* $P^\bullet : \mathsf{Set}^{\mathrm{op}} \to \mathsf{Set}$ is *self-adjoint*, i.e. $(P^\bullet)^{\mathrm{op}} : \mathsf{Set} \to \mathsf{Set}^{\mathrm{op}}$ is its left adjoint, as shown by Proposition II.2.5.2 (following) and by the natural bijection

$$\frac{X \xrightarrow{f} PY}{PX \xleftarrow{g} Y} ,$$

where f and g determine each other via $(x \in g(y) \iff y \in f(x))$ for all $x \in X$, $y \in Y$. Of course, this bijection stems from the self-dual category Rel for which $\mathsf{Rel}(X, Y) \cong \mathsf{Rel}(Y, X)$.

II.2.5.2 Proposition *A functor $G : \mathsf{A} \to \mathsf{X}$ is right adjoint if and only if there are a map $F :$ ob $\mathsf{X} \to$ ob A and an* ob X*-indexed family of natural isomorphisms*

$$\phi_X : \mathsf{A}(FX, -) \to \mathsf{X}(X, G-) = \mathsf{X}(X, -)G .$$

This last condition yields bijections

$$\phi_{X,A} : \mathsf{A}(FX, A) \to \mathsf{X}(X, GA)$$

for all $X \in \mathrm{ob}\,\mathsf{X}$, $A \in \mathrm{ob}\,\mathsf{A}$ that are "natural in A"; casually, one writes

$$\frac{FX \xrightarrow{g} A}{X \xrightarrow{f} GA}$$

with F appearing on the left, and G on the right. Note that one necessarily obtains that these bijections are also "natural in X." Any pair of arrows (g, f) related to each other by the bijection $\phi_{X,A}$ are called *mates* to each other under the adjunction.

Proof If $F \underset{\varepsilon}{\overset{\eta}{\dashv}} G$, then $\phi_{X,A}$ is defined by

$$\phi_{X,A}(g) = Gg \cdot \eta_X$$

for all $g : FX \to A$ in A, with its inverse given by

$$(f : X \to GA) \mapsto \varepsilon_A \cdot Ff .$$

Conversely, having the bijections $\phi_{X,A}$ (natural in A), one obtains morphisms

$$\eta_X = \phi_{X,FX}(1_{FX}) , \quad \varepsilon_A = \phi^{-1}_{GA,A}(1_{GA}) .$$

Then F, already defined on objects, can be made a functor via

$$(f : X \to Y) \xrightarrow{F} (\phi^{-1}_{X,FY}(\eta_Y \cdot f) : FX \to FY) ,$$

such that η and ε become natural transformations satisfying the triangular identities. □

II.2.5.3 Example Considering $\mathsf{2} = \{\bot, \top\}$ successively as a set, an ordered set, a meet-semilattice, a lattice, or a frame, one obtains right adjoint functors

$$\mathsf{C}(-, \mathsf{2}) : \mathsf{C}^{\mathrm{op}} \to \mathsf{Set} ,$$

with $\mathsf{C} = \mathsf{Set}, \mathsf{Ord}, \mathsf{SLat}, \mathsf{Lat}$, or Frm, respectively, with left adjoint $X \mapsto PX$. Of course, for $\mathsf{C} = \mathsf{Set}$, $\mathsf{C}(-, \mathsf{2}) \cong P^\bullet$ is isomorphic to the contravariant powerset functor of Example II.2.5.1(6), and the needed natural isomorphisms

$$\mathsf{Set}(X, \mathsf{C}(Y, \mathsf{2})) \cong \mathsf{C}(Y, PX)$$

(with $X \in \mathrm{ob}\,\mathsf{Set}$, $Y \in \mathrm{ob}\,\mathsf{C}$) are just restrictions of those described in Example II.2.5.1(6).

Returning to Proposition II.2.5.2, we note that naturality of $\phi_{X,A}$ in A gives, for $g : FX \to A$ in A, the commutative diagram

$$\begin{array}{ccc} \mathsf{A}(FX, FX) & \xrightarrow{\mathsf{A}(FX,g)} & \mathsf{A}(FX, A) \\ {\scriptstyle \phi_{X,FX}}\downarrow & & \downarrow{\scriptstyle \phi_{X,A}} \\ \mathsf{X}(X, GFX) & \xrightarrow{\mathsf{X}(X,Gg)} & \mathsf{X}(X, GA) \end{array}$$

so that with $\eta_X = \phi_{X,FX}(1_{FX})$, one has

$$\phi_{X,A}(g) = \phi_{X,A} \cdot \mathsf{A}(FX, g)(1_{FX}) = \mathsf{X}(X, Gg) \cdot \phi_{X,FX}(1_{FX}) = Gg \cdot \eta_X \ ;$$

in other words, the formula for $\phi_{X,A}$ in terms of the unit η is forced upon us by naturality in A. The condition that $\phi_{X,A}$ is bijective now says:

> for all $f : X \to GA$ in X (with $A \in \mathrm{ob}\,\mathsf{A}$), there is precisely one morphism $g : FX \to A$ in A with $Gg \cdot \eta_X = f$.

$$\begin{array}{ccc} X & \xrightarrow{\ \eta_X\ } & GFX \\ & {\scriptstyle f}\searrow & \downarrow{\scriptstyle Gg} \\ & & GA \end{array} \qquad\qquad \begin{array}{c} FX \\ \vdots\,{\scriptstyle g} \\ A\,. \end{array} \tag{II.2.5.i}$$

For a functor $G : \mathsf{A} \to \mathsf{X}$ and an object X in X, a morphism $u : X \to GU$ in X with an object U in A is called a *G-universal arrow* for X if it satisfies the *universal property* (II.2.5.i), with (U, u) in lieu of (FX, η_X). This property allows us to define $\phi_{X,A}$ as above whenever we have a chosen G-universal arrow for every $X \in \mathrm{ob}\,\mathsf{X}$. As a consequence, we get:

II.2.5.4 Theorem *A functor $G : \mathsf{A} \to \mathsf{X}$ is right adjoint if and only if there exist a map $F : \mathrm{ob}\,\mathsf{X} \to \mathrm{ob}\,\mathsf{A}$ and an $\mathrm{ob}\,\mathsf{X}$-indexed family of G-universal arrows $\eta_X : X \to GFX$. There is then a unique way of making F a functor such that η becomes a natural transformation.*

Of course, F will be left adjoint to G with unit η. There is a dual way of describing adjunctions in terms of their counit: for any functor $F : \mathsf{X} \to \mathsf{A}$ and object A in A, a morphism $\varepsilon_A : FGA \to A$ in A together with an object GA is an *F-couniversal arrow* for A if, as an arrow in A^{op}, it is F^{op}-universal, i.e.

> for all $g : FX \to A$ in A (with $X \in \mathrm{ob}\,\mathsf{X}$), there is precisely one morphism $f : X \to GA$ in X with $\varepsilon_A \cdot Ff = g$.

$$\begin{array}{c} GA \\ \vdots\,{\scriptstyle f} \\ X \end{array} \qquad\qquad \begin{array}{ccc} FGA & \xrightarrow{\ \varepsilon_A\ } & A\,. \\ \uparrow{\scriptstyle Ff} & \nearrow{\scriptstyle g} & \\ FX & & \end{array} \tag{II.2.5.ii}$$

Usually, one also refers to (II.2.5.ii) as a *universal property*, although, strictly speaking, it is a "couniversal" property. In summary, one obtains from Theorem II.2.5.4 the following statement by *dualization*:

II.2.5.4^{op} Theorem *A functor $F : \mathsf{X} \to \mathsf{A}$ is left adjoint if and only if there exist a map $G : \mathrm{ob}\,\mathsf{A} \to \mathrm{ob}\,\mathsf{X}$ and an $\mathrm{ob}\,\mathsf{A}$-indexed family of F-couniversal arrows $\varepsilon_X : FGX \to X$. There is then a unique way of making G a functor such that ε becomes a natural transformation.*

Proof Apply Theorem II.2.5.4 with F^{op} in lieu of G (recall from Section II.2.5 that $F \dashv G \iff G^{\mathrm{op}} \dashv F^{\mathrm{op}}$). □

In what follows, we rarely formulate explicitly such *dual statements*.

Adjunctions compose; more precisely, direct calculation based on the definitions shows:

II.2.5.5 Proposition *If* $F \overset{\eta}{\underset{\varepsilon}{\dashv}} G \ : \mathsf{A} \to \mathsf{X}$ *and* $H \overset{\gamma}{\underset{\delta}{\dashv}} J \ : \mathsf{C} \to \mathsf{A}$, *then* $HF \overset{\alpha}{\underset{\beta}{\dashv}} GJ \ : \mathsf{C} \to \mathsf{X}$, *with*

$$\alpha = (1_{\mathsf{X}} \xrightarrow{\ \eta\ } GF \xrightarrow{\ G\gamma F\ } GJHF)\ , \qquad \beta = (HFGJ \xrightarrow{\ H\varepsilon J\ } HJ \xrightarrow{\ \delta\ } 1_{\mathsf{C}})\,.$$

II.2.5.6 Corollary *If* $F \dashv G : \mathsf{A} \to \mathsf{X}$, $H \dashv J : \mathsf{C} \to \mathsf{A}$, *and* $L \dashv GJ$, *then* $L \cong HF$.

II.2.6 Reflective subcategories, equivalence of categories

Whether a right adjoint functor is fully faithful is detected by the counit.

II.2.6.1 Proposition *For an adjunction* $F \overset{\eta}{\underset{\varepsilon}{\dashv}} G \ : \mathsf{A} \to \mathsf{X}$, *the functor* G *is fully faithful if and only if* ε *is a natural isomorphism. Dually,* F *is fully faithful if and only if* η *is a natural isomorphism.*

Proof Since, in the notations of Section II.2.5, for all $g : A \to B$ in A,

$$\phi_{GA,B} \cdot \mathsf{A}(\varepsilon_A, B)(g) = G(g \cdot \varepsilon_A) \cdot \eta_{GA} = Gg\,,$$

the following diagram commutes for all $A, B \in \mathrm{ob}\,\mathsf{A}$:

$$\begin{array}{ccccc} & & \mathsf{A}(FGA, B) & & \\ & \overset{\mathsf{A}(\varepsilon_A, B)}{\nearrow} & & \overset{\phi_{GA,B}}{\searrow} & \\ \mathsf{A}(A, B) & & \xrightarrow{\qquad G_{A,B} \qquad} & & \mathsf{X}(GA, GB)\,. \end{array}$$

Since $\phi_{GA,B}$ is bijective, one therefore has that for all $A, B \in \mathrm{ob}\,\mathsf{A}$ the map $G_{A,B}$ is bijective precisely when every ε_A is an isomorphism in A. □

A full subcategory A of a category X is *reflective* in X if the inclusion functor $J : \mathsf{A} \hookrightarrow \mathsf{X}$ is right adjoint. A left adjoint $R : \mathsf{X} \to \mathsf{A}$ to J is called a *reflector* of X onto A, and the unit $\rho : 1_{\mathsf{X}} \to JR$ gives the A-*reflections* $\rho_X : X \to RX$ that are characterized as J-universal arrows for X. By Theorem II.2.5.4, A is reflective in X if and only if, for every object in X, there is a chosen A-reflection. By Proposition II.2.6.1, the counit of the adjunction is an isomorphism.

An example of a reflective subcategory encountered in Section II.1 is $\mathsf{Ord}_{\mathrm{sep}}$ in Ord, with the separated reflections of Section II.1.3 forming the unit. For a reflective subcategory A of X with unit ρ, an object X in X is isomorphic to an object in A if and only if ρ_X is an isomorphism. Hence, when A is *replete* in X, i.e. when $X \cong A \in \operatorname{ob}\mathsf{A}$ implies $X \in \operatorname{ob}\mathsf{A}$, an object of X lies in A if and only if its A-reflection is an isomorphism.

A full subcategory A is *coreflective* in X if A^{op} is reflective in X^{op}; equivalently, if the inclusion functor has a right adjoint which is the *coreflector* of X onto A. The counit of the adjunction gives the A-*coreflections*.

A functor $G : \mathsf{A} \to \mathsf{X}$ is an *equivalence* of categories if there is a functor $F : \mathsf{X} \to \mathsf{A}$ with $1_{\mathsf{X}} \cong GF$ and $FG \cong 1_{\mathsf{A}}$. A category A is *equivalent* to a category X if there is an equivalence $\mathsf{A} \to \mathsf{X}$. If G is an equivalence, it is fully faithful (see the proof of Proposition II.2.6.2), and therefore *reflects isomorphisms*, i.e. if Gf is an isomorphism then so is f (Exercise II.2.A). In fact, these last two notions may be used to verify that G is an equivalence.

II.2.6.2 Proposition *The following are equivalent for a functor $G : \mathsf{A} \to \mathsf{X}$:*

(i) *G is an equivalence;*
(ii) *G reflects isomorphisms and has a fully faithful left adjoint;*
(iii) *G is fully faithful and there are a map $F : \operatorname{ob}\mathsf{X} \to \operatorname{ob}\mathsf{A}$ and an $\operatorname{ob}\mathsf{X}$-indexed family of isomorphisms $\eta_X : X \to GFX$ $(X \in \operatorname{ob}\mathsf{X})$.*

Proof (i) $\Longrightarrow$ (iii): The hypothesis yields faithfulness of F and G. For fullness of G, consider $f : GA \to GB$ in X (with $A, B \in \operatorname{ob}\mathsf{A}$). In order to show $Gg = f$ for $g := \varepsilon_B \cdot Ff \cdot \varepsilon_A^{-1}$ (with $\varepsilon : FG \to 1_{\mathsf{A}}$ some isomorphism), it suffices to verify $FGg = Ff$. But this follows from

$$\varepsilon_B \cdot FGg \cdot \varepsilon_A^{-1} = g \cdot \varepsilon_A \cdot \varepsilon_A^{-1} = \varepsilon_B \cdot Ff \cdot \varepsilon_A^{-1} .$$

(iii) $\Longrightarrow$ (ii): As a fully faithful functor, G reflects isomorphisms, and the isomorphism η_X is easily seen to be G-universal arrow for $X \in \operatorname{ob}\mathsf{X}$. Hence (b) follows from Theorem II.2.5.4 and Proposition II.2.6.1.

(ii) $\Longrightarrow$ (i): By Proposition II.2.6.1, one has $F \dashv G$ with the unit η an isomorphism. Since $G\varepsilon \cdot \eta G = 1_G$, the natural transformation $G\varepsilon$ is an isomorphism, and so is ε whenever G reflects isomorphisms. □

II.2.6.3 Corollary *If $G : \mathsf{A} \to \mathsf{X}$ is fully faithful and right adjoint, A is equivalent to a replete reflective subcategory $\overline{\mathsf{A}}$ of X.*

Proof Take $\overline{\mathsf{A}}$ to be the full subcategory of objects X in X for which η_X is an isomorphism, with η the unit of an adjunction $F \dashv G$. □

II.2.7 Initial and terminal objects, comma categories

Let $\mathbf{1} = \{\star\}$ be the one-object category whose only morphism is $1_\star$. For every category C there is then a unique functor $! : \mathsf{C} \to \mathbf{1}$. A right adjoint to this functor is given by an object T in C such that

for all objects X in C, there is precisely one morphism $X \to T$ in C.

Any such object T is called *terminal* in C. As the value of a right adjoint, it is unique up to isomorphism (a fact that is easily established directly): *if S, T are both terminal in C, then $S \cong T$; moreover, for any isomorphic objects S, T in C, one is terminal if the other is.* A terminal object in a category is often denoted by 1.

An object I is *initial* in C if it is terminal in C^{op}, so that

for all objects X in C, there is precisely one morphism $I \longrightarrow X$ in C.

Equivalently, I is initial if the functor $\mathbf{1} \longrightarrow \mathsf{C}$ with $\star \longmapsto I$ is left adjoint to the functor !. An initial object in a category is often denoted by 0.

The empty set is initial in Set, and every one-element set is terminal; the same holds for Ord, Top, and Cat, for example. In Mon, the trivial (one-element) monoids are both initial and terminal; such objects are called *zero-objects*. An individual monoid M, considered as a one-object category, can have an initial or terminal object only if it is trivial. In an ordered set X considered as a category, a terminal object is given by a top element, and an initial object is given by a bottom element.

The process of defining initial and terminal objects via adjunctions may be reversed, as follows. For a functor $G : \mathsf{A} \longrightarrow \mathsf{X}$ and an object X in X, define the *comma category* $(X \downarrow G)$ to have as objects pairs (A, f), with $A \in \mathrm{ob}\,\mathsf{A}$ and $f : X \longrightarrow GA$ in X; a morphism $h : (A, f) \longrightarrow (B, g)$ in $(X \downarrow G)$ is given by a morphism $h : A \longrightarrow B$ in A that makes the diagram

$$\begin{array}{ccc} & X & \\ {\scriptstyle f}\swarrow & & \searrow{\scriptstyle g} \\ GA & \xrightarrow{\ Gh\ } & GB \end{array}$$

commute. Then

a G-universal arrow for X is simply an initial object in $(X \downarrow G)$,

and Theorem II.2.5.4 states that G is right adjoint if and only if there are chosen initial objects in every comma category $(X \downarrow G)$ (with X running through $\mathrm{ob}\,\mathsf{X}$).

For a functor $F : \mathsf{X} \longrightarrow \mathsf{A}$, and $A \in \mathrm{ob}\,\mathsf{A}$, one also has the *comma category*

$$(F \downarrow A) := (A \downarrow F^{\mathrm{op}})^{\mathrm{op}}$$

whose morphisms $h : (X, f) \longrightarrow (Y, g)$ are depicted by

$$\begin{array}{ccc} FX & \xrightarrow{\ Fh\ } & FY \\ {\scriptstyle f}\searrow & & \swarrow{\scriptstyle g} \\ & A\,. & \end{array}$$

Its terminal objects are precisely the F-couniversal arrows for A. In the special case where $F = 1_{\mathsf{X}}$, one calls

$$\mathsf{X}/A := (1_{\mathsf{X}} \downarrow A)$$

the *slice of* X *over* A. Note that X/A always has a terminal object given by 1_A:

$$\begin{array}{ccc} X & \overset{f}{\dashrightarrow} & A \\ & {\scriptstyle f}\searrow \quad \swarrow{\scriptstyle 1_A} & \\ & A\,. & \end{array}$$

We also denote the *slice* $(A \downarrow 1_{\mathsf{X}})$ *of* X *under* A by A/X.

II.2.8 Limits

A functor $D : \mathsf{J} \to \mathsf{C}$ is also called a *diagram* (*of shape* J) in C. If J is *discrete*, so that for all objects i, j in J

$$\mathsf{J}(i, j) = \begin{cases} \{1_i\} & \text{if } i = j, \\ \varnothing & \text{otherwise,} \end{cases}$$

then D is simply a family $(A_i)_{i \in \mathrm{ob}\,\mathsf{J}}$ of objects in C. If J has precisely two objects, and two non-identical arrows, as in

$$a \underset{y}{\overset{x}{\rightrightarrows}} b \ ,$$

then a diagram of shape J is given by a pair of (not necessarily distinct) parallel arrows in C.

For every object A in C and any J, one has the *constant diagram* of shape J in C given by

$$\Delta A : \mathsf{J} \to \mathsf{C}\,, \quad i \mapsto A\,, \quad (\delta : i \to j) \mapsto 1_A\,.$$

For a diagram $D : \mathsf{J} \to \mathsf{C}$, any natural transformation $\alpha : \Delta A \to D$ is called a *cone* over D in C with vertex A. Naturality of α amounts to the property that, for all $\delta : i \to j$ in J, the diagram

$$\begin{array}{ccc} & A & \\ {\scriptstyle \alpha_i}\swarrow & & \searrow{\scriptstyle \alpha_j} \\ Di & \xrightarrow{D\delta} & Dj \end{array}$$

commutes in C. Hence, a cone over D with vertex A is simply an object (A, α) in the comma category $(\Delta \downarrow D)$, where D is considered as an object of C^{J}, and where Δ has been made into a functor

$$\Delta : \mathsf{C} \to \mathsf{C}^{\mathsf{J}}\,, \quad (f : A \to B) \mapsto (\Delta f : \Delta A \to \Delta B)\,,$$

with the natural transformation Δf defined by $(\Delta f)_i = f$ for all $i \in \mathrm{ob}\, J$.

A *limit* of $D : \mathsf{J} \to \mathsf{C}$ is a terminal cone over D in C, or equivalently a Δ-couniversal arrow for $D \in \mathrm{ob}\,\mathsf{C}^{\mathsf{J}}$. Hence, a limit of D is a cone $\lambda : \Delta L \to D$, with $L \in \mathrm{ob}\,\mathsf{C}$ such that

> for every cone $\alpha : \Delta A \to D$ with $A \in \mathrm{ob}\,\mathsf{C}$, there is a unique morphism $f : A \to L$ in C with $\lambda \cdot \Delta f = \alpha$, i.e. $\lambda_i \cdot f = \alpha_i$ for all $i \in \mathrm{ob}\,\mathsf{J}$.

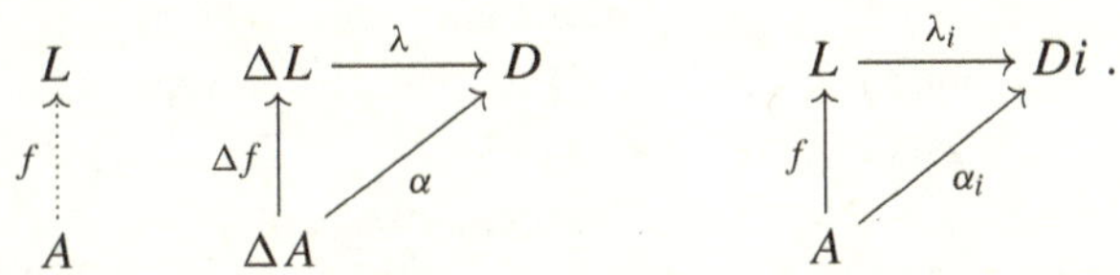

Limits are uniquely determined by D, up to isomorphism, so that, when (A, α) is also a limit of D, the unique morphism $f : A \to L$ is an isomorphism in C. We write $L \cong \lim D$ for any object that serves as the vertex of a limit for D in C.

Important instances of limits include: products, equalizers, and pullbacks (as well as certain special pullbacks), which we now proceed to describe.

Products

A *product* of a family of objects $(A_i)_{i \in I}$ in C is a limit of the diagram $\mathsf{J} \to \mathsf{C}$, $i \mapsto A_i$, with J discrete and $\mathrm{ob}\,\mathsf{J} = I$. Hence, a product is an object P in C together with a family of morphisms $p_i : P \to A_i$ $(i \in I)$ such that

> for every family $f_i : B \to A_i$ $(i \in I)$ of morphisms in C, there is a unique morphism $f : B \to P$ with $p_i \cdot f = f_i$ $(i \in I)$; this morphism is usually written
>
> $$f = \langle f_i \rangle_{i \in I} .$$

Any object P that serves as a product for $(A_i)_{i \in I}$ is denoted by $\prod_{i \in I} A_i$, and comes with *projections* p_i $(i \in I)$. Note that $I = \varnothing$ is permitted, in which case P is simply a terminal object of C. The case $I = \{\star\}$ is trivial: a limit of the one-object family A can be given by A, with projection 1_A. The product of a pair (A_1, A_2) of objects is denoted by $A_1 \times A_2$, with projections p_1, p_2, whose universal property may be depicted by

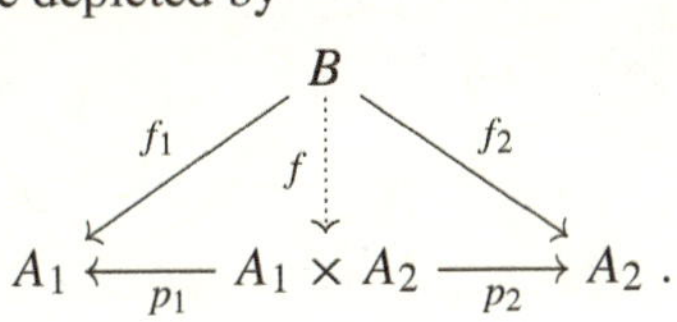

More casually, we may write

$$\frac{B \to A_1,\ B \to A_2}{B \to A_1 \times A_2 .}$$

In Set, small-indexed products can be constructed as Cartesian products. In many *concrete categories* (categories admitting a faithful functor to Set) like Ord and Mon, the Cartesian product is provided with an appropriate structure that makes the projections morphisms of the category; see Propositions II.5.8.3 and II.3.3.1, respectively, for specifications of this statement in two general contexts. In an ordered set X regarded as a category, a product of a family $(a_i)_{i \in I}$ of elements in X is simply an infimum of $\{a_i \mid i \in I\}$ in X.

Equalizers

An *equalizer* of a pair $(f, g : A \to B)$ of "parallel" morphisms in C is a limit (E, λ) of the corresponding diagram D of shape $\{\, a \overset{x}{\underset{y}{\rightrightarrows}} b \,\}$ in C, with $Dx = f$, $Dy = g$. Since $\lambda_b = Dx \cdot \lambda_a = Dy \cdot \lambda_a$, an equalizer of (f, g) is given by a morphism $u : E \to A$ with $f \cdot u = g \cdot u$ (namely, $u = \lambda_a$) such that

> for every morphism $h : C \to A$ with $f \cdot h = g \cdot h$, there is a unique morphism $t : C \to E$ with $u \cdot t = h$.

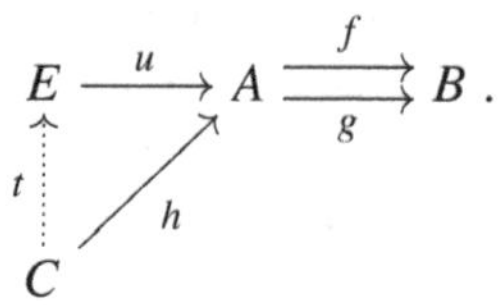

We sometimes write $u \cong \mathrm{equ}(f, g)$ in this case. If $f = g$, then u can be taken to be 1_A; conversely, if $\mathrm{equ}(f, g)$ exists and is an isomorphism, then $f = g$.

In Set, an equalizer $\mathrm{equ}(f, g)$ can be realized as the inclusion map

$$E = \{x \in A \mid f(x) = g(x)\} \hookrightarrow A\,,$$

and in concrete categories like Ord and Mon one can endow E with an obvious structure inherited from A. An ordered set X (regarded as a category) trivially has equalizers, since any two parallel arrows are equal. In a group regarded as a one-object category, an equalizer of two elements x, y can exist only if $x = y$.

Pullbacks

A pair of morphisms $f : A \to C$, $g : B \to C$ with common codomain can be viewed as a diagram D of shape J, where J is a three-object category with two non-identical arrows, as in

$$a \xrightarrow{\;x\;} c \xleftarrow{\;y\;} b\,.$$

A *pullback* of (f, g) in C is a limit (P, λ) of D in C. Since $\lambda_c = Dx \cdot \lambda_a = Dy \cdot \lambda_b$, a pullback of (f, g) is given by a pair $(p_1 : P \to A, p_2 : P \to B)$ with $f \cdot p_1 = g \cdot p_2$ (namely, $p_1 = \lambda_a$, $p_2 = \lambda_b$), such that

> for all morphisms $h_1 : X \to A$, $h_2 : X \to B$ with $f \cdot h_1 = g \cdot h_2$, there is a unique $t : X \to P$ with $p_1 \cdot t = h_1$, $p_2 \cdot t = h_2$. The morphism t is often denoted by $\langle h_1, h_2 \rangle$, and the square in the following diagram is called a *pullback diagram*:

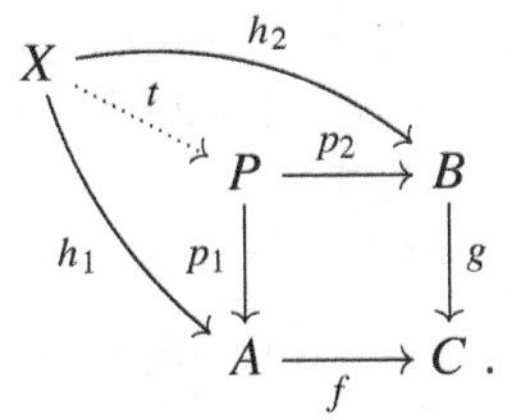

One often writes (somewhat casually) $P \cong A \times_C B$, and calls P a *fibered product* of A and B over C, thus unduly neglecting the morphisms f and g. But *pullbacks are in fact products*, since a pullback of (f, g) in C is nothing but the product of the pair of objects $(A, f), (B, g)$ in the comma category C/C. When $C = 1$ is terminal in C, one has $A \times_C B \cong A \times B$, i.e. the pullback is a product in C. *Multiple pullbacks* are simply products in C/C of families of objects in C (i.e. of families of morphisms in C with common codomain C).

In Set, a pullback of $(f : A \longrightarrow C, g : B \longrightarrow C)$ can be constructed as

$$P = \{(x, y) \in A \times B \mid f(x) = g(y)\} ,$$

with p_1, p_2 the restrictions of the product projections π_1, π_2. Hence, P is obtained as the equalizer of $f \cdot \pi_1, g \cdot \pi_2 : A \times B \longrightarrow C$. This construction works in general categories; see Corollary II.2.10.2.

Special pullbacks

A *kernel pair* of a morphism $f : A \longrightarrow B$ in C is a pullback of (f, f); hence, it is a pair $(p_1, p_2 : K \longrightarrow A)$ of morphisms in C with $f \cdot p_1 = f \cdot p_2$ satisfying the universal property of a pullback diagram. A morphism $f : A \longrightarrow B$ in C is a *monomorphism* (or is *monic*) if and only if $f \cdot x = f \cdot y$ implies $x = y$ whenever $x, y : X \longrightarrow A$; equivalently, if $(1_A, 1_A)$ serves as a kernel pair, or more generally if f has a kernel pair (p_1, p_2) with $p_1 = p_2$. For a monomorphism f, any commutative diagram

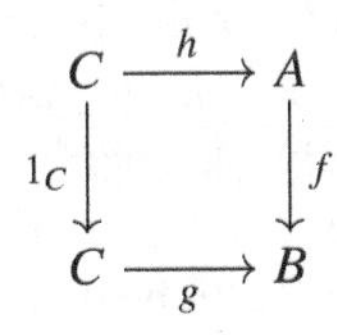

is a pullback diagram. Composites of monomorphisms are monic, and *monomorphisms are stable under pullback* in the following sense: for every pullback diagram

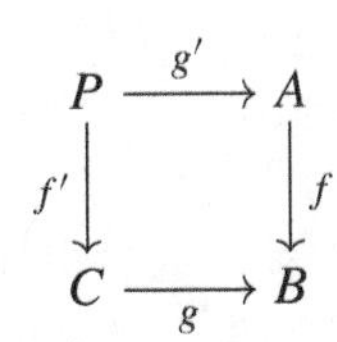

if f is monic, so is f'. One calls f' a *pullback* of f along g, alluding to the fact that in Set, when $f : A \hookrightarrow B$ is a subset inclusion, f' can be taken to be $g^{-1}(A) \hookrightarrow C$.

A frequently used property of pullback diagrams is:

II.2.8.1 Proposition *If in a commutative diagram in* C *of the form*

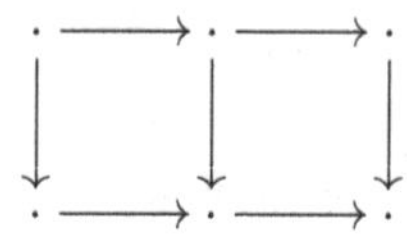

the outer rectangle and the right square are pullback diagrams, then so is the left square.

Proof The statement follows by routine diagram chasing. □

II.2.9 Colimits

A pair (K, γ) is a *colimit* of the diagram $D : \mathsf{J} \to \mathsf{C}$ in a category C if $(K, \gamma^{\mathrm{op}})$ is a limit of $D^{\mathrm{op}} : \mathsf{J}^{\mathrm{op}} \to \mathsf{C}^{\mathrm{op}}$ in C^{op}. Hence, $\gamma : D \to \Delta K$ is a *cocone* over D with vertex $K \in \mathrm{ob}\,\mathsf{C}$ such that

> for every cocone $\alpha : D \to \Delta A$ with $A \in \mathrm{ob}\,\mathsf{C}$, there is a unique morphism $f : K \to A$ in C with $\Delta f \cdot \gamma = \alpha$, i.e. $f \cdot \gamma_i = \alpha_i$ for all $i \in \mathrm{ob}\,\mathsf{J}$.

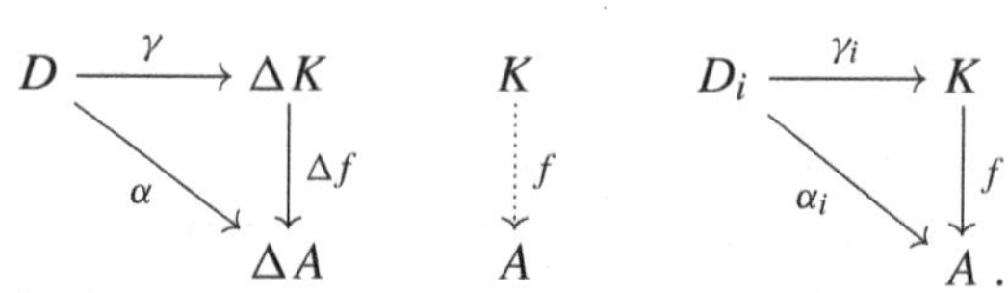

One writes $K \cong \mathrm{colim}\, D$ in this case. Dually to products, equalizers, pullbacks, kernel pairs, and monomorphisms (or morphisms that are monic), one obtains *coproducts*, *coequalizers*, *pushouts*, *cokernel pairs*, and *epimorphisms* (or morphisms that are *epic*), respectively. A coproduct $(A_i)_{i\in I}$ in C is denoted by $\coprod_{i\in I} A_i$, together with *injections* k_i $(i \in I)$. For $I = \varnothing$, the empty coproduct $\coprod_{i\in I} A_i \cong 0$ is an initial object of C. In the binary case $I = \{1, 2\}$, the coproduct is normally denoted by $A_1 + A_2$, and its universal property visualized by

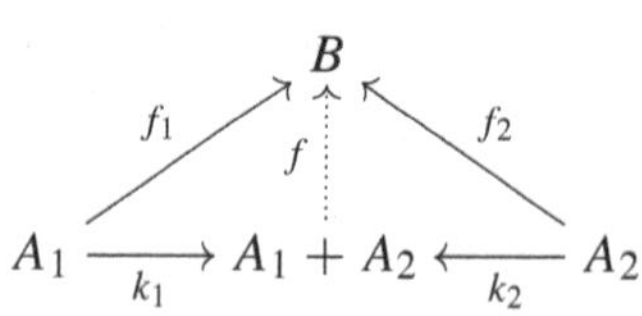

or

$$\frac{A_1 \to B,\ A_2 \to B}{A_1 + A_2 \to B}\ .$$

As for products, we often write $f = \langle f_1, f_2 \rangle$ for the arrow given by f_1, f_2. The universal properties of coequalizers and pushouts may be displayed as follows:

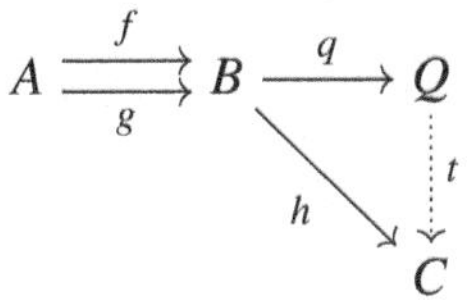

and

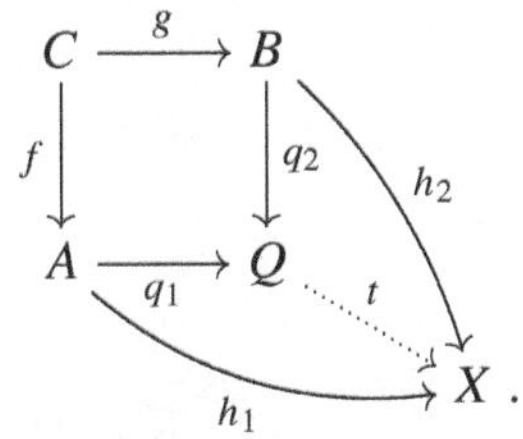

In **Set**, a coproduct of $(A_i)_{i \in I}$ may be constructed as its disjoint union $\bigcup_{i \in I} A_i \times \{i\}$. A coequalizer of $f, g : A \longrightarrow B$ may be constructed as the projection onto $Q = B/{\sim}$, where $\sim$ is the least equivalence relation on B with $f(x) \sim g(x)$ for all $x \in A$. For the pushout of $(f : C \longrightarrow A, g : C \longrightarrow B)$, form the coproduct $A + B$ with injections k_1, k_2 and then the coequalizer of $k_1 \cdot f, k_2 \cdot g$. In **Ord**, and **Top** for example, one may "lift" these constructions from **Set** by putting appropriate structures on them (see the dual statement of Proposition II.5.8.3). In an ordered set X considered as a category, coproducts are given by suprema, and coequalizers are, like equalizers, trivial.

II.2.10 Construction of limits and colimits

For categories C and J, one says that C is J-*complete* if the functor $\Delta : \mathsf{C} \longrightarrow \mathsf{C}^{\mathsf{J}}$ has a right adjoint, i.e. if all diagrams of shape J have a chosen limit in C; the right adjoint of Δ is usually denoted by lim. A category C is *finitely complete* if C is J-complete for every *finite category* J (categories for which ob J and all hom-sets of J are finite), and C is *small-complete* if C is J-complete for every small category J. A category C *has products* if C is J-complete for every small discrete category J, and C *has finite products* if this holds for every finite discrete category J. Similarly, $\{\cdot \rightrightarrows \cdot\}$-completeness and $\{\cdot \longrightarrow \cdot \longleftarrow \cdot\}$-completeness are phrased as "*has equalizers*," "*has pullbacks*," and so on. The dual notions are those of J-*cocompleteness* (with colim denoting the left adjoint of Δ), *finite cocompleteness, small-cocompleteness, has coproducts, has finite coproducts, has coequalizers, has pushouts, etc.*

General limits can be constructed in terms of products and equalizers, as follows. Given $D : \mathsf{J} \longrightarrow \mathsf{C}$, one forms the product $P = \prod_{i \in \mathrm{ob}\,\mathsf{J}} Di$ with projections p_i. Since in general the p_i do *not* form a cone (so that generally $D\delta \cdot p_j \neq p_k$ for $\delta : j \longrightarrow k$ in J), one also forms the product

$$Q = \prod_{\delta \in \mathrm{mor}\,\mathsf{J}} D(\mathrm{cod}\,\delta) \quad \text{with projections } q_\delta\,,$$

where $\mathrm{mor}\,\mathsf{J} = \bigcup_{j,k \in \mathrm{ob}\,\mathsf{J}} \mathsf{J}(j,k)$ (with the hom-sets of J assumed to be disjoint, without loss of generality). With

$$f = \langle p_{\mathrm{cod}\,\delta} \rangle_{\delta \in \mathrm{mor}\,\mathsf{J}}\,, \qquad g = \langle D\delta \cdot p_{\mathrm{dom}\,\delta} \rangle_{\delta \in \mathrm{mor}\,\mathsf{J}}\,,$$

one obtains for every $\delta : j \longrightarrow k$ in J the diagram

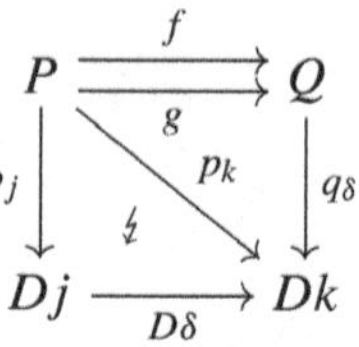

with $q_\delta \cdot f = p_k$, $q_\delta \cdot g = D\delta \cdot p_j$ (but generally $D\delta \cdot p_j \neq p_k$). Now, for all morphisms $h : A \to P$ in C,

$$f \cdot h = g \cdot h \iff \forall(\delta : j \to k) \text{ in } \mathsf{J}\ (D\delta \cdot (p_j \cdot h) = p_k \cdot h)\,.$$

Hence, when h satisfies $f \cdot h = g \cdot h$, then $\alpha_i = p_i \cdot h$ ($i \in \mathrm{ob}\,\mathsf{J}$) defines a cone $\alpha : \Delta A \to D$; conversely, for every cone $\alpha : \Delta A \to D$, the morphism $h = \langle \alpha_i \rangle_{i \in \mathrm{ob}\,\mathsf{J}}$ satisfies $f \cdot h = g \cdot h$. Consequently, an equalizer h of (f, g) gives a limit cone over D, and a limit cone over D gives an equalizer of (f, g). This sketches the proof of the following result:

II.2.10.1 Theorem *A category* C *is* J*-complete if it has equalizers as well as* ob J- *and* mor J*-indexed products. Hence,* C *is small-complete if and only if* C *has products and equalizers.*

II.2.10.2 Corollary *The following conditions are equivalent for a category* C*:*

(i) C *is finitely complete;*
(ii) C *has finite products and equalizers;*
(iii) C *has pullbacks and a terminal object.*

Proof The equivalence (i) $\iff$ (ii) follows from the Theorem.

(ii) $\Longrightarrow$ (iii): A terminal object exists as a product of the empty family of objects in C. To obtain the pullback (P, p_1, p_2) of $(f : A \to C, g : B \to C)$, let $u \cong \mathrm{equ}(f \cdot \pi_1, g \cdot \pi_2)$, with π_1, π_2 the projections of $A \times B$, and let $p_1 = \pi_1 \cdot u$, $p_2 = \pi_2 \cdot u$:

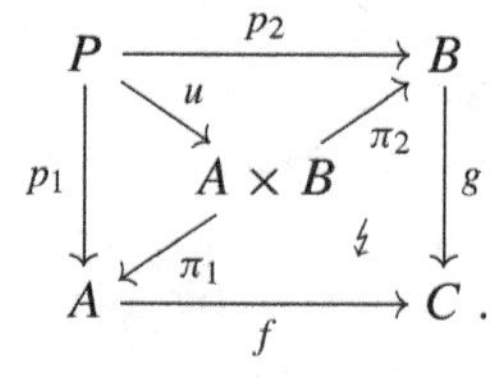

(iii) $\Longrightarrow$ (ii): Pullbacks over 1 give binary products; n-ary products can be constructed from binary products via $A_1 \times A_2 \times A_3 \cong (A_1 \times A_2) \times A_3$, etc. An equalizer u of $f, g : A \to B$ can be obtained as a pullback of $\delta = \langle 1_B, 1_B \rangle$ along $\langle f, g \rangle$:

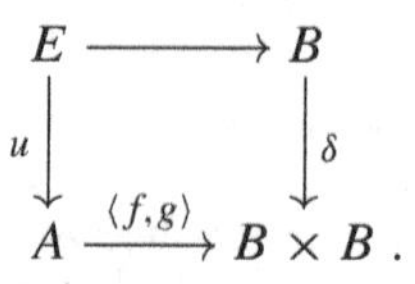

□

II.2.10.3 Example According to Theorem II.2.10.1, a limit (L, λ) of $D : \mathsf{J} \to \mathsf{Set}$ with J small is obtained as

$$L = \{(x_i)_{i \in \mathrm{ob}\,\mathsf{J}} \in \textstyle\prod_{i \in \mathrm{ob}\,\mathsf{J}} Di \mid \forall(\delta : j \to k) \text{ in } \mathsf{J}\ (D\delta(x_j) = x_k)\}\,,$$

with $\lambda_i : L \to Di$ the restriction of the product projection.

For the colimit (K, γ) of $D : \mathsf{J} \to \mathsf{Set}$ with J small, one considers dually the least equivalence relation $\sim$ on the disjoint union $\coprod_{i \in \mathrm{ob}\,\mathsf{J}} Di = \bigcup_{i \in \mathrm{ob}\,\mathsf{J}} Di \times \{i\}$ such that, for all $\delta : j \to k$ in J and $x \in Dj$,

$$(x, j) \sim (D\delta(x), k)\,.$$

Then $K = \coprod_{i \in \mathrm{ob}\,\mathsf{J}} Di/{\sim}$, and $\gamma_i(x)$ is the equivalence class of (x, i).

There is an important special case when this equivalence relation has an easy description, namely when J is an up-directed ordered set I, considered as a category J, with $\mathrm{ob}\,\mathsf{J} = I$. Then the diagram D is simply a family of sets and maps $f_{j,k} : X_j \to X_k$ (for $j \le k$ in I), with

$$f_{k,l} \cdot f_{j,k} = f_{j,l} \text{ if } j \le k \le l\,, \quad f_{j,j} = 1_{X_j}\,.$$

Now $\sim$ is given by

$$(x, j) \sim (y, k) \iff \exists l \ge j, k\ (f_{j,l}(x) = f_{k,l}(y))\,.$$

Classically, in any category C, *co*limits of diagrams in C of shape J, where J is an up-directed ordered set, are called *direct limits*, and limits of shape J^{op} are called *inverse limits* in C; we prefer to call them *directed colimits* and *directed limits*, respectively.

II.2.11 Preservation and reflection of limits and colimits

Let $F : \mathsf{C} \to \mathsf{D}$ be a functor and (L, λ) a limit of $D : \mathsf{J} \to \mathsf{C}$ in C. One says that *F preserves this limit* if $(FL, F\lambda)$ is a limit of FD in D. The functor *F preserves* J-*limits* (or is J-*continuous*) if F preserves all existing limits in C of diagrams of shape J; if C is J-complete, this property may be casually communicated by

$$F \lim \cong \lim F\,.$$

The functor *F reflects* J-*limits* if, for every cone $\lambda : \Delta L \to D$ (with $D : \mathsf{J} \to \mathsf{C}$) such that $(FL, F\lambda)$ is a limit of FD in D, then (L, λ) is a limit of D in C. This language is applied in various special contexts; for example, *F preserves monomorphisms* if Fm is monic in D whenever m is monic in C, and *F reflects monomorphisms* if m is monic whenever Fm is monic. This terminology is used similarly for colimits. For example, J-*cocontinuity* is *preservation of* J-*colimits*.

II.2.11.1 Proposition

(1) *A right adjoint functor preserves all (existing) limits, and a left adjoint functor preserves all (existing) colimits.*

(2) *A fully faithful functor reflects all (existing) limits and colimits.*

(3) *A replete reflective subcategory of a* J*-complete category is* J*-complete. Also, a replete reflective subcategory of a* J*-cocomplete category is* J*-cocomplete.*

Proof (1): If $F \dashv G : \mathsf{A} \to \mathsf{X}$ and $L \cong \lim D$, the following natural correspondences sketch the proof of $GL \cong \lim GD$:

$$\begin{array}{cl} \Delta X \longrightarrow GD & \\ \hline \Delta FX \longrightarrow D & (F \dashv G) \\ \hline FX \longrightarrow L & (L \cong \lim D) \\ \hline X \longrightarrow GL\,. & (F \dashv G) \end{array}$$

(2): If $G : \mathsf{A} \to \mathsf{X}$ is full and faithful and $GL \cong \lim GD$, one may sketch the proof for $L \cong \lim D$ as follows:

$$\begin{array}{cl} \Delta A \longrightarrow D & \\ \hline \Delta GA \longrightarrow GD & (G \text{ fully faithful}) \\ \hline GA \longrightarrow GL & (GL \cong \lim GD) \\ \hline A \longrightarrow L\,. & (G \text{ fully faithful}) \end{array}$$

Completeness part of (3): We show the stronger statement that A is *closed under limits* in X, i.e. whenever (L, λ) is a limit of JD in X (with $J : \mathsf{A} \hookrightarrow \mathsf{X}$), then $L \in \operatorname{ob}\mathsf{A}$. For this, it suffices to show that ρ_L is an isomorphism, with $\rho_L : L \to RL = JRL$ the A-reflection of L. But the inverse of ρ_L may be found as follows:

$$\begin{array}{cl} \Delta L \xrightarrow{\lambda} JD & \\ \hline \Delta RL \longrightarrow D & (R \dashv J) \\ \hline \Delta JRL \longrightarrow JD & (J \text{ fully faithful}) \\ \hline JRL \xrightarrow{\rho_L^{-1}} L\,. & (L \cong \lim JD) \end{array}$$

Cocompleteness part of (3): If (K, γ) is a colimit of JD in X, then $(RK, \Delta\rho_K \cdot \gamma)$ is a colimit of D in A:

$$\begin{array}{cl} D \longrightarrow \Delta A & \\ \hline JD \longrightarrow \Delta JA & (J \text{ fully faithful}) \\ \hline K \longrightarrow JA & (K \cong \operatorname{colim} JD) \\ \hline RK \longrightarrow A\,. & (R \dashv J) \end{array}$$

Hence, a left adjoint $\operatorname{colim}_{\mathsf{A}}$ to $\Delta_{\mathsf{A}} : \mathsf{A} \to \mathsf{A}^{\mathsf{J}}$ may be written as the composite functor

$$\mathsf{A}^{\mathsf{J}} \xrightarrow{J(-)} \mathsf{X}^{\mathsf{J}} \xrightarrow{\operatorname{colim}_{\mathsf{X}}} \mathsf{X} \xrightarrow{R} \mathsf{A}\,. \qquad \square$$

In addition to (3), note that, although the inclusion functor J preserves all limits, it generally fails to preserve colimits. Nonetheless, existence in A of colimits of a specified shape is guaranteed by their existence in X.

Limits and colimits in a functor category D^{C} can be formed "pointwise" in D. More precisely, for every $A \in \operatorname{ob}\mathsf{C}$, one has the *evaluation functor*

$$\mathrm{Ev}_A : \mathsf{D}^{\mathsf{C}} \to \mathsf{D}, \quad F \mapsto FA, \quad \alpha \mapsto \alpha_A .$$

A limit (L, λ) of $D : \mathsf{J} \to \mathsf{D}^{\mathsf{C}}$ can be formed by letting (LA, λ_A) be a limit of $\mathrm{Ev}_A D$ in D, if the latter exist for all $A \in \operatorname{ob}\mathsf{C}$; the same construction holds for colimits. This way one proves:

II.2.11.2 Proposition *The functor category D^{C} is J-complete if D is J-complete, and then the evaluation functors are J-continuous. Dually, D^{C} is J-cocomplete if D is J-cocomplete, in which case the evaluation functors are J-cocontinuous.*

II.2.11.3 Corollary *Every hom-functor $\mathsf{C}(A, -) : \mathsf{C} \to \mathsf{Set}$ of a locally small category C preserves all existing limits, and so does the Yoneda embedding $\mathbf{y} : \mathsf{C} \to \mathsf{Set}^{\mathsf{C}^{\mathrm{op}}}$.*

Proof When (L, λ) is a limit of D in C, the isomorphism $\mathsf{C}(A, L) \cong \lim \mathsf{C}(A, D(-))$ follows from

$$\frac{\dfrac{\dfrac{\Delta X \xrightarrow{\alpha} \mathsf{C}(A, D(-))}{\Delta A \xrightarrow{\alpha_x} D \ (x \in X)}}{A \xrightarrow{f_x} L \ (x \in X)}}{X \xrightarrow{f} \mathsf{C}(A, L)\,.} \qquad (L \cong \lim D)$$

To show that $\mathbf{y}L \cong \lim \mathbf{y}D$ in $\mathsf{Set}^{\mathsf{C}^{\mathrm{op}}}$, it suffices to show that $\mathrm{Ev}_A\,\mathbf{y}L \cong \lim \mathrm{Ev}_A\,\mathbf{y}D$ in Set for all $A \in \operatorname{ob}\mathsf{C}$. But $\mathrm{Ev}_A\,\mathbf{y} = \mathsf{C}(A, -)$, which preserves the limit, as we just saw. □

Note that, since Set is small-complete and small-cocomplete, every functor category of Set is too. Hence, every locally small category is fully embedded into a small-complete and small-cocomplete category by its Yoneda embedding.

II.2.12 Adjoint Functor Theorem

Preservation of limits by a functor $G : \mathsf{A} \to \mathsf{X}$ is a necessary condition for its right adjointness (see Proposition II.2.11.1). For ordered sets, completeness of A also makes this condition sufficient (see Corollary II.1.8.3). For general categories, however, small-completeness of A and preservation of limits by G do not generally suffice to make G right adjoint (see Exercise II.2.O). Here is a two-step procedure on how to construct a G-universal arrow for an object X in X.

Step 1

A *G-solution set* for X is a set $\mathcal{L}$ of objects in A with the property that, for all $f : X \to GA$ ($A \in \operatorname{ob}\mathsf{A}$), there exist $L \in \mathcal{L}$, $e : X \to GL$ in X, and $h : L \to A$ in A with $Gh \cdot e = f$. If X is locally small and A has products which are preserved by G, then there exists a *weakly G-universal arrow* for X, consisting of an object V in A and a morphism $v : X \to GV$ in X such that, for all $f : X \to GA$ ($A \in \mathrm{ob}\,\mathsf{A}$), there is a (not necessarily unique) morphism $g : V \to A$ with $Gg \cdot v = f$: indeed, simply take

$$V = \prod_{L\in\mathcal{L}} L^{\mathsf{X}(X,GL)}\,, \qquad v = \langle e\rangle_{e\in\mathsf{X}(X,GL),L\in\mathcal{L}}\,,$$

with $L^{\mathsf{X}(X,GL)} = \prod_{\mathsf{X}(X,GL)} L$, and with existence of v guaranteed by the product preservation by G.

Step 2

If A is locally small and has *generalized equalizers* (i.e. limits for non-empty sets – not just pairs – of parallel morphisms) that are preserved by G, then existence of a weakly G-universal arrow (V, v) for X forces existence of a G-universal arrow (U, u) for X as follows: let U be the limit of the diagram

$$\{t \in \mathsf{A}(V, V) \mid Gt \cdot v = v\} \longrightarrow \mathsf{A}\,,$$

with limit projection k; hence, $k : U \to V$ is universal with respect to the property $t \cdot k = k$ for all t with $Gt \cdot v = v$. Limit preservation by G gives the morphism $u : X \to GU$ with $Gk \cdot u = v$. Weak universality of (V, v) gives weak universality of (U, u). Furthermore, if $a, b : U \to A$ in A satisfy $Ga \cdot u = Gb \cdot u$, consider the equalizer $c : C \to U$ of a, b, and obtain $w : X \to GC$ with $Gc \cdot w = u$ from the limit preservation by G. Weak universality of (V, v) gives $d : V \to C$ with $Gd \cdot v = w$:

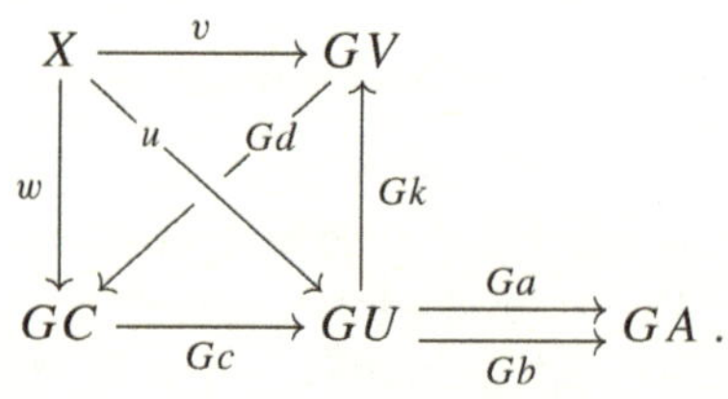

Since $G(k \cdot c \cdot d) \cdot v = G(k \cdot c) \cdot w = Gk \cdot u = v$, one has $k \cdot c \cdot d \cdot k = k$ by definition of k, and since k is monic (by the uniqueness part of the limit property) one obtains $c \cdot d \cdot k = 1_U$. Finally, since $a \cdot c = b \cdot c$, this implies $a = b$.

Hence, we have proved the non-trivial part of the following result:

II.2.12.1 Theorem *A functor $G : \mathsf{A} \to \mathsf{X}$ of locally small categories with A small-complete is right adjoint if and only if*

(1) *for every object X in X one has a chosen G-solution set, and*
(2) *G preserves small limits.*

Proof For the necessity of the conditions, if $F \dashv G$, then $\{FX\}$ is a G-solution set for $X \in \mathrm{ob}\,\mathsf{X}$, and limit preservation by G follows from Proposition II.2.11.1. □

II.2.12.2 Example With Grp denoting the category of groups, let us prove the existence of a left adjoint to the forgetful functor U : Grp $\longrightarrow$ Set, using Theorem II.2.12.1. For a set X and a map $f : X \longrightarrow G$ into a group G such that $f(X)$ generates G, the cardinality of G cannot exceed the cardinality of $(X+1)^{\mathbb{N}}$ (since there is a surjection $(X+1)^{\mathbb{N}} \longrightarrow G$). Hence, as a U-solution set for X, one may choose a representative system of non-isomorphic groups whose cardinality does not exceed the cardinality of $(X+1)^{\mathbb{N}}$. Since small limits in Grp may be formed by taking the limit in Set, and providing it with a group structure, small-limit preservation by U is clear. (Another argument for limit preservation would be $U \cong$ Grp$(\mathbb{Z}, -)$.) Hence, Theorem II.2.12.1 guarantees existence of the *free-group functor* F.

The argument just given extends to any algebraic structure defined by a set of operations and equations between them. That is, groups may be traded in this example for rings, R-modules, R-algebras, etc.

II.2.13 Kan extensions

For functors S : A $\longrightarrow$ B and T : A $\longrightarrow$ X, the *right Kan extension* of T along S is a F-couniversal arrow for T, where F is the whiskering functor

$$F = (-)S : \mathsf{X}^{\mathsf{B}} \longrightarrow \mathsf{X}^{\mathsf{A}}, \quad \alpha \mapsto \alpha S .$$

Hence, the right Kan extension is given by a functor K : B $\longrightarrow$ X and a natural transformation $\kappa : KS \longrightarrow T$ such that

> for all $\alpha : QS \longrightarrow T$ in A (with Q : B $\longrightarrow$ X), there is precisely one natural transformation $\sigma : Q \longrightarrow K$ with $\kappa \cdot \sigma S = \alpha$.

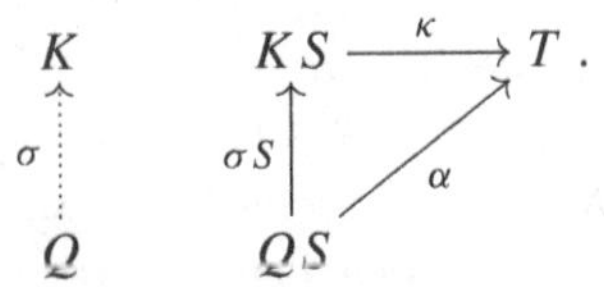

It is common to use the notation

$$K \cong \mathrm{Ran}_S\, T .$$

By Theorem II.2.5.4^{op}, if the right Kan extension of all functors T : A $\longrightarrow$ X along a fixed S exist, then $(-)S$ has a right adjoint. However, the Kan extension is useful without the existence of the entire right adjoint.

In the previous definition, one can factor in the categories to obtain the following picture for the right Kan extension (K, κ) of T along S:

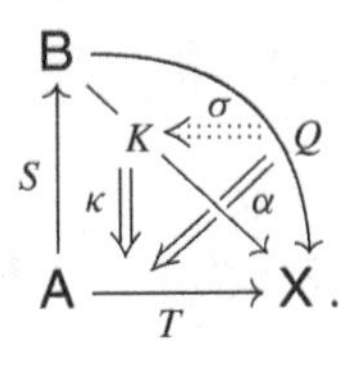

The terminology is explained by the case where A is a full subcategory of B, as K then does indeed *extend T along the full embedding S*. This extension is made to the *right* because one has a natural bijection

$$\mathsf{X}^{\mathsf{A}}(QS, T) \cong \mathsf{X}^{\mathsf{B}}(Q, \mathrm{Ran}_S\, T)\ .$$

To construct the right Kan extension, consider an object B in B. For any A in A admitting a B-morphism $f : B \to SA$, one expects to obtain an X-morphism $Kf : KB \to TA$. In fact, for any commutative diagram

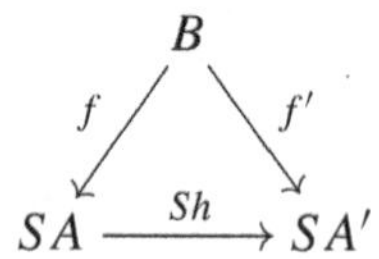

in B, one wants a similar commutative diagram in X, where B is replaced by KB and Sh is replaced by Th. Taking into consideration the couniversal property of the right Kan extension, one defines

$$KB \cong \lim D\ ,$$

where $D : (B \downarrow S) \to \mathsf{X}$ is the diagram sending an object $f : B \to SA$ to TA, and a morphism $h : (A, f) \to (A', f')$ to $Th : TA \to TA'$:

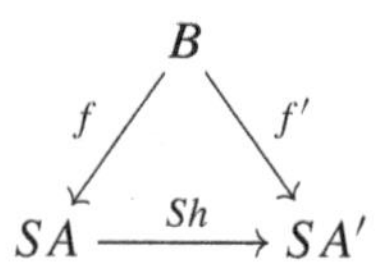

$$\longmapsto \qquad TA \xrightarrow{Th} TA'\ .$$

II.2.13.1 Theorem *Suppose that functors $S : \mathsf{A} \to \mathsf{B}$ and $T : \mathsf{A} \to \mathsf{X}$ are given such that the diagram $D : (B \downarrow S) \to \mathsf{X}$ has a limit for all objects B of B. Then the right Kan extension (K, κ) of T along S exists. When S is full and faithful, the natural transformation κ is an isomorphism.*

Proof The previous construction defined on objects of B extends to yield a functor $K : \mathsf{B} \to \mathsf{X}$ via the couniversal property of limits. The component at $A \in \mathsf{A}$ of the natural transformation $\kappa : KS \to T$ is then obtained as the component $\lambda_{1_{SA}} : KSA \to TA$ at 1_{SA} of the limit cone of $D : (SA \downarrow S) \to \mathsf{X}$. When S is full and faithful, every object $SA \to SB$ of the comma category $(SA \downarrow S)$ can be written as Sf for a unique A-morphism $f : A \to B$; hence, 1_{SA} is an initial object in this category, and the construction of KSA yields an object isomorphic to TA. □

If the right Kan extension (K, κ) of T along S is constructed as in Theorem II.2.13.1, so that

$$KB \cong \lim(\,(B \downarrow S) \longrightarrow \mathsf{A} \xrightarrow{\ T\ } \mathsf{X}\,)$$

for all $B \in \mathrm{ob}\,\mathsf{B}$, then the extension is called *pointwise*.

The *left Kan extension*

$$K \cong \mathrm{Lan}_S T : \mathsf{B} \to \mathsf{X}$$

of $T : \mathsf{A} \to \mathsf{X}$ along $S : \mathsf{A} \to \mathsf{B}$ is defined dually: K^{op} is the right Kan extension of T^{op} along S^{op}, so that K comes with a $(-)S$-universal natural transformation $\kappa : T \to KS$. This is briefly summarized in

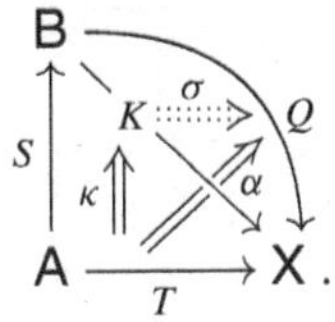

Again, the extension is *pointwise* if

$$KB \cong \mathrm{colim}(\,(S \downarrow B) \longrightarrow \mathsf{A} \xrightarrow{\;T\;} \mathsf{X}\,)$$

for all objects B in B, with the colimit cone constructed dually to Theorem II.2.13.1.

II.2.14 Dense functors

A functor $J : \mathsf{A} \to \mathsf{X}$ is *dense* if $(1_{\mathsf{X}}, 1_J)$ is a pointwise left Kan extension of J along J; i.e. if, for all $X \in \mathrm{ob}\,\mathsf{X}$, one has $\mathrm{colim}\, D_X \cong X$, with

$$D_X = (\,(J \downarrow X) \longrightarrow \mathsf{A} \xrightarrow{\;J\;} \mathsf{X}\,)$$

and the colimit cocone given by

$$\lambda_{(A,f)} = f : JA \to X$$

for all $(A, f) \in \mathrm{ob}(J \downarrow X)$. Arbitrary cocones

$$\alpha : D_X \to \Delta Y$$

correspond bijectively to natural transformations

$$\gamma : \mathsf{X}(J-, X) \to \mathsf{X}(J-, Y)$$

via

$$\gamma_A(f) = \alpha_{(A,f)}$$

for $A \in \mathrm{ob}\,\mathsf{A}$, $f : JA \to X$ in X, naturally in $Y \in \mathrm{ob}\,\mathsf{X}$.

II.2.14.1 Proposition *For a functor $J : \mathsf{A} \to \mathsf{X}$ of locally small categories, the following statements are equivalent:*

(i) *J is dense;*
(ii) *the maps $\mathsf{X}(X, Y) \to \widehat{\mathsf{A}}(\mathsf{X}(J-, X), \mathsf{X}(J-, Y))$, $f \mapsto \mathsf{X}(J-, f)$ are bijective for all $X, Y \in \mathrm{ob}\,\mathsf{X}$;*
(iii) *the functor $\widehat{J} : \widehat{\mathsf{X}} \to \widehat{\mathsf{A}}$, $\phi \mapsto \phi J^{\mathrm{op}}$ is fully faithful.*

Proof Condition (i) means that there is a natural bijective correspondence

$$\frac{X \longrightarrow Y}{D_X \longrightarrow \Delta Y},$$

and condition (ii) means that there is a natural bijective correspondence

$$\frac{X \longrightarrow Y}{\mathsf{X}(J-, X) \longrightarrow \mathsf{X}(J-, Y)}.$$

By the introductory observations, (i) and (ii) are therefore equivalent. Since $\mathsf{X}(X, Y) \cong \widehat{\mathsf{X}}(\mathsf{X}(-, X), \mathsf{X}(-, Y))$ by the Yoneda Lemma, (ii) and (iii) are equivalent as well. □

II.2.14.2 Theorem *For every locally small category* C, *the Yoneda embedding is dense.*

Proof The Yoneda Lemma facilitates a natural bijective correspondence

$$F \cong \widehat{C}(\mathbf{y}-, F)$$

for all $F : \mathsf{C}^{\mathrm{op}} \to \mathsf{Set}$, with $\mathbf{y} : \mathsf{C} \to \widehat{\mathsf{C}} = \mathsf{Set}^{\mathsf{C}^{\mathrm{op}}}$ denoting the Yoneda embedding. Hence, for $F, G \in \mathrm{ob}\,\widehat{\mathsf{C}}$, there is a natural bijective correspondence

$$\frac{F \longrightarrow G}{\widehat{\mathsf{C}}(\mathbf{y}-, F) \longrightarrow \widehat{\mathsf{C}}(\mathbf{y}-, G)},$$

showing that $\mathbf{y}$ satisfies condition (ii) of Proposition II.2.14.1. □

II.2.14.3 Corollary *Every functor* $F : \mathsf{C}^{\mathrm{op}} \to \mathsf{Set}$ *is the colimit of the diagram*

$$(1 \downarrow F)^{\mathrm{op}} \cong \mathrm{el}(F)^{\mathrm{op}} \longrightarrow \mathsf{C} \xrightarrow{\ \mathbf{y}\ } \widehat{\mathsf{C}}$$

(with 1 *the terminal object in* Set *and* $\mathrm{el}(F)$ *the category of "elements of F," see* Exercise II.2.P*).*

Proof By the Yoneda Lemma, $(1 \downarrow F)^{\mathrm{op}} \cong (\mathbf{y} \downarrow F)$, and the diagram becomes the same as the one used in the defining property of the density of $\mathbf{y}$. □

The Yoneda embedding of a locally small category C therefore embeds C into the category $\widehat{\mathsf{C}}$ with all small colimits, and every object in $\widehat{\mathsf{C}}$ is naturally a colimit of objects originating from C.

Exercises

II.2.A *Isomorphisms and functors.* If $f : A \to B$, $g : B \to C$ are isomorphisms in a category C, then so are $g \cdot f$ and f^{-1}, and $(g \cdot f)^{-1} = f^{-1} \cdot g^{-1}$, $(f^{-1})^{-1} = f$. Every functor $F : \mathsf{C} \to \mathsf{D}$ *preserves isomorphisms* but does not generally *reflect* them: if f is an isomorphism, then so is Ff, but not conversely in general; as an example, consider the forgetful functor $\mathsf{Ord} \to \mathsf{Set}$. However, if F is fully faithful, then F reflects isomorphisms.

II.2.B *Monomorphisms, epimorphisms, and functors.* Consider $f : A \to B$, $g : B \to C$ in C. If f and g are monic, then so is $g \cdot f$; if $g \cdot f$ is monic, then so is f. Establish the dual statements for epimorphisms. If $F : \mathsf{C} \to \mathsf{D}$ preserves kernel pairs, then F preserves monomorphisms; in particular, every right adjoint functor and every hom-functor does so. A faithful functor reflects both monomorphisms and epimorphisms. The monomorphisms of Set are precisely the injective maps, and the epimorphisms are the surjective maps. Not every epimorphism in Mon is surjective, and a morphism that is both monic and epic need not be an isomorphism.

II.2.C *Split monomorphisms and split epimorphisms.* A morphism $f : A \to B$ in C is a *split monomorphism* (or a *section*) if $g \cdot f = 1_A$ for some $g : B \to A$ in C; the dual notion is that of *split epimorphism* (or *retraction*). Every split monomorphism is monic, and it is an isomorphism precisely when it is epic. Composites of split epimorphisms are split epic, and split epimorphisms are stable under pullback. Dualize. In Set, every monomorphism $\varnothing \neq A \to B$ splits, and to say that every epimorphism splits is equivalent to the Axiom of Choice.

II.2.D *Regular epimorphisms.* A morphism $f : A \to B$ in C is a *regular epimorphism* in C if for any $g : A \to C$ with the property that $f \cdot x = f \cdot y$ always implies $g \cdot x = g \cdot y$, one has a uniquely determined morphism $h : B \to C$ with $h \cdot f = g$. Every split epimorphism is regular, and every regular epimorphism is in fact an epimorphism. If C has kernel pairs, then f is a regular epimorphism if and only if f is the coequalizer of some pair of morphisms $p, q : K \to A$; in fact, p, q may always be chosen to be the kernel pair of f. Dualize to *regular monomorphisms*.

II.2.E *Full or faithful right adjoints.* If $F \overset{\eta}{\underset{\varepsilon}{\dashv}} G : \mathsf{A} \to \mathsf{X}$, then G is faithful if and only if ε_A is epic for all $A \in \operatorname{ob}\mathsf{A}$, and G is full if and only if ε_A is a split monomorphism for all $A \in \operatorname{ob}\mathsf{A}$.

II.2.F *Limits and colimits in slices of categories.* For any object A of a category C, the comma category C/A is finitely complete when C has fibered products. If C is J-cocomplete, so is C/A, and the forgetful functor $\mathsf{C}/A \to \mathsf{C}$, $(X, f) \mapsto X$, is J-cocontinuous.

II.2.G *Products and coproducts in* Rel. Coproducts in Rel are formed as in Set, and also serve as products. In the category AbGrp of Abelian groups, finite coproducts and finite products may be formed using the same Abelian group.

II.2.H *Cartesian structure of categories with finite products.* In a category with finite products, establish natural isomorphisms $\alpha, \lambda, \rho, \sigma$ satisfying the same coherence conditions as in Section II.1.1.

II.2.I *Products as down-directed limits of finite products.* For a family $(A_i)_{i\in I}$ of objects in a category C with finite products, put $A_F = \prod_{i\in F} A_i$ for every finite set $F \subseteq I$. If $G \subseteq F$, there is then an obvious morphism $A_F \longrightarrow A_G$. With $\mathcal{F}$ the set of finite subsets of I ordered by inclusion, this defines a diagram $D : \mathcal{F}^{\mathrm{op}} \longrightarrow \mathsf{C}$, and the limit of D exists in C if and only if the product $\prod_{i \in I} A_i$ exists, and then both limits coincide up to isomorphism.

II.2.J *Hom-functors.* The metacategory CAT has products and coproducts. For every category C, there is a functor $\mathsf{C} : \mathsf{C}^{\mathrm{op}} \times \mathsf{C} \longrightarrow \mathsf{SET}$. Does C preserve products or coproducts?

II.2.K *Completeness of* Ord *and* Top. The categories Ord and Top are both small-complete and small-cocomplete. The respective forgetful functors to Set preserve all existing limits and colimits.

II.2.L *Representable functors.* A functor $H : \mathsf{C} \longrightarrow \mathsf{Set}$ (with C locally small) is *representable* if it is isomorphic to a hom-functor $\mathsf{C}(A, -)$. If H is right adjoint, it is representable. (*Hint.* If $F \dashv H$, then $\mathsf{Set}(1, H(-)) \cong \mathsf{C}(F1, -)$.) Conversely, a representable functor $H \cong \mathsf{C}(A, -)$ is right adjoint if and only if for every set X the coproduct $X \cdot A := \coprod_{x \in X} A_x$ with $A_x = A$ for all $x \in X$ exists in C.

II.2.M *Equivalence of categories, irrelevance of injectivity on objects.* Equivalence of categories is a reflexive, symmetric, and transitive property. A category equivalent to a J-complete category is J-complete itself. Every functor $F : \mathsf{C} \longrightarrow \mathsf{D}$ can be factored as

$$F = (\,\mathsf{C} \xrightarrow{\;\widetilde{F}\;} \widetilde{\mathsf{D}} \xrightarrow{\;U\;} \mathsf{D}\,)\,,$$

with $\widetilde{F}$ injective on objects and U an equivalence of categories. (*Hint.* When $\mathsf{C} \neq \varnothing$, let $\mathrm{ob}\,\widetilde{\mathsf{D}} = \mathrm{ob}\,\mathsf{C} \times \mathrm{ob}\,\mathsf{D}$ and $\widetilde{\mathsf{D}}((A, B), (A', B')) = \mathsf{D}(B, B')$.) Moreover, $\widetilde{F}$ is full or faithful precisely when F has the respective property. If F is faithful, C is isomorphic to a *subcategory* $\widetilde{\mathsf{C}}$ of $\widetilde{\mathsf{D}}$, so that there are inclusion functions both for objects and hom-sets forming a functor $\widetilde{\mathsf{C}} \longrightarrow \widetilde{\mathsf{D}}$.

II.2.N *Formal adjointness criterion.* A functor $G : \mathsf{A} \longrightarrow \mathsf{X}$ is right adjoint if and only if for every object X in X the (not necessarily small) diagram

$$D_X : (X \downarrow G) \longrightarrow \mathsf{A}\,, \quad (A, f) \longmapsto A$$

has a chosen limit in A that is preserved by G.
Hint. $FX \cong \lim D_X$.

II.2.O *Limit preservation does not suffice for right adjointness.* Let $\mathcal{S}$ be the class of all sets ordered by set inclusion. Then every non-empty subclass has a least element, but there is no top element. Hence, when considered as a category, $\mathcal{S}^{\mathrm{op}}$ is small-complete, and the only functor $\mathcal{S}^{\mathrm{op}} \longrightarrow \mathbf{1}$ trivially preserves all limits but has no left adjoint.

II.2.P *The element construction.* For a functor $H : \mathsf{C} \to \mathsf{Set}$ (with C locally small), one defines the category $\mathrm{el}(H)$ of "elements of H" to have objects (A, x) with $A \in \mathrm{ob}\,\mathsf{C}$, $x \in HA$; a morphism $f : (A, x) \to (B, y)$ is a C-morphism $f : A \to B$ with $Hf(x) = y$. Then H is representable (see Exercise II.2.L) if and only if $\mathrm{el}(H)$ has an initial object.

II.2.Q *Connectedness, colimits in* Set *revisited.* A category C is *connected* if $\mathsf{C} \neq \varnothing$ and, whenever $\mathsf{C} = \mathsf{A} + \mathsf{B}$ in CAT, $\mathsf{A} = \varnothing$ or $\mathsf{B} = \varnothing$; equivalently, if $\mathsf{C} \neq \varnothing$ and, for any $A, B \in \mathrm{ob}\,\mathsf{C}$, there is a finite "zigzag" $A \to \cdot \leftarrow \cdot \to \cdot \dots \cdot \to B$ of morphisms in C. Every category is a (not necessarily small) coproduct of its *connected components* (i.e. its maximal connected full subcategories). Show also that the colimit of a small diagram $D : \mathsf{J} \to \mathsf{Set}$ can be taken to be the set of connected components of the category $\mathrm{el}(D)$ (see Exercise II.2.P).

II.2.R *General comma categories.* For functors $G : \mathsf{A} \to \mathsf{X}$, $H : \mathsf{C} \to \mathsf{X}$, consider the category $(G \downarrow H)$ whose objects (A, f, C) are given by $A \in \mathrm{ob}\,\mathsf{A}$, $C \in \mathrm{ob}\,\mathsf{C}$, and $f : GA \to HC$ in X; a morphism $(h, j) : (A, f, C) \to (B, g, D)$ is formed by morphisms $h : A \to B$ in A and $j : C \to D$ in C that make

$$\begin{array}{ccc} GA & \xrightarrow{Gh} & GB \\ {\scriptstyle f}\downarrow & & \downarrow{\scriptstyle g} \\ HC & \xrightarrow{Hj} & HD \end{array}$$

commute in X. Composition is such that the projections

$$P : (G \downarrow H) \to \mathsf{A}, \quad (h, j) \mapsto h,$$
$$Q : (G \downarrow H) \to \mathsf{C}, \quad (h, j) \mapsto j,$$

become functors, and

$$\kappa : GP \to HQ, \quad \kappa_{(A,f,B)} = f$$

a natural transformation. Show that, given any functors $R : \mathsf{K} \to \mathsf{A}$, $S : \mathsf{K} \to \mathsf{C}$ and a natural transformation $\phi : GR \to HS$, there is a unique functor $T : \mathsf{K} \to (G \downarrow H)$ with

$$PT = R, \quad QT = S, \quad \kappa T = \phi.$$

Observe also that the comma categories discussed in Section II.2.7 are all special cases of the general type introduced here, up to isomorphism.

II.2.S *Testing limits via Yoneda.* A cone $\lambda : \Delta A \to D$ in a locally small category C is a limit if and only if, for all objects X in C, the cone $\mathsf{X}(X, \lambda_{(-)}) : \Delta\mathsf{X}(X, A) \to \mathsf{X}(X, D-)$ is a limit in Set.

II.2.T *Quantales freely generated by monoids.* Show that the forgetful functor $\mathsf{Qnt} \to \mathsf{Mon}$ has a left adjoint (see Exercise II.1.M).

II.3 Monads

II.3.1 Monads and adjunctions

A *monad* $\mathbb{T} = (T, m, e)$ on a category X is given by a functor $T : \mathsf{X} \to \mathsf{X}$ and two natural transformations, the *multiplication* and *unit* of the monad

$$m : TT \to T\ , \quad e : 1_{\mathsf{X}} \to T\ ,$$

satisfying the *multiplication law* and the *right* and *left unit laws*

$$m \cdot mT = m \cdot Tm\ , \quad m \cdot eT = 1_T = m \cdot Te\ ;$$

equivalently, these equalities mean that the diagrams

$$\begin{array}{ccc} TTT & \xrightarrow{Tm} & TT \\ {\scriptstyle mT}\downarrow & & \downarrow{\scriptstyle m} \\ TT & \xrightarrow{m} & T \end{array} \qquad\qquad \begin{array}{ccccc} T & \xrightarrow{eT} & TT & \xleftarrow{Te} & T \\ & {\scriptstyle 1_T}\searrow & \downarrow{\scriptstyle m} & \swarrow{\scriptstyle 1_T} & \\ & & T & & \end{array}$$

commute. A *morphism of monads* $\alpha : \mathbb{S} \to \mathbb{T}$ (where $\mathbb{S} = (S, n, d)$) is a natural transformation $\alpha : S \to T$ that preserves the monad structure:

$$\alpha \cdot n = m \cdot (\alpha \circ \alpha)\ , \quad \alpha \cdot d = e\ .$$

Hence, comparing with the definition of a monoid given in Section II.1.1, we observe that the set M has been replaced by a functor, the Cartesian product by horizontal composition, and maps by natural transformations. Since the natural isomorphisms α, λ, and β are identities, the coherence conditions are immediately verified, so that a monad is nothing but an object of the functor category X^{X} that carries a monoid structure with respect to the (horizontal) compositional structure.

Any adjunction $F \overset{\eta}{\underset{\varepsilon}{\dashv}} G : \mathsf{A} \to \mathsf{X}$ gives rise to the *associated monad* (or *induced monad*) $\mathbb{T} = (GF, G\varepsilon F, \eta)$ on X. Indeed, by whiskering the identities $\varepsilon \cdot \varepsilon FG = \varepsilon \cdot FG\varepsilon$, $G\varepsilon \cdot \eta G = 1_G$, and $\varepsilon F \cdot F\eta = 1_F$ appropriately by F on the right and G on the left, we obtain the corresponding multiplication and unit laws:

$$G\varepsilon F \cdot G\varepsilon FGF = G\varepsilon F \cdot GFG\varepsilon F\ , \quad G\varepsilon F \cdot \eta GF = 1_{GF} = G\varepsilon F \cdot GF\eta\ .$$

In fact, *any* monad can be obtained as a monad associated to an adjunction, although different adjunctions may yield the same monad. The "largest" of these adjunctions is described in Section II.3.2, and the "smallest" is treated in Section II.3.6.

II.3.1.1 Examples

(1) A closure operation on an ordered set X is a monad on X (considered as a category), and the closure operation $c = g \cdot f$ associated to an adjunction of ordered sets $f \dashv g : X \to Y$ is also the associated monad.

(2) The *list monad* (or the *free-monoid monad*) $\mathbb{L} = (L = GF, G\varepsilon F, \eta)$ on Set is the monad associated to the adjunction $F \dashv G : \mathsf{Mon} \to \mathsf{Set}$ described in Examples II.2.5.1. Similarly, from Example II.2.12.2, one obtains the free-group monad, the free-Abelian-group monad, the free-ring monad, etc.

(3) The covariant powerset functor $P : \mathsf{Set} \to \mathsf{Set}$, together with the union $m_X : PPX \to PX$ and singleton $e_X : X \to PX$ maps

$$m_X(\mathcal{A}) = \bigcup \mathcal{A}\,, \quad e_X(x) = \{x\}\,,$$

for all $x \in X$, $\mathcal{A} \in PPX$, form the *powerset monad* $\mathbb{P} = (P, m, e)$.

(4) The contravariant powerset adjunction $(P^\bullet)^{\mathrm{op}} \dashv P^\bullet : \mathsf{Set}^{\mathrm{op}} \to \mathsf{Set}$ (see Example II.2.5.1(6)) has the *double-powerset monad* $\mathbb{P}^2 = (P^2 = P^\bullet(P^\bullet)^{\mathrm{op}}, m, e)$ as its associated monad on Set, with $P^2 f(\mathfrak{x}) = f[\mathfrak{x}]$ defined by

$$B \in f[\mathfrak{x}] \iff f^{-1}(B) \in \mathfrak{x}$$

for $f : X \to Y$, and all $B \in PY$, $\mathfrak{x} \in P^2X$. The components $m_X : P^2P^2X \to P^2X$ are defined as in the filtered sum construction (see Section II.1.12):

$$m_X(\mathfrak{A}) = \sum \mathfrak{A} \quad (\mathfrak{A} \subseteq PPPX)\,,$$

with $(A \in \sum \mathfrak{A} \iff \{a \subseteq PX \mid A \in a\} \in \mathfrak{A})$ for all $A \subseteq X$, and the components $e_X : X \to P^2X$ are given by the principal filter

$$e_X(x) = \dot{x} \quad (x \in X)\,.$$

(5) The double-powerset functor $P^2 : \mathsf{Set} \to \mathsf{Set}$, along with its monad structure, can be restricted to the

- *up-set functor* U with $UX = \{a \subseteq PX \mid \uparrow_{PX} a = a\}$,
- *filter functor* F with $FX = \{a \subseteq PX \mid a \text{ a filter on } X\}$,
- *ultrafilter functor* β with $\beta X = \{a \subseteq PX \mid a \text{ an ultrafilter on } X\}$,
- *identity functor* 1_{Set} with $1_{\mathsf{Set}}X = X \cong \{\dot{x} \mid x \in X\}$,

and we obtain the *monads* $\mathbb{U} = (U, m, e)$, $\mathbb{F} = (F, m, e)$, $\mathbb{B} = (\beta, m, e)$, $\mathbb{I} = (1_{\mathsf{Set}}, 1, 1)$. These monads can be thought of as being induced by the adjunctions

$$\mathsf{C}^{\mathrm{op}} \underset{(P^\bullet)^{\mathrm{op}}}{\overset{\mathsf{C}(-,2)}{\underset{\longleftarrow}{\overset{\longrightarrow}{\top}}}} \mathsf{Set}$$

of Example II.2.5.3, with successively $\mathsf{C} = \mathsf{Ord}, \mathsf{SLat}, \mathsf{Lat}$, and Frm. We obtain a chain of monad morphisms

$$\mathbb{I} \to \mathbb{B} \to \mathbb{F} \to \mathbb{U} \to \mathbb{P}^2\,,$$

all given objectwise by inclusion maps.

Note that there is also for each set X a map $\alpha_X : PX \to FX$ sending a subset A of X to the principal filter $\dot{A}$, and extending to a monad morphism $\alpha : \mathbb{P} \to \mathbb{F}$.

(6) On every category X, there is a unique monad morphism from the identity monad $\mathbb{I} = (1_{\mathsf{X}}, 1, 1)$ to any monad $\mathbb{T} = (T, m, e)$ on X, and it is given by the unit $e : 1_{\mathsf{X}} \to T$ of $\mathbb{T}$. Hence, the identity monad is an initial object in the metacategory $\mathsf{MND}_{\mathsf{X}}$ of *monads* on X, with *monad morphisms*.

II.3.2 The Eilenberg–Moore category

Given a monad $\mathbb{T} = (T, m, e)$ on X, a $\mathbb{T}$*-algebra* (or *Eilenberg–Moore algebra*) is a pair (X, a), where X is an object of X, and the *structure morphism* $a : TX \to X$ satisfies

$$a \cdot Ta = a \cdot m_X \qquad \text{and} \qquad 1_X = a \cdot e_X \;;$$

diagrammatically:

$$\begin{array}{ccc} TTX & \xrightarrow{Ta} & TX \\ {\scriptstyle m_X}\downarrow & & \downarrow{\scriptstyle a} \\ TX & \xrightarrow{a} & X \end{array} \qquad\qquad \begin{array}{ccc} X & \xrightarrow{e_X} & TX \\ & {\scriptstyle 1_X}\searrow & \downarrow{\scriptstyle a} \\ & & X\,. \end{array}$$

A $\mathbb{T}$*-homomorphism* $f : (X, a) \to (Y, b)$ is an X-morphism $f : X \to Y$ such that

$$f \cdot a = b \cdot Tf \;,$$

which amounts to commutativity of the diagram

$$\begin{array}{ccc} TX & \xrightarrow{Tf} & TY \\ {\scriptstyle a}\downarrow & & \downarrow{\scriptstyle b} \\ X & \xrightarrow{f} & Y\,. \end{array}$$

The category of $\mathbb{T}$-algebras and $\mathbb{T}$-homomorphisms is denoted by $\mathsf{X}^{\mathbb{T}}$ and is also called the *Eilenberg–Moore category* of $\mathbb{T}$.

The forgetful functor $G^{\mathbb{T}} : \mathsf{X}^{\mathbb{T}} \to \mathsf{X}$ has a left adjoint $F^{\mathbb{T}} : \mathsf{X} \to \mathsf{X}^{\mathbb{T}}$:

$$X \mapsto (TX, m_X)\,, \quad (f : X \to Y) \mapsto (Tf : TX \to TY)\,,$$

where (TX, m_X) is the *free* $\mathbb{T}$*-algebra* on X. The unit $\eta^{\mathbb{T}} : 1_{\mathsf{X}} \to G^{\mathbb{T}}F^{\mathbb{T}} = T$ of this adjunction is e, and the counit is described by its components as

$$\varepsilon^{\mathbb{T}}_{(X,a)} = a : (TX, m_X) \to (X, a)\,.$$

Since the monad associated to $F^{\mathbb{T}} \dashv G^{\mathbb{T}}$ gives back the original $\mathbb{T} = (T, m, e)$, any monad may be obtained via an adjunction. Among such adjunctions, $F^{\mathbb{T}} \dashv G^{\mathbb{T}}$ is characterized as follows:

II.3.2.1 Proposition *If* $\mathbb{T} = (T, m, e)$ *is the monad associated to* $F \dashv_{\varepsilon}^{\eta} G : \mathsf{A} \longrightarrow \mathsf{X}$, *there exists a unique functor* $K : \mathsf{A} \to \mathsf{X}^{\mathbb{T}}$ *making both the inner and outer triangles of the following diagram commute:*

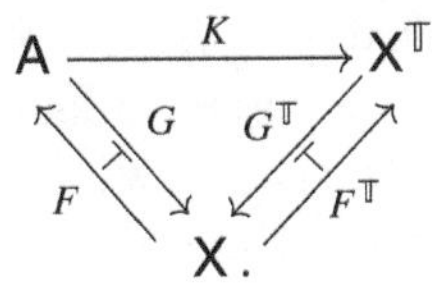

(II.3.2.i)

Proof The functor $K : \mathsf{A} \to \mathsf{X}^{\mathbb{T}}$ defined on objects and morphisms by

$$A \mapsto (GA, G\varepsilon_A)\,, \quad (f : A \to B) \mapsto (Gf : GA \to GB)$$

makes both triangles in (II.3.2.i) commute. If $K' : \mathsf{A} \to \mathsf{X}^{\mathbb{T}}$ is another such functor, commutativity of (II.3.2.i) yields $K'A = (GA, a)$ (for a certain $\mathbb{T}$-algebra structure $a : TGA \to GA$) and $K'f = Gf$; since $K'\varepsilon_A = G\varepsilon_A : K'FGA \to K'A$ is an $\mathsf{X}^{\mathbb{T}}$-morphism, the following diagram commutes:

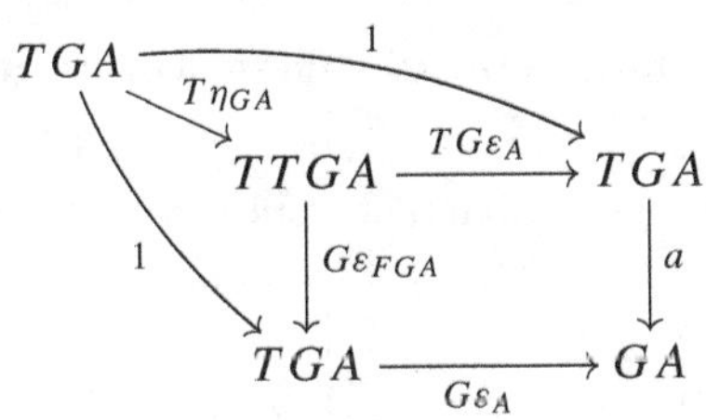

which tells us that $a = G\varepsilon_A$. □

A right adjoint functor $G : \mathsf{A} \to \mathsf{X}$ is *monadic* if the *comparison functor* $K : \mathsf{A} \to \mathsf{X}^{\mathbb{T}}$ of the previous proposition is an equivalence, and G is *strictly monadic* if K is an isomorphism; one then says that A is *monadic over* X, or *strictly monadic over* X, respectively. A monadic functor $G : \mathsf{A} \to \mathsf{X}$ can be considered as a measure of the "algebraic character" of A over X, where the Eilenberg–Moore category $\mathsf{X}^{\mathbb{T}}$ represents the "algebraic part" of A.

II.3.2.2 Examples

(1) The forgetful functor $G : \mathsf{Mon} \to \mathsf{Set}$ is monadic; in fact, the comparison functor is even an isomorphism:

$$\mathsf{Set}^{\mathbb{L}} \cong \mathsf{Mon}\,.$$

Similarly, the Eilenberg–Moore category of the free-group monad is isomorphic to Grp, that of the free-Abelian-group monad is isomorphic to AbGrp, etc.

(2) For the powerset monad $\mathbb{P}$, the structure map $a : PX \to X$ of a $\mathbb{P}$-algebra (X, a) defines a complete order on X via

$$\bigvee A := a(A) ,$$

and every $\mathbb{P}$-homomorphism $f : (X, a) \to (Y, b)$ is then a sup-map. On the other hand, when X is a complete lattice, the map $\bigvee : PX \to X$ is a $\mathbb{P}$-algebra structure, and sup-maps between complete lattices become $\mathbb{P}$-homomorphisms. Therefore, there is an isomorphism

$$\mathsf{Set}^{\mathbb{P}} \cong \mathsf{Sup} ,$$

and the forgetful functor $\mathsf{Sup} \to \mathsf{Set}$ is strictly monadic.

(3) Neither of the forgetful functors $\mathsf{Ord} \to \mathsf{Set}$ or $\mathsf{Top} \to \mathsf{Set}$ is monadic, since the Eilenberg–Moore category of each of these is equivalent to Set.

(4) Any full and faithful right adjoint functor is monadic (see also Exercise II.3.M).

II.3.3 Limits in the Eilenberg–Moore category

The forgetful functor $G^{\mathbb{T}} : \mathsf{X}^{\mathbb{T}} \to \mathsf{X}$ is very well behaved with respect to limits. Because $G^{\mathbb{T}}$ is right adjoint, it naturally preserves all the limits that exist in $\mathsf{X}^{\mathbb{T}}$ (Proposition II.2.11.1). It also reflects limits; in fact, it does better than that.

One says that a functor $F : \mathsf{C} \to \mathsf{D}$ *creates* J*-limits* if for every diagram $D : \mathsf{J} \to \mathsf{C}$, such that (A, α) is a limit of FD in D, there is a unique pair (L, λ) consisting of an object L of C and a cone $\lambda : \Delta L \to D$ satisfying $(FL, F\lambda) = (A, \alpha)$, and moreover (L, λ) is a limit of D. In particular, if F creates J-limits, then it reflects them. The dual notion is that of *creation of* J*-colimits*.

II.3.3.1 Proposition *The forgetful functor $G^{\mathbb{T}} : \mathsf{X}^{\mathbb{T}} \to \mathsf{X}$ creates all limits.*

Proof Let $D : \mathsf{J} \to \mathsf{X}^{\mathbb{T}}$ be a diagram and let (X, λ) be a limit of $G^{\mathbb{T}}D$ in X. There is then a unique morphism $a : TX \to X$ with

$$\lambda \cdot \Delta a = G^{\mathbb{T}}\varepsilon^{\mathbb{T}}D \cdot T\lambda .$$

All that needs to be shown is that (X, a) is a $\mathbb{T}$-algebra, and that $((X, a), \lambda)$ serves as a limit for D in $\mathsf{X}^{\mathbb{T}}$. For the $\mathbb{T}$-algebra laws, observe that

$$\begin{aligned} \lambda \cdot \Delta(a \cdot e_X) &= G^{\mathbb{T}}\varepsilon^{\mathbb{T}}D \cdot T\lambda \cdot \Delta e_X \\ &= G^{\mathbb{T}}\varepsilon^{\mathbb{T}}D \cdot eG^{\mathbb{T}}D \cdot \lambda \qquad \text{(naturality of } e = \eta^{\mathbb{T}}\text{)} \\ &= \lambda , \end{aligned}$$

and

$$\begin{aligned}
\lambda\cdot\Delta(a\cdot m_X) &= G^{\mathbb{T}}\varepsilon^{\mathbb{T}}D\cdot T\lambda\cdot\Delta m_X \\
&= G^{\mathbb{T}}\varepsilon^{\mathbb{T}}D\cdot mG^{\mathbb{T}}D\cdot TT\lambda \qquad (\text{naturality of } m=G^{\mathbb{T}}\varepsilon^{\mathbb{T}}F^{\mathbb{T}}) \\
&= G^{\mathbb{T}}\varepsilon^{\mathbb{T}}D\cdot TG^{\mathbb{T}}\varepsilon^{\mathbb{T}}D\cdot TT\lambda \quad (\text{naturality of } \varepsilon^{\mathbb{T}}) \\
&= G^{\mathbb{T}}\varepsilon^{\mathbb{T}}D\cdot T\lambda\cdot\Delta Ta \\
&= \lambda\cdot\Delta(a\cdot Ta)\,.
\end{aligned}$$

Obviously, $\lambda : \Delta(X,a) \to D$ is actually a cone in $\mathsf{X}^{\mathbb{T}}$, and for any cone $\alpha : \Delta(Y,b) \to D$ one obtains a unique morphism $f : Y \to X$ in X with

$$\lambda\cdot\Delta f = G^{\mathbb{T}}\alpha\,.$$

That $f : (Y,b) \to (X,a)$ is indeed a $\mathbb{T}$-homomorphism follows from

$$\begin{aligned}
\lambda\cdot\Delta(f\cdot b) &= G^{\mathbb{T}}\alpha\cdot\Delta G^{\mathbb{T}}\varepsilon^{\mathbb{T}}_{(Y,b)} \\
&= G^{\mathbb{T}}\varepsilon^{\mathbb{T}}D\cdot TG^{\mathbb{T}}\alpha \qquad (\text{naturality of } \varepsilon^{\mathbb{T}}) \\
&= G^{\mathbb{T}}\varepsilon^{\mathbb{T}}D\cdot T\lambda\cdot\Delta Tf \\
&= \lambda\cdot\Delta(a\cdot Tf)\,.
\end{aligned}$$

□

II.3.3.2 Corollary *For a monadic functor $G : \mathsf{A} \to \mathsf{X}$, if X is J-complete, then A is also J-complete, and G preserves and reflects J-limits.*

Proof This is an immediate consequence of Proposition II.3.3.1. □

II.3.4 Beck's monadicity criterion

In order to characterize monadic functors, let us first observe that for a monad $\mathbb{T}$ on X, and a $\mathbb{T}$-algebra (X,a), the diagram

$$TTX \underset{Ta}{\overset{m_X}{\rightrightarrows}} TX \xrightarrow{\ a\ } X\ , \qquad \text{(II.3.4.i)}$$

in addition to satisfying $a\cdot m_X = a\cdot Ta$, has the property that there are *splittings* η_X and η_{TX}: in general, one calls a diagram

$$X \underset{g}{\overset{f}{\rightrightarrows}} Y \xrightarrow{\ h\ } Z$$

in X with $h\cdot f = h\cdot g$ a *split fork* if there are morphisms $i : Z \to Y$ and $j : Y \to X$ (the aforementioned "splittings") with $h\cdot i = 1_Z$, $f\cdot j = 1_Y$, and $g\cdot j = i\cdot h$:

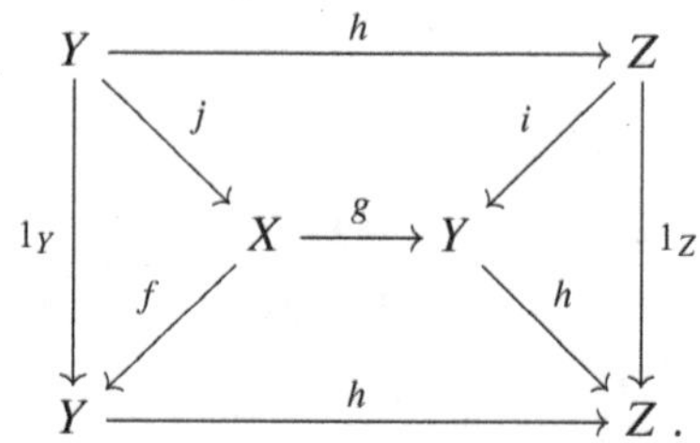

In this case, h is a coequalizer of (f, g) which, trivially, is *absolute*, in the sense that it is preserved by every functor defined on X.

Moreover, the coequalizer diagram (II.3.4.i) has the special property that it can be "lifted" along $G^{\mathbb{T}}$, i.e.

$$(TTX, m_{TX}) \underset{Ta}{\overset{m_X}{\rightrightarrows}} (TX, m_X) \xrightarrow{\ a\ } (X, a) \qquad \text{(II.3.4.ii)}$$

is a coequalizer in $\mathsf{X}^{\mathbb{T}}$, presenting the $\mathbb{T}$-algebra (X, a) as a quotient of the free $\mathbb{T}$-algebra $F^{\mathbb{T}}X$. In particular, the parallel pair of morphisms (m_X, Ta) in (II.3.4.ii) has the property that its $G^{\mathbb{T}}$-image is part of the split fork (II.3.4.i). This property turns out to be crucial, and for a functor $G : \mathsf{A} \to \mathsf{X}$ one therefore says that morphisms $s, t : A \to B$ in A form a *G-split pair* (s, t) if $f = Gs$, $g = Gt$ belong to a split fork in X; similarly, (s, t) is a *G-absolute pair* if (Gs, Gt) has a coequalizer in X that is absolute. Obviously, every G-split pair is G-absolute.

II.3.4.1 Theorem (Beck's monadicity criterion) *A functor $G : \mathsf{A} \to \mathsf{X}$ is monadic if and only if the following conditions hold:*

(1) *G is right adjoint;*
(2) *G reflects isomorphisms;*
(3) *A has coequalizers of G-split pairs, and G preserves them.*

Condition (2) *may be replaced by*

(2′) *G reflects coequalizers of G-split pairs.*

The criterion remains valid if "G-split" is replaced by "G-absolute" everywhere.

Proof For the necessity of the conditions, even in the formally stronger "G-absolute" version, we remark for (2) that isomorphisms are **1**-limits (then see Proposition II.3.3.1), and refer to Exercise II.3.D for (3) and (2′). For their sufficiency, consider the adjunction $F \overset{\eta}{\underset{\varepsilon}{\dashv}} G : \mathsf{A} \to \mathsf{X}$ with its induced monad $\mathbb{T} = (T, m, e)$. Since $G = G^{\mathbb{T}}K$, the comparison functor K also reflects isomorphisms, and because of Propositions II.2.6.2 and II.2.6.1 it suffices to establish an adjunction $L \overset{\kappa}{\underset{\lambda}{\dashv}} K : \mathsf{A} \to \mathsf{X}^{\mathbb{T}}$ with κ an isomorphism.

For a $\mathbb{T}$-algebra (X, a), the G-image of the pair (ε_{FX}, Fa) is (m_X, Ta). Hence, condition (3) guarantees the existence of a coequalizer diagram

$$FGFX \underset{Fa}{\overset{\varepsilon_{FX}}{\rightrightarrows}} FX \xrightarrow{\gamma_{(X,a)}} L(X, a) \tag{II.3.4.iii}$$

in A that is preserved by $G = G^{\mathbb{T}}K$. Since $G^{\mathbb{T}}$ reflects such coequalizers, K preserves the coequalizers (II.3.4.iii) as well. Consequently, the unique morphism $\kappa_{(X,a)}$ given by the coequalizer property of (II.3.4.ii) and making the following diagram commute, must actually be an isomorphism:

$$\begin{array}{ccccc}
F^{\mathbb{T}}TX & \underset{Ta}{\overset{m_X}{\rightrightarrows}} & F^{\mathbb{T}}X & \xrightarrow{\quad a \quad} & (X, a) \\
{\scriptstyle 1}\downarrow & & \downarrow{\scriptstyle 1} & & \vdots\ {\scriptstyle \kappa_{(X,a)}} \\
KFTX & \underset{KFa}{\overset{K\varepsilon_{FX}}{\rightrightarrows}} & KFX & \xrightarrow{K\gamma_{(X,a)}} & KL(X, a)\,.
\end{array}$$

That $\kappa_{(X,a)}$ is K-universal is easily seen with the natural correspondence

$$\frac{(X, a) \xrightarrow{f} KA}{L(X, a) \xrightarrow{g} A}$$

for all $A \in \mathrm{ob}\,\mathsf{A}$, which arises from the coequalizer property of (II.3.4.iii), as shown by

$$\begin{array}{ccccc}
FGFX & \underset{Fa}{\overset{\varepsilon_{FX}}{\rightrightarrows}} & FX & \xrightarrow{\gamma_{(X,a)}} & L(X, a) \\
{\scriptstyle FGFf}\downarrow & & \downarrow{\scriptstyle Ff} & & \downarrow{\scriptstyle g} \\
FGFGA & \underset{FG\varepsilon_A}{\overset{\varepsilon_{FGA}}{\rightrightarrows}} & FGA & \xrightarrow{\quad \varepsilon_A \quad} & A\,.
\end{array}$$

This completes the proof that (1)–(3) are sufficient for monadicity of G. If (2) is replaced by (2′), one should observe that the preceding diagram exhibits the counit $\lambda_A = g : LKA \to A$ if we let $(X, a) = KA$ and $f = 1_{KA}$. In this case, G maps the bottom row of the diagram to a coequalizer diagram (trivial case of (II.3.4.i)), so under assumption (2′) the bottom row is already a coequalizer in A, and the "comparison" morphism $\lambda_A = g$ must be an isomorphism. □

II.3.4.2 Example The contravariant powerset functor $P^{\bullet} : \mathsf{Set}^{\mathrm{op}} \to \mathsf{Set}$ (see Section II.2.2) is monadic. It certainly is right adjoint (see Example II.2.5.1(6)) and reflects isomorphisms. Moreover, one easily verifies that $P^{\bullet}$ transforms an equalizer diagram

$$E \hookrightarrow X \underset{g}{\overset{f}{\rightrightarrows}} Y$$

in **Set** into a coequalizer diagram

$$PY \underset{g^{-1}(-)}{\overset{f^{-1}(-)}{\rightrightarrows}} PX \xrightarrow{(-)\cap E} PE$$

in **Set**, provided that $f^{-1}(-) = P^{\bullet}f$ is surjective (i.e. if f is injective). Hence, the Beck criterion applies.

II.3.5 Duskin's monadicity criterion

Coequalizers are often difficult to construct explicitly, so the third condition in Theorem II.3.4.1 can be delicate to handle. In certain categories, however, coequalizers can be replaced by other, more practical structures. For example, in **Set** a kernel pair $(r, r' : R \to X)$ of a map $f : X \to Y$ yields the equivalence relation $R \subseteq X \times X$:

$$\forall x, y \in X \; ((x, y) \in R \iff f(x) = f(y)) \, ,$$

and a coequalizer of (r, r') is then simply given by the projection $\pi : X \to X/R$. In fact, every equivalence relation in **Set** is the kernel pair of a split epimorphism (namely the projection π); see Exercise II.2.C.

For a functor $G : \mathsf{A} \to \mathsf{X}$, a *G-kernel pair* is a pair $(f, g : A \to B)$ such that there is a diagram

$$X \underset{Gg}{\overset{Gf}{\rightrightarrows}} Y \xrightarrow{h} Z \; , \qquad \text{(II.3.5.i)}$$

where (Gf, Gg) is the kernel pair of h in X. In the case where h is a split epimorphism, this diagram becomes a split fork, so that h is a coequalizer of (Gf, Gg) (Exercise II.3.C).

A *joint kernel pair* of $(f, g : A \to B)$ in a category A is a pair $(s, s' : S \to A)$ with $f \cdot s = f \cdot s'$, $g \cdot s = g \cdot s'$, and such that if there is a pair $(t, t' : L \to A)$ with $f \cdot t = f \cdot t'$, $g \cdot t = g \cdot t'$, then there exists a unique A-morphism $u : L \to S$ with $s \cdot u = t$ and $s' \cdot u = t'$:

$$\begin{array}{ccccc} S & \underset{s'}{\overset{s}{\rightrightarrows}} & A & \underset{g}{\overset{f}{\rightrightarrows}} & B \, . \\ {\scriptstyle u}\uparrow & \nearrow_{t} \nearrow_{t'} & & & \\ L & & & & \end{array}$$

Since a joint kernel pair of (f, f) is simply a kernel pair of f, a category that has joint kernel pairs also has kernel pairs. Whenever the product $B \times B$ exists, the joint kernel pair of (f, g) is simply the kernel pair of $\langle f, g \rangle : A \to B \times B$.

II.3.5.1 Theorem (Duskin's monadicity criterion) *Suppose that* A *has joint kernel pairs and that* X *has kernel pairs of split epimorphisms. Then a functor* $G : \mathsf{A} \to \mathsf{X}$ *is monadic if and only if the following conditions hold:*

(1) *G is right adjoint;*
(2) *G reflects isomorphisms;*
(3) *every G-kernel pair of a split epimorphism has a coequalizer that is preserved by G.*

Proof The necessity of these conditions is immediate from Theorem II.3.4.1. Using this same result, the sufficiency obviously follows if one proves that every G-split pair

$$X \underset{g}{\overset{f}{\rightrightarrows}} Y$$

has a coequalizer that is preserved by G. Given hypothesis (3), it is enough to construct a G-kernel pair (f', g') of a split epimorphism whose coequalizer is the coequalizer of (f, g), and such that this property is preserved by G. Thus, consider the kernel pair (r, r') of g and the joint kernel pair (s, s') of $(f \cdot r, f \cdot r')$; in X, we obtain a diagram

$$\begin{array}{ccccccc} GS & \underset{Gs'}{\overset{Gs}{\rightrightarrows}} & GR & & K & & \\ & & {\scriptstyle Gr}\downarrow\downarrow{\scriptstyle Gr'} & & {\scriptstyle k}\downarrow\downarrow{\scriptstyle k'} & & \\ & & GA & \underset{Gg}{\overset{Gf}{\rightrightarrows}} & GB & \xrightarrow{\ h\ } & Z \end{array}$$

whose bottom row is a split fork with splittings $i : Z \to GB$ and $j : GB \to GA$ (so that $h \cdot i - 1_Z$, $Gf \cdot j = 1_{GB}$, and $Gg \cdot j = i \cdot h$), and where (k, k') is the kernel pair of h. Since

$$h \cdot (Gf \cdot Gr) = h \cdot Gg \cdot Gr = h \cdot Gg \cdot Gr' = h \cdot (Gf \cdot Gr'),$$

there is a uniquely determined X-morphism $v : GR \to K$ with $k \cdot v = Gf \cdot Gr$ and $k' \cdot v = Gf \cdot Gr'$. This morphism is split epic: as G is right adjoint, it preserves kernel pairs, so

$$Gg \cdot (j \cdot k) = i \cdot h \cdot k = i \cdot h \cdot k' = Gg \cdot (j \cdot k')$$

yields the existence of $w : K \to GR$ with $j \cdot k = Gr \cdot w$ and $j \cdot k' = Gr' \cdot w$; hence,

$$j \cdot k \cdot v \cdot w = j \cdot Gf \cdot Gr \cdot w = j \cdot Gf \cdot j \cdot k = j \cdot k$$

implies $k \cdot v \cdot w = k$ and similarly $k' \cdot v \cdot w = k'$, so $v \cdot w = 1_K$ by the universal property of (k, k'). Right adjointness of G also makes (Gs, Gs') into a joint kernel pair of $(Gf \cdot Gr, Gf \cdot Gr')$, and it follows that (Gs, Gs') is a kernel pair of the split epimorphism v. By hypothesis, (s, s') has a coequalizer $u : R \to Q$ that is preserved by G, and since $v : GR \to K$ is also a coequalizer of (Gs, Gs') (Exercise II.3.C), there is an isomorphism $\phi : K \to GQ$ with

$Gu = \phi \cdot v$. By definition of the joint kernel pair (s, s'), the universal property of u yields A-morphisms $q, q' : Q \to B$ such that $q \cdot u = f \cdot r$ and $q' \cdot u = f \cdot r'$:

$$\begin{array}{ccccc} S & \overset{s}{\underset{s'}{\rightrightarrows}} & R & \xrightarrow{\;u\;} & Q \\ & & {\scriptstyle r}\downarrow\downarrow{\scriptstyle r'} & & {\scriptstyle q}\downarrow\downarrow{\scriptstyle q'} \\ & & A & \overset{f}{\underset{g}{\rightrightarrows}} & B\,. \end{array}$$

One has $k \cdot v = G(f \cdot r) = Gq \cdot Gu = Gq \cdot \phi \cdot v$ so that $k = Gq \cdot \phi$ and similarly $k' = Gq' \cdot \phi$ because v is epic. Hence, (Gq, Gq') is a kernel pair of the split epimorphism h, and by hypothesis (q, q') has a coequalizer $c : B \to C$ that is preserved by G. Thus, one obtains a split fork

$$GA \overset{Gf}{\underset{Gg}{\rightrightarrows}} GB \xrightarrow{\;Gc\;} GC$$

and Gc is a coequalizer of (Gf, Gg). To see that c is a coequalizer of (f, g), consider an A-morphism $d : B \to D$ such that $d \cdot f = d \cdot g$. Since

$$d \cdot q \cdot u = d \cdot f \cdot r = d \cdot g \cdot r = d \cdot g \cdot r' = d \cdot q' \cdot u$$

and u – being a coequalizer – is epic, one has $d \cdot q = d \cdot q'$, and the universal property of (C, c) follows. □

II.3.6 The Kleisli category

The objects of the *Kleisli category* $\mathsf{X}_{\mathbb{T}}$ associated to the monad $\mathbb{T} = (T, m, e)$ on X are the objects of X, and a morphism $f : X \rightharpoonup Y$ in $\mathsf{X}_{\mathbb{T}}$ is simply an X-morphism $f : X \to TY$. The *Kleisli composition* of $f : X \rightharpoonup Y$ and $g : Y \rightharpoonup Z$ in $\mathsf{X}_{\mathbb{T}}$ is defined via the composition in X as

$$g \circ f := m_Z \cdot Tg \cdot f\,.$$

The identity $1_X : X \rightharpoonup X$ in this category is just the component $e_X : X \to TX$ of the unit e.

There is a functor $G_{\mathbb{T}} : \mathsf{X}_{\mathbb{T}} \to \mathsf{X}$ defined on objects and morphisms by

$$X \mapsto TX \qquad (f : X \rightharpoonup Y) \mapsto (m_Y \cdot Tf : TX \to TY)\,.$$

As for the Eilenberg–Moore category, this functor has a left adjoint $F_{\mathbb{T}} : \mathsf{X} \to \mathsf{X}_{\mathbb{T}}$:

$$X \mapsto X\,, \quad (f : X \to Y) \mapsto (e_Y \cdot f : X \rightharpoonup Y)\,.$$

The unit $\eta_{\mathbb{T}} : 1_{\mathsf{X}} \to G_{\mathbb{T}} F_{\mathbb{T}} = T$ of this adjunction is e, and the components of the counit $\varepsilon_{\mathbb{T}} : F_{\mathbb{T}} G_{\mathbb{T}} \to 1_{\mathsf{X}_{\mathbb{T}}}$ are simply the morphisms $1_{TX} : TX \to TX$ in X. Also in this case, the monad associated to this adjunction gives back the original monad $\mathbb{T} = (T, m, e)$, and the Kleisli category may be characterized dually to $\mathsf{X}^{\mathbb{T}}$:

II.3.6.1 Proposition *If* $\mathbb{T} = (T, m, e)$ *is the monad associated to* $F \overset{\eta}{\underset{\varepsilon}{\dashv}} G : \mathsf{A} \longrightarrow \mathsf{X}$, *then there exists a unique functor* $L : \mathsf{X}_{\mathbb{T}} \to \mathsf{A}$ *making both the inner and outer triangles of the following diagram commute:*

$$\begin{array}{ccc} \mathsf{X}_{\mathbb{T}} & \xrightarrow{\quad L \quad} & \mathsf{A} \\ {\scriptstyle F_{\mathbb{T}}}\nwarrow \searrow{\scriptstyle G_{\mathbb{T}}} & & {\scriptstyle G}\swarrow \nearrow{\scriptstyle F} \\ & \mathsf{X} . & \end{array}$$

Moreover, $L : \mathsf{X}_{\mathbb{T}} \to \mathsf{A}$ *is full and faithful.*

Proof The functor $L : \mathsf{X}_{\mathbb{T}} \overset{\cdot}{\to} \mathsf{A}$ maps objects like F does, and one has $Lf = \varepsilon_{FY} \cdot Ff$ on morphisms $f : X \rightharpoonup Y$. Uniqueness follows from the universal property of η, as does the fact that L is full and faithful. □

II.3.6.2 Example For $\mathbb{T} = \mathbb{P}$ the powerset monad, a $\mathsf{Set}_{\mathbb{P}}$-morphism from X to Y, is just a relation $r : X \to PY$, while the Kleisli composition is the ordinary relational composition of Section II.1.2:

$$s \cdot r = s \circ r \, ,$$

for any relations $r : X \to PY, s : Y \to PZ$. Therefore,

$$\mathsf{Set}_{\mathbb{P}} = \mathsf{Rel} \, .$$

II.3.7 Kleisli triples

There is an alternative presentation of monads which turns out to be very practical in verifying that given data describe a monad when there is no obvious adjunction inducing it. A *Kleisli triple* $(T, (-)^{\mathbb{T}}, e)$ on a category X consists of

- a function $T : \mathrm{ob}\,\mathsf{X} \to \mathrm{ob}\,\mathsf{X}$ sending X to TX,
- an *extension operation* $(-)^{\mathbb{T}}$ sending a morphism $f : X \to TY$ to a morphism $f^{\mathbb{T}} : TX \to TY$,
- a morphism $e_X : X \to TX$ for each $X \in \mathrm{ob}\,\mathsf{X}$,

subject to

$$(g^{\mathbb{T}} \cdot f)^{\mathbb{T}} = g^{\mathbb{T}} \cdot f^{\mathbb{T}} , \quad e_X^{\mathbb{T}} = 1_{TX} , \quad f^{\mathbb{T}} \cdot e_X = f \qquad \text{(II.3.7.i)}$$

for all $X \in \mathrm{ob}\,\mathsf{X}$, $f : X \to TY$, and $g : Y \to TZ$. One can set

$$g \circ f := g^{\mathbb{T}} \cdot f \, ,$$

so the previous conditions are equivalent to requiring that this "Kleisli composition" $\circ$ is associative, and that e_X acts as an identity. A *Kleisli triple morphism*

$\alpha : (S, (-)^{\mathbb{S}}, d) \to (T, (-)^{\mathbb{T}}, e)$ is given by a family of morphisms $\alpha_X : SX \to TX$ in X (with X running through $\mathrm{ob}\,\mathsf{X}$) such that

$$\alpha_Y \cdot f^{\mathbb{S}} = (\alpha_Y \cdot f)^{\mathbb{T}} \cdot \alpha_X\,, \quad \alpha_X \cdot d_X = e_X$$

for all $f : X \to SY$. That is, a Kleisli triple morphism is a family of morphisms that preserve the Kleisli composition together with its unit.

A Kleisli triple $(T, (-)^{\mathbb{T}}, e)$ gives rise to a monad $\mathbb{T} = (T, m, e)$ on X simply by setting

$$Tf := (e_Y \cdot f)^{\mathbb{T}}\,, \quad m_X := (1_{TX})^{\mathbb{T}}\,,$$

for all X-morphisms $f : X \to Y$. The facts that T defines a functor, e and m are natural transformations, and that the multiplication and unit laws are verified all follow from simple manipulations of the equalities (II.3.7.i). Similarly, a Kleisli triple morphism $\alpha : (S, (-)^{\mathbb{S}}, d) \to (T, (-)^{\mathbb{T}}, e)$ yields a natural transformation $\alpha : S \to T$, which turns out out to be a morphism $\alpha : \mathbb{S} \to \mathbb{T}$ between the corresponding monads.

Conversely, given a monad $\mathbb{T} = (T, m, e)$ on X, one obtains a Kleisli triple $(T, (-)^{\mathbb{T}}, e)$ via

$$f^{\mathbb{T}} := m_Y \cdot Tf\,,$$

for all X-morphisms $f : X \to Y$, and the conditions (II.3.7.i) readily follow by using naturality of e and m, as well as their multiplication and unit laws. A monad morphism $\alpha : \mathbb{S} \to \mathbb{T}$ also yields a morphism $\alpha : (S, (-)^{\mathbb{S}}, d) \to (T, (-)^{\mathbb{T}}, e)$ between the corresponding Kleisli triples.

Moreover, the passages from a Kleisli triple to a monad and from a monad to a Kleisli triple are mutually inverse, so that both definitions describe the same structure on X (and the two definitions of Kleisli composition correspond).

II.3.8 Distributive laws, liftings, and composite monads

Given monads $\mathbb{S} = (S, n, d)$ and $\mathbb{T} = (T, m, e)$ on X, one can consider the functor $ST : \mathsf{X} \to \mathsf{X}$ together with the natural transformation $d \circ e : 1_{\mathsf{X}} \to ST$, but there is no obvious choice in general for a natural transformation $w : STST \to ST$ making $\mathbb{S}\mathbb{T} = (ST, w, d \circ e)$ into a monad. A first step towards solving this problem is the introduction of a *distributive law* of $\mathbb{T}$ over $\mathbb{S}$, i.e. a natural transformation $\delta : TS \to ST$ making the following diagrams commute:

$$\begin{array}{ccccc} TSS & \xrightarrow{\delta S} & STS & \xrightarrow{S\delta} & SST \\ {\scriptstyle Tn}\downarrow & & & & \downarrow{\scriptstyle nT} \\ TS & & \xrightarrow{\quad\delta\quad} & & ST \\ {\scriptstyle mS}\uparrow & & & & \uparrow{\scriptstyle Sm} \\ TTS & \xrightarrow{T\delta} & TST & \xrightarrow{\delta T} & STT \end{array} \qquad \begin{array}{ccc} & T & \\ {\scriptstyle Td}\swarrow & & \searrow{\scriptstyle dT} \\ TS & \xrightarrow{\delta} & ST\,. \\ {\scriptstyle eS}\nwarrow & & \nearrow{\scriptstyle Se} \\ & S & \end{array} \tag{II.3.8.i}$$

Such a δ allows for a *lifting* of the monad $\mathbb{S}$ on X through $G^{\mathbb{T}} : \mathsf{X}^{\mathbb{T}} \to \mathsf{X}$, i.e. there exists a monad $\tilde{\mathbb{S}} = (\tilde{S}, \tilde{n}, \tilde{d})$ on $\mathsf{X}^{\mathbb{T}}$ such that

$$G^{\mathbb{T}}\tilde{S} = SG^{\mathbb{T}}\,, \quad G^{\mathbb{T}}\tilde{n} = nG^{\mathbb{T}}\,, \quad G^{\mathbb{T}}\tilde{d} = dG^{\mathbb{T}}\,.$$

The first of these conditions may be used to identify the domains and codomains of the natural transformations in the last two equalities. The latter state that the underlying X-morphisms of $\tilde{n}_{(X,a)}$, $\tilde{d}_{(X,a)}$ are n_X, d_X, respectively, for any $\mathbb{T}$-algebra (X, a), so the multiplication and unit laws of $\tilde{S}$ are automatically satisfied. Therefore, a lifting of $\mathbb{S}$ through $G^{\mathbb{T}} : \mathsf{X}^{\mathbb{T}} \to \mathsf{X}$ is simply provided by a functor $\tilde{S} : \mathsf{X}^{\mathbb{T}} \to \mathsf{X}^{\mathbb{T}}$ making

$$\begin{array}{ccc}
\mathsf{X}^{\mathbb{T}} & \xrightarrow{\tilde{S}} & \mathsf{X}^{\mathbb{T}} \\
{\scriptstyle G^{\mathbb{T}}}\downarrow & & \downarrow{\scriptstyle G^{\mathbb{T}}} \\
\mathsf{X} & \xrightarrow{S} & \mathsf{X}
\end{array}$$

commute, and such that $n_X : \tilde{S}\tilde{S}(X,a) \to \tilde{S}(X,a)$, $d_X : (X,a) \to \tilde{S}(X,a)$ are $\mathsf{X}^{\mathbb{T}}$-morphisms for all $(X,a) \in \mathrm{ob}(\mathsf{X}^{\mathbb{T}})$. Thus, if $\delta : TS \to ST$ is a distributive law of $\mathbb{T}$ over $\mathbb{S}$, a lifting $\tilde{S} : \mathsf{X}^{\mathbb{T}} \to \mathsf{X}^{\mathbb{T}}$ is obtained via

$$(X,a) \mapsto (SX, Sa \cdot \delta_X)\,, \quad (f : X \to Y) \mapsto (Sf : SX \to SY)\,.$$

The fact that the components of $\tilde{n} : \tilde{S}\tilde{S} \to \tilde{S}$ and $\tilde{d} : 1_{\mathsf{X}^{\mathbb{T}}} \to \tilde{S}$ have underlying $\mathsf{X}^{\mathbb{T}}$-morphisms then follows directly from commutativity of the diagrams in (II.3.8.i).

In turn, a lifting of $\mathbb{S}$ through $G^{\mathbb{T}} : \mathsf{X}^{\mathbb{T}} \to \mathsf{X}$ yields the following composite adjunction:

$$(\mathsf{X}^{\mathbb{T}})^{\tilde{\mathbb{S}}} \underset{F^{\tilde{\mathbb{S}}}}{\overset{G^{\tilde{\mathbb{S}}}}{\underset{\longleftarrow}{\overset{\longrightarrow}{\top}}}} \mathsf{X}^{\mathbb{T}} \underset{F^{\mathbb{T}}}{\overset{G^{\mathbb{T}}}{\underset{\longleftarrow}{\overset{\longrightarrow}{\top}}}} \mathsf{X}\,.$$

Since $ST = SG^{\mathbb{T}}F^{\mathbb{T}} = G^{\mathbb{T}}\tilde{S}F^{\mathbb{T}} = G^{\mathbb{T}}G^{\tilde{\mathbb{S}}}F^{\tilde{\mathbb{S}}}F^{\mathbb{T}}$, the monad on X associated to this adjunction is

$$\mathbb{S}\mathbb{T} = (ST, w, d \circ e)\,,$$

where $w = G^{\mathbb{T}}G^{\tilde{\mathbb{S}}}\varepsilon^{\tilde{\mathbb{S}}}F^{\tilde{\mathbb{S}}}F^{\mathbb{T}} \cdot G^{\mathbb{T}}G^{\tilde{\mathbb{S}}}F^{\tilde{\mathbb{S}}}\varepsilon^{\mathbb{T}}G^{\tilde{\mathbb{S}}}F^{\tilde{\mathbb{S}}}F^{\mathbb{T}}$ by Proposition II.2.5.5 and Section II.3.1, so that

$$w = G^{\mathbb{T}}\tilde{n}F^{\mathbb{T}} \cdot G^{\mathbb{T}}\tilde{S}\varepsilon^{\mathbb{T}}\tilde{S}F^{\mathbb{T}} = nT \cdot SG^{\mathbb{T}}\varepsilon^{\mathbb{T}}\tilde{S}F^{\mathbb{T}}\,.$$

By combining this expression for w with the lifting conditions, one easily verifies that the following equalities also hold:

(1) $w \cdot SeST = nT$;
(2) $w \cdot STdT = Sm$;
(3) $w \cdot STSm = Sm \cdot wT$;
(4) $w \cdot nTST = nT \cdot Sw$.

More generally, we say that a monad $\mathbb{ST} = (ST, w, d \circ e)$ is a *composite* of $\mathbb{S}$ and $\mathbb{T}$ if its multiplication w satisfies (1)–(4). These conditions are useful in proofs, but there is a more elegant and practical presentation, as Lemma II.3.8.1 shows. Given monads $\mathbb{S} = (S, n, d)$, $\mathbb{T} = (T, m, e)$ on X, one says that a monad $\mathbb{ST} = (ST, w, d \circ e)$ satisfies the *middle unit law* if the following diagram commutes:

$$\begin{array}{ccc} & ST & \\ {}^{Se\circ dT}\swarrow & & \searrow^{1_{ST}} \\ STST & \xrightarrow{\ w\ } & ST \end{array}$$

i.e. if $w \cdot (Se \circ dT) = 1_{ST}$.

II.3.8.1 Lemma *For monads* $\mathbb{S} = (S, n, d)$, $\mathbb{T} = (T, m, e)$, *and* $\mathbb{ST} = (ST, w, d \circ e)$ *on* X, *the following are equivalent:*

(i) $\mathbb{ST}$ *is a composite monad;*

(ii) *the natural transformations* $Se : S \to ST$ *and* $dT : T \to ST$ *are monad morphisms, and* $\mathbb{ST}$ *satisfies the middle unit law.*

Proof The proof that (i) $\Longrightarrow$ (ii) follows from routine verifications. To show (ii) $\Longrightarrow$ (i), we must verify the four composite monad conditions. For (1), observe that

$$\begin{aligned} w \cdot SeST &= w \cdot (STw \cdot STSTdT \cdot STSeT) \cdot SeST && \text{(middle unit law)} \\ &= w \cdot wST \cdot STSTdT \cdot STSeT \cdot SeST && \text{(multiplication law)} \\ &= w \cdot STdT \cdot wT \cdot STSeT \cdot SeST && \text{(naturality of } w\text{)} \\ &= w \cdot STdT \cdot SeT \cdot nT && (Se \text{ monad morphism}) \\ &= nT && \text{(middle unit law).} \end{aligned}$$

Condition (2) is proved in a similar way, by inserting the middle unit law $wST \cdot SeSTST \cdot SdTST = 1_{STST}$ into $w \cdot STdT$, and proceeding as above. To see (3), we write

$$\begin{aligned} w \cdot STSm &= w \cdot (STw \cdot STSTdT) && (ST \text{ applied to (2))} \\ &= w \cdot wST \cdot STSTdT && \text{(multiplication law)} \\ &= w \cdot STdT \cdot wT && \text{(naturality of } w\text{)} \\ &= Sm \cdot wT && \text{(by (2) again).} \end{aligned}$$

The final condition is proved similarly by exploiting (1) instead of (2). □

Given a monad $\mathbb{ST} = (ST, w, d \circ e)$ on X that is a composite of $\mathbb{S} = (S, n, d)$ and $\mathbb{T} = (T, m, e)$, there is a natural transformation

$$\delta := (TS \xrightarrow{dT \circ Se} STST \xrightarrow{\ w\ } ST)\,,$$

which turns out to be a distributive law of $\mathbb{T}$ over $\mathbb{S}$ (as can be checked routinely).

As these constructions suggest, distributive laws, liftings, and composite monads are equivalent concepts.

II.3.8.2 Proposition *For monads* $\mathbb{S}$ *and* $\mathbb{T}$ *on* X*, there is a bijective correspondence between*

(i) *distributive laws* δ *of* $\mathbb{T}$ *over* $\mathbb{S}$*;*
(ii) *liftings* $\tilde{\mathbb{S}}$ *of* $\mathbb{S}$ *through* $G^{\mathbb{T}} : \mathsf{X}^{\mathbb{T}} \to \mathsf{X}$*;*
(iii) *monads* $\mathbb{ST} = (ST, w, d \circ e)$ *that are composites of* $\mathbb{S}$ *and* $\mathbb{T}$*.*

Proof The previous discussion has already shown that these concepts are related as in:

$$\begin{array}{ccc} & \text{(i)} & \\ \nearrow\!\!\!\nearrow & & \searrow\!\!\!\searrow \\ \text{(iii)} & \Longleftarrow & \text{(ii)}\,. \end{array}$$

To verify the statement, it suffices to prove that any path of length 3 in this diagram is the identity. We check this for liftings, the other verifications being similar. By definition, a lifting is entirely determined by its behavior on the structure morphisms of $\mathsf{X}^{\mathbb{T}}$-objects (X, a). Let us denote by (SX, a') the $\tilde{S}$-image of (X, a), and by a'' the structure corresponding to the path (ii) $\Longrightarrow$ (iii) $\Longrightarrow$ (i) $\Longrightarrow$ (ii). By using the fact that $\tilde{S}\varepsilon^{\mathbb{T}}_{(X,a)} : (STX, m'_X) \to (SX, a')$ is an $\mathsf{X}^{\mathbb{T}}$-morphism, one observes

$$\begin{aligned} a'' &= Sa \cdot n_{TX} \cdot S(G^{\mathbb{T}}\varepsilon^{\mathbb{T}}_{\tilde{S}F^{\mathbb{T}}X} \cdot TSe_X) \cdot d_{TSX} && \\ &= Sa \cdot n_{TX} \cdot d_{STX} \cdot G^{\mathbb{T}}\varepsilon^{\mathbb{T}}_{(STX,m'_X)} \cdot TSe_X && \text{(naturality of } d\text{)} \\ &= Sa \cdot m'_X \cdot TSe_X && \text{(unit law for } \mathbb{S}\text{)} \\ &= a' \cdot TSa \cdot TSe_X && (\tilde{S}\varepsilon^{\mathbb{T}}_{(X,a)} \text{ a morphism}) \\ &= a'\,. \end{aligned}$$

□

II.3.8.3 Corollary *The monads* $\mathbb{ST}$ *that are composites of* $\mathbb{S} = (S, n, d)$ *and* $\mathbb{T} = (T, m, e)$ *are exactly those of the form*

$$\mathbb{ST} = (ST, (n \circ m) \cdot S\delta T, d \circ e)\,,$$

where δ *is a distributive law of* $\mathbb{T}$ *over* $\mathbb{S}$*. There are also monad morphisms*

$$Se : \mathbb{S} \to \mathbb{ST}\,, \quad dT : \mathbb{T} \to \mathbb{ST}\,.$$

Proof By Proposition II.3.8.2, any composite monad comes from a distributive law, and the resulting multiplication w is then easily seen to be $(n \circ m) \cdot S\delta T$. The fact that Se and dT are monad morphisms follows by a routine verification. □

II.3.8.4 Proposition *For a composite monad* $\mathbb{ST}$ *and the corresponding lifting* $\tilde{\mathbb{S}}$ *(see* Proposition II.3.8.2*), the comparison functor* $K : (\mathsf{X}^{\mathbb{T}})^{\tilde{\mathbb{S}}} \to \mathsf{X}^{\mathbb{ST}}$ *is an isomorphism.*

Proof The comparison functor sends an object $((X,t),\tilde{s})$ of $(\mathsf{X}^{\mathbb{T}})^{\tilde{\mathbb{S}}}$ to the $\mathbb{ST}$-algebra $(X,\tilde{s}\cdot St)$, and its inverse maps (X,a) to $((X,a\cdot d_{TX}),a\cdot Se_X)$. □

II.3.8.5 Examples

(1) For any monad $\mathbb{T}$ on X, there are trivial distributive laws $1_T : 1_{\mathsf{X}}T \longrightarrow T1_{\mathsf{X}}$ of the identity monad $\mathbb{I}$ over $\mathbb{T}$, and $1_T : T1_{\mathsf{X}} \longrightarrow 1_{\mathsf{X}}T$ of $\mathbb{T}$ over $\mathbb{I}$.

(2) A distributive law $\delta : LG \longrightarrow GL$ of the free-monoid monad $\mathbb{L}$ over the free-Abelian-group monad $\mathbb{G}$ (Example II.3.1.1(2)) is obtained by sending an element of LGX of the form

$$\Big(\big(\textstyle\sum_{i_1=1,\dots,n_1}\alpha_{1,i_1}\cdot x_{1,i_1}\big),\big(\textstyle\sum_{i_2=1,\dots,n_2}\alpha_{2,i_2}\cdot x_{k,i_2}\big),\dots,\\ \big(\textstyle\sum_{i_k=1,\dots,n_k}\alpha_{k,i_k}\cdot x_{k,i_k}\big)\Big)$$

to the element of GLX given by

$$\textstyle\sum_{i_1,i_2,\dots,i_k}\alpha_{1,i_1}\alpha_{2,i_2}\cdots\alpha_{k,i_k}\cdot(x_{1,i_1},x_{2,i_2},\dots,x_{k,i_k})\ .$$

The resulting category of $\mathbb{GL}$-algebras is the category Rng of unital rings and their homomorphisms (rings also occur as monoids in the category of Abelian groups, see Example II.4.2.1(2)). From this example, distributive laws may be seen as a generalization of the usual notion of distributivity (of multiplication over addition) that holds in rings.

II.3.9 Distributive laws and extensions

For monads $\mathbb{S}$ and $\mathbb{T}$ on X, an *extension* of $\mathbb{T}$ along $F_{\mathbb{S}} : \mathsf{X} \longrightarrow \mathsf{X}_{\mathbb{S}}$ is a monad $\hat{\mathbb{T}} = (\hat{T},\hat{m},\hat{e})$ on $\mathsf{X}_{\mathbb{S}}$ satisfying

$$F_{\mathbb{S}}T = \hat{T}F_{\mathbb{S}}\ ,\qquad F_{\mathbb{S}}m = \hat{m}F_{\mathbb{S}}\ ,\qquad F_{\mathbb{S}}e = \hat{e}F_{\mathbb{S}}\ .$$

As is the case for liftings, the first condition insures that the second two equalities make sense. Moreover, since $F_{\mathbb{S}}$ is identical on objects, one has $\hat{m}_X = F_{\mathbb{S}}m_X$, $\hat{e}_X = F_{\mathbb{S}}e_X$ for all X-objects X, so the multiplication and unit laws are immediate. An extension of $\mathbb{T}$ is therefore given by a functor $\hat{T} : \mathsf{X}_{\mathbb{S}} \longrightarrow \mathsf{X}_{\mathbb{S}}$ making

$$\begin{array}{ccc} \mathsf{X}_{\mathbb{S}} & \xrightarrow{\ \hat{T}\ } & \mathsf{X}_{\mathbb{S}} \\ {\scriptstyle F_{\mathbb{S}}}\uparrow & & \uparrow{\scriptstyle F_{\mathbb{S}}} \\ \mathsf{X} & \xrightarrow{\ T\ } & \mathsf{X} \end{array}$$

commute, and such that $\hat{m}_X = F_{\mathbb{S}}m_X : TTX \rightharpoonup TX$, $\hat{e}_X = F_{\mathbb{S}}e_X : X \rightharpoonup TX$ ($X \in \mathrm{ob}\,\mathsf{X}$) become the components of natural transformations $\hat{m} : \hat{T}\hat{T} \longrightarrow \hat{T}$, $\hat{e} : 1_{\mathsf{X}_{\mathbb{S}}} \longrightarrow \hat{T}$.

In the presence of a distributive law δ of $\mathbb{T}$ over $\mathbb{S}$, an extension $\hat{T} : \mathsf{X}_{\mathbb{S}} \to \mathsf{X}_{\mathbb{S}}$ is obtained via

$$X \mapsto TX \qquad (f : X \to SY) \mapsto (\delta_Y \cdot Tf : TX \to STY) .$$

Functoriality of $\hat{T}$ follows from naturality of δ, and commutativity of the lower diagrams in (II.3.8.i). To verify naturality of $\hat{m}$ and $\hat{e}$ in $\mathsf{X}_{\mathbb{S}}$, one uses the other two diagrams.

An extension of $\mathbb{T}$ along $F_{\mathbb{S}} : \mathsf{X} \to \mathsf{X}_{\mathbb{S}}$ yields the following composite adjunction:

$$(\mathsf{X}_{\mathbb{S}})_{\hat{\mathbb{T}}} \underset{F_{\hat{\mathbb{T}}}}{\overset{G_{\hat{\mathbb{T}}}}{\underset{\top}{\rightleftarrows}}} \mathsf{X}_{\mathbb{S}} \underset{F_{\mathbb{S}}}{\overset{G_{\mathbb{S}}}{\underset{\top}{\rightleftarrows}}} \mathsf{X} .$$

One can then consider the monad $\mathbb{S}\mathbb{T} = (ST, w, d \circ e)$ as being induced by $F_{\hat{\mathbb{T}}} F_{\mathbb{S}} \dashv G_{\mathbb{S}} G_{\hat{\mathbb{T}}}$, with

$$w = Sm \cdot G_{\mathbb{S}} \hat{T} \varepsilon_{\mathbb{S}} F_{\mathbb{S}} T .$$

The fact that this monad is a composite of $\mathbb{S}$ and $\mathbb{T}$ may then be checked directly. By Section II.3.8, we can get back a distributive law from an extension, and we have:

II.3.9.1 Proposition *For monads $\mathbb{S}$ and $\mathbb{T}$ on X, there is a bijective correspondence between*

(i) *distributive laws δ of $\mathbb{T}$ over $\mathbb{S}$;*
(ii) *extensions $\hat{\mathbb{T}}$ of $\mathbb{T}$ along $F_{\mathbb{S}} : \mathsf{X} \to \mathsf{X}_{\mathbb{S}}$.*

Proof An easy verification shows that, when an extension is obtained via a distributive law, the induced composite monad is the same as in Proposition II.3.8.2, so we need to check only that the path (ii) $\Longrightarrow$ (i) $\Longrightarrow$ (ii) returns the same extension. We denote by $\hat{T}$ the original extension, by $\hat{T}'$ the new one, and recall that $\varepsilon_{\mathbb{S}_Y} = 1_{SY}$. If $f : X \to SY$ is an $\mathsf{X}_{\mathbb{S}}$-morphism, then

$$\begin{aligned}
\hat{T}' f &= Sm_Y \cdot G_{\mathbb{S}} \hat{T} 1_{STY} \cdot G_{\mathbb{S}} F_{\mathbb{S}} T G_{\mathbb{S}} F_{\mathbb{S}} e_Y \cdot d_{TSY} \cdot Tf \\
&= Sm_Y \cdot G_{\mathbb{S}} \hat{T}(1_{STY} \circ F_{\mathbb{S}} G_{\mathbb{S}} F_{\mathbb{S}} e_Y) \cdot d_{TSY} \cdot Tf && \text{(functoriality of } \hat{T}) \\
&= Sm_Y \cdot G_{\mathbb{S}} \hat{T}(F_{\mathbb{S}} e_Y \circ 1_{SY}) \cdot d_{TSY} \cdot Tf && \text{(naturality of } \varepsilon_{\mathbb{S}}) \\
&= Sm_Y \cdot G_{\mathbb{S}} F_{\mathbb{S}} T e_Y \cdot G_{\mathbb{S}} \hat{T} 1_{SY} \cdot d_{TSY} \cdot Tf && (\hat{T} \text{ an extension}) \\
&= n_{TY} \cdot S\hat{T} 1_{SY} \cdot d_{TSY} \cdot Tf && \text{(unit law of } \mathbb{T}) \\
&= n_{TY} \cdot d_{STY} \cdot \hat{T} 1_{SY} \cdot Tf && \text{(naturality of } d) \\
&= \hat{T} 1_{SY} \circ F_{\mathbb{S}} Tf && \text{(unit law of } \mathbb{S}) \\
&= \hat{T}(1_{SY} \circ F_{\mathbb{S}} f) && (\hat{T} \text{ an extension}) \\
&= \hat{T} f .
\end{aligned}$$

□

II.3.9.2 Proposition *For a composite monad* $\mathbb{ST}$ *and the corresponding extension* $\hat{\mathbb{T}}$, *the functor* $L : (\mathsf{X}_{\mathbb{S}})_{\hat{\mathbb{T}}} \to \mathsf{X}_{\mathbb{ST}}$ *of* Proposition II.3.6.1 *is the identity.*

Proof The statement is immediate once the definition of the composition in $(\mathsf{X}_{\mathbb{S}})_{\hat{T}}$ has been unraveled. □

Exercises

II.3.A *Trivial monads on* Set. An object (X, a) of the category $\mathsf{Set}^{\mathbb{T}}$ is *trivial* if X is either the empty set $\varnothing$ or a singleton $1 = \{\star\}$. If there exists at least one non-trivial $\mathbb{T}$-algebra, then the unit $e_X : X \to TX$ is injective for all sets X. The only examples of *trivial monads* (monads that admit only trivial Eilenberg–Moore algebras) are the *terminal monad* $\mathbb{1}$ in $\mathsf{MND}_{\mathsf{Set}}$, whose functor sends all sets X to 1, and the monad whose functor sends the empty set to itself and all other sets to 1.
Hint. (TX, m_X) is a $\mathbb{T}$-algebra for any monad $\mathbb{T}$.

II.3.B *M-actions and equivariant maps.* Any monoid (M, m, e) (as in Section II.1.1) yields a monad $\mathbb{M} = (M \times (-), \overline{m}, \overline{e})$ on Set, where $\overline{m}_X(a, (b, x)) = (m(a, b), x)$ and $\overline{e}_X(x) = (e, x)$. The Eilenberg–Moore category $\mathsf{Set}^{\mathbb{M}}$ of this monad is the category M-Set of M-*actions* and *equivariant maps* which is isomorphic to the *functor category* Set^{M}, with M considered as a one-object category. Every monoid homomorphism $f : M \to N$ yields a monad morphism $\mathbb{M} \to \mathbb{N}$; in fact, this operation describes a functor $\mathsf{Mon} \to \mathsf{MND}_{\mathsf{Set}}$.

II.3.C *Split forks and coequalizers.* A *fork* in a category X is a diagram

$$X \underset{g}{\overset{f}{\rightrightarrows}} Y \xrightarrow{h} Z$$

such that $h \cdot f = h \cdot g$. Show that a fork splits if and only if h is a coequalizer of (f, g) and there is $j : Y \to X$ with $f \cdot j = 1_Y$ and $g \cdot j \cdot f = g \cdot j \cdot g$. If h is a split epimorphism and (f, g) is its kernel pair, then the fork splits.

II.3.D *Creation of colimits by $G^{\mathbb{T}}$.* If $\mathbb{T}$ is a monad on X, the functor $G^{\mathbb{T}} : \mathsf{X}^{\mathbb{T}} \to \mathsf{X}$ creates all colimits that exist in X and that are preserved by T and TT. Therefore, $G^{\mathbb{T}}$ creates coequalizers of G-absolute pairs.

II.3.E *Fullness and faithfulness of the comparison functor.* The comparison functor $K : \mathsf{A} \to \mathsf{X}^{\mathbb{T}}$ (see Proposition II.3.2.1) is full and faithful if and only if ε_A is a regular epimorphism for all $A \in \mathrm{ob}\,\mathsf{A}$; in that case, ε_A is coequalizer of $(\varepsilon_{FGA}, FG\varepsilon_A)$ for all $A \in \mathrm{ob}\,\mathsf{A}$.

II.3.F *Strict monadicity criterion.* For a functor $G : \mathsf{A} \to \mathsf{X}$, the following conditions are equivalent:

(i) G is strictly monadic;
(ii) G is right adjoint and creates coequalizers of G-absolute pairs;
(iii) G is right adjoint and creates coequalizers of G-split pairs.

II.3.G *Eilenberg–Moore algebras via Kleisli triples.* If $(T, (-)^{\mathbb{T}}, e)$ is a Kleisli triple on a category X, the Eilenberg–Moore algebras associated to the monad $\mathbb{T}$ are those pairs (X, a) with $X \in \operatorname{ob}\mathsf{X}$ and $a : TX \to X$ an X-morphism such that

$$\forall f, g \in \mathsf{X}(Y, TX)\ (a \cdot f = a \cdot g \implies a \cdot f^{\mathbb{T}} = a \cdot g^{\mathbb{T}}) \qquad \text{and} \qquad a \cdot e_X = 1_X .$$

II.3.H *Functor liftings and monad morphisms.* Let $F \overset{e}{\underset{\varepsilon}{\dashv}} G : \mathsf{A} \to \mathsf{X}$ be an adjunction with associated monad $\mathbb{T} = (T, m, e)$ on X. For a monad $\mathbb{S} = (S, n, d)$ on Y, a *lifting* of a functor $R : \mathsf{X} \to \mathsf{Y}$ through $(G, G^{\mathbb{S}})$ is a functor $\tilde{R} : \mathsf{A} \to \mathsf{Y}^{\mathbb{S}}$ that makes the diagram

$$\begin{array}{ccc} \mathsf{A} & \xrightarrow{\tilde{R}} & \mathsf{Y}^{\mathbb{S}} \\ {\scriptstyle G}\downarrow & & \downarrow{\scriptstyle G^{\mathbb{S}}} \\ \mathsf{X} & \xrightarrow{R} & \mathsf{Y} \end{array}$$

commute. Equivalently, a lifting of R through $(G, G^{\mathbb{S}})$ can be given by a natural transformation $\alpha : SR \to RT$ satisfying

$$Rm \cdot \alpha T \cdot S\alpha = \alpha \cdot nR \qquad \text{and} \qquad Re = \alpha \cdot dR . \tag{II.3.9.i}$$

Indeed, such an α yields a functor $\tilde{R} : \mathsf{A} \to \mathsf{Y}^{\mathbb{S}}$ via

$$\tilde{R}A := (RGA, RG\varepsilon_A \cdot \alpha_{GA}) ,$$

and conversely a lifting $\tilde{R}$ of R through $(G, G^{\mathbb{S}})$ returns a natural transformation α satisfying (II.3.9.i) via

$$\alpha_X := \tilde{m}_X \cdot SRe_X ,$$

where $\tilde{m}_X : SRTX \to RTX$ denotes the Y-morphism defined by $\tilde{R}FX = (RTX, \tilde{m}_X)$.

As a consequence, liftings of $1_{\mathsf{X}} : \mathsf{X} \to \mathsf{X}$ through $(G, G^{\mathbb{S}})$ are in one-to-one correspondence with monad morphisms $\alpha : \mathbb{S} \to \mathbb{T}$. A lifting of a monad $\mathbb{S}$ along $G^{\mathbb{T}}$ (in the sense of Section II.3.8) provides a lifting of the underlying functor $S : \mathsf{X} \to \mathsf{X}$ through $(G^{\mathbb{T}}, G^{\mathbb{T}})$.

II.3.I *Functor extensions and monad morphisms.* Consider an adjunction $F \overset{e}{\underset{\varepsilon}{\dashv}} G : \mathsf{A} \to \mathsf{X}$ with associated monad $\mathbb{T} = (T, m, e)$ on X. For a monad $\mathbb{S} = (S, n, d)$ on Y, an *extension* of a functor $R : \mathsf{Y} \to \mathsf{X}$ along $(F_{\mathbb{S}}, F)$ is a functor $\hat{R} : \mathsf{Y}_{\mathbb{S}} \to \mathsf{A}$ that makes the following diagram commute:

$$\begin{array}{ccc} \mathsf{Y}_{\mathbb{S}} & \xrightarrow{\hat{R}} & \mathsf{A} \\ {\scriptstyle F_{\mathbb{S}}}\uparrow & & \uparrow{\scriptstyle F} \\ \mathsf{Y} & \xrightarrow{R} & \mathsf{X}\,. \end{array}$$

Extensions of R along $(F_{\mathbb{S}}, F)$ are in one-to-one correspondence with natural transformations $\alpha : RS \to TR$ satisfying

$$mR \cdot T\alpha \cdot \alpha S = \alpha \cdot Rn \qquad \text{and} \qquad eR = \alpha \cdot Rd\,. \tag{II.3.9.ii}$$

Indeed, given such an α, one obtains a functor $\hat{R} : \mathsf{Y}_{\mathbb{S}} \to \mathsf{A}$ defined on Y-objects Y by $\hat{R}Y = FRY$, and on Y-morphisms $f : X \to SY$ by

$$\hat{R}f := \varepsilon_{FRY} \cdot F(\alpha_Y \cdot Rf)\,;$$

conversely, an extension $\hat{R}$ of R along $(F_{\mathbb{S}}, F)$ returns a natural transformation α satisfying (II.3.9.ii) via

$$\alpha_Y := G\hat{R}1_{SY} \cdot e_{RSY}\,.$$

In particular, an extension of $1_{\mathsf{X}} : \mathsf{X} \to \mathsf{X}$ along $(F_{\mathbb{S}}, F)$ is equivalently described by a monad morphism $\alpha : \mathbb{S} \to \mathbb{T}$, and an extension of a monad $\mathbb{S}$ along $F_{\mathbb{S}}$ (in the sense of Section II.3.9) yields an extension of $T : \mathsf{X} \to \mathsf{X}$ along $(F_{\mathbb{S}}, F_{\mathbb{S}})$. In fact, for monads $\mathbb{S}$ and $\mathbb{T}$ on X, a distributive law α of $\mathbb{T}$ over $\mathbb{S}$ is equivalent to a natural transformation α inducing both an extension of $T : \mathsf{X} \to \mathsf{X}$ along $(F_{\mathbb{S}}, F_{\mathbb{S}})$ and a lifting of $S : \mathsf{X} \to \mathsf{X}$ through $(G^{\mathbb{T}}, G^{\mathbb{T}})$ (as in Exercise II.3.H).

II.3.J *Distributive laws as monoids.* A lifting $\alpha : SR \to RT$ of a functor $R : \mathsf{X} \to \mathsf{A}$ through $(G^{\mathbb{T}}, G^{\mathbb{S}})$, where $\mathbb{S} = (S, n, d)$ is a monad on A and $\mathbb{T} = (T, m, e)$ a monad on X (see Exercise II.3.H), can be seen as an arrow $(R, \alpha) : \mathbb{S} \to \mathbb{T}$. Composition with $(L, \beta) : \mathbb{T} \to \mathbb{U}$ is given via

$$(L, \beta) \circ (R, \alpha) := (RL, R\beta \cdot \alpha L)\,.$$

Liftings from $\mathbb{S}$ to $\mathbb{T}$ form the objects of a metacategory $\mathsf{MNDLFT}(\mathbb{S}, \mathbb{T})$, whose morphisms $\lambda : (R, \alpha) \to (R', \alpha')$ are natural transformations $\lambda : R \to R'$ satisfying $\alpha' \cdot S\lambda = \lambda T \cdot \alpha$. In the same way that a monad is a monoid in a functor category (see Section II.3.1), a distributive law δ of $\mathbb{T}$ over $\mathbb{S}$ is equivalently described as a monoid in $\mathsf{MNDLFT}(\mathbb{T}, \mathbb{T})$ given by the lifting (S, δ) together with two morphisms

$$n : (S, \delta) \circ (S, \delta) \to (S, \delta) \qquad \text{and} \qquad d : (1_{\mathsf{X}}, 1_T) \to (S, \delta)\,.$$

A similar procedure using functor extensions (see Exercise II.3.I) leads to the description of a distributive law δ of $\mathbb{T}$ over $\mathbb{S}$ as a monoid $((T, \delta), m, e)$ in $\mathsf{MNDEXT}(\mathbb{S}, \mathbb{S})$, where $\mathsf{MNDEXT}(\mathbb{S}, \mathbb{T})$ denotes the metacategory of extensions from $\mathbb{S}$ to $\mathbb{T}$.

II.3.K *Adjoint triangle theorem.*

(1) Show that the following assertions are equivalent for an adjunction $F \overset{\eta}{\underset{\varepsilon}{\dashv}} G : \mathsf{A} \to \mathsf{X}$ with induced monad $\mathbb{T}$:
   (a) for every object A in A, ε_A is a coequalizer of ε_{FGA}, $FG\varepsilon_A$;
   (b) for every object A in A, ε_A is a regular epimorphism;
   (c) the comparison functor $K : \mathsf{A} \to \mathsf{X}^{\mathbb{T}}$ is full and faithful.

Adjunctions satisfying these equivalent conditions are said to be *of descent type*, in which case G is called *premonadic*.

(2) Consider a commutative triangle of functors

$$\begin{array}{ccc} \mathsf{A} & \xrightarrow{K} & \mathsf{B} \\ & {\scriptstyle G}\searrow \quad \swarrow{\scriptstyle J} & \\ & \mathsf{X} & \end{array}$$

with G, J right adjoint, so that one has $F \overset{\eta}{\underset{\varepsilon}{\dashv}} G$, $H \overset{\delta}{\underset{\gamma}{\dashv}} J$, and J premonadic. With $\mu : H \to KF$ determined by $J\mu \cdot \delta = \eta$, set

$$\alpha := FJ\gamma\,, \quad \beta := \varepsilon FJ \cdot FJ\mu J\,.$$

Prove the equivalence of the following assertions for $B \in \operatorname{ob}\mathsf{B}$:
   (a) there exists a K-universal arrow for B;
   (b) there exists a coequalizer of α_B, β_B in A.

(3) For functors $K : \mathsf{A} \to \mathsf{B}$, $J : \mathsf{B} \to \mathsf{X}$ with J premonadic and coequalizers existing in A, the functor K has a left adjoint if and only if JK has a left adjoint.

II.3.L *Cocompleteness of the Eilenberg–Moore category.* For a monadic functor $G : \mathsf{A} \to \mathsf{X}$, if A has coequalizers and X is J-cocomplete, then A is also J-cocomplete. In particular, for a monad $\mathbb{T}$ on a category X with coproducts, $\mathsf{X}^{\mathbb{T}}$ is small-cocomplete whenever it has coequalizers.
Hint. Consider the commutative diagram

$$\begin{array}{ccc} \mathsf{A} & \xrightarrow{\Delta} & \mathsf{A}^{\mathsf{J}} \\ {\scriptstyle G}\downarrow & & \downarrow{\scriptstyle G^{\mathsf{J}}} \\ \mathsf{X} & \xrightarrow{\Delta} & \mathsf{X}^{\mathsf{J}} \end{array}$$

and apply Exercise II.3.K(2).

II.3.M *Idempotent monads.* A monad $\mathbb{T} = (T, m, e)$ on X is *idempotent* if $m : TT \to T$ is an isomorphism. For any adjunction

$$F \overset{\eta}{\underset{\varepsilon}{\dashv}} G : \mathsf{A} \longrightarrow \mathsf{X}$$

inducing $\mathbb{T}$, the following statements are equivalent:

(i) $\mathbb{T}$ is idempotent;
(ii) $T\eta = \eta T$;
(iii) any one of $G\varepsilon$, ηG, εF, $F\eta$ is an isomorphism;
(iv) for all $\mathbb{T}$-algebras (X, a), the structure a is an X-isomorphism;
(v) $G^{\mathbb{T}} : \mathsf{X}^{\mathbb{T}} \to \mathsf{X}$ restricts to an isomorphism of $\mathsf{X}^{\mathbb{T}}$ with the full subcategory of X containing those X with η_X an isomorphism;
(vi) $FG^{\mathbb{T}}$ is left adjoint to the comparison functor $K : \mathsf{A} \to \mathsf{X}^{\mathbb{T}}$ and the unit κ of this adjunction satisfies $G^{\mathbb{T}}\kappa = \eta G^{\mathbb{T}}$.

These conditions hold in particular when G or F is full and faithful.

II.4 Monoidal and ordered categories

II.4.1 Monoidal categories

A *monoidal category* is a category C together with

(1) a functor $(-) \otimes (-) : \mathsf{C} \times \mathsf{C} \to \mathsf{C}$,
(2) a distinguished object E in C,
(3) natural isomorphisms

$$\alpha_{A,B,C} : A\otimes(B\otimes C) \to (A\otimes B)\otimes C\,,\ \lambda_A : E\otimes A \to A\,,\ \rho_A : A\otimes E \to A$$

making the following diagrams commute for all objects A, B, C, D:

$$\begin{array}{ccccc}
A\otimes(B\otimes(C\otimes D)) & \xrightarrow{\alpha_{A,B,C\otimes D}} & (A\otimes B)\otimes(C\otimes D) & \xrightarrow{\alpha_{A\otimes B,C,D}} & ((A\otimes B)\otimes C)\otimes D \\
{\scriptstyle 1_A\otimes\alpha_{B,C,D}}\downarrow & & & & \uparrow{\scriptstyle \alpha_{A,B,C}\otimes 1_D} \\
A\otimes((B\otimes C)\otimes D) & & \xrightarrow{\qquad\alpha_{A,B\otimes C,D}\qquad} & & (A\otimes(B\otimes C))\otimes D
\end{array}$$

$$\begin{array}{ccc}
A\otimes(E\otimes B) & \xrightarrow{\alpha_{A,E,B}} & (A\otimes E)\otimes B \\
& {\scriptstyle 1_A\otimes\lambda_B}\searrow \quad \swarrow{\scriptstyle \rho_A\otimes 1_B} & \\
& A\otimes B &
\end{array}$$

and satsifying $\lambda_E = \rho_E$.

The monoidal category C is *symmetric* if there are also

(4) natural isomorphisms $\sigma_{A,B} : A\otimes B \to B\otimes A$ with $\sigma_{B,A}\cdot\sigma_{A,B} = 1_{A\otimes B}$ such that the following diagrams commute:

$$\begin{array}{ccccc}
A\otimes(B\otimes C) & \xrightarrow{\alpha_{A,B,C}} & (A\otimes B)\otimes C & \xrightarrow{\sigma_{A\otimes B,C}} & C\otimes(A\otimes B) \\
{\scriptstyle 1_A\otimes\sigma_{B,C}}\downarrow & & & & \downarrow{\scriptstyle \alpha_{C,A,B}} \\
A\otimes(C\otimes B) & \xrightarrow{\alpha_{A,C,B}} & (A\otimes C)\otimes B & \xrightarrow{\sigma_{A,C}\otimes 1_B} & (C\otimes A)\otimes B
\end{array}
\qquad
\begin{array}{ccc}
E\otimes A & \xrightarrow{\sigma_{E,A}} & A\otimes E \\
& {\scriptstyle \lambda_A}\searrow \quad \swarrow{\scriptstyle \rho_A} & \\
& A\,. &
\end{array}$$

Since $\lambda_E = \rho_E$, the commutative triangle gives $\sigma_{E,E} = 1_{E,E}$.

This somewhat cumbersome definition is best illustrated in terms of examples.

II.4.1.1 Examples

(1) Set, with its Cartesian structure $\otimes = \times$ (see II.1.1) is symmetric monoidal. In fact, every category C with finite products can be considered a symmetric monoidal category (see Exercise II.2.H).

(2) The prototype of a symmetric monoidal category is the category AbGrp of Abelian groups with its usual *tensor product*. (For Abelian groups A, B, the Abelian group $A \otimes B$ comes with a bilinear map $\otimes : A \times B \to A \otimes B$ such that any bilinear map $f : A \times B \to C$ factors as $f = g \cdot \otimes$, with a uniquely determined homomorphism $g : A \otimes B \to C$.) One puts $E = \mathbb{Z}$ and produces α, λ, ρ, σ with the universal property of the tensor product. More generally, such a construction works for the category Mod_R of R-modules, where R is a commutative unital ring.

(3) Sup has a symmetric monoidal structure which may be constructed analogously to the tensor product of AbGrp. Indeed, for $X, Y, Z \in \mathrm{ob}\,\mathsf{Sup}$, call a mapping $f : X \times Y \to Z$ a *bimorphism* if $f(x, -) : Y \to Z$ and $f(-, y) : X \to Z$ are morphisms of Sup for all $x \in X$, $y \in Y$. Then one can construct the universal bimorphism $\otimes : X \times Y \to X \otimes Y$ (so that every bimorphism f factors as $f = g \cdot \otimes$, with a unique sup-map $g : X \otimes Y \to Z$) similarly to the tensor product of Abelian groups: on the free sup-lattice $P(X \times Y)$ (see Example II.2.5.1(5)), consider the least compatible congruence relation $\sim$ that makes the composite map

$$X \times Y \xrightarrow{e_{X \times Y}} P(X \times Y) \xrightarrow{\mathrm{proj}} P(X \times Y)/\!\sim$$

a bimorphism; this composite map is universal, hence $X \otimes Y = P(X \times Y)/\!\sim$.

Many of the monoidal categories considered in this book are *strict*; i.e. α, λ, and ρ can all be taken to be identity morphisms, so that the *coherence conditions* (expressed by the first two commutative diagrams) are trivially satisfied. These strict monoidal categories, however, typically fail to be symmetric.

(4) For a set X, consider the ordered set $\mathsf{Rel}(X, X)$ of relations on X, seen as a category. Relational composition defines a strict monoidal structure on $\mathsf{Rel}(X, X)$ (see Section II.1.2).

(5) More generally, let $\mathcal{V}$ be an ordered set, considered as a category. $\mathcal{V}$ becomes strict monoidal precisely when it carries a monoid structure whose multiplication $\mathcal{V} \times \mathcal{V} \to \mathcal{V}$ is monotone. In particular, every quantale $\mathcal{V}$ is a strict monoidal category.

(6) For a category C, the functor category C^{C} becomes a strict monoidal metacategory, with composition of functors as tensor product.

(7) If C is a monoidal category, then so is C^{op} (and similarly for a symmetric monoidal category C).

A *homomorphism of monoidal categories* C and D is a functor $F : \mathsf{C} \to \mathsf{D}$ which comes with natural isomorphisms

$$\delta_{A,B} : FA \otimes_{\mathsf{D}} FB \to F(A \otimes_{\mathsf{C}} B)\,, \quad \varepsilon : E_{\mathsf{D}} \to FE_{\mathsf{C}}$$

for all $A, B \in \operatorname{ob}\mathsf{C}$, which commute with the natural isomorphisms defining the structure of the monoidal categories C and D:

$$\begin{array}{ccc}
FA \otimes (FB \otimes FC) & \xrightarrow{\alpha} & (FA \otimes FB) \otimes FC \\
{\scriptstyle 1\otimes\delta}\downarrow & & \downarrow{\scriptstyle \delta\otimes 1} \\
FA \otimes F(B \otimes C) & & F(A \otimes B) \otimes FC \\
{\scriptstyle \delta}\downarrow & & \downarrow{\scriptstyle \delta} \\
F(A \otimes (B \otimes C)) & \xrightarrow{F\alpha} & F((A \otimes B) \otimes C)
\end{array}$$

$$\begin{array}{ccc}
E \otimes FA & \xrightarrow{\lambda} & FA \\
{\scriptstyle \varepsilon\otimes 1}\downarrow & & \uparrow{\scriptstyle F\lambda} \\
FE \otimes FA & \xrightarrow{\delta} & F(E \otimes A)
\end{array}
\qquad
\begin{array}{ccc}
FA \otimes E & \xrightarrow{\rho} & FA \\
{\scriptstyle 1\otimes\varepsilon}\downarrow & & \uparrow{\scriptstyle F\rho} \\
FA \otimes FE & \xrightarrow{\delta} & F(A \otimes E)\,.
\end{array}$$

Observe that naturality of δ makes F commute with $\otimes$, up to isomorphism, not just for objects but also for morphisms:

$$\begin{array}{ccc}
FA \otimes FB & \xrightarrow{Ff\otimes Fg} & FC \otimes FD \\
{\scriptstyle \delta}\downarrow & & \downarrow{\scriptstyle \delta} \\
F(A \otimes B) & \xrightarrow{F(f\otimes g)} & F(C \otimes D)\,.
\end{array}$$

II.4.2 Monoids

Let C be a monoidal category as in Section II.4.1. A *monoid M in* C is a C-object together with two morphisms

$$m : M \otimes M \to M\,, \quad e : E \to M$$

such that the diagrams

$$\begin{array}{ccccc}
M \otimes (M \otimes M) & \xrightarrow{\alpha_{M,M,M}} & (M \otimes M) \otimes M & \xrightarrow{m\otimes 1_M} & M \otimes M \\
{\scriptstyle 1_M\otimes m}\downarrow & & & & \downarrow{\scriptstyle m} \\
M \otimes M & & \xrightarrow{\quad\quad m \quad\quad} & & M
\end{array}$$

$$\begin{array}{ccccc}
E \otimes M & \xrightarrow{e\otimes 1_M} & M \otimes M & \xleftarrow{1_M\otimes e} & M \otimes E \\
& {\scriptstyle \lambda_M}\searrow & \downarrow{\scriptstyle m} & \swarrow{\scriptstyle \rho_M} & \\
& & M & &
\end{array}$$

commute. A *homomorphism of monoids* $f : (M, m, e) \to (N, n, d)$ is a C-morphism making the diagrams

$$\begin{array}{ccc} M \otimes M & \xrightarrow{f \otimes f} & N \otimes N \\ {\scriptstyle m}\downarrow & & \downarrow{\scriptstyle n} \\ M & \xrightarrow{f} & N \end{array} \qquad \begin{array}{ccc} & E & \\ {\scriptstyle e}\swarrow & & \searrow{\scriptstyle d} \\ M & \xrightarrow{f} & N \end{array}$$

commute. The resulting category is denoted by $\mathsf{Mon}_{\mathsf{C}}$. A *comonoid* M in C is simply a monoid M in C^{op}.

II.4.2.1 Examples

(1) For Set with its Cartesian structure, $\mathsf{Mon}_{\mathsf{Set}} = \mathsf{Mon}$.
(2) A unital ring R is an Abelian group that is also a monoid in which the distributive laws hold, i.e. the multiplication $R \times R \to R$ is $\mathbb{Z}$-bilinear and is therefore equivalently described as a homomorphism $R \otimes R \to R$. Hence, unital rings are precisely the monoids in AbGrp (with its usual tensor product), and $\mathsf{Mon}_{\mathsf{AbGrp}} = \mathsf{Rng}$, the category of unital rings and their homomorphisms.
(3) A quantale $\mathcal{V}$ is a complete lattice with a monoid operation $\otimes : \mathcal{V} \times \mathcal{V} \to \mathcal{V}$ that preserves suprema in each variable; with the tensor product in Sup (see Example II.4.1.1(3)), the monoid operation may equivalently be considered a morphism $\mathcal{V} \otimes \mathcal{V} \to \mathcal{V}$ in Sup. This way, one shows $\mathsf{Mon}_{\mathsf{Sup}} = \mathsf{Qnt}$.
(4) Monoids in $\mathsf{Rel}(X, X)$ (see Example II.4.1.1(4)) are precisely orders on the set X (see also Section II.1.3).
(5) A monoid in a quantale $\mathcal{V}$ (see Example II.4.1.1(5)) is simply an idempotent element $v \in \mathcal{V}$ (so that $v \otimes v = v$) with $k \leq v$. A comonoid in $\mathcal{V}$ is an idempotent element $v \in \mathcal{V}$ with $v \leq k$. A frame is thus an example of a commutative quantale in which every element is a comonoid.
(6) For a category C, monoids of the monoidal category C^{C} (see Example II.4.1.1(6)) are precisely monads on C, and homomorphisms of monoids are precisely morphisms of monads (see II.3.1).
(7) A topology on a set X may be defined as a monoid in $\mathsf{SLat}(PX, PX)$ (see Exercise II.1.F), where PX is considered as a join-semilattice.

II.4.3 Actions

Let C be a monoidal category and let $M = (M, m, e)$ be a monoid in C. A *left M-action* (or simply an *M-action*) is an object A in C that comes with a C-morphism

$$a : M \otimes A \to A$$

such that (in the notation of Sections II.4.1 and II.4.2) the following diagrams commute:

$$\begin{array}{ccc} M \otimes (M \otimes A) & \xrightarrow{1_M \otimes a} & M \otimes A \\ {\scriptstyle (m\otimes 1_A)\cdot \alpha_{M,M,A}}\downarrow & & \downarrow {\scriptstyle a} \\ M \otimes A & \xrightarrow{\quad a \quad} & A \end{array} \qquad \begin{array}{ccc} E \otimes A & \xrightarrow{e\otimes 1_M} & M \otimes A \\ & {\scriptstyle \lambda_A} \searrow & \downarrow {\scriptstyle a} \\ & & A\,. \end{array}$$

A C-morphism $f : A \to B$ between M-actions (A, a) and (B, b) is *equivariant* if

$$\begin{array}{ccc} M \otimes A & \xrightarrow{1_M \otimes f} & M \otimes B \\ {\scriptstyle a}\downarrow & & \downarrow {\scriptstyle b} \\ A & \xrightarrow{\quad f \quad} & B \end{array}$$

commutes. Similarly, a *right M-action* is an object A in C with a C-morphism $a : A \otimes M \to A$ making the corresponding diagrams commute.

Any monoid (M, m, e) in a monoidal category C gives rise to a monad $\mathbb{M} = (M \otimes (-), \overline{m}, \overline{e})$ on C, where

$$\overline{m}_A = (m \otimes 1_A) \cdot \alpha_{M,M,A} \qquad \text{and} \qquad \overline{e}_A = (e \otimes 1_A) \cdot \lambda_A^{-1}$$

for all $A \in \mathrm{ob}\,\mathsf{C}$ (using the notation of Section II.4.2). The Eilenberg–Moore category $\mathsf{C}^{\mathbb{M}}$ of this monad is the category of M-actions and equivariant C-morphisms.

II.4.3.1 Examples

(1) If C = Set, the category of M-actions is the category M-Set of M-actions and equivariant maps (see Exercise II.3.B).

(2) The monoidal structure of AbGrp is given by the tensor product over $\mathbb{Z}$ (Example II.4.1.1(2)), and a monoid R in AbGrp is a ring (Example II.4.2.1(2)). Hence, $\mathsf{AbGrp}^{\mathbb{R}}$ (with $\mathbb{R}$ the monad induced by the monoid R) is the usual category of left R-modules.

(3) Given a quantale $\mathcal{V} = (V, \otimes, k)$, i.e. a monoid in Sup (Example II.4.2.1(3)), the category $\mathsf{Sup}^{\mathcal{V}}$ is described as follows. A $\mathcal{V}$-action X in Sup is a complete lattice X together with a bimorphism $(-) \cdot (-) : \mathcal{V} \times X \to X$ in Sup (Example II.4.1.1(3)) such that

$$(u \otimes v) \cdot x = u \cdot (v \cdot x)\,, \qquad k \cdot x = x\,,$$

for all $v \in \mathcal{V}$, $x \in X$, and a sup-map $f : X \to Y$ is equivariant whenever

$$f(v \cdot x) = v \cdot f(x)$$

for all $v \in \mathcal{V}$, $x \in X$. Objects in $\mathsf{Sup}^{\mathcal{V}}$ are also known as $\mathcal{V}$-modules (see Kruml and Paseka [2008]), but here we reserve this name for a generalization of the term introduced in Section II.1.4 (see Section III.1.3).

II.4.4 Monoidal closed categories

An object A in a monoidal category C (as in Section II.4.1) is $\otimes$-*exponentiable* if the functors

$$A \otimes (-)\,,\; (-) \otimes A : \mathsf{C} \to \mathsf{C}$$

have right adjoints

$$A \multimap (-)\,,\; (-) \multimapinv A : \mathsf{C} \to \mathsf{C}\,,$$

respectively; i.e. if for all objects B, C in C there are bijections

$$\frac{A \otimes B \to C}{B \to (A \multimap C)} \qquad\qquad \frac{B \otimes A \to C}{B \to (C \multimapinv A)}$$

which are natural in B. The functor $A \multimap (-)$ is the *right internal hom-functor* of A, and $(-) \multimapinv A$ is the *left internal hom-functor*. Obviously, when C is symmetric, $(A \multimap (-)) \cong ((-) \multimapinv A)$, and there is no need for distinction between right and left, so either of the two notations may be used. The monoidal category C is *closed* if every object is $\otimes$-exponentiable.

If the monoidal structure of C is the Cartesian structure of C (given by finite direct products), then one says *exponentiable* instead of $\times$-exponentiable and usually writes

$$B^A$$

for the internal hom-object $A \multimap B \cong B \multimapinv A$, also called an *exponential*, and one says that C is *Cartesian closed* if it is closed.

II.4.4.1 Examples

(1) Set, Ord, and Cat are Cartesian closed. For Set, see Example II.2.5.1(4). Essentially the same proof can be used to prove Cartesian closure of Ord, with

$$B^A = \mathsf{Ord}(A, B)$$

provided with its pointwise order. In Cat, the functor category B^A serves as internal hom-object; see Exercise II.4.A.

(2) AbGrp, and more generally Mod_R, (for a commutative unital ring R), are monoidal closed, with

$$A \multimap B = \hom_R(A, B) = \mathsf{Mod}_R(A, B)\,.$$

(3) Sup is monoidal closed, with

$$X \multimap Y = \mathsf{Sup}(X, Y)\,,$$

provided with the pointwise structure.

(4) $\mathsf{Rel}(X, X)$, considered as a monoidal category as in Example II.4.1.1(4), is monoidal closed, and one has

$$x\ (r \multimap s)\ y \iff \forall z \in X\ (y\ r\ z \implies x\ s\ z)\,,$$
$$x\ (s \multimapinv r)\ y \iff \forall z \in X\ (z\ r\ x \implies z\ s\ y)\,,$$

for all $x, y \in X$.

(5) The internal hom-objects in a quantale $\mathcal{V}$ have been described in Section II.1.10; hence $\mathcal{V}$ is monoidal closed. As mentioned in Example II.4.1.1(5), any *ordered monoid* (i.e. any ordered set with a monotone monoid operation) can be considered a monoidal category. In particular, every monoid $\mathcal{V}$, provided with the discrete order, is a monoidal category, and it is closed precisely when it is a group. In that case, for $u, v \in \mathcal{V}$, one has

$$u \multimap v = u^{-1}v\,, \quad v \multimapinv u = vu^{-1}\,,$$

which provides some justification for the internal hom notation in general.

(6) For a category C, the monoidal category C^{C} of Example II.4.1.1(6) is rarely closed. For example, when C is a small discrete category, C^{C} is the monoid $\mathsf{Set}(X, X)$ with $X = \operatorname{ob}\mathsf{C}$, which, according to the preceding example, is monoidal closed only when it is a group, i.e. when $|X| \le 1$.

II.4.4.2 Proposition *In a locally small monoidal closed category* C, *one has natural isomorphisms*

(1) $(A \otimes B) \multimap C \cong B \multimap (A \multimap C)$;
(2) $C \multimapinv (A \otimes B) \cong (C \multimapinv B) \multimapinv A$;
(3) $(A \multimap C) \multimapinv B \cong A \multimap (C \multimapinv B)$.

If C *is Cartesian closed, then* $C^{A \times B} \cong (C^A)^B \cong (C^B)^A$.

Proof For all objects X, one has

$$\frac{\dfrac{\dfrac{\dfrac{X \to ((A \otimes B) \multimap C)}{(A \otimes B) \otimes X \to C}}{A \otimes (B \otimes X) \to C}}{B \otimes X \to (A \multimap C)}}{X \to (B \multimap (A \multimap C))}\,,$$

which implies (1) (because the Yoneda embedding reflects isomorphisms; see Corollary II.2.4.2 and Exercise II.2.A). Point (2) is proved analogously to (1). For (3), we note

$$\begin{array}{c}
X \to ((A \multimap C) \multimapinv B) \\ \hline
X \otimes B \to (A \multimap C) \\ \hline
A \otimes (X \otimes B) \to C \\ \hline
(A \otimes X) \otimes B \to C \\ \hline
A \otimes X \to (C \multimapinv B) \\ \hline
X \to (A \multimap (C \multimapinv B))\,.
\end{array}$$

□

Given an object C of a monoidal closed category C, there is a functor

$$(-) \multimap C : \mathsf{C}^{\mathrm{op}} \to \mathsf{C}\,, \quad (f : A \to B) \mapsto (f \multimap 1_C : (B \multimap C) \to (A \multimap C))\,,$$

with $f \multimap 1_C$ the unique C-morphism that makes the diagram

$$\begin{array}{ccc}
A \otimes (A \multimap C) & \xrightarrow{\mathrm{ev}^A_C} & C \\
{\scriptstyle 1_A \otimes (f \multimap 1_C)} \uparrow & \nearrow {\scriptstyle \mathrm{ev}^B_C \cdot (f \otimes 1_{B \multimap C})} & \\
A \otimes (B \multimap C) & &
\end{array}$$

commute (here $\mathrm{ev}^A : A \otimes (A \multimap (-)) \to 1_{\mathsf{C}}$ denotes the counit of the adjunction $A \otimes (-) \dashv (-)^A$). One can naturally define a functor $C \multimapinv (-)$ that will be isomorphic to $(-) \multimap C$ whenever C is symmetric. In this case, the functor is also right adjoint.

II.4.4.3 Proposition *For a symmetric monoidal closed category* C, *the functor* $(-) \multimap C : \mathsf{C}^{\mathrm{op}} \to \mathsf{C}$ *is self-adjoint (for all* C*-objects* C*).*

Proof By couniversality of the component at C of the counit $\mathrm{ev}^{A \multimap C} : (A \multimap C) \otimes ((A \multimap C) \multimap (-)) \to 1_{\mathsf{C}}$, there is a unique C-morphism $u_A : A \to ((A \multimap C) \multimap C)$ making the following diagram commute:

$$\begin{array}{ccc}
(A \multimap C) \otimes ((A \multimap C) \multimap C) & \xrightarrow{\mathrm{ev}^{A \multimap C}_C} & C\,. \\
{\scriptstyle 1_{A \multimap C} \otimes u_A} \uparrow & \nearrow {\scriptstyle \mathrm{ev}^A_C \cdot \sigma_{A \multimap C, A}} & \\
(A \multimap C) \otimes A & &
\end{array}$$

The morphism u_A is a $((-) \multimap C)$-universal arrow for A. Indeed, given a C-morphism $f : A \to (B \multimap C)$, we define $g : B \to (A \multimap C)$ as the unique C-morphism such that

$$\begin{array}{ccc}
A \otimes (A \multimap C) & \xrightarrow{\mathrm{ev}^A_C} & C \\
{\scriptstyle 1_A \otimes g} \uparrow & \nearrow {\scriptstyle \mathrm{ev}^B_C \cdot (1_B \otimes f) \cdot \sigma_{A,B}} & \\
A \otimes B & &
\end{array}$$

commutes; since

$$\begin{aligned}
\mathrm{ev}_C^B \cdot (1_B \otimes ((g \multimap 1_C) \cdot u_A)) &= \mathrm{ev}_C^B \cdot (1_B \otimes (g \multimap 1_C)) \cdot (1_B \otimes u_A) && ((-)\otimes(-) \text{ functor})\\
&= \mathrm{ev}_C^{A \multimap C} \cdot (g \otimes 1_{(A\multimap C)\multimap C}) \cdot (1_B \otimes u_A) && (\text{definition of } g \multimap 1_C)\\
&= \mathrm{ev}_C^{A \multimap C} \cdot (1_{A\multimap C} \otimes u_A) \cdot (g \otimes 1_A) && ((-)\otimes(-) \text{ functor})\\
&= \mathrm{ev}_C^{A} \cdot \sigma_{A\multimap C, A} \cdot (g \otimes 1_A) && (\text{definition of } u_A)\\
&= \mathrm{ev}_C^{A} \cdot (1_A \otimes g) \cdot \sigma_{B,A} && (\sigma \text{ natural})\\
&= \mathrm{ev}_C^{B} \cdot (1_B \otimes f) && (\text{definition of } g),
\end{aligned}$$

one can conclude that $(g \multimap 1_C) \cdot u_A = f$ by couniversality of ev_C^B. The displayed identities also show that g is uniquely determined, so that u_A is universal as claimed. To see that $((-) \multimap C)^{\mathrm{op}} : \mathsf{C} \to \mathsf{C}^{\mathrm{op}}$ is the required left adjoint, it suffices to verify that the morphisms u_A (with $A \in \mathsf{C}$) form a natural transformation $u : 1_{\mathsf{C}} \to (((-) \multimap B) \multimap B)$ (Theorem II.2.5.4); by couniversality of ev_C^B, this follows from the identities

$$\begin{aligned}
\mathrm{ev}_C^{B\multimap C} \cdot (1_{B\multimap C} \otimes (u_B \cdot f)) &= \mathrm{ev}_C^{B} \cdot \sigma_{B\multimap C, B} \cdot (1_{B\multimap C} \otimes f)\\
&= \mathrm{ev}_C^{B} \cdot (f \otimes 1_{B\multimap C}) \cdot \sigma_{B\multimap C, A}\\
&= \mathrm{ev}_C^{A} \cdot (1_A \otimes (f \multimap C)) \cdot \sigma_{B\multimap C, A}\\
&= \mathrm{ev}_C^{A} \cdot \sigma_{A\multimap C, A} \cdot ((f \multimap C) \otimes 1_A)\\
&= \mathrm{ev}_C^{A\multimap C} \cdot (1_{A\multimap C} \otimes u_A) \cdot ((f \multimap C) \otimes 1_A)\\
&= \mathrm{ev}_C^{A\multimap C} \cdot ((f \multimap C) \otimes 1_{(A\multimap C)\multimap C}) \cdot (1_{B\multimap C} \otimes u_A)\\
&= \mathrm{ev}_C^{B\multimap C} \cdot (1_{B\multimap C} \otimes (((f \multimap C) \multimap C) \cdot u_A))
\end{aligned}$$

(for all C-morphisms $f : A \to B$). □

II.4.4.4 Example When $\mathsf{C} = \mathsf{Set}$ with its Cartesian closed structure and $C = \mathsf{2}$, for every set X the internal hom-object $\mathsf{2}^X = X \multimap \mathsf{2}$ is the powerset PX. The functor $\mathsf{2}^{(-)} = (-) \multimap \mathsf{2}$ then describes the contravariant powerset functor $P^\bullet$ of Example II.2.5.1(6).

II.4.5 Ordered categories

An *ordered category* C is a category C with each hom-class $\mathsf{C}(A, B)$ carrying an order (i.e. a reflexive and transitive relation $\leq$), such that the composition maps

$$\mathsf{C}(A, B) \times \mathsf{C}(B, C) \to \mathsf{C}(A, C), \quad (f, g) \mapsto g \cdot f$$

are monotone. This is already guaranteed if composition from either side preserves the order, i.e. if, for all $f, f' : A \to B$, $h : B \to C$, $g : D \to A$, one has

$$f \leq f' \implies h \cdot f \cdot g \leq h \cdot f' \cdot g\,.$$

Hence, an ordered category is a special case of a *2-category* (see [Borceux, 1994a]), namely a 2-category whose local categories are *thin* (i.e. ordered classes). Alternatively, readers familiar with *enriched category theory* (see [Kelly, 1982]) may want to think of an ordered category as a category enriched over the metacategory ORD of ordered classes; see also Section II.4.10.

II.4.5.1 Examples

(1) Rel is an ordered category, with $\mathsf{Rel}(X, Y)$ ordered by inclusion for all sets X, Y.
(2) Mod is, like Rel, an ordered category.
(3) Ord is an ordered category, with $\mathsf{Ord}(X, Y)$ carrying the pointwise order.
(4) Sup is, like Ord, an ordered category.
(5) Every category can be considered an ordered category when provided with the discrete order.

Note that for any object A of an ordered category C, $\mathsf{C}(A, A)$ can be considered a strict monoidal category, with the tensor product given by composition in C (see Examples II.4.1.1(4), (5), and (6)).

If C is an ordered category, then its dual C^{op} (obtained by "turning arrows around") is also ordered. Moreover, one can form the *conjugate* ordered category C^{co} that leaves the arrows intact but turns around the order:

$$\mathsf{C}^{\mathrm{co}}(A, B) = (\mathsf{C}(A, B), \geq) = (\mathsf{C}(A, B))^{\mathrm{op}} .$$

There is only one combination of the two dualization processes, since

$$\mathsf{C}^{\mathrm{co\,op}} = \mathsf{C}^{\mathrm{op\,co}} .$$

II.4.6 Lax functors, pseudo-functors, 2-functors, and their transformations

A *lax functor* $F : \mathsf{C} \to \mathsf{D}$ of ordered categories is given by functions

$$F : \mathrm{ob}\,\mathsf{C} \to \mathrm{ob}\,\mathsf{D} \qquad \text{and} \qquad F_{A,B} : \mathsf{C}(A, B) \to \mathsf{D}(FA, FB)$$

for all $A, B \in \mathrm{ob}\,\mathsf{C}$ (with $F_{A,B}$ usually written as F), such that

(1) $F_{A,B}$ is monotone;
(2) $Fg \cdot Ff \leq F(g \cdot f)$;
(3) $1_{FA} \leq F1_A$,

for all $A, B, C \in \mathrm{ob}\,\mathsf{C}$, $f : A \to B$, $g : B \to C$ in C. The lax functor F is a *pseudo-functor* if the inequalities "$\leq$" in (2) and (3) may be replaced by "$\simeq$" (see Section II.1.3), and F is a *2-functor* if they may be replaced by "$=$".

An *oplax functor* $F : \mathsf{C} \to \mathsf{D}$ must satisfy condition (1) and, instead of (2) and (3),

(2*) $F(f \cdot g) \leq Ff \cdot Fg$;
(3*) $F1_A \leq 1_{FA}$.

Hence, pseudo-functors are precisely the simultaneously lax and oplax functors.

This terminology coincides with the one used in the more general context of 2-categories (see [Borceux, 1994a]). If one thinks of C and D as ORD-enriched categories, then 2-functors are precisely ORD-functors (see [Kelly, 1982]). For ordinary categories, considered as discrete ordered categories, all the functor notions just introduced coincide with the ordinary notion of functor.

For lax or oplax functors $F, G : \mathsf{C} \to \mathsf{D}$, a *lax transformation* $\alpha : F \to G$ is given by an ob C-indexed family of D-morphisms $\alpha_A : FA \to GA$ such that, for all $f : A \to B$ in C,

(4) $Gf \cdot \alpha_A \leq \alpha_B \cdot Ff,$

while an *oplax transformation* satisfies

(4*) $\alpha_B \cdot Ff \leq Gf \cdot \alpha_A$

instead. A *pseudo-natural transformation* is a simultaneously lax and oplax transformation, and the prefix "pseudo" may be dropped if the order of the hom-class of D is separated.

Lax transformations can be *vertically composed* and lead to a lax transformation again, via $(\beta \cdot \alpha)_A = \beta_A \cdot \alpha_A$ (for $\alpha : F \to G, \beta : G \to H$); oplax transformations compose similarly. *Horizontal composition* is more delicate. Here we mention only that, for a lax or oplax functor $S : \mathsf{B} \to \mathsf{C}$ and a lax transformation $\alpha : F \to G$, one obtains a lax transformation $\alpha S : FS \to GS$; one can proceed analogously with oplax transformations. However, "whiskering" from the other side with $T : \mathsf{D} \to \mathsf{E}$ requires a pseudo-functor in order to obtain from α again a lax or oplax transformation $T\alpha : TF \to TG$.

II.4.7 Maps

A morphism $f : A \to B$ in an ordered category C is a *map* if there is a morphism $g : B \to A$ in C with

$$1_A \leq g \cdot f \qquad \text{and} \qquad f \cdot g \leq 1_B\ ;$$

one writes $f \dashv g$ in this situation, and calls f the *left adjoint* and g the *right adjoint* of the *adjunction*.

For $\mathsf{C} = \mathsf{Ord}$, being a map means being left adjoint (see Section II.1.5). The map terminology is, however, better motivated by the example $\mathsf{C} = \mathsf{Rel}$: a relation $r : A \nrightarrow B$ of sets A, B is a map in Rel precisely when it is the graph of a morphism $r : A \to B$ in Set. Indeed, the existence of a relation $s : B \nrightarrow A$ with

$$1_A \leq s \cdot r \quad \text{means} \quad \forall x \in A\ \exists y \in B\ (x\ r\ y\ \&\ y\ s\ x)\ ,$$

while

$$r \cdot s \leq 1_B \quad \text{means} \quad \forall x \in A\ \forall y, y' \in B\ (y\ s\ x\ \&\ x\ r\ y' \implies y = y')\ .$$

Hence, the expression $(r(x) = y \iff x\ r\ y)$ defines a unique Set-map $r : A \to B$. One can also easily see that necessarily $s = r^\circ$ (see Exercise II.4.B).

Of course, in any ordered category (or more generally, in a 2-category), a right adjoint g of a map f as above is uniquely determined up to "$\simeq$"; we often write f^* (or $f^{\dashv}$) for g:

$$1_A \leq f^* \cdot f \qquad \text{and} \qquad f \cdot f^* \leq 1_B\,.$$

This notation is coherent with the fact that for a monotone map $f : A \to B$ of ordered sets one has $f_* \dashv f^*$ in Mod (see Section II.1.4). Adjunctions in C can be detected by "hom-ing" into ORD:

II.4.7.1 Proposition *For morphisms $f : A \to B$ and $g : B \to A$ in an ordered category C, one has $f \dashv g$ if and only if $\mathsf{C}(T, f) \dashv \mathsf{C}(T, g)$ in ORD for all $T \in \operatorname{ob}\mathsf{C}$.*

Proof $\mathsf{C}(T, f) \dashv \mathsf{C}(T, g)$ means (by Proposition II.1.5.1) that

$$\mathsf{C}(T, f)(x) \leq y \iff x \leq \mathsf{C}(T, g)(y)$$

for all $x \in \mathsf{C}(T, A)$, $y \in \mathsf{C}(T, B)$; i.e.

$$f \cdot x \leq y \iff x \leq g \cdot y\,.$$

But this equivalence (for all x, y) means exactly $f \dashv g$ in C (for which one may re-employ the proof of Proposition II.1.5.1). □

II.4.8 Quantaloids

A *quantaloid* is a category C with each hom-class being a complete lattice (although not necessarily small), and with composition-preserving suprema on either side:

$$g \cdot (\textstyle\bigvee_{i\in I} f_i) = \bigvee_{i\in I}(g \cdot f_i)\,, \qquad (\bigvee_{i\in I} g_i) \cdot f = \bigvee_{i\in I}(g_i \cdot f)\,,$$

for all C-morphisms $f, f_i : A \to B$, $g, g_i : B \to C$ $(i \in I)$. A quantaloid is therefore an ordered category. However, C is not just ORD-enriched, but SUP-enriched (where SUP denotes the metacategory of sup-complete classes and sup-maps). A *homomorphism of quantaloids* $F : \mathsf{C} \to \mathsf{D}$ is simply a functor that preserves suprema:

$$F(\textstyle\bigvee_{i\in I} f_i) = \bigvee_{i\in I} Ff_i\,.$$

II.4.8.1 Examples

(1) The ordered categories Rel and Sup are quantaloids, but Ord is not. An abstract category provided with the discrete order is not a quantaloid either, unless it is an ordered class (so that its hom-classes are at most singleton sets).

(2) Quantales are precisely the one-object quantaloids. Indeed, one-object quantaloids have a monoid structure that can be considered as the tensor product of the quantale (see Section II.1.10). A homomorphism of one-object quantaloids is precisely a homomorphism of quantales.
(3) For every quantaloid C, its opposite category C^{op} is a quantaloid as well, but C^{co} is generally not so: composition will preserve infima from either side, but not suprema in general.

A *biproduct* of a family $(A_i)_{i\in I}$ of objects in a quantaloid C is given by an object P together with morphisms

$$P \underset{e_i}{\overset{p_i}{\rightleftarrows}} A_i \quad (i \in I)$$

such that

$$p_i \cdot e_j = \begin{cases} 1_{A_i} & \text{if } i = j, \\ \bot & \text{otherwise,} \end{cases} \qquad \text{and} \qquad \textstyle\bigvee_{i\in I} e_i \cdot p_i = 1_P \ .$$

The terminology explains itself with the following statement.

II.4.8.2 Proposition *For a biproduct $(P, p_i, e_i)_{i\in I}$ of $(A_i)_{i\in I}$ in a quantaloid, $(P, p_i)_{i\in I}$ is a product and $(P, e_i)_{i\in I}$ is a coproduct of $(A_i)_{i\in I}$.*

Proof For $f_i : B \to A_i$, $(i \in I)$, $f := \bigvee_{i\in I} e_i \cdot f_i$ is the only morphism with $p_i \cdot f = f_i$ $(i \in I)$. Likewise, given $g_i : A_i \to B$ $(i \in I)$, $g := \bigvee_{i\in I} g_i \cdot p_i$ is the only morphism with $g \cdot e_i = g_i$ $(i \in I)$. □

II.4.8.3 Example In Rel, the biproduct of $(A_i)_{i\in I}$ can be taken to be the coproduct (i.e. the disjoint union in Set) $(e_i : A_i \to C)_{i\in I}$ with $p_i = e_i^\circ$ $(i \in I)$.

For a morphism $f : A \to B$ in a quantaloid C, the sup-map

$$\mathsf{C}(C, f) = f \cdot (-) : \mathsf{C}(C, A) \to \mathsf{C}(C, B)$$

has, for all objects C, a right adjoint, denoted by $f \multimap (-)$ and characterized by

$$\frac{f \cdot g \le h}{g \le f \multimap h}$$

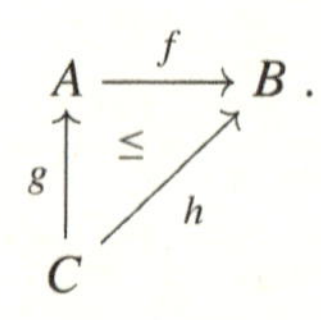

In this situation, $f \multimap h$ is called a *lifting* of h along f. Dually, for $f : B \to A$ in C, the sup-maps

$$\mathsf{C}(f, C) = (-) \cdot f : \mathsf{C}(A, C) \to \mathsf{C}(B, C)$$

have right adjoints, denoted by $(-) \mathrel{\bullet\!\!-} f$ and characterized by

$$\frac{g \cdot f \leq h}{g \leq h \mathrel{\bullet\!\!-} f}$$

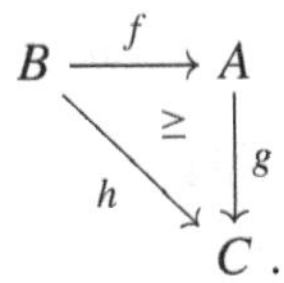

One calls $h \mathrel{\bullet\!\!-} f$ an *extension* of h along f.

For $A = B = C$, the notations used here coincide with those introduced in Section II.4.4 when the monoidal category is taken to be $\mathsf{C}(A, A)$.

Expanding on Proposition II.4.7.1, one can prove the following.

II.4.8.4 Proposition *For morphisms $f : A \to B$ and $g : B \to A$ in a quantaloid C, the following are equivalent:*

(i) $f \dashv g$ *in* C;
(ii) $\mathsf{C}(T, f) \dashv \mathsf{C}(T, g)$ *in* SUP *for all* $T \in \mathrm{ob}\,\mathsf{C}$;
(iii) $\mathsf{C}(g, T) \dashv \mathsf{C}(f, T)$ *in* SUP *for all* $T \in \mathrm{ob}\,\mathsf{C}$;
(iv) $\mathsf{C}(T, g) = f \mathrel{-\!\!\bullet} (-)$ *in* SUP *for all* $T \in \mathrm{ob}\,\mathsf{C}$;
(v) $\mathsf{C}(f, T) = (-) \mathrel{\bullet\!\!-} g$ *in* SUP *for all* $T \in \mathrm{ob}\,\mathsf{C}$.

Proof Note that, when deriving (i) from any of (ii)–(v), the choices $T = A$ and $T = B$ suffice. Further details are left to the reader. □

II.4.9 Kock–Zöberlein monads

A monad $\mathbb{T} = (T, m, e)$ on an ordered category C is *of Kock–Zöberlein type* (or simply *Kock–Zöberlein*) or *lax idempotent* if T is a 2-functor, and for every C-object X there is a chain of adjunctions:

$$Te_X \dashv m_X \dashv e_{TX} \,.$$

This condition can be replaced by any of the following equivalent expressions:

$$\begin{aligned} \forall X \in \mathrm{ob}\,\mathsf{C}\ (Te_X \leq e_{TX}) &\iff \forall X \in \mathrm{ob}\,\mathsf{C}\ (Te_X \dashv m_X) \\ &\iff \forall X \in \mathrm{ob}\,\mathsf{C}\ (m_X \dashv e_{TX}) \,. \end{aligned}$$

Indeed, if any one of the previous conditions holds, then the two others can be obtained by using the following implications:

$$\begin{aligned} Te_X \leq e_{TX} &\Longrightarrow Te_X \cdot m_X = m_{TX} \cdot TTe_X \leq m_{TX} \cdot Te_{TX} = 1_{TTX} \,, \\ Te_{TX} \cdot m_{TX} \leq 1_{TTTX} &\Longrightarrow 1_{TTX} = Tm_X \cdot Te_{TX} \cdot m_{TX} \cdot e_{TTX} \leq Tm_X \cdot e_{TTX} \\ &\qquad = e_{TX} \cdot m_X \,, \\ 1_{TTX} \leq e_{TX} \cdot m_X &\Longrightarrow Te_X \leq e_{TX} \cdot m_X \cdot Te_X = e_{TX} \,. \end{aligned}$$

The study of the Eilenberg–Moore algebras is facilitated in the case of a Kock–Zöberlein monad: if $a : TX \to X$ is a C-morphism with $a \cdot e_X \simeq 1_X$, then one has

$$1_{TX} \simeq Ta \cdot Te_X \leq Ta \cdot e_{TX} = e_X \cdot a\,,$$

so that $a \dashv e_X$. The adjunctions $Ta \dashv Te_X$, $a \dashv e_X$, and $m_X \dashv e_{TX}$ yield

$$a \cdot Ta \dashv Te_X \cdot e_X \qquad \text{and} \qquad a \cdot m_X \dashv e_{TX} \cdot e_X\,,$$

and therefore $a \cdot Ta \simeq a \cdot m_X$ because each side is a left adjoint of $Te_X \cdot e_X = e_{TX} \cdot e_X$.

II.4.9.1 Proposition *Let $\mathbb{T} = (T, m, e)$ be a monad on an ordered category* C*. The following are equivalent when T is a 2-functor:*

(i) *$\mathbb{T}$ is of Kock–Zöberlein type;*
(ii) *any* C*-morphism $a : TX \to X$ with $a \cdot e_X \simeq 1_X$ is left adjoint to e_X and satisfies $a \cdot Ta \simeq a \cdot m_X$.*

Proof The proof of (i) $\Longrightarrow$ (ii) is given in the previous discussion. For (ii) $\Longrightarrow$ (i), note that if (ii) holds, then $m_X \cdot e_{TX} = 1_{TX}$ implies $m_X \dashv e_{TX}$, so that $\mathbb{T}$ is Kock–Zöberlein. □

II.4.9.2 Corollary *The Eilenberg–Moore category $\mathsf{C}^{\mathbb{T}}$ of a Kock–Zöberlein monad $\mathbb{T} = (T, m, e)$ on a* separated *ordered category* C *may be described as follows: objects are those* C*-objects X for which e_X is a section; morphisms are those* C*-morphisms that commute in the usual way with the retractions of the unit.*

Proof When the order on the hom-sets of C is separated, "$\simeq$" becomes an equality in Proposition II.4.9.1, and left adjoints are uniquely determined. □

II.4.9.3 Example The down-set functor $\mathrm{Dn} : \mathsf{Ord} \to \mathsf{Ord}$ (see Section II.2.2) with the union maps $\bigvee_{\mathrm{Dn}\,X} = \bigvee : \mathrm{Dn}\,\mathrm{Dn}\,X \to \mathrm{Dn}\,X$ (where $\bigvee \mathcal{A} = \bigcup_{A \in \mathcal{A}} A$ for all $\mathcal{A} \in \mathrm{Dn}\,\mathrm{Dn}\,X$) and the down-set maps $\downarrow_X : X \to \mathrm{Dn}\,X$ form the *down-set monad*

$$\mathbb{D}\mathrm{n} = (\mathrm{Dn}, \textstyle\bigvee_{\mathrm{Dn}}, \downarrow)$$

on Ord. As $\bigvee_{\mathrm{Dn}\,X}$ is left adjoint to the down-set map $\downarrow_{\mathrm{Dn}\,X}$, and the down-set functor is monotone, the monad $\mathbb{D}\mathrm{n}$ is of Kock–Zöberlein type. The *up-set monad* $\mathbb{U}\mathrm{p} = (\mathrm{Up}, \bigwedge_{\mathrm{Up}}, \uparrow)$ is *dual Kock–Zöberlein* (i.e. *of dual Kock–Zöberlein type* or *colax idempotent*), as it is Kock–Zöberlein with the hom-sets equipped with their dual order (the infimum map is right adjoint to the up-set one); note that because $\mathrm{Up}\,X$ is ordered by reverse inclusion (see Section II.1.7), the map $\bigwedge_{\mathrm{Up}\,X} : \mathrm{Up}\,\mathrm{Up}\,X \to \mathrm{Up}\,X$ is given by set-theoretic union.

The down-set and the up-set monads on Ord restrict to separated ordered sets, so Proposition II.4.9.1 and Section II.1.7 yield 2-isomorphisms

$$\mathsf{Ord}_{\mathrm{sep}}^{\mathbb{D}\mathrm{n}} \cong \mathsf{Sup} \qquad \text{and} \qquad \mathsf{Ord}_{\mathrm{sep}}^{\mathbb{U}\mathrm{p}} \cong \mathsf{Inf}\,.$$

II.4.10 Enriched categories

A locally small ordered category C has its hom-sets and composition operation live in the Cartesian closed category Ord. More generally, one may consider any monoidal category V instead of Ord and, keeping the notations of Section II.4.1, define a V-*category* C to be given by a class ob C of objects, a *hom-object* $C(A, B) \in \mathrm{ob}\,V$ for each pair $A, B \in \mathrm{ob}\,C$, *composition* and *identity* operations in V

$$m_{A,B,C} : C(A, B) \otimes C(B, C) \to C(A, C) \qquad \text{and} \qquad e_A : E \to C(A, A)\,,$$

for all $A, B, C \in \mathrm{ob}\,C$, subject to the *associativity* and *identity laws* which require that the diagrams

$$\begin{array}{ccccc}
C(A,B)\otimes(C(B,C)\otimes C(C,D)) & \xrightarrow{\alpha} & (C(A,B)\otimes C(B,C))\otimes C(C,D) & \xrightarrow{m\otimes 1} & C(A,C)\otimes C(C,D) \\
\big\downarrow{\scriptstyle 1\otimes m} & & & & \big\downarrow{\scriptstyle m} \\
C(A,B)\otimes C(B,D) & & \xrightarrow{\qquad\qquad m \qquad\qquad} & & C(A,D)
\end{array}$$

$$\begin{array}{ccc}
E\otimes C(A,B) & \xrightarrow{e\otimes 1} & C(A,A)\otimes C(A,B) \\
& {\scriptstyle \lambda}\searrow & \big\downarrow{\scriptstyle m} \\
& & C(A,B)
\end{array}
\qquad
\begin{array}{ccc}
C(A,B)\otimes C(B,B) & \xleftarrow{1\otimes e} & C(A,B)\otimes E \\
\big\downarrow{\scriptstyle m} & \swarrow{\scriptstyle \rho} & \\
C(A,B) & &
\end{array}$$

commute in V.

For V = Set and V = Ord, the definition of a V-category returns the ordinary notions of locally small category and of locally small ordered category, respectively. A *2-category* is simply a CAT-category, where again the monoidal structure is given by the Cartesian product. Starting with Chapter III, a major part of this book is devoted to the study of V-categories, when V is merely a quantale, and generalizations thereof which involve the consideration of a monad on Set. If V is AbGrp, the prototype of a monoidal category, then V-categories are called *additive categories*.

The general theory of V-categories can be found in [Kelly, 1982]. For the reader's convenience, we mention here the definitions of V-functors and V-natural transformations. A V-*functor* $F : C \to D$ of V-categories C, D is given by a function $F : \mathrm{ob}\,C \to \mathrm{ob}\,D$ and morphisms

$$F_{A,B} : C(A, B) \to D(FA, FB)$$

in V for all $A, B \in \mathrm{ob}\,C$, subject to the commutativity of the diagrams

$$\begin{array}{ccc}
C(A,B)\otimes C(B,C) & \xrightarrow{F\otimes F} & D(FA,FB)\otimes D(FB,FC) \\
\big\downarrow{\scriptstyle m} & & \big\downarrow{\scriptstyle m} \\
C(A,C) & \xrightarrow{\qquad F \qquad} & D(FA,FC)\,. \\
& {\scriptstyle e}\nwarrow \quad \nearrow{\scriptstyle e} & \\
& E &
\end{array}$$

For V-functors $F, G : \mathsf{C} \to \mathsf{D}$, a V-*natural transformation* $\tau : F \to G$ is given by morphisms

$$\tau_A : E \to \mathsf{D}(FA, GA)$$

in V such that the diagrams

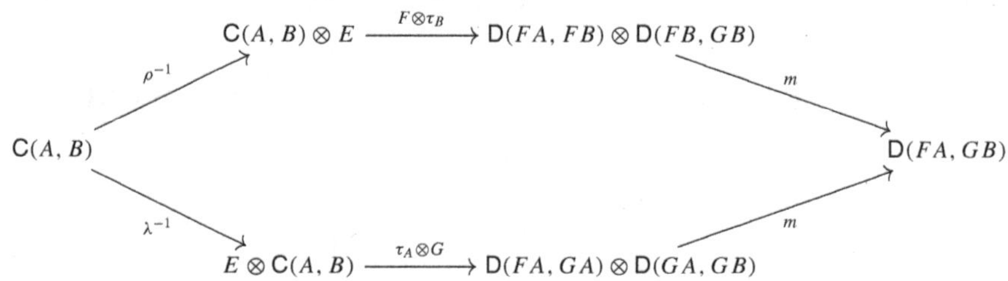

commute. Again, for $\mathsf{V} = \mathsf{Set}$, one obtains the ordinary categorical notions, and for $\mathsf{V} = \mathsf{Ord}$ a V-functor is a 2-functor of ordered categories (see Section II.4.6), while a V-natural transformation is simply a natural transformation.

Exercises

II.4.A *Cartesian closedness of* Cat. In Cat, the functor category B^{A} serves as internal hom-object.

II.4.B *Relational adjoints to maps.* In Rel, if $r : A \nrightarrow B$ is a map and $s : B \nrightarrow A$ is its right adjoint, then one has $s = r^\circ$.

II.4.C *Tensor commutes with coproducts.* In a closed monoidal category, one has the rules

$$A \otimes \coprod_{i\in I} B_i \cong \coprod_{i\in I} A \otimes B_i\,, \quad A \otimes 0 \cong 0\,,$$

whenever the coproduct on the left exists. Likewise,

$$A \multimap \prod_{i\in I} B_i \cong \prod_{i\in I}(A \multimap B_i)\,, \quad (A \multimap 1) \cong 1\,,$$

whenever the product on the left exists. Thus, in a Cartesian closed category, one has in particular

$$(A \times B)^C \cong A^C \times B^C\,, \quad 1^C \cong 1\,.$$

II.4.D *Further isomorphisms for the internal hom-functors.* If C is a monoidal closed category with $(-) \multimap B : \mathsf{C}^{\mathrm{op}} \to \mathsf{C}$ right adjoint (for example, if C is symmetric monoidal, see Proposition II.4.4.3), then, for all objects A, B, C, there are natural isomorphisms

$$(E \multimap A) \cong A \qquad \text{and} \qquad \prod_{i\in I}(A_i \multimap B) \cong (\coprod_{i\in I} A_i) \multimap B\,.$$

In the case where $B \mathrel{\bullet\!\!-} (-) : \mathsf{C}^{\mathrm{op}} \to \mathsf{C}$ is right adjoint, one has for all objects A, B, C

$$(A \mathrel{\bullet\!\!-} E) \cong A \qquad \text{and} \qquad \textstyle\prod_{i\in I}(B \mathrel{\bullet\!\!-} A_i) \cong B \mathrel{\bullet\!\!-} \coprod_{i\in I} A_i\ .$$

II.4.E *Adjunctions and equivalent maps.* If $f \dashv g : B \to A$ and $f' \dashv g' : B \to A$ in an ordered category, then $f \le f'$ if and only if $g' \le g$. In particular, if $f \le f'$ and $g \le g'$, then $f \simeq f'$ and $g \simeq g'$.

II.4.F *Trivial extensions and liftings.* For a morphism $f : A \to B$ in a quantaloid, one has

$$1_B \mathrel{-\!\!\bullet} f = f \qquad \text{and} \qquad f \mathrel{\bullet\!\!-} 1_A = f\ .$$

II.4.G *Adjoints in quantaloids.* A morphism $f : A \to B$ in a quantaloid has a right adjoint g if and only if $(f \mathrel{-\!\!\bullet} 1_B)\cdot f = (f \mathrel{-\!\!\bullet} f)$; in this case, $g = (f \mathrel{-\!\!\bullet} 1_B)$. Dually, $g : B \to A$ has a left adjoint f if and only if $g \cdot (1_B \mathrel{\bullet\!\!-} g) = (g \mathrel{\bullet\!\!-} g)$, and in this case, $f = (1_B \mathrel{\bullet\!\!-} g)$. Moreover, one has the following rules:

(1) $h \cdot (\phi \mathrel{\bullet\!\!-} \psi) = (h \cdot \phi) \mathrel{\bullet\!\!-} \psi$ if h is right adjoint;
(2) $(\phi \mathrel{\bullet\!\!-} \psi)\cdot f = \phi \mathrel{\bullet\!\!-} (g \cdot \psi)$ if $f \dashv g$;
(3) $(\phi \cdot p) \mathrel{\bullet\!\!-} \psi = \phi \mathrel{\bullet\!\!-} (\psi \cdot q)$ if $p \dashv q$.

II.4.H *Existence of biproducts.* For a family $(A_i)_{i\in I}$ of objects in a quantaloid C, the following are equivalent:

(i) the biproduct of $(A_i)_{i\in I}$ exists;
(ii) the product of $(A_i)_{i\in I}$ exists;
(iii) the coproduct of $(A_i)_{i\in I}$ exists.

II.4.I V *as a* V*-category.* For V a symmetric monoidal closed category, and all objects A, B, C in V, there are morphisms

$$(A \mathrel{-\!\!\bullet} B) \otimes (B \mathrel{-\!\!\bullet} C) \to (A \mathrel{-\!\!\bullet} C) \qquad \text{and} \qquad E \to (A \mathrel{-\!\!\bullet} A)$$

which render V a V-category, with the same object class and $(A \mathrel{-\!\!\bullet} B)$ as the internal hom.

Hint. For the first morphism, construct $A \otimes (A \mathrel{-\!\!\bullet} B) \to B$, and tensor with $(B \mathrel{-\!\!\bullet} C)$.

II.4.J *Categories as monoids*

(1) Let O be a set. A *directed graph* $G = (M, d : M \to O, c : M \to O)$ over O (the *vertices* of G) is given by a set M (the *edges* of G) and maps d, c (which assign to an edge its *domain* and *codomain*). A morphism $f : (M, d, c) \to (N, b, a)$ in the category $\mathsf{Gph}(O)$ of graphs over O is a map $f : M \to N$ with $b \cdot f = d$, $a \cdot f = c$. Show that $\mathsf{Gph}(O)$ becomes a monoidal category when one puts

$$(M, d, c) \otimes (N, b, a) = (M \times_O N, d \cdot b', a \cdot c')$$

using the pullback diagram

$$\begin{array}{ccc} M \times_O N & \xrightarrow{c'} & N \\ {\scriptstyle b'}\downarrow & & \downarrow{\scriptstyle b} \\ M & \xrightarrow{c} & O\,. \end{array}$$

However, the tensor product fails to be symmetric.

(2) Show that the category of monoids in $\mathsf{Gph}(O)$ is equivalent to the category $\mathsf{Cat}(O)$ of small categories with object set O and functors mapping O identically.

II.5 Factorizations, fibrations, and topological functors

II.5.1 Factorization systems for morphisms

We denote, respectively, by $\mathrm{Iso}\,\mathsf{C}$, $\mathrm{Epi}\,\mathsf{C}$, and $\mathrm{Mono}\,\mathsf{C}$ the classes of all isomorphisms, epimorphisms, and monomorphisms in C. An *orthogonal factorization system for morphisms* (or simply a *factorization system*) in a category C is given by a pair $(\mathcal{E}, \mathcal{M})$ of morphism classes in C such that

(1) $\mathrm{Iso}\,\mathsf{C} \cdot \mathcal{E} \subseteq \mathcal{E}$, $\mathcal{M} \cdot \mathrm{Iso}\,\mathsf{C} \subseteq \mathcal{M}$: i.e. $\mathcal{E}$ and $\mathcal{M}$ are closed under composition with isomorphisms from the left and the right, respectively;

(2) $\mathrm{mor}\,\mathsf{C} = \mathcal{M} \cdot \mathcal{E}$: every morphism factors into an $\mathcal{E}$-morphism followed by an $\mathcal{M}$-morphism; in fact, we tacitly assume that there is a fixed choice for these factorizations;

(3) $\mathcal{E} \perp \mathcal{M}$: every $\mathcal{E}$-morphism e is *orthogonal* to every $\mathcal{M}$-morphism m, so that for every commutative solid-arrow square

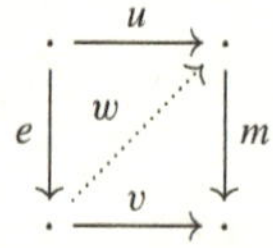

there is a unique morphism w with $w \cdot e = u$, $m \cdot w = v$; one writes $e \perp m$ in this situation.

The system $(\mathcal{E}, \mathcal{M})$ is called *proper* if $\mathcal{E} \subseteq \mathrm{Epi}\,\mathsf{C}$ and $\mathcal{M} \subseteq \mathrm{Mono}\,\mathsf{C}$; in that case, any w with $w \cdot e = u$ or $m \cdot w = v$ (where $m \cdot u = v \cdot e$) satisfies both equations and is trivially uniquely determined. The notion of factorization system is self-dual in the sense that $(\mathcal{M}, \mathcal{E})$ is a factorization system in C^{op} if $(\mathcal{E}, \mathcal{M})$ is one in C (and the same is true for proper factorization systems).

An immediate consequence of the *unique diagonalization property* (3) is that $(\mathcal{E}, \mathcal{M})$-factorizations of a morphism are unique, up to a unique isomorphism: if

$m \cdot e = m' \cdot e'$ with $e, e' \in \mathcal{E}$, $m, m' \in \mathcal{M}$, then $j \cdot e = e'$, $m' \cdot j = m$ for a unique isomorphism j. (This fact follows also from (1) and (2) in Proposition II.5.1.1.)

II.5.1.1 Proposition

(1) *For a factorization system* $(\mathcal{E}, \mathcal{M})$ *in* C*, one has*

$$\mathcal{E} = {}^{\perp}\mathcal{M} := \{e \in \operatorname{mor} \mathsf{C} \mid \forall m \in \mathcal{M} : e \perp m\} ,$$
$$\mathcal{M} = \mathcal{E}^{\perp} := \{m \in \operatorname{mor} \mathsf{C} \mid \forall e \in \mathcal{E} : e \perp m\} .$$

In particular, $\mathcal{E}$ *and* $\mathcal{M}$ *determine each other uniquely, and*

$$\mathcal{E} \cap \mathcal{M} = \operatorname{Iso} \mathsf{C} .$$

(2) *Any class* $\mathcal{M} = \mathcal{E}^{\perp}$ *(for some* $\mathcal{E} \subseteq \operatorname{mor} \mathsf{C}$*) satisfies*

(a) $\operatorname{Iso} \mathsf{C} \subseteq \mathcal{M}$ *and* $\mathcal{M} \cdot \mathcal{M} = \mathcal{M}$ *(*$\mathcal{M}$ *is closed under composition);*

(b) *if* $g \cdot f$, $g \in \mathcal{M}$*, then* $f \in \mathcal{M}$ *(*$\mathcal{M}$ *is* weakly left-cancelable*);*

(c) $\mathcal{M}$ *is* stable under pullbacks *(see* Section II.2.8*), so that for any pullback diagram*

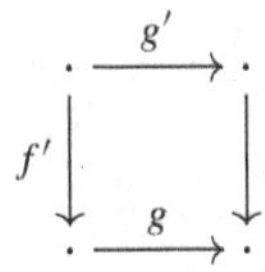

$f \in \mathcal{M}$ *implies* $f' \in \mathcal{M}$*;*

(d) $\mathcal{M}$ *is* stable under multiple pullbacks*, so that when* (A, f) *is a product of* (A_i, f_i) *in* C/B*, one has* $f \in \mathcal{M}$ *whenever all* $f_i \in \mathcal{M}$*;*

(e) $\mathcal{M}$ *is* closed under limits*, so that when* $\mu : D \to E$ *is a natural transformation that is componentwise in* $\mathcal{M}$ *(with* $D, E : \mathsf{J} \to \mathsf{C}$*), the induced morphism* $\lim \mu : \lim D \to \lim E$ *also lies in* $\mathcal{M}$ *(provided that the needed limits exist); in particular,* $\mathcal{M}$ *is* closed under products*:*

$$\forall i \in I \ (m_i \in \mathcal{M}) \implies \textstyle\prod_{i \in I} m_i \in \mathcal{M} .$$

Any class $\mathcal{E} = {}^{\perp}\mathcal{M}$ *(for some* $\mathcal{M} \subseteq \operatorname{mor} \mathsf{C}$*) satisfies the properties dual to* (a)–(e).

(3) *A factorization system* $(\mathcal{E}, \mathcal{M})$ *in* C *with* $\mathcal{E} \subseteq \operatorname{Epi} \mathsf{C}$ *satisfies*

(a) $\operatorname{ExtMono} \mathsf{C} \subseteq \mathcal{M}$*, where a monomorphism* f *is an* extremal monomorphism *in* C *if* $f = g \cdot e$ *with* e *epic only if* e *is an isomorphism;*

(b) $\operatorname{SplitMono} \mathsf{C} \subseteq \mathcal{M}$ *(see* Exercise II.2.C*);*

(c) $g \cdot f \in \mathcal{M}$ *implies* $f \in \mathcal{M}$ *(*$\mathcal{M}$ *is* left-cancelable*).*

Conversely, if C *has finite products, any of* (a), (b), (c) *implies* $\mathcal{E} \subseteq \operatorname{Epi} \mathsf{C}$*. The dual assertions hold as well.*

Proof (1) and (2) are shown by standard arguments. (It is less standard to show that properties (c) and (d) in (2) actually follow from (e); see [Dikranjan and Tholen, 1995].)

(3): First assume $\mathcal{E} \subseteq \mathrm{Epi}\,\mathsf{C}$. For (a), $(\mathcal{E}, \mathcal{M})$-factoring $f \in \mathrm{ExtMono}\,\mathsf{C}$ gives $f = m \cdot e$ with e epic, so that e must be an isomorphism and $f \in \mathcal{M}$. (b) follows trivially since $\mathrm{SplitMono}\,\mathsf{C} \subseteq \mathrm{ExtMono}\,\mathsf{C}$. For (c), let $g \cdot f \in \mathcal{M}$. In order to show $f \in \mathcal{M} = \mathcal{E}^{\perp}$, assume $f \cdot u = v \cdot e$ with $e \in \mathcal{E}$. Since $(g \cdot f) \cdot u = (g \cdot v) \cdot e$ with $g \cdot f \in \mathcal{M}$, there is w with $w \cdot e = u$. Thus, $\mathcal{E} \subseteq \mathrm{Epi}\,\mathsf{C}$ implies that w is unique and also satisfies $f \cdot w = v$.

Conversely, in order to show $\mathcal{E} \subseteq \mathrm{Epi}\,\mathsf{C}$ under hypotheses (a), (b), or (c), assume $e \in \mathcal{E}$ and $f \cdot e = g \cdot e$. With $A = \mathrm{cod}\, f = \mathrm{dom}\, f$, $h = \langle f, g \rangle$, $\delta = \langle 1_A, 1_A \rangle$, one obtains the commutative diagram

$$\begin{array}{ccc} \cdot & \xrightarrow{f \cdot e} & A \\ {\scriptstyle e}\big\downarrow & & \big\downarrow{\scriptstyle \delta} \\ \cdot & \xrightarrow{h} & A \times A\,. \end{array}$$

Since $p_i \cdot \delta = 1_A \in \mathcal{M}$ (with p_1, p_2 the projections of $A \times A$), $\delta \in \mathcal{M}$ follows under any of hypotheses (a), (b), or (c), so that $e \perp \delta$ gives w with $\delta \cdot w = h$, hence $f = p_1 \cdot h = w = p_2 \cdot h = g$. $\square$

Because of (1), one calls $\mathcal{E}$ a *left factorization class* and $\mathcal{M}$ a *right factorization class* if they belong to a factorization system $(\mathcal{E}, \mathcal{M})$, and then $\mathcal{E}$ is the *left companion* of $\mathcal{M}$ and $\mathcal{M}$ the *right companion* of $\mathcal{E}$. Of course, one talks about *extremal epimorphisms* in the situation dual to (3)(a).

II.5.1.2 Examples Every category C has both *trivial factorization systems* $(\mathrm{Iso}\,\mathsf{C}, \mathrm{mor}\,\mathsf{C})$ and $(\mathrm{mor}\,\mathsf{C}, \mathrm{Iso}\,\mathsf{C})$. The class of monomorphisms is a right factorization class in the categories Set, Mon, Grp, Ord, Sup, and Top; its left companion $\mathcal{E}$ is given by the surjective morphisms in Set, Mon, Grp and Sup, and by the *quotient morphisms* in Ord and Top (see Exercise II.5.N); $\mathcal{E}$ is precisely the class of regular epimorphisms in all six categories, which only in Set and Grp coincide with the class of all epimorphisms. The class of epimorphisms is a left factorization class with an easily determined right companion $\mathcal{M}$ both in Ord and Top, given by the fully faithful injections (see Section II.1.3) and the subspace embeddings, respectively; $\mathcal{M}$ is the class of regular monomorphisms in both categories. The following two diagrams depict these two fundamental factorizations in Ord and Top:

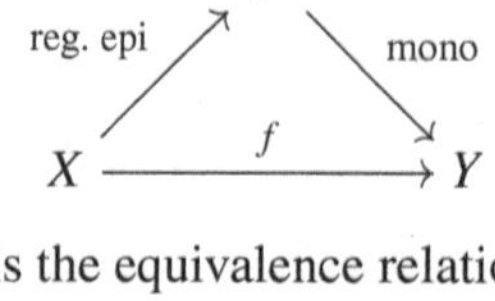

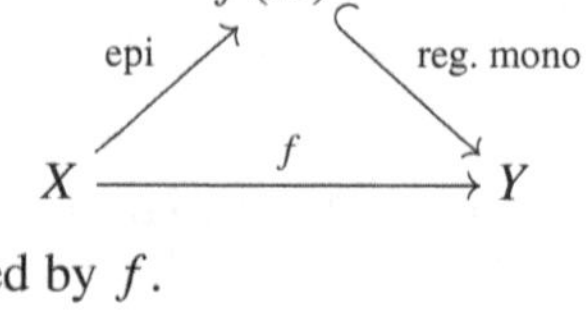

where $\sim$ is the equivalence relation induced by f.

For a class $\mathcal{E}$ of morphisms in C, a subcategory B of C is called *$\mathcal{E}$-reflective* if all B-reflections lie in $\mathcal{E}$; accordingly, *epi-reflective*, *regular epi-reflective*, etc., refer to the cases $\mathcal{E} = \{\text{epimorphisms}\}$, $\{\text{regular epimorphisms}\}$, etc. *Bireflective* means epi-reflective and mono-reflective. The dual notions are *$\mathcal{M}$-coreflective*, *mono-coreflective*, *regular mono-coreflective*, *bicoreflective*, etc.

II.5.1.3 Proposition *For a factorization system $(\mathcal{E}, \mathcal{M})$ in C, a replete reflective subcategory B of C is $\mathcal{E}$-reflective if and only if for all $m : A \to B$ in $\mathcal{M}$, one has $A \in \operatorname{ob}\mathsf{B}$ when $B \in \operatorname{ob}\mathsf{B}$.*

Proof For the necessity of the condition, factor $m : A \to B$ in $\mathcal{M}$ with $B \in \operatorname{ob}\mathsf{B}$ through the B-reflection $\rho_A : A \to RA$ and obtain $t : RA \to A$ with $t \cdot \rho_A = 1_A$ from $\rho_A \perp m$. Now, $(\rho_A \cdot t) \cdot \rho_A = \rho_A$ given $\rho_A \cdot t = 1_{RA}$, by universality of ρ_A. For the sufficiency of the condition, consider an $(\mathcal{E}, \mathcal{M})$-factorization $\rho_A : (A \xrightarrow{e} B \xrightarrow{n} RA)$ of the B-reflection of $A \in \operatorname{ob}\mathsf{C}$. Since $B \in \operatorname{ob}\mathsf{B}$, there is $s : RA \to B$ with $s \cdot \rho_A = e$, hence $(n \cdot s) \cdot \rho_A = \rho_A$ and $n \cdot s = 1_{RA}$. Now $e \perp n$ implies $s \cdot n = 1_B$. □

II.5.2 Subobjects, images, and inverse images

For a class $\mathcal{M} \subseteq \operatorname{Mono}\mathsf{C}$ containing all isomorphisms and being closed under composition with them, and for every $A \in \operatorname{ob}\mathsf{C}$, one forms the full subcategory

$$\operatorname{sub} A = \operatorname{sub}_{\mathcal{M}} A = \mathcal{M}/A$$

of the comma category C/A, given by the objects $m : M \to A$ with $m \in \mathcal{M}$. (If $\mathcal{M}$ is closed under composition and is weakly left-cancelable, one may also consider $\mathcal{M}/A$ as a comma category of the *category* $\mathcal{M}$ with $\operatorname{ob}\mathcal{M} = \operatorname{ob}\mathsf{C}$.) Of course, $\operatorname{sub} A$ is just an ordered class (with $m \leq n$ if $n \cdot l = m$ for some morphism l). Its objects are called *$\mathcal{M}$-subobjects* (or simply *subobjects*) of A. Most authors reserve this term for isomorphism classes of objects in $\mathcal{M}/A$. The category C is said to have *inverse $\mathcal{M}$-images* (or *inverse images*) if for all $f : A \to B$ in C and all $n : N \to B$ in $\mathcal{M}$ there is a pullback diagram

$$\begin{array}{ccc} f^{-1}(N) & \xrightarrow{f'} & N \\ {\scriptstyle f^{-1}(n)}\downarrow & & \downarrow{\scriptstyle n} \\ A & \xrightarrow{f} & B \end{array} \tag{II.5.2.i}$$

with $f^{-1}(n)$ in $\mathcal{M}$ (for convenience, the choice of such a diagram is supposed to be fixed). One then has the *inverse-image functor*

$$f^{-1}(-) : \operatorname{sub} B \to \operatorname{sub} A\ .$$

II.5.2.1 Proposition *Let C have inverse $\mathcal{M}$-images. Then $\mathcal{M}$ is a right factorization class of C if and only if $\mathcal{M}$ is closed under composition and $f^{-1}(-)$*

has a left adjoint $f(-)$, for every morphism f in C. *In that case, the* $\mathcal{M}$-image $f(m)$ *of* $m : M \to A$ *in* $\mathcal{M}$ *under* $f : A \to B$ *may be constructed by the* $(\mathcal{E}, \mathcal{M})$*-factorization of* $f \cdot m$ *(with* $\mathcal{E}$ *the left companion of* $\mathcal{M}$*):*

$$\begin{array}{ccc} M & \xrightarrow{e\in\mathcal{E}} & f(M) \\ {\scriptstyle m}\downarrow & & \downarrow{\scriptstyle f(m)} \\ A & \xrightarrow{f} & B\,. \end{array}$$

One has

(1) $m \le f^{-1}(f(m))$ *for all* $m \in \operatorname{sub} A$, *with* $m \cong f^{-1}(f(m))$ *when* $f \in \mathcal{M}$;
(2) f *lies in* $\mathcal{E}$ *if and only if* $f(1_A) \cong 1_B$;
(3) $f(f^{-1}(n)) \le n$ *for all* $n \in \operatorname{sub} B$, *while* $n \cong f(f^{-1}(n))$ *for all* n *precisely when* f *lies* $\mathcal{M}$-hereditarily *in* $\mathcal{E}$, *i.e. when* $f' \in \mathcal{E}$ *in every pullback diagram* (II.5.2.i) *with* $n \in \mathcal{M}$.

Proof All verifications are routine, except the proof that $\mathcal{M}$ is a right factorization class when $\mathcal{M} \cdot \mathcal{M} \subseteq \mathcal{M}$ and there is $f(-) \dashv f^{-1}(-)$ for all f. To establish a factorization of f, one puts $m := f(1_A)$ and lets e be the upper horizontal composite arrow in

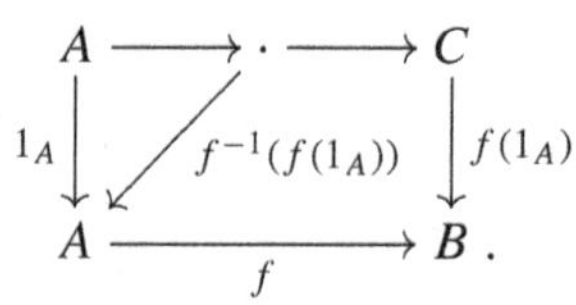

The universal property of the adjunction amounts to the fact that for every commutative solid-arrow diagram

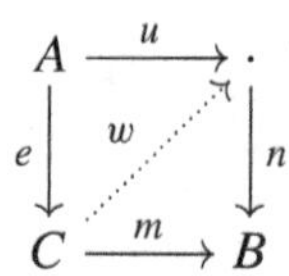

with $n \in \mathcal{M}$ there is a unique "diagonal" w. Applying the same factorization procedure to e in lieu of f gives $e = m' \cdot e'$ and the above solid arrow diagram with $n = m \cdot m'$ and $u = e'$. Since $m \cdot m' \in \mathcal{M}$, by hypothesis one obtains w with $m \cdot m' \cdot w = m$, hence $m' \cdot w = 1_C$. Consequently, m' is an isomorphism so that e must have the universal property symbolized by

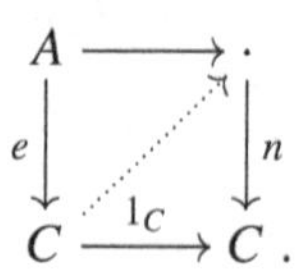

This is sufficient to verify $e \in {}^{\perp}\mathcal{M}$, since the search for a diagonal for an arbitrary square

$$\begin{array}{ccc} \cdot & \xrightarrow{u} & \cdot \\ {\scriptstyle e}\downarrow & & \downarrow{\scriptstyle n} \\ \cdot & \xrightarrow{v} & \cdot \end{array}$$

can be reduced to the case $v = 1$ by pulling n back along v. □

It follows that the image and inverse-image functors associated with a factorization system $(\mathcal{E}, \mathcal{M})$ with $\mathcal{M} \subseteq \mathrm{Mono}\,\mathsf{C}$ preserve existing joins and meets, respectively:

II.5.2.2 Corollary

$$f(\bigvee_{i\in I} m_i) \simeq \bigvee_{i\in I} f(m_i)\,, \quad f^{-1}\big(\bigwedge_{i\in I} m_i\big) \simeq \bigwedge_{i\in I} f^{-1}(m_i)\,.$$

Proof Left adjoints preserve colimits, and right adjoints preserve limits. □

II.5.3 Factorization systems for sinks and sources

A *source* in a category C is simply a discrete cone in C; hence, it is given by an object A and a family $(g_i : A \to B_i)_{i\in I}$ of morphisms in C (I may be empty and is not required to be small); the dual notion is *sink*. One says that the sink $(f_i : A_i \to B)_{i\in I}$ is *orthogonal* to the morphism $m : C \to D$ if for every solid-arrow diagram

$$\begin{array}{ccc} A_i & \xrightarrow{u_i} & C \\ {\scriptstyle f_i}\downarrow & \overset{w}{\nearrow} & \downarrow{\scriptstyle m} \\ B & \xrightarrow{v} & D \end{array} \qquad (i \in I)$$

there is a unique morphism w with $w \cdot f_i = u_i$ $(i \in I)$ and $m \cdot v = v$; one writes $(f_i)_{i\in I} \perp m$ in this case. A *factorization system for sinks* (also called an *orthogonal factorization system for sinks*) is a pair $(\mathbb{E}, \mathcal{M})$ consisting of a collection $\mathbb{E}$ of sinks and a class $\mathcal{M}$ of morphisms in C such that

(1) $\mathbb{E}$ and $\mathcal{M}$ are closed under composition with isomorphisms from the left and right, respectively;
(2) every sink in C factors into an $\mathbb{E}$-sink followed by an $\mathcal{M}$-morphism;
(3) $\mathbb{E} \perp \mathcal{M}$.

We leave it to the reader to formulate the appropriate sink generalizations of the statements of Proposition II.5.1.1. For example, closure of $\mathbb{E}$ under composition means that, when all $(f_{i,j} : A_{ij} \to B_i)_{j\in J_i}$ and $(g_i : B_i \to C)_{i\in I}$ lie in $\mathbb{E}$, the composite sink $(g_i \cdot f_{i,j})_{i\in I,\, j\in J_i}$ also lies in $\mathbb{E}$. Conversely, *weak right-cancelation* of $\mathbb{E}$ stipulates that, when the composite sink and all $(f_{i,j})_{j\in J_i}$ lie in $\mathbb{E}$, then $(g_i)_{i\in I}$ also lies in $\mathbb{E}$.

Of course, a factorization system $(\mathbb{E}, \mathcal{M})$ for sinks gives in particular a factorization system $(\mathcal{E}, \mathcal{M})$ for morphisms, but not vice versa. For example, there cannot be a factorization system $(\mathbb{E}, \operatorname{mor}\mathsf{C})$ for sinks in any category C containing morphisms that are not monic, as we show next.

II.5.3.1 Lemma *For a factorization system $(\mathbb{E}, \mathcal{M})$ for sinks, one has $\mathcal{M} \subseteq \operatorname{Mono}\mathsf{C}$.*

Proof For $f : A \to B$ in $\mathcal{M}$, assume $f \cdot x = f \cdot y = h$ with $x, y : D \to A$, and consider the constant sink $(h)_{i \in I}$, with $I := \operatorname{ob}(\mathsf{C}/A)$. We obtain an $(\mathbb{E}, \mathcal{M})$-factorization $h = m \cdot e_i$ $(i \in I)$ and consider

$$J := \{w \mid \forall i \in I \;:\; w \cdot e_i \in \{x, y\}\} .$$

The diagrams

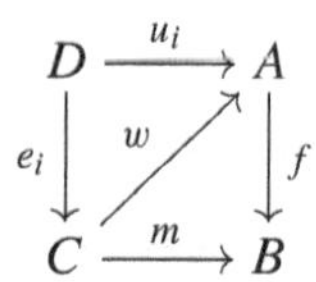

show $J \neq \varnothing$ (take $u_i := x$ for all i); therefore, the inclusion map $\tau : J \hookrightarrow I$ has a retraction $\sigma : I \to J$ (so that $\sigma \cdot \tau = 1_J$). For $i \in I$, consider

$$u_i := \begin{cases} x & \text{if } \sigma(i) \cdot e_i = y, \\ y & \text{if } \sigma(i) \cdot e_i = x, \end{cases}$$

and obtain $w \in J$ as above. Then, with $\sigma(i_0) = w$, one has

$$\sigma(i_0) \cdot e_{i_0} = x \iff \sigma(i_0) \cdot e_{i_0} = y ,$$

which is possible only if $x = y$. □

II.5.3.2 Theorem *A class of morphisms $\mathcal{M}$ belongs to a factorization system $(\mathbb{E}, \mathcal{M})$ for sinks in C if and only if*

(1) $\operatorname{Iso}\mathsf{C} \subseteq \mathcal{M} \subseteq \operatorname{Mono}\mathsf{C}$ *and* $\mathcal{M} \cdot \mathcal{M} \subseteq \mathcal{M}$,
(2) C *has inverse* $\mathcal{M}$*-images,*
(3) C *has* $\mathcal{M}$-intersections*; i.e. multiple pullbacks of families of* $\mathcal{M}$*-morphisms exist in* C*, and* $\mathcal{M}$ *is stable under them.*

Proof Lemma II.5.3.1 takes care of the only delicate part in the proof of the necessity of condition (1). For the necessity of (2), let $f : A \to B$ be in C, $n : N \to B$ in $\mathcal{M}$ and consider the class

$$J := \{k \in \mathcal{M} \mid \exists x_k \;:\; f \cdot k = n \cdot x_k\} .$$

The sink $(k)_{k\in J}$ has an $(\mathbb{E}, \mathcal{M})$-factorization $m \cdot e_k = k$, with $m \in \mathcal{M}$ and $(e_k)_{k\in J}$ in $\mathbb{E}$. Since $(e_k) \perp n$, one finds f' making

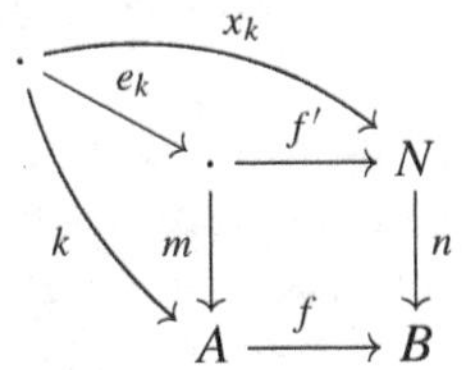

commute for all $k \in J$. Using the fact that we have in particular an $(\mathbb{E}, \mathcal{M})$-factorization system for morphisms, one easily verifies that the square of the diagram is a pullback. Condition (3) is shown similarly.

For the sufficiency of conditions (1)–(3), let us consider a sink $(f_i : A_i \to B)_{i\in I}$, and form

$$J := \{k \in \mathcal{M} \mid \forall i \in I\ \exists x_{i,k} : k \cdot x_{i,k} = f_i\}.$$

The universal property of the multiple pullback

$$m = \textstyle\bigwedge_{k\in J} k$$

shows $m \in J$; i.e. there are morphisms e_i with $f_i = m \cdot e_i$ $(i \in I)$. It remains to be shown that $(e_i) \perp t$ for all $t \in \mathcal{M}$. But if

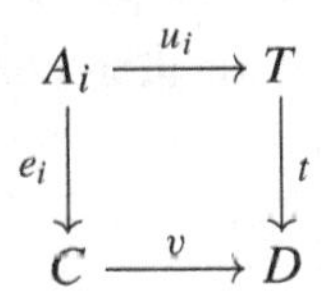

for all $i \in I$, then $t' = v^{-1}(t)$ is a factor of each e_i, so that $m \cdot t' \in \mathcal{M}$ is a factor of each f_i. Consequently, $m \cdot t' \in J$, and one has that t' is an isomorphism by construction of m. Hence, $w = v' \cdot (t')^{-1}$ is the needed diagonal for the above square, with v' the pullback of v along t. □

With $\mathcal{M} \subseteq \text{Mono}\,\mathsf{C}$, a category C is $\mathcal{M}$-*wellpowered* if, for every $A \in \text{ob}\,\mathsf{C}$, the separated reflection of $\text{sub}_{\mathcal{M}} A$ (see Section II.1.3) is small; i.e. every object has only a set of isomorphism types of $\mathcal{M}$-subobjects. The prefix $\mathcal{M}$ is omitted when $\mathcal{M} = \text{Mono}\,\mathsf{C}$.

II.5.3.3 Corollary *Let* C *be small-complete and* $\mathcal{M}$*-wellpowered, for a class* $\mathcal{M}$ *with* $\text{Iso}\,\mathsf{C} \subseteq \mathcal{M} \subseteq \text{Mono}\,\mathsf{C}$. *The following are equivalent:*

(i) $\mathcal{M}$ *is a right factorization class for morphisms;*
(ii) $\mathcal{M}$ *is a right factorization class for sinks;*
(iii) $\mathcal{M}$ *is closed under composition, and stable under pullbacks, as well as under multiple pullbacks.*

Proof The equivalence follows from Proposition II.5.1.1 and Theorem II.5.3.2. □

A pair $(\mathcal{E}, \mathbb{M})$ is a *factorization system for sources* in C if $(\mathbb{M}, \mathcal{E})$ is a factorization system for sinks in C^{op}. Similarly, C is $\mathcal{E}$*-cowellpowered* if C^{op} is $\mathcal{E}$-wellpowered. We leave the formulation of the statements dual to II.5.3.1–II.5.3.3 to the reader.

For C, any of the categories Set, Mon, Grp, Ord, Sup, and Top, $\mathrm{Mono}\,\mathsf{C}$ is not just a right factorization class for morphisms, but also for sinks. For $\mathsf{C} = \mathsf{Set}$, the left companion $\mathbb{E}$ contains precisely the *jointly surjective sinks* $(f_i : A_i \to B)_{i \in I}$, so that $B = \bigcup_{i \in I} f_i(A_i)$. In Mon, Grp, and Sup, B is only *generated* by $\bigcup_{i \in I} f_i(A_i)$ for $(f_i)_{i \in I} \in \mathbb{E}$. In Ord and Top, sinks in $\mathbb{E}$ are still jointly surjective but must carry the appropriate structure, which will be discussed more generally in Examples II.5.6.1.

$\mathrm{Epi}\,\mathsf{C}$ is a left factorization class for sources, with an easily described companion $\mathbb{M}$ for $\mathsf{C} = \mathsf{Set}$ or Grp: $\mathbb{M}$ contains precisely the *point-separating sources* $(f_i : A \to B_i)_{i \in I}$, so that for $x, y : X \to A$ one has $x = y$ when $f_i \cdot x = f_i \cdot y$ for all $i \in I$. In a general category, sources with this property are called *mono-sources* (the dual notion is that of an *epi-sink*). In Ord and Top, the domain A must carry the appropriate structure, the description of which will appear more generally in Examples II.5.6.1.

II.5.4 Closure operators

Let $\mathcal{M} \subseteq \mathrm{Mono}\,\mathsf{C}$ be a right factorization class, hence part of a factorization system $(\mathcal{E}, \mathcal{M})$ for morphisms. As in Section II.5.2, for every object A in C, we consider the ordered class $\mathrm{sub}\,A = \mathcal{M}/A$ and have, for every morphism $f : A \to B$, the image function

$$f(-) : \mathrm{sub}\,A \to \mathrm{sub}\,B$$

(constructed as in Proposition II.5.2.1) which, if C has inverse $\mathcal{M}$-images, is left adjoint to $f^{-1}(-)$. An $\mathcal{M}$*-closure operator* (or simply *closure operator*) on C is a family

$$c = (c_A : \mathrm{sub}\,A \to \mathrm{sub}\,A)_{A \in \mathrm{ob}\,\mathsf{C}}$$

of functions which are

(1) *extensive*: $m \leq c_A(m)$,
(2) *monotone*: if $m \leq m'$, then $c_A(m) \leq c_A(m')$,
(3) and which satisfy the *continuity condition*: $f(c_A(m)) \leq c_B(f(m))$,

for all $f : A \to B$ in C and $m, m' \in \mathrm{sub}\,A$. In the presence of inverse images, it is easy to see that the continuity condition may be equivalently formulated as

(3′) $c_A(f^{-1}(n)) \leq f^{-1}(c_B(n))$,

for all $f : A \to B$ in C and $n \in \mathrm{sub}\,B$. Extensivity yields for every $m : M \to A$ in $\mathcal{M}$ a factorization

$$\begin{array}{ccccc} & & c_A(M) & & \\ & {\scriptstyle j_m}\nearrow & & \searrow{\scriptstyle c_A(m)} & \\ M & & \xrightarrow{\quad m \quad} & & A \end{array}$$

with $c(m) = c_A(m) \in \mathcal{M}$ and a uniquely determined morphism $j_m \in \mathcal{M}$. Monotonicity and continuity ensure that the passage from m to its *c-closure* $c_A(m)$ is *functorial* as follows: for every commutative diagram

$$\begin{array}{ccc} M & \xrightarrow{f'} & N \\ {\scriptstyle m}\downarrow & & \downarrow{\scriptstyle n} \\ A & \xrightarrow{f} & B \end{array}$$

in C, with $m, n \in \mathsf{C}$, there is a unique morphism f'' making the following diagram:

$$\begin{array}{ccc} M & \xrightarrow{\quad f' \quad} & N \\ {\scriptstyle j_m}\downarrow & & \downarrow{\scriptstyle j_n} \\ c_A(M) & \xrightarrow{\quad f'' \quad} & c_B(N) \\ {\scriptstyle c_A(m)}\downarrow & & \downarrow{\scriptstyle c_B(n)} \\ A & \xrightarrow{\quad f \quad} & B \end{array}$$

commute (see Exercise II.5.G for a specification of the functoriality claim).

A subobject m is *c-closed* if j_m is an isomorphism, and it is *c-dense* if $c(m)$ is an isomorphism. More generally, a morphism $f : A \to B$ is *c-dense* if $f(1_A) : f(A) \to B$ is c-dense. The closure operator c is *idempotent* if $c(m)$ is c-closed (so that $c(c(m)) \cong c(m)$ for all $m \in \mathcal{M}$), and c is *weakly hereditary* if j_m is c-dense for all $m \in \mathcal{M}$.

II.5.4.1 Proposition *The following conditions on a closure operator c are equivalent:*

(i) *c is idempotent and weakly hereditary;*
(ii) *c is idempotent, and the class $\mathcal{M}^c$ of c-closed subobjects in C is closed under composition;*
(iii) *the class $\mathcal{M}^c$ is a right factorization class for morphisms in C.*

If these conditions hold, the left companion of $\mathcal{M}^c$ is the class $\mathcal{E}^c$ of all c-dense morphisms in C. The $(\mathcal{E}^c, \mathcal{M}^c)$-system factors a morphism $f : A \to B$ through $c_B(f(A))$.

Proof The statement is easily proved by using functoriality as previously described. □

It is also a straightforward exercise to show that for any closure operator c the class $\mathcal{M}^c$ is closed under limits, in particular it is stable under pullbacks and under multiple pullbacks.

These latter properties are characteristic in the following sense:

II.5.4.2 Theorem *Let $\mathcal{M}$ be a right factorization class for sinks in* C*, and consider a subclass $\mathcal{K} \subseteq \mathcal{M}$. Then $\mathcal{K} = \mathcal{M}^c$ for an idempotent closure operator c if and only if $\mathcal{K}$ is stable under pullbacks and multiple pullbacks. In fact, the closure operator c is uniquely determined by $\mathcal{K}$.*

Moreover, c is weakly hereditary if and only if $\mathcal{K}$ is closed under composition.

Proof See Exercise II.5.H. □

A closure operator is called *hereditary* if for all $m : M \to A$, $k : A \to B$ in $\mathcal{M}$, the lower rectangle in

$$\begin{array}{ccc} M & \xrightarrow{1_M} & M \\ {\scriptstyle j_m}\downarrow & & \downarrow{\scriptstyle j_{k\cdot m}} \\ c_A(M) & \longrightarrow & c_B(M) \\ {\scriptstyle c_A(m)}\downarrow & & \downarrow{\scriptstyle c_B(k\cdot m)} \\ A & \xrightarrow{k} & B \end{array} \qquad \text{(II.5.4.i)}$$

is a pullback diagram; i.e. if $c_A(m) \cong k^{-1}(c_B(k{\cdot}m))$. Exploitation of this property with $m = j_n$ and $k = c_B(n)$ for any $N \to B$ in $\mathcal{M}$ yields

$$c_{c_B(N)}(j_n) \cong c_B(n)^{-1}(c_B(n)) \cong j_n \ ,$$

so that j_n is c-dense. Hence, heredity implies weak heredity. The following result describes the extent to which heredity is stronger than weak heredity.

II.5.4.3 Proposition *A closure operator c is hereditary if and only if c is weakly hereditary and satisfies the following cancelation condition: for all $m, k \in \mathcal{M}$, if $k \cdot m$ is c-dense, m is also c-dense.*

Proof The cancelation condition is certainly necessary for heredity of c since, when $c_B(k \cdot m)$ is an isomorphism, any pullback of it is also an isomorphism. Conversely, let the weakly hereditary closure operator c satisfy the cancelation condition. We want to see that the canonical morphism $t : c_A(M) \to k^{-1}(c_B(M))$ induced by the commutative lower rectangle of (II.5.4.i) is an isomorphism. But weak heredity makes

$$j_{k\cdot m} = k^{-1}(c_B(k \cdot m)) \cdot t \cdot j_m$$

c-dense, so that $s := t \cdot j_m : M \to k^{-1}(c_B(M))$ is c-dense by hypothesis. Functoriality of c then yields the inverse of t:

$$\begin{array}{ccc}
M & \xrightarrow{\;1_M\;} & M \\
{\scriptstyle j_s}\big\downarrow & & \big\downarrow{\scriptstyle j_m} \\
\cdot & \cdots\cdots\cdots\rightarrow & c_A(M) \\
{\scriptstyle \cong}\big\downarrow & & \big\downarrow{\scriptstyle c_A(m)} \\
k^{-1}(c_B(M)) & \longrightarrow & A\,.
\end{array}$$

□

II.5.4.4 Examples

(1) We may think of an idempotent closure operator c as a family $(c_A)_{A \in \mathrm{ob}\,\mathsf{C}}$ of closure operations c_A on sub A (in the sense of Section II.1.6) which collectively must satisfy the continuity condition. (The notion of c-closedness defined here then coincides with the one defined in Section II.1.6.)

(2) The down- and up-closure for subsets of ordered sets define idempotent and hereditary closure operators of Ord (with its (Epi, RegMono)-factorization system, see Section II.1.7, where RegMono refers to RegMono C, the class of all regular monomorphisms in C). Likewise, Kuratowski closure defines an idempotent and hereditary closure operator of Top (with its (Epi, RegMono)-factorization system, see Exercise II.1.F).

(3) In terms of Theorem II.5.4.2, the Kuratowski closure operator of Top corresponds to the class $\mathcal{K}$ of closed subspace injections in Top. The class $\mathcal{O}$ of open subspace injections in Top still induces a closure operator θ via

$$\begin{aligned}\theta_X(M) &= \bigcap \{O \subseteq X \mid M \subseteq O \text{ and } O \text{ open}\} \\ &= \{x \in X \mid \forall V \text{ neighborhood of } x \,:\, M \cap \overline{V} \neq \varnothing\}\end{aligned}$$

(with $\overline{V}$ the Kuratowski closure of V in X). But failure of $\mathcal{O}$ to be closed under multiple pullbacks (i.e. intersections) makes θ fail to be idempotent, and θ is not weakly hereditary either.

II.5.5 Generators and cogenerators

A class $\mathcal{G}$ of objects in a category C is *generating* if for every object A in C the family $\mathsf{C}(\mathcal{G}, A)$ of all morphisms with codomain A and domain in $\mathcal{G}$ forms an epi-sink. Hence, for all $f, g : A \to B$ in C, one has $f = g$ whenever $f \cdot x = g \cdot x$ for all $x : G \to A$, $G \in \mathcal{G}$. Equivalently, $\mathcal{G}$ is generating if the generalized hom-functor

$$\mathsf{C}(\mathcal{G}, -) : \mathsf{C} \to \mathsf{Set}^{\mathcal{G}}\,, \quad A \mapsto (\mathsf{C}(G, A))_{G \in \mathcal{G}}$$

is faithful (where C is assumed to be locally small). A class $\mathcal{G}$ is *strongly generating* in C if $\mathsf{C}(\mathcal{G}, A)$ is an *extremal epi-sink* for all objects A, so that one has the additional property that, for a monomorphism $m : B \to A$, one can have $x = m \cdot h_x$ for all x in $\mathsf{C}(\mathcal{G}, A)$ only if m is an isomorphism (the dual notion is that of an *extremal mono-source*). Equivalently, $\mathcal{G}$ is strongly generating if $\mathsf{C}(\mathcal{G}, -)$ is faithful and reflects isomorphisms. We call a small generating class a *generator* of C, and a single object G in C a *generator* of C, if $\{G\}$ is one. Similarly, one says that a class is a *strong generator* if it is a small strongly generating class, and a single object G can similarly be called a *strong generator*. Note that one commonly uses *separator* as an alternative name for generator. The terms *cogenerating*, *cogenerator*, or *coseparator* are used in the dual situation.

When $\mathcal{G}$ is small, the functor $\mathsf{C}(\mathcal{G}, -)$ has a left adjoint F if and only if all coproducts

$$FX = \coprod_{G \in \mathcal{G}} X_G \cdot G$$

(with $X = (X_G)_{G \in \mathcal{G}}$ an object in $\mathsf{Set}^{\mathcal{G}}$) exist; here $X_G \cdot G$ denotes the coproduct of X_G-many copies of G in C. The counits

$$\varepsilon_A : \coprod_{G \in \mathcal{G}} \mathsf{C}(G, A) \cdot G \to A$$

are the canonical morphisms with $\varepsilon_A \cdot i_x = x$ for all $x : G \to A$, $G \in \mathcal{G}$, with i_x denoting a coproduct injection.

II.5.5.1 Proposition *The following conditions on a set $\mathcal{G}$ of objects in a locally small category C with coproducts are equivalent:*

(i) *$\mathcal{G}$ is generating;*
(ii) *for all objects A, the canonical morphisms ε_A are epimorphisms;*
(iii) *for every object A, there is some epimorphism*

$$\coprod_{i \in I} G_i \to A$$

with a small family $(G_i)_{i \in I}$ of objects in $\mathcal{G}$.

The same equivalence holds if one specializes to a strongly *generating set $\mathcal{G}$ in* (i), *and* extremal *epimorphisms in* (ii) *and* (iii).

Proof The equivalence follows from Exercise II.5.I. □

A singleton set (in fact, every non-empty set) is a single-object strong generator of Set. The terminal object is also a generator in Ord or Top, but it is not strong. The free algebra over a singleton set is a strong generator in every Eilenberg–Moore category over Set (see Section II.3.2). Hence, the additive group $\mathbb{Z}$ is a generator of both Grp and AbGrp.

A two-element set is a single-object strong cogenerator of Set. Provided with the indiscrete structure, it is a cogenerator in both Ord and Top, but it is not strong. The "rational circle" $\mathbb{Q}/\mathbb{Z}$ is a strong cogenerator in AbGrp, but Grp has no cogenerator at all.

II.5.6 *U*-initial morphisms and sources

For a functor $U : \mathsf{A} \to \mathsf{X}$, a source $(g_i : A \to B_i)_{i \in I}$ of A-morphisms is *U-initial* if, for every source $(h_i : C \to B_i)_{i \in I}$ in A and every X-morphism $s : UC \to UA$ with $Ug_i \cdot s = Uh_i$, there is exactly one morphism $t : C \to A$ in A with $Ut = s$ and $g_i \cdot t = h_i$ for all $i \in I$:

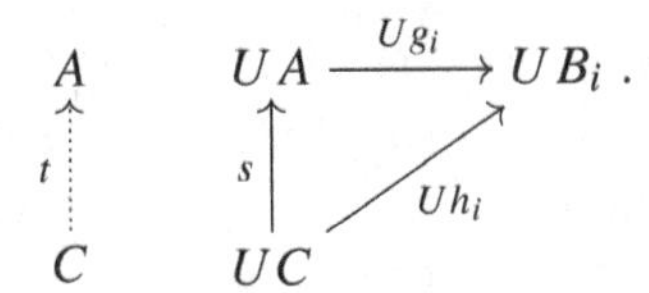

Of course, uniqueness of t as well as $g_i \cdot t = h_i$ $(i \in I)$ follow from $Ut = s$ when U is faithful. Hence, for U faithful, U-initiality of $(g_i)_{i \in I}$ simply means that any X-morphism $s : UC \to UA$ can be lifted to an A-morphism $t : C \to A$ along U whenever all $Ug_i \cdot s : UC \to UB_i$ can be lifted to A-morphisms $h_i : C \to B_i$ along U.

If the given source consists of a single morphism $f : A \to B$ (hence, if $|I| = 1$), it is more customary to say that f is *U-Cartesian* instead of U-initial, and we shall do so especially when U is not necessarily faithful.

In the case where $I = \emptyset$, the source $(g_i)_{i \in I}$ is given by an object A, and we say that A is *U-indiscrete* when A is U-initial as an empty source. The dual notions for sinks are those of *U-final sink*, *U-co-Cartesian morphism*, and *U-discrete* object, with the universal property depicted by

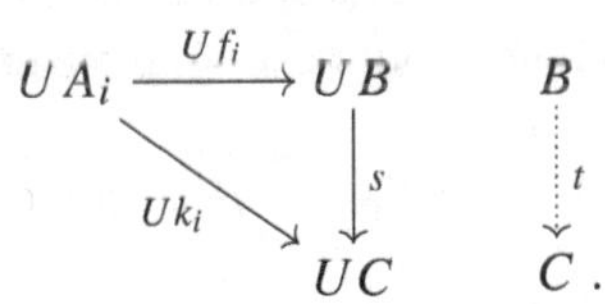

(We note a certain contradiction in the common terminology here, since U-final sinks with domain $(A_i)_{i \in I}$ are characterized as *initial* objects in a certain category; but Examples II.5.6.1 (1) and (2) give some justification.)

II.5.6.1 Examples

(1) For the forgetful functor $U : \mathsf{Ord} \to \mathsf{Set}$, a source $(g_i : A \to B_i)_{i \in I}$ is U-initial precisely when

$$x \leq y \iff \forall i \in I \; (g_i(x) \leq g_i(y))$$

for all $x, y \in A$. A sink $(f_i : A_i \to B)_{i \in I}$ is U-final precisely when, for all $z \neq w$ in A,

$$\begin{aligned} z \leq w \iff & \exists i_0, \ldots, i_n \in I \; \exists x_0 \leq y_0 \text{ in } A_{i_0}, x_1 \leq y_1 \text{ in } \\ & A_{i_1} \ldots, x_n \leq y_n \text{ in } A_{i_n}: \\ & z = f_{i_0}(x_0), f_{i_0}(y_0) = f_{i_1}(x_1), \ldots, f_{i_{n-1}}(y_{n-1}) \\ & = f_{i_n}(x_n), f_{i_n}(y_n) = w \; . \end{aligned}$$

Briefly, B carries the least order making all f_i monotone.

(2) For the forgetful functor $U : \mathsf{Top} \to \mathsf{Set}$, a source $(g_i : A \to B_i)_{i \in I}$ is U-initial precisely when $\{g_i^{-1}(V) \mid i \in I, V \subseteq B_i \text{ open}\}$ is a generating system of open sets for A. A sink $(f_i : A_i \to B)_{i \in I}$ is U-final precisely if $V \subseteq B$ is open whenever all $f_i^{-1}(V) \subseteq A_i$ are open.

(3) Let $\mathbb{T}$ be a monad on a category X. Then every mono-source in $\mathsf{X}^{\mathbb{T}}$ is initial with respect to the forgetful functor $G^{\mathbb{T}} : \mathsf{X}^{\mathbb{T}} \to \mathsf{X}$.

(4) For a category C, the functor category C^{2} (with $\mathsf{2} = \{\bot, \top\}$ considered as a category) can be thought of as having the morphisms of C as its objects, and a morphism $(u, v) : f \to g$ in C^{2} is given by a commutative square

$$\begin{array}{ccc} \cdot & \xrightarrow{u} & \cdot \\ {\scriptstyle f}\big\downarrow & & \big\downarrow{\scriptstyle g} \\ \cdot & \xrightarrow[v]{} & \cdot \end{array} \tag{II.5.6.i}$$

in C. The evaluation functor at $\top \in \mathrm{ob}\,\mathsf{2}$ now appears as the *codomain functor*

$$\mathrm{cod} : \mathsf{C}^{\mathsf{2}} \to \mathsf{C}\,, \quad f \mapsto \mathrm{cod}\, f\,, \quad (u, v) \mapsto v\,.$$

The morphism $(u, v) : f \to g$ is cod-Cartesian if and only if (II.5.6.i) is a *Cartesian square* in C, i.e. a pullback diagram in C.

For $U : \mathsf{A} \to \mathsf{X}$, we denote the classes of U-initial and U-final morphisms in A by

$$\mathrm{Ini}\, U \qquad \text{and} \qquad \mathrm{Fin}\, U\,,$$

respectively, and list some easily established properties for them.

II.5.6.2 Proposition

(1) $\mathrm{Ini}\, U$ *contains all isomorphisms, is closed under composition, and is weakly left-cancelable; it is even left-cancelable when U is faithful.*

(2) $\mathrm{Ini}\, U$ *is stable under pullbacks and multiple pullbacks when U preserves them, and is closed under those limits in A that are preserved by U.*

Proof The statements follow by routine verifications. □

We leave it to the reader to formulate the corresponding generalized statements for sources, as well as their dualizations for morphisms and sinks.

II.5.7 Fibrations and cofibrations

A functor $U : \mathsf{A} \to \mathsf{X}$ is a *fibration* (more precisely, a *cloven fibration*) when, for all $f : X \to UB$ in X with $B \in \mathrm{ob}\,\mathsf{A}$, there is a (tacitly chosen) *U-Cartesian lifting*, i.e. a morphism $g \in \mathrm{Ini}\,U$ with $Ug = f$:

$$\begin{array}{ccc} A & \xrightarrow{\;g\;} & B \\ \mathrel{\rotatebox[origin=c]{-90}{$\mapsto$}} & & \big\downarrow U \\ X & \xrightarrow[f]{} & UB\,. \end{array}$$

It is easy to show that U is a fibration if and only if the functors

$$U_B : \mathsf{A}/B \to \mathsf{X}/UB\,, \quad (A, g) \to (UA, Ug)\,,$$

have right adjoints Γ_B such that the counits are identity morphisms. Hence, $U_B \Gamma_B = 1_{\mathsf{X}\downarrow UB}$ and Γ_B is a full embedding for all $B \in \mathrm{ob}\,\mathsf{A}$.

A functor $U : \mathsf{A} \to \mathsf{X}$ is a *cofibration* (also called an *opfibration*) when $U^{\mathrm{op}} : \mathsf{A}^{\mathrm{op}} \to \mathsf{X}^{\mathrm{op}}$ is a fibration; explicitly, when for all $f : UA \to Y$ in X with $A \in \mathrm{ob}\,\mathsf{A}$ there is a *U-co-Cartesian lifting* (again, tacitly chosen) $g : A \to B$ in $\mathrm{Fin}\,U$:

$$\begin{array}{ccc} A & \xrightarrow{\;g\;} & B \\ \mathrel{\rotatebox[origin=c]{-90}{$\mapsto$}} & & \big\downarrow U \\ UA & \xrightarrow[f]{} & Y\,. \end{array}$$

For $X \in \mathrm{ob}\,\mathsf{X}$, let $U^{-1}X$ denote the *fiber* of U over X: its objects are the A-objects A with $UA = X$, and a morphism $t : A \to A'$ in $U^{-1}X$ is an A-morphism t with $Ut = 1_X$ (and composition as in A). When U is a fibration, any $f : X \to Y$ in X gives rise to a functor $f^* : U^{-1}Y \to U^{-1}X$ which assigns to $B \in \mathrm{ob}(U^{-1}Y)$ the domain A of the chosen U-Cartesian lifting of $f : X \to UB$. Dually, when U is a cofibration, there is a functor $f_* : U^{-1}X \to U^{-1}Y$ which assigns to $A \in \mathrm{ob}(U^{-1}X)$ the codomain B of the chosen U-co-Cartesian lifting of $f : UA \to Y$. One easily checks that there is an adjunction

$$U^{-1}Y \underset{f_*}{\overset{f^*}{\rightleftarrows}} U^{-1}X\,, \quad (\top)$$

when U is a fibration *and* a cofibration.

This adjunction describes best the "transport of structure along f." For example, for the forgetful functor $U : \mathsf{Top} \to \mathsf{Set}$ and a set X, the fiber $U^{-1}X$ is the lattice of topologies on X, and for a map $f : X \to Y$ of sets and a topology $\tau \in U^{-1}X$, the topology $\sigma = f_*(\tau)$ is described by $(V \in \sigma \iff$

$f^{-1}(V) \in \tau$). Likewise, for $\sigma \in U^{-1}Y$, the topology $\tau = f^*(\sigma)$ is given by $(V \in \tau \iff \exists W \in \sigma : V = f^{-1}(W))$. This way, one sees that U is a fibration and a cofibration.

The forgetful functor $\mathsf{Ord} \to \mathsf{Set}$ is another example of a fibration and a cofibration. The (non-faithful) object-functor ob : $\mathsf{Cat} \to \mathsf{Set}$ is a fibration, but not a cofibration (see Exercise II.5.R).

A useful generalization of the notions of fibration and cofibration is that of an $\mathcal{M}_0$-*fibration* and $\mathcal{E}_0$-*fibration*, for classes of morphisms $\mathcal{M}_0$ and $\mathcal{E}_0$ in X. For an $\mathcal{M}_0$-fibration, one requires the existence of a U-Cartesian lifting of $f : X \to UB$ in X only when $f \in \mathcal{M}_0$; dually, for an $\mathcal{E}_0$-fibration, U-co-Cartesian liftings are only required for morphisms in $\mathcal{E}_0$.

For example, the forgetful functor $\mathsf{Ord}_{\mathrm{sep}} \to \mathsf{Set}$ of the category of separated ordered sets is not a fibration, but it is a *mono-fibration* (i.e. a Mono Set-fibration). Similarly, the forgetful functor $\mathsf{Haus} \to \mathsf{Set}$ from the category Haus of Hausdorff topological spaces is a mono-fibration. (A topological space X is *Hausdorff* if, for all $x, y \in X$ with $x \neq y$, there exist open subsets $A, B \subseteq X$ with $x \in A$, $y \in B$ and $A \cap B = \varnothing$.) The following result offers a general explanation of these examples.

II.5.7.1 Proposition *For a functor $U : \mathsf{A} \to \mathsf{X}$ and a factorization system $(\mathcal{E}_0, \mathcal{M}_0)$ in X, let*

$$\mathcal{E} = U^{-1}\mathcal{E}_0 = \{e \in \operatorname{mor} \mathsf{A} \mid Ue \in \mathcal{E}_0\}, \quad \mathcal{M} = U^{-1}\mathcal{M}_0 \cap \operatorname{Ini} U .$$

(1) *If U is an $\mathcal{M}_0$-fibration, then $(\mathcal{E}, \mathcal{M})$ is a factorization system in A.*
(2) *If U is a fibration and B is an $\mathcal{E}$-reflective subcategory of A, then $U|_{\mathsf{B}}$ is an $\mathcal{M}_0$-fibration.*

Proof (1): In order to $(\mathcal{E}, \mathcal{M})$-factorize $f : A \to B$ in A, let $m_0 \cdot e_0 = Uf$ be an $(\mathcal{E}_0, \mathcal{M}_0)$-factorization in X, and consider a U-Cartesian lifting $m : C \to B$ of $m_0 : Z \to UB$. Then $e : A \to C$ is the only morphism with $Ue = e_o$ and $m \cdot e = f$. The fact that $\mathcal{E} \perp \mathcal{M}$ follows from routine diagram chasing. (2) follows from (1) and Proposition II.5.1.3. □

II.5.7.2 Corollary *If $U : \mathsf{A} \to \mathsf{X}$ is a fibration, then $U^{-1}(\operatorname{Iso} \mathsf{X})$ is a left factorization class in A.*

Proof Apply Proposition II.5.7.1 with $\mathcal{E}_0 = \operatorname{Iso} \mathsf{X}$. □

II.5.8 Topological functors

A functor $U : \mathsf{A} \to \mathsf{X}$ is *topological* if every source $(f_i : X \to UB_i)_{i \in I}$ with a family $(B_i)_{i \in I}$ of A-objects admits a U-*initial lifting* (once more, assumed to be chosen), i.e. a U-initial source $(g_i : A \to B_i)_{i \in I}$ with $UA = X$ and $Ug_i = f_i$ for all $i \in I$:

$$\begin{array}{ccc} A & \xrightarrow{g_i} & B_i \\ \downarrow & & \downarrow U \\ X & \xrightarrow[f_i]{} & UB_i . \end{array}$$

Hence, in the generalization of the corresponding statement for fibrations, U is topological if, for every discrete category I and every object $B = (B_i)_{i \in I} \in \mathrm{ob}\,\mathsf{A}^I$, the induced functor

$$U_B : (\Delta_{\mathsf{A}} \downarrow B) \to (\Delta_{\mathsf{X}} \downarrow UB)$$

has a right adjoint Γ_B such that the counits are identity morphisms; here $\Delta_{\mathsf{A}} : \mathsf{A} \to \mathsf{A}^I$ is as in Section II.2.8.

Obviously, *every topological functor is a fibration* ($|I| = 1$) *and has a full and faithful right adjoint* ($I = \varnothing$; in this case, for the only object B of $\mathsf{A}^{\varnothing}$, one has $(\Delta_{\mathsf{A}} \downarrow B) \cong \mathsf{A}$ and $(\Delta_{\mathsf{X}} \downarrow UB) \cong \mathsf{X}$).

The forgetful functor $U : \mathsf{Ord} \to \mathsf{Set}$ is topological. For a source $(g_i : X \to B_i)_{i \in I}$ with ordered sets B_i, make the set into an ordered set A by providing it with the appropriate order described in Example II.5.6.1(1). Similarly, $U : \mathsf{Top} \to \mathsf{Set}$ is topological: provide X with the topology generated by the sets $g_i^{-1}(V)$ ($i \in I$, $V \subseteq B_i$ open), see Example II.5.6.1(2).

Topological functors possess extremely good lifting properties. We have already seen that, as fibrations, they lift factorization systems for morphisms (see Proposition II.5.7.1(1)). In fact, for a collection $\mathbb{M}_0$ of sources in X, we can say that $U : \mathsf{A} \to \mathsf{X}$ is $\mathbb{M}_0$-*topological* by requiring U-initial liftings only for sources $(f_i : X \to UB_i)_{i \in I}$ in $\mathbb{M}_0$. One then has, analogously to Proposition II.5.7.1, the following:

II.5.8.1 Proposition *For a functor $U : \mathsf{A} \to \mathsf{X}$ and a factorization system $(\mathcal{E}_0, \mathbb{M}_0)$ for sources in X, let*

$$\mathcal{E} = U^{-1}\mathcal{E}_0 , \quad \mathbb{M} = \{(f_i)_{i \in I} \mid (f_i)_{i \in I}\ U\text{-initial } \& \ (Uf_i)_{i \in I} \in \mathbb{M}_0\} .$$

(1) *If U is $\mathbb{M}_0$-topological, then $(\mathcal{E}, \mathbb{M})$ is a factorization system for sources in A.*
(2) *If U is topological and B is an $\mathcal{E}$-reflective subcategory of A, then $U|_{\mathsf{B}}$ is $\mathbb{M}_0$-topological.*

Proof The proof of Proposition II.5.7.1 can be adapted to the present situation. □

Next, we show how to "lift" limits. For that, one first proves a lemma similar to II.5.3.1. (In fact, there is a common generalization for both lemmata; see [Börger and Tholen, 1978].)

II.5.8.2 Lemma *A topological functor $U : \mathsf{A} \to \mathsf{X}$ is faithful.*

Proof For $x, y : A \to D$ in A, assume $Ux = Uy = h$, and consider the constant source $(h)_{i\in I}$ with $I := \operatorname{mor} U^{-1}(UA)$. Let $(e_i : C \to D)_{i\in I}$ be a U-initial lifting of $(h)_{i\in I}$, and consider

$$J := \{w \in I \mid \forall i \in I : e_i \cdot w \in \{x, y\}\}.$$

With a retraction $\sigma : I \to J$, consider for $i \in I$

$$u_i := \begin{cases} x & \text{if } e_i \cdot \sigma(i) = y, \\ y & \text{if } e_i \cdot \sigma(i) = x, \end{cases}$$

and derive $x = y$ as in Lemma II.5.3.1. □

II.5.8.3 Proposition *For a faithful functor $U : \mathsf{A} \to \mathsf{X}$, the U-initial lifting of a limit cone $\lambda : \Delta L \to UD$ in X (with $D : \mathsf{J} \to \mathsf{A}$) yields a limit cone $\alpha : \Delta A \to D$ in A.*

Proof Faithfulness of U makes sure that the U-initial lifting $(\alpha_i : A \to D_i)_{i\in \operatorname{ob}\mathsf{J}}$ of the source $(\lambda_i : L \to UD_i)_{i\in\operatorname{ob}\mathsf{J}}$ does give a cone in A. Its limit property follows routinely. □

II.5.8.4 Corollary *For a topological functor $U : \mathsf{A} \to \mathsf{X}$, if X is J-complete, then so is A, and U preserves J-limits.*

Proof This follows immediately from Proposition II.5.8.3. □

We note that a topological functor in fact preserves all limits since it has a left adjoint (see Theorem II.5.9.1).

Proposition II.5.8.3 and its Corollary fully explain how to construct limits in Top. Indeed, one just needs to provide the limit of the underlying sets with the U-initial structure with respect to the limit projections.

The previous assertions hold analogously for colimits since topologicity of a functor is a self-dual concept, as we show next.

II.5.9 Self-dual characterization of topological functors

A functor $U : \mathsf{A} \to \mathsf{X}$ is *transportable* if, for every isomorphism $f : X \to UB$ in X with $B \in \operatorname{ob}\mathsf{A}$, there is a (chosen) isomorphism $g : A \to B$ in A with $Ug = f$. Since U-Cartesian liftings of isomorphisms are isomorphisms, transportability of U means precisely that U is an $\operatorname{Iso}\mathsf{X}$-fibration.

When U is faithful, all fibers $U^{-1}X$ ($X \in \operatorname{ob}\mathsf{X}$) are ordered classes. We call them *large-complete* if the infimum (or supremum) of any subclass exists.

II.5.9.1 Theorem *The following conditions are equivalent for a functor $U : \mathsf{A} \to \mathsf{X}$:*

(i) *U is topological;*
(ii) *every sink $(f_i : UA_i \to Y)_{i\in I}$ admits a U-final lifting $(g_i : A_i \to B)_{i\in I}$;*

(iii) *U is a faithful fibration and a cofibration with large-complete fibers;*
(iv) *U is faithful and transportable with a fully faithful left adjoint, and $U^{-1}(\mathrm{Iso}\,\mathsf{X})$ is a left factorization class for sources in A;*
(v) *U is faithful and transportable with a fully faithful right adjoint, and $U^{-1}(\mathrm{Iso}\,\mathsf{X})$ is a right factorization class for sinks in A.*

Proof (i) $\Longrightarrow$ (iii): An infimum of a family $(B_i)_{i\in I}$ in $U^{-1}Y$ is obtained from a U-initial lifting of the source $(1_Y : Y \longrightarrow UB_i)_{i\in I}$. Hence, the fibers of U are large-complete. It remains to be shown that U is a cofibration. For $f : UA \longrightarrow Y$ in X with $A \in \mathrm{ob}\,\mathsf{A}$, one considers all morphisms $g_i : A \longrightarrow B_i$ in A with $Ug_i = f$ $(i \in I)$, and then a U-initial lifting $(e_i : B \longrightarrow B_i)_{i\in I}$ of the source $(1_Y : Y \longrightarrow UB_i)_{i\in I}$. There is then a morphism $g : A \longrightarrow B$ with $Ug = f$, and we should check that it is U-final. Hence, let $k : A \longrightarrow C$ in A and $s : UB \longrightarrow UC$ in X with $s \cdot Ug = Uk$. With a U-initial lifting $t : D \longrightarrow C$ of s, one obtains $g' : A \longrightarrow D$ with $Ug' = f$, so that $g' = g_i$ for some $i \in I$. Now, $t \cdot e_i : B \longrightarrow C$ satisfies $U(t \cdot e_i) = s$, as desired.

(iii) $\Longrightarrow$ (i): Given a source $(f_i : X \longrightarrow UB_i)_{i\in I}$ in X with $B_i \in \mathrm{ob}\,\mathsf{A}$ for every $i \in I$, let $g_i : A_i \longrightarrow B_i$ be a U-initial lifting of $f_i : X \longrightarrow UB_i$, and let $(e_i : A \longrightarrow A_i)_{i\in I}$ represent an infimum in $U^{-1}X$. We need to show that $(g_i \cdot e_i : A \longrightarrow B_i)_{i\in I}$ is U-initial. Hence, we consider $s : UC \longrightarrow X$ in X and $k_i : C \longrightarrow B_i$ in A with $f_i \cdot s = Uk_i$ for all $i \in I$. Now, U-initiality of every g_i gives $t_i : C \longrightarrow A_i$ in A with $Ut_i = s$, and every t_i factors through the U-final lifting $t : C \longrightarrow D$ of $s : UC \longrightarrow X$. The infimum property of A in $U^{-1}X$ then yields a morphism $j : D \longrightarrow A$ in $U^{-1}X$, and we have $U(j \cdot t) = s$, as desired.

Since (iii)^{op} = (iii) and (i)^{op} = (ii), we have established the equivalence of (i), (ii), (iii). In particular, a topological functor has both a fully faithful left adjoint and a fully faithful right adjoint, and it is faithful and transportable. Furthermore, for a source $(f_i : A \longrightarrow B_i)_{i\in I}$ in A, one obtains a $(U^{-1}(\mathrm{Iso}), \{U\text{-initial sources}\})$-factorization by U-initially lifting the source $(Uf_i : UA \longrightarrow UB_i)_{i\in I}$. The unique diagonalization property follows as in the morphism case. Hence, (i) $\Longrightarrow$ (iv) is shown.

(iv) $\Longrightarrow$ (i): Let $D \dashv U$ with unit η an isomorphism. A source $(f_i : X \longrightarrow UB_i)_{i\in I}$ in X gives rise to a source $(g_i : DX \longrightarrow B_i)_{i\in I}$, for which there is then a morphism $e : DX \longrightarrow C$ in $U^{-1}(\mathrm{Iso}\,\mathsf{X})$ and a U-initial source $(m_i : C \longrightarrow B_i)_{i\in I}$ with $m_i \cdot e = g_i$ for all $i \in I$. The X-isomorphism $Ue \cdot \eta_X : X \longrightarrow UC$ may be lifted to an A-isomorphism $j : A \longrightarrow C$ with $Uj = Ue \cdot \eta_X$. Hence, $(m_i \cdot j)_{i\in I}$ is the desired U-initial lifting of $(f_i)_{i\in I}$. Since (iv)^{op} = (v), this completes the proof. □

The factorization needed in (v) may be constructed using Theorem II.5.3.2. For this, we say that X has small *connected limits* if every diagram $D : \mathsf{J} \longrightarrow \mathsf{X}$ with J small and connected (see Exercise II.2.Q) has a chosen limit in X.

© **II.5.9.2 Corollary** *Let* X *have small connected limits, and let* $U : \mathsf{A} \to \mathsf{X}$ *have small fibers. The functor* U *is topological if and only if the following conditions hold:*

(1) U *is faithful and transportable;*
(2) A *has small connected limits, and* U *preserves them;*
(3) U *has a fully faithful right adjoint.*

Proof For the necessity of the conditions, see Proposition II.5.8.3 and Theorem II.5.9.1. For their sufficiency, after Theorem II.5.9.1(v) we must show only that $\mathcal{M} = U^{-1}(\mathrm{Iso}\,\mathsf{X})$ satisfies the conditions of Theorem II.5.3.2. Certainly, $\mathcal{M}$ satisfies $\mathrm{Iso}\,\mathsf{A} \subseteq \mathcal{M} \subseteq \mathrm{Mono}\,\mathsf{A}$, is closed under composition, and is also stable under pullbacks since U preserves them. The only delicate point is the existence of (not necessarily small) intersections of morphisms in $\mathcal{M}$. Hence, consider $m_i : A_i \to B$ in $\mathcal{M}$ $(i \in I)$. Transportability gives isomorphisms $f_i : A_i \to B_i$ with $Uf_i = Um_i$ $(i \in I)$, and

$$\{B_i \in \mathrm{ob}\,\mathsf{X} \mid i \in I\} \subseteq U^{-1}(UB)$$

is just a set. Furthermore, if $B_i = B_j$, then $(A_i, m_i) \cong (A_j, m_j)$ in A/B for all $i, j \in I$. Consequently, in order to form the multiple pullback of $(m_i)_{i \in I}$ in A, it
© suffices to form the multiple pullback of a small subfamily, which exists and is preserved by U, so that it lies in $\mathcal{M}$ again. □

II.5.10 Epireflective subcategories

In Proposition II.5.1.3, we noted that a replete reflective subcategory B of a category C with a factorization system $(\mathcal{E}, \mathcal{M})$ for morphisms has its reflection morphisms in $\mathcal{E}$ if and only if B is *closed under* $\mathcal{M}$*-morphisms in* C, i.e. if $m : A \to B$ in $\mathcal{M}$ with $B \in \mathrm{ob}\,\mathsf{B}$ implies $A \in \mathsf{B}$. More generally, for a collection $\mathbb{M}$ of sources in C, one says that B is *closed under* $\mathbb{M}$*-sources in* C if $(m_i : A \to B_i)_{i \in I}$ in $\mathbb{M}$ with $B_i \in \mathrm{ob}\,\mathsf{B}$ implies $A \in \mathrm{ob}\,\mathsf{B}$, and one proves the following result.

II.5.10.1 Proposition *In a category* C *with a factorization system* $(\mathcal{E}, \mathbb{M})$ *for sources, a full replete subcategory* B *of* C *is* $\mathcal{E}$*-reflective if and only if* B *is closed under* $\mathbb{M}$*-sources in* C.

Proof For the "only if" part, one proceeds as in Proposition II.5.1.3. For the "if" part, given $C \in \mathrm{ob}\,\mathsf{C}$, one considers the source $(f_i : C \to A_i)_{i \in I}$ of all morphisms with domain C and codomain in B, and one $(\mathcal{E}, \mathbb{M})$-factors it as $f_i = g_i \cdot e$ with $e : C \to B$ in $\mathcal{E}$. Since $B \in \mathrm{ob}\,\mathsf{B}$ by hypothesis and $\mathcal{E} \subseteq \mathrm{Epi}\,\mathsf{C}$ (by the dual of Lemma II.5.3.1), e is a B-reflection morphism for C. □

© **II.5.10.2 Corollary** *Let* C *have products and a factorization system* $(\mathcal{E}, \mathcal{M})$ *for morphisms, with* $\mathcal{E} \subseteq \mathrm{Epi}\,\mathsf{C}$*, and suppose that* C *is* $\mathcal{E}$*-cowellpowered. Then a full* *replete subcategory* B *of* C *is* $\mathcal{E}$*-reflective if and only if* B *is closed under products and* $\mathcal{M}$*-morphisms in* C.

Proof After Proposition II.5.10.1, it suffices to guarantee the existence of a factorization system $(\mathcal{E}, \mathbb{M})$ for sources. Given a source $(f_i : A \to B_i)_{i \in I}$ in C, one may $(\mathcal{E}, \mathcal{M})$-factor each $f_i = m_i \cdot e_i$, and then choose a small representative family $(e_j : A \to C_j)_{j \in J}$ (with $J \subseteq I$) among the morphisms $e_i \in \mathcal{E}$ $(i \in I)$. By $(\mathcal{E}, \mathcal{M})$-factoring the induced morphism $g : A \to \prod_{j \in J} C_j$ with $p_j \cdot g = e_j$ (for $j \in J$) as $g = m \cdot e$, one obtains

$$f_j = (m_j \cdot p_j \cdot m) \cdot e \qquad (j \in J)$$

with $e \in \mathcal{E}$ and $d \perp (m_j \cdot p_j \cdot m)_{j \in J}$ for all $d \in \mathcal{E}$. Since each e_i $(i \in I)$ is isomorphic to some e_j $(j \in J)$, this factorization extends to the original source. □

In the presence of a topological functor $U : \mathsf{C} \to \mathsf{X}$, we may apply Proposition II.5.10.1 in particular to the $(U^{-1}(\mathrm{Iso}\,\mathsf{X}), \mathbb{M})$-factorization system for sources in C, where now $\mathbb{M}$ consists of all U-initial sources, and obtain at once the equivalence (i) $\Longleftrightarrow$ (ii) in the following Theorem.

II.5.10.3 Theorem *Let* $U : \mathsf{C} \to \mathsf{X}$ *be a topological functor. The following assertions for a full replete subcategory* B *of* C *are equivalent:*

(i) B *is* $U^{-1}(\mathrm{Iso}\,\mathsf{X})$*-reflective in* C*;*
(ii) B *is closed under* U*-initial sources in* C*;*
(iii) $U|_{\mathsf{B}}$ *is topological, and* $U|_{\mathsf{B}}$ *initial sources in* B *are also* U*-initial in* C.

Proof From the comment preceding Theorem II.5.10.3, we are left to verify (ii) $\Longleftrightarrow$ (iii), but this is immediate. □

Finally, let us discuss an important sufficient condition for the inclusion functor $\mathsf{B} \hookrightarrow \mathsf{C}$ to preserve initiality, as in (iii). One says that a full replete subcategory B is *U-finally dense* (or simply *finally dense*) if every object C in C is the codomain of some U-final sink $(f_i : A_i \to C)_{i \in I}$ with all $A_i \in \mathrm{ob}\,\mathsf{B}$; the dual notion is that of a *U-initially dense* (or just *initially dense*) subcategory. Without the topologicity assumption on U, one can still prove:

II.5.10.4 Proposition *Let* $U : \mathsf{C} \to \mathsf{X}$ *be a functor and let* B *be a full replete subcategory of* C *that is finally dense in* C*. Then* $U|_{\mathsf{B}}$*-initial sources in* B *are* U*-initial in* C*, and when* U *is faithful and* $U|_{\mathsf{B}}$ *is topological,* B *is* $U^{-1}(\mathrm{Iso}\,\mathsf{X})$*-reflective in* C.

Proof In order to show U-initiality of a $U|_{\mathsf{B}}$-initial source $(f_i : A \to B)_{i \in I}$ in B, consider $(g_i : C \to B_i)_{i \in I}$ in C and $s : UC \to UA$ in X with $Uf_i \cdot s = Ug_i$ for all $i \in I$, and let $(h_j : A_j \to C)_{j \in J}$ be U-final with all $A_j \in \mathrm{ob}\,\mathsf{B}$. $U|_{\mathsf{B}}$-initiality then gives, for every $j \in J$, a unique $t_j : C \to A_j$ with $Ut_j = s \cdot Uh_j$ and $f_i \cdot t_j = g_i \cdot h_j$ for all $i, j \in I$, and U-finality yields a unique $t : C \to A$ in B with $Ut = s$ and $t \cdot h_j = t_j$ for all $j \in J$. By $U|_{\mathsf{B}}$-initiality again, t also satisfies $f_i \cdot t = g_i$ for all $i \in I$, and one easily sees that t is the only morphism "over s" satisfying these equations:

$$\begin{array}{ccccc} & & A & \xrightarrow{f_i} & B_i \\ & {\scriptstyle t_j}\nearrow & \uparrow{\scriptstyle t} & \nearrow{\scriptstyle g_i} & \\ A_j & \xrightarrow[h_j]{} & C & & \end{array} \qquad\qquad \begin{array}{c} UA \\ \uparrow{\scriptstyle s=Ut} \\ UC\,. \end{array}$$

For the reflexivity assertion, one proceeds as in the proof of Proposition II.5.10.1, and, given $C \in \operatorname{ob}\mathsf{C}$, one considers the source $(f_i : C \to A_i)_{i \in I}$ of all morphisms with domain C and codomain in B. Let $(g_i : B \to A_i)_{i \in I}$ be a $U|_{\mathsf{B}}$-initial lifting of $(Uf_i : UC \to UA_i)_{i \in I}$. Since $(g_i)_{i \in I}$ must be even U-initial, there is $e : C \to B$ in C with $Ue = 1_{UC}$, which is easily seen to serve as the B-reflection morphism for C. □

II.5.10.5 Example There is a full coreflective embedding $E : \mathsf{Ord} \to \mathsf{Top}$ which provides an ordered set $(X, \leq)$ with the topology of open sets generated by the down-sets $\downarrow x$, for $x \in X$. Its right adjoint S provides a topological space with its underlying order (see Section II.1.9). A topological space X is in the image of E precisely when it is an *Alexandroff space*, i.e. when arbitrary intersections of open sets in X are open. By the dual of Theorem II.5.10.3, Ord is closed under U-final sinks in Top (with the forgetful functor $U : \mathsf{Top} \to \mathsf{Set}$), and $U|_{\mathsf{Ord}}$-final sinks in Ord are U-final in Top. In fact, Ord is even initially dense in Top: for a topological space X, consider the source of characteristic functions of all open sets of X into the two-element chain. But Ord is not closed under U-initial sources, not even under products, as no infinite power in Top of the two-element chain is Alexandroff.

II.5.11 Taut Lift Theorem

We consider a commutative diagram of functors

$$\begin{array}{ccc} \mathsf{A} & \xrightarrow{G} & \mathsf{B} \\ {\scriptstyle U}\downarrow & & \downarrow{\scriptstyle V} \\ \mathsf{X} & \xrightarrow{J} & \mathsf{Y} \end{array}$$

such that J has a left adjoint $H \underset{\delta}{\overset{\gamma}{\dashv}} J$. Our goal is to find a left adjoint $F \underset{\varepsilon}{\overset{\eta}{\dashv}} G$ when U has good lifting properties.

II.5.11.1 Theorem (Taut Lift Theorem) *Let U be a topological functor. Then, in the preceding diagram, G has a left adjoint F with $UF = HV$ if and only if G maps U-initial sources to V-initial sources.*

Proof For the necessity of the stated condition, see Exercise II.5.T. For its sufficiency, we consider an object B in B and the source of all morphisms $t : B \to GA_t$ with $A_t \in \mathsf{A}$. The source of all

$$HVB \xrightarrow{HVt} HVGA_t = HJUA_t \xrightarrow{\delta_{UA_t}} UA_t$$

has a U-initial lifting

$$FB \xrightarrow{f_t} A_t$$

so that $Uf_t = \delta_{UA_t} \cdot HVt$. As $JHVB = JUFB$, the codomain of γ_{VB} is the domain of JUf_t, and

$$JUf_t \cdot \gamma_{VB} = J\delta_{UA_t} \cdot JHVt \cdot \gamma_{VB} = J\delta_{UA_t} \cdot \gamma_{JUA_t} \cdot Vt = Vt\,.$$

By hypothesis, G transforms the source into a V-initial source, so the diagram

$$\begin{array}{ccc} JUFB = VGFB & \xrightarrow{JUf_t = VGf_t} & JUA_t = VGA_t \\ \uparrow{\scriptstyle \gamma_{VB}} & \nearrow{\scriptstyle Vt} & \\ VB & & \end{array}$$

produces a morphism $\eta_B : B \to GFB$ with $V\eta_B = \gamma_{VB}$ and $Gf_t \cdot \eta_B = t$ for all t. By adjointness of J and faithfulness of U, the factorization $Vt = JUf_t \cdot \gamma_{VB}$ determines f_t uniquely. □

II.5.11.2 Example Consider the diagram of forgetful functors

$$\begin{array}{ccc} \mathsf{TopGrp} & \longrightarrow & \mathsf{Top} \\ \downarrow & & \downarrow \\ \mathsf{Grp} & \longrightarrow & \mathsf{Set} \end{array}$$

where TopGrp is the category of *topological groups* (i.e. groups G with a topology that makes the group operations $G \times G \to G$, $(x, y) \mapsto x \cdot y$, and $G \to G$, $x \mapsto x^{-1}$ continuous), and of continuous group homomorphisms. It is easy to see that $\mathsf{TopGrp} \to \mathsf{Grp}$ is topological with initial structures formed as for $\mathsf{Top} \to \mathsf{Set}$. Hence, the free-group functor $F : \mathsf{Set} \to \mathsf{Grp}$ may be lifted to a left adjoint functor $\mathsf{Top} \to \mathsf{TopGrp}$. In lieu of groups, the example generalizes to any algebraic structures defined by a set of operations and equations between them, and that admit free functors (see Example II.2.12.2).

Exercises

II.5.A *More cancelation rules.* Let $\mathcal{M}$ be a pullback-stable class of morphisms in C. Then $\mathcal{M}$ satisfies the cancelation property: if $g \cdot f \in \mathcal{M}$ with g monic implies $f \in \mathcal{M}$. If $\mathcal{M}$ belongs to a factorization system $(\mathcal{E}, \mathcal{M})$, then $g \cdot f \in \mathcal{M}$

with a split epimorphism f implies $g \in \mathcal{M}$. The family $(f_i : A \to B_i)_{i \in I}$ is a mono-source if $(g_{ij} \cdot f_i)_{i \in I, j \in J}$ is a mono-source for sources $(g_{ij} : B_i \to C_{ij})_{j \in J_i}$, $i \in I$.

II.5.B *More on proper systems.* For a factorization system $(\mathcal{E}, \mathcal{M})$ in a category C with equalizers, one has $\mathcal{E} \subseteq \mathrm{Epi}\,\mathsf{C}$ if and only if $\mathrm{RegMono}\,\mathsf{C} \subseteq \mathcal{M}$.

II.5.C *Extremal, strong, and regular epimorphisms.* The class of *strong epimorphisms* in C is

$$\mathrm{StrongEpi}\,\mathsf{C} := \mathrm{Epi}\,\mathsf{C} \cap {}^{\perp}(\mathrm{Mono}\,\mathsf{C})\,.$$

One has

$$\mathrm{RegEpi}\,\mathsf{C} \subseteq \mathrm{StrongEpi}\,\mathsf{C} \subseteq \mathrm{ExtEpi}\,\mathsf{C}\,,$$

where $\mathrm{RegEpi}\,\mathsf{C}$ and $\mathrm{ExtEpi}\,\mathsf{C}$ are the classes of regular and extremal epimorphisms, respectively (see Section II.5.1 for the latter). If C has pullbacks, then $\mathrm{StrongEpi}\,\mathsf{C} = \mathrm{ExtEpi}\,\mathsf{C}$. Of course, the dual statements hold for *strong monomorphisms*. Furthermore, each of the following statements implies the next, and all are equivalent when C has kernel pairs and coequalizers of kernel pairs:

(i) $\mathrm{Mono}\,\mathsf{C} \cdot \mathrm{RegEpi}\,\mathsf{C} = \mathrm{mor}\,\mathsf{C}$;
(ii) $(\mathrm{RegEpi}\,\mathsf{C}, \mathrm{Mono}\,\mathsf{C})$ is a factorization system in C;
(iii) $\mathrm{RegEpi}\,\mathsf{C} = \mathrm{ExtEpi}\,\mathsf{C}$;
(iv) $\mathrm{RegEpi}\,\mathsf{C}$ is closed under composition.

II.5.D *Regular monomorphisms in a topological category.* If $U : \mathsf{A} \to \mathsf{X}$ is topological (or just a faithful functor with both a left and a right adjoint), an A-morphism f is a regular monomorphism if and only if it is U-initial and Uf is a regular monomorphism. The same statement holds if "regular" is replaced by "extremal."

II.5.E *Functoriality of factorizations.* For every factorization system $(\mathcal{E}, \mathcal{M})$, there is a functor $F : \mathsf{C}^2 \to \mathsf{C}$ which assigns to a morphism f in C the object $\mathrm{dom}\, m = \mathrm{cod}\, e$ for a chosen $(\mathcal{E}, \mathcal{M})$-factorization $f = m \cdot e$.

II.5.F *Wellpowered and cowellpowered categories over* Set. The category Set and every topological category over Set is wellpowered and cowellpowered. Every monadic category over Set is wellpowered and $\mathcal{E}$-cowellpowered with $\mathcal{E}$ the class of regular epimorphisms. However, a monadic category over Set is not necessarily cowellpowered: Frm is not cowellpowered (see [Johnstone, 1982]).

II.5.G *Closure operators as functors.* A class $\mathcal{M}$ of morphisms in C can be considered as a full subcategory of C^2 (see Example II.5.6.1(4)). Show that an $\mathcal{M}$-closure operator c defines a functor $c : \mathcal{M} \to \mathcal{M}$, together with a natural transformation $j : 1_{\mathcal{M}} \to c$. If there is a mono-fibration $U : \mathsf{C} \to \mathsf{Set}$ such that $\mathcal{M} = U^{-1}\,\mathrm{Mono} \cap \mathrm{Ini}\, U$ (see Section II.5.9) and C is $\mathcal{M}$-wellpowered, then c defines a functor

$$c : \mathsf{C} \to \mathsf{Cls}\,, \quad X \mapsto (UX, c_X)\,.$$

II.5.H *Closure operators and closed subobjects.* Let $\mathcal{M}$ be a right factorization class for sinks in C, and let $\mathcal{K} \subseteq \mathcal{M}$ be stable under pullbacks. Then

$$c_A(m) = \bigwedge\{k \in \mathrm{sub}\,A \mid k \in \mathcal{K}, k \leq m\}$$

defines an idempotent $\mathcal{M}$-closure operator of C, and one has $\mathcal{M}^c = \mathcal{K}$ if and only if $\mathcal{K}$ is stable under multiple pullbacks. Furthermore, as an idempotent closure operator, c is uniquely determined by the condition $\mathcal{M} \subseteq \mathcal{K}$, and c is weakly hereditary if and only if $\mathcal{K}$ is closed under composition.

II.5.I *Adjunctions, epimorphisms, and generating classes.* For an adjunction $F \xrightarrow[\varepsilon]{\eta}\!\!\dashv G : \mathsf{A} \longrightarrow \mathsf{X}$, one has that, if G is faithful and $\mathcal{H}$ is generating in X, then

$$F\mathcal{H} = \{FH \mid H \in \mathcal{H}\}$$

is generating in A. Furthermore, the following conditions are equivalent:

(i) G is faithful;
(ii) G reflects epimorphisms;
(iii) the counits ε_A are epimorphisms;
(iv) the class $\{FX \mid X \in \mathsf{X}\}$ is generating in A.

The same equivalence holds if one specializes to a faithful functor that *reflects isomorphisms* in (i), *extremal* epimorphisms in (ii) and (iii), and a *strongly* generating class in (iv).

II.5.J *Special Adjoint Functor Theorem.* Suppose that A is locally small, small-complete, has a cogenerator $\mathcal{G}$, and is wellpowered, so that for every $A \in \mathrm{ob}\,\mathsf{A}$ there is a chosen set $\mathcal{J}_A \subseteq \mathrm{ob}\,\mathsf{A}$ such that

if $m : B \to A$ is a monomorphism, then there is $J \in \mathcal{J}_A$ with $J \cong B$.

Then a functor $G : \mathsf{A} \to \mathsf{X}$ into a locally small category X is right adjoint if and only if it preserves small limits.
Hint. Use Proposition II.5.5.1^{op} and the sets $\mathcal{J}_A$ to construct a G-solution set for every $X \in \mathrm{ob}\,\mathsf{X}$.

II.5.K *Special Adjoint Functor Theorem for a class of monomorphisms.* For some class $\mathcal{M}$ of monomorphisms closed under composition with isomorphisms, let the locally small category A have pullbacks of morphisms in $\mathcal{M}$ (along arbitrary morphisms) and intersections of arbitrarily large families of morphisms in $\mathcal{M}$, and suppose that both belong to $\mathcal{M}$ again. Furthermore, let A have an $\mathcal{M}$-cogenerator, i.e. a set $\mathcal{G}$ of objects in A such that the product

$P_A = \prod_{G\in\mathcal{G}} G^{\mathsf{A}(G,A)}$ exists for all $A \in \mathrm{ob}\,\mathsf{A}$, and the canonical morphism $P_A \to A$ lies in $\mathcal{M}$. Then a functor $G : \mathsf{A} \to \mathsf{X}$ has a left adjoint if and only if G preserves all limits whose existence is guaranteed by the hypotheses.

II.5.L *Dense generators.* A class $\mathcal{G}$ of objects in a locally small category C is *densely generating* in C if $\mathsf{C}(\mathcal{G}, A)$ is a *strict epi-sink* for all objects A, i.e., whenever a sink $(h_x : G_x \to B)_{x\in\mathsf{C}(\mathcal{G},A)}$ has the property

$$x \cdot a = y \cdot b \implies h_x \cdot a = h_y \cdot b$$

for all $a : D \to G_x$, $b : D \to G_y$, $x, y \in \mathsf{C}(\mathcal{G}, A)$, then $h_x = f \cdot x$ for all x, with a uniquely determined morphism $f : A \to B$. Show that $\mathcal{G}$ is densely generating if and only if the functor

$$\mathbf{y}_{\mathcal{G}} : \mathsf{C} \to \mathsf{Set}^{\mathcal{G}^{\mathrm{op}}}, \quad A \mapsto (\mathsf{C}(-, A) : \mathcal{G}^{\mathrm{op}} \to \mathsf{Set})$$

(which has the Yoneda embedding as a factor) is full and faithful; here, and differently from Section II.5.5, we consider $\mathcal{G}$ as a full subcategory of C. A densely generating set is strongly generating, but not vice versa: a one-point space is a strong generator of $\mathsf{CompHaus}$, the category of compact Hausdorff spaces and continuous maps, but it is not dense ($\mathsf{CompHaus}$ in fact has no dense generator; see [Gabriel and Ulmer, 1971]).

II.5.M *$\mathcal{M}$-injective objects.* An object C in a category C is *$\mathcal{M}$-injective* for a class $\mathcal{M}$ of morphisms in C if, for any $m : A \to B$ with $m \in \mathcal{M}$ and $f : A \to C$, there exists a map $g : B \to C$ extending f, i.e. such that the following diagram commutes:

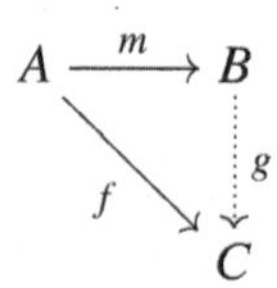

(the dual notion is that of an *$\mathcal{E}$-projective* object in C). If $r : C \to D$ is a retraction to an $\mathcal{M}$-injective object C, then D is also $\mathcal{M}$-injective. Also, a product of $\mathcal{M}$-injective objects is an $\mathcal{M}$-injective object.

II.5.N *U-co-Cartesian liftings of regular epimorphisms.* Let A have kernel pairs, and suppose that $U : \mathsf{A} \to \mathsf{X}$ preserves them. Then a U-co-Cartesian lifting of a regular epimorphism $f : UA \to Y$ in X is a regular epimorphism in A. Morphisms in $\mathrm{Fin}\,U \cap U^{-1}(\mathrm{RegEpi}\,\mathsf{X})$ are also called *quotient morphisms* of A (with respect to U).

II.5.O *Fibrations and equivalences.* A fibration $U : \mathsf{A} \to \mathsf{X}$ which reflects isomorphisms is an equivalence, provided that A has a terminal object and U preserves it. The provision is essential: for any category C and $T \in \mathrm{ob}\,\mathsf{C}$, the *domain* *functor* dom $= \mathrm{dom}_T : \mathsf{C}/T \longrightarrow \mathsf{C}$ is a fibration and reflects isomorphisms, but it is not an equivalence, unless T is terminal in C.

II.5.P *The codomain functor as a fibration.* The functor cod : $\mathsf{C}^2 \longrightarrow \mathsf{C}$ of Example II.5.6.1(4) is a fibration if and only if C has pullbacks. It is topological if and only if C has multiple pullbacks (of any size).

II.5.Q *The domain functor is not necessarily topological.* Let C be locally small and small-complete. For any $T \in \mathrm{ob}\,\mathsf{C}$, the domain functor dom : $\mathsf{C}/T \longrightarrow \mathsf{C}$ satisfies all conditions of Corollary II.5.9.2, except that the right adjoint of dom generally fails to be fully faithful. Thus, in general, dom fails to be topological.

II.5.R *The set-of-objects functor as a fibration.* The *set-of-objects functor* ob : $\mathsf{Cat} \longrightarrow \mathsf{Set}$ is a fibration, but not a cofibration. Hence, it is not topological.

II.5.S *Topological restrictions of topological functors.* Let $U : \mathsf{C} \longrightarrow \mathsf{X}$ be topological and let A be a full replete subcategory of C, and set $\mathcal{J} := U^{-1}(\mathrm{Iso}\,\mathsf{X})$. The following conditions are equivalent:

(i) $U|_{\mathsf{A}}$ is topological;
(ii) there is a full $\mathcal{J}$-reflective subcategory B of C which contains A as a $\mathcal{J}$-coreflective subcategory;
(iii) there is a full $\mathcal{J}$-coreflective subcategory B of C which contains A as a $\mathcal{J}$-reflective subcategory;
(iv) there is a functor $R : \mathsf{C} \longrightarrow \mathsf{A}$ with $(U|_{\mathsf{A}})R = U$ and $R|_{\mathsf{A}} = 1_{\mathsf{A}}$.

Hint. For (i) $\Longrightarrow$ (ii), take for ob B all C-objects C for which there exist a U-initial source with domain C and all codomains in A.

II.5.T *Preservation of initiality.* For a commutative diagram of functors

$$\begin{array}{ccc} \mathsf{A} & \xrightarrow{G} & \mathsf{B} \\ {\scriptstyle U}\downarrow & & \downarrow{\scriptstyle V} \\ \mathsf{X} & \xrightarrow{J} & \mathsf{Y} \end{array}$$

such that there are adjunctions $F \underset{\varepsilon}{\overset{\eta}{\dashv}} G$ and $H \underset{\delta}{\overset{\gamma}{\dashv}} J$, with the canonical transformation $HV \longrightarrow UF$ an isomorphism, show that G maps U-initial sources to V-initial sources. (Note that no topologicity assumption for U or V is required.)

II.5.U *Generalized Taut Lift Theorem.* In the commutative diagram of functors

$$\begin{array}{ccc} \mathsf{A} & \xrightarrow{G} & \mathsf{B} \\ {\scriptstyle U}\downarrow & & \downarrow{\scriptstyle V} \\ \mathsf{X} & \xrightarrow{J} & \mathsf{Y} \end{array}$$

let A have an $(\mathcal{E}, \mathbb{M})$-factorization system for sources such that G maps sources in $\mathbb{M}$ to V-initial sources. If U and J have left adjoints, then G has also a left adjoint.

II.5.V *Grothendieck construction versus faithful fibrations and topological functors.*

(1) For a faithful cloven fibration $U : \mathsf{A} \to \mathsf{X}$, one obtains (in the notation of Section II.5.7) a pseudo-functor

$$U^{-1} : \mathsf{X}^{\mathrm{op}} \to \mathsf{ORD}$$

into the ordered metacategory of ordered classes; it assigns to $f : X \to Y$ in X the monotone function $f^* : U^{-1}Y \to U^{-1}X$. If U is topological, ORD may be replaced by INF, where objects have all infima and whose morphisms preserve them.

(2) For a pseudo-functor $T : \mathsf{X}^{\mathrm{op}} \to \mathsf{ORD}$, let the objects of the category X_T be pairs (X, τ) with $X \in \mathrm{ob}\,\mathsf{X}$, $\tau \in TX$; a morphism $f : (X, \tau) \to (Y, \sigma)$ is an X-morphism $f : X \to Y$ with $\tau \le Tf(\sigma)$. The forgetful functor

$$U_T : \mathsf{X}_T \to \mathsf{X}\,, \quad (X, \tau) \to X$$

is a faithful fibration. When T takes values in INF, the functor U_T is topological.

(3) For a faithful fibration $U : \mathsf{A} \to \mathsf{X}$ and $T = U^{-1}$, there is an isomorphism $G : \mathsf{A} \to \mathsf{X}_T$ with $U_T G = U$.

(4) For a pseudo-functor $T : \mathsf{X}^{\mathrm{op}} \to \mathsf{ORD}$, there is a pseudo-natural isomorphism $\gamma : T \to (U_T)^{-1}$.

II.5.W *Equivalence of topological functors and* INF*-valued pseudo-functors.*

(1) Let the objects of the metacategory FFIB be faithful fibrations, and let morphisms $(F, G) : U \to V$ be commutative diagrams

$$\begin{array}{ccc} \mathsf{A} & \xrightarrow{G} & \mathsf{B} \\ {\scriptstyle U}\downarrow & & \downarrow{\scriptstyle V} \\ \mathsf{X} & \xrightarrow{F} & \mathsf{Y} \end{array}$$

in CAT, such that G transforms U-initial morphisms into V-initial morphisms. The objects of the metacategory $\mathsf{CAT}/\!/\ \mathsf{ORD}$ are contravariant pseudo-functors with values in ORD; a morphism

$$(F, \gamma) : T \to S$$

is given by a diagram

$$\begin{array}{ccc} \mathsf{X}^{\mathrm{op}} & \xrightarrow{\ F^{\mathrm{op}}\ } & \mathsf{Y}^{\mathrm{op}} \\ & \overset{\gamma}{\Rightarrow} & \\ {\scriptstyle T}\searrow & & \swarrow{\scriptstyle S} \\ & \mathsf{ORD} & \end{array}$$

with a functor $F : \mathsf{X} \to \mathsf{Y}$ and a pseudo-natural transformation $\gamma : T \to SF^{\mathrm{op}}$, and its composite with $(H, \delta) : S \to R$ is given by

$$(H, \delta)(F, \gamma) = (HF, \delta F^{\mathrm{op}} \cdot \gamma)\,.$$

Show that FFIB and $\mathsf{CAT}/\!/\mathsf{ORD}$ are equivalent metacategories.

(2) Let TOPFUN be the (non-full) subcategory of FFIB formed by those $(F, G) : U \to V$ with U, V topological and G transforming U-initial sources into V-initial sources. Prove that TOPFUN is equivalent to $\mathsf{CAT}/\!/\mathsf{INF}$.

(3) Formulate (1) and (2) for cofibrations instead of fibrations.

Notes on Chapter II

Most of the topics presented in this chapter may be found in more elaborate form in the standard books on category theory: [Mac Lane, 1971], [Adámek, Herrlich, and Strecker, 1990], [Borceux, 1994a,b,c]. Each of these books contains remarks or brief sections on the foundations of category theory and its connections with logic or set theory, a topic that has been addressed in many articles, including [Lawvere, 1966], [Mac Lane, 1969], [Feferman, 1969, 1977], [Bénabou, 1985], as well as in books on topos theory, such as [Mac Lane and Moerdijk, 1994]. Beginners in category theory will enjoy [Lawvere and Rosebrugh, 2003] and [Awodey, 2006], and advanced readers are referred to [Kelly, 1982] as the standard text on enriched category theory, and to [Johnstone, 2002a,b] as a rich resource for a broad range of categorical topics. Readers looking for further reading on order and quantale theory as it pertains to Section II.1 are referred to [Johnstone, 1982], [Rosenthal, 1990], and [Wood, 2004].

We highlight some particular aspects that distinguish this chapter from the literature mentioned thus far and give some additional references. The chapter covers some aspects of monad theory in greater detail than the standard texts on category theory, as it includes Duskin's monadicity criterion (see Theorem II.3.5.1, originally established in [Duskin, 1969]) and a discussion of Beck's distributive laws [Beck, 1969]; see also [Manes and Mulry, 2007]), as well as a treatment of Kock–Zöberlein monads (see [Kock, 1995] and references therein), albeit in the simplified context of ordered categories. For further reading on monads and their connection with algebraic theories, we refer the reader to [Manes, 1976], [Barr and Wells, 1985], [MacDonald and Sobral, 2004], and [Adámek, Rosický, and

Vitale, 2011]. In a higher-order context, they are treated in [Street, 1974] and [Lack and Street, 2002]. In this chapter, however, we touch upon higher-categorical structures only to a minimal level, with 2-cells most often given simply by order, the notion of quantaloid being an important example as treated in Stubbe's articles [2005; 2006; 2007].

Another non-standard emphasis of this chapter concerns factorization systems and topological functors which, unlike in [Adámek *et al.*, 1990], are presented in concert with fibrations (see [Grothendieck, Verdier, and Deligne, 1972], [Bénabou, 1985], [Streicher, 1998–2012]), as highlighted by Theorem II.5.9.1. A predecessor of its Corollary II.5.9.2 first appeared in Hoffmann's thesis [1972]. Wyler's Taut Lift Theorem II.5.11.1 was first proved in [Wyler, 1971], and presented in a general categorical context in [Tholen, 1978]. The key existence theorem on factorization systems (Theorem II.5.3.2) appeared in generalized form in [Tholen, 1979]. For a categorical treatment of closure operators, the reader is referred to [Dikranjan and Tholen, 1995].

III

Lax algebras

Dirk Hofmann, Gavin J. Seal, and Walter Tholen

For a quantale $\mathcal{V}$ and a monad $\mathbb{T}$ on Set, laxly extended to the category $\mathcal{V}$-Rel of sets and $\mathcal{V}$-valued relations, this chapter introduces the key category of interest to this book, the category $(\mathbb{T}, \mathcal{V})$-$\mathsf{Cat}$, whose objects, depending on context, may be called $(\mathbb{T}, \mathcal{V})$-*algebras*, $(\mathbb{T}, \mathcal{V})$-*spaces*, or $(\mathbb{T}, \mathcal{V})$-*categories*. After a first introduction to the $\mathcal{V}$-relational setting and the required lax monad extension, the guiding examples (ordered sets, metric spaces, topological spaces, approach spaces) are presented in full detail, followed by the basic properties of the category $(\mathbb{T}, \mathcal{V})$-$\mathsf{Cat}$, such as its topologicity over Set and its embeddability into the quasitopos of $(\mathbb{T}, \mathcal{V})$-graphs. The seemingly "technical" lax extendability of $\mathbb{T}$ to the "syntactical" category $\mathcal{V}$-Rel permits us to consider $\mathbb{T}$ as a monad on $\mathcal{V}$-Cat and even $(\mathbb{T}, \mathcal{V})$-$\mathsf{Cat}$, the Eilenberg–Moore algebras of which lead to the consideration of objects that combine relevant ordered, topological, and metric structures in a very natural way.

III.1 Basic concepts

III.1.1 $\mathcal{V}$-relations

Recall from II.1.2 that a relation r from a set X to a set Y associates to every pair $(x, y) \in X \times Y$ a truth value in $\mathsf{2} = \{\text{false}, \text{true}\}$ which tells us whether x and y are r-related or not. In order to model situations where quantitative information is available, r can be allowed to take values in any quantale $\mathcal{V} = (\mathcal{V}, \otimes, k)$ rather than just in $\mathsf{2} = (\mathsf{2}, \wedge, \top)$. (The quantale $\mathcal{V}$ is associative and unital, as defined in II.1.10.) A $\mathcal{V}$-*relation* $r : X \nrightarrow Y$ from X to Y is therefore presented by a

Monoidal Topology: A Categorical Approach to Order, Metric, and Topology, eds. Dirk Hofmann, Gavin J. Seal, and Walter Tholen. Published by Cambridge University Press.

map $r : X \times Y \longrightarrow \mathcal{V}$. As for ordinary relations, a $\mathcal{V}$-relation $r : X \nrightarrow Y$ can be composed with another $\mathcal{V}$-relation $s : Y \nrightarrow Z$ via "matrix multiplication"

$$(s \cdot r)(x, z) = \bigvee_{y \in Y} r(x, y) \otimes s(y, z)$$

(for all $x \in X$, $z \in Z$) to yield a $\mathcal{V}$-relation $s \cdot r : X \nrightarrow Z$. This composition is associative (see Exercise III.1.C), and the $\mathcal{V}$-relation $1_X : X \nrightarrow X$ that sends every diagonal element (x, x) to k, and all other elements to the bottom element $\bot$ of $\mathcal{V}$, serves as the identity morphism on X. Thus, sets and $\mathcal{V}$-relations form a category, denoted by

$$\mathcal{V}\text{-}\mathsf{Rel}\,.$$

The set $\mathcal{V}\text{-}\mathsf{Rel}(X, Y)$ of all $\mathcal{V}$-relations from X to Y inherits the pointwise order induced by $\mathcal{V}$: given $r : X \nrightarrow Y$ and $r' : X \nrightarrow Y$, we have

$$r \leq r' \iff \forall (x, y) \in X \times Y \; (r(x, y) \leq r'(x, y))\,.$$

Since the order on $\mathcal{V}$ is complete, so is the pointwise order on $\mathcal{V}\text{-}\mathsf{Rel}(X, Y)$, and since the tensor in $\mathcal{V}$ distributes over suprema, $\mathcal{V}$-relational composition preserves suprema in each variable:

$$s \cdot \bigvee_{i \in I} r_i = \bigvee_{i \in I} (s \cdot r_i) \qquad \text{and} \qquad \bigvee_{i \in I} r_i \cdot t = \bigvee_{i \in I} (r_i \cdot t)$$

for $\mathcal{V}$-relations $r_i : X \nrightarrow Y$ $(i \in I)$, $s : Y \nrightarrow Z$, and $t : W \nrightarrow X$. *Thus,* $\mathcal{V}\text{-}\mathsf{Rel}$ *is not just an ordered category, but a quantaloid* (see II.4.5 and II.4.8).

The canonical isomorphism $X \times Y \cong Y \times X$ induces a bijection between $\mathcal{V}\text{-}\mathsf{Rel}(X, Y)$ and $\mathcal{V}\text{-}\mathsf{Rel}(Y, X)$, so that for every $\mathcal{V}$-relation $r : X \nrightarrow Y$ one has the *opposite* (or *dual*) $\mathcal{V}$-relation $r^\circ : Y \nrightarrow X$ defined by

$$r^\circ(x, y) = r(y, x)$$

for all $x \in X$, $y \in Y$. This operation preserves the order on $\mathcal{V}\text{-}\mathsf{Rel}(X, Y)$:

$$r \leq r' \implies r^\circ \leq (r')^\circ\,,$$

and one has $1_X^\circ = 1_X$ as well as $r^{\circ\circ} = r$. Let us also note that the equality

$$(s \cdot r)^\circ = r^\circ \cdot s^\circ$$

holds whenever $\mathcal{V}$ is commutative.

III.1.1.1 Examples

(1) As already mentioned, a $\mathsf{2}$-relation is just an ordinary relation. For relations $r : X \nrightarrow Y$ and $s : Y \nrightarrow Z$, the previous "matrix multiplication" formula specializes to the usual relational composition:

$$x\,(s \cdot r)\,z \iff \exists y \in Y\; (x\,r\,y \;\&\; y\,s\,z)\,.$$

Therefore, the category of 2-relations is just the category of relations:

$$\mathsf{2\text{-}Rel} \cong \mathsf{Rel}\,.$$

(2) For $\mathcal{V} = \mathsf{P}_+$ (see II.1.10), a P_+-relation is a "distance function" $r : X \times Y \to \mathsf{P}_+$, and composition with $s : Y \times Z \to \mathsf{P}_+$ yields

$$s \cdot r(x,z) = \inf\{r(x,y) + s(y,z) \mid y \in Y\}$$

for all $x \in X$ and $z \in Z$. Therefore, P_+-Rel can be seen as the category of sets and *metric relations*.

(3) For $\mathsf{2}^2 = \{\bot, u, v, \top\}$ the diamond lattice of Exercise II.1.H, a $\mathsf{2}^2$-relation is a "choice relation" that chooses between the truth values u and v, taking value $\bot$ if none is selected and $\top$ if both are. Each $\mathsf{2}^2$-relation r can therefore be considered as a pair of relations (r_u, r_v), and $\mathsf{2}^2$-Rel as the category of sets and *birelations*:

$$\frac{X \times Y \xrightarrow{\;r\;} \mathsf{2}^2 \cong \mathsf{2} \times \mathsf{2}}{X \times Y \xrightarrow{\;r_u\;} \mathsf{2}\,,\;\; X \times Y \xrightarrow{\;r_v\;} \mathsf{2}\,.}$$

III.1.2 Maps in $\mathcal{V}$-Rel

There is a functor from Set to $\mathcal{V}$-Rel that interprets the graph of a Set-map $f : X \to Y$ as the $\mathcal{V}$-relation $f_\circ : X \nrightarrow Y$ given by

$$f_\circ(x,y) = \begin{cases} k & \text{if } f(x) = y, \\ \bot & \text{otherwise.} \end{cases}$$

The functor

$$(-)_\circ : \mathsf{Set} \to \mathcal{V}\text{-}\mathsf{Rel}$$

is faithful if and only if $\bot < k$ in $\mathcal{V}$. Therefore, *from now on, the quantale $\mathcal{V}$ is assumed to be non-trivial*, so that $\mathcal{V}$ is not reduced to a singleton (see Exercise III.1.A). To keep notation simple, we usually write $f : X \to Y$ instead of $f_\circ : X \nrightarrow Y$ to designate a $\mathcal{V}$-relation induced by a map.

The formula for $\mathcal{V}$-relational composition becomes considerably easier if one of the $\mathcal{V}$-relations comes from a Set-map:

$$s \cdot f(x,z) = s(f(x),z)\,, \qquad h^\circ \cdot s(y,w) = s(y,h(w))\,,$$
$$g \cdot r(x,z) = \bigvee\nolimits_{y \in g^{-1}(z)} r(x,y)\,, \qquad t \cdot f^\circ(y,z) = \bigvee\nolimits_{x \in f^{-1}(y)} t(x,z)\,,$$

for all maps $f : X \to Y$, $g : Y \to Z$, $h : W \to Z$, $\mathcal{V}$-relations $r : X \nrightarrow Y$, $s : Y \nrightarrow Z$, $t : X \nrightarrow Z$, and elements $w \in W$, $x \in X$, $y \in Y$, $z \in Z$. In particular, one should keep in mind that the pointwise expression of a relation of the form $h^\circ \cdot s \cdot f$ is

$$h^\circ \cdot s \cdot f(x,w) = s(f(x),h(w))\,,$$

as this formula will be used systematically from now on. We note that, without any commutativity assumption on $\mathcal{V}$, composition of $\mathcal{V}$-relations with Set-maps is also compatible with the involution $(-)^\circ$:

$$(s \cdot f)^\circ = f^\circ \cdot s^\circ \qquad \text{and} \qquad (g \cdot r)^\circ = r^\circ \cdot g^\circ .$$

The formulas for "relation-with-map composition" imply at once that every Set-map $f : X \to Y$ satisfies the inequalities

$$1_X \leq f^\circ \cdot f_\circ \qquad \text{and} \qquad f_\circ \cdot f^\circ \leq 1_Y$$

in $\mathcal{V}$-Rel, so that $f_\circ$ is a map in the sense of II.4.7, thus providing further justification for the identification $f_\circ = f$. Given Set-maps $f : X \to Y$ and $g : Y \to Z$, we therefore obtain the *adjunction rules*:

$$g \cdot r \leq t \iff r \leq g^\circ \cdot t \qquad \text{and} \qquad t \cdot f^\circ \leq s \iff t \leq s \cdot f \tag{III.1.2.i}$$

for $\mathcal{V}$-relations $r : X \nrightarrow Y$, $s : Y \nrightarrow Z$, and $t : X \nrightarrow Z$ (see Proposition II.4.7.1).

For $\mathcal{V} = 2$, every map (i.e. every left-adjoint morphism) in $\mathcal{V}$-Rel $=$ Rel is given by a Set-map: whenever $r \dashv s : Y \nrightarrow X$, one has $r = f_\circ$, $s = f^\circ$ for a uniquely determined map $f : X \to Y$ (see II.4.7). We briefly discuss to what extent this fact holds more generally and call the quantale $\mathcal{V}$ *integral* if $k = \top$, i.e. if the top element is the neutral element of the tensor. In an integral quantale $\mathcal{V}$, one has $u \otimes v \leq u \wedge v$ for all $u, v \in \mathcal{V}$ (since $u \otimes v \leq u \otimes \top = u \otimes k = u$ and likewise $u \otimes v \leq v$). We say that $\mathcal{V}$ is *lean* if

$$(u \vee v = \top \text{ and } u \otimes v = \bot) \implies (u = \top \text{ or } v = \top)$$

for all $u, v \in \mathcal{V}$. In an integral and lean quantale, $\top$ and $\bot$ are the only *complemented* elements (i.e. elements u for which there is v with $u \vee v = \top, u \wedge v = \bot$). The quantales 2, P_+, $\mathsf{P}_{\max}$ are integral and lean, 3 and $\mathsf{P}_\times$ are lean but not integral, and 2^2 is integral but not lean.

III.1.2.1 Proposition *For an integral quantale $\mathcal{V}$, all left-adjoint $\mathcal{V}$-relations are* Set-*maps if and only if $\mathcal{V}$ is lean.*

Proof Let $\mathcal{V}$ be integral and lean, and assume $r \dashv s : Y \nrightarrow X$ in $\mathcal{V}$-Rel. If $X = \varnothing$, then $r : \varnothing \hookrightarrow Y$ is the inclusion map. Hence, one can consider $x \in X$. Since

$$\bot < k = (s \cdot r)(x, x) = \bigvee\nolimits_{y' \in Y} r(x, y') \otimes s(y', x) ,$$

there is some $y \in Y$ with $u := r(x, y) \otimes s(y, x) > \bot$, and we can write $k = \top = u \vee v$, where $v := \bigvee_{y' \neq y} r(x, y') \otimes s(y', x)$. From $r \cdot s(y, y') \leq \bot$ one obtains $s(y, x) \otimes r(x, y') = \bot$ for all $y' \neq y$, and therefore

$$u \otimes v = \bigvee\nolimits_{y' \neq y} r(x, y) \otimes (s(y, x) \otimes r(x, y')) \otimes s(y', x) = \bot .$$

Consequently, $u = r(x, y) \otimes s(y, x) = \top$ (since $v = \top = k$ would force $u = u \otimes v = \bot$), so that $r(x, y) = \top = s(y, x)$ because $\top = r(x, y) \otimes s(y, x) \leq r(x, y) \wedge s(y, x)$. Since $u = \top$ and $v = \bot$, we have shown that for every $x \in X$ there is precisely one $y := f(x)$ in Y with $r(x, y) \otimes s(y, x) > \bot$, and then $r(x, y) = s(y, x) = k$. Consequently, $f \leq r$ and $f^\circ \leq s$, which actually forces $r = f$:

$$r = r \cdot 1_X \leq r \cdot f^\circ \cdot f \leq r \cdot s \cdot f \leq 1_Y \cdot f = f \,.$$

Conversely, for maps in $\mathcal{V}$-Rel to be Set-maps we must show that $\mathcal{V}$ is necessarily lean. If $u \vee v = \top$ and $u \otimes v = \bot$, with $X := \{u, v\}$ we may define $r : 1 = \{\star\} \nrightarrow X$ by $r(\star, x) = x$ and claim $r \dashv r^\circ$. Indeed, since $k = \top$ and $(r \cdot r^\circ)(u, v) = u \otimes v = \bot$, one has $r \cdot r^\circ \leq 1_Y$; for $1_1 \leq r^\circ \cdot r$ we first observe

$$u = u \otimes k = u \otimes (u \vee v) = (u \otimes u) \vee (u \otimes v) = u \otimes u$$

and likewise $v = v \otimes v$, which implies

$$(r^\circ \cdot r)(\star, \star) = (u \otimes u) \vee (v \otimes v) = u \vee v = k \,.$$

By hypothesis, r is then given by the maps $\star \mapsto u$ or $\star \mapsto v$, which means $u = k$ or $v = k$. □

The adjunction $f \dashv f^\circ$ detects injectivity and surjectivity of the Set-map f by the equivalent conditions $f^\circ \cdot f = 1_X$ and $f \cdot f^\circ = 1_Y$, respectively. In fact, one can easily prove a more general fact:

III.1.2.2 Proposition

(1) *The* Set-*maps* $f_i : X \to Y_i$ $(i \in I \neq \varnothing)$ *form a mono-source if and only if* $\bigwedge_{i \in I} f_i^\circ \cdot f_i = 1_X$.

(2) *The* Set-*maps* $g : X_i \to Y$ $(i \in I)$ *form an epi-sink if and only if* $\bigvee_{i \in I} g_i \cdot g_i^\circ = 1_Y$.

Proof The statements follow from

$$(\textstyle\bigwedge_{i \in I} f_i^\circ \cdot f_i)(x, x') = \bigwedge_{i \in I} 1_Y(f_i(x), f_i(x')) = k \iff \forall i \in I\ (f_i(x) = f_i(x'))$$

for all $x, x' \in X$, and

$$(\textstyle\bigvee_{i \in I} g_i \cdot g_i^\circ)(x, x') = \bigvee_{i \in I} \bigvee_{x \in g_i^{-1} y} k = k \iff \bigcup_{i \in I} g_i^{-1} y \neq \varnothing$$

for all $y \in Y$. □

III.1.2.3 Remarks

(1) The empty source with domain X is a mono-source in Set if and only if $|X| \leq 1$. Hence, the assertion of Proposition III.1.2.2(1) remains valid also in the case $I = \varnothing$ precisely when either $X = \varnothing$, or when $|X| = 1$ and $\top = k$. Consequently, for an integral quantale $\mathcal{V}$, the restriction $I \neq \varnothing$ of Proposition III.1.2.2(1) may be dropped.

(2) The functor $(-)_\circ : \mathsf{Set} \to \mathcal{V}\text{-}\mathsf{Rel}$ has a right adjoint $\mathcal{V}\text{-}\mathsf{Rel} \to \mathsf{Set}$ that sends a set X to the set $\mathcal{V}^X$ of all maps $\phi : X \to \mathcal{V}$, and a $\mathcal{V}$-relation $r : X \nrightarrow Y$ to the map $r^{\mathbb{P}_\mathcal{V}} : \mathcal{V}^X \to \mathcal{V}^Y$ defined by

$$r^{\mathbb{P}_\mathcal{V}}(\phi)(y) = \bigvee_{x\in X} \phi(x) \otimes r(x, y)$$

for all $\phi \in \mathcal{V}^X$ and $y \in Y$. The monad $\mathbb{P}_\mathcal{V}$ induced by this adjunction is the $\mathcal{V}$-*powerset monad* – whose Kleisli category is $\mathcal{V}\text{-}\mathsf{Rel}$ (see Exercise III.1.D).

III.1.3 $\mathcal{V}$-categories, $\mathcal{V}$-functors, and $\mathcal{V}$-modules

We introduced $\mathcal{V}$-categories and $\mathcal{V}$-functors for a monoidal category $\mathcal{V}$ in II.4.10. Here we recall the definition in the highly simplified case where $\mathcal{V} = (V, \otimes, k)$ is a quantale.

A $\mathcal{V}$-relation $a : X \nrightarrow X$ is *transitive* if $a \cdot a \le a$ and *reflexive* if $1_X \le a$. A $\mathcal{V}$-*category* (X, a) is a set X with a transitive and reflexive $\mathcal{V}$-relation a. A $\mathcal{V}$-*functor* $f : (X, a) \to (Y, b)$ of $\mathcal{V}$-categories is given by a map $f : X \to Y$ with $f \cdot a \le b \cdot f$ or, equivalently, $a \le f^\circ \cdot b \cdot f$. Hence, in pointwise notation, the characteristic conditions for a $\mathcal{V}$-category read as

$$a(x, y) \otimes a(y, z) \le a(x, z) \qquad \text{and} \qquad k \le a(x, x)\ ,$$

and for a $\mathcal{V}$-functor as

$$a(x, y) \le b(f(x), f(y))$$

for all $x, y, z \in X$. Since identity maps and composites of $\mathcal{V}$-functors are $\mathcal{V}$-functors, $\mathcal{V}$-categories and $\mathcal{V}$-functors form a category

$$\mathcal{V}\text{-}\mathsf{Cat}\,.$$

III.1.3.1 Examples

(1) For $\mathcal{V} = \mathsf{2} = \{\text{true}, \text{false}\}$, writing $x \le y$ for $a(x, y) = \text{true}$, the transitivity and reflexivity conditions read as expected:

$$(x \le y \;\&\; y \le z \implies x \le z) \qquad \text{and} \qquad x \le x$$

for all $x, y, z \in X$. Thus, a $\mathsf{2}$-category $(X, \le)$ is just an ordered set. (Recall from II.1.3 that we do not require an order to be antisymmetric.) A $\mathsf{2}$-functor $f : (X, \le) \to (Y, \le)$ is a map $f : X \to Y$ with

$$x \le y \implies f(x) \le f(y)$$

(for all $x, y \in X$), so the category of $\mathsf{2}$-categories is just the category of ordered sets:

$$\mathsf{2}\text{-}\mathsf{Cat} = \mathsf{Ord}\,.$$

(2) For $\mathcal{V} = \mathsf{P}_+$ (see Example II.1.10.1(3)), a transitive and reflexive P_+-relation is equivalently described as a *metric* on X, i.e. a map $a : X \times X \to \mathsf{P}_+$ such that

$$a(x, y) + a(y, z) \geq a(x, z) \qquad \text{and} \qquad 0 = a(x, x) ,$$

for all $x, y, z \in X$, and we say that X is a *metric space*. Whenever we require any of the other traditionally assumed conditions, namely *symmetry* ($a(x, y) = a(y, x)$), *separation* ($a(x, y) = 0 = a(y, x) \implies x = y$), and *finiteness* ($a(x, y) < \infty$), we will say so explicitly, thus calling X a *symmetric*, *separated*, or *finitary metric space*, respectively. A P_+-functor $f : (X, a) \to (X, b)$ is a map $f : X \to Y$ that is *non-expansive*:

$$a(x, y) \geq b(f(x), f(y))$$

(for all $x, y \in X$). Hence, the category of P_+-categories is equivalently described as the category Met of metric spaces in the generalized sense as specified above:

$$\mathsf{P}_+\text{-}\mathsf{Cat} = \mathsf{Met} .$$

It contains the full subcategories $\mathsf{Met}_{\mathrm{sym}}$ and $\mathsf{Met}_{\mathrm{sep}}$ of symmetric and separated metric spaces, respectively.

We sketch some general procedures for creating $\mathcal{V}$-categories, starting with $\mathcal{V}$ itself: the $\mathcal{V}$-valued binary operation $\multimap$ on $\mathcal{V}$ with

$$v \otimes t \leq w \iff t \leq v \multimap w$$

defines a $\mathcal{V}$-relation, and one obtains:

III.1.3.2 Proposition *The quantale $\mathcal{V}$ endowed with the $\mathcal{V}$-relation $\multimap$ is a $\mathcal{V}$-category.*

Proof Since $v \otimes k \leq v$, one has $k \leq v \multimap v$, and from $v \multimap w \leq v \multimap w$ one obtains $v \otimes (v \multimap w) \leq w$. Consequently,

$$v \otimes (v \multimap w) \otimes (w \multimap z) \leq w \otimes (w \multimap z) \leq z ,$$

which yields transitivity: $(v \multimap w) \otimes (w \multimap z) \leq v \multimap z$. □

Note that the more general situation of this Proposition, where $\mathcal{V}$ is not just a quantale but a monoidal category, was sketched in Exercise II.4.I.

For $\mathcal{V} = \mathsf{2}$, $\multimap$ returns the order of $\mathsf{2}$ (false < true), while for $\mathcal{V} = \mathsf{P}_+$ $\multimap$ is the truncated difference: $v \multimap w = w - v$ if $v \leq w < \infty$, $v \multimap w = 0$ if $w \leq v$, and $v \multimap \infty = \infty$ if $v < \infty$ (see II.1.10).

Now let $X = (X, a)$ and $Y = (Y, b)$ be $\mathcal{V}$-categories. One defines a $\mathcal{V}$-relation $[-, -]$ on $\mathcal{V}\text{-}\mathsf{Cat}(X, Y) = \{f : X \to Y \mid f \text{ is a } \mathcal{V}\text{-functor}\}$ by

$$[f, g] = \textstyle\bigwedge_{x \in X} b(f(x), g(x)) ,$$

and a $\mathcal{V}$-relation $a \otimes b$ on $X \times Y$ by

$$(a \otimes b)((x, y), (x', y')) = a(x, x') \otimes b(y, y')$$

for all $x, x' \in X$, $y, y' \in Y$.

III.1.3.3 Proposition *Let $X = (X, a)$, $Y = (Y, b)$ be $\mathcal{V}$-categories.*

(1) *$[X, Y] = (\mathcal{V}\text{-}\mathsf{Cat}(X, Y), [-, -])$ is a $\mathcal{V}$-category.*
(2) *If $\mathcal{V}$ is commutative, then $X \otimes Y := (X \times Y, a \otimes b)$ is a $\mathcal{V}$-category.*

Proof Straightforward verifications. □

III.1.3.4 Example For $\mathcal{V} = \mathsf{P}_+$, $[f, g] = \sup_{x\in X} b(f(x), g(x))$ is the usual "sup-metric" on the function space $[X, Y]$, and $(a \otimes b)((x, y), (x', y')) = a(x, x') + b(y, y')$ endows $X \times Y$ with the usual "+-metric."

III.1.3.5 Remark The structure on $[X, Y]$ may be rewritten as

$$[f, g] = \bigwedge_{x,x'\in X} a(x, x') \multimap b(f(x), g(x')) \,. \tag{III.1.3.i}$$

Indeed, since $g : X \to Y$ is a $\mathcal{V}$-functor, for all $x, x' \in X$, one has

$$b(f(x), g(x)) \otimes a(x, x') \le b(f(x), g(x)) \otimes b(g(x), g(x')) \le b(f(x), g(x')),$$

and then $b(f(x), g(x)) \le a(x, x') \multimap b(f(x), g(x'))$, which proves "$\le$" of (III.1.3.i). "$\ge$" follows from

$$a(x, x) \multimap b(f(x), g(x)) \le k \multimap b(f(x), g(x)) \le b(f(x), g(x))$$

for all $x \in X$.

III.1.3.6 Theorem *For a commutative quantale $\mathcal{V}$, $\mathcal{V}$-Cat is a symmetric monoidal closed category.*

Proof We prove that

$$\frac{Z \xrightarrow{\phi} [X, Y]}{Z \otimes X \xrightarrow{\tilde{\phi}} Y}$$

with $\phi(z)(x) = \tilde{\phi}(z, x)$ for all $z \in Z$, $x \in X$ establishes a bijective correspondence of $\mathcal{V}$-functors ϕ and $\tilde{\phi}$ when $X = (X, a)$, $Y = (Y, b)$, $Z = (Z, c)$ are $\mathcal{V}$-categories. In fact, by Remark III.1.3.5, $\mathcal{V}$-functoriality of ϕ means equivalently

$$c(z, z') \le a(x, x') \multimap b(\phi(z)(x), \phi(z)(x'))$$

for all $z, z' \in Z$, $x, x' \in X$, which may be equivalently rewritten as

$$(c \otimes a)((z, x), (z', x')) = c(z, z') \otimes a(x, x') \le b(\tilde{\phi}(z, x), \tilde{\phi}(z, x')) \,,$$

and this last inequality describes the $\mathcal{V}$-functoriality of $\tilde{\phi}$. We note that this inequality also entails $\mathcal{V}$-functoriality of $\phi(z) : X \to Y$ for all $z \in Z$, since

$$a(x,x') = k \otimes a(x,x') \leq c(z,z) \otimes a(x,x') \leq b(\phi(z)(x), \phi(z)(x'))\,.$$

The correspondence is obviously bijective and natural in Z, which shows monoidal closure, and symmetry ($X \otimes Y \cong Y \otimes X$) holds trivially. □

III.1.3.7 Examples For $\mathcal{V} = 2$, Theorem III.1.3.6 confirms that Ord is Cartesian closed and, for $\mathcal{V} = \mathsf{P}_+$, that Met is monoidal closed. But note that the tensor product $X \otimes Y$ in Met must not be confused with the product $X \times Y$: while the structure of $X \otimes Y$ is given by the "+-metric," $X \times Y$ carries the "max-metric"; see Exercise III.1.G.

The notion of module for ordered sets (see II.1.4) extends naturally from the case $\mathcal{V} = 2$ to the arbitrary case: for $\mathcal{V}$-categories (X,a), (Y,b) one calls a $\mathcal{V}$-relation $r : X \nrightarrow Y$ a *$\mathcal{V}$-module* (also *$\mathcal{V}$-bimodule*, *$\mathcal{V}$-profunctor*, or *$\mathcal{V}$-distributor*) if

$$r \cdot a \leq r \qquad \text{and} \qquad b \cdot r \leq r\,.$$

Since the reversed inequalities always hold, these are in fact equalities:

$$r \cdot a = r \qquad \text{and} \qquad b \cdot r = r\,.$$

We write

$$r : (X,a) \nrightarrow (Y,b)$$

if the $\mathcal{V}$-relation r is a $\mathcal{V}$-module. The module inequalities are stable under $\mathcal{V}$-relational composition, and

$$a : (X,a) \nrightarrow (X,a)$$

serves as an identity morphism in the category

$$\mathcal{V}\text{-Mod}$$

whose objects are $\mathcal{V}$-categories and morphisms are $\mathcal{V}$-modules. This category is *ordered*, with the order inherited from $\mathcal{V}$-Rel; in fact, $\mathcal{V}$-Mod *is a quantaloid*, with suprema in its hom-sets formed as in $\mathcal{V}$-Rel.

There is now a structured version of the functors

$$\mathsf{Set} \xrightarrow{(-)_\circ} \mathcal{V}\text{-}\mathsf{Rel} \xleftarrow{(-)^\circ} \mathsf{Set}^{\mathrm{op}}$$

as follows. For a $\mathcal{V}$-functor $f : (X,a) \to (Y,b)$, one defines $\mathcal{V}$-modules

$$f_* : (X,a) \nrightarrow (Y,b) \qquad \text{and} \qquad f^* : (Y,b) \nrightarrow (X,a)$$

by

$$f_* := b \cdot f \qquad \text{and} \qquad f^* := f^\circ \cdot b\,,$$

i.e.

$$f_*(x,y) = b(f(x),y) \qquad \text{and} \qquad f^*(y,x) = b(y,f(x))$$

for all $x \in X$, $y \in Y$. One easily verifies the $\mathcal{V}$-module conditions and functoriality:

$$\mathcal{V}\text{-}\mathsf{Cat} \xrightarrow{(-)_*} \mathcal{V}\text{-}\mathsf{Mod} \xleftarrow{(-)^*} (\mathcal{V}\text{-}\mathsf{Cat})^{\mathrm{op}} .$$

In particular,

$$1_{(X,a)} = a = (1_X)^* = (1_X)_* ,$$

so that we can simply write 1_X^* for the identity $\mathcal{V}$-module on (X, a). Moreover, there is also a structured version of the adjunction $f_\circ \dashv f^\circ$:

III.1.3.8 Proposition *For a $\mathcal{V}$-functor $f : (X, a) \to (Y, b)$, one has $f_* \dashv f^*$ in* $\mathcal{V}$-Mod.

Proof One writes

$$f_* \cdot f^* = b \cdot f \cdot f^\circ \cdot b \leq b \cdot 1_Y \cdot b \leq b = 1_Y^* ,$$

and

$$1_X^* = a \leq a \cdot 1_X \cdot a \leq a \cdot f^\circ \cdot f \cdot a \leq f^\circ \cdot b \cdot b \cdot f = f^* \cdot f_* ,$$

which proves the claim. □

Let us finally observe that *if $\mathcal{V}$ is commutative*, then, for every $\mathcal{V}$-category $X = (X, a)$, the pair

$$X^{\mathrm{op}} := (X, a^\circ)$$

is also a $\mathcal{V}$-category, called the *dual* of X. For every $\mathcal{V}$-functor $f : (X, a) \to (Y, b)$ one has a $\mathcal{V}$-functor $f^{\mathrm{op}} : X^{\mathrm{op}} \to Y^{\mathrm{op}}$ given by f, so there is a functor

$$(-)^{\mathrm{op}} : \mathcal{V}\text{-}\mathsf{Cat} \to \mathcal{V}\text{-}\mathsf{Cat} .$$

Furthermore,

$$(f^{\mathrm{op}})_* = (f^*)^\circ \qquad \text{and} \qquad (f^{\mathrm{op}})^* = (f_*)^\circ .$$

III.1.4 Lax extensions of functors

Section III.1.2 shows that the category $\mathcal{V}$-Rel of $\mathcal{V}$-relations can be seen as an extension of Set. For a given monad $\mathbb{T} = (T, m, e)$ on Set, we now consider extensions of $\mathbb{T}$ to $\mathcal{V}$-Rel. For this, we first concentrate on the underlying Set-functor T; the natural transformations e and m will be considered in Section III.1.5.

III.1.4.1 Definition For a quantale $\mathcal{V}$ and a functor $T : \mathsf{Set} \to \mathsf{Set}$, a *lax extension* $\hat{T} : \mathcal{V}\text{-}\mathsf{Rel} \to \mathcal{V}\text{-}\mathsf{Rel}$ of T to $\mathcal{V}\text{-}\mathsf{Rel}$ is given by functions

$$\hat{T}_{X,Y} : \mathcal{V}\text{-}\mathsf{Rel}(X, Y) \to \mathcal{V}\text{-}\mathsf{Rel}(TX, TY)$$

for all sets X, Y (with $\hat{T}_{X,Y}$ simply written as $\hat{T}$), such that

(1) $r \le r' \implies \hat{T}r \le \hat{T}r'$,
(2) $\hat{T}s \cdot \hat{T}r \le \hat{T}(s \cdot r)$,
(3) $Tf \le \hat{T}f$ and $(Tf)^\circ \le \hat{T}(f^\circ)$,

for all sets X, Y, Z, $\mathcal{V}$-relations $r, r' : X \nrightarrow Y$, $s : Y \nrightarrow Z$, and maps $f : X \to Y$. By setting $\hat{T}X = TX$ for all sets X, and observing that condition (3) yields $1_{TX} \le \hat{T}1_X$, one can define a lax extension of a Set-functor T equivalently as a lax functor $\hat{T} : \mathcal{V}\text{-}\mathsf{Rel} \to \mathcal{V}\text{-}\mathsf{Rel}$ (see II.4.6) that agrees with T on objects of $\mathcal{V}\text{-}\mathsf{Rel}$ and satisfies the extension condition (3).

III.1.4.2 Examples

(1) The identity functor on Set has a lax extension given by the identity functor on $\mathcal{V}\text{-}\mathsf{Rel}$.
(2) For $\mathcal{V} = 2$, the covariant powerset functor $P : \mathsf{Set} \to \mathsf{Set}$ has lax extensions $\check{P}, \hat{P} : \mathsf{Rel} \to \mathsf{Rel}$ given by

$$A\,(\check{P}r)\,B \iff A \subseteq r^\circ(B) \iff \forall x \in A\, \exists y \in B\ (x\ r\ y)\,,$$
$$A\,(\hat{P}r)\,B \iff B \subseteq r(A) \iff \forall y \in B\, \exists x \in A\ (x\ r\ y)\,,$$

for every relation $r : X \nrightarrow Y$, and all $A \subseteq X$, $B \subseteq Y$.
(3) Every functor T on Set admits a largest lax extension to $\mathcal{V}\text{-}\mathsf{Rel}$ given by

$$T^\top r : TX \times TY \to V\,, \quad (\mathfrak{x}, \mathfrak{y}) \mapsto \top$$

for all $\mathcal{V}$-relations $r : X \nrightarrow Y$.

Although a lax extension $\hat{T}$ preserves composition of $\mathcal{V}$-relations only up to inequality, it operates more strictly on composites of $\mathcal{V}$-relations with Set-maps, as the Corollary to the following Proposition shows.

III.1.4.3 Proposition *Given functions* $\hat{T}_{X,Y} : \mathcal{V}\text{-}\mathsf{Rel}(X, Y) \to \mathcal{V}\text{-}\mathsf{Rel}(TX, TY)$ *that satisfy conditions* (1) *and* (2) *of* Definition III.1.4.1, *the following are equivalent:*

(i) $Tf \le \hat{T}f$ *and* $(Tf)^\circ \le \hat{T}(f^\circ)$ *for all* $f : X \to Y$ *(this is condition* III.1.4.1(3)*);*
(ii) $Tf \le \hat{T}f$ *and* $\hat{T}(s \cdot f) = \hat{T}s \cdot Tf$ *for all* $f : X \to Y$ *and* $s : Y \nrightarrow Z$*;*
(iii) $(Tf)^\circ \le \hat{T}(f^\circ)$ *and* $\hat{T}(f^\circ \cdot r) = (Tf)^\circ \cdot \hat{T}r$ *for all* $f : X \to Y$ *and* $r : Z \nrightarrow Y$.

The next condition is a consequence of any of the previous ones, and is equivalent to each of them if $\hat{T}$ also satisfies $1_{TX} \leq \hat{T}1_X$:

(iv) $\hat{T}(s \cdot f) = \hat{T}s \cdot Tf$ *and* $\hat{T}(f^\circ \cdot r) = (Tf)^\circ \cdot \hat{T}r$ *for all* $f : X \to Y$ *and* $r : Z \nrightarrow Y$, $s : Y \nrightarrow Z$.

Proof For (i) $\Longrightarrow$ (ii), we observe

$$\begin{aligned}\hat{T}s \cdot Tf &\leq \hat{T}s \cdot \hat{T}f \leq \hat{T}(s \cdot f) \leq \hat{T}(s \cdot f) \cdot (Tf)^\circ \cdot Tf \\ &\leq \hat{T}(s \cdot f) \cdot \hat{T}(f^\circ) \cdot Tf \leq \hat{T}(s \cdot f \cdot f^\circ) \cdot Tf \leq \hat{T}s \cdot Tf\,,\end{aligned}$$

so these inequalities are all equalities, and (i) $\Longrightarrow$ (iii) is shown in the same way.

For (ii) $\Longrightarrow$ (iv), we see that $1_{TX} \leq \hat{T}1_X$ follows immediately from $Tf \leq \hat{T}f$; moreover, one observes

$$Tf \cdot \hat{T}(f^\circ \cdot r) \leq \hat{T}f \cdot \hat{T}(f^\circ \cdot r) \leq \hat{T}(f \cdot f^\circ \cdot r) \leq \hat{T}r\,,$$

so $\hat{T}(f^\circ \cdot r) \leq (Tf)^\circ \cdot \hat{T}r$ follows by the first adjunction rule in (III.1.2.i); for the other inequality, we apply $\hat{T}$ to $1_X \leq f^\circ \cdot f$ to obtain $1_X \leq \hat{T}(f^\circ) \cdot Tf$ thanks to the hypothesis, and $(Tf)^\circ \leq \hat{T}(f^\circ)$ by the second adjunction rule, so that

$$(Tf)^\circ \cdot \hat{T}r \leq \hat{T}(f^\circ) \cdot \hat{T}r \leq \hat{T}(f^\circ \cdot r)\,.$$

The implication (iii) $\Longrightarrow$ (iv) is proved similarly. Finally, for (iv) $\Longrightarrow$ (i), we observe that $1_X \leq f^\circ \cdot f$ yields

$$1_{TX} \leq \hat{T}(1_X) \leq \hat{T}(f^\circ \cdot f) = \hat{T}(f^\circ) \cdot Tf\,,$$

which implies $(Tf)^\circ \leq \hat{T}(f^\circ)$, and $Tf \leq \hat{T}f$ follows in a similar way. □

III.1.4.4 Corollary *For a lax extension* $\hat{T} : \mathcal{V}\text{-}\mathsf{Rel} \to \mathcal{V}\text{-}\mathsf{Rel}$ *of a* Set*-functor* T *one has*

$$\hat{T}(s \cdot f) = \hat{T}s \cdot \hat{T}f = \hat{T}s \cdot Tf\,, \quad \hat{T}(f^\circ \cdot r) = \hat{T}(f^\circ) \cdot \hat{T}r = (Tf)^\circ \cdot \hat{T}r$$

for all maps $f : X \to Y$ *and* $\mathcal{V}$*-relations* $r : Z \nrightarrow Y$, $s : Y \nrightarrow Z$.

Proof This follows immediately from Proposition III.1.4.3 since a lax extension is a lax functor, and $\hat{T}s \cdot \hat{T}f \leq \hat{T}(s \cdot f) = \hat{T}s \cdot Tf \leq \hat{T}s \cdot \hat{T}f$; the other equalities are obtained in the same way. □

A lax extension $\hat{T}$ of T is *flat* if

$$\hat{T}1_X = T1_X = 1_{TX}\,,$$

i.e. if both diagrams

$$\begin{array}{ccc} \mathcal{V}\text{-}\mathsf{Rel} & \xrightarrow{\hat{T}} & \mathcal{V}\text{-}\mathsf{Rel} \\ {\scriptstyle (-)_\circ}\uparrow & & \uparrow{\scriptstyle (-)_\circ} \\ \mathsf{Set} & \xrightarrow{T} & \mathsf{Set} \end{array} \qquad\qquad \begin{array}{ccc} \mathcal{V}\text{-}\mathsf{Rel} & \xrightarrow{\hat{T}} & \mathcal{V}\text{-}\mathsf{Rel} \\ {\scriptstyle (-)^\circ}\uparrow & & \uparrow{\scriptstyle (-)^\circ} \\ \mathsf{Set}^{\mathrm{op}} & \xrightarrow{T^{\mathrm{op}}} & \mathsf{Set}^{\mathrm{op}} \end{array}$$

commute. Indeed, if $\hat{T}$ is flat, by Proposition III.1.4.3 one obtains

$$\hat{T}f = \hat{T}1_Y \cdot Tf = Tf \qquad \text{and} \qquad \hat{T}(f^\circ) = (Tf)^\circ \cdot \hat{T}1_X = (Tf)^\circ$$

for all $f : X \to Y$ in Set. Note that, of all the Examples III.1.4.2, only the given lax extension of the identity functor is flat.

III.1.5 Lax extensions of monads

Let us now turn our attention to the natural transformations e and m that we wish to extend from Set to $\mathcal{V}$-Rel together with the functor T.

III.1.5.1 Definition A triple $\hat{\mathbb{T}} = (\hat{T}, m, e)$ is a *lax extension of the monad* $\mathbb{T} = (T, m, e)$ if $\hat{T}$ is a lax extension of T which makes both $m : \hat{T}\hat{T} \to \hat{T}$ and $e : 1_{\mathcal{V}\text{-Rel}} \to \hat{T}$ oplax (see II.4.6), i.e.

(4) $m_Y \cdot \hat{T}\hat{T}r \leq \hat{T}r \cdot m_X$,
(5) $e_Y \cdot r \leq \hat{T}r \cdot e_X$,

for all $\mathcal{V}$-relations $r : X \nrightarrow Y$.

By using both adjunction rules (III.1.2.i) for the maps m_X and e_X, we obtain the following equivalent formulations of (4) and (5):

(4°) $\hat{T}\hat{T}r \cdot m_X^\circ \leq m_Y^\circ \cdot \hat{T}r$,
(5°) $r \cdot e_X^\circ \leq e_Y^\circ \cdot \hat{T}r$.

Similarly, these conditions are equivalent to

($4'$) $\hat{T}\hat{T}r \leq m_Y^\circ \cdot \hat{T}r \cdot m_X$,
($5'$) $r \leq e_Y^\circ \cdot \hat{T}r \cdot e_X$.

These inequalities then yield the following pointwise expressions:

(4^*) $\hat{T}\hat{T}r(\mathfrak{X}, \mathfrak{Y}) \leq \hat{T}r(m_X(\mathfrak{X}), m_Y(\mathfrak{Y}))$,
(5^*) $r(x, y) \leq \hat{T}r(e_X(x), e_Y(y))$,

for all $x \in X$, $y \in Y$, $\mathfrak{X} \in TTX$, $\mathfrak{Y} \in TTY$, and $\mathcal{V}$-relations $r : X \nrightarrow Y$.

One says that a lax extension $\hat{\mathbb{T}} = (\hat{T}, m, e)$ of the monad $\mathbb{T}$ is *flat* if the lax extension $\hat{T}$ of the functor T is flat.

The construction of lax extensions in a general setting can be rather technical, and is postponed until Chapter IV. In this chapter, however, we will describe especially important extensions in Sections III.1.10 and III.2.4. Here, we restrict ourselves to reconsidering the easy examples of III.1.4.2. These demonstrate in particular that a monad on Set may generally admit more than one lax extension to $\mathcal{V}$-Rel.

III.1.5.2 Examples

(1) The identity monad $\mathbb{I}$ on Set can be extended to the identity monad $\mathbb{I}$ on $\mathcal{V}$-Rel, and, unless otherwise stated, this is the flat lax extension that will be used from now on for this monad.
(2) The lax extensions $\check{P}$, $\hat{P}$ of Examples III.1.4.2 provide non-flat lax extensions $\check{\mathbb{P}}$, $\hat{\mathbb{P}}$ of the powerset monad $\mathbb{P}$ (see Example II.3.1.1(3)) to Rel.
(3) Every monad $\mathbb{T}$ on Set admits a largest lax extension $\mathbb{T}^{\top}$ to $\mathcal{V}$-Rel. It fails to be flat.

III.1.6 $(\mathbb{T}, \mathcal{V})$-categories and $(\mathbb{T}, \mathcal{V})$-functors

Let $\mathcal{V}$ be a quantale and let $\hat{\mathbb{T}} = (\hat{T}, m, e)$ be a lax extension to $\mathcal{V}$-Rel of a monad $\mathbb{T} = (T, m, e)$ on Set. A $(\mathbb{T}, \mathcal{V})$-relation $a : TX \nrightarrow X$ is *transitive* if it satisfies

$$a \cdot \hat{T}a \cdot m_X^{\circ} \leq a \qquad \text{or equivalently} \qquad a \cdot \hat{T}a \leq a \cdot m_X$$

by adjunction (see Section III.1.2). In pointwise notation, this transitivity condition becomes

$$\hat{T}a(\mathfrak{X}, \mathfrak{y}) \otimes a(\mathfrak{y}, z) \leq a(m_X(\mathfrak{X}), z)$$

for all $\mathfrak{X} \in TTX$, $\mathfrak{y} \in TX$, and $z \in X$. A $(\mathbb{T}, \mathcal{V})$-relation $a : TX \nrightarrow X$ is *reflexive* if it satisfies

$$e_X^{\circ} \leq a \qquad \text{or equivalently} \qquad 1_X \leq a \cdot e_X\,.$$

In pointwise notation, $a : TX \nrightarrow X$ is reflexive if and only if

$$k \leq a(e_X(x), x)$$

holds for all $x \in X$.

III.1.6.1 Definition A $(\mathbb{T}, \mathcal{V})$-*category*, depending on context also referred to as a *lax algebra*, a $(\mathbb{T}, \mathcal{V})$-*algebra*, or a $(\mathbb{T}, \mathcal{V})$-*space*, is a pair (X, a) consisting of a set X and a transitive and reflexive $(\mathbb{T}, \mathcal{V})$-relation $a : TX \nrightarrow X$; i.e. it is a set X with a $\mathcal{V}$-relation $a : TX \nrightarrow X$ satisfying the two laws for an Eilenberg–Moore algebra laxly:

$$\begin{array}{ccc} TTX & \xrightarrow{\hat{T}a} & TX \\ {\scriptstyle m_X}\downarrow & \geq & \downarrow{\scriptstyle a} \\ TX & \xrightarrow[a]{} & X \end{array} \qquad\qquad \begin{array}{ccc} X & \xrightarrow{e_X} & TX \\ & {\scriptstyle 1_X}\searrow \;\; \leq & \downarrow{\scriptstyle a} \\ & & X\,. \end{array}$$

Note that the notion depends in fact not just on $\mathbb{T}$ but also on $\hat{\mathbb{T}}$; hence, whenever needed, we will refer to a $(\mathbb{T}, \mathcal{V})$-category more precisely as a $(\mathbb{T}, \mathcal{V}, \hat{\mathbb{T}})$-category.

We already considered an important special type of $(\mathbb{T}, \mathcal{V})$-categories in Section III.1.3. When $\mathbb{T}$ is the identity monad $\mathbb{I}$ identically extended to $\mathcal{V}$-**Rel**, an $(\mathbb{I}, \mathcal{V})$-category is simply a $\mathcal{V}$-category. Hence, in what follows we consider easy examples with other choices of $\mathbb{T}$. Further examples will follow in Section III.2.

III.1.6.2 Examples

(1) With $\mathcal{V} = \mathsf{2}$ and $T = P$ laxly extended by $\check{P}$ (Example III.1.5.2(2)), a transitive and reflexive relation $a : PX \nrightarrow X$ must satisfy the conditions

$$(\mathcal{A} \subseteq a^\circ(B)\ \&\ B\ a\ z \implies (\bigcup \mathcal{A})\ a\ z) \qquad \text{and} \qquad \{x\}\ a\ x$$

for all $x \in X$, $B \subseteq X$, $\mathcal{A} \subseteq PX$. Since $\{x\}\ a\ y$ may be re-written as $\big\{\{x\}\big\} \subseteq a^\circ(\{y\})$, by

$$x \le y \iff \{x\}\ a\ y$$

one defines an order on X. We claim that this order completely determines a, since

$$A\ a\ y \iff \forall x \in A\ (\{x\}\ a\ y) \iff A \subseteq \downarrow y\,. \qquad \text{(III.1.6.i)}$$

Indeed, when $A\ a\ y$ and $x \in A$ one has $\big\{\{x\}\big\} \subseteq a^\circ(A)$, hence $\{x\}\ a\ y$ by transitivity; when $\{x\}\ a\ y$ for all $x \in A$ one uses $\big\{\{x\} \mid x \in A\big\} \subseteq a^\circ(\{y\})$ to obtain $A\ a\ y$ by transitivity.

Conversely, starting with an order $\le$ on X, (III.1.6.i) defines a $(\mathbb{P}, \mathsf{2}, \check{\mathbb{P}})$-category structure a on X which reproduces the original order.

(2) Trading $\check{P}$ for $\hat{P}$ (Example III.1.5.2(2)), for a transitive and reflexive relation $a : PX \nrightarrow X$ we may define a closure operation c on PX by

$$x \in c(A) \iff A\ a\ x\,.$$

(For idempotency of c, consider $\mathcal{A} = \big\{c(\{x\}) \mid x \in A\big\}$, where $A \subseteq X$.) Conversely, given c, this definition yields a $(\mathbb{P}, \mathsf{2}, \check{\mathbb{P}})$-category structure a on X.

(3) For arbitrary $\mathcal{V}$ and the largest lax extension $\mathbb{T}^\top$ of a monad $\mathbb{T}$ (Example III.1.5.2(3)), the only $(\mathbb{T}, \mathcal{V}, \mathbb{T}^\top)$-category structure t on a set X is given by $t(\mathfrak{x}, y) = \top$ for all $\mathfrak{x} \in TX$, $y \in X$. Indeed, for any $(\mathbb{T}, \mathcal{V}, \mathbb{T}^\top)$-category structure a on X, one has

$$a = a \cdot e_{TX}^\circ \cdot m_X^\circ \le e_X^\circ \cdot T^\top a \cdot m_X^\circ \le a \cdot T^\top a \cdot m_X^\circ \le a\,,$$

so

$$a(\mathfrak{x}, y) = e_X^\circ \cdot T^\top a \cdot m_X^\circ(\mathfrak{x}, y) = \bigvee_{\mathfrak{X} \in m_X^{-1}(\mathfrak{x})} T^\top a(\mathfrak{X}, e_X(y)) = \top$$

since $e_{TX}(\mathfrak{x}) \in m_X^{-1}(\mathfrak{x}) \neq \varnothing$.

III.1.6.3 Definition A map $f : X \to Y$ between $(\mathbb{T}, \mathcal{V})$-categories (X, a) and (Y, b) is a $(\mathbb{T}, \mathcal{V})$-*functor* if it satisfies

$$f \cdot a \leq b \cdot Tf \,.$$

Diagrammatically, this means that f is a *lax homomorphism* of lax algebras:

$$\begin{array}{ccc} TX & \xrightarrow{Tf} & TY \\ {\scriptstyle a}\big\downarrow\!\!\!\!\!\!\;\mid & \leq & \big\downarrow\!\!\!\!\!\!\;\mid{\scriptstyle b} \\ X & \xrightarrow[f]{} & Y\,. \end{array}$$

We can transcribe this condition equivalently as $a \leq f^{\circ} \cdot b \cdot Tf$, which, in pointwise notation, reads as

$$a(\mathfrak{x}, x) \leq b(Tf(\mathfrak{x}), f(x))$$

for all $\mathfrak{x} \in TX$ and $x \in X$.

The identity map $1_X : (X, a) \to (X, a)$ is a $(\mathbb{T}, \mathcal{V})$-functor, and so is the composite of $(\mathbb{T}, \mathcal{V})$-functors. Hence, $(\mathbb{T}, \mathcal{V})$-categories and $(\mathbb{T}, \mathcal{V})$-functors form a category, denoted by

$$(\mathbb{T}, \mathcal{V})\text{-}\mathsf{Cat}\,.$$

Of course, this category depends on the lax extension $\hat{T}$ of T, but we will always assume that such an extension is given beforehand and will therefore write more precisely $(\mathbb{T}, \mathcal{V}, \hat{\mathbb{T}})$-Cat in lieu of $(\mathbb{T}, \mathcal{V})$-Cat only if there is a danger of ambiguity. When $\mathbb{T} = \mathbb{I}$ is identically extended to $\mathcal{V}$-Rel, an $(\mathbb{I}, \mathcal{V})$-functor is simply a $\mathcal{V}$-functor. Hence,

$$(\mathbb{I}, \mathcal{V})\text{-}\mathsf{Cat} = \mathcal{V}\text{-}\mathsf{Cat}\,.$$

III.1.6.4 Examples

(1) For a map between $(\mathbb{P}, 2, \check{\mathbb{P}})$-categories to be a $(\mathbb{P}, 2)$-functor means equivalently that the map must be monotone with respect to the induced orders (Example III.1.6.2(1)). As a consequence, one obtains an isomorphism

$$(\mathbb{P}, 2, \check{\mathbb{P}})\text{-}\mathsf{Cat} \cong \mathsf{Ord}$$

which leaves underlying sets invariant.

(2) Similarly, with Example III.1.6.2(2) one has an isomorphism

$$(\mathbb{P}, 2, \hat{\mathbb{P}})\text{-}\mathsf{Cat} \cong \mathsf{Cls}\,.$$

(3) Because of Example III.1.6.2(3), there is an isomorphism

$$(\mathbb{T}, \mathcal{V}, \mathbb{T}^{\top})\text{-}\mathsf{Cat} \cong \mathsf{Set}$$

for every monad $\mathbb{T}$ on Set and every quantale $\mathcal{V}$.

A $\mathbb{T}$-algebra (X, a) (where a is a map satisfying $a \cdot e_X = 1_X$ and $a \cdot Ta = a \cdot m_X$) is generally not a $(\mathbb{T}, \mathcal{V})$-algebra (X, a) (which requires $a \cdot \hat{T}a \leq a \cdot m_X$), as one may see already in the case $\mathbb{T} = \mathbb{P}$ with its extensions $\hat{\mathbb{P}}$ or $\check{\mathbb{P}}$. There is, of course, no problem when $\hat{T}$ is flat.

III.1.6.5 Proposition *If $\hat{T}$ is a flat lax extension of T to $\mathcal{V}$-*Rel*, then a $\mathbb{T}$-algebra $(X, a : TX \to X)$ is also a $(\mathbb{T}, \mathcal{V})$-category. In this case, morphisms of $\mathbb{T}$-algebras yield $(\mathbb{T}, \mathcal{V})$-functors between the corresponding $(\mathbb{T}, \mathcal{V})$-categories, and there is a full embedding*

$$\mathsf{Set}^{\mathbb{T}} \hookrightarrow (\mathbb{T}, \mathcal{V})\text{-}\mathsf{Cat}\,.$$

Proof The only non-obvious fact in the statement of the Proposition is that the embedding of $\mathsf{Set}^{\mathbb{T}}$ in $(\mathbb{T}, \mathcal{V})$-Cat is full. To see this, consider $\mathbb{T}$-algebras (X, a) and (Y, b) with a $(\mathbb{T}, \mathcal{V})$-functor $f : (X, a) \to (Y, b)$, i.e. a map $f : X \to Y$ satisfying

$$f \cdot a \leq b \cdot Tf$$

in $\mathcal{V}$-Rel. As $f \cdot a$ and $b \cdot Tf$ are really Set-maps, the inequality means that the graph of the first is contained in the second, but an inclusion of graphs of Set-maps with the same domain is an equality. □

III.1.7 Kleisli convolution

The relations representing $(\mathbb{T}, \mathcal{V})$-category structures are of the form $a : TX \nrightarrow X$. More generally, a $(\mathbb{T}, \mathcal{V})$-*relation* is a $\mathcal{V}$-relation $r : TX \nrightarrow Y$, also denoted by $r : X \rightharpoondown Y$. In order to compose such relations, we introduce the *Kleisli convolution* of $(\mathbb{T}, \mathcal{V})$-relations as a variation of the Kleisli composition presented in II.3.6. Let us emphasize that *associativity* of this operation turns out to depend on the monad lax extension, so that sets with $\mathcal{V}$-relations $r : TX \nrightarrow Y$ only form a category in particular cases. In Section III.1.8, we will provide a context in which the Kleisli convolution allows for identity morphisms, and in Section III.1.9 we will introduce a category that has the Kleisli convolution as its composition.

III.1.7.1 Definition Given a lax extension $\hat{\mathbb{T}} = (\hat{T}, m, e)$ of a Set-monad $\mathbb{T} = (T, m, e)$, the *Kleisli convolution* $s \circ r : X \rightharpoondown Z$ of $(\mathbb{T}, \mathcal{V})$-relations $r : X \rightharpoondown Y$ and $s : Y \rightharpoondown Z$ is the $(\mathbb{T}, \mathcal{V})$-relation defined by

$$s \circ r := s \cdot \hat{T}r \cdot m_X^\circ\,,$$

an operation that may be depicted as

$$(\,TX \xrightarrow{r} Y\,,\; TY \xrightarrow{s} Z\,) \quad \longmapsto \quad (\,TX \xrightarrow{m_X^\circ} TTX \xrightarrow{\hat{T}r} TY \xrightarrow{s} Z\,)\,.$$

When $\mathbb{T} = \mathbb{1}$,

$$s \circ r = s \cdot r$$

is just the relational composition of $\mathcal{V}$-relations.

The set of all $(\mathbb{T}, \mathcal{V})$-relations from X to Y inherits the order of $\mathcal{V}\text{-}\mathsf{Rel}(TX, Y)$:

$$r \leq r' \iff \forall(\mathfrak{x}, y) \in TX \times Y\ (r(\mathfrak{x}, y) \leq r'(\mathfrak{x}, y)) ,$$

and the Kleisli convolution preserves this order in each variable:

$$r \leq r', s \leq s' \implies r \circ s \leq r' \circ s' ,$$

for all $r, r' : X \rightharpoonup Y$ and $s, s' : Y \rightharpoonup Z$. The $(\mathbb{T}, \mathcal{V})$-relation $e_X^\circ : X \rightharpoonup X$ is a *lax identity* for this composition: one has

$$e_Y^\circ \circ r = e_Y^\circ \cdot \hat{T}r \cdot m_X^\circ \geq r \cdot e_{TX}^\circ \cdot m_X^\circ = r ,$$

with equality holding if $e^\circ = (e_X^\circ)_X : \hat{T} \longrightarrow 1$ is a natural transformation, and

$$r \circ e_X^\circ = r \cdot \hat{T}(e_X^\circ) \cdot m_X^\circ \geq r \cdot (Te_X)^\circ \cdot m_X^\circ = r ,$$

with equality holding if $\hat{T}$ is flat. In particular, $e_X^\circ \circ e_X^\circ \geq e_X^\circ$, but generally this inequality is strict when $\hat{T}$ fails to be flat. It turns out that e_X° can be modified to become idempotent without loss of its lax identity properties. To this end, we first prove the following result.

III.1.7.2 Lemma *For a lax extension* $\hat{\mathbb{T}} = (\hat{T}, m, e)$ *to* $\mathcal{V}\text{-}\mathsf{Rel}$ *of a monad* $\mathbb{T} = (T, m, e)$ *on* Set, *one has*

$$\hat{T}1_X = \hat{T}(e_X^\circ) \cdot m_X^\circ .$$

Proof On one hand, we can exploit $1_{TX} = 1_{TX}^\circ = (m_X \cdot Te_X)^\circ = (Te_X)^\circ \cdot m_X^\circ$ to obtain

$$\begin{aligned}
\hat{T}1_X &= \hat{T}1_X \cdot 1_{TX}^\circ && (1_{TX} = 1_{TX}^\circ)\\
&= \hat{T}1_X \cdot (Te_X)^\circ \cdot m_X^\circ && (1_{TX}^\circ = (Te_X)^\circ \cdot m_X^\circ)\\
&\leq \hat{T}(1_X) \cdot \hat{T}(e_X^\circ) \cdot m_X^\circ && ((Te_X)^\circ \leq \hat{T}(e_X^\circ))\\
&\leq \hat{T}(e_X^\circ) \cdot m_X^\circ && (\hat{T} \text{ lax functor}).
\end{aligned}$$

On the other hand,

$$\begin{aligned}
\hat{T}(e_X^\circ) \cdot m_X^\circ &\leq \hat{T}(e_X^\circ \cdot \hat{T}1_X) \cdot m_X^\circ && (1_{TX} \leq \hat{T}1_X)\\
&= (Te_X)^\circ \cdot \hat{T}\hat{T}1_X \cdot m_X^\circ && (\text{Corollary III.1.4.4})\\
&\leq (Te_X)^\circ \cdot m_X^\circ \cdot \hat{T}1_X && (m \text{ oplax})\\
&= \hat{T}1_X && (1_{TX} = (Te_X)^\circ \cdot m_X^\circ),
\end{aligned}$$

which concludes the proof. □

We set

$$1_X^\sharp := e_X^\circ \circ e_X^\circ ,$$

hence

$$1_X^\sharp = e_X^\circ \cdot \hat{T}1_X$$

by Lemma III.1.7.2, and we can prove:

III.1.7.3 Proposition *If* $\hat{\mathbb{T}} = (\hat{T}, m, e)$ *is a lax extension of the monad* $\mathbb{T} = (T, m, e)$ *to* $\mathcal{V}$-Rel, *then*

$$r \circ e_X^\circ = r \cdot \hat{T}1_X = r \circ 1_X^\sharp \qquad \textit{and} \qquad e_Y^\circ \circ r = 1_Y^\sharp \circ r$$

for all $(\mathbb{T}, \mathcal{V})$*-relations* $r : X \nrightarrow Y$. *In particular,* $1_X^\sharp \circ 1_X^\sharp = 1_X^\sharp$, *so that* $(X, 1_X^\sharp)$ *is a* $(\mathbb{T}, \mathcal{V})$*-algebra.*

Proof We first observe that

$$\begin{aligned} r \circ 1_X^\sharp &= r \cdot \hat{T}(e_X^\circ \cdot \hat{T}1_X) \cdot m_X^\circ \\ &= r \cdot (Te_X)^\circ \cdot \hat{T}\hat{T}1_X \cdot m_X^\circ && \text{(Corollary III.1.4.4)} \\ &\leq r \cdot (Te_X)^\circ \cdot m_X^\circ \cdot \hat{T}1_X && (m \text{ oplax}) \\ &= r \cdot \hat{T}1_X && (1_X = (Te_X)^\circ \cdot m_X^\circ) \\ &= r \cdot \hat{T}(e_X^\circ) \cdot m_X^\circ && \text{(Lemma III.1.7.2)} \\ &= r \circ e_X^\circ \,. \end{aligned}$$

This inequality suffices to prove the first set of equalities, since $e_X^\circ \leq 1_X^\sharp$ implies $r \circ e_X^\circ \leq r \circ 1_X^\sharp$. The other equality follows directly from Corollary III.1.4.4, as

$$1_Y^\sharp \circ r = e_Y^\circ \cdot \hat{T}(1_Y^\circ) \cdot \hat{T}r \cdot m_X^\circ = e_Y^\circ \cdot \hat{T}r \cdot m_X^\circ = e_Y^\circ \circ r \,.$$

Finally,

$$1_X^\sharp \circ 1_X^\sharp = 1_X^\sharp \cdot \hat{T}1_X = e_X^\circ \cdot \hat{T}1_X \cdot \hat{T}1_X = e_X^\circ \cdot \hat{T}1_X = 1_X^\sharp \,. \qquad \square$$

We call $(X, 1_X^\sharp)$ the *discrete* $(\mathbb{T}, \mathcal{V})$*-category* over X; see Section III.3.2.

III.1.8 Unitary $(\mathbb{T}, \mathcal{V})$-relations

Our candidates for the identities of the Kleisli convolution are the $(\mathbb{T}, \mathcal{V})$-relations $1_X^\sharp$, but the array of $(\mathbb{T}, \mathcal{V})$-relations from X to Y that are left invariant by composition with these identities must still be determined.

III.1.8.1 Definition A $(\mathbb{T}, \mathcal{V})$-relation $r : X \nrightarrow Y$ is *right unitary* if it satisfies

$$r \circ e_X^\circ \leq r \,,$$

and it is *left unitary* if

$$e_Y^\circ \circ r \leq r$$

holds. In terms of the relational composition, these conditions amount to

$$r \cdot \hat{T}1_X \leq r \qquad \text{and} \qquad e_Y^\circ \cdot \hat{T}r \cdot m_X^\circ \leq r\,,$$

respectively. The $(\mathbb{T}, \mathcal{V})$-relation r is *unitary* if it is both left and right unitary.

The $(\mathbb{T}, \mathcal{V})$-relation e_X° itself is not unitary in general, but Proposition III.1.7.3 shows that we can replace it in the previous definitions by $1_X^\sharp$. It also follows from the discussion preceding Lemma III.1.7.2 that the inequalities appearing in the left and right unitary conditions are in fact equalities. Hence, a $(\mathbb{T}, \mathcal{V})$-relation r is right unitary, respectively left unitary, if

$$r \circ 1_X^\sharp = r\,, \quad \text{respectively} \quad 1_Y^\sharp \circ r = r\,.$$

Let us examine $(\mathbb{T}, \mathcal{V})$-categories and $(\mathbb{T}, \mathcal{V})$-functors in the light of Kleisli convolution and unitary $(\mathbb{T}, \mathcal{V})$-relations. By definition, a $(\mathbb{T}, \mathcal{V})$-category structure is a relation $a : TX \nrightarrow X$ such that

$$a \circ a \leq a \qquad \text{and} \qquad e_X^\circ \leq a\,.$$

These conditions imply that such a $(\mathbb{T}, \mathcal{V})$-relation $a : X \nrightarrow X$ is always unitary:

$$a \circ 1_X^\sharp = a = 1_X^\sharp \circ a\,,$$

since $a \circ e_X^\circ \leq a \circ a \leq a$ and $e_X^\circ \circ a \leq a \circ a \leq a$. As a consequence, $a : TX \nrightarrow X$ is a $(\mathbb{T}, \mathcal{V})$-category structure if and only if

$$a \circ a = a \qquad \text{and} \qquad 1_X^\sharp \leq a\,. \tag{III.1.8.i}$$

Indeed, the first condition follows from transitivity: $a \leq a \circ e_X^\circ \leq a \circ a \leq a$, and the second condition follows from reflexivity: $1_X^\sharp = e_X^\circ \circ e_X^\circ \leq a \circ e_X^\circ \leq a$ (the converse resulting from $e_X^\circ \leq 1_X^\sharp$). Hence, a $(\mathbb{T}, \mathcal{V})$-algebra structure a can also be seen as a monoid in the ordered set of unitary $(\mathbb{T}, \mathcal{V})$-relations from X to X, considered as a category that is provided with the $\circ$-operation. But we recall that associativity of $\circ$ is guaranteed only under additional hypotheses (see Section III.1.9), a property that is needed to consider $\circ$ as a monoidal structure.

By definition, a $(\mathbb{T}, \mathcal{V})$-functor $f : (X, a) \to (Y, b)$ is a map $f : X \to Y$ satisfying

$$f \cdot a \leq b \cdot Tf\,.$$

Since a $(\mathbb{T}, \mathcal{V})$-category structure b is right unitary, it satisfies $b \cdot \hat{T}1_Y = b \circ e_Y^\circ = b$ by Proposition III.1.7.3; one then obtains by Proposition III.1.4.3 the equalities $b \cdot \hat{T}f = b \cdot \hat{T}1_Y \cdot Tf = b \cdot Tf$. Hence, the $(\mathbb{T}, \mathcal{V})$-functor condition can equivalently be expressed by using the lax extension of T:

$$f \cdot a \leq b \cdot \hat{T}f\,.$$

Setting

$$f^\sharp := f^\circ \cdot 1_Y^\sharp : Y \nrightarrow X\,,$$

one can also express $(\mathbb{T}, \mathcal{V})$-functoriality of a map $f : X \to Y$ via Kleisli convolution as

$$a \circ f^\sharp \leq f^\sharp \circ b\,;$$

see Exercise III.1.M.

III.1.9 Associativity of unitary $(\mathbb{T}, \mathcal{V})$-relations

With respect to the Kleisli convolution, the unitary $(\mathbb{T}, \mathcal{V})$-relation $1_X^\sharp$ serves as an identity for all unitary $(\mathbb{T}, \mathcal{V})$-relations composable with $1_X^\sharp$. In general, however, unitary $(\mathbb{T}, \mathcal{V})$-relations do not compose associatively, even when $\mathbb{T}$ is the identity monad (see Proposition III.1.9.7).

III.1.9.1 Definition A lax extension $\hat{\mathbb{T}}$ to $\mathcal{V}$-Rel of a monad $\mathbb{T} = (T, m, e)$ on Set is *associative* whenever the Kleisli convolution of unitary $(\hat{\mathbb{T}}, \mathcal{V})$-relations is associative. Explicitly, a lax extension $\hat{\mathbb{T}}$ is associative whenever

$$t \circ (s \circ r) = (t \circ s) \circ r\,,$$

or, equivalently,

$$t \cdot \hat{T}(s \cdot \hat{T}r \cdot m_X^\circ) \cdot m_X^\circ = t \cdot \hat{T}s \cdot m_Y^\circ \cdot \hat{T}r \cdot m_X^\circ \qquad \text{(III.1.9.i)}$$

for all unitary $(\mathbb{T}, \mathcal{V})$-relations $r : X \nrightarrow Y$, $s : Y \nrightarrow Z$, and $t : Z \nrightarrow W$.

For $\hat{\mathbb{T}}$ associative, unitary $(\mathbb{T}, \mathcal{V})$-relations are closed under Kleisli convolution:

$$(s \circ r) \circ 1_X^\sharp = s \circ (r \circ 1_X^\sharp) = s \circ r \qquad \text{and} \qquad 1_X^\sharp \circ (s \circ r) = (1_X^\sharp \circ s) \circ r = s \circ r\,.$$

Hence, in the presence of an associative lax extension $\hat{\mathbb{T}}$, we can form the category

$$(\mathbb{T}, \mathcal{V})\text{-URel}$$

whose objects are sets, and whose morphisms are unitary $(\mathbb{T}, \mathcal{V})$-relations that compose via Kleisli convolution. We note that, like $(\mathbb{T}, \mathcal{V})$-Cat, $(\mathbb{T}, \mathcal{V})$-URel depends on the lax extension $\hat{\mathbb{T}}$; we write $(\mathbb{T}, \mathcal{V}, \hat{\mathbb{T}})$-URel whenever this dependency needs to be emphasized. When the hom-sets $(\mathbb{T}, \mathcal{V})\text{-URel}(X, Y)$ are equipped with the pointwise order induced by $\mathcal{V}$,

$$r \leq r' \iff \forall (\mathfrak{x}, y) \in TX \times Y\ (r(\mathfrak{x}, y) \leq r'(\mathfrak{x}, y))\,,$$

$(\mathbb{T}, \mathcal{V})$-URel becomes an ordered category.

Condition (III.1.9.i) appears daunting to verify directly, so we postpone examples of associative lax extensions until after Proposition III.1.9.4, which presents more practical conditions. To this end, we introduce the unitary $(\mathbb{T}, \mathcal{V})$-relation

$$r_\sharp := e_Y^\circ \cdot \hat{T}r : TX \nrightarrow Y$$

associated to a $\mathcal{V}$-relation $r : X \nrightarrow Y$ (see Exercise III.1.N). Note that $(1_X)_\sharp = 1_X^\sharp$ as defined in Section III.1.7.

III.1.9.2 Lemma *Let $\hat{\mathbb{T}}$ be a lax extension to $\mathcal{V}$-*Rel *of a monad $\mathbb{T} = (T, m, e)$ on* Set. *Then*

$$\hat{T}(s_\sharp \cdot \hat{T}r) \cdot m_X^\circ = \hat{T}(s \cdot r)$$

for all $\mathcal{V}$-relations $r : X \nrightarrow Y$ and $s : Y \nrightarrow Z$. In particular,

$$\hat{T}(s_\sharp) \cdot m_Y^\circ = \hat{T}s$$

for all $\mathcal{V}$-relations $s : Y \nrightarrow Z$.

Proof The first stated equality follows from

$$\begin{aligned}
\hat{T}(s \cdot r) &= \hat{T}(s \cdot r) \cdot \hat{T}(e_X^\circ) \cdot m_X^\circ && \text{(Lemma III.1.7.2)}\\
&\le \hat{T}(s \cdot r \cdot e_X^\circ) \cdot m_X^\circ && (\hat{T} \text{ lax functor})\\
&\le \hat{T}(e_Z^\circ \cdot \hat{T}s \cdot \hat{T}r) \cdot m_X^\circ = \hat{T}(s_\sharp \cdot \hat{T}r) \cdot m_X^\circ && (e^\circ \text{ lax natural})\\
&\le \hat{T}(e_Z^\circ \cdot \hat{T}(s \cdot r)) \cdot m_X^\circ && (\hat{T} \text{ lax functor})\\
&= (Te_Z)^\circ \cdot \hat{T}\hat{T}(s \cdot r) \cdot m_X^\circ && \text{(Corollary III.1.4.4)}\\
&\le (Te_Z)^\circ \cdot m_Z^\circ \cdot \hat{T}(s \cdot r) && (m^\circ \text{ lax natural})\\
&= \hat{T}(s \cdot r)\,.
\end{aligned}$$

The particular case is obtained by setting $r = 1_Y$. □

III.1.9.3 Lemma *Let $\hat{\mathbb{T}}$ be a lax extension to $\mathcal{V}$-*Rel *of a monad $\mathbb{T} = (T, m, e)$ on* Set. *Then*

$$m_X^\circ \cdot \hat{T}1_X = \hat{T}1_{TX} \cdot m_X^\circ \cdot \hat{T}1_X = \hat{T}\hat{T}1_X \cdot m_X^\circ \cdot \hat{T}1_X\,.$$

Proof Since $1_{TTX} \le \hat{T}1_{TX} \le \hat{T}\hat{T}1_X$, we have

$$m_X^\circ \cdot \hat{T}1_X \le \hat{T}1_{TX} \cdot m_X^\circ \cdot \hat{T}1_X \le \hat{T}\hat{T}1_X \cdot m_X^\circ \cdot \hat{T}1_X \le m_X^\circ \cdot \hat{T}1_X$$

by lax naturality of m°. □

III.1.9.4 Proposition *Let $\hat{\mathbb{T}}$ be a lax extension to $\mathcal{V}$-*Rel *of a monad $\mathbb{T} = (T, m, e)$ on* Set. *The following are equivalent:*

(i) *$\hat{\mathbb{T}}$ is associative;*
(ii) *$\hat{T} : \mathcal{V}$-*Rel *$\to \mathcal{V}$-*Rel *preserves composition and $m^\circ : \hat{T} \to \hat{T}\hat{T}$ is natural;*

(iii) $t \circ (s \circ r) = (t \circ s) \circ r$ *for all* $\mathcal{V}$*-relations* $t : TZ \nrightarrow W$, $s : TY \nrightarrow Z$ *and right unitary* $\mathcal{V}$*-relations* $r : TX \nrightarrow Y$.

Proof For (i) $\Longrightarrow$ (ii), consider $\mathcal{V}$-relations $r : X \nrightarrow Y$ and $s : Y \nrightarrow Z$. We first prove that an associative lax extension preserves composition. Since all of $r_\sharp$, $s_\sharp$, and $\hat{T}1_Z$ are unitary (Exercise III.1.N), one has

$$\hat{T}1_Z \circ (s_\sharp \circ r_\sharp) = (\hat{T}1_Z \circ s_\sharp) \circ r_\sharp .$$

This identity is equivalent to $\hat{T}(s \cdot r) = \hat{T}s \cdot \hat{T}r$: indeed,

$$\hat{T}1_Z \circ (s_\sharp \circ r_\sharp) = \hat{T}(s_\sharp \cdot \hat{T}(r_\sharp) \cdot m_X^\circ) \cdot m_X^\circ = \hat{T}(s_\sharp \cdot \hat{T}r) \cdot m_X^\circ = \hat{T}(s \cdot r)$$

by using Lemma III.1.9.2 twice, and

$$(\hat{T}1_Z \circ s_\sharp) \circ r_\sharp = \hat{T}(s_\sharp) \cdot m_Y^\circ \cdot \hat{T}(r_\sharp) \cdot m_X^\circ = \hat{T}s \cdot \hat{T}r$$

by Lemma III.1.9.2 again. To see that m° is natural, we compute

$$\hat{T}1_Y \circ (\hat{T}1_Y \circ r_\sharp) = \hat{T}1_Y \cdot \hat{T}(\hat{T}1_Y \cdot \hat{T}(r_\sharp) \cdot m_X^\circ) \cdot m_X^\circ = \hat{T}1_Y \cdot \hat{T}\hat{T}r \cdot m_X^\circ = \hat{T}\hat{T}r \cdot m_X^\circ$$

and

$$(\hat{T}1_Y \circ \hat{T}1_Y) \circ r_\sharp = \hat{T}1_Y \cdot \hat{T}\hat{T}1_Y \cdot m_Y^\circ \cdot \hat{T}(r_\sharp) \cdot m_X^\circ = m_Y^\circ \cdot \hat{T}r$$

using Lemmata III.1.9.2 and III.1.9.3. Since Kleisli convolution is associative on unitary relations, we obtain $\hat{T}\hat{T}r \cdot m_X^\circ = m_Y^\circ \cdot \hat{T}r$.

For (ii) $\Longrightarrow$ (iii), we use right unitariness of r to write

$$\begin{aligned}
t \circ (s \circ r) &= t \cdot \hat{T}(s \cdot \hat{T}(r \cdot \hat{T}1_X) \cdot m_X^\circ) \cdot m_X^\circ \\
&= t \cdot \hat{T}(s \cdot \hat{T}r \cdot \hat{T}\hat{T}1_X \cdot m_X^\circ) \cdot m_X^\circ \\
&= t \cdot \hat{T}(s \cdot \hat{T}r \cdot m_X^\circ \cdot \hat{T}1_X) \cdot m_X^\circ \\
&= t \cdot \hat{T}s \cdot \hat{T}\hat{T}r \cdot \hat{T}(m_X^\circ) \cdot \hat{T}\hat{T}1_X \cdot m_X^\circ \\
&= t \cdot \hat{T}s \cdot \hat{T}\hat{T}r \cdot (Tm_X)^\circ \cdot \hat{T}\hat{T}1_X \cdot m_X^\circ \\
&= t \cdot \hat{T}s \cdot \hat{T}\hat{T}r \cdot m_{TX}^\circ \cdot m_X^\circ \cdot \hat{T}1_X \\
&= t \cdot \hat{T}s \cdot m_Y^\circ \cdot \hat{T}r \cdot m_X^\circ \cdot \hat{T}1_X \\
&= t \cdot \hat{T}s \cdot m_Y^\circ \cdot \hat{T}(r \cdot \hat{T}1_X) \cdot m_X^\circ = (t \circ s) \circ r .
\end{aligned}$$

(iii) $\Longrightarrow$ (i) is immediate by definition of an associative lax extension. □

III.1.9.5 Remark In the case where the lax extension $\hat{T}$ to V-Rel satisfies $\hat{T}(r^\circ) = (\hat{T}r)^\circ$ for all $\mathcal{V}$-relations r, the naturality condition for m° in Proposition III.1.9.4(ii) can equivalently be expressed as naturality of $m : \hat{T}\hat{T} \to \hat{T}$ (see Exercise III.1.J).

III.1.9.6 Examples

(1) The identity extension $\mathbb{I}$ to $\mathcal{V}$-Rel of the identity monad is associative. Indeed, the Kleisli convolution of $\mathcal{V}$-relations is their usual composition, which is associative (Exercise III.1.C).

(2) The lax extensions $\check{\mathbb{P}}$, $\hat{\mathbb{P}}$ of the powerset monad $\mathbb{P}$ (Example III.1.5.2(2)) to Rel are both associative. Let us verify (ii) of Proposition III.1.9.4 for $\check{\mathbb{P}}$ (the verifications for $\hat{\mathbb{P}}$ are similar). Since $A\ (\hat{P}r)\ B$ is defined as $A \subseteq r^\circ(B)$, the equivalence

$$A \subseteq (s \cdot r)^\circ(C) \iff \exists B \subseteq Y\ (A \subseteq r^\circ(B)\ \&\ B \subseteq s^\circ(C))$$

(for all $A \subseteq X$, $C \subseteq Y$, and relations $r : X \nrightarrow Y$, $s : Y \nrightarrow X$) shows that the lax extension $\check{P}$ preserves relational composition. As the monad multiplication $\bigcup$ yields a lax natural transformation $\bigcup^\circ$ in Rel, one needs only to verify that $\bigcup_Y^\circ \cdot \check{P}r \subseteq \check{P}\check{P}r \cdot \bigcup_X^\circ$: suppose that, for $A \subseteq X$ and $\mathcal{B} \subseteq PY$, one has $A \subseteq r^\circ(\bigcup \mathcal{B})$; then the subset $\mathcal{A} = \{\{x\} \mid x \in A\} \subseteq PX$ is such that $\bigcup \mathcal{A} = A$ and, for all $A' \in \mathcal{A}$, there is $B \in \mathcal{B}$ such that $A' \subseteq r^\circ(B)$.

(3) The largest lax extension $\mathbb{T}^\top$ to $\mathcal{V}$-Rel of a monad $\mathbb{T}$ on Set is associative.

Let us exhibit now a non-associative lax extension. To this end, we consider the five-element frame C depicted by

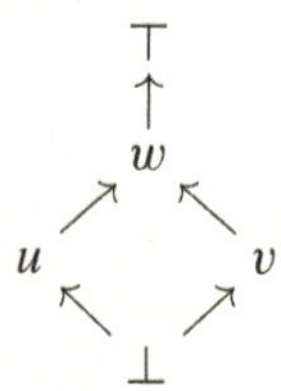

(considered as a quantale with $\otimes = \wedge$). The identity monad $\mathbb{I} = (I = 1_{\mathsf{Set}}, 1, 1)$ on Set may be extended non-identically to C-Rel by

$$\hat{I}r(x,y) = \begin{cases} \top & \text{if } w \le r(x,y), \\ r(x,y) & \text{otherwise,} \end{cases}$$

for all $x \in X$, $y \in Y$, and C-relations $r : X \nrightarrow Y$.

III.1.9.7 Proposition *C is a commutative, integral, and lean quantale, and $\hat{\mathbb{I}} = (\hat{I}, m = 1, e = 1)$ is a flat lax extension of the identity monad $\mathbb{I}$ to C-Rel, which makes $m^\circ : \hat{I} \to \hat{I}\hat{I}$ a natural transformation. But $\hat{I}$ fails to preserve Kleisli convolution, so $\hat{\mathbb{I}}$ is not associative.*

Proof The claims about C are immediate. One also easily sees that

$$r \le \hat{I}r = \hat{I}\hat{I}r \qquad \text{and} \qquad \hat{I}f = f\,, \quad \hat{I}(f^\circ) = f^\circ$$

for all $r : X \nrightarrow Y$ in C-Rel, and $f : X \to Y$ in Set. Hence, all claims but the last will follow from

$$\hat{I}s \cdot \hat{I}r \le \hat{I}(s \cdot r)$$

(with $s : Y \nrightarrow Z$). To see this, suppose first that, for $x \in X$ and $z \in Z$, we have $\hat{I}s \cdot \hat{I}r(x,z) = \top$. Then necessarily $\hat{I}r(x,y) = \top = \hat{I}s(y,z)$ for some $y \in Y$, i.e. $w \le r(x,y)$ and $w \le s(y,z)$. In this case, $w \le (s \cdot r)(x,z)$ and $\hat{I}(s \cdot r)(x,z) = \top$ follows. Suppose now that

$$(\hat{I}s \cdot \hat{I}r)(x,z) = \bigvee_{y \in Y} \hat{I}r(x,y) \wedge \hat{I}s(y,z) \le w < \top\,,$$

and consider any $y \in Y$. We may assume $\hat{I}r(x,y) < w$ (since $\hat{I}r(x,y)$ and $\hat{I}s(y,z)$ cannot both have value w), so that $\hat{I}r(x,y) = r(x,y)$. But then, regardless of whether $\hat{I}s(y,z) = \top$ (in which case $w \le s(y,z)$) or $\hat{I}s(y,z) < \top$ (in which case $\hat{I}s(y,z) = s(y,z)$), we obtain

$$\hat{I}r(x,y) \wedge \hat{I}s(y,z) = r(x,y) \wedge s(y,z)\,.$$

Consequently,

$$(\hat{I}s \cdot \hat{I}r)(x,z) = \bigvee_{y \in Y} \hat{I}r(x,y) \wedge \hat{I}s(y,z) = (s \cdot r)(x,z) < \top\,,$$

and $(\hat{I}s \cdot \hat{I}r)(x,z) \le \hat{I}(s \cdot r)(x,z)$ follows.

For the last claim, consider $X = \{0,1,2,3\}$ and $r : X \nrightarrow X$ with

$$r(0,1) = \top\,, \quad r(1,3) = u\,, \quad r(0,2) = v\,, \quad r(2,3) = \top\,,$$

while $r(x,y) = \bot$ for all other pairs (x,y). Then $\hat{I}r = r$, and therefore

$$(\hat{I}r \cdot \hat{I}r)(0,3) = (r \cdot r)(0,3) = \bigvee_{y \in X} r(0,y) \wedge r(y,3) = u \vee v = w\,,$$

which, however, implies $(\hat{I}r \cdot \hat{I}r)(0,3) < \top = \hat{I}(r \cdot r)(0,3)$. □

III.1.10 The Barr extension

Finding a lax extension of a functor T : Set $\to$ Set to $\mathcal{V}$-Rel requires some effort in general, but the problem is much simpler for $\mathcal{V} = 2$. Recall that 2-Rel $\cong$ Rel. Given a relation $r : X \times Y \to 2$, we denote its representation as a subset of $X \times Y$ by R (compare with II.1.2). With $\pi_1 : R \to X$ and $\pi_2 : R \to Y$ the respective projections, r is represented as a *span*, i.e. as a diagram of the form

$$\begin{array}{ccc} & R & \\ {\scriptstyle\pi_1}\swarrow & & \searrow{\scriptstyle\pi_2} \\ X & & Y \end{array}$$

and we have

$$r = \pi_2 \cdot \pi_1^\circ$$

in Rel.

III.1.10.1 Definition The *Barr extension* of a functor $T : \mathsf{Set} \to \mathsf{Set}$ to Rel is defined by

$$\overline{T}r := T\pi_2 \cdot (T\pi_1)^\circ .$$

Elementwise, for $\mathfrak{x} \in TX$ and $\mathfrak{y} \in TY$ the Barr extension is given by

$$\mathfrak{x}\, \overline{T}r\, \mathfrak{y} \iff \exists \mathfrak{w} \in TR\, (T\pi_1(\mathfrak{w}) = \mathfrak{x}\ \&\ T\pi_2(\mathfrak{w}) = \mathfrak{y}) .$$

III.1.10.2 Remarks

(1) The Barr extension $\overline{T}$ preserves the order on hom-sets. Indeed, if $s \leq r$, then we may assume $S \subseteq R$ with S the domain of the span representing s; hence, with $i : S \hookrightarrow R$ denoting the inclusion map,

$$\overline{T}s = T\pi_2 \cdot Ti \cdot (Ti)^\circ \cdot (T\pi_1)^\circ \leq T\pi_2 \cdot (T\pi_1)^\circ = \overline{T}r .$$

(2) One easily verifies that

$$\overline{T}(r^\circ) = (\overline{T}r)^\circ \qquad \text{and} \qquad \overline{T}f = Tf$$

for all relations $r : X \nrightarrow Y$ and Set-maps $f : X \to Y$. Moreover, given Set-maps $f : A \to X$ and $g : Y \to B$, one has, by definition,

$$\overline{T}(g \cdot r) = Tg \cdot \overline{T}r \qquad \text{and} \qquad \overline{T}(r \cdot f^\circ) = \overline{T}r \cdot (Tf)^\circ .$$

(3) In the definition of $\overline{T}r$, the pair (π_1, π_2) can be replaced by any other mono-source representing r, or even by any other source (p, q) with $r = q \cdot p^\circ$ if T sends surjections to surjections. (Recall that, in the presence of the Axiom of Choice, every Set-functor preserves surjections, since every epimorphism in Set splits if and only if the Axiom of Choice holds; see Exercise II.2.C.)

Given any other factorization $r = q \cdot p^\circ$ via maps $p : P \to X$ and $q : P \to Y$, the equation $r = q \cdot p^\circ$ says precisely that the canonical map $P \to X \times Y$ has image R and therefore defines a surjection $l : P \to R$. Moreover, this map l is a bijection if (p, q) forms a mono-source: l is monic by a trivial cancelation rule for mono-sources (see Section II.5.3 in the dual situation, and Exercise II.5.A).

In general, for a factorization $r = q \cdot p^\circ$, one has

$$Tq \cdot (Tp)^\circ = T(\pi_2 \cdot l) \cdot (T(\pi_1 \cdot l))^\circ = T\pi_2 \cdot Tl \cdot (Tl)^\circ \cdot (T\pi_1)^\circ \leq T\pi_2 \cdot (T\pi_1)^\circ,$$

with equality holding if $Tl \cdot (Tl)^\circ = 1_{TX}$, i.e. if Tl is surjective (Proposition III.1.2.2).

III.1.10.3 Examples

(1) The Barr extension $\overline{1}_{\mathsf{Set}}$ of the identity functor 1_{Set} on Set is simply the identity functor 1_{Rel} on Rel.

(2) For the filter functor $F : \mathsf{Set} \to \mathsf{Set}$, the Barr extension $\overline{F}$ is obtained as follows. First, note that, for filters $\mathfrak{a} \in FX$, $\mathfrak{b} \in FY$, and a relation $r : X \nrightarrow Y$,

$$\mathfrak{a}\,(\overline{F}r)\,\mathfrak{b} \iff \exists \mathfrak{c} \in FR\,(\pi_1[\mathfrak{c}] = \mathfrak{a}\ \&\ \pi_2[\mathfrak{c}] = \mathfrak{b})\,.$$

If such a filter $\mathfrak{c}$ exists, then, for all $A \in \mathfrak{a}$, one has: $C := \pi_1^{-1}(A) \in \mathfrak{c}$, and the set

$$r(A) := \{y \in Y \mid \exists x \in A\,(x\ r\ y)\}$$

must be in $\mathfrak{b}$, as it contains $\pi_2(C)$ and $\pi_2(C) \in \pi_2[\mathfrak{c}] = \mathfrak{b}$. Similarly, one observes that $r^\circ(B) \in \mathfrak{a}$ for all $B \in \mathfrak{b}$. Conversely, if $r(A) \in \mathfrak{b}$ and $r^\circ(B) \in \mathfrak{a}$ for all $A \in \mathfrak{a}$ and $B \in \mathfrak{b}$, the sets $C_{A,B} = \pi_1^{-1}(A) \cap \pi_2^{-1}(B)$ (with A running through $\mathfrak{a}$, and B through $\mathfrak{b}$) form a filter base for $\mathfrak{c} \in FR$ such that $\pi_1[\mathfrak{c}] = \mathfrak{a}$ and $\pi_2[\mathfrak{c}] = \mathfrak{b}$.

Therefore, the Barr extension of the filter functor is given by

$$\mathfrak{a}\,(\overline{F}r)\,\mathfrak{b} \iff r[\mathfrak{a}] \subseteq \mathfrak{b}\ \&\ r^\circ[\mathfrak{b}] \subseteq \mathfrak{a}$$

for all $\mathfrak{a} \in FX$ and $\mathfrak{b} \in FY$, and relations $r : X \nrightarrow Y$, where $r[\mathfrak{a}]$ is the filter generated by the filter base $\{r(A) \mid A \in \mathfrak{a}\}$ (this notation coincides with the image-filter notation for maps of II.1.12).

(3) In the previous example, if both $\mathfrak{a}$ and $\mathfrak{b}$ are ultrafilters, then

$$r[\mathfrak{a}] \subseteq \mathfrak{b} \iff r^\circ[\mathfrak{b}] \subseteq \mathfrak{a}\,.$$

Indeed, for an ultrafilter $\mathfrak{b} \in \beta Y$ and $A' \subseteq X$, one has

$$A' \in \mathfrak{b} \iff \forall B \in \mathfrak{b}\,(A' \cap B \neq \varnothing)\,.$$

Hence, $r[\mathfrak{a}] \subseteq \mathfrak{b}$ means that, for all $A \in \mathfrak{a}$ and $B \in \mathfrak{b}$, one has $r(A) \cap B \neq \varnothing$, i.e. $A \cap r^\circ(B) \neq \varnothing$, and one obtains $r^\circ[\mathfrak{b}] \subseteq \mathfrak{a}$ (the other implication also follows). The Barr extension of the ultrafilter functor β is therefore described by

$$\mathfrak{a}\,(\overline{\beta}r)\,\mathfrak{b} \iff r[\mathfrak{a}] \subseteq \mathfrak{b} \iff r^\circ[\mathfrak{b}] \subseteq \mathfrak{a}\,,$$

for all $\mathfrak{a} \in \beta X$, $\mathfrak{b} \in \beta Y$, and relations $r : X \nrightarrow Y$, or equivalently by

$$\mathfrak{a}\,(\overline{\beta}r)\,\mathfrak{b} \iff \forall A \in \mathfrak{a}, B \in \mathfrak{b}\ \exists x \in A, y \in B\,(x\ r\ y)\,.$$

(4) The equivalent descriptions of the Barr extension of the ultrafilter monad lead to distinct extensions when we consider the filter monad instead. Similarly to the extensions $\check{P}$, $\hat{P}$ of the powerset functor (Example III.1.4.2), one obtains two lax extensions of the filter functor, neither of which is the Barr extension of F. First, by setting

$$\begin{aligned}\mathfrak{a}\,(\check{F}r)\,\mathfrak{b} &\iff \mathfrak{a} \supseteq r^\circ[\mathfrak{b}]\\ &\iff \forall B \in \mathfrak{b}\ \exists A \in \mathfrak{a}\ \forall x \in A\ \exists y \in B\,(x\ r\ y)\end{aligned}$$

for all relations $r : X \nrightarrow Y$, and filters $\mathfrak{a} \in FX$, $\mathfrak{b} \in FY$, one obtains a non-flat lax extension whose lax algebras, similarly to the Barr extension of β, provide a convergence description of the category of topological spaces (see Theorem III.2.2.5 and Corollary IV.1.5.4). By contrast, the lax algebras with respect to the lax extension given by

$$\begin{aligned} \mathfrak{a}\,(\hat{F}r)\,\mathfrak{b} &\iff \mathfrak{b} \supseteq r[\mathfrak{a}] \\ &\iff \forall A \in \mathfrak{a}\, \exists B \in \mathfrak{b}\, \forall y \in B\, \exists x \in A\, (x\; r\; y) \end{aligned}$$

and their lax homomorphisms form a category isomorphic to the category of closure spaces (see Exercise III.1.O).

We still have to address the question of *whether the Barr extension is actually a lax extension of the functor T or, even better, of the monad $\mathbb{T}$*. To this end, the functor needs to satisfy an important additional condition that we now proceed to describe. Reassuringly, all the functors presented so far satisfy this condition, and therefore their Barr extensions are lax extensions (see Examples III.1.12.3).

III.1.11 The Beck–Chevalley condition

Consider Set-maps $f : X \to Z$ and $g : Y \to Z$, and let

$$\begin{array}{ccc} P & \xrightarrow{p_2} & Y \\ {\scriptstyle p_1}\downarrow & & \downarrow{\scriptstyle g} \\ X & \xrightarrow{f} & Z \end{array}$$

be a pullback diagram. The inequality $p_2 \cdot p_1^\circ \leq g^\circ \cdot f$ holds by commutativity of the diagram, while the pullback property forces equality: $p_2 \cdot p_1^\circ = g^\circ \cdot f$. More generally, a commutative diagram

$$\begin{array}{ccc} W & \xrightarrow{h_2} & Y \\ {\scriptstyle h_1}\downarrow & & \downarrow{\scriptstyle g} \\ X & \xrightarrow{f} & Z \end{array} \qquad \text{(III.1.11.i)}$$

is a *Beck–Chevalley square*, or simply a *BC-square*, if the maps involved satisfy $h_2 \cdot h_1^\circ = g^\circ \cdot f$, or equivalently if $h_1 \cdot h_2^\circ = f^\circ \cdot g$, i.e. if

$$\begin{array}{ccc} W & \xrightarrow{h_2} & Y \\ {\scriptstyle h_1^\circ}\uparrow & & \uparrow{\scriptstyle g^\circ} \\ X & \xrightarrow{f} & Z \end{array} \qquad \text{or equivalently} \qquad \begin{array}{ccc} W & \xleftarrow{h_2^\circ} & Y \\ {\scriptstyle h_1}\downarrow & & \downarrow{\scriptstyle g} \\ X & \xleftarrow{f^\circ} & Z \end{array}$$

commutes in Rel. In fact, we may trade 2 in Rel $=$ 2-Rel with any non-trivial quantale $\mathcal{V}$.

III.1.11.1 Lemma *The following conditions are equivalent for the commutative square* (III.1.11.i)*:*

(i) $h_2 \cdot h_1^\circ = g^\circ \cdot f$ *in* Rel;
(ii) $h_2 \cdot h_1^\circ = g^\circ \cdot f$ *in* $\mathcal{V}$-Rel, *for every quantale* $\mathcal{V}$;
(iii) $h_2 \cdot h_1^\circ = g^\circ \cdot f$ *in* $\mathcal{V}$-Rel, *for some non-trivial quantale* $\mathcal{V}$;
(iv) (III.1.11.i) *is a* weak pullback diagram *in* Set, *i.e. the canonical map* $c : W \to X \times_Z Y$ *is surjective.*

Proof The unique quantale homomorphism $\iota : 2 \to \mathcal{V}$ (with $\bot \mapsto \bot, \top \mapsto k$) induces a faithful functor ι : Rel $\to$ $\mathcal{V}$-Rel which sends $r : X \nrightarrow Y$ to $\iota r : X \nrightarrow Y$ (with $(\iota r)(x, y) = \iota(r(x, y))$); moreover, $\iota(f_\circ) = f_\circ$ and $\iota(f^\circ) = f^\circ$ for all maps $f = f_\circ$. Hence, ι facilitates the proof that (i), (ii), (iii) are equivalent. For (i) $\iff$ (iv), we just note that $c(w) = (h_1(w), h_2(w))$ for all $w \in W$, and

$$(x\ (g^\circ \cdot f)\ y \iff f(x) = g(y))\ ,$$
$$(x\ (h_2 \cdot h_1^\circ)\ y \iff \exists w \in W\ (h_1(w) = x\ \&\ h_2(w) = y))$$

for all $x \in X$, $y \in Y$. □

III.1.11.2 Definitions

(1) A Set-functor T satisfies the *Beck–Chevalley condition*, or *BC* for short, if it sends BC-squares to BC-squares:

$$h_2 \cdot h_1^\circ = g^\circ \cdot f \implies Th_2 \cdot (Th_1)^\circ = (Tg)^\circ \cdot Tf$$

for all maps f, g, h_1, h_2 with $h_1 \cdot f = g \cdot h_2$.

(2) A natural transformation $\alpha : S \to T$ between Set-functors S and T satisfies *BC* if its naturality diagrams

$$\begin{array}{ccc} SX & \xrightarrow{\alpha_X} & TX \\ {\scriptstyle Sf}\downarrow & & \downarrow{\scriptstyle Tf} \\ SY & \xrightarrow{\alpha_Y} & TY \end{array}$$

are BC-squares for all maps $f : X \to Y$, i.e. if $\alpha_X \cdot (Sf)^\circ = (Tf)^\circ \cdot \alpha_Y$, or equivalently if $Sf \cdot \alpha_X^\circ = \alpha_Y^\circ \cdot Tf$.

Note that, by Lemma III.1.11.1, it does not matter whether we read the equational conditions appearing in this definition in Rel or $\mathcal{V}$-Rel (for a non-trivial $\mathcal{V}$). Furthermore, we can easily prove the following characterization.

III.1.11.3 Proposition *The following statements are equivalent for a* Set-*functor* T*:*

(i) T *satisfies BC;*
(ii) T *preserves weak pullback diagrams;*
(iii) T *transforms pullbacks into weak pullbacks and preserves the surjectivity of maps.*

Proof (i) $\iff$ (ii) follows from Lemma III.1.11.1. The first assertion of (iii) follows trivially from (ii), and the second from (i) and Proposition III.1.2.2: $f : X \to Y$ is surjective precisely when $f \cdot f^\circ = 1_Y$, i.e. if and only if

$$\begin{array}{ccc} X & \xrightarrow{f} & Y \\ {\scriptstyle f}\downarrow & & \downarrow{\scriptstyle 1_Y} \\ Y & \xrightarrow{1_Y} & Y \end{array}$$

is a BC-square. Finally, for (iii) $\Longrightarrow$ (i), with the BC-square (III.1.11.i) and the canonical surjection $c : W \to X \times_Z Y$, the canonical map

$$TW \xrightarrow{Tc} T(X \times_Z Y) \xrightarrow{t} TX \times_{TZ} TY$$

is by hypothesis the composite of two surjectives, hence surjective. □

III.1.11.4 Remarks

(1) Assuming the Axiom of Choice, so that epimorphisms in Set split and are preserved by functors, we may add

   (iv) *T transforms pullbacks into weak pullbacks.*

   to the list of equivalent statements in Proposition III.1.11.3.

(2) This list may be extended further by the following statement, which is equivalent to (iv):

   (v) *for all Set-maps $f : X \to Z$, $g : Y \to Z$, the monotone map*

$$(-) \cdot t : \mathsf{Rel}(Q, W) \to \mathsf{Rel}(TP, W)$$

   is fully faithful for all sets W; here, $P = X \times_Z Y$, $Q = TX \times_{TZ} TY$, and $t : TP \to Q$ is the comparison map.

   In fact, for (iv) $\Longrightarrow$ (v), one considers $r, s : Q \nrightarrow W$ with $s \cdot t \leq r \cdot t$, and concludes $s = s \cdot t \cdot t^\circ \leq r \cdot t \cdot t^\circ = r$ since t is surjective. For (v) $\Longrightarrow$ (iv), one observes that t, like any map, satisfies $t = t \cdot t^\circ \cdot t$, in particular $1_Y \cdot t \leq t \cdot t^\circ \cdot t$. By hypothesis, one concludes $1_Y \leq t \cdot t^\circ$, so that t must be surjective.

We can now return to our primary purpose of introducing BC.

III.1.11.5 Theorem *For a functor $T : \mathsf{Set} \to \mathsf{Set}$, the following assertions are equivalent:*

(i) *the functor T satisfies BC;*
(ii) *the Barr extension $\overline{T}$ is a flat lax extension of T to Rel and a functor $\overline{T} : \mathsf{Rel} \to \mathsf{Rel}$;*
(iii) *there is some functor $\hat{T} : \mathsf{Rel} \to \mathsf{Rel}$ which is a lax extension of T to Rel.*

Moreover, any functor $\hat{T} : \mathsf{Rel} \to \mathsf{Rel}$ *as in* (iii) *is uniquely determined, i.e.* $\hat{T} = \overline{T}$.

Proof (i) $\Longrightarrow$ (ii): To see that $\overline{T}$ is a flat lax extension, the only issue lies in verifying $\overline{T}s \cdot \overline{T}r = \overline{T}(s \cdot r)$ for relations $r : X \nrightarrow Y$ and $s : Y \nrightarrow Z$ with respective factorizations $r = \pi_2 \cdot \pi_1^\circ$ and $s = \rho_2 \cdot \rho_1^\circ$. As the pullback (p_1, p_2) of $R \xrightarrow{\pi_2} Y \xleftarrow{\rho_1} S$ yields a mono-source that moreover forms a factorization $p_2 \cdot p_1^\circ$ of the relation $\rho_1^\circ \cdot \pi_2$,

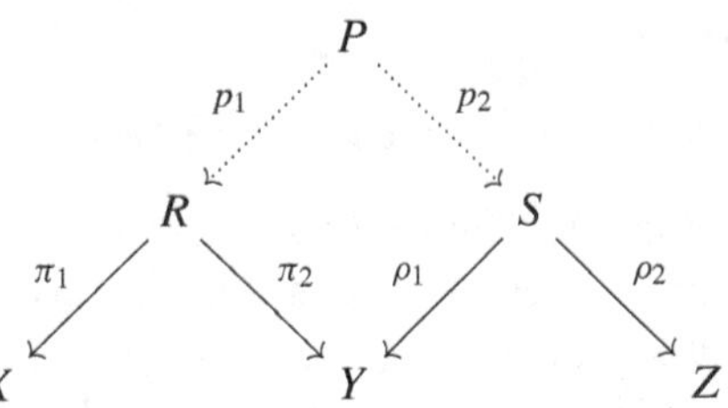

one has $\overline{T}(\rho_1^\circ \cdot \pi_2) = Tp_2 \cdot (Tp_1)^\circ$ (see Remark III.1.10.2(3)), and $Tp_2 \cdot (Tp_1)^\circ = (T\rho_1)^\circ \cdot T\pi_2$ since T satisfies BC. Consequently, with Remark III.1.10.2(2) one obtains

$$\begin{aligned}\overline{T}(s) \cdot \overline{T}(r) = T\rho_2 \cdot (T\rho_1)^\circ \cdot T\pi_2 \cdot (T\pi_1)^\circ &= T\rho_2 \cdot \overline{T}(\rho_1^\circ \cdot \pi_2) \cdot (T\pi_1)^\circ \\ &= \overline{T}(\rho_2 \cdot \rho_1^\circ \cdot \pi_2 \cdot \pi_1^\circ) = \overline{T}(s \cdot r)\,.\end{aligned}$$

(ii) $\Longrightarrow$ (iii): This is trivial.

(iii) $\Longrightarrow$ (i): Let $h_2 \cdot h_1^\circ = g^\circ \cdot f$ as in III.1.11.2(1). Since functoriality makes $\hat{T}$ also flat, one obtains, with Corollary III.1.4.4,

$$Th_2 \cdot (Th_1)^\circ = \hat{T}(h_2 \cdot h_1^\circ) = \hat{T}(g^\circ \cdot f) = (Tg)^\circ \cdot Tf\,.$$

For the same reasons, one has $\hat{T}r = T\pi_2 \cdot (T\pi_1)^\circ = \overline{T}r$, for $r = \pi_2 \cdot \pi_1^\circ$. □

III.1.12 The Barr extension of a monad

Theorem III.1.11.5 proves that if T satisfies BC, then the Barr extension $\overline{T}$ is a lax extension of T to Rel. It does not require much more effort to show that, under the same assumption, the Barr extension yields a lax extension of the monad $\mathbb{T} = (T, m, e)$.

Consider first a natural transformation $\alpha : S \to T$ between functors $S, T : \mathsf{Set} \to \mathsf{Set}$ provided with their lax extensions $\overline{S}, \overline{T}$. Then, for a relation $r : X \nrightarrow Y$ with $r = \pi_2 \cdot \pi_1^\circ$, we have

$$\alpha_Y \cdot \overline{S}r = \alpha_Y \cdot S\pi_2 \cdot (S\pi_1)^\circ = T\pi_2 \cdot \alpha_R \cdot (S\pi_1)^\circ \le T\pi_2 \cdot (T\pi_1)^\circ \cdot \alpha_X = \overline{T}r \cdot \alpha_X\,,$$

i.e. $\alpha : \overline{S} \to \overline{T}$ is oplax. From the same computation, we observe that $\alpha : \overline{S} \to \overline{T}$ is a natural transformation if all naturality diagrams

$$\begin{array}{ccc} SX & \xrightarrow{\alpha_X} & TX \\ {\scriptstyle Sf}\downarrow & & \downarrow{\scriptstyle Tf} \\ SY & \xrightarrow[\alpha_Y]{} & TY \end{array}$$

form BC-squares.

Therefore, if T belongs to a monad $\mathbb{T} = (T, m, e)$, then m and e become oplax natural transformations in Rel:

$$m : \overline{TT} \to \overline{T} \qquad \text{and} \qquad e : 1_{\mathsf{Rel}} \to \overline{T}\,.$$

An issue remains with the domain of the multiplication, which should be $\overline{T}\,\overline{T}$ rather than $\overline{TT}$. Hence, in order to obtain a lax extension of the monad $\mathbb{T}$ to Rel, we show that the identities 1_{TTX} are the components of an oplax natural transformation $\overline{T}\,\overline{T} \to \overline{TT}$. It follows from Remark III.1.10.2(3) and the equality $\overline{T}r = T\pi_2 \cdot (T\pi_1)^\circ$ that

$$\overline{TT}r = TT\pi_2 \cdot (TT\pi_1)^\circ \leq \overline{T}(\overline{T}r)$$

for any relation $r = \pi_2 \cdot \pi_1^\circ$, with equality holding if T preserves surjections. Thus, the Barr extension $\hat{\mathbb{T}} = (\overline{T}, m, e)$ is a lax extension of the Set-monad $\mathbb{T} = (T, m, e)$ to Rel provided that T satisfies BC.

III.1.12.1 Theorem *For a monad $\mathbb{T} = (T, m, e)$ on* Set*, the following assertions are equivalent:*

(i) *the functor T satisfies BC;*
(ii) *the Barr extension yields a flat lax extension $\overline{\mathbb{T}} = (\overline{T}, m, e)$ of $\mathbb{T}$ to* Rel.

Proof The implication (i) $\Longrightarrow$ (ii) follows from the preceding discussion since one knows from Theorem III.1.11.5 that if T satisfies BC then $\overline{T}$ is a flat lax extension of T. The converse implication is also an immediate consequence of the same result, since a monad is a flat lax extension exactly when its underlying functor is one. □

III.1.12.2 Corollary *Suppose that $\mathbb{T} = (T, m, e)$ is a monad on* Set *such that T and m satisfy BC. Then $\overline{\mathbb{T}}$ is an associative lax extension of $\mathbb{T}$ to* Rel.

Proof The proof of (i) $\Longrightarrow$ (ii) in Theorem III.1.12.1 yields that $\overline{T}s \cdot \overline{T}r = \overline{T}(s \cdot r)$ if T satisfies BC; the discussion preceding Theorem III.1.12.1 shows that if T preserves epimorphisms (as do functors that satisfy BC, see Section III.1.11) and naturality diagrams of m are BC-squares, then $m : \overline{T}\,\overline{T} \to \overline{T}$ is a natural transformation. The fact that the Barr extension is flat allows us to conclude that the Barr extension is associative by Proposition III.1.9.4. □

III.1.12.3 Examples

(1) The identity functor 1_{Set} on Set obviously satisfies BC; it is also immediate that the Barr extension 1_{Rel} is a lax extension.

(2) The filter functor $F : \mathsf{Set} \to \mathsf{Set}$ satisfies BC. Indeed, suppose that

$$\begin{array}{ccc} W & \xrightarrow{h_2} & Y \\ {\scriptstyle h_1}\downarrow & & \downarrow{\scriptstyle g} \\ X & \xrightarrow[f]{} & Z \end{array}$$

is a BC-square. Since the square commutes, one immediately obtains the inequality $Fh_2 \cdot (Fh_1)^\circ \leq (Fg)^\circ \cdot Ff$. For the other direction, we must show that, for all filters $\mathfrak{a} \in FX$ and $\mathfrak{b} \in FY$,

$$f[\mathfrak{a}] = g[\mathfrak{b}] \implies \exists \mathfrak{c} \in FW\ (h_1[\mathfrak{c}] = \mathfrak{a}\ \&\ h_2[\mathfrak{c}] = \mathfrak{b})\,.$$

But the sets $h_1^{-1}(A) \cap h_2^{-1}(B)$ (for $A \in \mathfrak{a}$, and $B \in \mathfrak{b}$) form a base for a filter $\mathfrak{c}$ satisfying $h_1[\mathfrak{c}] = \mathfrak{a}$ and $h_2[\mathfrak{c}] = \mathfrak{b}$; indeed, $g^\circ \cdot f \leq h_2 \cdot h_1^\circ$ means that, for any pair $(x, y) \in A \times B$ with $f(x) = g(y)$, there is an element $w \in W$ satisfying $h_1(w) = x$ and $h_2(w) = y$. Thus, the Barr extension $\overline{F}$ described in Example III.1.10.3(2) is a lax extension of F to Rel.

(3) The ultrafilter functor satisfies BC for similar reasons. In this case, to see that $(\beta g)^\circ \cdot \beta f \leq \beta h_2 \cdot (\beta h_1)^\circ$, one obtains a filter $\mathfrak{c}$ from ultrafilters $\mathfrak{a} \in FX$, $\mathfrak{b} \in FY$ as above. By Proposition II.1.13.2, there is an ultrafilter $\mathfrak{x}$ on W with $\mathfrak{c} \subseteq \mathfrak{x}$, so that $\mathfrak{a} \subseteq h_1[\mathfrak{x}]$ and $\mathfrak{b} \subseteq h_2[\mathfrak{x}]$, and maximality of ultrafilters yields the required equalities. Thus, as in the filter case, the Barr extension $\overline{\beta}$ of Example III.1.10.3(2) is a flat lax extension of β.

We now show that the multiplication m of the ultrafilter monad $\mathbb{\beta}$ satisfies BC. For any map $f : X \to Y$ and all $\mathfrak{x} \in \beta X$ and $\mathcal{Y} \in \beta\beta Y$ with $m_Y(\mathcal{Y}) = \beta f(\mathfrak{x})$, we must find $\mathcal{X} \in \beta\beta X$ with

$$\beta\beta f(\mathcal{X}) = \mathcal{Y} \qquad \text{and} \qquad m_X(\mathcal{X}) = \mathfrak{x}\,.$$

By hypothesis, $f(A)^{\mathbb{\beta}} \cap \mathcal{B} \neq \varnothing$, for all $A \in \mathfrak{x}$ and $\mathcal{B} \in \mathcal{Y}$. One easily verifies $f(A)^{\mathbb{\beta}} = \beta f(A^{\mathbb{\beta}})$, and from $\beta f(A^{\mathbb{\beta}}) \cap \mathcal{B} \neq \varnothing$ one obtains $A^{\mathbb{\beta}} \cap (\beta f)^{-1}(\mathcal{B}) \neq \varnothing$. Therefore,

$$\{A^{\mathbb{\beta}} \mid A \in \mathfrak{x}\} \cup \{(\beta f)^{-1}(\mathcal{B}) \mid \mathcal{B} \in \mathcal{Y}\}$$

is a filter base, and any ultrafilter $\mathcal{X}$ containing it has the desired property. Consequently, by Corollary III.1.12.2, $\overline{\mathbb{\beta}}$ is associative.

As in Section III.1.9, one can now consider the category $(\mathbb{\beta}, 2)$-URel of sets and unitary $(\mathbb{\beta}, 2)$-relations. (The same statement holds for the filter monad with its Barr extension, but we will not consider this particular instance any further.)

The unit of neither the filter nor the ultrafilter monad satisfies BC (see Exercise III.1.Q). In the case of the ultrafilter monad, there is a general reason for this claim, as follows.

III.1.12.4 Proposition *Any monad* $\mathbb{T} = (T, m, e)$ *with* $T1 \cong 1$ *and* e *satisfying BC must be isomorphic to the identity monad.*

Proof Since e satisfies BC, the diagram

$$\begin{array}{ccc} TX & \xrightarrow{T!_X} & T1 \\ {\scriptstyle e_X^\circ}\big\downarrow & & \big\downarrow{\scriptstyle e_1^\circ} \\ X & \xrightarrow{!_X} & 1 \end{array}$$

commutes in Rel for every set X. Expressed elementwise (with $1 = \{\star\}$), this reads as

$$\forall \chi \in TX \; (T!_X(\chi) = e_1(\star) \iff \exists x \in X \; (e_X(x) = \chi)) \, ,$$

with $T!_X(\chi) = e_1(\star)$ holding since $T1 \cong 1$. Hence, e_X must be surjective, or even bijective if $\mathbb{T}$ is non-trivial (see Exercise II.3.A). But neither of the two trivial monads on Set has a unit satisfying BC (Exercise III.1.Q). □

III.1.13 A double-categorical presentation of lax extensions

The notion of a lax extension $\hat{T} : \mathcal{V}\text{-Rel} \to \mathcal{V}\text{-Rel}$ of a functor $T : \mathsf{Set} \to \mathsf{Set}$ as given in Definition III.1.4.1 allows for a very natural double-categorical interpretation. We briefly describe this presentation here without formally introducing double categories, their functors, and natural transformations. To this end, we consider diagrams

$$\begin{array}{ccc} X & \xrightarrow{f} & Y \\ {\scriptstyle r}\big\downarrow & \leq & \big\downarrow{\scriptstyle s} \\ U & \xrightarrow[g]{} & V \end{array}$$

consisting of Set-maps f, g, and $\mathcal{V}$-relations r, s, such that

$$g \cdot r \leq s \cdot f \, ,$$

or equivalently $r \leq g^\circ \cdot s \cdot f$, i.e. $r(x, u) \leq s(f(x), g(u))$ for all $x \in X, u \in U$. We call these diagrams *cells*. The point is that such a cell may be considered as a morphism $r \to s$ horizontally as well as a morphism $f \to g$ vertically. With map composition used horizontally and $\mathcal{V}$-relational composition used vertically, one obtains two intertwined category structures whose main interaction is captured by the *middle-interchange law*:

$$\begin{array}{|c|c|}\hline \alpha & \beta \\ \hline \gamma & \delta \\ \hline \end{array}$$

For cells $\alpha, \beta, \gamma, \delta$ that fit together as indicated above, one has

$$(\delta \cdot \gamma) \circ (\beta \cdot \alpha) = (\delta \circ \beta) \cdot (\gamma \circ \alpha)$$

(with vertical composition denoted by $\circ$). Cells and their compositions form a *double category* $\mathcal{V}\text{-}\mathcal{R}el$. A *double category functor* $\mathcal{T} : \mathcal{V}\text{-}\mathcal{R}el \to \mathcal{V}\text{-}\mathcal{R}el$ returns for every cell α a cell $\mathcal{T}\alpha$, preserves horizontal composition and identity morphisms strictly, and vertical composition and identity morphisms laxly. Hence $\mathcal{T}$ is in fact given by a functor $T : \mathsf{Set} \to \mathsf{Set}$ and a lax functor $\hat{T} : \mathcal{V}\text{-}\mathsf{Rel} \to \mathcal{V}\text{-}\mathsf{Rel}$ such that

$$\begin{array}{ccc} X & \xrightarrow{f} & Y \\ {\scriptstyle r}\downarrow & \leq & \downarrow{\scriptstyle s} \\ U & \xrightarrow[g]{} & V \end{array} \quad \overset{\mathcal{T}}{\mapsto} \quad \begin{array}{ccc} TX & \xrightarrow{Tf} & TY \\ {\scriptstyle \hat{T}r}\downarrow & \leq & \downarrow{\scriptstyle \hat{T}s} \\ TU & \xrightarrow[Tg]{} & TV \end{array}$$

i.e. $g \cdot r \leq s \cdot f$ implies $Tg \cdot \hat{T}r \leq \hat{T}s \cdot Tf$.

III.1.13.1 Proposition *Double category functors* $\mathcal{T} = (T, \hat{T}) : \mathcal{V}\text{-}\mathcal{R}el \to \mathcal{V}\text{-}\mathcal{R}el$ *are precisely the lax extensions* $\hat{T} : \mathcal{V}\text{-}\mathsf{Rel} \to \mathcal{V}\text{-}\mathsf{Rel}$ *of functors* $T : \mathsf{Set} \to \mathsf{Set}$.

Proof It suffices to show that the cell preservation condition for $\mathcal{T} = (T, \hat{T})$ is equivalent to the lax extension conditions $Tf \leq \hat{T}f$ and $(Tf)^\circ \leq \hat{T}(f^\circ)$, given that T is a functor and $\hat{T}$ a lax functor. When one exploits the preservation conditions for the cells

$$\begin{array}{ccc} X & \xrightarrow{1_X} & Y \\ {\scriptstyle 1_X}\downarrow & \leq & \downarrow{\scriptstyle f} \\ X & \xrightarrow[f]{} & X \end{array} \quad \text{and} \quad \begin{array}{ccc} X & \xrightarrow{f} & Y \\ {\scriptstyle 1_X}\downarrow & \leq & \downarrow{\scriptstyle f^\circ} \\ X & \xrightarrow[1_X]{} & X \end{array}$$

one obtains

$$Tf = Tf \cdot T1_X \leq Tf \cdot \hat{T}1_X \leq \hat{T}f \cdot T1_X = \hat{T}f$$

and

$$(Tf)^\circ = 1_{TX} \cdot (Tf)^\circ \leq \hat{T}1_X \cdot (Tf)^\circ \leq \hat{T}(f^\circ) \cdot Tf \cdot (Tf)^\circ \leq \hat{T}(f^\circ),$$

respectively. Conversely, by Corollary III.1.4.4, a lax extension satisfies the preservation condition: expressing $g \cdot r \leq s \cdot f$ equivalently by $r \leq g^\circ \cdot s \cdot f$, one obtains $\hat{T}r \leq (Tg)^\circ \cdot \hat{T}s \cdot Tf$, or $Tg \cdot \hat{T}r \leq \hat{T}s \cdot Tf$. □

One can now proceed to consider an appropriate monad structure on $\mathcal{T}$, by suitable natural transformations $m : \mathcal{T}\mathcal{T} \to \mathcal{T}$ and $e : 1 \to \mathcal{T}$: we require these transformations to be given by *horizontal* natural transformations $m : TT \to T$ and $e : 1 \to T$ (in the ordinary sense) that are compatible with the vertical structure, so that there are cells

$$\begin{array}{ccc} TTX & \xrightarrow{m_X} & TX \\ {\scriptstyle \hat{T}\hat{T}r}\big\downarrow & \leq & \big\downarrow{\scriptstyle \hat{T}r} \\ TTY & \xrightarrow[m_Y]{} & TY \end{array} \quad \text{and} \quad \begin{array}{ccc} X & \xrightarrow{e_X} & TX \\ {\scriptstyle r}\big\downarrow & \leq & \big\downarrow{\scriptstyle \hat{T}r} \\ Y & \xrightarrow[e_Y]{} & TY \end{array}$$

for every $\mathcal{V}$-relation $r : X \nrightarrow Y$. But the existence requirement for these cells gives precisely the condition that $m : \hat{T}\hat{T} \to \hat{T}$ and $e : 1 \to \hat{T}$ be op-lax (see Definition III.1.5.1). We therefore have:

III.1.13.2 Corollary *Monads of the double category $\mathcal{V}$-$\mathcal{R}el$ are precisely lax extensions to $\mathcal{V}$-*Rel *of monads on* Set.

Exercises

III.1.A *The trivial and integral quantales.* Show that a quantale $\mathcal{V}$ is trivial (i.e. $|\mathcal{V}| = 1$) if and only if $\bot = k$, where k denotes the neutral element of $\mathcal{V}$. Furthermore, $\mathcal{V}$ is integral (i.e. $k = \top$) if and only if the terminal object of $\mathcal{V}$-Cat is a generator.

III.1.B *Lean but not integral.* For a monoid M, the powerset PM has a quantale structure as in Exercise II.1.M. Then PM is lean, but integral only if M is trivial.

III.1.C *Associativity of $\mathcal{V}$-relational composition.* Verify that

$$t \cdot (s \cdot r) = (t \cdot s) \cdot r$$

holds for all $\mathcal{V}$-relations $r : X \nrightarrow Y$, $s : Y \nrightarrow Z$, and $t : Z \nrightarrow A$.

III.1.D *A $\mathcal{V}$-powerset monad.* Given a quantale $\mathcal{V}$, the *$\mathcal{V}$-powerset functor* $P_\mathcal{V}$ sends a set X to its *$\mathcal{V}$-powerset* $\mathcal{V}^X$, and a map $f : X \to Y$ to $P_\mathcal{V} f : \mathcal{V}^X \to \mathcal{V}^Y$, where

$$P_\mathcal{V} f(\phi)(y) := \textstyle\bigvee_{x \in f^{-1}(y)} \phi(x)\,,$$

for all $\phi \in \mathcal{V}^Y$, $y \in Y$. The multiplication $\mu : P_\mathcal{V} P_\mathcal{V} \to P_\mathcal{V}$ and unit $\delta : 1_{\mathsf{Set}} \to P_\mathcal{V}$ of the *$\mathcal{V}$-powerset monad* $\mathbb{P}_\mathcal{V}$ are defined, respectively, by

$$\mu_X(\Phi)(y) := \bigvee_{\phi \in \mathcal{V}^X} \Phi(\phi) \otimes \phi(y) \qquad \text{and} \qquad \delta_X(x)(y) := \begin{cases} k & \text{if } x = y, \\ \bot & \text{otherwise,} \end{cases}$$

for all $x, y \in X$, $\Phi \in \mathcal{V}^{\mathcal{V}^X}$. The extension operation $(-)^{\mathbb{P}_\mathcal{V}}$ of the corresponding Kleisli triple (see II.3.7) is given for any $f : X \to P_\mathcal{V} Y$ by

$$f^{\mathbb{P}_\mathcal{V}}(\phi)(y) = \bigvee_{x \in X} \phi(x) \otimes f(x)(y)$$

for all $\phi \in \mathcal{V}^X$, $y \in Y$. The $\mathsf{2}$-powerset monad is simply the powerset monad $\mathbb{P}$, and the Kleisli category returns $\mathcal{V}$-Rel:

$$\mathsf{Set}_{\mathbb{P}_\mathcal{V}} = \mathcal{V}\text{-}\mathsf{Rel}\,.$$

III.1.E *Extensions and liftings in* $\mathcal{V}$-Rel. For $r : X \nrightarrow Y$ and $s : X \nrightarrow Z$, show that the extension $s \multimapdotinv r$ in the quantaloid $\mathcal{V}$-Rel (see II.4.8) may be described as

$$(s \multimapdotinv r)(y, z) = \bigwedge_{x \in X} r(x, y) \multimapdotinv s(x, z)\,.$$

Likewise, for $t : Z \nrightarrow Y$, the lifting $r \multimapdot t$ is described by

$$(r \multimapdot t)(z, x) = \bigwedge_{y \in Y} t(z, y) \multimapdotinv r(x, y)\,.$$

III.1.F *Symmetric and separated metric spaces.* The category $\mathsf{Met}_{\mathrm{sym}}$ is bicoreflective in Met, and $\mathsf{Met}_{\mathrm{sep}}$ is strongly epireflective in Met (see Example III.1.3.1(2)). The notion of symmetry and separation can be introduced for $\mathcal{V}$-categories and the previous statements generalized to $\mathcal{V}$-Cat, for arbitrary $\mathcal{V}$ in lieu of P_+.

III.1.G *Limits and colimits in* $\mathcal{V}$-Cat

(1) Products and coproducts of $(X_i, a_i)_{i \in I}$ in $\mathcal{V}$-Cat are formed by endowing the sets $\prod_{i \in I} X_i$ and $\coprod_{i \in I} X_i$ with the structures

$$p((x_i)_{i \in I}, (y_i)_{i \in I}) = \bigwedge_{i \in I} a_i(x_i, y_i) \qquad \text{and}$$

$$s((x, i), (y, j)) = \begin{cases} a_i(x, y) & \text{if } i = j, \\ \bot & \text{otherwise,} \end{cases}$$

respectively.

(2) Show that the forgetful functor $O : \mathcal{V}$-$\mathsf{Cat} \to \mathsf{Set}$ is topological. Conclude that $\mathcal{V}$-Cat is small-complete and small-cocomplete, with all limits and colimits in $\mathcal{V}$-Cat preserved by O.

(3) For a surjective map of sets $f : X \to Y$ and $z, w \in Y$, a tuple $(x_0, y_0, x_1, y_1, \dots, x_n, y_n)$ (with $0 \le n$) is f*-admissible* for z, w if $z = f(x_0)$, $f(y_n) = w$ and

$$f(y_0) = f(x_1),\ f(y_1) = f(x_2), \dots, f(y_{n-1}) = f(x_n)\,.$$

For $f : (X, a) \to (Y, b)$ in $\mathcal{V}$-Cat, prove that f is O-final if and only if

$$b(z, w) = \bigvee_{\substack{(x_0, y_0, \ldots, x_n, y_n) \text{ is} \\ f\text{-admissible for } z,\, w}} a(x_0, y_0) \otimes \ldots \otimes a(x_n, y_n)$$

for all $z, w \in Y$.

(4) Describe equalizers and coequalizers in $\mathcal{V}$-Cat and apply the description to Met.

III.1.H *Yoneda functor, Yoneda Lemma, initial density of* $\mathcal{V}$ *in* $\mathcal{V}$-Cat. Let $\mathcal{V}$ be a commutative quantale, considered as a $\mathcal{V}$-category $(\mathcal{V}, \multimap)$, and for a $\mathcal{V}$-category $X = (X, a)$, set $\widehat{X} = [X^{\mathrm{op}}, \mathcal{V}]$ (see Proposition III.1.3.3).

(1) Show that $\mathbf{y} : X \to \widehat{X}$ with

$$\mathbf{y}(x) = a(-, x) : X^{\mathrm{op}} \to \mathcal{V}$$

defines a $\mathcal{V}$-functor.

(2) Prove $\hat{a}(\mathbf{y}(x), \phi) = \phi(x)$ for all $x \in X$, $\phi \in \widehat{X}$, where $\hat{a}$ denotes the $\mathcal{V}$-category structure of $\widehat{X}$.

(3) Conclude that $\mathbf{y}$ is O-initial with respect to the forgetful functor $O : \mathcal{V}\text{-Cat} \to \text{Set}$ (see Exercise III.1.G).

(4) With ev_x denoting the evaluation $\mathcal{V}$-functor at $x \in X$, show that the source

$$(\, X \xrightarrow{\;\mathbf{y}\;} \widehat{X} \xrightarrow{\;\mathrm{ev}_x\;} \mathcal{V} \,)_{x \in X}$$

is O-initial, and conclude that $\mathcal{V}$ is O-initially dense in $\mathcal{V}$-Cat.

III.1.I *Lax distributive laws and lax monad extensions.* A *lax distributive law* of the $\mathcal{V}$-powerset monad $\mathbb{P}_{\mathcal{V}}$ (Exercise III.1.D) over the monad $\mathbb{T}$ on Set is a natural transformation $\lambda_X : TP_{\mathcal{V}}X \to P_{\mathcal{V}}TX$, such that the diagrams

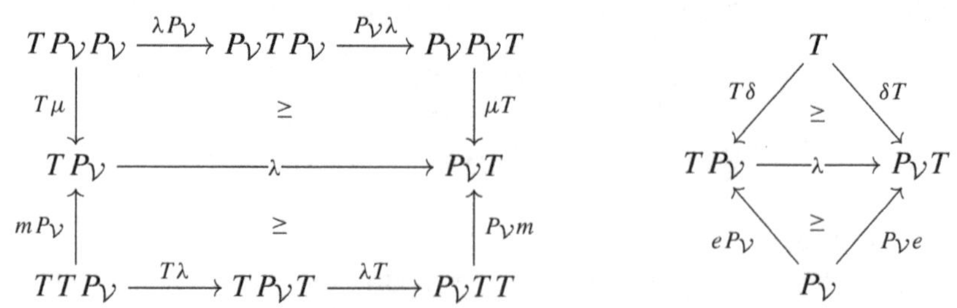

commute up to "$\geq$" as indicated. There is a bijective correspondence between lax extensions $\hat{\mathbb{T}}$ of $\mathbb{T}$ to $\mathcal{V}$-Rel and lax distributive laws λ of $\mathbb{P}_{\mathcal{V}}$ over $\mathbb{T}$ satisfying

$$r \leq r' \implies \lambda_Y \cdot Tr \leq \lambda_Y \cdot Tr'$$

for all $\mathcal{V}$-relations $r, r' : X \nrightarrow Y$ presented as maps $r, r' : X \to P_{\mathcal{V}}Y$: a lax extension $\hat{T} : \mathcal{V}\text{-Rel} \to \mathcal{V}\text{-Rel}$ yields a lax distributive law $\lambda = (\hat{T}1_{P_{\mathcal{V}}X} :$

$TP_{\mathcal{V}}X \to P_{\mathcal{V}}TX)_{X\in \mathrm{ob}\,\mathsf{Set}}$, and a lax distributive law $\lambda : TP_{\mathcal{V}} \to P_{\mathcal{V}}T$ determines a lax extension $\hat{T}$ that sends a $\mathcal{V}$-relation $r : X \to P_{\mathcal{V}}Y$ to the $\mathcal{V}$-relation $\lambda_Y \cdot Tr : TX \to P_{\mathcal{V}}TY$.

III.1.J *The dual lax extension.* Let $\mathcal{V}$ be a commutative quantale. If $\hat{T}$ is a lax extension to $\mathcal{V}$-Rel of a Set-functor T, then, for any $\mathcal{V}$-relation $r : X \nrightarrow Y$, its *dual*

$$\hat{T}^{\circ}r := (\hat{T}(r^{\circ}))^{\circ}$$

is also a lax extension to $\mathcal{V}$-Rel of T, and one has $(\hat{T}^{\circ})^{\circ} = \hat{T}$. If $\hat{\mathbb{T}} = (\hat{T}, m, e)$ is a lax extension of a monad $\mathbb{T}$, then so is $\hat{\mathbb{T}}^{\circ} = (\hat{T}^{\circ}, m, e)$. In this case, $\hat{\mathbb{T}}^{\circ}$ is associative if and only if $\hat{T}$ preserves composition of $\mathcal{V}$-relations and $m : \hat{T}\hat{T} \to \hat{T}$ is a natural transformation.

For the lax extensions of the powerset and filter monads coming from Examples III.1.4.2(2) and III.1.10.3(4), one has

$$\check{P} = \hat{P}^{\circ} \qquad \text{and} \qquad \check{F} = \hat{F}^{\circ} .$$

The Barr extension of a functor T to Rel is always self-dual:

$$\overline{T}^{\circ} = \overline{T} .$$

III.1.K *Checking* $(\mathbb{T}, \mathcal{V})$*-functoriality.* Prove that the following equivalences hold:

$$f \cdot a \leq b \cdot Tf \iff a \leq f^{\circ} \cdot b \cdot Tf \iff f \cdot a \cdot (Tf)^{\circ} \leq b$$

for all Set-maps $f : X \to Y$, and $\mathcal{V}$-relations $a : TX \nrightarrow X$, $b : TY \nrightarrow Y$.

III.1.L *Associativity of the Kleisli convolution.* Consider a lax extension $\hat{T}$ to $\mathcal{V}$-Rel of the underlying functor of $\mathbb{T} = (T, m, e)$, and $(\mathbb{T}, \mathcal{V})$-relations $r : X \nrightarrow Y$, $s : Y \nrightarrow Z$, $t : Z \nrightarrow W$.

(1) If $m_Y^{\circ} \cdot \hat{T}r \leq \hat{T}\hat{T}r \cdot m_{TX}^{\circ}$, in particular if $m^{\circ} : \hat{T} \to \hat{T}\hat{T}$ is a natural transformation, then $(t \circ s) \circ r \leq t \circ (s \circ r)$.
(2) If $\hat{T}r \cdot m_X^{\circ} \cdot \hat{T}1_X \leq \hat{T}r \cdot m_X^{\circ}$ and $\hat{T}$ preserves the composition of $\mathcal{V}$-relations, then $t \circ (s \circ r) \leq (t \circ s) \circ r$.

III.1.M *Kleisli convolution for* $(\mathbb{T}, \mathcal{V})$*-functors.* For a monad $\mathbb{T}$ with a lax extension $\hat{\mathbb{T}}$, one defines for a map $f : X \to Y$ the unitary $\mathcal{V}$-relation $f^{\sharp} = (e_Y \cdot f)^{\circ} \cdot \hat{T}1_Y : TY \nrightarrow X$. The condition

$$a \cdot (Tf)^{\circ} \leq f^{\circ} \cdot b ,$$

for unitary $\mathcal{V}$-relations $a : TX \nrightarrow X$ and $b : TY \nrightarrow Y$, is then equivalent to

$$a \circ f^{\sharp} \leq f^{\sharp} \circ b .$$

Briefly put, $(\mathbb{T}, \mathcal{V})$-functoriality of a map $f : X \to Y$ between $(\mathbb{T}, \mathcal{V})$-categories (X, a) and (Y, b) can be expressed in terms of the Kleisli convolution.

III.1.N *Unitary* $(\mathbb{T}, \mathcal{V})$*-relations.* Let $(\hat{T}, m, e)$ be a lax extension of a monad $\mathbb{T} = (T, m, e)$ on Set. Show that $\hat{T}r : X \nrightarrow TY$ and $r_\sharp = e_Y^\circ \cdot \hat{T}r$ are unitary $(\mathbb{T}, \mathcal{V})$-relations for every $\mathcal{V}$-relation $r : X \nrightarrow Y$. Moreover, if $(\mathbb{T}, \mathcal{V})$-relations $r : X \nrightarrow Y$ and $s : Y \nrightarrow Z$ are respectively right and left unitary, then $s \circ r$ is unitary.

III.1.O *Closure spaces via the filter monad.* A lax extension of the filter monad $\mathbb{F}$ to Rel is obtained via

$$\mathfrak{a}\ \hat{F}r\ \mathfrak{b} \iff \mathfrak{b} \supseteq r[\mathfrak{a}]$$

for all $\mathfrak{a} \in FX$, $\mathfrak{b} \in FY$, and relations $r : X \nrightarrow Y$ (Example III.1.10.3(4)). Verify that, for a map $c : PX \to PX$ and relation $r : FX \nrightarrow X$, the relation $r_c : FX \nrightarrow X$ and map $c_r : PX \to PX$ given by

$$\mathfrak{a}\ r_c\ x \iff x \in \textstyle\bigcap_{A \in \mathfrak{a}} c(A) \qquad \text{and} \qquad x \in c_r(A) \iff \dot{A}\ r\ x$$

(where $\dot{A}$ denotes the principal filter over A) determine an isomorphism between the categories of $(\mathbb{F}, 2)$-categories and of closure spaces:

$$(\mathbb{F}, 2, \hat{\mathbb{F}})\text{-}\mathsf{Cat} \cong \mathsf{Cls}\,.$$

III.1.P *Functors on* Set *preserving monomorphisms.*

(1) The following statements are equivalent for a functor $T : \mathsf{Set} \to \mathsf{Set}$:
   (a) T preserves monomorphisms;
   (b) T preserves the monomorphism $\varnothing \to T\varnothing$;
   (c) T preserves the pullback

$$\begin{array}{ccc} \varnothing & \longrightarrow & \varnothing \\ \big\downarrow & & \big\downarrow \\ \varnothing & \longrightarrow & T\varnothing \end{array}\ ;$$

   (d) T preserves the pullback of (c) *weakly*, i.e. the canonical map $T\varnothing \to T\varnothing \times_{TT\varnothing} T\varnothing$ is surjective.

(2) Each of the following conditions implies (a)–(d) of (1):
   (a) $T\varnothing = \varnothing$;
   (b) T preserves the disjointness of some binary coproduct, that is, $TX \times_{T(X+Y)} TY = \varnothing$ for some X, Y;
   (c) T preserves some binary coproduct;
   (d) T preserves kernel pairs;
   (e) T preserves kernel pairs weakly;
   (f) T satisfies BC;
   (g) T is the functor of a monad $\mathbb{T} = (T, m, e)$.

(3) Show that there is a functor $T : \mathsf{Set} \to \mathsf{Set}$, with $T\varnothing = \{\star\} + \{\star\}$ and $TX = X + \{\star\}$ for $X \neq \varnothing$, which does not preserve monomorphisms. But there are natural transformations $m : TT \to T$ and $e : 1_{\mathsf{Set}} \to T$ with

$$m \cdot mT = m \cdot Tm \qquad \text{and} \qquad m \cdot eT = 1\,.$$

III.1.Q *Units of monads and BC.*

(1) The units of the two trivial monads on Set (see Exercise II.3.A) do not satisfy BC.
(2) The units of the powerset and the filter monad do not satisfy BC.
Hint. Consider the map $!_X : X \to 1$ for $|X| \geq 2$.
(3) The list monad (see Example II.3.1.1(2)) is *Cartesian*, i.e. its functor L preserves pullbacks, and every naturality square of the unit and of the multiplication is a pullback. The functor L also preserves surjectivity of maps. In particular, L and its monad multiplication satisfy BC.

III.2 Fundamental examples

In this section, we present some of the motivating examples of lax algebras that will accompany us throughout this book. Further examples appear in Exercises III.2.B and III.2.D.

III.2.1 Ordered sets, metric spaces, and probabilistic metric spaces

In Example III.1.3.1(1), we saw that $\mathsf{2}$-categories with $\mathsf{2}$-functors (i.e. $(\mathbb{I}, \mathsf{2})$-categories with $(\mathbb{I}, \mathsf{2})$-functors, where $\mathbb{I}$ is identically extended to Rel; see Section III.1.6) were equivalenty described as ordered sets with monotone maps:

$$(\mathbb{I}, \mathsf{2})\text{-}\mathsf{Cat} = \mathsf{2}\text{-}\mathsf{Cat} = \mathsf{Ord}\,.$$

Similarly, P_+-categories and P_+-functors are the metric spaces with non-expansive maps of Example III.1.3.1(2):

$$(\mathbb{I}, \mathsf{P}_+)\text{-}\mathsf{Cat} = \mathsf{P}_+\text{-}\mathsf{Cat} = \mathsf{Met}\,.$$

The quantale isomorphism between P_+ and the unit interval with its multiplication:

$$([0, \infty]^{\mathrm{op}}, +, 0) \cong ([0, 1], \cdot, 1)$$

(see Exercise III.2.A) allows for a "probabilistic" interpretation of metric spaces. Say that a map $\phi : [0, \infty] \to [0, 1]$ is a *distance distribution* if

$$\phi(v) = \bigvee_{w<v} \phi(w)$$

for all $v \in [0, \infty]$; the *convolution* of two distance distribution maps $\phi, \psi : [0, \infty] \to [0, 1]$ yields a distance distribution map $\phi \otimes \psi : [0, \infty] \to [0, 1]$ via

$$(\phi \otimes \psi)(u) = \bigvee_{v+w \leq u} \phi(v) \cdot \psi(w)$$

for all $u \in [0, \infty]$; with $\kappa : [0, \infty] \to [0, 1]$ defined by $\kappa(0) = 0$ and $\kappa(u) = 1$ if $0 < u$, the set D of all distribution maps ordered pointwise forms a quantale

$$\mathsf{D} = (\mathsf{D}, \otimes, \kappa) \,.$$

A *probabilistic metric* is then a map $a : X \times X \to \mathsf{D}$ such that

$$a(x, x) = \kappa \qquad \text{and} \qquad a(x, y) \otimes a(y, z) \leq a(x, z)$$

for all $x, y, z \in X$. Here, the value $p = a(x, y)(u)$ can be loosely interpreted as the "probability" that a given randomized metric $\tilde{a} : X \times X \to [0, \infty]$ satisfies $\tilde{a}(x, y) < u$. A set X with a probabilistic metric $a : X \times X \to \mathsf{D}$ forms a *probabilistic metric space* (X, a), and a map $f : X \to Y$ between probabilistic metric spaces (X, a) and (X, b) is *probabilistically non-expansive* if

$$a(x, y) \leq b(f(x), f(y))$$

for all $x, y \in X$. Probabilistic metric spaces with probabilistically non-expansive maps are the objects and morphisms of the category $\mathsf{ProbMet}$, and one observes

$$(\mathbb{I}, \mathsf{D})\text{-}\mathsf{Cat} = \mathsf{D}\text{-}\mathsf{Cat} = \mathsf{ProbMet} \,.$$

In Example III.3.5.2(2) we exhibit full embeddings

$$\mathsf{Ord} \hookrightarrow \mathsf{Met} \hookrightarrow \mathsf{ProbMet}$$

which make explicit how the structures discussed here generalize each other.

III.2.2 Topological spaces

Our paradigmatic example of a $(\mathbb{T}, \mathcal{V})$-category comes from [Barr, 1970], in which topological spaces are presented as so-called *relational algebras* for the ultrafilter monad; in our context, this result reads as

© $$(\mathbb{\beta}, 2)\text{-}\mathsf{Cat} \cong \mathsf{Top} \,.$$

Recall from Examples II.3.1.1 that $\mathbb{\beta}$ stands for the ultrafilter monad, and the required lax extension $\overline{\beta}$ of $\beta : \mathsf{Set} \to \mathsf{Set}$ to $\mathsf{Rel} \cong 2\text{-}\mathsf{Rel}$ is described in Example III.1.10.3(3). In the following, we will freely use the notations introduced in II.1.12 and II.1.13. There are many ways to prove the isomorphism mentioned above, depending in particular on the choice of the standard presentation of topological spaces: open or closed sets, interior or closure operations, or neighborhood systems. Here we work with closure operations, while in Chapter IV, with a more developed theory at our disposal, we will present another approach. The idea of the isomorphism between $(\beta, \mathsf{2})$-categories and topological spaces is that a relation $r : \beta X \nrightarrow X$ represents convergence and specifies which ultrafilters converge to which points of X. We can then associate with r a finitely additive closure operation $c : PX \to PX$, and, conversely, show that every finitely additive closure operation c determines a convergence relation $r : \beta X \nrightarrow X$.

In II.1.6, closure spaces were introduced as pairs (X, c) consisting of a set X and a closure operation $c : PX \to PX$ on the powerset PX of X. We observed that such a closure operation is a monotone map (equivalently an element of $\mathsf{Ord}(PX, PX)$) carrying a monoid structure with respect to the compositional structure, i.e.

$$c \cdot c(A) \subseteq c(A)\ , \quad A \subseteq c(A)$$

for all $A \subseteq X$. Furthermore, $c : PX \to PX$ defines a topology on X if and only if c is finitely additive:

$$c(A \cup B) = c(A) \cup c(B)\ , \quad c(\varnothing) = \varnothing$$

for all $A, B \subseteq X$ (see Exercises II.1.F and II.1.G). To any relation $r : \beta X \nrightarrow X$, we can associate a finitely additive map $\mathrm{clos}(r) : PX \to PX$ given by

$$\mathrm{clos}(r)(A) = \{y \in X \mid \exists \mathfrak{x} \in \beta X\ (A \in \mathfrak{x}\ \&\ \mathfrak{x}\ r\ y)\}\ .$$

Conversely, given a map $c : PX \to PX$, we define a relation $\mathrm{conv}(c) : \beta X \nrightarrow X$ by setting

$$\mathfrak{x}\ \mathrm{conv}(c)\ y \iff \forall A \in PX\ (A \in \mathfrak{x} \implies y \in c(A))\ .$$

Note that we have the identities

$$\mathrm{conv}(c)(A^{\beta}) = c(A) \qquad \text{and} \qquad \mathrm{clos}(r)(A) = r(A^{\beta})\ , \tag{III.2.2.i}$$

for all $A \subseteq X$, where $A^{\beta} = \{\mathfrak{x} \in \beta X \mid A \in \mathfrak{x}\}$, and

$$r(A^{\beta}) = \{x \in X \mid \exists \mathfrak{x} \in A^{\beta}\ (\mathfrak{x}\ r\ x)\}\ .$$

III.2.2.1 Lemma *If* $\mathsf{Set}(PX, PX)$ *is equipped with the pointwise order, then the maps* ©

$$\mathrm{clos} : \mathsf{Rel}(\beta X, X) \to \mathsf{Set}(PX, PX) \quad \textit{and}$$
$$\mathrm{conv} : \mathsf{Set}(PX, PX) \to \mathsf{Rel}(\beta X, X)$$

form an adjunction clos $\dashv$ conv. *The fixpoints of* clos · conv *are the maps* $c : PX \to PX$ *that are finitely additive.*

Proof Monotonicity of clos and conv follows immediately from the definitions. One also has

$$1 \le \mathrm{conv} \cdot \mathrm{clos} \qquad \text{and} \qquad \mathrm{clos} \cdot \mathrm{conv} \le 1\ .$$

Indeed, for $\mathfrak{x} \in \beta X$ and $y \in X$,

$$\mathfrak{x}\ r\ y \implies \forall A \in PX\ (A \in \mathfrak{x} \implies \mathfrak{x}\ r\ y) \implies \mathfrak{x}\ \mathrm{conv}(\mathrm{clos}(r))\ y,$$

and $\mathfrak{x}$ conv(c) y implies that $y \in c(A)$ for all $A \in \mathfrak{x}$; thus, for any $A \in PX$, the set clos(conv(c))(A) is given by

$$\{y \in X \mid \exists \mathfrak{x} \in \beta X\ (A \in \mathfrak{x}\ \&\ \forall B \in PX\ (B \in \mathfrak{x} \implies y \in c(B)))\} \subseteq c(A)\ . \tag{III.2.2.ii}$$

In particular, for $c : PX \to PX$ one necessarily has $\mathrm{clos} \cdot \mathrm{conv}(c)(\varnothing) = \varnothing$ (since no ultrafilter $\mathfrak{x} \in \beta X$ contains $\varnothing$), and

$$\mathrm{clos} \cdot \mathrm{conv}(c)(A \cup B) = \mathrm{clos} \cdot \mathrm{conv}(c)(A) \cup \mathrm{clos} \cdot \mathrm{conv}(c)(B)\ ,$$

as $A \cup B \in \mathfrak{x}$ implies that either $A \in \mathfrak{x}$ or $B \in \mathfrak{x}$ (Lemma II.1.13.1), and the converse holds because $\mathfrak{x}$ is an up-set. Thus, the fixpoints of clos · conv must be finitely additive.

Consider now a map $c : PX \to PX$ that is finitely additive (and therefore monotone) and $A \in PX$ with $y \in c(A)$. The set

$$j := \{J \in PX \mid y \notin c(J)\}$$

© is an ideal on X. By Corollary II.1.13.5 (with the principal filter $a = \dot{A}$), one obtains the existence of an ultrafilter $\mathfrak{x} \in \beta X$ with $\mathfrak{x} \supseteq \dot{A}$ and $\mathfrak{x} \cap j = \varnothing$; in other words, $A \in \mathfrak{x}$ and $y \in c(B)$ for all $B \in \mathfrak{x}$, so one can conclude that $c(A) = \mathrm{clos} \cdot \mathrm{conv}(c)(A)$ by (III.2.2.ii). □

A crucial observation is that both maps, clos and conv, are homomorphisms of the monoids $(\beta, 2)$-URel(X, X) and SLat(PX, PX), whose operations are given by Kleisli convolution and map composition, respectively.

© **III.2.2.2 Proposition** *The maps* conv *and* clos *satisfy*

$$\mathrm{clos}(s \circ r) = \mathrm{clos}(s) \cdot \mathrm{clos}(r)\ , \qquad \mathrm{conv}(d \cdot c) = \mathrm{conv}(d) \circ \mathrm{conv}(c)\ ,$$
$$\mathrm{clos}(e_X^\circ) = 1_X\ , \qquad \mathrm{conv}(1_{PX}) = e_X^\circ\ ,$$

for all $(\beta, 2)$*-relations* $r, s : X \relbar\!\!\!\rightharpoonup X$ *and finitely additive maps* $c, d : PX \to PX$.

Proof For $\mathfrak{x} \in \beta X$ and $x \in X$, if $\mathfrak{x}$ conv(1_{PX}) x, then $x \in A$ for all $A \in \mathfrak{x}$; thus, $\mathrm{conv}(1_{PX}) = e_X^\circ$. Since 1_{PX} is finitely additive, it is a fixpoint of clos · conv, so the previous equality yields $1_{PX} = \mathrm{clos} \cdot \mathrm{conv}(1_{PX}) = \mathrm{clos}(e_X^\circ)$.

Consider now finitely additive maps $c, d : PX \to PX$, and let $\mathfrak{x} \in \beta X$, $z \in X$. Set also

$$a := \uparrow_{PX}\{c(A) \mid A \in \mathfrak{x}\}\ ,$$

which is a filter since c is monotone. Then (III.2.2.i), together with Corollary II.1.13.3, tells us

$$a \subseteq y \iff \exists \mathcal{X} \in \beta\beta X \ (m_X(\mathcal{X}) = \chi \ \& \ \mathcal{X} \ \overline{\beta}(\mathrm{conv}(c)) \ y)$$

for all $y \in \beta X$. Assume first that $\chi \ \mathrm{conv}(d \cdot c) \ z$. Then a is disjoint from the ideal

$$j = \{B \subseteq X \mid z \notin d(B)\} .$$

Applying Corollary II.1.13.5, we see that there is an ultrafilter $y \in \beta X$ containing a and disjoint from j. Hence $y \ \mathrm{conv}(d) \ z$, and there exists $\mathcal{X} \in \beta\beta X$ with $\mathcal{X} \ \overline{\beta}(\mathrm{conv}(c)) \ y$ and $m_X(\mathcal{X}) = \chi$. We conclude that $\chi \ (\mathrm{conv}(d) \circ \mathrm{conv}(c)) \ z$. Assume now that $\chi \ (\mathrm{conv}(d) \circ \mathrm{conv}(c)) \ z$, so there is a $y \in \beta X$ such that $a \subseteq y$ and $y \ \mathrm{conv}(d) \ z$. From this we obtain $z \in d \cdot c(A)$ for every $A \in \chi$, i.e. $\chi \ \mathrm{conv}(d \cdot c) \ z$.

Finally, let $r, s : X \nrightarrow X$ be $(\mathbb{\beta}, 2)$-relations, and consider $A \subseteq X$, $z \in X$. If $z \in \mathrm{clos}(s \circ r)(A)$, then we have $\mathcal{X} \in \beta\beta X$ and $y \in \beta X$ with

$$A^{\mathbb{\beta}} \in \mathcal{X} , \quad \mathcal{X} \ \overline{\beta} r \ y , \quad y \ s \ z ;$$

and it follows that $z \in \mathrm{clos}(s)(r(A^{\mathbb{\beta}})) = \mathrm{clos}(s)(\mathrm{clos}(r)(A))$. Thus, $\mathrm{clos}(s \circ r) \leq \mathrm{clos}(s) \cdot \mathrm{clos}(r)$. For the other inequality, we use that $\mathrm{conv}(d \cdot c) = \mathrm{conv}(d) \circ \mathrm{conv}(c)$ for fixpoints c, d of $\mathrm{clos} \cdot \mathrm{conv}$, and in particular for $c = \mathrm{clos}(r)$, $d = \mathrm{clos}(s)$, and $c \cdot d$:

$$\begin{aligned} \mathrm{clos}(s) \cdot \mathrm{clos}(r) &= \mathrm{clos} \cdot \mathrm{conv}(\mathrm{clos}(s) \cdot \mathrm{clos}(r)) \\ &= \mathrm{clos}(\mathrm{conv}(\mathrm{clos}(s)) \circ \mathrm{conv}(\mathrm{clos}(r))) \leq \mathrm{clos}(s \circ r) \end{aligned}$$

because $\mathrm{conv} \cdot \mathrm{clos} \leq 1$ (by Lemma III.2.2.1). □

III.2.2.3 Lemma *Every* $(\mathbb{\beta}, 2)$*-relation* $r : X \nrightarrow X$ *satisfies* $\mathrm{conv} \cdot \mathrm{clos}(r) = e_X^\circ \circ r$.

Proof On one hand,

$$\mathrm{conv}(\mathrm{clos}(r)) = \mathrm{conv}(1_{PX} \cdot \mathrm{clos}(r)) = e_X^\circ \circ \mathrm{conv}(\mathrm{clos}(r)) \geq e_X^\circ \circ r .$$

On the other hand, consider $\chi \in \beta X$, $y \in X$, and suppose that $\chi \ \mathrm{conv}(\mathrm{clos}(r)) \ y$. Then for all $A \in \chi$ we have $y \in \mathrm{clos}(r)(A) = r(A^{\mathbb{\beta}})$, which implies that there is an $\mathcal{X} \in \beta\beta X$ with $m_X(\mathcal{X}) = \chi$ and $\mathcal{X} \ (\overline{\beta} r) \ e_X(y)$. Hence, $\chi \ (e_X^\circ \circ r) \ y$, which shows that $\mathrm{conv}(\mathrm{clos}(r)) \leq e_X^\circ \circ r$. □

As a consequence, we obtain $r = \mathrm{conv} \cdot \mathrm{clos}(r)$ for every left unitary $(\mathbb{\beta}, 2)$-relation $r : X \nrightarrow X$, and in particular for every reflexive and transitive $(\mathbb{\beta}, 2)$-relation.

© **III.2.2.4 Proposition** *The fixpoints of the adjunction* clos $\dashv$ conv *are on one hand the finitely additive maps* $c : PX \to PX$, *and on the other hand the left unitary relations* $r : \beta X \nrightarrow X$.

Proof This follows from Lemmata III.2.2.1 and III.2.2.3. □

© **III.2.2.5 Theorem** *There is an isomorphism*

$$(\mathbb{B}, 2)\text{-}\mathsf{Cat} \cong \mathsf{Top}$$

that commutes with the underlying-set functors.

Proof By Proposition III.2.2.4, clos and conv define a one-to-one correspondence between finitely additive maps $c : PX \to PX$ and left unitary $(\mathbb{B}, 2)$-relations $a : X \nrightarrow X$. If c is a topological closure operation, so that $1_{PX} \leq c$ and $c \cdot c = c$, then $a := \mathrm{conv}(c)$ satisfies

$$e_X^\circ = \mathrm{conv}(1_{PX}) \leq \mathrm{conv}(c) = a \qquad \text{and}$$
$$a \circ a = \mathrm{conv}(c) \circ \mathrm{conv}(c) = \mathrm{conv}(c \cdot c) = \mathrm{conv}(c) = a$$

by Proposition III.2.2.2, i.e. $(X, a : \beta X \nrightarrow X)$ is a $(\mathbb{B}, 2)$-category. Likewise, from $e_X^\circ \leq a$ and $a \circ a = a$ one gets $1_{PX} \leq c$ and $c \cdot c = c$ for $c := \mathrm{clos}(a)$, and therefore clos and conv actually define a bijective correspondence between $(\mathbb{B}, 2)$-categorical structures and topological closure operations on X. An easy verification shows that a $(\mathbb{B}, 2)$-functor preserves the corresponding closure operation – and is therefore continuous – while a continuous map preserves the corresponding $(\mathbb{B}, 2)$-categorical structure – and is therefore a $(\mathbb{B}, 2)$-functor. □

Spelled out, the previous theorem states that a topological space $(X, \mathcal{O}X)$ can be equivalently described as a pair (X, a), with $a : \beta X \nrightarrow X$ a relation representing convergence which, when we denote both a and $\overline{\beta}a$ by $\longrightarrow$, satisfies

$$\mathfrak{X} \longrightarrow \mathfrak{y} \;\&\; \mathfrak{y} \longrightarrow z \implies \sum \mathfrak{X} \longrightarrow z \qquad \text{and} \qquad \dot{x} \longrightarrow x\,,$$

for all $z \in X$, $\mathfrak{y} \in \beta X$, and $\mathfrak{X} \in \beta\beta X$; here $\mathfrak{X} \longrightarrow \mathfrak{y} \iff \mathfrak{X} \supseteq a^\circ[\mathfrak{y}]$, and $\sum$ is the Kowalsky sum restricted to ultrafilters (see Section II.1.12 and Example II.3.1.1(5)). In this context, the continuous maps $f : (X, a) \to (X, b)$ are exactly the convergence-preserving maps, i.e. the maps $f : X \to Y$ such that

$$\chi \longrightarrow y \implies f[\chi] \longrightarrow f(y)$$

for all $y \in X$, and $\chi \in \beta X$.

III.2.3 Compact Hausdorff spaces

Because the Barr extension of $\mathbb{B}$ to Rel is flat, it is possible to exploit Theorem III.2.2.5 to obtain an elegant description of the category of $\mathbb{B}$-algebras (i.e. of Eilenberg–Moore algebras associated to $\mathbb{B}$; see Section II.3.2).

Let us recall that a topological space X is *compact* if every open cover of X has a finite subcover, i.e. for every open cover $\mathcal{A}$ there exists a finite subset $\mathcal{F} \subseteq \mathcal{A}$ with $\bigcup \mathcal{F} = X$.

In the following results we freely exploit that a topological space X can equivalently be described via a set $\mathcal{O}X$ of open sets, a closure operation $c : PX \longrightarrow PX$, or a convergence relation $a : \beta X \nrightarrow X$ (Exercises II.1.F and II.1.G, and Theorem III.2.2.5). In fact, we can avoid the closure operation by translating the definition of the convergence relation given in Section III.2.2 as

$$\mathfrak{x} \longrightarrow x \iff \forall A \in \mathcal{O}X \; (x \in A \implies A \in \mathfrak{x}) \qquad \text{(III.2.3.i)}$$

(this easily follows from Lemma II.1.13.1 and $\mathcal{O}X = \{(c(B))^{\complement} \mid B \in PX\}$).

III.2.3.1 Proposition *The following statements are equivalent for a topological space X:* ©

(i) *X is compact;*
(ii) *every ultrafilter on X converges, i.e. $1_{\beta X} \leq a^{\circ} \cdot a$.*

Proof Assume first that X is compact. Let $\mathfrak{x} \in \beta X$ and set

$$\mathcal{A} = \{A \subseteq X \mid A \text{ open}, A \notin \mathfrak{x}\}\,.$$

The set $\mathcal{A}$ cannot cover X since otherwise there would exist $A_1, \ldots, A_n \in \mathcal{A}$ with $A_1 \cup \ldots \cup A_n = X \in \mathfrak{x}$, and therefore $A_k \in \mathfrak{x}$ for some $k \in \{1, \ldots, n\}$ by Lemma II.1.13.1, contradicting the definition of $\mathcal{A}$. Thus, there exists $x \in X$ such that every open set A with $x \in A$ belongs to $\mathfrak{x}$. This implies that $\mathfrak{x} \longrightarrow x$ by (III.2.3.i).

Assume now that any ultrafilter $\mathfrak{x} \in \beta X$ converges. Let $\mathcal{A}$ be a set of open subsets of X with the property that no finite subset of $\mathcal{A}$ covers X. Then

$$\mathfrak{f} = \uparrow_{PX}\{(\textstyle\bigcup \mathcal{F})^{\complement} \mid \mathcal{F} \subseteq \mathcal{A}, \mathcal{F} \text{ finite}\}$$

is a proper filter on X, and therefore it is contained in an ultrafilter $\mathfrak{x}$ (Proposition II.1.13.2) that converges to some $x \in X$ by hypothesis. Hence $x \notin A$ for any $A \in \mathcal{A}$ (by (III.2.3.i) and the definition of $\mathfrak{x} \supseteq \mathfrak{f}$), i.e. $\mathcal{A}$ does not cover X. □

III.2.3.2 Proposition *The following statements are equivalent for a topological space X:* ©

(i) *X is Hausdorff;*
(ii) *every ultrafilter on X has at most one convergence point, i.e. $a \cdot a^{\circ} \leq 1_X$.*

Proof Assume that X is Hausdorff and let $\mathfrak{x} \in \beta X$ and $x, y \in X$ with $\mathfrak{x} \longrightarrow x$ and $\mathfrak{x} \longrightarrow y$. By (III.2.3.i), for all open subsets $A \subseteq X$ and $B \subseteq X$ with $x \in A$ and $y \in B$ one has $A \cap B \neq \varnothing$, which is possible only if $x = y$.

Assume now that every ultrafilter on X converges to at most one point, and let $x, y \in X$. If every open neighborhood of x intersects every open neighborhood of y, there exists an ultrafilter $\mathfrak{x}$ containing all neighborhoods of x and all neighborhoods of y, hence converging to both x and y, which, by hypothesis, implies $x = y$. □

© **III.2.3.3 Theorem** *There is an isomorphism*

$$\mathsf{Set}^{\mathbb{B}} \cong \mathsf{CompHaus}$$

that commutes with the respective forgetful functors to Set.

Proof Combining both the preceding propositions, we see that a topological space X is compact Hausdorff if and only if its convergence relation $a : \beta X \nrightarrow X$ is actually a map $a : \beta X \longrightarrow X$, which therefore satisfies $a \cdot \beta a = a \cdot m_X$. (Since $\overline{\beta}a = \beta a$ because a is a map, the condition $a \cdot \overline{\beta}a \leq a \cdot m_X$ is an inclusion of graphs of functions with same domain, hence necessarily an equality.) Furthermore, a continuous map $f : X \longrightarrow Y$ between compact Hausdorff spaces with convergence structures $a : \beta X \longrightarrow X$ and $b : \beta Y \longrightarrow Y$, respectively, satisfies $f \cdot a \leq b \cdot Tf$, but, since all relations involved are functions, this condition is actually an equality. □

III.2.4 Approach spaces

Approach spaces provide a common framework for both topological and metric structures. More precisely, an *approach space* is a pair (X, δ) consisting of a set X and a function $\delta : X \times PX \longrightarrow [0, \infty]$ subject to the conditions:

(1) $\delta(x, \{x\}) = 0$,
(2) $\delta(x, \varnothing) = \infty$,
(3) $\delta(x, A \cup B) = \min\{\delta(x, A), \delta(x, B)\}$,
(4) $\delta(x, A) \leq \delta(x, A^{(u)}) + u$,

for all $x \in X$, $A, B \subseteq X$, $u \in [0, \infty]$, and where

$$A^{(u)} = A^{(u)}_{\delta} = \{x \in X \mid \delta(x, A) \leq u\} .$$

A function $\delta : X \times PX \longrightarrow [0, \infty]$ satisfying the axioms (1)–(4) is called an *approach distance* on X. A morphism $f : (X, \delta) \longrightarrow (Y, \delta')$ of approach spaces is given by a *non-expansive map* $f : X \longrightarrow Y$, so that f satisfies

$$\delta'(f(x), f(A)) \leq \delta(x, A)$$

for all $x \in X$ and $A \subseteq X$. Approach spaces and non-expansive maps are the objects and morphisms of the category App.

III.2.4.1 Examples

(1) Every metric space (X, a) (in the sense of Example III.1.3.1(2)) becomes an approach space (X, δ) when one puts

$$\delta(y, A) := \inf\{a(x, y) \mid x \in A\} .$$

In fact, this defines a full embedding

$$\mathsf{Met} \hookrightarrow \mathsf{App}$$

that exhibits Met as a coreflective subcategory of App (this follows from Proposition III.3.4.2 together with the following Theorem III.2.4.5).

(2) Every topological space (X, c) (as in Exercises II.1.F and II.1.G) becomes an approach space (X, δ) when one sets

$$\delta(y, A) := \begin{cases} 0 & \text{if } y \in c(A), \\ \infty & \text{otherwise.} \end{cases}$$

In this way, one obtains a full embedding

$$\mathsf{Top} \hookrightarrow \mathsf{App}$$

that exhibits Top as a coreflective subcategory of App (see Section III.3.6 together with Theorems III.2.2.5 and III.2.4.5).

We now show that approach spaces can be seen as "numerified topological spaces," i.e. we prove the existence of an isomorphism

$$\mathsf{App} \cong (\mathbb{\beta}, \mathsf{P}_+)\text{-}\mathsf{Cat}$$

for a suitable extension $\overline{\beta} : \mathsf{P}_+\text{-}\mathsf{Rel} \longrightarrow \mathsf{P}_+\text{-}\mathsf{Rel}$ of $\beta : \mathsf{Set} \longrightarrow \mathsf{Set}$ (defined in the following). In order to keep the proof similar to the one for topological spaces as presented in Section III.2.2, we think of maps

$$\delta : X \times PX \longrightarrow [0, \infty]$$

as of morphisms "from X to X." Given also $\gamma : X \times PX \longrightarrow [0, \infty]$, we define the composite $\gamma \cdot \delta : X \times PX \longrightarrow [0, \infty]$ by

$$\gamma \cdot \delta(z, A) := \inf\{\gamma(z, A_\delta^{(u)}) + u \mid u \in [0, \infty]\},$$

for all $A \subseteq X$ and $z \in X$. One easily verifies that this composition preserves the order in both variables, and that the map $\varepsilon_X : X \times PX \longrightarrow [0, \infty]$, sending (x, A) to 0 if $x \in A$ and to ∞ otherwise, is neutral with respect to this composition. We say that $\delta : X \times PX \longrightarrow \mathsf{P}_+$ is *finitely additive* if

$$\delta(y, \varnothing) = \infty \qquad \text{and} \qquad \delta(y, A \cup B) = \min\{\delta(y, A), \delta(y, B)\},$$

for all $y \in X$ and $A, B \in PX$. With this notation, $\delta : X \times PX \longrightarrow [0, \infty]$ is an approach distance if and only if δ is finitely additive and

$$\delta \cdot \delta \geq \delta, \quad \varepsilon_X \geq \delta.$$

The former inequality is actually an equality thanks to the latter.

We saw in Section III.2.2 that $\beta : \mathsf{Set} \longrightarrow \mathsf{Set}$ can be extended to a 2-functor $\overline{\beta} : \mathsf{Rel} \longrightarrow \mathsf{Rel}$ by setting $\mathfrak{x}\ (\overline{\beta}r)\ \mathfrak{y} \iff r(\mathfrak{x}) \subseteq \mathfrak{y}$ for every relation

$r : X \nrightarrow Y$. We now go one step further and extend the ultrafilter monad $\mathbb{\beta}$ to a lax monad $(\overline{\beta}, m, e)$ on P_+-Rel as follows. Let us recall from II.1.10 that $\leq$ always refers to the natural order of $[0, \infty]$. For $r : X \times Y \to \mathsf{P}_+$ and $v \in \mathsf{P}_+$, we consider the relation $r_v : X \nrightarrow Y$ defined by

$$x \, r_v \, y \iff r(x, y) \leq v \, .$$

Given another P_+-relation $s : Y \nrightarrow Z$ and $u \in \mathsf{P}_+$, one easily verifies $s_u \cdot r_v \leq (s \cdot r)_{u+v}$. We now consider the *Barr extension* of β to P_+-Rel:

$$\begin{aligned} \overline{\beta} r := \overline{\beta}_{\mathsf{P}_+} r : \beta X \times \beta Y &\to \mathsf{P}_+ \\ (\mathfrak{x}, \mathfrak{y}) &\mapsto \inf\{v \in \mathsf{P}_+ \mid \mathfrak{x} \; \overline{\beta}(r_v) \; \mathfrak{y}\} \, , \end{aligned}$$

which can be equivalently expressed by

$$\overline{\beta} r(\mathfrak{x}, \mathfrak{y}) = \sup\nolimits_{A \in \mathfrak{x}, B \in \mathfrak{y}} \inf\nolimits_{x \in A, y \in B} r(x, y) \, ,$$

for all P_+-relations $r : X \nrightarrow Y$, $\mathfrak{x} \in \beta X$, and $\mathfrak{y} \in \beta Y$ (see Exercise III.2.E). Furthermore, $\overline{\beta}(r_u) \leq (\overline{\beta} r)_u$ for any $u \in \mathsf{P}_+$; and $(\overline{\beta} r)_u \leq \overline{\beta}(r_v)$ whenever $u < v$ in P_+.

© **III.2.4.2 Lemma** *Let $r : X \nrightarrow Y$ be a P_+-relation, $\mathfrak{f}$ a filter on X, and $\mathfrak{y} \in \beta Y$. Then there exists an ultrafilter $\mathfrak{x} \in \beta X$ such that $\mathfrak{f} \subseteq \mathfrak{x}$ and*

$$\overline{\beta} r(\mathfrak{x}, \mathfrak{y}) = \sup\nolimits_{A \in \mathfrak{f}, B \in \mathfrak{y}} \inf\nolimits_{x \in A, y \in B} r(x, y) \, .$$

Proof Certainly, for every ultrafilter $\mathfrak{x} \in \beta X$ with $\mathfrak{f} \subseteq \mathfrak{x}$ we have

$$\overline{\beta} r(\mathfrak{x}, \mathfrak{y}) \geq \sup\nolimits_{A \in \mathfrak{f}, B \in \mathfrak{y}} \inf\nolimits_{x \in A, y \in B} r(x, y) \, .$$

Let us set $v := \sup_{A \in \mathfrak{f}, B \in \mathfrak{y}} \inf_{x \in A, y \in B} r(x, y)$. If $v = \infty$, then any $\mathfrak{x} \supseteq \mathfrak{f}$ has the desired property. If $v < \infty$, we consider

$$\mathfrak{j} = \{A \subseteq X \mid \sup\nolimits_{B \in \mathfrak{y}} \inf\nolimits_{x \in A, y \in B} r(x, y) > v\} \, .$$

Of course, $\mathfrak{f} \cap \mathfrak{j} = \varnothing$, and it is not hard to see that $\mathfrak{j}$ is an ideal. By Corollary II.1.13.5, there exists an ultrafilter $\mathfrak{x} \in \beta X$ with $\mathfrak{f} \subseteq \mathfrak{x}$ and $\mathfrak{x} \cap \mathfrak{j} = \varnothing$, and therefore $\overline{\beta} r(\mathfrak{x}, \mathfrak{y}) \leq v$. □

© **III.2.4.3 Proposition** *The Barr extension $\overline{\mathbb{\beta}} = (\overline{\beta}, m, e)$ is a flat associative lax extension to P_+-Rel of the ultrafilter monad $\mathbb{\beta} = (\beta, m, e)$. Moreover, $\overline{\beta}(r^\circ) = (\overline{\beta} r)^\circ$ for every P_+-relation r.*

Proof We show only

$$\overline{\beta} s \cdot \overline{\beta} r = \overline{\beta}(s \cdot r) \qquad \text{and} \qquad m_Y \cdot \overline{\beta}\, \overline{\beta} r = \overline{\beta} r \cdot m_X \, ,$$

for all P_+-relations $r : X \nrightarrow Y$ and $s : Y \nrightarrow Z$, as the other verifications are straightforward.

In order to show the inequality $\overline{\beta}s \cdot \overline{\beta}r \geq \overline{\beta}(s \cdot r)$, assume $\overline{\beta}s \cdot \overline{\beta}r(\mathfrak{x}, \mathfrak{z}) < u$ for $\mathfrak{x} \in \beta X$ and $\mathfrak{z} \in \beta Z$. Therefore, there is some $\mathfrak{y} \in \beta Y$ satisfying

$$\overline{\beta}r(\mathfrak{x}, \mathfrak{y}) + \overline{\beta}s(\mathfrak{y}, \mathfrak{z}) < u \ .$$

Consider $u_1, u_2 \in \mathsf{P}_+$ such that

$$\overline{\beta}r(\mathfrak{x}, \mathfrak{y}) < u_1 \ , \quad \overline{\beta}s(\mathfrak{y}, \mathfrak{z}) < u_2 \ , \quad u_1 + u_2 = u \ .$$

Hence, we have $\mathfrak{x} \ \overline{\beta}(r_{u_1}) \ \mathfrak{y}$ and $\mathfrak{y} \ \overline{\beta}(s_{u_2}) \ \mathfrak{z}$, so that $\mathfrak{x} \ \overline{\beta}(s_{u_2} \cdot r_{u_1}) \ \mathfrak{z}$. Since $s_{u_2} \cdot r_{u_1} \leq (s \cdot r)_{u_1+u_2}$, we conclude $\overline{\beta}(s \cdot r)(\mathfrak{x}, \mathfrak{z}) \leq u_1 + u_2 = u$.

To see $\overline{\beta}s \cdot \overline{\beta}r \leq \overline{\beta}(s \cdot r)$, let $\mathfrak{x} \in \beta X$, $\mathfrak{z} \in \beta Z$, and $u \in \mathsf{P}_+$, with

$$u > \overline{\beta}(s \cdot r)(\mathfrak{x}, \mathfrak{z}) = \sup_{A \in \mathfrak{x}, C \in \mathfrak{z}} \inf_{x \in A, z \in C} s \cdot r(x, z) \ .$$

Hence, for every $A \in \mathfrak{x}$ and every $C \in \mathfrak{z}$, there exist $x \in A$, $y \in Y$, and $z \in C$, with $r(x, y) + s(y, z) \leq u$, i.e.

$$B_{A,C} := \{y \in Y \mid \exists x \in A, \ z \in Z \ : \ r(x, y) + s(y, z) \leq u\} \neq \varnothing \ .$$

Since $B_{A \cap A', C \cap C'} \subseteq B_{A,C} \cap B_{A',C'}$, the set

$$\{B_{A,C} \mid A \in \mathfrak{x}, \ C \in \mathfrak{z}\}$$

is a filter base; let $\mathfrak{y} \in \beta Y$ be any ultrafilter containing it. Then

$$\overline{\beta}r(\mathfrak{x}, \mathfrak{y}) = \sup_{A \in \mathfrak{x}, B \in \mathfrak{y}} \inf_{x \in A, y \in B} \leq u$$

because for any $A \in \mathfrak{x}$ and $B \in \mathfrak{y}$ one has $B \cap B_{A,Z} \neq \varnothing$. Similarly, $\overline{\beta}s(\mathfrak{y}, \mathfrak{z}) \leq u$. Consider $\varepsilon > 0$ and set $u_0 = \overline{\beta}r(\mathfrak{x}, \mathfrak{y})$; if $u_0 = 0$, we are done, so we can assume $u_0 > 0$. Thus, there is some $A \in \mathfrak{x}$ with $r_{u_0 - \varepsilon}[A] \notin \mathfrak{y}$, and therefore its complement

$$B_0 := \{y \in Y \mid \forall x \in A \ (r(x, y) > u_0 - \varepsilon)\}$$

belongs to $\mathfrak{y}$. We show that $\overline{\beta}s(\mathfrak{y}, \mathfrak{z}) \leq (u - u_0) + \varepsilon$. To this end, let $B \in \mathfrak{y}$ and $C \in \mathfrak{z}$. Then $B \cap B_0 \cap B_{A,C} \neq \varnothing$, which implies that there are $x \in A$, $y \in B$, and $z \in C$, with

$$r(x, y) + s(y, z) \leq u \qquad \text{and} \qquad r(x, y) > u_0 - \varepsilon \ ,$$

therefore $s(y, z) \leq (u - u_0) + \varepsilon$. Consequently,

$$\overline{\beta}s(\mathfrak{y}, \mathfrak{z}) \leq \sup_{B \in \mathfrak{y}, C \in \mathfrak{z}} \inf_{y \in B, z \in C} s(y, z) \leq (u - u_0) + \varepsilon \ .$$

To verify $m_Y \cdot \overline{\beta}\,\overline{\beta}r \geq \overline{\beta}r \cdot m_X$, let $u, u' \in \mathsf{P}_+$ and $\mathfrak{X} \in \beta\beta X$, $\mathfrak{Y} \in \beta\beta Y$ be such that

$$\overline{\beta}\,\overline{\beta}r(\mathfrak{X}, \mathfrak{Y}) < u < u' \ .$$

Hence, $\mathfrak{X} \ \overline{\beta}\,\overline{\beta}(r_u) \ \mathfrak{Y}$, and therefore $\mathfrak{X} \ \overline{\beta}\,\overline{\beta}(r_{u'}) \ \mathfrak{Y}$. This implies $m_X(\mathfrak{X}) \ \overline{\beta}(r_{u'}) \ m_Y(\mathfrak{Y})$, i.e. $\overline{\beta}r(m_X(\mathfrak{X}), m_Y(\mathfrak{Y})) \leq u'$.

Finally, we show $m_Y \cdot \overline{\beta}\,\overline{\beta}r \leq \overline{\beta}r \cdot m_X$. Let $\mathfrak{X} \in \beta\beta X$ and $\mathfrak{y} \in \beta Y$, and assume

$$\overline{\beta}r(m_X(\mathfrak{X}), \mathfrak{y}) < u'' < u' < u\,.$$

Since the Barr extension of β to Rel is associative, there is some $\mathcal{Y} \in \beta\beta Y$ with $\mathfrak{X}\ (\overline{\beta}\,\overline{\beta}r_{u''})\ \mathcal{Y}$, so that one has $\mathfrak{X}\ \overline{\beta}(\overline{\beta}r)_{u'}\ \mathcal{Y}$ and $\mathfrak{X}\ (\overline{\beta}\,\overline{\beta}r)_u\ \mathcal{Y}$, i.e. $\overline{\beta}\,\overline{\beta}r(\mathfrak{X}, \mathcal{Y}) \leq u$.

Finally, since $\overline{\beta}$ commutes with the involution on P_+-Rel, $m^\circ : \overline{\beta} \to \overline{\beta}\,\overline{\beta}$ is a natural transformation too (see also Exercise III.1.J). Proposition III.1.9.4 then yields that $\overline{\mathbb{B}}$ is associative. □

Every P_+-relation $r : \beta X \times X \to \mathsf{P}_+$ defines a finitely additive function

$$\begin{aligned} \mathrm{clos}(r) : X \times PX &\longrightarrow [0, \infty] \\ (y, A) &\longmapsto \inf\{r(\mathfrak{x}, y) \mid \mathfrak{x} \in \beta A\}\,, \end{aligned}$$

and every function $\delta : X \times PX \longrightarrow [0, \infty]$ yields a P_+-relation

$$\begin{aligned} \mathrm{conv}(\delta) : \beta X \times X &\longrightarrow \mathsf{P}_+ \\ (\mathfrak{x}, y) &\longmapsto \sup\{\delta(y, A) \mid A \in \mathfrak{x}\}\,. \end{aligned}$$

As in Section III.2.2, the following proposition contains the key facts about clos and conv.

© **III.2.4.4 Proposition** *The operations* conv *and* clos *satisfy*

$$\mathrm{clos}(s \circ r) = \mathrm{clos}(s) \cdot \mathrm{clos}(r)\,, \qquad \mathrm{conv}(\gamma \cdot \delta) = \mathrm{conv}(\gamma) \circ \mathrm{conv}(\delta)\,,$$
$$\mathrm{clos}(e_X^\circ) = \varepsilon_X\,, \qquad \mathrm{conv}(\varepsilon_X) = e_X^\circ\,,$$

for all finitely additive functions $\delta, \gamma : X \times PX \to [0, \infty]$, and $(\mathbb{B}, \mathsf{P}_+)$-relations $r, s : X \nrightarrow X$.

Proof We show only $\mathrm{clos}(s \circ r) = \mathrm{clos}(s) \cdot \mathrm{clos}(r)$, as the verifications of the other statements are either very similar or straightforward.

Assume first that $\mathrm{clos}(s) \cdot \mathrm{clos}(r)(z, A) < v$, with $z \in Z$, $A \subseteq X$, and $v \in [0, \infty]$. Then, for some $u \in [0, \infty]$,

$$\mathrm{clos}(s)(z, A^{(u)}_{\mathrm{clos}(r)}) + u < v\,.$$

Consider $u' \in \mathsf{P}_+$ such that

$$\mathrm{clos}(s)(z, A^{(u)}_{\mathrm{clos}(r)}) < u' \qquad \text{and} \qquad u' + u = v\,.$$

Hence, there exists $\mathfrak{y} \in \beta X$ with $A^{(u)}_{\mathrm{clos}(r)} \in \mathfrak{y}$ and $r(\mathfrak{y}, z) < u'$. For every $B \in \mathfrak{y}$, there is an element $y \in B$ satisfying

$$\inf_{\mathfrak{y} \in A^{\mathbb{B}}} r(\mathfrak{y}, y) = \mathrm{clos}(r)(y, A) \leq u\,.$$

Therefore, we obtain

$$\sup_{B\in y}\inf_{y\in A^{\beta}}\inf_{y\in B} r(y, y) \le u ,$$

and Lemma III.2.4.2 guarantees the existence of some $\mathcal{X} \in \beta\beta X$ with $A^{\beta} = \{\chi \in \beta X \mid A \in \chi\} \in \mathcal{X}$ and $\overline{\beta}r(\mathcal{X}, y) \le u$. Since

$$A \in m_X(\mathcal{X}) \qquad \text{and} \qquad s \circ r(m_X(\mathcal{X}), z) \le u' + u = v ,$$

we deduce $\mathrm{clos}(s \circ r)(z, A) \le v$.

Assume now $\mathrm{clos}(s \circ r)(z, A) < v$. Hence, for some $\chi \in \beta A$, we have $s \circ r(\chi, z) < v$, and there exist $\mathcal{X} \in \beta\beta X$, $y \in \beta X$ with

$$m_X(\mathcal{X}) = \chi \qquad \text{and} \qquad \overline{\beta}r(\mathcal{X}, y) + s(y, z) < v .$$

Let $u \in [0, \infty]$ with $\overline{\beta}r(\mathcal{X}, y) < u$ and $u + s(y, z) = v$. We have

$$A^{(u)}_{\mathrm{clos}(r)} \supseteq r_u(A^{\beta}) \in y ,$$

and therefore $\mathrm{clos}(s)(z, A^{(u)}_{\mathrm{clos}(r)}) + u \le v$. □

From the identities

$$\mathrm{clos}(\mathrm{conv}(\delta))(y, A) = \inf_{\chi\in\beta A}\mathrm{conv}(\delta)(\chi, y) = \inf_{\chi\in\beta A}\sup_{B\in\chi}\delta(y, B) ,$$

we obtain $\delta(x, A) \le \mathrm{clos}(\mathrm{conv}(\delta))(x, A)$, for all $\delta : X \times PX \to [0, \infty]$, $A \subseteq X$, and $y \in X$. Moreover, equality holds if $\delta(y, A) = \infty$. If $\delta(y, A) < \infty$, we can consider

$$j = \{B \subseteq X \mid \delta(y, B) > \delta(y, A)\} ;$$

when δ is finitely additive, j is an ideal with $A \notin j$, and there exists an ultrafilter $\chi \in \beta X$ with $A \in \chi$ and $\chi \cap j = \varnothing$. Therefore, $\delta = \mathrm{clos}(\mathrm{conv}(\delta))$ if and only if δ is finitely additive. On the other hand,

$$\mathrm{conv}(\mathrm{clos}(r))(\chi, y) = \sup_{A\in\chi}\mathrm{clos}(r)(y, A) = \sup_{A\in\chi}\inf_{y\in\beta A} r(y, y) ,$$

so that $\mathrm{conv}(\mathrm{clos}(r)) = e_X^{\circ} \circ r$ (see Exercise III.2.F), for every P_+-relation $r : \beta X \nrightarrow X$. Hence, $r = \mathrm{conv}(\mathrm{clos}(r))$ if and only if r is unitary.

III.2.4.5 Theorem *There is an isomorphism* $(\beta, \mathsf{P}_+)\text{-}\mathsf{Cat} \cong \mathsf{App}$ *that commutes with the underlying-set functors.* ©

Proof To every approach space (X, δ) we associate the (β, P_+)-category $(X, \mathrm{conv}(\delta))$, and to every (β, P_+)-category (X, r) corresponds the approach space $(X, \mathrm{clos}(r))$. Then non-expansive maps preserve the corresponding (β, P_+)-categorical structure, and (β, P_+)-functors become non-expansive. All said, we obtain functors

$$\mathsf{App} \to (\beta, \mathsf{P}_+)\text{-}\mathsf{Cat} \qquad \text{and} \qquad (\beta, \mathsf{P}_+)\text{-}\mathsf{Cat} \to \mathsf{App}$$

that are inverse to one another. □

Thus, as in the case of topological spaces, we obtain an alternative description of approach spaces. Here, the relational arrow for a convergence relation $\longrightarrow$ is replaced by a numerified "degree of convergence," so that an approach space can be seen as a pair (X, a), with $a : \beta X \times X \to [0, \infty]$ a map satisfying

$$\overline{\beta}a(\mathcal{X}, \mathfrak{y}) + a(\mathfrak{y}, z) \geq a(\sum \mathcal{X}, z) \qquad \text{and} \qquad a(\dot{x}, x) = 0\,,$$

for all $z \in X$, $\mathfrak{y} \in \beta X$, and $\mathcal{X} \in \beta\beta X$, and where

$$\overline{\beta}a(\mathcal{X}, \mathfrak{y}) = \sup_{\mathcal{A}\in\mathcal{X}, B\in\mathfrak{y}} \inf_{\mathfrak{x}\in\mathcal{A}, y\in B} a(\mathfrak{x}, y)\,.$$

With this description, the non-expansive maps $f : (X, a) \to (Y, b)$ between approach spaces are those satisfying

$$a(\mathfrak{x}, y) \geq b(f[\mathfrak{x}], f(y))$$

for all $y \in X$, and $\mathfrak{x} \in \beta X$, i.e. the maps that improve the "degree of convergence." The embedding $\mathsf{Top} \hookrightarrow \mathsf{App}$ of Example III.2.4.1(2) now describes topological spaces as those approach spaces whose measure of convergence $a : \beta X \times X \to [0, \infty]$ takes its values in $\{0, \infty\}$. Hence, ultrafilter convergence in X is defined by

$$\mathfrak{x} \longrightarrow y \iff a(\mathfrak{x}, y) = 0$$

for all $\mathfrak{x} \in \beta X$, $y \in X$.

III.2.5 Closure spaces

Example III.1.6.4(2) shows that when the powerset monad $\mathbb{P}$ is equipped with the lax extension $\hat{P} : \mathsf{Rel} \to \mathsf{Rel}$

$$A\ (\hat{P}r)\ B \iff B \subseteq r(A) \qquad \text{(III.2.5.i)}$$

(for all $A \in PX$, $B \in PY$, and relations $r : X \nrightarrow Y$), $(\mathbb{P}, \mathsf{2})$-category structures $a : PX \nrightarrow X$ are in one-to-one correspondence with closure operations $c : PX \to PX$ via

$$x \in c(A) \iff A\ a\ x$$

for all $A \in PX$, $x \in X$. Under this correspondence, a $(\mathbb{P}, \mathsf{2})$-functor $f : (X, a) \to (X, b)$ is equivalently described as a continuous map $f : (X, c_X) \to (Y, c_Y)$, and one obtains an isomorphism

$$(\mathbb{P}, \mathsf{2})\text{-}\mathsf{Cat} \cong \mathsf{Cls}$$

that commutes with the underlying-set functors. The extension formula (III.2.5.i) can also be used with the *finite-powerset monad* $\mathbb{P}_{\mathrm{fin}} = (P_{\mathrm{fin}}, \bigcup, \{-\})$, whose functor $P_{\mathrm{fin}} : \mathsf{Set} \to \mathsf{Set}$ sends a set X to the set $P_{\mathrm{fin}}X$ of its finite subsets; the multiplication and unit of the monad are just the appropriate restrictions of those of $\mathbb{P}$. By the same procedure as in the powerset case, one can identify a $(\mathbb{P}_{\text{fin}}, 2)$-category with a *finitary closure space* (also called an *algebraic closure space*), i.e. a closure space (X, c) whose closure operation $c : PX \longrightarrow PX$ is *finitary*:

$$c(A) = \bigcup_{B \in P_{\text{fin}} A} c(B) .$$

Denoting by $\mathsf{Cls}_{\text{fin}}$ the full subcategory of Cls whose objects are finitary closure spaces, one obtains an isomorphism

$$(\mathbb{P}_{\text{fin}}, 2)\text{-}\mathsf{Cat} \cong \mathsf{Cls}_{\text{fin}}$$

that commutes with the underlying-set functors.

Exercises

III.2.A *The probabilistic and metric quantales.* The unit interval $[0, 1]$ with its natural order and multiplication yields a quantale

$$([0, 1], \cdot, 1)$$

that is isomorphic to the quantale $\mathsf{P}_{+} = ([0, \infty]^{\text{op}}, +, 0)$.

III.2.B *Ultrametric spaces.* An *ultrametric* is a map $a : X \times X \longrightarrow [0, \infty]$ satisfying

$$\max\{a(x, y), a(y, z)\} \geq a(x, z) \qquad \text{and} \qquad 0 = a(x, x)$$

for all $x, y, z \in X$, and an ultrametric space is a pair (X, a) composed of a set X and an ultrametric $a : X \times X \longrightarrow [0, \infty]$; a non-expansive map $f : (X, a) \longrightarrow (X, b)$ is, as in Example III.1.3.1(2), a map $f : X \longrightarrow Y$ such that

$$a(x, y) \geq b(f(x), f(y)) .$$

With the quantale $\mathsf{P}_{\max} = ([0, \infty]^{\text{op}}, \max, 0)$ (see Example II.1.10.1(3)), the category $\mathsf{UltraMet}$ of *ultrametric spaces* with *non-expansive maps* is the category of $\mathsf{P}_{\max}$-categories and $\mathsf{P}_{\max}$-functors:

$$(\mathbb{I}, \mathsf{P}_{\max})\text{-}\mathsf{Cat} = \mathsf{P}_{\max}\text{-}\mathsf{Cat} = \mathsf{UltraMet} .$$

III.2.C *The underlying order via ultrafilter convergence.* Use Exercise II.1.F © and the correspondence between convergence of ultrafilters and closure operations of Theorem III.2.4.5 to show that the underlying order of a $(\beta, 2)$-category (X, a) is given by

$$x \leq y \iff \dot{x} \longrightarrow y$$

for all $x, y \in X$.

© **III.2.D** *Bitopological spaces.* When $\mathcal{V} = 2^2$ is the diamond lattice of Exercise II.1.H, the *Barr extension* to 2^2-Rel of the ultrafilter functor β is defined by

$$\overline{\beta}r(\mathfrak{x}, \mathfrak{y}) = \bigvee \{w \in 2^2 \mid \mathfrak{x}\ \overline{\beta}(r_w)\ \mathfrak{y}\} = \bigwedge_{A\in\mathfrak{x}, B\in\mathfrak{y}} \bigvee_{x\in A, y\in B} r(x, y)\,,$$

for all 2^2-relations $r : X \nrightarrow Y$, $\mathfrak{x} \in \beta X$, and $\mathfrak{y} \in \beta Y$. Verify that this yields a flat lax extension of the ultrafilter monad $\mathbb{B}$ from Set to 2^2-Rel.

A *bitopological space* is a triple $(X, \mathcal{O}_1, \mathcal{O}_2)$, with $(X, \mathcal{O}_1)$ and $(X, \mathcal{O}_2)$ topological spaces, and a *bicontinuous map* $f : (X, \mathcal{O}_1, \mathcal{O}_2) \to (Y, \mathcal{O}_1', \mathcal{O}_2')$ between bitopological spaces is a map $f : X \to Y$ that is continuous as $f : (X, \mathcal{O}_1) \to (Y, \mathcal{O}_1')$ and $f : (X, \mathcal{O}_2) \to (Y, \mathcal{O}_2')$. Show that the category of $(\mathbb{B}, 2^2)$-categories obtained from the previous Barr extension is isomorphic to the category BiTop of bitopological spaces with bicontinuous maps:

© $$(\mathbb{B}, 2^2)\text{-Cat} \cong \text{BiTop}\,.$$

III.2.E *An alternative description of $\overline{\beta}$: P$_+$-Rel $\to$ P$_+$-Rel.* The Barr extension to P$_+$-Rel of the ultrafilter functor (see Proposition III.2.4.3) can be equivalently expressed as

$$\overline{\beta}r(\mathfrak{x}, \mathfrak{y}) = \sup\nolimits_{A\in\mathfrak{x}, B\in\mathfrak{y}} \inf\nolimits_{x\in A, y\in B} r(x, y)\,,$$

for every P$_+$-relation $r : X \nrightarrow Y$, $\mathfrak{x} \in \beta X$, and $\mathfrak{y} \in \beta Y$.

© **III.2.F** *Unitary* $(\mathbb{B}, \mathsf{P}_+)$*-relations.* Show

$$e_X \circ r(\mathfrak{x}, y) = \sup\nolimits_{A\in\mathfrak{x}} \inf\nolimits_{\mathfrak{y}\in A^{\mathbb{B}}} r(\mathfrak{y}, y)\,,$$

for every $(\mathbb{B}, \mathsf{P}_+)$-relation $r : X \nrightarrow X$ (recall from Section III.2.2 that $A^{\mathbb{B}} = \{\mathfrak{y} \in \beta X \mid A \in \mathfrak{y}\}$).

III.2.G *Metric closure spaces.* A *metric closure space* is a pair (X, c) composed of a set X and a *metric closure operation* c, i.e. a map $c : X \times PX \to [0, \infty]$ such that

(1) $c(x, A) = 0$,
(2) $A \subseteq B \implies c(x, B) \le c(x, A)$,
(3) $c(x, A) \le c(x, A^{(u)}) + u$,

for all $x \in X$, $A, B \in PX$, $u \in [0, \infty]$, and where $A^{(u)} = \{x \in X \mid c(x, A) \le u\}$. A morphism $f : (X, c) \to (Y, c')$ of metric closure spaces is a *non-expansive map*, i.e. a map $f : X \to Y$ that satisfies

$$c'(f(x), f(A)) \le c(x, A)$$

for all $x \in X$ and $A \subseteq X$. Metric closure spaces with non-expansive maps form the category MetCls. By equipping the powerset monad $\mathbb{P}$ with the lax extension given by

$$\hat{P}r : PX \times PY \to \mathsf{P}_+ , \quad (A, B) \mapsto \sup_{x\in A} \inf_{y\in B} r(x, y)$$

(for all $x \in X$, $y \in Y$, and $\mathcal{V}$-relations $r : X \nrightarrow Y$), one obtains an isomorphism

$$(\mathbb{P}, \mathsf{P}_+)\text{-}\mathsf{Cat} \cong \mathsf{MetCls} .$$

III.2.H *An alternative description of approach spaces.* For a map $\delta : X \times PX \to [0, \infty]$ satisfying conditions (1)–(3) of Section III.2.4, condition (4) is equivalent to

$$\sup_{x\in X} |\delta(x, A) - \delta(x, B)| \le \min\{\sup_{y\in A} \delta(y, B), \sup_{z\in B} \delta(z, A)\}$$

for all $A, B \subseteq X$.

III.3 Categories of lax algebras

In this section we prove topologicity of the underlying-set functor $(\mathbb{T}, \mathcal{V})\text{-}\mathsf{Cat} \to \mathsf{Set}$ and describe how $(\mathbb{T}, \mathcal{V})\text{-}\mathsf{Cat}$ varies functorially under appropriate changes of the parameters $\mathbb{T}$ and $\mathcal{V}$, where $\mathcal{V}$ is a quantale and $\mathbb{T}$ is a monad on Set with a lax extension $\hat{\mathbb{T}}$ to $\mathcal{V}\text{-}\mathsf{Rel}$.

III.3.1 Initial structures

Our first goal is to show that the forgetful functor $O : (\mathbb{T}, \mathcal{V})\text{-}\mathsf{Cat} \to \mathsf{Set}$ is topological (see II.5.8). It is in fact easy to give an explicit description of the O-initial lifting of sources, as follows.

III.3.1.1 Proposition *The O-initial lifting of a source $(f_i : X \to Y_i)_{i\in I}$, where (Y_i, b_i) is a family of $(\mathbb{T}, \mathcal{V})$-categories, is provided by the structure $a := \bigwedge_{i\in I} f_i^\circ \cdot b_i \cdot Tf_i$ on the set X or, in pointwise notation, by*

$$a(\mathfrak{x}, y) = \bigwedge_{i\in I} b_i(Tf_i(\mathfrak{x}), f_i(y))$$

for all $\mathfrak{x} \in TX$ and $y \in X$.

Proof Let us first verify that the $\mathcal{V}$-relation $a : TX \nrightarrow X$ is reflexive and transitive. Since each b_i is reflexive, one has

$$1_X \le f_i^\circ \cdot f_i \le f_i^\circ \cdot b_i \cdot e_{Y_i} \cdot f_i = f_i^\circ \cdot b_i \cdot Tf_i \cdot e_X ,$$

so that $1_X \le a \cdot e_X$ by taking the infimum on all $i \in I$. One also observes

$$\begin{aligned}
a \cdot \hat{T}a &\le \textstyle\bigwedge_{i\in I}(f_i^\circ \cdot b_i \cdot Tf_i) \cdot \bigwedge_{j\in J} \hat{T}(f_j^\circ \cdot b_j \cdot Tf_j) && (\hat{T} \text{ monotone})\\
&\le \textstyle\bigwedge_{i\in I, j\in J} f_i^\circ \cdot b_i \cdot Tf_i \cdot (Tf_j)^\circ \cdot \hat{T}b_j \cdot TTf_j && (\text{Corollary III.1.4.4})\\
&\le \textstyle\bigwedge_{i\in I} f_i^\circ \cdot b_i \cdot Tf_i \cdot (Tf_i)^\circ \cdot \hat{T}b_i \cdot TTf_i && \\
&\le \textstyle\bigwedge_{i\in I} f_i^\circ \cdot b_i \cdot \hat{T}b_i \cdot TTf_i && (Tf_i \cdot (Tf_i)^\circ \le 1_{TY_i})\\
&\le \textstyle\bigwedge_{i\in I} f_i^\circ \cdot b_i \cdot m_{Y_i} \cdot TTf_i && (b_i \text{ transitive})\\
&= a \cdot m_X && (m \text{ natural}).
\end{aligned}$$

Thus, (X, a) is a $(\mathbb{T}, \mathcal{V})$-category, and every $f_i : X \to Y_i$ is a $(\mathbb{T}, \mathcal{V})$-functor because, by definition of a, one has

$$a(\mathfrak{x}, y) \leq b_i(Tf_i(\mathfrak{x}), f_i(y))$$

for all $\mathfrak{x} \in TX$ and $y \in X$. To prove that the source $(f_i : (X, a) \to (Y_i, b_i))_{i \in I}$ is O-initial, consider a source $(h_i : (Z, c) \to (Y_i, b_i))$ with a map $g : Z \to X$ satisfiying $h_i = f_i \cdot g$ for all $i \in I$. One then has

$$g \cdot c \leq \bigwedge_{i \in I} f_i^{\circ} \cdot f_i \cdot g \cdot c \leq \bigwedge_{i \in I} f_i^{\circ} \cdot b_i \cdot Tf_i \cdot Tg = a \cdot Tg\,,$$

as desired. □

III.3.1.2 Examples Rewriting the formula of Proposition III.3.1.1 for the main example categories considered previously, we obtain:

(1) $x \leq y \iff \forall i \in I\ (f_i(x) \leq f_i(y))$ in $\mathsf{Ord} = (\mathbb{I}, \mathsf{2})\text{-}\mathsf{Cat}$;
(2) $a(x, y) = \sup_{i \in I} b_i(f_i(x), f_i(y))$ in $\mathsf{Met} = (\mathbb{I}, \mathsf{P}_+)\text{-}\mathsf{Cat}$;
(3) $\mathfrak{x} \longrightarrow y \iff \forall i \in I(f_i[\mathfrak{x}] \longrightarrow f_i(y))$ in $\mathsf{Top} \cong (\beta, \mathsf{2})\text{-}\mathsf{Cat}$;
(4) $a(\mathfrak{x}, y) = \sup_{i \in I} b_i(f_i[\mathfrak{x}], f_i(y))$ in $\mathsf{App} \cong (\beta, \mathsf{P}_+)\text{-}\mathsf{Cat}$.

III.3.1.3 Theorem *For a quantale $\mathcal{V}$ and a lax extension $\hat{T}$ of a* Set*-functor T to $\mathcal{V}$-*Rel*, the forgetful functor $O : (\mathbb{T}, \mathcal{V})\text{-}\mathsf{Cat} \to \mathsf{Set}$ is topological. It therefore admits initial and final liftings, is transportable, has both a fully faithful left and a fully faithful right adjoint, and makes $(\mathbb{T}, \mathcal{V})$-*Cat *a small-complete, small-cocomplete, well-powered, and cowell-powered category with a generator and a cogenerator.*

Proof Proposition III.3.1.1 shows that every source $(f_i : X \to O(Y_i, b_i))$ admits an O-initial lifting, so that O is topological. Theorem II.5.9.1 takes care of the final liftings, transportability, and the existence of adjoints. Corollary II.5.8.4 and its dual prove the statements on completeness and cocompleteness, since Set is both small-complete and small-cocomplete. Well- and cowell-poweredness of $(\mathbb{T}, \mathcal{V})$-Cat are also consequences of topologicity; see Exercise II.5.F. To obtain a generator in $(\mathbb{T}, \mathcal{V})$-Cat, one can apply the left adjoint of O to a generator of Set, and proceed dually for a cogenerator. □

III.3.1.4 Remark The proof of Proposition III.3.1.1 makes it easy to describe explicitly the limit (L, a) of a diagram $D : \mathsf{J} \to (\mathbb{T}, \mathcal{V})\text{-}\mathsf{Cat}$: just form the limit L of OD in Set and then use the formula for a as given in Proposition III.3.1.1, where the f_i ($i \in \operatorname{ob}\mathsf{J}$) are the limit projections in Set. There is unfortunately no easy general formula for the construction of O-final liftings, and consequently for colimits in $(\mathbb{T}, \mathcal{V})$-Cat. Under additional hypotheses, coproducts may be described easily; see Section III.4.3.

III.3.2 Discrete and indiscrete lax algebras

The topological functor $O : (\mathbb{T}, \mathcal{V})\text{-}\mathsf{Cat} \to \mathsf{Set}$ has both a left and right adjoint (Theorem III.3.1.3), which we now proceed to describe explicitly. We first consider the functor $(-)_d : \mathsf{Set} \to (\mathbb{T}, \mathcal{V})\text{-}\mathsf{Cat}$ that is the identity on maps, and sends a set X to $X_d = (X, 1_X^\sharp)$, where the *discrete structure* $1_X^\sharp$ on X is described by

$$1_X^\sharp = e_X^\circ \cdot \hat{T} 1_X$$

(see Section III.1.7). The reflexivity condition for a $(\mathbb{T}, \mathcal{V})$-category structure a is equivalent to $1_X^\sharp \leq a$ by (III.1.8.i), so one obtains a $(\mathbb{T}, \mathcal{V})$-functor $\varepsilon_X : (X, 1_X^\sharp) \to (X, a)$ whose underlying map is the identity on X. Thus, the natural transformations $1 : 1_{\mathsf{Set}} \to O(-)_d = 1_{\mathsf{Set}}$ and $\varepsilon : (-)_d O \to 1_{(\mathbb{T},\mathcal{V})\text{-}\mathsf{Cat}}$ trivially satisfy the triangular identities of an adjunction (see II.2.5), and one concludes that $(-)_d$ is left adjoint to O:

$$(-)_d \underset{\varepsilon}{\overset{1}{\dashv}} O : (\mathbb{T}, \mathcal{V})\text{-}\mathsf{Cat} \longrightarrow \mathsf{Set} \, .$$

As is the case for every topological functor (see Theorem II.5.9.1), the left adjoint $(-)_d$ embeds Set as a full coreflective category of $(\mathbb{T}, \mathcal{V})\text{-}\mathsf{Cat}$.

The right adjoint $(-)_i : \mathsf{Set} \to (\mathbb{T}, \mathcal{V})\text{-}\mathsf{Cat}$ of O provides a set X with the O-initial structure of the empty source with domain X (see II.5.8). Hence, $(-)_i$ sends a set X to the $(\mathbb{T}, \mathcal{V})$-category $X_i = (X, \top_X)$, where $\top_X : TX \to\!\!\!\!\!\!\mapsto X$ is the *indiscrete structure* on X given by

$$\top_X(\mathfrak{x}, y) = \top$$

for all $\mathfrak{x} \in TX$, $y \in X$. This describes a full reflective embedding of Set into $(\mathbb{T}, \mathcal{V})\text{-}\mathsf{Cat}$.

As the described discrete and indiscrete structures are determined by adjunctions, they correspond to the respective structures described for $\mathsf{Ord} = \mathsf{2}\text{-}\mathsf{Cat}$ and $\mathsf{Top} \cong (\beta, \mathsf{2})\text{-}\mathsf{Cat}$ in Example II.2.5.1(3).

III.3.3 Induced orders

Given a lax extension $\hat{\mathbb{T}}$ to $\mathcal{V}\text{-}\mathsf{Rel}$ of a monad $\mathbb{T}$ on Set, the structure of a $(\mathbb{T}, \mathcal{V})$-category is a reflexive and transitive $\mathcal{V}$-relation. This terminology not only extends the usual concept used for ordinary relations, but also suggests that the structure induces a natural order on the underlying set. Indeed, since a $(\mathbb{T}, \mathcal{V})$-category structure $a : TX \to\!\!\!\!\!\!\mapsto X$ is left unitary, one has

$$e_X^\circ \cdot \hat{T}a \cdot e_{TX} \leq e_X^\circ \cdot \hat{T}a \cdot m_X^\circ = e_X^\circ \circ a = a \, .$$

The inequality in the other direction is just the expression of oplaxness of the unit $e : 1_{\mathcal{V}\text{-}\mathsf{Rel}} \to \hat{T}$, so we have

$$a = e_X^\circ \cdot \hat{T}a \cdot e_{TX} \tag{III.3.3.i}$$

for any $(\mathbb{T}, \mathcal{V})$-category structure $a : TX \nrightarrow X$. This identity is used to prove the following result.

III.3.3.1 Proposition *Let $\hat{\mathbb{T}}$ be a lax extension of a monad $\mathbb{T}$ on* Set *to* $\mathcal{V}$-Rel. *If $a : TX \nrightarrow X$ is a $(\mathbb{T}, \mathcal{V})$-category structure, then the relation*

$$x \leq y \iff k \leq a(e_X(x), y)$$

(for all $x, y \in X$) defines an order on X, called the underlying order *induced by a (or sometimes simply the* induced order*). The structure a is then monotone in its second variable with respect to this order:*

$$x \leq y \implies a(\mathfrak{x}, x) \leq a(\mathfrak{x}, y)$$

for all $x, y \in X$, $\mathfrak{x} \in TX$.

Proof For the given relation $\leq$ on X, one immediately has $x \leq x$ since $k \leq a(e_X(x), x)$ by reflexivity of a. By transitivity of a and the identity (III.3.3.i), if $x \leq y$ and $y \leq z$, then

$$\begin{aligned} k &\leq a(e_X(x), y) \otimes a(e_X(y), z)) \\ &= \hat{T}a(e_{TX}(e_X(x)), e_X(y)) \otimes a(e_X(y), z) \leq a(e_X(x), z) , \end{aligned}$$

i.e. $x \leq z$, so the relation $\leq$ is also transitive. Finally, if $x \leq y$, then transitivity of a also yields

$$a(\mathfrak{x}, x) = \hat{T}a(e_{TX}(\mathfrak{x}), e_X(x)) \leq \hat{T}a(e_{TX}(\mathfrak{x}), e_X(x)) \otimes a(e_X(x), y) \leq a(\mathfrak{x}, y) ,$$

so that a is monotone with respect to this order. □

III.3.3.2 Corollary *The order induced by a $(\mathbb{T}, \mathcal{V})$-category structure yields a functor*

$$(\mathbb{T}, \mathcal{V})\text{-}\mathsf{Cat} \to \mathsf{Ord}$$

*that makes $(\mathbb{T}, \mathcal{V})$-*Cat *an ordered category. Hence, as in* Ord, *hom-sets in $(\mathbb{T}, \mathcal{V})$-*Cat *are ordered pointwise:*

$$f \leq g \iff \forall x \in X \, (f(x) \leq g(x))$$

for $f, g : (X, a) \to (Y, b)$.

Proof The order given by Proposition III.3.3.1 makes every $(\mathbb{T}, \mathcal{V})$-functor $f : (X, a) \to (Y, b)$ monotone since, for $x, y \in X$, $x \leq y$ implies

$$k \leq a(e_X(x), y) \leq b(Tf \cdot e_X(x), f(y)) = b(e_Y \cdot f(x), f(y)) ,$$

i.e. $f(x) \leq f(y)$. Since the hom-sets in the ordered category Ord are ordered pointwise, the order on the hom-sets in $(\mathbb{T}, \mathcal{V})$-Cat induced by $(\mathbb{T}, \mathcal{V})$-Cat $\to$ Ord is also pointwise. □

III.3.3.3 Corollary *The following conditions are equivalent for $(\mathbb{T}, \mathcal{V})$-functors $f, g : (X, a) \to (Y, b)$:*

(i) $f \le g$;
(ii) $\forall y \in TY, x \in X\ (b(y, f(x)) \le b(y, g(x)))$;
(iii) $\forall \chi \in TX, x \in X\ (a(\chi, x) \le b(Tf(\chi), g(x)))$.

Proof (i) $\Longrightarrow$ (ii) follows from Proposition III.3.3.1.

For (ii) $\Longrightarrow$ (iii), observe that $a(\chi, x) \le b(Tf(\chi), f(x)) \le b(Tf(\chi), g(x))$.

Finally, $k \le a(e_X(x), x) \le b(Tf(e_X(x)), g(x)) = b(e_Y(f(x)), g(x))$ proves (iii) $\Longrightarrow$ (i). □

III.3.3.4 Remark If $\hat{\mathbb{T}}$ is associative, then the following condition is also equivalent to (i)–(iii) of Corollary III.3.3.3:

(iv) $\forall \chi \in TX, y \in Y\ (b(Tg(\chi), y) \le b(Tf(\chi), y))$.

A proof using Kleisli convolution is indicated in Exercise III.3.E.

III.3.3.5 Examples

(1) For the identity lax extension of $\mathbb{I}$ to 2-Rel, we have 2-Cat = Ord (Example III.1.3.1(1)), and the underlying order on an ordered set (X, a) induced by a returns the original order on X. For the identity lax extension of $\mathbb{I}$ to P_+-Rel, we had P_+-Cat = Met (Example III.1.3.1(2)), and the underlying order induced on a metric space (X, a) is given by

$$x \le y \iff a(x, y) = 0$$

for all $x, y \in X$.

(2) For the Barr extension of the ultrafilter monad β to 2-Rel, Theorem III.2.2.5 states that $(\beta, 2)$-Cat $\cong$ Top. Here, the underlying order on (X, a) is given by (when we write a as $\longrightarrow$)

$$x \le y \iff \dot{x} \longrightarrow y\,.$$

By Exercise III.2.C, this is precisely the underlying order of topological spaces II.1.9. The underlying order on an approach space (X, a) in the guise of a (β, P_+)-category (see Theorem III.2.4.5) is defined by

$$x \le y \iff a(\dot{x}, y) = 0$$

for all $x, y \in X$. In terms of the set-point distance $\delta : X \times PX \to \mathsf{P}_+$, this means

$$x \le y \iff \delta(y, \{x\}) = 0\,.$$

For a $(\mathbb{T}, \mathcal{V})$-category (X, a), we wish to define an order on the set TX such that – as one would expect of a hom-functor – a becomes order reversing in its first variable. The strategy should be to provide TX with a "natural"

$(\mathbb{T}, \mathcal{V})$-structure and then to consider the order induced via Proposition III.3.3.1. Unfortunately, the free $\mathbb{T}$-algebra structure m_X on TX will generally not provide a $(\mathbb{T}, \mathcal{V})$-category structure (unless $\hat{T}$ is flat, see Proposition III.1.6.5), but there is a $(\mathbb{T}, \mathcal{V})$-category structure $\tilde{m}_X \geq m_X$ on TX satisfying our requirements, namely $\tilde{m}_X := \hat{T}1_X \cdot m_X$, as we show next.

III.3.3.6 Proposition *There is a functor* $\tilde{T} : \mathsf{Set} \to (\mathbb{T}, \mathcal{V})\text{-}\mathsf{Cat}$ *sending* X *to* $(TX, \tilde{m}_X)$ *that makes*

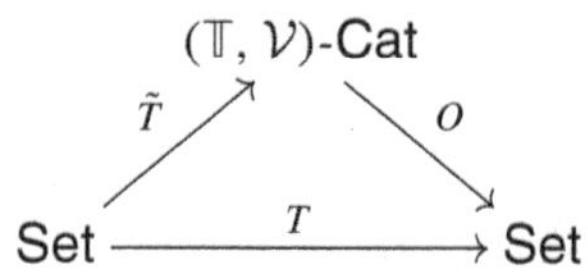

commute. The order on TX *induced by* $\tilde{m}_X$ *is given by*

$$\mathfrak{x} \leq \mathfrak{y} \iff k \leq \hat{T}1_X(\mathfrak{x}, \mathfrak{y}),$$

for all $\mathfrak{x}, \mathfrak{y} \in TX$. *Any* $(\mathbb{T}, \mathcal{V})$*-category structure* a *on* X *reverses this order in its first variable:*

$$\mathfrak{x} \leq \mathfrak{y} \implies a(\mathfrak{y}, z) \leq a(\mathfrak{x}, z)$$

for all $\mathfrak{x}, \mathfrak{y} \in TX$, $z \in X$, *so that* $a : TX \nrightarrow X$ *becomes a module of ordered sets.*

Proof Reflexivity and transitivity of $\tilde{m}_X$ are easily verified:

$$1_{TX} \leq \hat{T}1_X = \hat{T}1_X \cdot m_X \cdot e_{TX} = \tilde{m}_X \cdot e_{TX},$$

and

$$\begin{aligned} \tilde{m}_X \cdot \hat{T}\tilde{m}_X &= \hat{T}1_X \cdot m_X \cdot \hat{T}\hat{T}1_X \cdot Tm_X && \text{(Corollary III.1.4.4)} \\ &\leq \hat{T}1_X \cdot m_X \cdot Tm_X && (m \text{ oplax}) \\ &= \tilde{m}_X \cdot m_{TX}. \end{aligned}$$

Also, for any map $f : X \to Y$, $Tf : \tilde{T}X \to \tilde{T}Y$ is a $(\mathbb{T}, \mathcal{V})$-functor:

$$\begin{aligned} Tf \cdot \tilde{m}_X &= Tf \cdot \hat{T}(e_X^\circ) \cdot m_X^\circ \cdot m_X && \text{(Lemma III.1.7.2)} \\ &\leq \hat{T}(f \cdot e_X^\circ) \cdot m_X^\circ \cdot m_X \\ &\leq \hat{T}(e_Y^\circ \cdot Tf) \cdot m_X^\circ \cdot m_X \\ &= \hat{T}(e_Y^\circ) \cdot TTf \cdot m_X^\circ \cdot m_X && \text{(Corollary III.1.4.4)} \\ &\leq \hat{T}(e_Y^\circ) \cdot m_Y^\circ \cdot Tf \cdot m_X \\ &= \hat{T}1_Y \cdot m_Y \cdot TTf = \tilde{m}_X \cdot TTf && \text{(Lemma III.1.7.2).} \end{aligned}$$

By Proposition III.3.3.1, the order induced by $\tilde{m}_X$ on TX is given by the $\mathcal{V}$-relation $\hat{T}1_X \cdot m_X \cdot e_{TX} = \hat{T}1_X$, which gives the description in the claim.

Finally, if a is a $(\mathbb{T}, \mathcal{V})$-category structure, then it is right unitary, so that $k \le \hat{T}1_X(\mathfrak{x}, y)$ yields

$$a(y, z) \le \hat{T}1_X(\mathfrak{x}, y) \otimes a(y, z) \le a \cdot \hat{T}1_X(\mathfrak{x}, z) = a(\mathfrak{x}, z)\,,$$

which means that $\mathfrak{x} \le y$ implies $a(y, z) \le a(\mathfrak{x}, z)$, as claimed. □

We will see in Section III.5.4 that $\tilde{T}$ factors through the left adjoint of O; see Exercise III.5.K. Also, alternative orders on TX will be considered (see, in particular, Examples III.5.3.7).

III.3.4 Algebraic functors

Consider lax extensions $\hat{\mathbb{S}}$, $\hat{\mathbb{T}}$ to $\mathcal{V}$-**Rel** of monads $\mathbb{S} = (S, n, d)$ and $\mathbb{T} = (T, m, e)$ on **Set**. A *morphism of lax extensions* $\alpha : (S, \hat{S}) \to (T, \hat{T})$ is a natural transformation $\alpha : S \to T$ that becomes an oplax transformation $\hat{S} \to \hat{T}$, so that

$$\alpha_Y \cdot \hat{S}r \le \hat{T}r \cdot \alpha_X$$

for all $\mathcal{V}$-relations $r : X \nrightarrow Y$. A monad morphism $\alpha : \mathbb{S} \to \mathbb{T}$ which is also a morphism of lax extensions $\alpha : \hat{S} \to \hat{T}$ is denoted by $\alpha : (\mathbb{S}, \hat{S}) \to (\mathbb{T}, \hat{T})$. Any such natural transformation α induces a functor

$$A_\alpha : (\mathbb{T}, \mathcal{V})\text{-}\mathsf{Cat} \to (\mathbb{S}, \mathcal{V})\text{-}\mathsf{Cat}\,,$$

sending (X, a) to $(X, a \cdot \alpha_X)$, and mapping morphisms identically. Indeed, one has $1_X \le a \cdot e_X = a \cdot \alpha_X \cdot d_X$, and

$$\begin{aligned} a \cdot \alpha_X \cdot \hat{S}(a \cdot \alpha_X) &= a \cdot \alpha_X \cdot \hat{S}a \cdot S\alpha_X \\ &\le a \cdot \hat{T}a \cdot \alpha_{TX} \cdot S\alpha_X \\ &\le a \cdot m_X \cdot \alpha_{TX} \cdot S\alpha_X \\ &= a \cdot \alpha_X \cdot n_X\,. \end{aligned}$$

Moreover, a $(\mathbb{T}, \mathcal{V})$-functor $f : (X, a) \to (Y, b)$ is an $(\mathbb{S}, \mathcal{V})$-functor $f : (X, a \cdot \alpha_X) \to (Y, b \cdot \alpha_Y)$:

$$f \cdot a \cdot \alpha_X \le b \cdot Tf \cdot \alpha_X = b \cdot \alpha_Y \cdot Sf\,.$$

The functor A_α is called the *algebraic functor* associated with α.

Given a lax extension $\hat{\mathbb{T}}$ of $\mathbb{T} = (T, m, e)$ to $\mathcal{V}$-**Rel**, the unit $e : 1_{\mathsf{Set}} \to T$ immediately yields an algebraic functor. Indeed, oplaxness of e means precisely that there is a morphism of lax extensions $e : (1_{\mathsf{Set}}, 1_{\mathcal{V}\text{-}\mathsf{Rel}}) \to (T, \hat{T})$. As e is also a monad morphism $e : \mathbb{I} \to \mathbb{T}$, one obtains a functor

$$A_e : (\mathbb{T}, \mathcal{V})\text{-}\mathsf{Cat} \to \mathcal{V}\text{-}\mathsf{Cat}$$

that sends (X, a) to its *underlying $\mathcal{V}$-category* $(X, a \cdot e_X)$ and commutes with the underlying-set functors (recall from Section III.1.6 that we write $\mathcal{V}$-**Cat** rather than $(\mathbb{I}, \mathcal{V})$-**Cat**). This functor has a left adjoint that we now proceed to describe. In Section III.1.9, we defined the unitary $(\mathbb{T}, \mathcal{V})$-relation

$$r_\sharp = e_Y^\circ \cdot \hat{T}r$$

associated to a $\mathcal{V}$-relation $r : X \nrightarrow Y$.

III.3.4.1 Lemma *The $(-)_\sharp$ transformation defined above satisfies*

$$r \le r' \implies r_\sharp \le r'_\sharp \qquad \textit{and} \qquad s_\sharp \circ r_\sharp \le (s \cdot r)_\sharp$$

*for all $\mathcal{V}$-relations $r, r' : X \nrightarrow Y$ and $s : Y \nrightarrow Z$. Moreover, if $\hat{\mathbb{T}}$ is associative, then the second inequality is actually an equality, and $(-)_\sharp$ defines a 2-functor $(-)_\sharp : \mathcal{V}$-**Rel** $\to (\mathbb{T}, \mathcal{V})$-**URel**.*

Proof The first implication is straightforward, since composition and $\hat{T}$ both preserve the order on hom-sets. For the second expression, we write

$$\begin{aligned} s_\sharp \circ r_\sharp &= e_Z^\circ \cdot \hat{T}s \cdot \hat{T}(e_Y^\circ \cdot \hat{T}r) \cdot m_X^\circ \\ &= e_Z^\circ \cdot \hat{T}s \cdot (Te_Y)^\circ \cdot \hat{T}\hat{T}r \cdot m_X^\circ \\ &\le e_Z^\circ \cdot \hat{T}s \cdot (Te_Y)^\circ \cdot m_Y^\circ \cdot \hat{T}r \\ &= e_Z^\circ \cdot \hat{T}s \cdot \hat{T}r \\ &\le e_Z^\circ \cdot \hat{T}(s \cdot r) \\ &= (s \cdot r)_\sharp \,. \end{aligned}$$

If $\hat{\mathbb{T}}$ is associative, then $\hat{T}$ preserves composition and m° is a natural transformation (Proposition III.1.9.4), so the inequalities in the previous displayed formulas become equalities. Since $(1_X)_\sharp = 1_X^\sharp$ is the identity in $(\mathbb{T}, \mathcal{V})$-**URel**, the $(-)_\sharp$ transformation defines a functor $\mathcal{V}$-**Rel** $\to (\mathbb{T}, \mathcal{V})$-**URel**. □

III.3.4.2 Proposition *The algebraic functor A_e has a left adjoint*

$$A^\circ : \mathcal{V}\text{-}\mathsf{Cat} \to (\mathbb{T}, \mathcal{V})\text{-}\mathsf{Cat}$$

which associates to a $\mathcal{V}$-category (X, r) the $(\mathbb{T}, \mathcal{V})$-category $(X, r_\sharp)$.

Proof Let (X, r) be a $\mathcal{V}$-category. Since $1_X \le r$ and $r \cdot r \le r$, we deduce from Lemma III.3.4.1

$$1_X^\sharp \le r_\sharp \qquad \text{and} \qquad r_\sharp \circ r_\sharp \le (r \cdot r)_\sharp \le r_\sharp \,,$$

so that $r_\sharp$ is both reflexive and transitive if r is. For a $\mathcal{V}$-functor $f : (X, r) \to (Y, s)$, we also have

$$f \cdot r_\sharp = f \cdot e_X^\circ \cdot \hat{T}r \le e_X^\circ \cdot \hat{T}(f \cdot r) \le e_X^\circ \cdot \hat{T}(s \cdot f) = e_X^\circ \cdot \hat{T}s \cdot Tf = s_\sharp \cdot Tf \,,$$

which yields $(\mathbb{T}, \mathcal{V})$-functoriality. This functor is left adjoint to A_e: for a $\mathcal{V}$-category (X, r),

$$r \leq e_X^\circ \cdot \hat{T}r \cdot e_X = r_\sharp \cdot e_X \ ,$$

and for a $(\mathbb{T}, \mathcal{V})$-category (X, a),

$$(a \cdot e_X)_\sharp = e_X^\circ \cdot \hat{T}(a \cdot e_X) = e_X^\circ \cdot \hat{T}a \cdot Te_X \leq e_X^\circ \cdot \hat{T}a \cdot m_X^\circ = a \ . \qquad \square$$

III.3.4.3 Examples

(1) For $\mathcal{V} = 2$ and $\mathbb{T} = \mathbb{\beta}$, we have $2\text{-}\mathsf{Cat} = \mathsf{Ord}$ and $(\mathbb{\beta}, 2)\text{-}\mathsf{Cat} = \mathsf{Top}$, and the functor $A_e : \mathsf{Top} \to \mathsf{Ord}$ sends a topological space (X, a) to the ordered set $(X, a \cdot e_X)$ whose order is the underlying order of the former; i.e. A_e is the forgetful functor to Ord of Corollary III.3.3.2. The left adjoint $\mathsf{Ord} \hookrightarrow \mathsf{Top}$ provides an ordered set $(X, \leq)$ with the Alexandroff topology, i.e. the topology whose open sets are generated by the down-sets $\downarrow x$, for $x \in X$ (Example II.5.10.5).

(2) For $\mathcal{V} = \mathsf{P}_+$ and $\mathbb{T} = \mathbb{\beta}$, we have $\mathsf{P}_+\text{-}\mathsf{Cat} = \mathsf{Met}$ and $(\mathbb{\beta}, \mathsf{P}_+)\text{-}\mathsf{Cat} = \mathsf{App}$, so $A_e : \mathsf{App} \to \mathsf{Met}$ sends an approach space (X, a) to the metric space $(X, a \cdot e_X)$. The left adjoint $\mathsf{Met} \hookrightarrow \mathsf{App}$ sends a metric space (X, r) to the approach space whose structure is given by

$$r_\sharp(\mathfrak{x}, y) = \sup_{A \in \mathfrak{x}} \inf_{x \in A} r(x, y)$$

for all $\mathfrak{x} \in \beta X$ and $y \in X$. This coreflective embedding has been described in terms of point-set distance in Example III.2.4.1(1).

III.3.5 Change-of-base functors

The algebraic functors deal with monads, i.e. with the first variable in $(\mathbb{T}, \mathcal{V})$-Cat. The change-of-base functors deal with the second. Consider lax extensions $\hat{\mathbb{T}}$ and $\check{\mathbb{T}}$ of the monad $\mathbb{T} = (T, m, e)$ to $\mathcal{V}$-Rel and $\mathcal{W}$-Rel, respectively.

Let $\varphi : \mathcal{V} \to \mathcal{W}$ be a lax homomorphism of quantales (see Section II.1.10), so that φ is order preserving, and

$$\varphi(u) \otimes \varphi(v) \leq \varphi(u \otimes v) \ , \quad l \leq \varphi(k)$$

for all $u, v \in \mathcal{V}$, and where k, l are the units of $\mathcal{V}, \mathcal{W}$, respectively. Then φ induces a lax functor

$$\varphi : \mathcal{V}\text{-}\mathsf{Rel} \to \mathcal{W}\text{-}\mathsf{Rel}$$

which leaves objects unchanged and sends $r : X \times Y \to \mathcal{V}$ to $\varphi r : X \times Y \to \mathcal{W}$. Clearly, for any Set-map f we have

$$f \leq \varphi f \qquad \text{and} \qquad f^\circ \leq \varphi(f^\circ) \ ,$$

where f and f° are considered as $\mathcal{W}$-relations when appearing on the left of the inequality sign, and as $\mathcal{V}$-relations on the right. Furthermore, we assume that φ is *compatible* with the respective lax extensions $\hat{\mathbb{T}}$ and $\check{\mathbb{T}}$ of $\mathbb{T}$ to $\mathcal{V}$-Rel and $\mathcal{W}$-Rel, i.e. $\check{T}(\varphi r) \leq \varphi(\hat{T} r)$ for all $\mathcal{V}$-relations r:

$$\begin{array}{ccc} \mathcal{V}\text{-}\mathsf{Rel} & \xrightarrow{\hat{T}} & \mathcal{V}\text{-}\mathsf{Rel} \\ {\scriptstyle\varphi}\downarrow & \leq & \downarrow{\scriptstyle\varphi} \\ \mathcal{W}\text{-}\mathsf{Rel} & \xrightarrow[\check{T}]{} & \mathcal{W}\text{-}\mathsf{Rel} \,. \end{array}$$

Under these conditions, φ induces a functor

$$B_\varphi : (\mathbb{T}, \mathcal{V})\text{-}\mathsf{Cat} \to (\mathbb{T}, \mathcal{W})\text{-}\mathsf{Cat}\,,$$

called the *change-of-base functor* associated to φ, sending (X, a) to $(X, \varphi a)$ and leaving maps unchanged. Indeed, we observe that $e_X^\circ \leq \varphi(e_X^\circ) \leq \varphi a$ holds, as well as

$$\begin{aligned} \varphi a \circ \varphi a &= \varphi a \cdot \check{T}(\varphi a) \cdot m_X^\circ \\ &\leq \varphi a \cdot \varphi(\hat{T} a) \cdot \varphi(m_X^\circ) \\ &\leq \varphi(a \cdot \hat{T} a \cdot m_X^\circ) \\ &= \varphi a \,. \end{aligned}$$

Moreover, given a $(\mathbb{T}, \mathcal{V})$-functor $f : (X, a) \to (Y, b)$, one can adapt Corollary III.1.4.4 to φ to obtain the last equality in

$$f \cdot \varphi a \leq \varphi f \cdot \varphi a \leq \varphi(f \cdot a) \leq \varphi(b \cdot Tf) = \varphi b \cdot Tf \,.$$

III.3.5.1 Proposition *Adjunctions of maps become adjunctions of the corresponding change-of-base functors. More precisely, suppose that $\varphi : \mathcal{V} \to \mathcal{W}$ and $\psi : \mathcal{W} \to \mathcal{V}$ are lax homomorphisms of quantales that are compatible with the respective lax extensions of $\mathbb{T}$ to $\mathcal{V}$-*Rel *and $\mathcal{W}$-*Rel*. Then one has*

$$\varphi \dashv \psi \implies B_\varphi \dashv B_\psi \,.$$

Proof The composite $B_\psi B_\varphi$ sends a $(\mathbb{T}, \mathcal{V})$-category (X, a) to $(X, \psi\varphi a)$. If a is seen as a map $a : TX \times X \to \mathcal{V}$, then $\psi\varphi a$ is the map $\psi \cdot \varphi \cdot a$ with $a \leq \psi \cdot \varphi \cdot a$ because $1_\mathcal{V} \leq \psi \cdot \varphi$; therefore, the identity $1_X : (X, a) \to (X, \psi\varphi a)$ is a $(\mathbb{T}, \mathcal{V})$-functor. Dually, the identity $1_X : (X, \varphi\psi b) \to (X, b)$ is a $(\mathbb{T}, \mathcal{W})$-functor for a $(\mathbb{T}, \mathcal{W})$-category (X, b). These maps then yield the respective components of the unit and counit of an adjunction $B_\varphi \dashv B_\psi$ since the triangular identities are trivially satisfied.

III.3.5.2 Examples

(1) The quantale homomorphism $\iota : \mathsf{2} \to \mathcal{V}$ always has a right adjoint $p : \mathcal{V} \to \mathsf{2}$ that is a lax homomorphism of quantales (with $p(v) = \top$ if $k \le v$ and $p(v) = \bot$ otherwise; see Exercise II.1.I). In fact, p is a quantale homomorphism if and only if

$$k \le u \otimes v \implies k \le u \text{ and } k \le v$$

for all $u, v \in \mathcal{V}$. As a monotone map, ι has a left adjoint $o : \mathcal{V} \to \mathsf{2}$ if and only if $k = \top$, given then by $o(v) = \top$ if and only if $\bot < v$; furthermore, o is a quantale homomorphism if and only if

$$u \otimes v = \bot \implies u = \bot \text{ or } v = \bot$$

for all $u, v \in \mathcal{V}$. These maps are all obviously compatible with the identical lax extensions of the identity monad $\mathbb{I}$ to Rel and $\mathcal{V}$-Rel, and Proposition III.3.5.1 yields adjunctions

$$\mathsf{Ord} \underset{B_p}{\overset{B_o}{\leftarrow}} \hookrightarrow \mathcal{V}\text{-}\mathsf{Cat}\,, \qquad B_o \dashv B_\iota \dashv B_p .$$

In this diagram, $B_\iota : \mathsf{2}\text{-}\mathsf{Cat} \to \mathcal{V}\text{-}\mathsf{Cat}$ is the embedding $\mathsf{Ord} \hookrightarrow \mathcal{V}\text{-}\mathsf{Cat}$, and B_p is the forgetful functor $\mathcal{V}\text{-}\mathsf{Cat} \to \mathsf{Ord}$ from Corollary III.3.3.2 (in the $\mathbb{T} = \mathbb{I}$ case).

(2) By (1) there is in particular a full reflective and coreflective embedding $\mathsf{Ord} \hookrightarrow \mathsf{Met}$ which provides an ordered set $(X, \le)$ with the metric d given by $d(x, y) = 0$ if $x \le y$ and $d(x, y) = \infty$ otherwise. There is also a full embedding

$$B_\delta : \mathsf{Met} \hookrightarrow \mathsf{ProbMet}$$

(see Section III.2.1) which is induced by the *Dirac morphism*

$$\delta : \mathsf{P}_+ = ([0, \infty]^{\mathrm{op}}, +, 0) \to \mathsf{D} = (\mathsf{D}, \otimes, \kappa)\,, \qquad w \mapsto \delta_w\,,$$

with

$$\delta_w(v) = \begin{cases} 0 & \text{if } v \le w, \\ 1 & \text{if } v > w, \end{cases}$$

for all $v, w \in [0, \infty]$. (That δ is indeed a quantale homomorphism is easy to verify.) Indeed, B_δ sends the metric space (X, d) to the probabilistic metric space (X, a) with

$$a(x, y)(v) = \begin{cases} 0 & \text{if } d(x, y) \ge v, \\ 1 & \text{if } d(x, y) < v \end{cases}$$

$(x, y \in X, v \in [0, \infty])$. For another metric space (Y, c) with $B_\delta(Y, c) = (Y, b)$, a map $f : (X, a) \longrightarrow (Y, b)$ is probabilistically non-expansive if and only if

$$\forall x, y \in X, v \in [0, \infty]\ (a(x, y)(v) \leq b(f(x), f(y))(v))\ ;$$

this means that $d(x, y) < v$ always implies $c(f(x), f(y)) < v$, or equivalently that $f : (X, d) \longrightarrow (Y, c)$ is non-expansive. Consequently, B_δ is indeed full.

The Dirac morphism δ has a left adjoint

$$\omega : \mathsf{D} \longrightarrow \mathsf{P}_+\,, \quad \phi \longmapsto \sup\{v \in [0, \infty] \mid \phi(v) \leq 0\}$$

that is a quantale homomorphism. As a monotone map, δ also has a right adjoint

$$\rho : \mathsf{D} \longrightarrow \mathsf{P}_+\,, \quad \phi \longmapsto \inf\{v \in [0, \infty] \mid 1 \leq \phi(v)\}$$

that, although it satisfies $\rho(\kappa) = 0$ and $\rho(\phi \otimes \psi) = \phi + \psi$, does not preserve arbitrary suprema, and is therefore only a lax homomorphism of quantales. Proposition III.3.5.1 then yields adjunctions

$$\mathsf{Met} \underset{B_\rho}{\overset{B_\omega}{\leftrightarrows}} \mathsf{ProbMet}\,.$$

III.3.6 Fundamental adjunctions

Set, Ord, $\mathcal{V}$-Cat, *and* $(\mathbb{T}, \mathcal{V})$-Cat

In the general setting of $(\mathbb{T}, \mathcal{V})$-categories (as in Section III.1.6), the composite of the algebraic functor $A_e : (\mathbb{T}, \mathcal{V})\text{-}\mathsf{Cat} \longrightarrow \mathcal{V}\text{-}\mathsf{Cat}$ (Proposition III.3.4.2) with the underlying-order functor $B_p : \mathcal{V}\text{-}\mathsf{Cat} \longrightarrow \mathsf{Ord}$ (see Example III.3.5.2(1)) is precisely the induced-order functor of Corollary III.3.3.2. This functor has a left adjoint, since both B_p and A_e have left adjoints. We may further compose this left adjoint with the discrete-order functor $\mathsf{Set} \longrightarrow \mathsf{Ord}$, which then gives the following decomposition of the adjunction $(-)_d \dashv O : (\mathbb{T}, \mathcal{V})\text{-}\mathsf{Cat} \longrightarrow \mathsf{Set}$ described in III.3.2:

$$\mathsf{Set} \underset{O}{\leftrightarrows} \mathsf{Ord} \underset{B_p}{\overset{B_\iota}{\leftrightarrows}} \mathcal{V}\text{-}\mathsf{Cat} \underset{A_e}{\overset{A^\circ}{\leftrightarrows}} (\mathbb{T}, \mathcal{V})\text{-}\mathsf{Cat}\,.$$

Ord *and* Met

In the case where $\mathcal{V} = \mathsf{P}_+$, Example III.3.5.2(1) yields the embedding $B_\iota : \mathsf{Ord} \hookrightarrow \mathsf{Met}$, together with its right and left adjoints B_p and B_o, respectively. The functor B_ι has been described in Example III.3.5.2(2), and its adjoints provide a metric space (X, a) with the orders given by

$$B_o : x \leq y \iff a(x, y) < \infty \qquad \text{and} \qquad B_p : x \leq y \iff a(x, y) = 0\,.$$

Top *and* App

The homomorphism ι is compatible with the lax extensions of the ultrafilter monad to Rel and P_+-Rel, and the induced change-of-base functor is the embedding Top $\hookrightarrow$ App described at the end of III.2.4.

The lax homomorphism $p : P_+ \to 2$ is also compatible with the lax extensions, and provides the embedding with a right adjoint B_p : App $\to$ Top. This adjoint sends an approach space (X, a) to a topological space in which an ultrafilter χ converges to a point x precisely when $a(\chi, x) = 0$.

Unfortunately, the lax homomorphism $o : P_+ \to 2$ is not compatible with the ultrafilter lax extensions. Nevertheless, given an approach space (X, a), one can still consider the pair (X, oa) that has an ultrafilter χ converging to x precisely when $a(\chi, x) < \infty$. This structure satisfies the reflexivity but not the transitivity condition for topologies defined via convergence. In other words, (X, oa) is just a pseudotopological space (see Exercise III.3.D). But we may apply the left adjoint of the full reflective embedding Top $\hookrightarrow$ PsTop to (X, oa) to obtain a topological space and thereby a left adjoint L : App $\to$ Top to the embedding Top $\hookrightarrow$ App.

Ord, Met, Top, *and* App

The following diagram relates the functors described in the preceding paragraphs with the adjunction of Proposition III.3.4.2. This diagram commutes with respect to both the solid and the dotted arrows (but not the dashed arrows); moreover, the two full embeddings Ord $\hookrightarrow$ App described by it coincide.

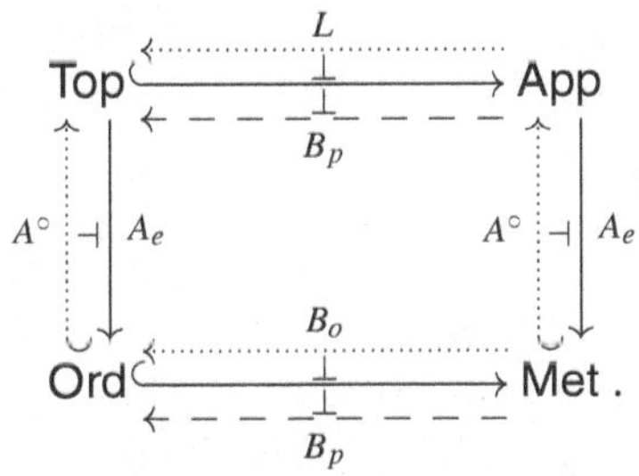

Exercises

III.3.A *The initial lax extension.* The pair $(1_{\mathsf{Set}}, 1_{\mathcal{V}\text{-}\mathsf{Rel}})$ is the initial object in the metacategory $\mathcal{V}$-LXT of lax extensions of Set-functors to $\mathcal{V}$-Rel and morphisms of lax extensions.

III.3.B *Functoriality of the lax-extension-to-lax-algebras transformation.* The correspondence

$$(\alpha : (\mathbb{S}, \hat{S}) \to (\mathbb{T}, \hat{T})) \mapsto (A_\alpha : (\mathbb{T}, \mathcal{V})\text{-}\mathsf{Cat} \to (\mathbb{S}, \mathcal{V})\text{-}\mathsf{Cat})$$

of Section III.3.4 defines a functor $(\mathcal{V}\text{-}\mathsf{LXT})^{\mathrm{op}} \to$ CAT from the dual of $\mathcal{V}$-LXT (see Exercise III.3.A) to the metacategory CAT of categories and functors.

III.3.C *Functoriality of the change-of-base transformation.* Given a monad $\mathbb{T}$ on Set, the correspondence

$$(\varphi : (\mathcal{V}, \hat{\mathbb{T}}) \to (\mathcal{W}, \check{\mathbb{T}})) \mapsto (B_\varphi : (\mathbb{T}, \mathcal{V})\text{-}\mathsf{Cat} \to (\mathbb{T}, \mathcal{W})\text{-}\mathsf{Cat})$$

of Section III.3.5 defines a 2-functor $\mathsf{QNT}(\mathbb{T}) \to \mathsf{CAT}$ from the metacategory $\mathsf{QNT}(\mathbb{T})$ of quantales with lax extension of $\mathbb{T}$ and compatible lax homomorphisms to the metacategory CAT of categories and functors. As a consequence, this functor preserves adjoint pairs (see also Proposition III.3.5.1).

© **III.3.D** *Pseudotopological spaces.* The category PsTop of pseudotopological spaces is defined as follows: its objects are sets equipped with a reflexive relation $a : \beta X \nrightarrow X$ representing convergence of ultrafilters to points, and its morphisms are the convergence-preserving maps. Theorem III.2.2.5 shows in particular that any topological space can be regarded as a pseudotopological space. In fact, one has a full reflective embedding

$$\mathsf{Top} \hookrightarrow \mathsf{PsTop}\,.$$

Hint. Given a reflexive relation $a : \beta X \nrightarrow X$, define $A \subseteq X$ to be open precisely when

$$\forall x \in X, \mathfrak{y} \in \beta X\ (\mathfrak{y}\, a\, x\ \&\ x \in A \implies A \in \mathfrak{y})\,.$$

III.3.E *Order and Kleisli convolution.* For a $(\mathbb{T}, \mathcal{V})$-functor $f : (X, a) \to (X, b)$, define $(\mathbb{T}, \mathcal{V})$-relations $f_* : X \rightharpoonup\!\!\!\!\!+\ Y$ and $f^* : Y \rightharpoonup\!\!\!\!\!+\ X$ by

$$f_* := b \cdot Tf \qquad \text{and} \qquad f^* := f^\circ \cdot b\,.$$

Then

(1) $f_* \circ f^* \le b$;
(2) $a \le f^* \circ f_*$ if m satisfies BC;
(3) $f \le g$ in $(\mathbb{T}, \mathcal{V})$-$\mathsf{Cat}$ if and only if $f^* \le g^*$ in $\mathcal{V}$-Rel;
(4) condition (iv) in Remark III.3.3.4 holds if and only if $g_* \le f_*$ in $\mathcal{V}$-Rel;
(5) if $\hat{\mathbb{T}}$ is associative, then

$$f^* \le g^* \iff g_* \le f_*\,,$$

and condition III.3.3.4(iv) is equivalent to conditions III.3.3.3(i)–(iii).
Hint. Use Kleisli convolution and statements (1) and (2).

III.3.F *Adjunctions in* $(\mathbb{T}, \mathcal{V})$-$\mathsf{Cat}$. Use Exercise III.3.E(5) to prove that if $\hat{\mathbb{T}}$ is associative, then $f : (X, a) \to (Y, b)$ is left adjoint to $g : (Y, b) \to (X, a)$ in the ordered category $(\mathbb{T}, \mathcal{V})$-$\mathsf{Cat}$ if and only if

$$a(\mathfrak{x}, g(y)) = b(Tf(\mathfrak{x}), y)$$

for all $\mathfrak{x} \in TX$, $y \in Y$.

III.3.G *Tensored $\mathcal{V}$-categories.* A $\mathcal{V}$-category (X, a) is called *tensored* if for all $x \in X$ and $u \in \mathcal{V}$ there exists $z \in X$ such that

$$\forall y \in X \; (a(z, y) = (u \multimap a(x, y)))\,. \tag{III.3.6.i}$$

(1) Show that (X, a) is tensored if and only if, for all $x \in X$, the $\mathcal{V}$-functor

$$a(x, -) : X \to \mathcal{V}$$

has a left adjoint in $\mathcal{V}$-**Cat**. Conclude that any element z satisfying (III.3.6.i) is, up to order-equivalence in X, uniquely determined by x and u; one writes $z = x \otimes u$.

(2) Show that the $\mathcal{V}$-category $\mathcal{V}$ is tensored, and so is $\widehat{X} = [X^{\mathrm{op}}, \mathcal{V}]$ (see Exercise III.1.H), for every $\mathcal{V}$-category X.

III.3.H *Change-of-base needs lax homomorphisms.* For quantales $\mathcal{V}$ and $\mathcal{W}$, let $\varphi : \mathcal{V} \to \mathcal{W}$ be monotone. The following assertions are equivalent:

(i) for every $\mathcal{V}$-category (X, a), the pair $(X, \varphi a)$ forms a $\mathcal{W}$-category;
(ii) $(\mathcal{V}, \varphi h)$ is a $\mathcal{W}$-category, with $h = (-) \multimap (-)$;
(iii) φ is a lax homomorphism of quantales.

III.3.I *Characterizing change-of-base functors.* Let $\mathcal{V}$ and $\mathcal{W}$ be quantales. We write $(\mathcal{V}, h) = (\mathcal{V}, \multimap)$ (see Exercise III.3.H).

(1) For a lax homomorphism $\varphi : \mathcal{V} \to \mathcal{W}$, show that the change-of-base functor $B_\varphi : \mathcal{V}$-**Cat** $\to \mathcal{W}$-**Cat** is a 2-functor that preserves initial morphisms with respect to the underlying-**Set** functors.

(2) Let $F : \mathcal{V}$-**Cat** $\to \mathcal{W}$-**Cat** be a 2-functor preserving underlying sets and initial morphisms. Writing $(X, \tilde{a})$ for $F(X, a)$, show the following.

  (a) $\tilde{h}(k, -) : \mathcal{V} \to \mathcal{W}$ is monotone.

  (b) For every tensored $\mathcal{V}$-category (X, a),

$$\forall u \in \mathcal{V} \; \forall x, y \in X \; (\tilde{a}(x \otimes u, y) = \tilde{h}(u, a(x, y)))\;;$$

  in particular,

$$\forall x, y \in X \; (\tilde{a}(x, y) = \tilde{h}(k, a(x, y)))\,. \tag{III.3.6.ii}$$

  (c) Formula (III.3.6.ii) holds for every $\mathcal{V}$-category, thanks to initiality of the Yoneda functor (see Exercise III.1.H(3)).

  (d) $F = B_\varphi$, for a unique lax homomorphism $\varphi : \mathcal{V} \to \mathcal{W}$ of quantales.

III.3.J *Many structures on* 1. For the topological functor $O : \mathcal{V}$-**Cat** $\to$ **Set**, the complete lattice $O^{-1}1$ (see Theorem II.5.9.1) of $\mathcal{V}$-category structures on a singleton set is order-isomorphic to $\{v \in \mathcal{V} \mid k \le v, v \otimes v \le v\}$, which is closed under infima in $\mathcal{V}$ but not under suprema (unless $\mathcal{V}$ is trivial).

III.3.K *Lax extensions of the same monad.* Let $\check{\mathbb{T}}$ and $\hat{\mathbb{T}}$ be lax extensions to $\mathcal{V}$-Rel of the monad $\mathbb{T} = (T, m, e)$ on Set with $\check{T}r \leq \hat{T}r$ for all $\mathcal{V}$-relations r. Then there is a full and faithful algebraic functor $(\mathbb{T}, \mathcal{V}, \check{\mathbb{T}})\text{-Cat} \to (\mathbb{T}, \mathcal{V}, \hat{\mathbb{T}})\text{-Cat}$.

III.3.L *Restricting a lax extension from* $\mathcal{V}$-Rel *to* Rel. Let $\mathbb{T}$ come with a lax extension $\hat{\mathbb{T}}$ to $\mathcal{V}$-Rel, and assume

$$k = \top \qquad \text{and} \qquad (u \otimes v = \bot \implies u = \bot \text{ or } v = \bot)$$

(for all $u, v \in \mathcal{V}$), so that the left adjoint o to $\iota : 2 \to \mathcal{V}$ becomes a quantale homomorphism. For $r : X \nrightarrow Y$ in Rel $=$ 2-Rel, define $\tilde{T}r := o\hat{T}(\iota r)$ and assume that

$$\hat{T}(\iota r) = \iota(\tilde{T}r)$$

holds for all relations r. Then $\tilde{\mathbb{T}}$ is a lax extension to Rel of $\mathbb{T}$, and ι induces the change-of-base functor

$$B_\iota : (\mathbb{T}, 2, \tilde{\mathbb{T}})\text{-Cat} \to (\mathbb{T}, \mathcal{V}, \hat{\mathbb{T}})\text{-Cat}\,.$$

In particular, for $\mathcal{V} = \mathsf{P}_+$ and $\hat{\mathbb{T}} = \overline{\beta}$ the Barr extension of β to P_+-Rel, $\tilde{\mathbb{T}}$ is the Barr extension of β to Rel.

III.4 Embedding lax algebras into a quasitopos

In Theorem III.3.1.3 we showed that the forgetful functor $O : (\mathbb{T}, \mathcal{V})\text{-Cat} \to \mathsf{Set}$ is topological, giving an explicit description of O-initial structures and consequently of limits (see Proposition III.3.1.1 and Remark III.3.1.4). In this section, we introduce a supercategory of $(\mathbb{T}, \mathcal{V})$-Cat which remains topological over Set but, unlike $(\mathbb{T}, \mathcal{V})$-Cat, also allows for an easy description of colimits. In addition, this supercategory turns out to be a quasitopos – a term that we will explain in Section III.4.8 below – and therefore Cartesian closed. For $(\mathbb{T}, \mathcal{V}) = (\beta, 2)$, this quasitopos is the category PsTop of pseudotopological spaces. We also discuss in general terms the role of the intermediate category PrTop of pretopological spaces, which still enjoys an important property of quasitopoi (existence of a partial-map classifier), although it fails to be Cartesian closed.

We continue to work with a quantale $\mathcal{V}$ which has a $\otimes$-neutral element k, and with a monad $\mathbb{T} = (T, m, e)$ on Set which comes with a lax extension $\hat{\mathbb{T}}$ to $\mathcal{V}$-Rel (see Section III.1.5).

III.4.1 $(\mathbb{T}, \mathcal{V})$-graphs

A $(\mathbb{T}, \mathcal{V})$-*graph* (X, a) is a set X equipped with a reflexive $(\mathbb{T}, \mathcal{V})$-relation $a : X \rightharpoonup X$ (see Section III.1.6), i.e. a $\mathcal{V}$-relation $a : TX \nrightarrow X$ with $e_X^\circ \leq a$.

A morphism $f : (X, a) \to (X, b)$ of $(\mathbb{T}, \mathcal{V})$-graphs is defined as for $(\mathbb{T}, \mathcal{V})$-categories $(f \cdot a \le b \cdot Tf)$ and is therefore also called a $(\mathbb{T}, \mathcal{V})$-*functor* in the more general context of the category

$$(\mathbb{T}, \mathcal{V})\text{-}\mathsf{Gph}$$

of $(\mathbb{T}, \mathcal{V})$-graphs. There is a string of full subcategories

$$(\mathbb{T}, \mathcal{V})\text{-}\mathsf{Cat} \hookrightarrow (\mathbb{T}, \mathcal{V})\text{-}\mathsf{UGph} \hookrightarrow (\mathbb{T}, \mathcal{V})\text{-}\mathsf{RGph} \hookrightarrow (\mathbb{T}, \mathcal{V})\text{-}\mathsf{Gph}\ ,$$

with $(\mathbb{T}, \mathcal{V})$-$\mathsf{RGph}$ the category of *right unitary* $(\mathbb{T}, \mathcal{V})$-graphs (X, a) that must satisfy $a \circ e_X^\circ \le a$, and with $(\mathbb{T}, \mathcal{V})$-$\mathsf{UGph}$ the category of *unitary* $(\mathbb{T}, \mathcal{V})$-graphs (X, a) for which a is also left unitary: $e_X^\circ \circ a \le a$. Recall from Proposition III.1.7.3 that right-unitariness of a $(\mathbb{T}, \mathcal{V})$-relation a is equivalently expressed by $a \cdot \hat{T}1_X \le a$, so it comes for free when $\hat{T}$ is flat. For a map $f : X \to Y$ between $(\mathbb{T}, \mathcal{V})$-graphs (X, a), (Y, b), one therefore has

$$b \cdot Tf = b \cdot \hat{T}1_X \cdot Tf = b \cdot \hat{T}f$$

when (Y, b) is right unitary, so that the condition $f \cdot a \le b \cdot \hat{T}f$ then suffices to make f a $(\mathbb{T}, \mathcal{V})$-functor.

Before presenting examples, let us state these definitions elementwise.

III.4.1.1 Lemma *A set X with a $\mathcal{V}$-relation $a : TX \nrightarrow X$ is a $(\mathbb{T}, \mathcal{V})$-graph if*

$$k \le a(e_X(x), x)$$

for all $X \in X$. It is right unitary if

$$\hat{T}1_X(\mathfrak{x}, \mathfrak{y}) \otimes a(\mathfrak{y}, y) \le a(\mathfrak{x}, y)$$

for all $y \in X$, $\mathfrak{x}, \mathfrak{y} \in TX$, and unitary if, in addition,

$$\hat{T}a(\mathfrak{X}, e_X(y)) \le a(m_X(\mathfrak{X}), y)$$

for all $\mathfrak{X} \in TTX$, $y \in X$.

Proof By definition,

$$(a \cdot \hat{T}1_X)(\mathfrak{x}, y) = \bigvee_{\mathfrak{y} \in TX} \hat{T}1_X(\mathfrak{x}, \mathfrak{y}) \otimes a(\mathfrak{y}, y)$$

for all $\mathfrak{x} \in TX$, $y \in X$. Furthermore, $e_X^\circ \circ a \le a$ means $e_X^\circ \cdot \hat{T}a \le a \cdot m_X$, which, when expressed elementwise, gives the stated condition for (X, a) being unitary. □

Note also that with $1_X^\sharp = e_X^\circ \cdot \hat{T}1_X$, (X, a) is a left unitary $(\mathbb{T}, \mathcal{V})$-graph if and only if

$$1_X^\sharp \le a = 1_X^\sharp \circ a\ ,$$

and that it becomes unitary precisely under the additional condition

$$a = a \circ 1_X^\sharp$$

(see Proposition III.1.7.3). With $f^\sharp = f^\circ \cdot 1_Y^\sharp$, $(\mathbb{T}, \mathcal{V})$-functors are equivalently described by

$$a \circ f^\sharp \leq f^\sharp \circ b$$

(see Section III.1.8).

III.4.1.2 Remark The definition of a $(\mathbb{T}, \mathcal{V})$-graph does not depend on the monad multiplication m of $\mathbb{T} = (T, m, e)$ or the lax extension $\hat{T}$ of the functor T. Right-unitariness depends on $\hat{T}$ but not on m, whereas unitariness depends on both m and $\hat{T}$.

III.4.1.3 Examples

(1) $(\mathbb{I}, \mathcal{V})$-graphs, with $\mathbb{I}$ identically extended to $\mathcal{V}$-Rel, are simply sets X with a reflexive $\mathcal{V}$-relation $a : X \nrightarrow X$. For $\mathcal{V} = \mathsf{2}$ one obtains sets with a reflexive relation, and for $\mathcal{V} = \mathsf{P}_+$ sets with a function $a : X \times X \to [0, \infty]$, which is 0 on the diagonal of $X \times X$. $(\mathbb{I}, \mathcal{V})$-graphs are automatically unitary.

(2) $(\mathbb{\beta}, \mathsf{2})$-graphs, where $\mathbb{\beta}$ is equipped with its Barr extension, are sets X with a relation $a : \beta X \nrightarrow X$ that is only required to satisfy $\dot{x} \longrightarrow x$ for all $x \in X$. (Here we write $\chi \longrightarrow x$ instead of $\chi\ a\ x$.) These are precisely the *pseudotopological spaces* of Exercise III.3.D. Indeed, since the Barr extension is flat, being right unitary comes for free, and the only remaining condition ($\dot{x} \longrightarrow x$ for all $x \in X$) is characteristic for pseudotopologicity; hence,

$$(\mathbb{\beta}, \mathsf{2})\text{-}\mathsf{Gph} = (\mathbb{\beta}, \mathsf{2})\text{-}\mathsf{RGph} = \mathsf{PsTop}\,.$$

Unitary $(\mathbb{\beta}, \mathsf{2})$-graphs must satisfy the additional condition

$$(\mathcal{X} \longrightarrow \dot{y}) \implies (\textstyle\sum \mathcal{X} \longrightarrow y)\,, \qquad \text{(III.4.1.i)}$$

where $\sum \mathcal{X}$ is the Kowalsky sum of $\mathcal{X} \in \beta\beta X$, and where $\mathcal{X} \longrightarrow \dot{y}$ amounts to

$$\forall \mathcal{A} \in \mathcal{X}\ \exists \chi \in \mathcal{A}\ (\chi \longrightarrow y)\,.$$

Hence, (III.4.1.i) means equivalently (see Example III.1.10.3(3))

$$(\{\chi \in \beta X \mid \chi \longrightarrow y\} \in \mathcal{X}) \implies (\textstyle\sum \mathcal{X} \longrightarrow y) \qquad \text{(III.4.1.ii)}$$

for all $\mathcal{X} \in \beta\beta X$, $y \in X$. From Proposition III.2.2.4 we obtain immediately that unitary $(\mathbb{\beta}, \mathsf{2})$-graphs on a set X correspond bijectively to maps $c : PX \to PX$ with

$$A \subseteq c(A)\,, \quad c(\varnothing) = \varnothing\,, \quad c(A \cup B) = c(A) \cup c(B)\,,$$

for all $A, B \subseteq X$. A set X equipped with such a map c is called a *pretopological space*. A morphism $f : X \to Y$ in the category PrTop must satisfy $f(c_X(A)) \subseteq c_Y(f(A))$ for all $A \subseteq X$, and it is now easy to see that there is an isomorphism

$$(\mathbb{\beta}, \mathsf{2})\text{-}\mathsf{UGph} \cong \mathsf{PrTop}$$ ©

which leaves underlying sets unchanged.

(3) For the filter monad $\mathbb{F}$ extended by

$$\mathfrak{a}\ (\check{F}r)\ \mathfrak{b} \iff \mathfrak{a} \supseteq r^\circ[\mathfrak{b}]$$

(for all relations $r : X \nrightarrow Y$, $\mathfrak{a} \in FX$, $\mathfrak{b} \in FY$) as in Example III.1.10.3(4), a right unitary $(\mathbb{F}, \mathsf{2})$-graph $(X, \longrightarrow)$ must satisfy

$$\dot{x} \longrightarrow x \qquad \text{and} \qquad (\mathfrak{a} \supseteq \mathfrak{b}\ \&\ \mathfrak{b} \longrightarrow y \implies \mathfrak{a} \longrightarrow y)\,, \qquad \text{(III.4.1.iii)}$$

and $(X, \longrightarrow)$ is unitary if

$$(\{\mathfrak{a} \in FX \mid \mathfrak{a} \longrightarrow y\} \in \mathcal{A}) \implies (\textstyle\sum \mathcal{A} \longrightarrow y) \qquad \text{(III.4.1.iv)}$$

for all $\mathcal{A} \in FFX$, $x, y \in X$. The category $(\mathbb{F}, \mathsf{2})\text{-}\mathsf{RGph}$ contains the full subcategory $(\mathbb{F}, \mathsf{2})\text{-}\mathsf{RGph}_{\mathrm{Ps}}$ whose objects $(X, \longrightarrow)$ satisfy (III.4.1.iii) and the condition

(Ps) if $\mathfrak{a}$ is a filter such that for every proper filter $\mathfrak{b} \supseteq \mathfrak{a}$ there exists a proper filter $\mathfrak{c} \supseteq \mathfrak{b}$ with $\mathfrak{c} \longrightarrow y$, then $\mathfrak{a} \longrightarrow y$

for all $\mathfrak{a} \in FX$, $y \in X$. It is not difficult to see that $(\mathbb{F}, \mathsf{2})\text{-}\mathsf{RGph}_{\mathrm{Ps}}$ is isomorphic to PsTop (Exercise III.4.I). Moreover, the full subcategory ©
$(\mathbb{F}, \mathsf{2})\text{-}\mathsf{UGph}$ of $(\mathbb{F}, \mathsf{2})\text{-}\mathsf{RGph}_{\mathrm{Ps}}$ is isomorphic to PrTop (Exercise III.4.L), so using (2) one obtains the following diagram of full embeddings and isomorphisms:

$$\begin{array}{ccccccc}
(\mathbb{F},\mathsf{2})\text{-}\mathsf{UGph} & \xrightarrow{\cong} & \mathsf{PrTop} & \underset{©}{\hookrightarrow} & \mathsf{PsTop} & \underset{©}{\xrightarrow{\cong}} & (\mathbb{F},\mathsf{2})\text{-}\mathsf{RGph}_{\mathrm{Ps}} \\
 & & {\scriptstyle\cong}\downarrow{\scriptstyle ©} & & {\scriptstyle =}\downarrow & & \downarrow \\
 & & (\mathbb{\beta},\mathsf{2})\text{-}\mathsf{UGph} & \hookrightarrow & (\mathbb{\beta},\mathsf{2})\text{-}\mathsf{RGph} & \hookrightarrow & (\mathbb{F},\mathsf{2})\text{-}\mathsf{RGph} \\
 & & & & {\scriptstyle =}\downarrow & & \downarrow \\
 & & & & (\mathbb{\beta},\mathsf{2})\text{-}\mathsf{Gph} & \hookrightarrow & (\mathbb{F},\mathsf{2})\text{-}\mathsf{Gph}\,.
\end{array}$$

The horizontal arrows on the right are obtained by extension of the convergence relation: for $\mathfrak{a} \in FX$ and $y \in X$,

$$(\mathfrak{a} \longrightarrow y) \iff \forall \mathfrak{x} \in \beta X\ (\mathfrak{x} \supseteq \mathfrak{a} \implies \mathfrak{x} \longrightarrow y)\,.$$

(4) We write

$$\mathsf{PsApp} := (\mathbb{\beta}, \mathsf{P}_+)\text{-}\mathsf{Gph} = (\mathbb{\beta}, \mathsf{P}_+)\text{-}\mathsf{RGph}$$

(with the Barr extension of β to P_+-Rel of Section III.2.4) and calls its objects *pseudo-approach spaces*. PsApp contains the full subcategory

$$\mathsf{PrApp} := (\beta, \mathsf{P}_+)\text{-}\mathsf{UGph}$$

of *pre-approach spaces* whose objects may be described equivalently as sets equipped with a finitely additive distance function $\delta : X \times PX \to [0, \infty]$ satisfying $\delta(x, \{x\}) = 0$ for all $x \in X$. The needed isomorphism is established as in Theorem III.2.4.5, with Proposition III.2.4.4 providing the key ingredient to the proof. (See also Exercise III.4.O.)

III.4.1.4 Proposition *The forgetful functor* $O : (\mathbb{T}, \mathcal{V})\text{-}\mathsf{Gph} \to \mathsf{Set}$ *is topological. O-initial liftings of sources may be formed as for $(\mathbb{T}, \mathcal{V})$-categories (see* Proposition III.3.1.1*), while a sink* $(f_i : (X_i, a_i) \to (Y, b))_{i\in I}$ *in* $(\mathbb{T}, \mathcal{V})\text{-}\mathsf{Gph}$ *is O-final precisely when*

$$b = e_Y^\circ \vee \bigvee_{i\in I} f_i \cdot a_i \cdot (Tf_i)^\circ . \qquad \text{(III.4.1.v)}$$

For an epi-sink $(f_i)_{i\in I}$*, this formula simplifies to* $b = \bigvee_{i\in I} f_i \cdot a_i \cdot (Tf_i)^\circ$.

Proof For a family (X_i, a_i) of $(\mathbb{T}, \mathcal{V})$-graphs and Set-maps $f_i : X_i \to Y$ $(i \in I)$, define b by (III.4.1.v). Trivially, b is reflexive, and every $(f_i : (X_i, a_i) \to (Y, b)$ has trivially become a $(\mathbb{T}, \mathcal{V})$-functor. For O-finality of $(f_i)_{i\in I}$, consider a $(\mathbb{T}, \mathcal{V})$-graph (Z, c) and a map $h : Y \to Z$ such that

$$h \cdot f_i \cdot a_i \le c \cdot T(h \cdot f_i)$$

for all $i \in I$. Then

$$h \cdot f_i \cdot a_i \cdot (Tf_i)^\circ \le c \cdot Th \cdot Tf_i \cdot (Tf_i)^\circ \le c \cdot Th ,$$

and therefore $f_i \cdot a_i \cdot (Tf_i)^\circ \le h^\circ \cdot c \cdot Th$ by adjunction (for all $i \in I$). Since

$$e_Y^\circ \le e_Y^\circ \cdot (Th)^\circ \cdot Th = h^\circ \cdot e_Z^\circ \cdot Th \le h^\circ \cdot c \cdot Th ,$$

one has $b \le h^\circ \cdot c \cdot Th$, and h is a $(\mathbb{T}, \mathcal{V})$-functor. This concludes the proof that O is topological, with O-final sinks characterized by (III.4.1.v). That O-initial sources may be described as in Proposition III.3.1.1 follows from a direct verification (see Exercise III.4.A). Finally, if $(f_i)_{i\in I}$ is epic, with Proposition III.1.2.2 one obtains $1_Y \le \bigvee_{i\in I} f_i \cdot f_i^\circ$. Consequently,

$$e_Y^\circ \le \bigvee_{i\in I} f_i \cdot f_i^\circ \cdot e_Y^\circ = \bigvee_{i\in I} f_i \cdot e_{X_i}^\circ \cdot (Tf_i)^\circ \le \bigvee_{i\in I} f_i \cdot a_i \cdot (Tf_i)^\circ . \qquad \square$$

The category $(\mathbb{T}, \mathcal{V})\text{-}\mathsf{RGph}$ is reflective in $(\mathbb{T}, \mathcal{V})\text{-}\mathsf{Gph}$, with the reflection morphisms given by $(X, a) \to (X, a \cdot \hat{T}1_X)$. Indeed, any $(\mathbb{T}, \mathcal{V})$-functor $f : (X, a) \to (Y, b)$ with (Y, b) right unitary is a $(\mathbb{T}, \mathcal{V})$-functor $f : (X, a \cdot \hat{T}1_X) \to (Y, b)$:

$$f \cdot a \cdot \hat{T}1_X \le b \cdot Tf \cdot \hat{T}1_X \le b \cdot \hat{T}f = b \cdot \hat{T}1_Y \cdot Tf = b \cdot Tf .$$

III.4.1.5 Corollary *The forgetful functor* $O : (\mathbb{T},\mathcal{V})\text{-}\mathsf{RGph} \to \mathsf{Set}$ *is topological. O-initial liftings of sources may be formed as for $(\mathbb{T},\mathcal{V})$-categories (see* Proposition III.3.1.1*), while a sink $(f_i : (X_i, a_i) \to (Y, b))_{i\in I}$ in $(\mathbb{T},\mathcal{V})\text{-}\mathsf{RGph}$ is O-final precisely when*

$$b = 1_Y^\sharp \vee \bigvee_{i\in I} f_i \cdot a_i \cdot \hat{T}(f_i^\circ)\,. \tag{III.4.1.vi}$$

For an epi-sink $(f_i)_{i\in I}$, this formula simplifies to $b = \bigvee_{i\in I} f_i \cdot a_i \cdot \hat{T}(f_i^\circ)$.

Proof We may apply Theorem II.5.10.3. O-final liftings in $(\mathbb{T},\mathcal{V})\text{-}\mathsf{RGph}$ are obtained by reflecting the O-final lifting in $(\mathbb{T},\mathcal{V})\text{-}\mathsf{Gph}$. With b_0 denoting the codomain structure of the O-final lifting in $(\mathbb{T},\mathcal{V})\text{-}\mathsf{Gph}$, an easy computation in the quantaloid $\mathcal{V}\text{-}\mathsf{Rel}$ shows

$$\begin{aligned} b = b_0 \cdot \hat{T}1_Y = (e_Y^\circ \cdot \hat{T}1_Y) \vee \textstyle\bigvee_{i\in I}(f_i \cdot a_i \cdot (Tf_i)^\circ \cdot \hat{T}1_Y) \\ = 1_Y^\sharp \vee \textstyle\bigvee_{i\in I} f_i \cdot a_i \cdot \hat{T}(f_i^\circ)\,. \quad \square \end{aligned}$$

III.4.1.6 Remarks

(1) Since $(\mathbb{T},\mathcal{V})\text{-}\mathsf{Cat} \hookrightarrow (\mathbb{T},\mathcal{V})\text{-}\mathsf{Gph}$ preserves initial sources, one may use the Taut Lift Theorem II.5.11.1 to obtain its reflectivity; likewise for $(\mathbb{T},\mathcal{V})\text{-}\mathsf{Cat} \hookrightarrow (\mathbb{T},\mathcal{V})\text{-}\mathsf{RGph}$. The reflector provides a $(\mathbb{T},\mathcal{V})$-graph (X, a) with the $(\mathbb{T},\mathcal{V})$-category structure

$$\bigwedge\{c : X \rightharpoonup X \mid a \leq c,\ c \circ c \leq c\}\,.$$

In Section III.4.2, using ordinal recursion we give an alternative description of the reflector $(\mathbb{T},\mathcal{V})\text{-}\mathsf{RGph} \to (\mathbb{T},\mathcal{V})\text{-}\mathsf{Cat}$ that turns out to be essential in Section III.4.3 when describing coproducts in $(\mathbb{T},\mathcal{V})\text{-}\mathsf{Cat}$.

(2) A surjective morphism $f : X \to Y$ in $(\mathbb{T},\mathcal{V})\text{-}\mathsf{RGph}$ with X in $(\mathbb{T},\mathcal{V})\text{-}\mathsf{Cat}$ that is final with respect to $O : (\mathbb{T},\mathcal{V})\text{-}\mathsf{RGph} \to \mathsf{Set}$ will generally fail to make Y into a $(\mathbb{T},\mathcal{V})$-category and be final with respect to the forgetful functor $(\mathbb{T},\mathcal{V})\text{-}\mathsf{Cat} \to \mathsf{Set}$, even when $\mathbb{T} = \mathbb{I}$ and $\mathcal{V} = 2$ (Exercise III.1.G(3)). Nevertheless, the explicit description of O-final sinks in $(\mathbb{T},\mathcal{V})\text{-}\mathsf{RGph}$ turns out to be useful for the computation of colimits in $(\mathbb{T},\mathcal{V})\text{-}\mathsf{Cat}$ in special instances (see Section III.4.2).

III.4.2 Reflecting $(\mathbb{T},\mathcal{V})$-RGph into $(\mathbb{T},\mathcal{V})$-Cat

For a right unitary $(\mathbb{T},\mathcal{V})$-graph (X, a) we define recursively (see Section II.1.14) an ascending chain of $(\mathbb{T},\mathcal{V})$-relations a_ν, for every ordinal number ν, as follows:

$$\begin{aligned} a_0 &:= e_X^\circ\,, \\ a_{\nu+1} &:= a \circ a_\nu\,, \\ a_\lambda &:= \textstyle\bigvee_{\nu<\lambda} a_\nu \qquad (\lambda \text{ a limit ordinal}). \end{aligned}$$

Since a is right unitary, $a_1 = a \circ e_X^\circ = a$, and reflexivity makes the chain ascending: $a_\nu \leq e_X^\circ \circ a_\nu \leq a \circ a_\nu = a_{\nu+1}$. Consequently, this chain in the set $\mathcal{V}$-Rel(TX, X) must become stationary, i.e.

$$a_\mu = a_{\mu+1}$$

for some ordinal number μ. Next we will show that under a mild additional assumption on the lax extension of $\mathbb{T}$, the pair $(X, a_\infty) := (X, a_\mu)$ provides a reflection of (X, a) into $(\mathbb{T}, \mathcal{V})$-Cat.

III.4.2.1 Proposition *For the lax extension $\hat{\mathbb{T}} = (\hat{T}, m, e)$ of $\mathbb{T}$, suppose that $m^\circ : \hat{T} \to \hat{T}\hat{T}$ is a natural transformation. Then $(\mathbb{T}, \mathcal{V})$-Cat is reflective in $(\mathbb{T}, \mathcal{V})$-RGph, with the reflection morphisms obtained by ordinal recursion, as described previously.*

Proof When m° is a natural transformation, $(t \circ s) \circ r \leq t \circ (s \circ r)$ for all $(\mathbb{T}, \mathcal{V})$-relations $r : X \nrightarrow Y$, $s : Y \nrightarrow Z$, $t : Z \nrightarrow W$ (Exercise III.1.L). In particular,

$$(a \circ a_\nu) \circ a_\mu \leq a \circ (a_\nu \circ a_\mu)$$

for all ordinals ν, μ. Consequently, if $a_\mu = a_{\mu+1}$, one can easily show $a_\nu \circ a_\mu \leq a_\mu$ for all ν by ordinal recursion, and $a_\mu \circ a_\mu \leq a_\mu$ follows. Since $e_X^\circ \leq a \leq a_\mu$, one has that (X, a_μ) is a $(\mathbb{T}, \mathcal{V})$-category, and $1_X : (X, a) \to (X, a_\mu)$ is a $(\mathbb{T}, \mathcal{V})$-functor. It remains to be shown that every $(\mathbb{T}, \mathcal{V})$-functor $f : (X, a) \to (Y, b)$ with (Y, b) a $(\mathbb{T}, \mathcal{V})$-category gives a $(\mathbb{T}, \mathcal{V})$-functor $f : (X, a_\mu) \to (Y, b)$, and for that is suffices to show that $a \leq f^\circ \cdot b \cdot Tf$ implies $a_\nu \leq f^\circ \cdot b \cdot Tf$ for all ordinals ν. We show the successor step of the ordinal recursion:

$$\begin{aligned} a_{\nu+1} = a \circ a_\nu &\leq (f^\circ \cdot b \cdot Tf) \circ (f^\circ \cdot b \cdot Tf) \\ &\leq f^\circ \cdot b \cdot Tf \cdot (Tf)^\circ \cdot \hat{T}b \cdot TTf \cdot m_X^\circ \\ &\leq f^\circ \cdot b \cdot \hat{T}b \cdot m_Y^\circ \cdot Tf \\ &= f^\circ \cdot (b \circ b) \cdot Tf = f^\circ \cdot b \cdot Tf \,. \end{aligned}$$ □

It follows from Proposition III.4.2.1 that $(\mathbb{T}, \mathcal{V})$-Cat is reflective in $(\mathbb{T}, \mathcal{V})$-UGph, but there is no obvious simplification in this proof when applying the ordinal recursion to a *unitary* $(\mathbb{T}, \mathcal{V})$-graph. However, under additional hypotheses on the lax extension of $\mathbb{T}$, there is an easy one-step construction for a reflector $(\mathbb{T}, \mathcal{V})$-RGph $\to$ $(\mathbb{T}, \mathcal{V})$-UGph, as follows.

III.4.2.2 Proposition *Suppose that the lax extension $\hat{\mathbb{T}} = (\hat{T}, m, e)$ of $\mathbb{T}$ to $\mathcal{V}$-Rel is associative. Then*

$$1_X : (X, a) \to (X, e_X^\circ \circ a)$$

is a reflection of the right unitary $(\mathbb{T}, \mathcal{V})$-graph (X, a) into $(\mathbb{T}, \mathcal{V})$-UGph.

Proof By Proposition III.1.9.4, $e_X^\circ \circ a$ is right unitary because a is:

$$(e_X^\circ \circ a) \circ 1_X^\sharp = e_X^\circ \circ (a \circ 1_X^\sharp) = e_X^\circ \circ a \, .$$

Similarly, $e_X^\circ \circ a = 1_X^\sharp \circ a$ is also left unitary:

$$1_X^\sharp \circ (1_X^\sharp \circ a) = (1_X^\sharp \circ 1_X^\sharp) \circ a = 1_X^\sharp \circ a \, .$$

Furthermore, if $f : (X, a) \to (Y, b)$ is a $(\mathbb{T}, \mathcal{V})$-functor with (Y, b) unitary, $f : (X, e_X^\circ \circ a) \to (Y, b)$ is also a $(\mathbb{T}, \mathcal{V})$-functor:

$$\begin{aligned} e_X^\circ \circ a &\le e_X \circ (f^\circ \cdot b \cdot Tf) \\ &= e_X^\circ \cdot \hat{T}(f^\circ \cdot b \cdot Tf) \cdot m_X^\circ \\ &= e_X^\circ \cdot (Tf)^\circ \cdot (\hat{T}b) \cdot TTf \cdot m_X^\circ \\ &\le f^\circ \cdot e_Y^\circ \cdot (\hat{T}b) \cdot m_Y^\circ \cdot Tf \\ &\le f^\circ \cdot (e_Y^\circ \circ b) \cdot Tf = f^\circ \cdot b \cdot Tf \, . \end{aligned}$$

□

III.4.3 Coproducts of $(\mathbb{T}, \mathcal{V})$-categories

In the topological category $(\mathbb{T}, \mathcal{V})$-RGph over Set, the coproduct of a family (X_i, a_i) of right unitary $(\mathbb{T}, \mathcal{V})$-graphs is given by the disjoint union

$$t_i : X_i \to X = \coprod_{i \in I} X_i \quad (i \in I)$$

endowed with the final structure $a = \bigvee_{i \in I} t_i \cdot a_i \cdot \hat{T}(t_i^\circ)$ (see Corollary III.4.1.5). Since

$$t_j^\circ \cdot t_i = \begin{cases} 1_{X_j} & \text{if } j = i, \\ \bot & \text{otherwise,} \end{cases}$$

each t_j is not just a $(\mathbb{T}, \mathcal{V})$-functor, but also each satisfies the equation

$$t_j^\circ \cdot a = \bigvee_{i \in I} t_j^\circ \cdot t_i \cdot a_i \cdot \hat{T}(t_i^\circ) = a_j \cdot \hat{T}(t_j^\circ) \, .$$

In what follows we use the following terminology, on which we will elaborate further in Chapter V.

III.4.3.1 Definition A morphism $f : (X, a) \to (Y, b)$ in $(\mathbb{T}, \mathcal{V})$-RGph is *nearly open* if $f^\circ \cdot b \le a \cdot \hat{T}(f^\circ)$, and *open* if $f^\circ \cdot b \le a \cdot (Tf)^\circ$.

Since every $(\mathbb{T}, \mathcal{V})$-functor satisfies $a \cdot (Tf)^\circ \le f^\circ \cdot b$, near openness of f means equivalently $f^\circ \cdot b = a \cdot \hat{T}(f^\circ)$, and openness $f^\circ \cdot b = a \cdot (Tf)^\circ$. Clearly, openness implies near openness, and the converse statement holds if

$$\hat{T}(f^\circ) = \hat{T}1_X \cdot (Tf)^\circ \tag{III.4.3.i}$$

for all maps $f : X \to Y$.

III.4.3.2 Lemma *Let $f : (X, a) \to (Y, b)$ be a $(\mathbb{T}, \mathcal{V})$-functor from a $(\mathbb{T}, \mathcal{V})$-category to a right unitary $(\mathbb{T}, \mathcal{V})$-graph. Then near openness of f implies near openness of $f : (X, a) \to (Y, b_\infty)$ with $b_\infty = \bigvee_\nu b_\nu$ (as in* Section III.4.2*), provided that $\hat{T}$ preserves $\mathcal{V}$-relational composition.*

Proof Proceeding to show $f^\circ \cdot b_\infty \leq a \cdot \hat{T}(f^\circ)$ by ordinal recursion, we consider the successor step and assume $f^\circ \cdot b_\nu \leq a \cdot \hat{T}(f^\circ)$. If $\hat{T}$ preserves composition of $\mathcal{V}$-relations, we then obtain

$$\begin{aligned} f^\circ \cdot b_{\nu+1} &= f^\circ \cdot (b \circ b_\nu) \\ &= f^\circ \cdot b \cdot \hat{T} b_\nu \cdot m_Y^\circ \\ &\leq a \cdot \hat{T}(f^\circ) \cdot \hat{T} b_\nu \cdot m_Y^\circ \\ &\leq a \cdot \hat{T}(f^\circ \cdot b_\nu) \cdot m_Y^\circ \\ &\leq a \cdot \hat{T}(a \cdot \hat{T}(f^\circ)) \cdot m_Y^\circ \\ &= a \cdot \hat{T} a \cdot \hat{T}\hat{T}(f^\circ) \cdot m_Y^\circ \\ &\leq a \cdot m_X \cdot m_X^\circ \cdot \hat{T}(f^\circ) = a \cdot \hat{T}(f^\circ) . \end{aligned}$$

□

III.4.3.3 Theorem *If $\hat{\mathbb{T}}$ is associative, then $(\mathbb{T}, \mathcal{V})$-*Cat *is closed under coproducts in $(\mathbb{T}, \mathcal{V})$-*RGph.

Proof For $(\mathbb{T}, \mathcal{V})$-categories (X_i, a_i), one forms the coproduct $(t_i : (X_i, a_i) \to (X, a))_{i \in I}$ in $(\mathbb{T}, \mathcal{V})$-RGph, with $a = \bigvee_{i \in I} t_i \cdot a_i \cdot \hat{T}(t_i^\circ)$, and one obtains the coproduct in $(\mathbb{T}, \mathcal{V})$-Cat by applying the reflector $(X, a) \to (X, a_\infty)$. Hence, we must show $a_\infty = a$. But, by Lemma III.4.3.2, every $t_i : (X_i, a_i) \to (X, a_\infty)$ is nearly open. Since $(t_i)_{i \in I}$ is epic, near openness yields

$$a_\infty = \bigvee_{i \in I} t_i \cdot t_i^\circ \cdot a_\infty = \bigvee_{i \in I} t_i \cdot a_i \cdot \hat{T}(t_i^\circ) = a .$$

□

III.4.3.4 Remarks

(1) Near openness of every $t_i : (X_i, a_i) \to (X, a)$ $(i \in I)$ is characteristic for $\mathcal{O}$-finality of the sink $(t_i)_{i \in I}$ (see Exercise III.4.B).

(2) Since T preserves injections (see Exercise III.1.P), every open injection $f : (X, a) \to (Y, b)$ in $(\mathbb{T}, \mathcal{V})$-Gph is $\mathcal{O}$-initial. Indeed,

$$f^\circ \cdot b \cdot Tf = a \cdot (Tf)^\circ \cdot Tf = a$$

since Tf is injective. We refer to such $(\mathbb{T}, \mathcal{V})$-functors as *open embeddings*.

(3) With $\hat{\mathbb{T}}$ an associative lax extension to $\mathcal{V}$-Rel satisfying (III.4.3.i), a coproduct $(t_i : (X_i, a_i) \to (X, a))_{i \in I}$ in $(\mathbb{T}, \mathcal{V})$-Cat satisfies

$$a = \bigvee_{i \in I} t_i \cdot a_i \cdot (Tt_i)^\circ .$$

Hence, for all $\chi \in TX$, $y \in X$, one has

$$a(\chi, y) = \begin{cases} a_i(\chi, y) & \text{if } \chi \in TX_i,\ y \in X_i \text{ for some } i \in I, \\ \bot & \text{otherwise;} \end{cases}$$

here we write $\chi \in TX_i$ instead of $\chi = Tt_i(\chi_i)$ for some $\chi_i \in TX_i$.

(4) In Lemma III.4.3.2, the hypotheses that $\hat{T}$ preserves $\mathcal{V}$-relational composition may be traded for the assumption that $\hat{T}(r \cdot g^\circ) = \hat{T}r \cdot (Tg)^\circ$ for all $r : X \nrightarrow Y$ and $g : Z \to X$. In fact, in this case a nearly open map $f : (X, a) \to (Y, b)$ is open, since $a \cdot \hat{T}(f^\circ) = a \cdot \hat{T}1_X \cdot (Tf)^\circ = a \cdot (Tf)^\circ$.

III.4.3.5 Definition A functor $T : \mathsf{Set} \to \mathsf{Set}$ is *taut* if it sends every BC-square

$$\begin{array}{ccc} M & \xrightarrow{g} & N \\ {\scriptstyle m}\downarrow & & \downarrow{\scriptstyle n} \\ X & \xrightarrow{f} & Y \end{array} \tag{III.4.3.ii}$$

with n monic to a BC-square; i.e. $n^\circ \cdot f = g \cdot m^\circ$ with n monic implies $(Tn)^\circ \cdot Tf = Tg \cdot (Tm)^\circ$.

Clearly, *a BC-square* (III.4.3.ii) *with n monic is actually a pullback diagram*: since the canonical morphism $c : M \to X \times_Y N$ factors through the monomorphism m, it is injective, but also surjective (by Lemma III.1.11.1).

Trivially, a functor satisfying BC is taut, and every taut functor preserves monomorphisms, since m is a monomorphism if and only if

$$\begin{array}{ccc} M & \xrightarrow{1} & M \\ {\scriptstyle 1}\downarrow & & \downarrow{\scriptstyle m} \\ M & \xrightarrow{m} & X \end{array}$$

is a BC-square (and then necessarily a pullback square).

III.4.3.6 Corollary *A* Set*-functor T is taut if and only if T preserves pullback squares* (III.4.3.ii) *with n monic.*

Proof This is immediate from the preceding discussion. □

In what follows, in addition to tautness of T, we also need that the underlying ordered set of the quantale $\mathcal{V}$ is *Cartesian closed*, so that both $\otimes$ and $\wedge$ distribute over suprema in $\mathcal{V}$. The status of this assumption for our purposes is clarified by the following Lemma.

III.4.3.7 Lemma *The following conditions are equivalent for a quantale* $\mathcal{V}$*:*

(i) $\mathcal{V}$ *is Cartesian closed;*

(ii) *the underlying ordered set of* $\mathcal{V}$ *is a frame;*

(iii) *the* right Frobenius law

$$(r \wedge s \cdot f) \cdot f^\circ = r \cdot f^\circ \wedge s$$

holds in $\mathcal{V}$-Rel *for all* $f : X \to Y,\ r : X \nrightarrow Z,\ s : Y \nrightarrow Z$;

(iv) *the* left Frobenius law

$$f \cdot (r \wedge f^\circ \cdot s) = f \cdot r \wedge s$$

holds in $\mathcal{V}$-Rel *for all* $f : X \to Y,\ r : Z \nrightarrow X,\ s : Z \nrightarrow Y$.

Proof Straightforward verifications. □

III.4.3.8 Proposition *Let* $\mathcal{V}$ *be a Cartesian closed quantale. If* T *is taut, then open embeddings are stable under pullback in* $(\mathbb{T}, \mathcal{V})$-$\mathsf{RGph}$, *and if* T *satisfies BC, all open maps are stable under pullback.*

Proof Consider a pullback diagram

$$\begin{array}{ccc} (P,d) & \xrightarrow{q} & (Y,b) \\ {\scriptstyle p}\downarrow & & \downarrow{\scriptstyle g} \\ (X,a) & \xrightarrow{f} & (Z,c) \end{array}$$

in $(\mathbb{T}, \mathcal{V})$-$\mathsf{RGph}$ with g open. Then d is the O-initial structure with respect to (p, q), i.e. $d = (p^\circ \cdot a \cdot Tp) \wedge (q^\circ \cdot b \cdot Tq)$, and $g^\circ \cdot c = b \cdot (Tg)^\circ$. If T satisfies BC, with Lemma III.4.3.7 one obtains openness of p as follows:

$$\begin{aligned} p^\circ \cdot a &= p^\circ \cdot (a \wedge a) \\ &\leq p^\circ \cdot (a \wedge (f^\circ \cdot c \cdot Tf)) \\ &\leq p^\circ \cdot a \wedge (p^\circ \cdot f^\circ \cdot c \cdot Tf) \\ &= p^\circ \cdot a \wedge (q^\circ \cdot g^\circ \cdot c \cdot Tf) \\ &= p^\circ \cdot a \wedge (q^\circ \cdot b \cdot (Tg)^\circ \cdot Tf) \\ &= p^\circ \cdot a \wedge (q^\circ \cdot b \cdot Tq \cdot (Tp)^\circ) && \text{(BC)} \\ &= \big((p^\circ \cdot a \cdot Tp) \wedge (q^\circ \cdot b \cdot Tq)\big) \cdot (Tp)^\circ && (\mathcal{V}\text{ Cartesian closed}) \\ &= d \cdot (Tp)^\circ . \end{aligned}$$

Tautness of T suffices in lieu of BC when g is injective. □

One says that a coproduct $(t_i : X_i \to X)_{i \in I}$ in a category is *universal* (or *stable under pullback*) if for all morphisms $f : Y \to X$ and pullback diagrams

$$\begin{array}{ccc} Y_i & \xrightarrow{s_i} & Y \\ {\scriptstyle f_i}\downarrow & & \downarrow{\scriptstyle f} \\ X_i & \xrightarrow{t_i} & X , \end{array}$$

the family $(s_i : Y_i \to Y)_{i \in I}$ is a coproduct.

III.4.3.9 Theorem *Suppose that $\mathcal{V}$ is Cartesian closed and T is taut. Then coproducts in $(\mathbb{T},\mathcal{V})$-RGph are universal if $\hat{T}(f^\circ) = \hat{T}1_X \cdot (Tf)^\circ$ for all $f : X \to Y$. If moreover $\hat{\mathbb{T}}$ is associative, coproducts are also universal in $(\mathbb{T},\mathcal{V})$-Cat.*

Proof Under the stated hypotheses, coproduct injections in $(\mathbb{T},\mathcal{V})$-RGph are open and stable under pullback. This implies universality of coproducts by Remark III.4.3.4(1). The additional hypothesis makes sure that $(\mathbb{T},\mathcal{V})$-Cat is closed under coproducts in $(\mathbb{T},\mathcal{V})$-RGph. □

A coproduct $(t_i : X_i \to X)_{i\in I}$ in a category is *disjoint* if for all $i \neq j$, the pullback of t_i and t_j is given by an initial object in the category. This is certainly true in Set and, since there is only one structure on $\varnothing$ in $(\mathbb{T},\mathcal{V})$-RGph and $(\mathbb{T},\mathcal{V})$-Cat, likewise in these categories. A category with coproducts (of small families of objects) and pullbacks is *extensive* if its coproducts are universal and disjoint (see also Exercise III.4.E and Section V.5.1).

III.4.3.10 Corollary *Suppose that $\mathcal{V}$ is Cartesian closed, that T is taut, and that $\hat{\mathbb{T}}$ is associative. If $\hat{T}(f^\circ) = \hat{T}1_X \cdot (Tf)^\circ$ for all $f : X \to Y$, then $(\mathbb{T},\mathcal{V})$-RGph and $(\mathbb{T},\mathcal{V})$-Cat are extensive categories.*

Proof The result restates Theorem III.4.3.9. □

III.4.4 Interlude on partial products and local Cartesian closedness

A finitely complete category C is *locally cartesian closed* if its comma categories C/Z are Cartesian closed for all $Z \in \operatorname{ob} \mathsf{C}$. By definition of Cartesian closedness, this means that for all $f : X \to Z$ in C the functor

$$(-)\times_Z X : \mathsf{C}/Z \to \mathsf{C}/Z\,, \quad (Y,g) \mapsto (Y\times_Z X, g\cdot\pi_1 = f\cdot\pi_2)$$

(with π_1 and π_2 the pullback projections) has a right adjoint. Since the domain functor $\mathrm{dom}_Z : \mathsf{C}/Z \to \mathsf{C}$ always has a right adjoint, given by

$$W \mapsto (W\times Z, W\times Z \to Z)\,,$$

local Cartesian closedness of C makes the composite functor

$$\mathrm{dom}_Z((-)\times_Z X) : \mathsf{C}/Z \to \mathsf{C}\,, \quad (Y,g)\mapsto Y\times_Z X$$

have a right adjoint.

Let us spell out the condition for a C-object Y to admit a $(\mathrm{dom}_Z((-)\times_Z X))$-couniversal arrrow. Such an arrow is given by an object $(P, p : P \to Z)$ in C/Z and a morphism $\varepsilon : P\times_Z X \to Y$ in C such that, for any $(Q, q : Q \to Z)$ in C/Z and any $g : Q\times_Z X \to Y$ in C, there is a uniquely determined morphism $h : Q \to P$ with $p\cdot h = q$ and $\varepsilon\cdot(h\times_Z 1_X) = g$ in C:

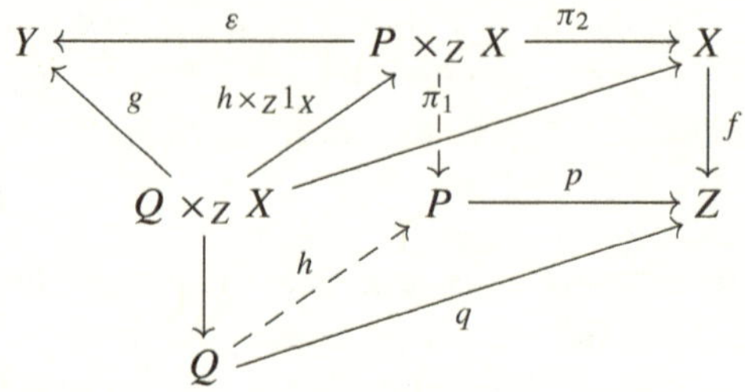

One calls $P = P(Y, f)$ together with the *projection* p and the *evaluation* ε a *partial product* of Y over f. The category C *has partial products* if $P(Y, f)$ exists for all (Y, f).

Note that, for $f = 1_X$, $P(Y, 1_X) \cong Y \times X$ is simply a product. If Z is a terminal object, the existence of $P(Y, ! : X \to Z)$ means precisely the existence of the exponential Y^X in C, i.e. the existence of a $((-) \times X)$-couniversal arrow for Y in C (see II.4.4). In particular, *local Cartesian closedness of* C *implies Cartesian closedness of* C.

III.4.4.1 Examples

(1) The existence of $P = P(Y, f)$ means, by definition, the existence of a natural bijective correspondence

$$\frac{Q \longrightarrow P}{Q \longrightarrow Z\,,\; Q \times_Z X \longrightarrow Y\,.}$$

Hence, when Q is a terminal object, we see that in $\mathsf{C} = \mathsf{Set}$ the partial product P must be, up to isomorphism,

$$P = \{(j, z) \mid z \in Z,\, j : f^{-1}z \to Y\} \cong \textstyle\coprod_{z \in Z} Y^{f^{-1}z}\,,$$

with $p : (j, z) \mapsto z$ and ε the evaluation map. In fact, one can write $P \times_Z X$ as

$$P \times_Z X = \{(j, x) \mid x \in X,\, j : f^{-1}f(x) \mapsto Y\} \cong \textstyle\coprod_{x \in X} Y^{f^{-1}f(x)}$$

with $\varepsilon : (j, x) \mapsto j(x)$. In this way one sees that Set has partial products.

(2) Ord is Cartesian closed but fails to have partial products (see Exercise III.4.F).

(3) Top fails to be Cartesian closed (see Exercise III.4.G).

III.4.4.2 Lemma *For a category* A *with equalizers, a functor* $F : \mathsf{A} \to \mathsf{C}/X$ *has a right adjoint if* $\mathrm{dom}_X F : \mathsf{A} \to \mathsf{C}$ *has a right adjoint.*

Proof Let $\kappa : \mathrm{dom}_X \to \Delta X$ be the natural transformation with $\kappa_{(W,t)} = t$ for every object (W, t) in C/X, and let J be right adjoint to $\mathrm{dom}_Z F$. Then, by adjunction, $\kappa F : \mathrm{dom}_X \to \Delta X$ corresponds to a natural transformation $\sigma : 1_{\mathsf{A}} \to \Delta J X$. Given (W, t), one forms the equalizer

$$G(W,t) \longrightarrow JW \underset{\sigma_{JW}}{\overset{Jt}{\rightrightarrows}} JX$$

whose domain defines a right adjoint G of F. Indeed, for every A-object A, one obtains the equalizer diagram in Set given by the top row of

$$\begin{array}{ccccc} \mathsf{A}(A, G(W,t)) & \longrightarrow & \mathsf{A}(A, JW) & \underset{\mathsf{A}(A,\sigma_{JW})}{\overset{\mathsf{A}(A,Jt)}{\rightrightarrows}} & \mathsf{A}(A, JX) \\ & & \downarrow\cong & & \downarrow\cong \\ & & \mathsf{C}(\mathrm{dom}_X FA, W) & \underset{\phi}{\overset{\mathsf{C}(\mathrm{dom}_X FA, t)}{\rightrightarrows}} & \mathsf{C}(\mathrm{dom}_X FA, X) \end{array}$$

in which the map $\mathsf{A}(A, \sigma_{JW})$ has constant value σ_A by naturality of σ. Consequently, the map ϕ that makes the diagram commute (in the obvious sense) must send every C-morphism to κ_{FA}, which corresponds to σ_A by adjunction. The equalizer of the two lower maps in Set is

$$\{h : \mathrm{dom}_X FA \to W \mid t \cdot h = \kappa_{FA}\} \cong \mathsf{C}/X(FA, (W,t)) ,$$

which is therefore isomorphic to $\mathsf{A}(A, G(W,t))$, naturally in A. □

III.4.4.3 Proposition *For a morphism $f : X \to Z$ of a finitely complete category C, the following assertions are equivalent and characterize f as an exponentiable object of C/Z:*

(i) *f is exponentiable in C/Z, i.e. $(-) \times_Z X : \mathsf{C}/Z \to \mathsf{C}/Z$ has a right adjoint;*
(ii) *for all C-objects Y, the partial product $P(Y, f)$ exists in C;*
(iii) *the pullback functor*

$$f^* : \mathsf{C}/Z \to \mathsf{C}/X , \quad (Y, g) \mapsto (Y \times_Z X, \pi_2 : Y \times_Z X \to X)$$

has a right adjoint.

Proof (i) $\Longrightarrow$ (ii) follows from the definitions. Lemma III.4.4.2 yields (ii) $\Longrightarrow$ (iii) by commutativity of the diagram

$$\begin{array}{ccccc} & & \mathsf{C}/X & & \\ & {}^{f^*}\nearrow & & \searrow^{\mathrm{dom}_X} & \\ \mathsf{C}/Z & & \xrightarrow[\mathrm{dom}_Z((-)\times_Z X)]{} & & \mathsf{C} . \end{array}$$

For (iii) $\Longrightarrow$ (i), consider the commutative diagram

$$\begin{array}{ccc} & \mathsf{C}/X & \\ {\scriptstyle f^*}\nearrow & & \searrow{\scriptstyle f_!} \\ \mathsf{C}/Z & \xrightarrow[(-)\times_Z X]{} & \mathsf{C}/Z \end{array}$$

with $f_! : (W, t) \mapsto (W, f \cdot t)$. Actually, one easily shows $f_! \dashv f^*$ (so that $(-) \times_Z X$ is the functor part of a comonad on C/Z). Hence, if f^* has a right adjoint, the composite functor $f_! f^*$ has a right adjoint. $\square$

III.4.4.4 Example In $\mathsf{C} = \mathsf{Set}$, the right adjoint $f_* : \mathsf{C}/X \to \mathsf{C}/Z$ of f^* sends (W, t) in C/X to (P, p), with

$$P = \{(j, z) \mid z \in Z, j : f^{-1}z \to W, t \cdot j = (f^{-1} \hookrightarrow X)\}$$

and $p : (j, z) \mapsto z$. One may write

$$P \times_Z X = \{(j, x) \mid x \in X, j : f^{-1}f(x) \to W, t \cdot j = (f^{-1}f(x) \hookrightarrow X)\},$$

with evalutation map $\varepsilon : (j, x) \mapsto j(x)$; this ε is a C/X-morphism since $t \cdot \varepsilon = \pi_2$. When setting $(W, t) = (Y \times X, Y \times X \to X)$, one retrieves the partial product $P(Y, f)$ from this construction.

III.4.5 Local Cartesian closedness of $(\mathbb{T}, \mathcal{V})$-Gph

Throughout this section we assume that

- *$\mathcal{V}$ is Cartesian closed and integral;*
- *T satisfies the Beck–Chevalley condition.*

(In fact, the tensor product of the quantale $\mathcal{V}$ and the multiplication m of the monad $\mathbb{T}$ play no role in this section: it suffices that $\mathcal{V}$ is a frame and T an endofunctor equipped with a natural transformation $e : 1_{\mathsf{Set}} \to T$. In particular, *no lax extension of T to $\mathcal{V}$-*Rel *is needed.*)

Our goal is to construct partial products in $(\mathbb{T}, \mathcal{V})$-$\mathsf{Gph}$ and thereby show local Cartesian closedness of this category. For that, it is useful to point out that the ordered set $\mathcal{V}$ (considered as a category) is locally Cartesian closed. Indeed, for $\alpha, \beta, \gamma, \upsilon \in \mathcal{V}$, one has the equivalence

$$\frac{\upsilon \leq \gamma \wedge (\alpha \to \beta)}{\upsilon \leq \gamma\,,\ \upsilon \wedge \alpha \leq \beta\,,}$$

where $\alpha \to \beta$ denotes the internal hom on $\mathcal{V}$ with its Cartesian structure $\wedge$. Hence, if $\alpha \leq \gamma$, the partial product $P(\beta, \alpha \leq \gamma)$ exists in $\mathcal{V}$ and is given by $\gamma \wedge (\alpha \to \beta)$.

Let us now consider $f : (X, a) \to (Z, c)$ and (Y, b) in $(\mathbb{T}, \mathcal{V})$-Gph, and form the set

$$P = \{(j, z) \mid z \in Z,\ j : (X_z, a_z) \to (Y, b) \text{ a } (\mathbb{T}, \mathcal{V})\text{-functor}\}\ ,$$

where $X_z = f^{-1}z$ and a_z is the restriction of a. We can write

$$W = P \times_Z X = \{(j, x) \mid x \in X,\ j : (X_{f(x)}, a_{f(x)}) \to (Y, b)\}$$

and define $p : P \to X$ and $\varepsilon : W \to Y$ as at the level of sets (see Example III.4.4.1(1); but note that P is a *subset* of the partial product of Y over f in Set). The $\mathcal{V}$-relation $d : TP \nrightarrow P$ is defined by

$$d(\mathfrak{p}, (j, z)) = c(Tp(\mathfrak{p}), z) \wedge \bigwedge\nolimits_{\mathfrak{w},x}(a(T\pi_2(\mathfrak{w}), x) \to b(T\varepsilon(\mathfrak{w}), j(x))) \tag{III.4.5.i}$$

(for $\mathfrak{p} \in TP$, $(j, z) \in P$), with the meet ranging over all $\mathfrak{w} \in TW$ with $T\pi_1(\mathfrak{w}) = \mathfrak{p}$ and $x \in X$ with $f(x) = z$.

III.4.5.1 Theorem *The pair* (P, d) *is a partial product of* (Y, b) *over* $f : (X, a) \to (Z, c)$ *in* $(\mathbb{T}, \mathcal{V})$-Gph.

Proof We show reflexivity of d, $(\mathbb{T}, \mathcal{V})$-functoriality of p and ε, and verify the universal property of (P, d).

d is reflexive: For $(j, z) \in P$ one has

$$k \le c(e_Z(z), z) = c(e_Z \cdot p(j, z), z) = c(Tp(e_P(j, z)), z)\ .$$

We now consider the pullback diagrams

$$\begin{array}{ccccc} X_z & \xrightarrow{s} & W & \xrightarrow{\pi_2} & X \\ {\scriptstyle !}\downarrow & & \downarrow{\scriptstyle \pi_1} & & \downarrow{\scriptstyle f} \\ 1 & \xrightarrow{r} & P & \xrightarrow{p} & Z \end{array}$$

in Set, with r the constant map to (j, z) and s the map $(x' \mapsto (j, x'))$. Since for every $\mathfrak{w} \in TW$ with $T\pi_1(\mathfrak{w}) = e_P(j, z)$ and every $x \in X$ with $f(x) = z$ one has

$$T\pi_1(\mathfrak{w}) = e_P \cdot r \cdot !(x) = Tr(e_1(!(X)))\ ,$$

and since T satisfies BC, there is $\chi \in TX_z$ with $T!(\chi) = e_1(!(x))$ and $Ts(\chi) = \mathfrak{w}$. Considering $T(\pi_2 \cdot s)$ as a subset inclusion, one obtains by $(\mathbb{T}, \mathcal{V})$-functoriality of $j : (X_z, a_z) \to (Y, b)$

$$\begin{aligned} a(T\pi_2(\mathfrak{w}), x) &= a(T\pi_2 \cdot Ts(\chi), x) \\ &= a(\chi, x) \\ &\le b(Tj(\chi), j(x)) \\ &= b(T(\varepsilon \cdot s)(\chi), j(x)) = b(T\varepsilon(\mathfrak{w}), j(x))\ . \end{aligned}$$

Consequently, $k \leq a(T\pi_2(w), x) \to b(T\varepsilon(w), j(x))$, which implies $k \leq d(e_P(j, z), (j, z))$.

p and ε are $(\mathbb{T}, \mathcal{V})$-functors: Since $d(p, (j, z)) \leq c(Tp(p), z)$ by definition of d, p is a morphism in $(\mathbb{T}, \mathcal{V})$-Gph. Denoting the structure of W by m, for all $w \in TW$ and $(j, x) \in W$, one has, by definition of d,

$$\begin{aligned} m(w, (j, x)) &= d(T\pi_1(w), (j, f(x))) \wedge a(T\pi_2(w), x) \\ &\leq (a(T\pi_2(w), x) \to b(T\varepsilon(w), j(x))) \wedge a(T\pi_2(w), x) \\ &\leq b(T\varepsilon(w), j(x)) = b(T\varepsilon(w), \varepsilon(j, x)) . \end{aligned}$$

(P, d) satisfies the universal property: For morphisms $q : (Q, l) \to (Z, c)$, $g : (Q \times_Z X, n) \to (Y, b)$ in $(\mathbb{T}, \mathcal{V})$-Gph, we must show that the map

$$\begin{aligned} h : Q &\to P \\ t &\mapsto (j_t, q(t)) \end{aligned} \qquad \text{with} \qquad \begin{aligned} j_t : X_{q(t)} &\to Y \\ x &\mapsto g(t, x) \end{aligned}$$

is a well-defined $(\mathbb{T}, \mathcal{V})$-functor. For $t \in Q$ consider the pullback diagrams

$$\begin{array}{ccccc} X_{q(t)} & \xrightarrow{u} & U & \xrightarrow{\tilde{\pi}_2} & X \\ {\scriptstyle !}\downarrow & & \downarrow{\scriptstyle \tilde{\pi}_1} & & \downarrow{\scriptstyle f} \\ 1 & \xrightarrow{t} & Q & \xrightarrow{q} & Z \end{array}$$

in Set, with $U = Q \times_Z X$ and u the map $(x \mapsto (t, x))$. Since $k = \top$ one has $t : (1, \top) \to (Q, l)$. Consequently, $u : (X_{q(t)}, a_{q(t)}) \to (U, n)$ and

$$j = g \cdot u : (X_{q(t)}, a_{q(t)}) \to (Y, b)$$

are $(\mathbb{T}, \mathcal{V})$-functors. Hence, $(j_t, q(t)) \in P$. Finally, for $\mathfrak{q} \in TQ$, $t \in Q$, we first note

$$l(\mathfrak{q}, t) \leq c(Tq(\mathfrak{q}), q(t)) = c(Tp(Th(\mathfrak{q})), q(t)) .$$

Exploiting BC for the left pullback diagram in

$$\begin{array}{ccccc} U & \xrightarrow{h \times 1_X} & W & \xrightarrow{\pi_2} & X \\ {\scriptstyle \tilde{\pi}_1}\downarrow & & \downarrow{\scriptstyle \pi_1} & & \downarrow{\scriptstyle f} \\ Q & \xrightarrow{h} & P & \xrightarrow{p} & Z \end{array}$$

for every $w \in TW$ with $T\pi_1(w) = Th(\mathfrak{q})$ we obtain $\mathfrak{u} \in TU$ with $T\tilde{\pi}_1(\mathfrak{u}) = \mathfrak{q}$ and $T(h \times 1_X)(\mathfrak{u}) = w$, hence $T\tilde{\pi}_2(\mathfrak{u}) = T\pi_2(w)$. Now for every $x \in X_{q(t)}$ we have

$$\begin{aligned} l(\mathfrak{q}, t) \wedge a(T\pi_2(w), x) &= l(T\tilde{\pi}_1(\mathfrak{u}), t) \wedge a(T\tilde{\pi}_2(\mathfrak{u}), x) \\ &= n(\mathfrak{u}, (t, x)) \\ &\leq b(Tg(\mathfrak{u}), g(t, x)) \\ &= b(T\varepsilon(T(h \times 1_X)(\mathfrak{u})), g(t, x)) = b(T\varepsilon(w), j_t(x)) . \end{aligned}$$

Consequently, $l(q,t) \leq a(T\pi_2(\mathfrak{w}),x) \rightarrow b(T\varepsilon(\mathfrak{w}), j_t(x)$ and $l(q,t) \leq d(Th(q),h(t))$ follows. □

III.4.5.2 Corollary *The category* $(\mathbb{T},\mathcal{V})$-Gph *is locally Cartesian closed.*

III.4.5.3 Examples

(1) For $\mathbb{T} = \mathbb{I}$ identically extended to $\mathcal{V}$-Rel, every $(\mathbb{T},\mathcal{V})$-graph is unitary, Corollary III.4.5.2 gives that *the category*

$$\mathcal{V}\text{-Gph} := (\mathbb{I},\mathcal{V})\text{-Gph} = (\mathbb{I},\mathcal{V})\text{-UGph}$$

is locally Cartesian closed whenever $\mathcal{V}$ *is Cartesian closed with* $k = \top$. For $\mathcal{V} = 2$, this is the category

$$\mathsf{RRel} = 2\text{-Gph}$$

of *reflexive relations*, i.e. of sets endowed with a reflexive relation which, even in the absence of transitivity, we write as $\leq$; morphisms are then "monotone" maps. Keeping the notation of Section III.4.4, by Theorem III.4.5.1 one can form the partial product $P = P(Y,f)$ in RRel by endowing the set

$$P = \{(j,z) \mid z \in Z,\ j : X_z \to Y \text{ monotone}\}$$

with the relation

$$(j,z) \leq (j',z') \iff z \leq z' \ \&\ \forall x \in X_z, x' \in X_{z'}\ (x \leq x' \implies j(x) \leq j'(x'))\,.$$

While p and ε are defined as for sets (see Example III.4.4.1(1)), we note that $O : \mathsf{RRel} \to \mathsf{Set}$ does not preserve partial products, not even exponentials.

(2) For general $\mathcal{V}$ (as in (1) above), Theorem III.4.5.1 constructs partial products in $\mathcal{V}$-Gph as follows. For $f : (X,a) \to (Z,c)$ and (Y,b) in $\mathcal{V}$-Gph and

$$P = \{(j,z) \mid z \in Z,\ j : (X_z,a_z) \to (Y,b) \text{ a } \mathcal{V}\text{-functor}\}\,,$$

provide $(P,d) = P(Y,f)$ with the structure

$$d((j,z),(j',z')) = c(z,z') \wedge \bigwedge\nolimits_{x\in X_z, x'\in X_{z'}} (a(x,x') \to b(j(x),j'(x'))\,.$$

In particular, in the category

$$\mathsf{RNRel} := \mathsf{P}_{\max}\text{-Gph}$$

of *reflexive numerical relations* and non-expansive maps, the "metric" d on the partial product P of f over Y is given by

$$d((z,j),(z',j')) = \max\{c(z,z'), \sup\nolimits_{x\in X_z, x'\in X_{z'}} (a(x,x') \to b(j(x),j'(x')\}$$

(see Example II.1.10.1(5)).

(3) Theorem III.4.5.1 shows that

$$\mathsf{PsTop} = (\mathbb{\beta}, 2)\text{-}\mathsf{Gph} = (\mathbb{\beta}, 2)\text{-}\mathsf{RGph}$$

is locally Cartesian closed, with the ultrafilter convergence $\longrightarrow$ on

$$P = P(Y, f) = \{(j, z) \mid z \in Z,\ j : X_z \to Y \text{ continuous}\}$$

described as follows: $\mathfrak{p} \longrightarrow (j, z) \in P$ if and only if

- $p(\mathfrak{p}) \longrightarrow z$ in Z (with $p : P \to Z$ the projection);
- for all $x \in X_z$, $\mathfrak{w} \in \beta(P \times_Z X)$ with $\pi_1[\mathfrak{w}] = \mathfrak{p}$, $\pi_2[\mathfrak{w}] \longrightarrow x$ in X implies $\varepsilon[\mathfrak{w}] \longrightarrow j(x)$ in Y (with $\varepsilon : P \times_Z X \to Y$ the evaluation map). With ultrafilters traded for filters, the same condition describes partial products in $(\mathbb{F}, 2)$-Gph (see Example III.4.1.3(3)).

III.4.6 Local Cartesian closedness of subcategories of $(\mathbb{T}, \mathcal{V})$-$\mathsf{Gph}$

Throughout this section we assume that

- *$\mathcal{V}$ is Cartesian closed and integral;*
- *T satisfies the Beck–Chevalley condition.*

Let us first give sufficient conditions for Cartesian and local Cartesian closedness of reflective or coreflective full subcategories of a Cartesian or locally Cartesian closed category.

III.4.6.1 Proposition *Let A be a full replete subcategory of a Cartesian closed category C with finite products.*

(1) *If A is reflective, with the reflector preserving binary products, then A is Cartesian closed with* exponentials *(that is, internal hom-objects) formed as in X.*
(2) *If A is coreflective and closed under binary products in X, then A is Cartesian closed with exponentials in A obtained by coreflection of exponentials formed in X.*

Proof (1): For A-objects A and B, let B^A be the exponential formed in X with counit $\varepsilon : B^A \times A \to B$, and let $\rho : B^A \to R(B^A)$ be the reflection into A. Since $R(B^A \times A) \cong R(B^A) \times A$ by hypothesis, the morphism $\rho \times 1_A : B^A \times A \to R(B^A) \times A$ serves as a reflection into A. Hence, ε factors uniquely as $\varepsilon = \overline{\varepsilon} \cdot (\rho \times 1_A)$. The mate s of $\overline{\varepsilon}$ that makes

$$\begin{array}{ccc} A \times B^A & \xrightarrow{\ \varepsilon\ } & B \\ {\scriptstyle 1_A \times s}\uparrow & \nearrow_{\overline{\varepsilon}} & \\ A \times R(B^A) & & \end{array}$$

commute must be inverse to ρ. Consequently, $B^A \cong R(B^A)$ lies in A and serves there in the same capacity as in X.

(2): The claim is straightforward. □

III.4.6.2 Corollary *Let* A *be a full replete subcategory of a locally Cartesian closed category* C *with pullbacks.*

(1) *If* A *is reflective, with the reflector preserving pullbacks, then* A *is locally Cartesian closed with partial products formed as in* X.
(2) *If* A *is coreflective and closed under pullbacks in* X, *then* A *is locally Cartesian closed with partial products in* A *obtained by coreflection of partial products formed in* X.

Proof Apply Proposition III.4.6.1 to the reflective or coreflective subcategory A/B of X/B (for $B \in \mathrm{ob}\,\mathsf{A}$). □

III.4.6.3 Corollary *If the functor* $R : (\mathbb{T}, \mathcal{V})\text{-}\mathsf{Gph} \to (\mathbb{T}, \mathcal{V})\text{-}\mathsf{RGph}$, *sending* (X, a) *to* $(X, a \cdot \hat{T}1_X)$, *preserves binary products, then* $(\mathbb{T}, \mathcal{V})\text{-}\mathsf{RGph}$ *is Cartesian closed with exponentials formed as in* $(\mathbb{T}, \mathcal{V})\text{-}\mathsf{Gph}$. *Similarly, if* $R : (\mathbb{T}, \mathcal{V})\text{-}\mathsf{Gph} \to (\mathbb{T}, \mathcal{V})\text{-}\mathsf{RGph}$ *preserves pullbacks, then* $(\mathbb{T}, \mathcal{V})\text{-}\mathsf{RGph}$ *is locally Cartesian closed with partial products formed as in* $(\mathbb{T}, \mathcal{V})\text{-}\mathsf{Gph}$.

Proof By Corollary III.4.1.5, R is the reflector of $(\mathbb{T}, \mathcal{V})\text{-}\mathsf{Gph}$ onto $(\mathbb{T}, \mathcal{V})\text{-}\mathsf{RGph}$. Hence, Proposition III.4.6.1 and Corollary III.4.6.2 may be applied. □

III.4.6.4 Example With the filter monad $\mathbb{F}$ extended by $\check{F}$ (as in Example III.1.10.3.(4)), the reflector $R : (\mathbb{F}, 2)\text{-}\mathsf{Gph} \to (\mathbb{F}, 2)\text{-}\mathsf{RGph}$ extends the reflexive filter convergence relation $\longrightarrow$ on a set X to

$$\mathfrak{a} \longrightarrow x \iff \exists \mathfrak{b} \in FX\ (\mathfrak{a} \supseteq \mathfrak{b}\ \&\ \mathfrak{b} \longrightarrow x)$$

for all $x \in X$, $\mathfrak{a} \in FX$. It is not difficult to show that R preserves finite limits. Hence, $(\mathbb{F}, 2)\text{-}\mathsf{RGph}$ is locally Cartesian closed with partial products formed as in $(\mathbb{F}, 2)\text{-}\mathsf{Gph}$.

In addition to the two general hypotheses on $\hat{\mathbb{T}}$ and $\mathcal{V}$ stated at the beginning of Section III.4.6, we now assume that

- $\hat{\mathbb{T}}$ *is associative;*
- $\hat{T}(f^\circ) = \hat{T}1_X \cdot (Tf)^\circ$ *for all* $f : X \to Y$ *(see* (III.4.3.i)*).*

Under these hypotheses, not only coproduct sinks in $(\mathbb{T}, \mathcal{V})\text{-}\mathsf{RGph}$ are universal (as proved in Theorem III.4.3.9), but also so are all O-final epi-sinks, where $O : (\mathbb{T}, \mathcal{V})\text{-}\mathsf{RGph} \to \mathsf{Set}$ is the forgetful functor:

III.4.6.5 Proposition *O-final epi-sinks are stable under pullback in* $(\mathbb{T}, \mathcal{V})\text{-}\mathsf{RGph}$.

Proof For an O-final epi-sink $(g_i : (Y_i, b_i) \to (Z, c))_{i\in I}$ in $(\mathbb{T}, \mathcal{V})$-RGph, the formula given in Corollary III.4.1.5 simplifies under the additional hypotheses to $c = \bigvee_{i\in I} g_i \cdot b_i \cdot (Tg_i)^\circ$. When for $f : (X, a) \to (Z, c)$ we form the pullbacks

$$\begin{array}{ccc} (P_i, d_i) & \xrightarrow{p_i} & (X, a) \\ {\scriptstyle q_i}\downarrow & & \downarrow{\scriptstyle f} \\ (Y_i, b_i) & \xrightarrow{g_i} & (Z, c) \end{array}$$

we have $d_i = (p_i^\circ \cdot a \cdot Tp_i) \wedge (q_i^\circ \cdot b_i \cdot Tq_i)$ for all $i \in I$ and need to show $a = \bigvee_{i\in I} p_i \cdot d_i \cdot (Tp_i)^\circ$:

$$\begin{aligned} &\textstyle\bigvee_{i\in I} p_i \cdot d_i \cdot (Tp_i)^\circ \\ &\quad = \textstyle\bigvee_{i\in I} p_i \cdot \big((p_i^\circ \cdot a \cdot Tp_i) \wedge (q_i^\circ \cdot b_i \cdot Tq_i)\big) \cdot (Tp_i)^\circ \\ &\quad = \textstyle\bigvee_{i\in I} a \wedge (p_i \cdot q_i^\circ \cdot b_i \cdot Tq_i \cdot (Tp_i)^\circ) && \text{(Lemma III.4.3.7)} \\ &\quad = a \wedge \textstyle\bigvee_{i\in I} f^\circ \cdot g_i \cdot b_i \cdot (Tg_i)^\circ \cdot Tf && \text{(BC)} \\ &\quad = a \wedge f^\circ \cdot (\textstyle\bigvee_{i\in I} g_i \cdot b_i \cdot (Tg_i)^\circ) \cdot Tf \\ &\quad = a \wedge (f^\circ \cdot c \cdot Tf) = a\,. \end{aligned}$$

□

III.4.6.6 Corollary *Colimits are stable under pullback in* $(\mathbb{T}, \mathcal{V})$-RGph.

Proof Colimit sinks in $(\mathbb{T}, \mathcal{V})$-RGph are those O-final epi-sinks which are colimits in Set. □

For $f : (X, a) \to (Z, c)$ in $(\mathbb{T}, \mathcal{V})$-RGph, we now consider the commutative diagram

$$\begin{array}{ccccc} (\mathbb{T}, \mathcal{V})\text{-RGph}/(Z, c) & \xrightarrow{f^*} & (\mathbb{T}, \mathcal{V})\text{-RGph}/(X, a) & \xrightarrow{\mathrm{dom}_X} & (\mathbb{T}, \mathcal{V})\text{-RGph} \\ {\scriptstyle O_Z}\downarrow & & {\scriptstyle O_X}\downarrow & & \downarrow{\scriptstyle O} \\ \mathsf{Set}/Z & \xrightarrow{f^*} & \mathsf{Set}/X & \xrightarrow{\mathrm{dom}_X} & \mathsf{Set}\,. \end{array} \tag{III.4.6.i}$$

Since O is topological, O_X is also topological (and so is O_Z), with O_X-final structures formed like O-final structures, i.e. the upper row domain functor dom_X (see Section III.4.4) transforms O_X-final structures into O-final sinks (see Exercise III.4.H). More importantly, by Proposition III.4.6.5, the upper row pullback functor f^* (see Proposition III.4.4.3) sends O_Z-final epi-sinks to O_X-final epi-sinks. In particular, f^* preserves all colimits. In fact, f^* has a right adjoint by virtue of the Special Adjoint Functor Theorem (see Exercise II.5.J).

III.4.6.7 Theorem *Suppose that* $\hat{\mathbb{T}}$ *is associative and* $\hat{T}(f^\circ) = \hat{T}1_X \cdot (Tf)^\circ$ *for all* $f : X \to Y$. *Then the category* $(\mathbb{T}, \mathcal{V})$-RGph *is locally Cartesian closed and therefore has all partial products.*

Proof We must verify that the sufficient conditions of the dual of the Special Adjoint Functor Theorem are satisfied by the functor f^*. But the topological category $(\mathbb{T}, \mathcal{V})$-$\mathsf{RGph}$ over Set inherits the relevant properties from Set and passes them on to its comma categories. In fact, for every category C and C-object X, one easily verifies the following facts:

(1) if C is locally small, then C/X is locally small (because $\mathrm{dom}_X : \mathsf{C}/X \to \mathsf{C}$ is faithful);
(2) if C is wellpowered, then C/X is wellpowered (because dom_X preserves and reflects monomorphisms);
(3) if C is small-complete, then C/X is small-complete (see Exercise II.2.J);
(4) if C is locally small and $\mathcal{G}$ is generating in C, then the class of arrows $\bigcup_{G\in\mathcal{G}} \mathsf{C}(G, X)$ is generating in C/X. □

III.4.6.8 Remark It is important to notice that, in general, f^* will *not* transform *all* O_Z-final sinks into O_X-final sinks, but only those that are epic. Otherwise, we could have applied the Taut Lift Theorem II.5.11 to (III.4.6.i), with the result that the lower-row right adjoints would have been lifted along the vertical topological functors. But a partial product in $(\mathbb{T}, \mathcal{V})$-$\mathsf{Gph}$ is generally *not* formed by endowing the partial product of the underlying Set-data with a $(\mathbb{T}, \mathcal{V})$-graph structure. However, one can apply the Generalized Taut Lift Theorem (Exercise II.5.U) as an alternative method of proof of Theorem III.4.6.7.

III.4.7 Interlude on subobject classifiers and partial-map classifiers

Throughout this section we consider a class $\mathcal{M}$ of morphisms in a category C with pullbacks such that:

- $\mathrm{SplitMono}\,\mathsf{C} \subseteq \mathcal{M} \subseteq \mathrm{Mono}\,\mathsf{C}$;
- *$\mathcal{M}$ is closed under composition with isomorphisms;*
- *$\mathcal{M}$ is stable under pullback.*

We adopt the notation of II.5.2, and write $\mathrm{sub}\,X = \mathrm{sub}_{\mathcal{M}}\,X := \mathcal{M}/X$ for the full subcategory of C/X whose objects lie in $\mathcal{M}$, and assume throughout that the separated reflection (see II.1.3)

$$\mathrm{sub}\,X \to \mathrm{sub}^{\simeq}\,X := \mathrm{sub}\,X/\simeq$$

comes with a section in SET, so that it is a split epimorphism in SET. Since the pullback functors $f^* : \mathsf{C}/Y \to \mathsf{C}/X$ restrict to $f^* : \mathrm{sub}\,Y \to \mathrm{sub}\,X$ and satisfy $(1_X)^* \cong 1_{\mathrm{sub}\,X}$ and $(g\cdot f)^* \cong f^*g^*$ (for $f : X \to Y$, $g : Y \to Z$ in C), one has a functor

$$\mathrm{sub}^{\simeq} : \mathsf{C}^{\mathrm{op}} \to \mathsf{SET}\,.$$

An $\mathcal{M}$-*subobject classifier* is a C-object Ω that represents $\mathrm{sub}^{\simeq}$, so that $\mathsf{C}(-,\Omega) \cong \mathrm{sub}^{\simeq}$. Hence, Ω is characterized by the existence of maps

$$\psi_X : \mathsf{C}(X,\Omega) \longrightarrow \mathrm{sub}\, X$$

that are pseudo-natural in X (with respect to the order of $\mathrm{sub}\, X$) and are such that for every subobject $m : M \to X$ there is a unique morphism $f : X \to \Omega$ with $\psi_X(f) \simeq m$; one calls f the *characteristic morphism* of m. Here is the hands-on characterization of $\mathcal{M}$-subobject classifiers:

III.4.7.1 Proposition *An object Ω is an $\mathcal{M}$-subobject classifier in C if and only if there is an $\mathcal{M}$-subobject $t : 1 \to \Omega$ (with 1 a terminal object of C) such that, for any $\mathcal{M}$-subobject $m : M \to X$, there is a unique morphism $f : X \to \Omega$ making*

$$\begin{array}{ccc} M & \longrightarrow & 1 \\ {\scriptstyle m}\downarrow & & \downarrow{\scriptstyle t} \\ X & \xrightarrow{\;f\;} & \Omega \end{array} \qquad \text{(III.4.7.i)}$$

a pullback diagram.

Proof For an $\mathcal{M}$-subobject classifier Ω, one puts $t := \psi_\Omega(1_\Omega) : Z \to \Omega$. Then, using pseudo-naturality of ψ, to obtain the characteristic morphism f of $m : M \to X$ in $\mathcal{M}$ we can chase $1_\Omega \in \mathsf{C}(\Omega,\Omega)$ in two ways around the diagram

$$\begin{array}{ccc} \mathsf{C}(\Omega,\Omega) & \xrightarrow{\;\psi_\Omega\;} & \mathrm{sub}\,\Omega \\ {\scriptstyle \mathsf{C}(f,\Omega)}\downarrow & & \downarrow{\scriptstyle f^*} \\ \mathsf{C}(X,\Omega) & \xrightarrow{\;\psi_X\;} & \Omega \end{array}$$

yielding $f^*(t) \simeq m$, so that (III.4.7.i) is a pullback diagram with 1 replaced by Z. We now show that Z must necessarily be terminal in C. Considering $m = 1_X$ for any object X, we obtain a pullback diagram

$$\begin{array}{ccc} X & \xrightarrow{\;u\;} & Z \\ {\scriptstyle 1_X}\downarrow & & \downarrow{\scriptstyle t} \\ X & \xrightarrow{\;f\;} & \Omega\,. \end{array}$$

For any other $v : X \to Z$,

$$\begin{array}{ccc} X & \xrightarrow{\;v\;} & Z \\ {\scriptstyle 1_X}\downarrow & & \downarrow{\scriptstyle t} \\ X & \xrightarrow{\;t\cdot v\;} & \Omega \end{array}$$

is a pullback diagram since t is a monomorphism. Hence, $t \cdot v$ is a characteristic morphism for 1_X, and we must have $t \cdot v = f = t \cdot u$, which implies $u = v$. The sufficiency condition is left to the reader. □

III.4.7.2 Remarks

(1) The proof of Proposition III.4.7.1 shows that a category C with an $\mathcal{M}$-subobject classifier necessarily has a terminal object and, in the presence of pullbacks, is therefore finitely complete. Furthermore, if C is locally small, C must be $\mathcal{M}$-wellpowered.

(2) Since m in (III.4.7.i) is a pullback of the split monomorphism $1 \longrightarrow \Omega$, one necessarily has $\mathcal{M} \subseteq \mathrm{RegMono}\,\mathsf{C}$ if C has an $\mathcal{M}$-subobject classifier; we speak of a *regular-subobject classifier* if $\mathcal{M} = \mathrm{RegMono}\,\mathsf{C}$, while *subobject classifier* refers to the case $\mathcal{M} = \mathrm{Mono}\,\mathsf{C}$. Hence, the existence of a subobject classifier in C forces $\mathrm{Mono}\,\mathsf{C} = \mathrm{RegMono}\,\mathsf{C}$.

III.4.7.3 Examples

(1) Every two-element set 2 is a subobject classifier in Set, since 2 represents the powerset functor P of Set. Less trivially, every functor category $\mathsf{Set}^{\mathsf{D}^{\mathrm{op}}}$ has a subobject classifier for a small category D (see [Johnstone, 1977] and [Mac Lane and Moerdijk, 1994]).

(2) If $U : \mathsf{A} \longrightarrow \mathsf{X}$ is topological with right adjoint I, and Ω is a regular-subobject classifier in X, then $I\Omega$ is a regular-subobject classifier in A; see Exercise III.4.J. In particular, a two-element set will assume that role both in Ord and in Top when provided with its respective indiscrete structure.

An $\mathcal{M}$*-partial map* from X to Z in C consists of an $\mathcal{M}$-subobject $m : M \longrightarrow X$ and a morphism $g : M \longrightarrow Z$. The (possibly large) class

$$\mathrm{part}(X, Z) = \mathrm{part}_{\mathcal{M}}(X, Z)$$

of all $\mathcal{M}$-partial maps from X to Z is ordered by

$$(m, g) \leq (m', g') \iff \exists h\ (m' \cdot h = m \ \&\ g' \cdot h = g)\ .$$

For $f : X \longrightarrow Y$ in C there is now a pullback functor

$$f^* : \mathrm{part}(Y, Z) \longrightarrow \mathrm{part}(X, Z)\ , \quad (n, g) \mapsto (f^*(n), g \cdot f_n)\ ,$$

(with $n : N \longrightarrow Y$ and $g : N \longrightarrow Z$) defined by the pullback diagram

$$\begin{array}{ccc} \cdot & \xrightarrow{\ f_n\ } & N \\ {\scriptstyle f^*(n)}\downarrow & & \downarrow{\scriptstyle n} \\ X & \xrightarrow{\ f\ } & Y\ . \end{array}$$

As for sub, when taking isomorphism classes of $\mathcal{M}$-partial maps, part$(-, Z)$ becomes a functor

$$\mathrm{part}^{\simeq}(-, Z) : \mathsf{C}^{\mathrm{op}} \longrightarrow \mathsf{SET}\,.$$

Note that, for a terminal object $Z = 1$, the functor $\mathrm{part}^{\simeq}(-, Z)$ is isomorphic to $\mathrm{sub}^{\simeq}$. An *$\mathcal{M}$-partial-map classifier* for Z is an object Z^* that represents $\mathrm{part}^{\simeq}(-, Z)$, so that $\mathsf{C}(-, Z^*) \cong \mathrm{part}^{\simeq}(-, Z)$. Hence, Z^* is characterized by the existence of maps

$$\phi_X : \mathsf{C}(X, Z^*) \longrightarrow \mathrm{part}(X, Z)$$

that are pseudo-natural in X and are such that, for all $(m, g) \in \mathrm{part}(X, Z)$, there is a unique morphism $f : X \longrightarrow Z^*$ with $\phi_X(f) \simeq (m, g)$.

III.4.7.4 Proposition *Let Z be an object in C. An object Z^* is an $\mathcal{M}$-partial-map classifier for Z if and only if there is a morphism $t_Z : Z \longrightarrow Z^*$ in $\mathcal{M}$ such that, for any $\mathcal{M}$-partial map (m, g) from X to Z, there is a unique morphism $f : X \longrightarrow Z^*$ making*

$$\begin{array}{ccc} M & \xrightarrow{\;g\;} & Z \\ {\scriptstyle m}\big\downarrow & & \big\downarrow{\scriptstyle t_Z} \\ X & \xrightarrow{\;f\;} & Z^* \end{array} \qquad \text{(III.4.7.ii)}$$

a pullback diagram.

Proof If Z^* is an $\mathcal{M}$-partial-map classifier for Z, one sets

$$(t_Z : W_Z \longrightarrow Z^*, s_Z : W_Z \longrightarrow Z) := \phi_{Z^*}(1_{Z^*})\,.$$

Given (m, g) from X to Z, let $f : X \longrightarrow Z^*$ be such that $\phi_X \simeq (m, g)$. Chasing 1_{Z^*} around the pseudo-naturality diagram

$$\begin{array}{ccc} \mathsf{C}(Z^*, Z^*) & \xrightarrow{\;\phi_{Z^*}\;} & \mathrm{part}(Z^*, Z) \\ {\scriptstyle \mathsf{C}(f, Z^*)}\big\downarrow & & \big\downarrow{\scriptstyle f^*} \\ \mathsf{C}(X, Z^*) & \xrightarrow{\;\phi_X\;} & \mathrm{part}(X, Z)\,, \end{array}$$

one obtains $(m, g) \simeq (f^*(t_Z), s_Z \cdot f_{t_Z})$, as in the diagram

$$\begin{array}{ccccc} \cdot & \xrightarrow{\;f_{t_Z}\;} & W_Z & \xrightarrow{\;s_Z\;} & Z \\ {\scriptstyle f^*(t_Z)}\big\downarrow & & \big\downarrow{\scriptstyle t_Z} & & \\ X & \xrightarrow{\;f\;} & Z^*\,. & & \end{array}$$

A comparison with (III.4.7.ii) shows that, if s_Z is an isomorphism, the necessity part of the statement is proved. When we specialize $(m, g) = (1_Z, 1_Z)$, one obtains $s_Z \cdot f_{t_Z}$ for the corresponding morphism $f : Z \longrightarrow Z^*$. Furthermore, chasing $f \in \mathsf{C}(Z, Z^*)$ both ways through

$$\begin{array}{ccc} \mathsf{C}(Z^*, Z^*) & \xrightarrow{\phi_Z} & \mathrm{part}(Z, Z) \\ {\scriptstyle \mathsf{C}(s_Z, Z^*)}\downarrow & & \downarrow{\scriptstyle s_Z^*} \\ \mathsf{C}(W_Z, Z^*) & \xrightarrow{\phi_{W_Z}} & \mathrm{part}(W_Z, Z)\ , \end{array}$$

we obtain $\phi_{W_Z}(f \cdot s_Z) \simeq (1_{W_Z}, s_Z)$. Since $\phi_{W_Z}(t_Z) \simeq (1_{W_Z}, s_Z)$ trivially, we conclude $f \cdot s_Z \simeq t_Z$. Hence, the split epimorphism s_Z is a monomorphism, and therefore an isomorphism.

The proof of the sufficiency of the stated condition is left to the reader. □

As for $\mathcal{M}$-subobject classifiers, the term *partial-map classifier* refers to the case $\mathcal{M} = \mathrm{Mono}\,\mathsf{C}$, and the term *regular-partial-map classifier* to the case $\mathcal{M} = \mathrm{RegMono}\,\mathsf{C}$. Actually, as we shall see in Corollary III.4.7.6, the general hypotheses on $\mathcal{M}$ leave no choice for $\mathcal{M}$ when there is an $\mathcal{M}$-partial-map classifier.

III.4.7.5 Examples

(1) A partial-map classifier for Z in Set is $Z^* = Z + 1$. More generally, such classifiers exist in every functor category $\mathsf{Set}^{\mathsf{C}^{\mathrm{op}}}$ with C small (see [Johnstone, 1977] and [Mac Lane and Moerdijk, 1994]).

(2) Unlike subobject classifiers, partial-map classifiers may not be lifted along topological functors. Indeed, the category Ord fails to have a regular-partial-map classifier for the two-chain $\mathsf{2} = \{0 < 1\}$; see Exercise III.4.K. A similar statement holds for Top. Top does have $\mathcal{M}$-partial-map classifiers for all spaces when $\mathcal{M}$ is the class of open embeddings or the class of closed embeddings. However, for these choices of $\mathcal{M}$ the general hypothesis that $\mathcal{M}$ contain all split monomorphisms fails, while all others are satisfied.

III.4.7.6 Corollary *In a finitely complete category* C *with an* $\mathcal{M}$*-partial-map classifier,* $\mathcal{M}$ *is precisely the class of equalizers in* C.

Proof Setting $g = m \in \mathcal{M}$ in (III.4.7.ii), one sees that m is an equalizer of t_X and f for some $f : X \to X^*$. Conversely, if $m : M \to X$ is the equalizer of some pair $g, h : X \to Y$ of morphisms, then m is a pullback of the split monomorphism $\langle 1_X, h\rangle$ along $\langle 1_X, g\rangle$ and must therefore be a morphism in $\mathcal{M}$:

$$\begin{array}{ccc} M & \xrightarrow{m} & X \\ {\scriptstyle m}\downarrow & & \downarrow{\scriptstyle \langle 1_X, h\rangle} \\ X & \xrightarrow{\langle 1_X, g\rangle} & X \times Y\ . \end{array}$$

□

Consequently, when C has an $\mathcal{M}$-partial-map classifer and cokernel pairs of regular monomorphisms, so that every regular monomorphism is an equalizer (Exercise II.2.D), then $\mathcal{M} = \mathrm{RegMono}\,\mathsf{C}$.

III.4.7.7 Theorem *For a finitely complete category* C, *the following assertions are equivalent:*

(i) C *has partial products and an* $\mathcal{M}$*-subobject classifier;*
(ii) C *is locally Cartesian closed and has an* $\mathcal{M}$*-subobject classifier;*
(iii) C *is Cartesian closed and has* $\mathcal{M}$*-partial-map classifiers for all objects.*

Proof The equivalence of (i) and (ii) follows from Proposition III.4.4.3.

For (i) $\Longrightarrow$ (iii) one forms the partial product $P = P(Z, t : 1 \to \Omega)$ and considers the diagram

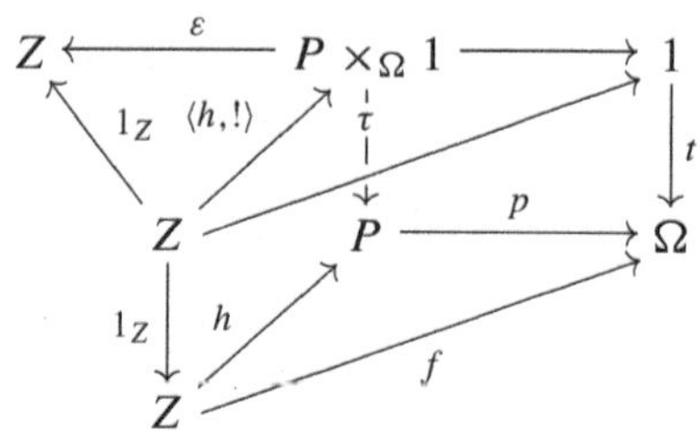

where f is the characteristic morphism for 1_Z. With the induced morphism h, the evaluation morphism ε becomes a split epimorphism. In addition, the partial-product property guarantees $h \cdot \varepsilon = \tau$, which makes ε a monomorphism since τ is one. Hence ε is an isomorphism, and we may assume $\varepsilon = 1_Z$. Now it is easy to see that $\tau : Z \to P$ classifies $\mathcal{M}$-partial maps to Z.

For (iii) $\Longrightarrow$ (i) we prove that $f^* : \mathsf{C}/Z \to \mathsf{C}/X$ has a right adjoint (for every $f : X \to Z$ in C; see Proposition III.4.4.3) by factoring f as

$$f = (\, X \xrightarrow{\langle f, 1_X \rangle} Z \times X \xrightarrow{\pi_1} Z \,)$$

and showing that both π_1^* and $\langle f, 1_X\rangle^*$ have right adjoints. Since π_1^* makes the diagram

$$\begin{array}{ccc} \mathsf{C}/Z & \xrightarrow{\pi_1^*} & \mathsf{C}/Z \times X \\ {\scriptstyle \mathrm{dom}_Z}\downarrow & & \downarrow{\scriptstyle \mathrm{dom}_{Z\times X}} \\ \mathsf{C} & \xrightarrow{(-)\times X} & \mathsf{C} \end{array}$$

(with $\pi_1^*(W, u) = (W \times X, u \times 1_X)$) commute, and $(-) \times X$ is left adjoint by Cartesian closedness of C, the functor π_1^* is also left adjoint by Proposition III.4.4.3.

The split monomorphism $m := \langle f, 1_X\rangle : X \to Z \times X$ lies in $\mathcal{M}$ by our general hypothesis on $\mathcal{M}$, so the $\mathcal{M}$-partial morphism $(m, 1_X)$ is represented by $h : Z \times X \to X^*$. For (W, w) in C/X we construct the value $m_*(w)$, to obtain an adjunction $m^* \dashv m_* : \mathsf{C}/X \to \mathsf{C}/Z \times X$, as follows. For $g : W^* \to X^*$ representing the $\mathcal{M}$-partial morphism (t_W, w), one obtains $m_*(w)$ as a pullback of g along h, and the counit $\varepsilon_w : m^*(m_*(w)) \to w$ by the pullback property of $W = X \times_{X^*} W^*$:

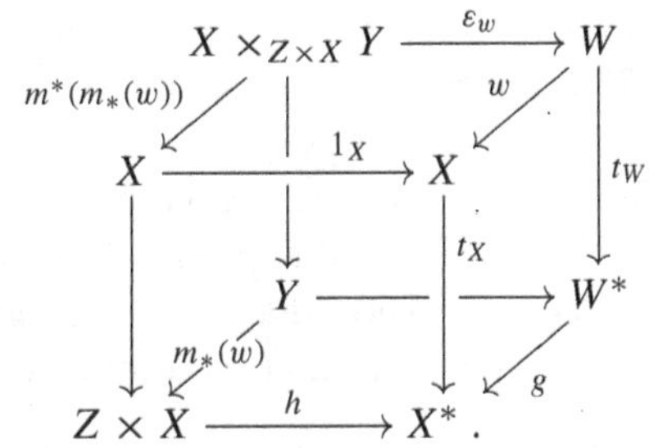

That ε_w assumes the alleged role as a counit is an easy diagrammatic exercise. □

III.4.7.8 Example Set satisfies the equivalent conditions of Theorem III.4.7.7. Ord is Cartesian closed and has a regular-subobject classifier, but not all regular-partial-map classifiers. The category Set_* of *pointed sets* and *pointed maps* has all partial-map classifiers, but fails to be Cartesian closed (Exercise III.4.K), where a pointed set (X, x_0) is a set X with $x_0 \in X$, and a pointed map $f : (X, x_0) \to (Y, y_0)$ is a map $f : X \to Y$ with $f(x_0) = y_0$. Top has a regular-subobject classifier, but none of the other properties of Theorem III.4.7.7, and Cat is Cartesian closed but not locally so, and does not have a regular-subobject classifier.

The following theorem gives in particular a necessary and sufficient condition for a topological category over Set to have regular-partial-map classifiers with underlying sets as in Set.

III.4.7.9 Theorem *Let $U : \mathsf{A} \to \mathsf{X}$ be a topological functor, with* X *finitely complete. The following assertions are equivalent:*

(i) A *has regular-partial-map classifiers and U preserves them;*
(ii) X *has regular-partial-map classifiers, and U-final sinks in* A *are stable under pullback along regular monomorphisms.*

Proof (ii) $\Longrightarrow$ (i): In order to construct a regular-partial-map classifier for B in A, one takes a regular-partial-map classifier $t_Z : Z \to Z^*$ for $Z = UB$ in X and considers the (possibly large) family of all regular partial maps

$$(m_i : M_i \to A_i,\ g_i : M_i \to B)_{i \in I} \qquad \text{(III.4.7.iii)}$$

in A with codomain B. Since a morphism in A is a regular monomorphism if and only if it is U-initial and its U-image is a regular monomorphism in X (Exercise II.5.D), for every $i \in I$ one obtains a unique morphism $h_i : UA_i \to Z^*$ in X which makes

$$\begin{array}{ccc} UM_i & \xrightarrow{Ug_i} & Z \\ {\scriptstyle Um_i}\downarrow & & \downarrow{\scriptstyle t_Z} \\ UA_i & \xrightarrow{h_i} & Z^* \end{array} \qquad \text{(III.4.7.iv)}$$

a pullback diagram. We let $(f_i : A_i \to B^*)_{i\in I}$ be a U-final lifting of $(h_i)_{i\in I}$. Since the family (III.4.7.iii) contains in particular $(1_B, 1_B)$, there is $j \in I$ with $m_j = g_j = 1_B$ and $h_j = t_Z$. We claim that $s_B := f_j : B \to B^*$ is a regular-partial-map classifier. Since $Us_B = t_Z$ is a regular mono, we must show that s_B is U-initial. Let $n : C \to B^*$ be a U-initial lifting of $t_Z : Z \to UB^*$. Then the diagrams (III.4.7.iv) with $h_i = Uf_i$ show that there are morphisms $\tilde{g}_i : M_i \to C$ with $U\tilde{g}_i = Ug_i$, by U-initiality of n, and the U-initiality of every m_i shows that all

$$\begin{array}{ccc} M_i & \xrightarrow{\tilde{g}_i} & C \\ {\scriptstyle m_i}\downarrow & & \downarrow{\scriptstyle n} \\ A_i & \xrightarrow{f_i} & B^* \end{array} \qquad \text{(III.4.7.v)}$$

are pullback diagrams in A. By hypothesis (ii), the sink $(\tilde{g}_i)_{i\in I}$ is U-final, so that $U\tilde{g}_i = Ug_i$ for all $i \in I$ yields $C \le B$ in $U^{-1}Z$, with $B \le C$ holding trivially by U-initiality of n. It follows that s_B is a regular monomorphism, and the diagrams (III.4.7.v) show that it has the required universal property in A. Furthermore, $Us_B = t_Z$ is a regular-partial-map classifier in X.

(i) $\Longrightarrow$ (ii): For the existence of regular-partial-map classifiers in X, see Exercise III.4.N. Now consider a family of pullback diagrams

$$\begin{array}{ccc} M_i & \xrightarrow{g_i} & N \\ {\scriptstyle m_i}\downarrow & & \downarrow{\scriptstyle n} \\ A_i & \xrightarrow{f_i} & B \end{array}$$

$(i \in I)$ in A, with n a regular monomorphism and $(f_i)_{i\in I}$ a U-final morphism, and consider $k : UN \to UC$ in X such that $k \cdot Ug_i = Uh_i$ with $h_i : M_i \to C$ in A for all $i \in I$. In order to confirm U-finality of $(g_i)_{i\in I}$, we must lift k to a morphism $N \to C$ in A. But with s_C a regular-partial-map classifier for C in A, for every $i \in I$ there is a unique morphism $t_i : A_i \to C^*$ in A exhibiting m_i as a pullback of s_C along t_i which restricts to h_i. Furthermore, since U preserves this classifier, there is a unique morphism $j : UB \to UC^*$ in X exhibiting Un as pullback of Us_C along j which restricts to k. The diagram

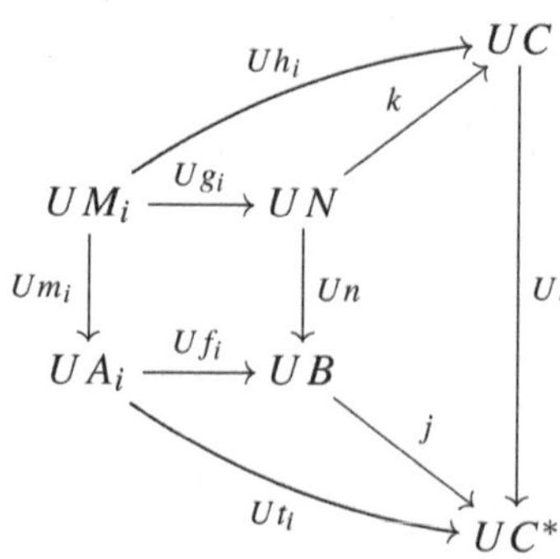

makes it obvious that j lifts to an A-morphism $B \longrightarrow C^*$ by U-finality of $(f_i)_{i\in I}$, and that k then lifts to an A-morphism $N \longrightarrow C$ by U-initiality of s_C. □

III.4.8 The quasitopos $(\mathbb{T}, \mathcal{V})$-Gph

A category C is a *quasitopos* if

- C *is finitely complete and finitely cocomplete;*
- C *is locally Cartesian closed;*
- C *has a regular-subobject classifier.*

A quasitopos is a *topos* if it has a subobject classifier. Hence, a topos is a quasitopos in which all monomorphisms are regular. Set and, more generally, every functor category $\mathsf{Set}^{\mathsf{D}^{\mathrm{op}}}$ (with D small) is a topos.

As a topological category over Set, the category $(\mathbb{T}, \mathcal{V})$-Gph is small-complete and small-cocomplete, and Theorem III.4.5.1 gives sufficient conditions on $\mathbb{T}$ and $\mathcal{V}$ for $(\mathbb{T}, \mathcal{V})$-Gph to be locally Cartesian closed. The existence of a regular-subobject classifier is easily established, not only in $(\mathbb{T}, \mathcal{V})$-Gph, but also in its relevant subcategories.

III.4.8.1 Proposition *The categories* $(\mathbb{T}, \mathcal{V})$-Gph, $(\mathbb{T}, \mathcal{V})$-RGph, $(\mathbb{T}, \mathcal{V})$-UGph, *and* $(\mathbb{T}, \mathcal{V})$-Cat *all have a regular-subobject classifier.*

Proof Simply put the indiscrete structure on the subobject classifier 2 of Set (see Exercise III.4.J). □

As a consequence, $(2, \top)$ is a regular subobject classifier in all four categories.

III.4.8.2 Corollary *The category* $(\mathbb{T}, \mathcal{V})$-Gph *is a quasitopos provided that $\mathcal{V}$ is Cartesian closed and integral, and T satisfies BC. If, in addition, $\hat{\mathbb{T}}$ is associative and $\hat{T}(f^\circ) = \hat{T}1_X \cdot (Tf)^\circ$ for all $f : X \longrightarrow Y$, then* $(\mathbb{T}, \mathcal{V})$-RGph *is also a quasitopos.*

Proof The statement follows immediately from the preceding discussion, with Theorems III.4.5.1 and III.4.6.7. □

Theorem III.4.7.9 applied to $O : (\mathbb{T}, \mathcal{V})\text{-}\mathsf{RGph} \longrightarrow \mathsf{Set}$ provides simpler conditions for $(\mathbb{T}, \mathcal{V})$-RGph to have regular-partial-map classifiers.

III.4.8.3 Proposition *For T taut, let $\hat{T}$ satisfy $\hat{T}t = Tt \cdot \hat{T}1_Z$ for every injective function $t : Y \longrightarrow Z$. Then* $(\mathbb{T}, \mathcal{V})$-RGph *has regular-partial-map classifiers preserved by O.*

Proof We must show that O-final sinks in $(\mathbb{T}, \mathcal{V})$-RGph are *hereditary*, i.e. stable under pullback along embeddings. Hence, consider a sink $(f_i : (X_i, a_i) \longrightarrow (Y, b))$ in $(\mathbb{T}, \mathcal{V})$-RGph with

$$b = 1_Y^\sharp \vee \bigvee_{i\in I} f_i \cdot a_i \cdot \hat{T}(f_i^\circ) = e_Y^\circ \cdot \hat{T}1_Y \vee \bigvee_{i\in I} f_i \cdot a_i \cdot (Tf_i)^\circ \cdot \hat{T}1_Y$$

(see Proposition III.4.1.4), an embedding $t : (Z, c) \hookrightarrow (Y, b)$ with $c = t^\circ \cdot b \cdot Tt$, and let $\tilde{f}_i : (f_i^{-1}(Z), d_i) \to (Z, c)$ be the restriction of f_i with $d_i = s_i^\circ \cdot a_i \cdot Ts_i$, $s_i \cdot f_i^{-1}(Z) \hookrightarrow X_i$. From $t^\circ \cdot e_Y^\circ = e_Z^\circ \cdot (Tt)^\circ$ one obtains $e_Z^\circ = t^\circ \cdot e_Y^\circ \cdot Tt$ since Tt is injective. With the hypothesis on T and $\hat{T}$,

$$\begin{aligned} c &= t^\circ \cdot e_Y^\circ \cdot \hat{T}1_Y \cdot Tt \vee \textstyle\bigvee_{i\in I} t^\circ \cdot f_i \cdot a_i \cdot (Tf_i)^\circ \cdot \hat{T}1_Y \cdot Tt \\ &= t^\circ \cdot e_Y^\circ \cdot Tt \cdot \hat{T}1_Z \vee \textstyle\bigvee_{i\in I} t^\circ \cdot f_i \cdot a_i \cdot (Tf_i)^\circ \cdot Tt \cdot \hat{T}1_Z \\ &= e_Z^\circ \cdot \hat{T}1_Z \vee \textstyle\bigvee_{i\in I} \tilde{f}_i \cdot s_i^\circ \cdot a_i \cdot Ts_i \cdot (T\tilde{f}_i)^\circ \cdot \hat{T}1_Z \\ &= 1_Z^\sharp \vee \textstyle\bigvee_{i\in I} \tilde{f}_i \cdot d_i \cdot \hat{T}(\tilde{f}_i^\circ)\,, \end{aligned}$$

so that $(\tilde{f}_i)_{i\in I}$ is O-final, as claimed. □

III.4.8.4 Remark Examining the proof of Theorem III.4.7.9, one sees that the hypothesis $\hat{T}t = Tt \cdot \hat{T}1_Z$ is used only for $t : Z \hookrightarrow Z^* = Z \cup \{\star\}$ with $\star \notin Z$. Since $\hat{T}t = \hat{T}1_{Z^*} \cdot Tt$ always holds, one can in fact show that it suffices that $\hat{T}$ satisfy

$$\mathfrak{x} \in TZ\,,\ \bot < \hat{T}1_{Z^*}(\mathfrak{x}, \mathfrak{y}) \implies \mathfrak{y} \in TZ \qquad \text{(III.4.8.i)}$$

for Proposition III.4.8.3 to hold true.

III.4.8.5 Examples

(1) The categories RRel, RNRel, PsTop, PsApp are quasitopoi (see Examples III.4.1.3 and III.4.5.3), but none of them is a topos (since monomorphisms need not be regular in these categories).

(2) Condition (III.4.8.i) fails for the lax extension $\hat{\mathbb{F}}$ of the monad $\mathbb{F}$, but it is satisfied for the lax extension $\check{\mathbb{F}}$ of Example III.1.10.3(4). Since $(\mathbb{F}, \check{\mathbb{F}}, 2)$-RGph is Cartesian closed (Example III.4.6.4), it is a quasitopos.

(3) PrTop $\cong$ $(\mathbb{F}, 2)$-UGph (see Example III.4.1.3(3)) has a regular-partial-map classifier which may be constructed via Theorem III.4.7.9 or Proposition III.4.8.3, but PrTop is not Cartesian closed (see [Herrlich, Colebunders, and Schwarz, 1991]).

III.4.9 Final density of $(\mathbb{T}, \mathcal{V})$-Cat in $(\mathbb{T}, \mathcal{V})$-Gph

Let us finally show that $(\mathbb{T}, \mathcal{V})$-Gph is "not too big" an extension of $(\mathbb{T}, \mathcal{V})$-Cat, in the sense that every $(\mathbb{T}, \mathcal{V})$-graph is the codomain of a small O-final sink with domains in $(\mathbb{T}, \mathcal{V})$-Cat, where $O : (\mathbb{T}, \mathcal{V})\text{-Gph} \to \mathsf{Set}$ is the forgetful functor. Here we assume that the functor T of the monad $\mathbb{T}$ *preserves disjointness*, so that $X \cap Y = \varnothing$ implies $TX \cap TY = \varnothing$; more precisely,

$$\begin{array}{ccc} \varnothing & \longrightarrow & TX \\ \big\downarrow & & \big\downarrow \\ TY & \longrightarrow & T(X+Y) \end{array}$$

is a pullback diagram for all sets X and Y. This is certainly the case when $T\varnothing = \varnothing$ and T is taut, or when T preserves binary coproducts.

We commence by constructing a $(\mathbb{T}, \mathcal{V})$-category structure c on the set

$$Z = X + 1$$

depending on a given set X and elements $\chi \in TX$ and $v \in \mathcal{V}$. Writing $1 = \{\star\}$ and using the disjointness hypothesis we define $c = c_{\chi,v}$ by

$$c(\mathfrak{z}, z) = \begin{cases} k & \text{if } \mathfrak{z} = e_Z(z), \\ v & \text{if } \mathfrak{z} = \chi \text{ and } z = \star, \\ \bot & \text{otherwise,} \end{cases}$$

for all $z \in Z$, $\mathfrak{z} \in TZ$. We observe that the $\mathcal{V}$-relation $c : TZ \nrightarrow Z$ *has finite fibers*, i.e.

$$c^\circ(z) = \{\mathfrak{z} \in TZ \mid \bot < c(\mathfrak{z}, z)\}$$

is finite for all $z \in Z$. We say that the lax natural transformation $e^\circ : \hat{T} \nrightarrow 1$ is *finitely strict* if

$$\begin{array}{ccc} TX & \xrightarrow{\hat{T}r} & TY \\ {\scriptstyle e_X^\circ}\downarrow & & \downarrow{\scriptstyle e_Y^\circ} \\ X & \xrightarrow{r} & Y \end{array}$$

commutes for every $\mathcal{V}$ relation r with finite fibers (i.e. for those $r : X \nrightarrow Y$ with $r^\circ(y) = \{x \in X \mid \bot < r(x, y)\}$ finite for all $y \in Y$).

III.4.9.1 Proposition *Suppose that T preserves disjointness, $\hat{T}$ is flat, and e° is finitely strict. Then*

$$(X + 1, c_{\chi,v})$$

is a $(\mathbb{T}, \mathcal{V})$-category for any set X and all $\chi \in TX$, $v \in \mathcal{V}$.

Proof As c is reflexive by definition, we must only show transitivity:

$$\hat{T}c(\mathfrak{Z}, \mathfrak{z}) \otimes c(\mathfrak{z}, z) \leq c(m_Z(\mathfrak{Z}), z)$$

for all $z \in Z = X + 1$, $\mathfrak{z} \in TZ$, $\mathfrak{Z} \in TTZ$. With $i : X \hookrightarrow Z$ denoting the inclusion, one trivially has $i^\circ \cdot c = i^\circ \cdot e_Z^\circ$, hence $(Ti)^\circ \cdot \hat{T}c = (Ti)^\circ \cdot (Te_Z)^\circ$ since $\hat{T}$ is flat. This means in pointwise terms

$$\hat{T}x(\mathfrak{Z}, \mathfrak{z}) = \begin{cases} k & \text{if } \mathfrak{Z} = Te_Z(\mathfrak{z}), \\ \bot & \text{otherwise,} \end{cases}$$

whenever $\mathfrak{z} \in TX$. In proving transitivity of c, we can disregard cases where $c(\mathfrak{z}, z) = \bot$ or $\hat{T}c(\mathfrak{Z}, \mathfrak{z}) = \bot$; hence, we are left to consider the following cases.

Case 1: $z \in X$, $z = e_Z(z)$, $\mathfrak{Z} = Te_Z(\mathfrak{z})$. Then all three terms appearing in the transitivity condition are equal to k, and the inequality is actually an equality.

Case 2: $z = \star$, $\mathfrak{z} = \chi$, $\mathfrak{Z} = Te_Z(\mathfrak{z})$. Then, similarly to the previous case, one obtains

$$\hat{T}c(\mathfrak{Z}, \mathfrak{z}) \otimes c(\mathfrak{z}, z) = k \otimes v = v = c(m_Z(\mathfrak{Z}), z) .$$

Case 3: $z = \star$, $\mathfrak{z} = e_Z(\star)$. Since e° is finitely strict, one has $e_Z^\circ \cdot \hat{T}c = c \cdot e_{TZ}^\circ$, so that

$$\hat{T}c(\mathfrak{Z}, \mathfrak{z}) = \bigvee_{w \in e_{TZ}^\circ(\mathfrak{Z})} c(w, \star) .$$

As this value is assumed not to be $\bot$, there is $w \in TZ$ with $\bot < c(w, \star)$ and $e_{TZ}(w) = \mathfrak{Z}$, and for every such w one has $c(m_Z(\mathfrak{Z}), z) = c(w, z)$. Consequently,

$$\hat{T}c(\mathfrak{Z}, \mathfrak{z}) \otimes c(\mathfrak{z}, z) = c(m_Z(\mathfrak{Z}, z)) \otimes k = c(m_Z(\mathfrak{Z}), z) ,$$

which concludes the last case. □

III.4.9.2 Theorem *Suppose that T preserves disjointness, $\hat{T}$ is flat, and e° is finitely strict. Then $(\mathbb{T}, \mathcal{V})$-*$\mathsf{Cat}$ *is finally dense in $(\mathbb{T}, \mathcal{V})$-*$\mathsf{Gph}$.

Proof For (X, a) a $(\mathbb{T}, \mathcal{V})$-graph, $x \in X$, and $\chi \in TX$, we claim that

$$f_{\chi,x} : (X + 1, c_{\chi,a(\chi,x)}) \to (X, a)$$

is a $(\mathbb{T}, \mathcal{V})$-functor, with the map $f_{\chi,x}$ induced by $1_X : X \to X$ and $x : 1 \to X$. Indeed, for $z \in Z = X + 1$ and $\mathfrak{z} \in TZ$ with $\bot < c(\mathfrak{z}, z)$ (where $c = c_{\chi,a(\chi,x)}$), one has $\mathfrak{z} = e_Z(z)$ or $\mathfrak{z} = \chi$, $z = \star \in 1$. In the former case,

$$c(\mathfrak{z}, z) = k \leq a(e_X(f_{\chi,x}(z)), f_{\chi,x}(z)) = a(Tf_{\chi,x}(\mathfrak{z}), f_{\chi,x}(z))$$

since a is reflexive, and in the latter case

$$c(\mathfrak{z}, z) = a(\chi, x) = a(Tf_{\chi,x}(\mathfrak{z}), f_{\chi,x}(z))$$

since $f_{\chi,x}|X = 1_X$. Furthermore, this last equation also shows

$$a(\chi, x) = (f_{\chi,x} \cdot c_{\chi,a(\chi,x)}) \cdot (Tf_{\chi,x})^\circ)(\chi, x) ,$$

so that the epi-sink $(f_{\chi,x})_{\chi \in TX, x \in X}$ must be O-final for the forgetful functor $O : (\mathbb{T}, \mathcal{V})$-$\mathsf{Gph} \to \mathsf{Set}$ (see Proposition III.4.1.4). □

Since the O-final sink just considered is small, we may state Theorem III.4.9.2 more strongly as follows.

III.4.9.3 Corollary *Suppose that T preserves disjointness, $\hat{T}$ is flat, and e° is finitely strict. Then every $(\mathbb{T}, \mathcal{V})$-graph is a quotient of a coproduct of $(\mathbb{T}, \mathcal{V})$-categories. If, moreover, $\hat{\mathbb{T}}$ is associative, then every $(\mathbb{T}, \mathcal{V})$-graph is a quotient of a $(\mathbb{T}, \mathcal{V})$-category.*

Proof The small O-final epi-sink used in the proof of Theorem III.4.9.2 factors through a coproduct of $(\mathbb{T}, \mathcal{V})$-categories followed by a quotient map. That coproduct is again a $(\mathbb{T}, \mathcal{V})$-category in case $\hat{\mathbb{T}}$ is associative and flat by Theorem III.4.3.3. □

This corollary can be applied to Examples III.4.1.3 with the following consequence.

III.4.9.4 Corollary *Every pseudotopological space is a quotient of a topological space, and every pseudo-approach space is a quotient of an approach space.*

Exercises

III.4.A *Initial sources in* $(\mathbb{T}, \mathcal{V})$-$\mathsf{RGph}$ *and* $(\mathbb{T}, \mathcal{V})$-$\mathsf{UGph}$. For the forgetful functor $O : (\mathbb{T}, \mathcal{V})\text{-}\mathsf{RGph} \to \mathsf{Set}$, O-initial sources are characterized as in Proposition III.3.1.1. Likewise for $(\mathbb{T}, \mathcal{V})\text{-}\mathsf{UGph} \to \mathsf{Set}$.

III.4.B *Sinks of nearly open morphisms.* Let $(f_i : (X_i, a_i) \to (Y, b))_{i \in I}$ be an epi-sink of morphisms in $(\mathbb{T}, \mathcal{V})$-$\mathsf{RGph}$, with all f_i nearly open for all $i \in I$. Then $(f_i)_{i \in I}$ is O-final, with $O : (\mathbb{T}, \mathcal{V})\text{-}\mathsf{RGph} \to \mathsf{Set}$.

III.4.C *Taut functors*

(1) The Set-functor T with $T\varnothing = \varnothing$, and $TX = 1$ if $X \neq \varnothing$, preserves monomorphisms but is not taut.
(2) The filter functor F is *weakly terminal* amongst all taut functors, i.e. for every taut functor $T : \mathsf{Set} \to \mathsf{Set}$, there is a natural transformation $\alpha : T \to F$.

Hint. Since T preserves monomorphisms, one may assume $TA \subset TX$ for $A \subseteq X$, and define $\alpha_X(\chi) = \{A \subseteq X \mid \chi \in TA\}$ for $\chi \in TX$.

III.4.D *Powerset graphs.* With respect to both lax extensions of the powerset monad $\check{\mathbb{P}}$ and $\hat{\mathbb{P}}$ (see Example III.1.4.2(2)), show

$$(\mathbb{P}, 2)\text{-}\mathsf{Gph} = (\mathbb{P}, 2)\text{-}\mathsf{RGph}$$

and identify $(\mathbb{P}, 2)$-UGph as a subcategory of Ord and Cls, respectively (see Examples III.1.6.4).

III.4.E *Extensive categories.* For a category C with small-indexed coproducts and pullbacks, the following statements are equivalent.

(i) C is extensive (i.e. coproducts are universal and disjoint);
(ii) coproducts are universal, and for all small families $(f_i : X_i \to Y_i)_{i \in I}$ the diagrams

$$\begin{array}{ccc} X_j & \longrightarrow & \coprod_{i\in I} X_i \\ {\scriptstyle f_j}\big\downarrow & & \big\downarrow{\scriptstyle \coprod_i f_i} \\ Y_j & \longrightarrow & \coprod_{i\in I} Y_i \end{array}$$

are pullback diagrams;

(iii) for all small families $(Y_i)_{i\in I}$ of C-objects, the functor

$$\textstyle\prod_{i\in I} \mathsf{C}/Y_i \to \mathsf{C}/\coprod_{i\in I} Y_i\,, \quad (f_i : X_i \to Y_i)_{i\in I} \mapsto \coprod_{i\in I} f_i$$

is an equivalence of categories.

III.4.F Ord *is not locally Cartesian closed.* Ord is Cartesian closed but not locally so: the partial product $P = (\{0,1\}, \{0,1\} \hookrightarrow \{0, \frac{1}{2}, 1\})$ does not exist in Ord.

III.4.G Top *is not Cartesian closed.*

(1) Consider the topological spaces X, Y, W, Z whose non-empty open sets are illustrated by the diagram

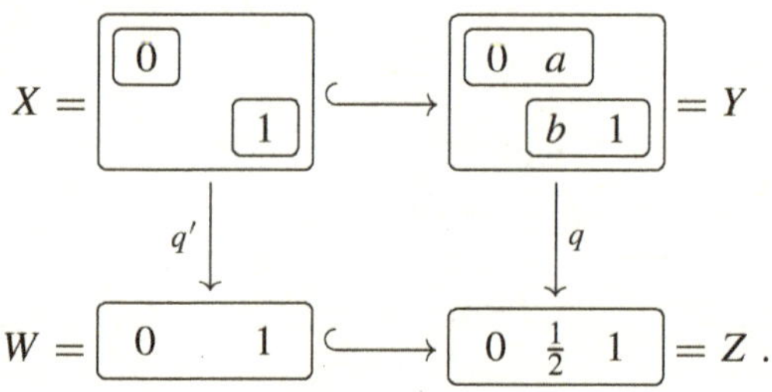

Conclude that quotient maps fail to be stable under pullback in Top, and that Top fails to be locally Cartesian closed.

(2) Find a topological space X and a quotient map $q : Y \to Z$ such that $q \times 1_X : Y \times X \to Z \times X$ fails to be a quotient map. Conclude that X is not exponentiable.

Hint. Consider $X = \mathbb{Q}$, the rational numbers, as a subspace of $\mathbb{R}$.

III.4.H *Slicing topological functors.* Consider a topological functor $U : \mathsf{A} \to \mathsf{X}$ and an A-object A. The induced functor $U_A : \mathsf{A}/A \to \mathsf{X}/UA$ is also topological, with the domain functor $\mathrm{dom}_A : \mathsf{A}/A \to \mathsf{A}$ sending U_A-final sinks to U-final sinks.

© **III.4.I** *Pseudotopological spaces via filter convergence.* With the help of the Axiom of Choice, condition (Ps) of Example III.4.1.3(3) can equivalently be expressed as

$$(\forall \chi \in \beta X\,(\chi \supseteq a \implies \chi \longrightarrow y)) \implies (a \longrightarrow y)$$

for all $a \in FX$, $y \in X$. Use this to show that the categories PsTop and $(\mathbb{F}, 2)$-$\mathsf{RGph}_{\mathsf{Ps}}$ of Example III.4.1.3(3) are isomorphic. Describe the reflector $\mathsf{PsTop} \to \mathsf{PrTop}$.

III.4.J *Lifting regular-subobject classifiers.* If I is right adjoint to the topological functor $U : \mathsf{A} \to \mathsf{X}$, and if Ω is a regular-subobject classifier of X, then $I\Omega$ is a regular-subobject classifier of A (see Example III.4.7.3(2) and Exercise II.5.D).

III.4.K *Partial-map classifiers.* The categories Ord and Top fail to have partial-map classifiers. The category Set_* of pointed sets has partial-map classifiers but fails to be Cartesian closed.

III.4.L *Pretopological spaces via filter convergence.* The multiplication and unit of the filter monad $\mathbb{F}$ satisfy

$$\sum \dot{\mathcal{A}} = \bigcap \mathcal{A} \qquad \text{and} \qquad (\mathcal{X} \supseteq \dot{\mathcal{A}}) \implies (\sum \mathcal{X} \supseteq \bigcap \mathcal{A})$$

for all $\mathcal{A} \subseteq FX$ and $\mathcal{X} \in FFX$. Hence, a $(\mathbb{F}, 2)$-graph (X, a) is unitary if and only if it is right unitary and

$$\nu(y) \longrightarrow y$$

holds for all $y \in X$, where $\nu(y) = \bigcap\{a \subset FX \mid a \longrightarrow y\}$. Deduce that unitary graphs on a set X are in bijective correspondence with maps $c : PX \to PX$ satisfying

$$c(\varnothing) = \varnothing\,, \qquad A \subseteq c(A)\,, \qquad c(A \cup B) = c(A) \cup c(B)\,,$$

for all $A, B \subseteq X$. The categories PrTop and $(\mathbb{F}, 2)$-UGph (see Example III.4.1.3(3)) are therefore isomorphic. Describe the reflector $\mathsf{PrTop} \to \mathsf{Top}$.

Hint. Consider the correspondence between maps $\nu : X \to PPX$ and $c : PX \to PX$ given by

$$c(A) = \{y \in X \mid B \in \nu(y) \implies A \cap B \neq \varnothing\} \qquad \text{and}$$
$$\nu(y) = \{B \in PX \mid y \in c(B^{\complement})^{\complement}\}\,,$$

where $B^{\complement} = X \setminus B$ denotes the complement of B in X.

III.4.M *Partial-map classifiers in* $\mathcal{V}$-Cat. If $k = \top$ in $\mathcal{V}$, then an object (Z, c) in $\mathcal{V}$-Cat may have a regular-partial-map classifier only when Z is indiscrete, i.e. when $c(x, y) = \top$ for all $x, y \in Z$. Hence, $\mathcal{V}$-Cat is not a quasitopos, even when $\otimes = \wedge$ in $\mathcal{V}$.

III.4.N *Inheriting regular-partial-map classifiers.* Let A be a finitely complete category with regular-partial-map classifiers $s_A : A \to A^*$.

(1) If B is a full, replete, and reflective subcategory of A with monic reflection morphisms $r_A : A \to RA$ and the reflector R preserving regular monomorphisms, then $r_{B^*} \cdot s_B$ is a regular-partial-map classifier for B in B.

(2) If $U : \mathsf{A} \to \mathsf{X}$ is topological, then X has regular-partial-map classifiers.

Hint. Recall Exercise II.5.D and apply (1) to the subcategory of indiscrete objects in A.

III.4.O Top *versus* App. There are full embeddings

$$\begin{array}{ccccc} \mathsf{Top} & \hookrightarrow & \mathsf{PrTop} & \hookrightarrow & \mathsf{PsTop} \\ \downarrow & & \downarrow & & \downarrow \\ \mathsf{App} & \hookrightarrow & \mathsf{PrApp} & \hookrightarrow & \mathsf{PsApp}\,, \end{array}$$

all of which are reflective, and the vertical ones are also coreflective (via $a(\mathfrak{x}, x) = 0$ if and only if $\mathfrak{x} \longrightarrow x$, as in Section III.3.6). Describe the reflectors and coreflectors.

III.5 Representable lax algebras

In this section we show that the Set-monad $\mathbb{T}$ with its lax extension $\hat{\mathbb{T}}$ to $\mathcal{V}$-Rel may be lifted, first to $\mathcal{V}$-Cat and then to $(\mathbb{T}, \mathcal{V})$-$\mathsf{Cat}$. The respective Eilenberg–Moore categories over $\mathcal{V}$-Cat and $(\mathbb{T}, \mathcal{V})$-$\mathsf{Cat}$ are topological over the Eilenberg–Moore category $\mathsf{Set}^{\mathbb{T}}$ and, somewhat surprisingly, turn out to be isomorphic. These categories lead us to interesting classes of so-called ordered and metric compact Hausdorff spaces, respectively.

III.5.1 The monad $\mathbb{T}$ on $\mathcal{V}$-Cat

We continue to work with a quantale $\mathcal{V}$ that is assumed to be non-trivial, as in Section III.1.2. As shown in Section III.3.3, every lax extension $\hat{\mathbb{T}} = (\hat{T}, m, e)$ of a monad $\mathbb{T} = (T, m, e)$ from Set to $\mathcal{V}$-Rel induces an order relation on TX, namely

$$\mathfrak{x} \le \mathfrak{y} \iff k \le \hat{T}1_X(\mathfrak{x}, \mathfrak{y})\,, \tag{III.5.1.i}$$

for $\mathfrak{x}, \mathfrak{y} \in TX$. In fact, the $\mathcal{V}$-relation $\hat{T}1_X : TX \nrightarrow TX$ makes TX into a $\mathcal{V}$-category $(TX, \hat{T}1_X)$ since

$$1_{TX} = T(1_X) \le \hat{T}1_X \qquad \text{and} \qquad \hat{T}1_X \cdot \hat{T}1_X \le \hat{T}(1_X \cdot 1_X) = \hat{T}1_X\,,$$

and its underlying order is given by (III.5.1.i). More generally, the same computation shows that, for every $\mathcal{V}$-category (X, a_0),

$$T(X, a_0) = (TX, \hat{T}a_0)\,, \tag{III.5.1.ii}$$

is a $\mathcal{V}$-category. This construction is the object part of a functor

$$T : \mathcal{V}\text{-}\mathsf{Cat} \to \mathcal{V}\text{-}\mathsf{Cat}$$

as Tf is a $\mathcal{V}$-functor $Tf : T(X, a_0) \to T(Y, b_0)$, for every $f : (X, a_0) \to (Y, b_0)$. Furthermore, oplaxness of m and e with respect to $\hat{T}$ imply that

$$m_X : TT(X, a_0) \to T(X, a_0) \qquad \text{and} \qquad e_X : (X, a_0) \to T(X, a_0)$$

are $\mathcal{V}$-functors, for every $\mathcal{V}$-category (X, a_0). Hence, the monad $\mathbb{T} = (T, m, e)$ on Set lifts to a monad on $\mathcal{V}$-Cat, which we denote by $\mathbb{T} = (T, m, e)$ again.

III.5.1.1 Proposition *A lax extension $\hat{\mathbb{T}}$ of a monad $\mathbb{T}$ on* Set *induces a 2-monad $\mathbb{T}$ on* $\mathcal{V}$-Cat *via* (III.5.1.ii)*, i.e. a monad $\mathbb{T} = (T, m, e)$ with T a 2-functor.*

Proof We are left to show that $T : \mathcal{V}\text{-}\mathsf{Cat} \to \mathcal{V}\text{-}\mathsf{Cat}$ is a 2-functor. To this end, let $f, g : (X, a_0) \to (Y, b_0)$ be $\mathcal{V}$-functors with $f \leq g$, i.e. $1_X \leq g^\circ \cdot b_0 \cdot f$. Then

$$1_{TX} \leq \hat{T} 1_X \leq \hat{T}(g^\circ \cdot b_0 \cdot f) = (Tg)^\circ \cdot \hat{T} b_0 \cdot Tf ,$$

i.e. $Tf \leq Tg$. □

III.5.1.2 Examples

(1) The identity monad on Set extended to the identity monad on $\mathcal{V}$-Rel yields the identity monad on $\mathcal{V}$-Cat.

(2) For the powerset monad $\mathbb{P} = (P, \{-\}, \bigcup)$ with its extensions $\check{\mathbb{P}}$, $\hat{\mathbb{P}}$, and $\overline{\mathbb{P}}$ to Rel (see Example III.1.5.2(2) and Section III.1.12), for an ordered set $(X, \leq)$ and $A, B \subseteq X$ we find

$$\begin{aligned} A\ (\check{P} \leq)\ B &\iff A \subseteq {\downarrow} B , \\ A\ (\hat{P} \leq)\ B &\iff B \subseteq {\uparrow} A , \\ A\ (\overline{P} \leq)\ B &\iff A \subseteq {\downarrow} B \ \&\ B \subseteq {\uparrow} A . \end{aligned}$$

Less formally, $A\ (\check{P} \leq)\ B$ whenever every element of A is covered by an element of B, $A\ (\hat{P} \leq)\ B$ whenever every element of B covers an element of A, and $A\ (\overline{P} \leq)\ B$ whenever both conditions are satisfied. Note that in all three cases the order relation on PX need not be antisymmetric even if the order on X is so. For instance, for $X = \mathbb{N}$ with its natural order $\leq$, and E and O the sets of even and odd numbers, respectively, one has

$$O\ (\check{P} \leq)\ E \qquad \text{and} \qquad E\ (\check{P} \leq)\ O .$$

(3) For the list monad $\mathbb{L} = (L, m, e)$ and its Barr extension $\overline{\mathbb{L}}$ to Rel, given by

$$(x_1, \ldots, x_n)\ \overline{L}r\ (y_1, \ldots, y_m) \iff n = m \ \&\ x_1\ r\ y_1 \ \&\ldots\ \&\ x_n\ r\ y_n$$

for all $(x_1, \ldots, x_n) \in LX$, $(y_1, \ldots, y_m) \in LY$, and relations $r : X \nrightarrow Y$ (see also V.1.4), one has

$$L(X, \leq) \cong \coprod_{n \in \mathbb{N}} (X, \leq)^n$$

for every ordered set $(X, \leq)$. Under this identification Lf corresponds to $\coprod_{n \in \mathbb{N}} f^n$.

(4) With the lax extension $\hat{\mathbb{P}}$ of $\mathbb{P}$ to P_+-Rel given in Exercise III.2.G, the powerset monad induces a 2-monad $\mathbb{P}$ on P_+-$\mathsf{Cat} \cong \mathsf{Met}$. In particular, for a metric space (X, d), the sets $(PX, \hat{P}d)$ become metric spaces with the (non-symmetric) *Hausdorff metric*

$$\hat{P}d(A, B) = \sup_{x \in A} \inf_{y \in B} d(x, y) \,,$$

for all $A, B \subseteq X$.

(5) We now consider the ultrafilter monad β on Set, together with the Barr-extension $\overline{\beta}$ to $\mathsf{Rel} \cong 2$-Rel (see Examples III.1.10.3). Then, for an ordered set $(X, \leq)$, the order relation on βX is given by

$$\mathfrak{x} \leq \mathfrak{y} \iff \forall A \in \mathfrak{x}, B \in \mathfrak{y} \ \exists x \in A, y \in B \ (x \leq y) \,,$$

for $\mathfrak{x}, \mathfrak{y} \in \beta X$.

Finally, we consider the quantale P_+ and the extension of the ultrafilter monad β to P_+-Rel described in Section III.2.4. Here, for a metric space (X, d) and ultrafilters $\mathfrak{x}, \mathfrak{y} \in \beta X$, the distance between $\mathfrak{x}$ and $\mathfrak{y}$ is

$$\sup_{A \in \mathfrak{x}, B \in \mathfrak{y}} \inf_{x \in A, y \in B} d(x, y) \,.$$

III.5.2 $\mathbb{T}$-algebras in $\mathcal{V}$-Cat

Let $\mathbb{T}$ be a monad lifted from Set to $\mathcal{V}$-Cat as in Section III.5.1. Following the notation of II.3.2, we denote the category of $\mathbb{T}$-algebras and $\mathbb{T}$-homomorphisms by

$$(\mathcal{V}\text{-}\mathsf{Cat})^{\mathbb{T}} \,.$$

We also recall that there is an adjunction

$$(\mathcal{V}\text{-}\mathsf{Cat})^{\mathbb{T}} \underset{F^{\mathbb{T}}}{\overset{G^{\mathbb{T}}}{\underset{\longleftarrow}{\overset{\longrightarrow}{\top}}}} \mathcal{V}\text{-}\mathsf{Cat} \,,$$

that $(\mathcal{V}\text{-}\mathsf{Cat})^{\mathbb{T}}$ is complete, and that $G^{\mathbb{T}} : (\mathcal{V}\text{-}\mathsf{Cat})^{\mathbb{T}} \to \mathcal{V}\text{-}\mathsf{Cat}$ preserves and creates limits.

An object of $(\mathcal{V}\text{-}\mathsf{Cat})^{\mathbb{T}}$ can be described as a triple (X, a_0, α), where (X, a_0) is a $\mathcal{V}$-category and $\alpha : TX \to X$ is simultaneously a $\mathbb{T}$-algebra structure on the set X and a $\mathcal{V}$-functor $\alpha : T(X, a_0) \to (X, a_0)$. For $\mathbb{T}$-algebras (X, a_0, α) and (Y, b_0, β), a map $f : X \to Y$ is a $\mathbb{T}$-homomorphism $f : (X, a_0, \alpha) \to (Y, b_0, \beta)$ if and only if $f : (X, a_0) \to (Y, b_0)$ is a $\mathcal{V}$-functor and $f : (X, \alpha) \to (Y, \beta)$ is a morphism of $\mathbb{T}$-algebras.

III.5.2.1 Examples

(1) For the monad $\mathbb{P}$ on $\mathsf{Ord} = \mathsf{2\text{-}Cat}$ obtained from the extension $\check{\mathbb{P}}$ of $\mathbb{P}$ to Rel, an Eilenberg–Moore algebra is an ordered set $(X, \leq)$ together with a monotone map $\alpha : PX \to X$ satisfying

$$\alpha(\{\alpha(A) \mid A \in \mathcal{A}\}) = \alpha(\bigcup \mathcal{A}) \qquad \text{and} \qquad \alpha(\{x\}) = x\,, \qquad \text{(III.5.2.i)}$$

for all $x \in X$ and $\mathcal{A} \subseteq PX$. We show that $\alpha(A)$ is a supremum of $A \subseteq X$ in $(X, \leq)$. Clearly, $\{x\} \leq A$ in $(PX, \leq)$ for any $x \in A$, so that $x \leq \alpha(A)$. Furthermore, for $y \in X$ with $x \leq y$ for all $x \in A$, one has $A \leq \{y\}$ and therefore $\alpha(A) \leq y$. Hence, $\mathsf{Ord}^{\mathbb{P}}$ can be equivalently described as the category of complete ordered sets with chosen suprema $\alpha : PX \to X$ satisfying (III.5.2.i), and a morphism is a monotone map which preserves these chosen suprema. Of course, suprema are unique if X is separated, and the full subcategory of $\mathsf{Ord}^{\mathbb{P}}$ defined by the separated complete ordered sets is precisely the category Sup (see II.2.1); moreover, Sup is a reflective subcategory of $\mathsf{Ord}^{\mathbb{P}}$.

Dually, if we consider the extension $\hat{\mathbb{P}}$ of $\mathbb{P}$ to Rel, the same reasoning applies with $\alpha : PX \to X$ now representing chosen infima of an order relation on X. Hence, the Eilenberg–Moore category $\mathsf{Ord}^{\mathbb{P}}$ has as objects complete ordered sets with chosen infima $\alpha : PX \to X$ satisfying (III.5.2.i), and as morphisms those monotone maps which preserve the chosen infima. Furthermore, the category Inf of separated complete ordered sets and inf-maps is a full reflective subcategory of $\mathsf{Ord}^{\mathbb{P}}$.

(2) For the list monad $\mathbb{L} = (L, m, e)$ with its Barr extension to Rel (see Example III.5.1.2(3)), the category $\mathsf{Ord}^{\mathbb{L}}$ is isomorphic to the category of *ordered monoids*, i.e. of monoids in the Cartesian closed category Ord.

(3) An object in Ord^{β} (here we consider the Barr extension $\overline{\beta}$ of β to Rel) is a triple $(X, \leq, \alpha)$, where $(X, \leq)$ is an ordered set and $\alpha : \beta X \to X$ is the convergence relation of a compact Hausdorff topology on X (see Section III.2.3); moreover, $\alpha : \beta(X, \leq) \to (X, \leq)$ is monotone. We write $R \subseteq X \times X$ for the graph of the order relation $\leq$ and $\pi_1 : R \to X$, $\pi_2 : R \to X$ for the projection maps. Recall from Section III.1.10 that, for $\mathfrak{x}, \mathfrak{y} \in \beta X$,

$$\mathfrak{x}\ (\overline{\beta}(\leq))\ \mathfrak{y} \iff \exists\, \mathfrak{w} \in \beta R\ (\beta\pi_1(\mathfrak{w}) = \mathfrak{x}\ \&\ \beta\pi_2(\mathfrak{w}) = \mathfrak{y})\,.$$

Therefore, $\alpha : \beta X \to X$ is monotone if and only if $(\alpha(\beta\pi_1(\mathfrak{w})), \alpha(\beta\pi_2(\mathfrak{w}))) \in R$ for every ultrafilter $\mathfrak{w} \in \beta R$; i.e. if and only if R is a closed subset of $X \times X$ with respect to the product topology.

Hence, an Eilenberg–Moore algebra for the monad β on Ord can be equivalently described as a compact Hausdorff space X equipped with an order relation whose graph is a closed subspace of $X \times X$; in other words, X

is an "internal order" in $\mathsf{Set}^{\mathbb{\beta}}$. These spaces are known as *ordered compact Hausdorff spaces*. We write

$$\mathsf{OrdCompHaus}$$

for the category of ordered compact Hausdorff spaces and morphisms. Every finite ordered set is an ordered compact Hausdorff space with the discrete topology. In particular, the ordered set $\mathsf{2} = \{0 \leq 1\}$ can be seen as an ordered compact Hausdorff space.

© (4) Considering the extension $\overline{\beta}$ of β to P_{+}-Rel of Section III.2.4, one obtains a monad $\mathbb{\beta}$ on $\mathsf{Met} = \mathsf{P}_{+}$-$\mathsf{Cat}$. The objects of $\mathsf{Met}^{\mathbb{\beta}}$ are triples (X, d, α), where (X, d) is a metric space and α is the convergence relation of a compact Hausdorff topology on X such that $\alpha : \beta(X, d) \to (X, d)$ is non-expansive. We call these spaces *metric compact Hausdorff spaces*, and denote the category of metric compact Hausdorff spaces and their morphisms by

$$\mathsf{MetCompHaus}\,.$$

The metric space $[0, \infty]$ with distance $\mu(u, v) = v \ominus u$ becomes a metric compact Hausdorff space with $\xi : \beta[0, \infty] \to [0, \infty]$ sending u to $\sup_{A \in u} \inf_{u \in A} u$ (see Exercise III.5.J).

Besides sitting over $\mathcal{V}$-Cat, the category $(\mathcal{V}\text{-}\mathsf{Cat})^{\mathbb{T}}$ also admits a forgetful functor

$$\tilde{O}\colon (\mathcal{V}\text{-}\mathsf{Cat})^{\mathbb{T}} \to \mathsf{Set}^{\mathbb{T}},\ (f\colon (X, a_0, \alpha) \to (Y, b_0, \beta)) \mapsto (f\colon (X, \alpha) \to (Y, \beta))$$

to $\mathsf{Set}^{\mathbb{T}}$, which can be seen as a lifting of the forgetful functor $O : \mathcal{V}\text{-}\mathsf{Cat} \to \mathsf{Set}$. Recall from Example II.5.6.1(3) that every mono-source in $\mathsf{Set}^{\mathbb{T}}$ is initial with respect to the forgetful functor $G^{\mathbb{T}} : \mathsf{Set}^{\mathbb{T}} \to \mathsf{Set}$, and therefore a mono-source $(f_i : (X, a_0, \alpha) \to (X_i, a_{0i}, \alpha_i))_{i \in I}$ in $(\mathcal{V}\text{-}\mathsf{Cat})^{\mathbb{T}}$ is initial with respect to the forgetful functor $(\mathcal{V}\text{-}\mathsf{Cat})^{\mathbb{T}} \to \mathsf{Set}$ if and only if $(f_i : (X, a_0) \to (X_i, a_{0i}))_{i \in I}$ is initial with respect to the forgetful functor $O : \mathcal{V}\text{-}\mathsf{Cat} \to \mathsf{Set}$.

III.5.2.2 Proposition *The functor $\tilde{O} : (\mathcal{V}\text{-}\mathsf{Cat})^{\mathbb{T}} \to \mathsf{Set}^{\mathbb{T}}$ is topological.*

Proof For a family $(f_i : (X, \alpha) \to (X_i, \alpha_i))_{i \in I}$ of morphisms of $\mathbb{T}$-algebras where (X_i, a_{0i}, α_i) is in $(\mathcal{V}\text{-}\mathsf{Cat})^{\mathbb{T}}$, let (X, a_0) be the initial lift of $(f : X \to X_i)_{i \in I}$ in $\mathcal{V}$-Cat (see Theorem III.3.1.3). Then (X, a_0, α) is indeed an object of $(\mathcal{V}\text{-}\mathsf{Cat})^{\mathbb{T}}$, since

$$\begin{aligned}
\alpha \cdot \hat{T} a_0 &= \alpha \cdot \hat{T}(\textstyle\bigwedge_{i \in I} f_i^{\circ} \cdot a_{0i} \cdot f_i) \\
&\leq \textstyle\bigwedge_{i \in I} (\alpha \cdot \hat{T}(f_i^{\circ} \cdot a_{0i} \cdot f_i)) \\
&= \textstyle\bigwedge_{i \in I} (\alpha \cdot (T f_i)^{\circ} \cdot \hat{T} a_{0i} \cdot T f_i)
\end{aligned}$$

$$\begin{aligned}
&\leq \bigwedge_{i\in I}(f_i^\circ \cdot \alpha_i \cdot \hat{T}a_{0i} \cdot Tf_i) \\
&\leq \bigwedge_{i\in I}(f_i^\circ \cdot a_{0i} \cdot f_i \cdot \alpha) \\
&= (\bigwedge_{i\in I} f_i^\circ \cdot a_{0i} \cdot f_i) \cdot \alpha = a_0 \cdot \alpha\,,
\end{aligned}$$

with the penultimate equality holding by right adjointness of $(-)\cdot\alpha$ (recall that $\alpha \dashv \alpha^\circ$ in $\mathcal{V}$-Rel). □

As a consequence, the functor $\tilde{O} : (\mathcal{V}\text{-Cat})^{\mathbb{T}} \to \mathsf{Set}^{\mathbb{T}}$ has both a right adjoint and a left adjoint given by the $\tilde{O}$-initial liftings of the empty family and the "all-family," respectively (see Theorem II.5.9.1). In summary, in the commutative diagram

$$\begin{array}{ccc}
(\mathcal{V}\text{-Cat})^{\mathbb{T}} & \xrightarrow{G^{\mathbb{T}}} & \mathcal{V}\text{-Cat} \\
{\scriptstyle \tilde{o}}\downarrow & & \downarrow{\scriptstyle o} \\
\mathsf{Set}^{\mathbb{T}} & \xrightarrow{G^{\mathbb{T}}} & \mathsf{Set}
\end{array}$$

the vertical arrows are topological and therefore have left and right adjoints; the horizontal arrows are monadic and therefore have left adjoints.

For a flat extension $\hat{\mathbb{T}}$ of $\mathbb{T}$, the left adjoint $(-)_d : \mathsf{Set}^{\mathbb{T}} \to (\mathcal{V}\text{-Cat})^{\mathbb{T}}$ has a simple description as $(X,\alpha)_d = (X, 1_X, \alpha)$. If, moreover, $\mathcal{V}$ is integral so that $k = \top$ is the top element of $\mathcal{V}$, then $(-)_d : \mathsf{Set}^{\mathbb{T}} \to (\mathcal{V}\text{-Cat})^{\mathbb{T}}$ preserves limits and therefore has a left adjoint $\pi_0 : (\mathcal{V}\text{-Cat})^{\mathbb{T}} \to \mathsf{Set}^{\mathbb{T}}$ (see Exercise II.3.K and also Exercises III.5.A and III.5.B).

III.5.3 Comparison with lax algebras

The two structures $a_0 : X \nrightarrow X$ and $\alpha : TX \to X$ of an object (X, a_0, α) in $(\mathcal{V}\text{-Cat})^{\mathbb{T}}$ can be combined to yield a single structure $a = a_0 \cdot \alpha : TX \nrightarrow X$ turning X into a $(\mathbb{T},\mathcal{V})$*-category* (X, a). In fact, one has

$$a \cdot e_X = a_0 \cdot \alpha \cdot e_X = a_0 \geq 1_X$$

and

$$a\cdot\hat{T}a = a_0\cdot\alpha\cdot\hat{T}(a_0\cdot\alpha) = a_0\cdot\alpha\cdot\hat{T}(a_0)\cdot T(\alpha) \leq a_0\cdot a_0\cdot\alpha\cdot T(\alpha) = a_0\cdot\alpha\cdot m_X = a\cdot m_X\,.$$

Note also that the underlying $\mathcal{V}$-category of (X,a) is (X,a_0). Furthermore, every morphism $f : (X, a_0, \alpha) \to (Y, b_0, \beta)$ in $(\mathcal{V}\text{-Cat})^{\mathbb{T}}$ becomes a $(\mathbb{T},\mathcal{V})$-functor $f : (X,a) \to (Y,b)$ (where $b = \beta \cdot b_0$). More generally, one has the following result.

III.5.3.1 Lemma *Let (X, a_0, α), (Y, b_0, β) be objects in $(\mathcal{V}\text{-Cat})^{\mathbb{T}}$ with corresponding $(\mathbb{T},\mathcal{V})$-categories (X,a) and (Y,b), and let $f : (X,a_0) \to (Y,b_0)$ be a $\mathcal{V}$-functor. Then f is a $(\mathbb{T},\mathcal{V})$-functor $f : (X,a) \to (Y,b)$ if and only if*

*$\beta \cdot Tf \leq f \cdot \alpha$ in $\mathcal{V}$-*Cat*, i.e. $\beta \cdot Tf(\chi) \leq f \cdot \alpha(\chi)$ in the $\mathcal{V}$-category (Y, b_0), for all $\chi \in TX$.*

Proof First note that $\beta \cdot Tf \leq f \cdot \alpha$ in $\mathcal{V}$-Cat is equivalent to $b_0 \cdot f \cdot \alpha \leq b_0 \cdot \beta \cdot Tf$ in $\mathcal{V}$-Rel. Since $f \cdot a_0 \leq b_0 \cdot f$, one obtains $f \cdot a = f \cdot a_0 \cdot \alpha \leq b_0 \cdot f \cdot \alpha \leq b_0 \cdot \beta \cdot Tf = b \cdot Tf$. Conversely, assuming $f \cdot a \leq b \cdot Tf$, one derives $f \cdot \alpha \leq b_0 \cdot \beta \cdot Tf$, and therefore $b_0 \cdot f \cdot \alpha \leq b_0 \cdot b_0 \cdot \beta \cdot Tf \leq b_0 \cdot \beta \cdot Tf$. □

The previous constructions define a functor

$$K : (\mathcal{V}\text{-Cat})^{\mathbb{T}} \to (\mathbb{T}, \mathcal{V})\text{-Cat}, \quad (X, a_0, \alpha) \mapsto (X, a_0 \cdot \alpha).$$

It is worth noting that K is a 2-functor.

III.5.3.2 Example We apply the functor K to Examples III.5.2.1(3) and (4). The ordered compact Hausdorff space $\mathsf{2}$ gives rise to the *Sierpiński space* $\mathsf{2} = \{0, 1\}$ having $\{0\}$ as its only non-trivial open subset. The metric compact Hausdorff space $([0, \infty], \mu, \xi)$ (see Example III.5.2.1(4)) induces the approach space $[0, \infty]$ with convergence P_+-relation $\lambda : \beta[0, \infty] \nrightarrow [0, \infty]$ defined by

$$\lambda(u, u) = u \ominus (\sup_{A \in u} \inf_{v \in A} v).$$

III.5.3.3 Proposition *The functor $K : (\mathcal{V}\text{-Cat})^{\mathbb{T}} \to (\mathbb{T}, \mathcal{V})\text{-Cat}$ sends initial sources with respect to $\tilde{O} : (\mathcal{V}\text{-Cat})^{\mathbb{T}} \to \mathsf{Set}^{\mathbb{T}}$ to initial sources with respect to $O : (\mathbb{T}, \mathcal{V})\text{-Cat} \to \mathsf{Set}$.*

Proof Assume that $(f_i : (X, a_0, \alpha) \to (X_i, a_{0i}, \alpha_i))_{i \in I}$ in $(\mathcal{V}\text{-Cat})^{\mathbb{T}}$ is initial with respect to $\tilde{O} : (\mathcal{V}\text{-Cat})^{\mathbb{T}} \to \mathsf{Set}^{\mathbb{T}}$. Then

$$\begin{aligned} a = a_0 \cdot \alpha &= (\textstyle\bigwedge_{i \in I} f_i^\circ \cdot a_{0i} \cdot f_i) \cdot \alpha \\ &= \textstyle\bigwedge_{i \in I} (f_i^\circ \cdot a_{0i} \cdot f_i \cdot \alpha) \\ &= \textstyle\bigwedge_{i \in I} (f_i^\circ \cdot a_{0i} \cdot \alpha_i \cdot Tf_i) = \bigwedge_{i \in I} (f_i^\circ \cdot a_i \cdot Tf_i). \end{aligned}$$ □

Consequently, applying the Taut Lift Theorem II.5.11.1 to the diagram

$$\begin{array}{ccc} (\mathcal{V}\text{-Cat})^{\mathbb{T}} & \xrightarrow{K} & (\mathbb{T}, \mathcal{V})\text{-Cat} \\ {\scriptstyle \tilde{O}} \downarrow & & \downarrow {\scriptstyle O} \\ \mathsf{Set}^{\mathbb{T}} & \xrightarrow[G^{\mathbb{T}}]{} & \mathsf{Set} \end{array}$$

we obtain the following corollary.

III.5.3.4 Corollary *The functor $K : (\mathcal{V}\text{-Cat})^{\mathbb{T}} \to (\mathbb{T}, \mathcal{V})\text{-Cat}$ has a left adjoint $M : (\mathbb{T}, \mathcal{V})\text{-Cat} \to (\mathcal{V}\text{-Cat})^{\mathbb{T}}$. In particular, K preserves limits.*

Analyzing the proof of the Taut Lift Theorem II.5.11.1, we find that the underlying $\mathsf{Set}^{\mathbb{T}}$-object of $M(X, a)$ can be chosen as (TX, m_X), for a $(\mathbb{T}, \mathcal{V})$-category $X = (X, a)$, and then $Mf = Tf$ for a $(\mathbb{T}, \mathcal{V})$-functor f. The unit $X \to KM(X)$

is given by $e_X : X \to TX$ and, for every $Y = (Y, b_0, \beta)$ in $(\mathcal{V}\text{-}\mathsf{Cat})^{\mathbb{T}}$, the counit $MK(Y) \to Y$ is given by $\beta : TY \to Y$. We will now give an explicit description of the $\mathcal{V}$-category structure of $M(X, a)$, under some conditions on the lax extension $\hat{T}$.

III.5.3.5 Theorem *Assume that the lax extension $\hat{\mathbb{T}}$ of $\mathbb{T}$ is associative. Then the left adjoint of $K : (\mathcal{V}\text{-}\mathsf{Cat})^{\mathbb{T}} \to (\mathbb{T}, \mathcal{V})\text{-}\mathsf{Cat}$ is given by*

$$M : (\mathbb{T}, \mathcal{V})\text{-}\mathsf{Cat} \to (\mathcal{V}\text{-}\mathsf{Cat})^{\mathbb{T}}, \quad (X, a) \mapsto (TX, \hat{T}a \cdot m_X^\circ, m_X), \quad f \mapsto Tf .$$

Proof According to Theorem II.5.11.1 and Proposition III.5.2.2, one has $M(X, a) = (TX, \hat{a}, m_X)$, where $\hat{a}$ is the initial $\mathcal{V}$-category structure on TX with respect to the family of maps

$$TX \xrightarrow{Tf} TY \xrightarrow{\beta} (Y, b_0)$$

running over all $(\mathbb{T}, \mathcal{V})$-functors $f : (X, a) \to (Y, b_0 \cdot \beta)$ with (Y, b_0, β) in $(\mathcal{V}\text{-}\mathsf{Cat})^{\mathbb{T}}$. We show that

$$\hat{a} = \hat{T}a \cdot m_X^\circ .$$

Let (X, a) be a $(\mathbb{T}, \mathcal{V})$-category. First, $(TX, \hat{T}a \cdot m_X^\circ)$ is a $\mathcal{V}$-category since

$$\hat{T}a \cdot m_X^\circ \geq \hat{T}e_X^\circ \cdot m_X^\circ \geq 1_{TX}$$

and

$$\begin{aligned} \hat{T}a \cdot m_X^\circ \cdot \hat{T}a \cdot m_X^\circ &= \hat{T}a \cdot \hat{T}\hat{T}a \cdot m_{TX}^\circ \cdot m_X^\circ \\ &= \hat{T}a \cdot \hat{T}\hat{T}a \cdot (Tm_X)^\circ \cdot m_X^\circ \\ &\leq \hat{T}a \cdot \hat{T}\hat{T}a \cdot \hat{T}(m_X^\circ) \cdot m_X^\circ \\ &\leq \hat{T}(a \cdot \hat{T}a \cdot m_X^\circ) \cdot m_X^\circ = \hat{T}a \cdot m_X^\circ . \end{aligned}$$

Furthermore,

$$\begin{aligned} m_X \cdot \hat{T}(\hat{T}a \cdot m_X^\circ) &\leq m_X \cdot \hat{T}\hat{T}a \cdot Tm_X^\circ \cdot \hat{T}\hat{T}1_X \\ &\leq \hat{T}a \cdot m_{TX} \cdot Tm_X^\circ \cdot \hat{T}\hat{T}1_X \\ &\leq \hat{T}a \cdot m_X^\circ \cdot m_X \cdot \hat{T}\hat{T}1_X \\ &\leq \hat{T}(a \cdot \hat{T}1_X) \cdot m_X^\circ \cdot m_X = \hat{T}a \cdot m_X^\circ \cdot m_X \end{aligned}$$

since $m_X \cdot m_{TX} = m_X \cdot Tm_X$ implies $m_{TX} \cdot Tm_X^\circ \leq m_X^\circ \cdot m_X$; therefore $(TX, \hat{T}a \cdot m_X^\circ, m_X)$ is indeed an object of $(\mathcal{V}\text{-}\mathsf{Cat})^{\mathbb{T}}$. The map $e_X : X \to TX$ is actually a $(\mathbb{T}, \mathcal{V})$-functor

$$e_X : (X, a) \to (TX, \hat{T}a \cdot m_X^\circ \cdot m_X) ,$$

since $\hat{T}a \cdot m_X^\circ \cdot m_X \cdot Te_X = \hat{T}a \cdot m_X^\circ \geq \hat{T}a \cdot e_{TX} \geq e_X \cdot a$; hence, the identity map $1_{TX} = m_X \cdot Te_X$ on TX is a $\mathcal{V}$-functor $(TX, \hat{a}) \to (TX, \hat{T}a \cdot m_X^\circ)$, so that

$\hat{a} \leq \hat{T}a \cdot m_X^\circ$. Secondly, for any $(\mathbb{T}, \mathcal{V})$-functor $f : (X, a) \to (Y, b_0 \cdot \beta)$ with (Y, b_0, β) in $(\mathcal{V}\text{-}\mathsf{Cat})^{\mathbb{T}}$, we find that

$$\begin{aligned}\beta \cdot Tf \cdot \hat{T}a \cdot m_X^\circ &\leq \beta \cdot \hat{T}(f \cdot a) \cdot m_X^\circ \\ &\leq \beta \cdot \hat{T}b_0 \cdot T\beta \cdot TTf \cdot m_X^\circ \\ &\leq b_0 \cdot \beta \cdot T\beta \cdot TTf \cdot m_X^\circ \\ &= b_0 \cdot \beta \cdot m_Y \cdot TTf \cdot m_X^\circ \\ &= b_0 \cdot \beta \cdot Tf \cdot m_X \cdot m_X^\circ \leq b_0 \cdot \beta \cdot Tf \,.\end{aligned}$$

Consequently, $\beta \cdot Tf : (TX, \hat{T}a \cdot m_X^\circ) \to (Y, b_0)$ is a $\mathcal{V}$-functor, and we conclude that $\hat{a} \geq \hat{T}a \cdot m_X^\circ$. □

III.5.3.6 Remark Under the conditions of Theorem III.5.3.5, the unit $e_X : (X, a) \to KM(X, a)$ of $M \dashv K$ at $(X, a) \in (\mathbb{T}, \mathcal{V})\text{-}\mathsf{Cat}$ is even O-initial (for the forgetful functor $O : (\mathbb{T}, \mathcal{V})\text{-}\mathsf{Cat} \to \mathsf{Set}$) since a is left unitary:

$$e_X^\circ \cdot \hat{T}a \cdot m_X^\circ \cdot m_X \cdot Te_X = e_X^\circ \cdot \hat{T}a \cdot m_X^\circ = e_X^\circ \circ a = a \,.$$

The same computation shows that the $(\mathbb{T}, \mathcal{V})$-structure a on X may be recovered from $\hat{a} = \hat{T}a \cdot m_X^\circ$:

$$e_X^\circ \cdot \hat{a} = e_X^\circ \cdot \hat{T}a \cdot m_X^\circ = a \,.$$

III.5.3.7 Examples

© (1) For the Barr extension $\overline{\beta}$ of the ultrafilter monad β to Rel and a topological space X with convergence relation $a : \beta X \rightarrow\!\!\!\!\!\rightarrow X$ (see Section III.2.2), the order $(\leq) = (\overline{\beta}a \cdot m_X^\circ)$ on βX is given by

$$\begin{aligned}\mathfrak{x} \leq \mathfrak{x}' &\iff \text{every closed set } A \in \mathfrak{x} \text{ belongs to } \mathfrak{x}' \\ &\iff \text{every open set } A \in \mathfrak{x}' \text{ belongs to } \mathfrak{x} \,.\end{aligned}$$

In fact, $\mathfrak{x} \leq \mathfrak{x}'$ if and only if there is some $\mathfrak{X} \in \beta\beta X$ with $m_X(\mathfrak{X}) = \mathfrak{x}$ and $\mathfrak{X}\ (\overline{\beta}a)\ \mathfrak{x}'$, i.e.

$$\forall A \in \mathfrak{x}\ (A^{\beta} \in \mathfrak{X}) \qquad \text{and} \qquad \forall \mathcal{A} \in \mathfrak{X}\ (a(\mathcal{A}) \in \mathfrak{x}') \,,$$

where $A^{\beta} = \{\mathfrak{a} \in \beta X \mid A \in \mathfrak{a}\}$ (see Example III.1.10.3(3)). Since $a(A^{\beta}) = \overline{A}$ is the closure of A, from $\mathfrak{x} \leq \mathfrak{x}'$ it follows that every closed set $A \in \mathfrak{x}$ belongs to $\mathfrak{x}'$. Conversely, this condition guarantees that the filter base

$$\{A^{\beta} \mid A \in \mathfrak{x}\}$$

is disjoint from the ideal

$$\{\mathcal{B} \subseteq \beta X \mid a[\mathcal{B}] \notin \mathfrak{x}'\} \,,$$

so that, by Corollary II.1.13.5, there is an ultrafilter $\mathfrak{X} \in \beta\beta X$ with $m_X(\mathfrak{X}) = \mathfrak{x}$ and $\mathfrak{X}\ (\overline{\beta}a)\ \mathfrak{x}'$.

(2) Similarly, for the Barr extension $\overline{\beta}$ of β to P_+-Rel and an approach space X with convergence P_+-relation $a : \beta X \nrightarrow X$ and corresponding approach distance $\delta : X \nrightarrow PX$ (see Section III.2.4), the metric $\hat{a} = \overline{\beta}a \cdot m_X^\circ$ on βX can be written in terms of the approach distance δ as

$$\hat{a}(\chi, \chi') = \inf\{u \in [0, \infty] \mid \forall A \in \chi \; (A^{(u)} \in \chi')\} .$$

To see this, let

$$v := \hat{a}(\chi, \chi') = \inf\{\overline{\beta}a(\mathcal{X}, \chi') \mid \mathcal{X} \in \beta\beta X \; (m_X(\mathcal{X}) = \chi)\}$$

and

$$w := \inf\{u \in [0, \infty] \mid \forall A \in \chi \; (A^{(u)} \in \chi')\} .$$

Since

$$v \geq \sup_{\substack{A\in\chi \\ B\in\chi'}} \inf_{\substack{a\in A^{\beta} \\ y\in B}} a(a, y) ,$$

for every $\varepsilon > 0$, $A \in \chi$, and $B \in \chi'$, there exist $a \in A^{\beta}$ and $y \in B$ with $a(a, y) \leq v + \varepsilon$; hence, $\delta(y, A) \leq v + \varepsilon$. Therefore, $A^{(v+\varepsilon)} \cap B \neq \varnothing$, and we conclude that $A^{(v+\varepsilon)} \in \chi'$ for every $A \in \chi$ and $\varepsilon > 0$. This proves $w \leq v$. For the reverse inequality, note that $A^{(w+\varepsilon)} \cap B \neq \varnothing$, for every $\varepsilon > 0$, $A \in \chi$ and $B \in \chi'$; this implies that

$$\sup_{\substack{A\in\chi \\ B\in\chi'}} \inf_{\substack{a\in A^{\beta} \\ y\in B}} a(a, y) \leq w + \varepsilon .$$

Hence, by Lemma III.2.4.2, there is some $\mathcal{X} \in \beta\beta X$ with

$$\{A^{\beta} \mid A \in \chi\} \subseteq \mathcal{X} \qquad \text{and} \qquad \overline{\beta}a(\mathcal{X}, \chi') \leq w + \varepsilon ,$$

so that $v \leq w$.

III.5.4 The monad $\mathbb{T}$ on $(\mathbb{T}, \mathcal{V})$-Cat

In order to be able to apply Theorem III.5.3.5, throughout this section we assume that

- *the lax extension $\hat{\mathbb{T}}$ of $\mathbb{T}$ is associative.*

The adjunction $M \dashv K$ of Section III.5.3 induces a monad on $(\mathbb{T}, \mathcal{V})$-Cat, whose functor sends a $(\mathbb{T}, \mathcal{V})$-category (X, a) to $(TX, \hat{T}a \cdot m_X^\circ \cdot m_X)$ and a $(\mathbb{T}, \mathcal{V})$-functor f to Tf. Since the multiplication and unit are given by m and e, respectively, this monad constitutes a lifting of the Set-monad $\mathbb{T}$ to $(\mathbb{T}, \mathcal{V})$-Cat. We therefore denote this monad again by $\mathbb{T} = (T, m, e)$.

The induced comparison functor is denoted by

$$\tilde{K} : (\mathcal{V}\text{-Cat})^{\mathbb{T}} \to ((\mathbb{T}, \mathcal{V})\text{-Cat})^{\mathbb{T}} ;$$

here $\tilde{K}(X, a_0, \alpha) = (X, a_0 \cdot \alpha, \alpha)$ and $\tilde{K}f = f$.

The algebraic functor $A_e : (\mathbb{T}, \mathcal{V})\text{-}\mathsf{Cat} \to \mathcal{V}\text{-}\mathsf{Cat}$ (defined by $(X, a) \mapsto (X, a \cdot e_X)$, see Section III.3.4) has a left adjoint

$$A^\circ : \mathcal{V}\text{-}\mathsf{Cat} \to (\mathbb{T}, \mathcal{V})\text{-}\mathsf{Cat}\,, \quad (X, a_0) \mapsto (X, e_X^\circ \cdot \hat{T} a_0)\,,$$

and composing the adjunction $A^\circ \dashv A_e$ with

$$((\mathbb{T}, \mathcal{V})\text{-}\mathsf{Cat})^{\mathbb{T}} \underset{F^{\mathbb{T}}}{\overset{G^{\mathbb{T}}}{\underset{\longleftarrow}{\overset{\longrightarrow}{\top}}}} (\mathbb{T}, \mathcal{V})\text{-}\mathsf{Cat}$$

yields a new adjunction

$$((\mathbb{T}, \mathcal{V})\text{-}\mathsf{Cat})^{\mathbb{T}} \underset{F_0^{\mathbb{T}}}{\overset{G_0^{\mathbb{T}}}{\underset{\longleftarrow}{\overset{\longrightarrow}{\top}}}} \mathcal{V}\text{-}\mathsf{Cat}\,.$$

A direct computation shows that $F_0^{\mathbb{T}} \dashv G_0^{\mathbb{T}}$ induces the monad $\mathbb{T}$ on $\mathcal{V}$-Cat, so we obtain the comparison functor

$$\tilde{A}_e : ((\mathbb{T}, \mathcal{V})\text{-}\mathsf{Cat})^{\mathbb{T}} \to (\mathcal{V}\text{-}\mathsf{Cat})^{\mathbb{T}}$$

that sends (X, a, α) to $(X, a \cdot e_X, \alpha)$. Clearly, $\tilde{A}_e \tilde{K} = 1$, and that $\tilde{A}_e$ and $\tilde{K}$ are in fact inverse to each other follows from a very pleasant property of the monad $\mathbb{T}$ on $(\mathbb{T}, \mathcal{V})$-Cat, as we show next.

III.5.4.1 Theorem *Assume that the lax extension $\hat{\mathbb{T}}$ of $\mathbb{T}$ is associative. Then the monad $\mathbb{T}$ on $(\mathbb{T}, \mathcal{V})$-Cat is of Kock–Zöberlein type.*

Proof First, if $f \le g$ for $f, g : (X, a) \to (Y, b)$ in $(\mathbb{T}, \mathcal{V})$-Cat, then $Tf \le Tg$; indeed, it follows from $f^\circ \cdot b \le g^\circ \cdot b$ that

$$(Tf)^\circ \cdot \hat{T} b \cdot m_Y^\circ \cdot m_Y = \hat{T}(f^\circ \cdot b) \cdot m_Y^\circ \cdot m_Y \le \hat{T}(g^\circ \cdot b) \cdot m_Y^\circ \cdot m_Y = (Tg)^\circ \cdot \hat{T} b \cdot m_Y^\circ \cdot m_Y\,.$$

Secondly, we show that $e_{TTX} : TX \to TTX$ is right adjoint to $m_X : TTX \to TX$, for every $(\mathbb{T}, \mathcal{V})$-category $X = (X, a)$. Clearly, $m_X \cdot e_{TX} = 1_{TX}$. To see $1_{TTX} \le e_{TX} \cdot m_X$, we recall that the $(\mathbb{T}, \mathcal{V})$-category structure on TX is given by $\hat{T} a \cdot m_X^\circ \cdot m_X$, and the $\mathcal{V}$-category structure on TTX is given by $c = \hat{T}\hat{a} \cdot Tm_X \cdot m_{TX}^\circ$, where $\hat{a} = \hat{T} a \cdot m_X^\circ$. Using the fact that $\hat{a} \cdot m_X$ is unitary, we then compute

$$m_X^\circ \cdot e_{TX}^\circ \cdot c = m_X^\circ \cdot e_{TX}^\circ \cdot \hat{T}(\hat{a} \cdot m_X) \cdot m_{TX}^\circ = m_X^\circ \cdot \hat{a} \cdot m_X \ge m_X^\circ \cdot m_X \ge 1_{TTX}\,.$$

Hence, for all $\mathcal{X} \in TTX$, we have $c(\mathcal{X}, e_{TX} \cdot m_X(\mathcal{X})) \ge k$; and therefore $1_{TTX} \le e_{TX} \cdot m_X$ in $(\mathbb{T}, \mathcal{V})$-Cat. $\square$

For $X = (X, a)$ in $(\mathbb{T}, \mathcal{V})$-Cat, an Eilenberg–Moore structure $\alpha : TX \to X$ in $(\mathbb{T}, \mathcal{V})$-Cat gives rise to an adjunction $\alpha \dashv e_X$ in $(\mathbb{T}, \mathcal{V})$-Cat (see Proposition II.4.9.1). Since the algebraic functor $A_e : (\mathbb{T}, \mathcal{V})\text{-}\mathsf{Cat} \to \mathcal{V}\text{-}\mathsf{Cat}$ is a 2-functor, the underlying $\mathcal{V}$-functor

$$e_X : A_e(X, a) = (X, a_0) \to A_e T(X, a) = (TX, \hat{T}a \cdot m_X^\circ \cdot m_X \cdot e_{TX})$$
$$= (TX, \hat{T}a \cdot m_X^\circ)$$

is right adjoint to $\alpha : (TX, \hat{T}a \cdot m_X^\circ) \to (X, a_0)$ as well. Hence, with Exercise III.3.F applied to $\mathcal{V}$-Cat, we obtain from the adjunction $\alpha \dashv e_X$ in $\mathcal{V}$-Cat the equation

$$a_0(\alpha(\mathfrak{x}), x) = \hat{T}a \cdot m_X^\circ(\mathfrak{x}, e_X(x)) \qquad \text{(III.5.4.i)}$$

for all $\mathfrak{x} \in TX$ and $x \in X$; this means $a_0 \cdot \alpha = e_X^\circ \cdot \hat{T}a \cdot m_X^\circ = e_X^\circ \circ a = a$. As a consequence, $\tilde{K}$ and $\tilde{A}_e$ are inverse to each other:

III.5.4.2 Corollary *If the lax extension* $\hat{\mathbb{T}}$ *of* $\mathbb{T}$ *is associative, then* $((\mathbb{T}, \mathcal{V})\text{-Cat})^{\mathbb{T}} \simeq (\mathcal{V}\text{-Cat})^{\mathbb{T}}$.

The following diagram summarizes the situation exhibited so far:

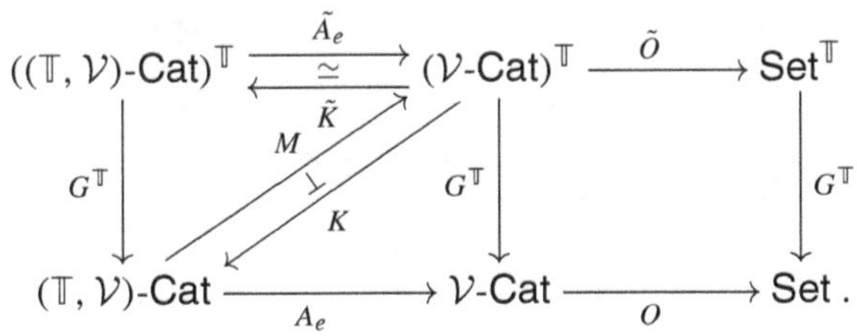

(III.5.4.ii)

One important consequence of Theorem III.5.4.1 is that a $(\mathbb{T}, \mathcal{V})$-category X admits up to equivalence at most one $\mathbb{T}$-algebra structure $\alpha : TX \to X$, since necessarily $\alpha \dashv e_X$ in $(\mathbb{T}, \mathcal{V})$-Cat (see Proposition II.4.9.1).

III.5.4.3 Definitions

(1) A $(\mathbb{T}, \mathcal{V})$-category X is *representable* if $e_X : X \to TX$ has a left adjoint in $(\mathbb{T}, \mathcal{V})$-Cat. Since $\mathbb{T}$ is of Kock–Zöberlein type, a $(\mathbb{T}, \mathcal{V})$-functor $\alpha : TX \to X$ is a left adjoint of e_X if and only if $\alpha \cdot e_X \simeq 1_X$. We hasten to remark that a representable $(\mathbb{T}, \mathcal{V})$-category does not need to be a $\mathbb{T}$-algebra since $\alpha \dashv e_X$ only implies $\alpha \cdot e_X \simeq 1_X$ and $\alpha \cdot T\alpha \simeq \alpha \cdot m_X$. Of course, if X is *separated*, in the sense that its underlying order is separated (see also V.2.1), then α is a $\mathbb{T}$-algebra structure.

(2) A $(\mathbb{T}, \mathcal{V})$-functor $f : X \to Y$ between representable $(\mathbb{T}, \mathcal{V})$-categories X and Y, with left adjoints $\alpha : TX \to X$ and $\beta : TY \to Y$, respectively, is a *pseudo-homomorphism* whenever

$$\beta \cdot Tf \simeq f \cdot \alpha .$$

As before, if Y is separated, then we have equality above. We note that this condition does not depend on the particular choice of the left adjoints α and β.

(3) The category of representable $(\mathbb{T},\mathcal{V})$-categories and pseudo-homomorphism will be denoted as

$$(\mathbb{T},\mathcal{V})\text{-}\mathsf{RepCat}\,.$$

(4) A $(\mathbb{T},\mathcal{V})$-category X is $\mathbb{T}$-*cocomplete* if the $\mathcal{V}$-functor $e_X : A_eX \to A_eTX$ has a left adjoint in $\mathcal{V}$-Cat. While we will not elaborate on this notion in this book, here we note that, by (III.5.4.i), $X = (X,a)$ is $\mathbb{T}$-cocomplete if and only if a can be written as $a = a_0 \cdot \alpha$ (with $a_0 = a \cdot e_X$), for some map $\alpha : TX \to X$. Put differently, for every $\chi \in TX$ there must exist a tacitly chosen *generic point* $x_0 \in X$ so that

$$a(\chi, x) = a_0(x_0, x)$$

for all $x \in X$, and such a generic point is unique up to equivalence.

III.5.4.4 Proposition *Assume that the lax extension $\hat{\mathbb{T}}$ of $\mathbb{T}$ is associative. Then a $(\mathbb{T},\mathcal{V})$-category $X = (X,a)$ is representable if and only if X is $\mathbb{T}$-cocomplete and $a \cdot \hat{T}a = a \cdot m_X$.*

Proof Clearly, a representable $(\mathbb{T},\mathcal{V})$-category $X = (X,a)$ is $\mathbb{T}$-cocomplete and, with $a = a_0 \cdot \alpha$,

$$a \cdot \hat{T}a = a_0 \cdot \alpha \cdot \hat{T}a_0 \cdot T\alpha \geq a_0 \cdot a_0 \cdot \alpha \cdot T\alpha = a_0 \cdot \alpha \cdot m_X = a \cdot m_X\,.$$

Assume now that $X = (X,a)$ is $\mathbb{T}$-cocomplete and that $a \cdot \hat{T}a = a \cdot m_X$. Since $\mathbb{T}$ is of Kock–Zöberlein type, it is enough to verify that the map $\alpha : TX \to X$ is a $(\mathbb{T},\mathcal{V})$-functor. In fact,

$$\begin{aligned}\alpha \cdot \hat{T}a \cdot m_X^\circ \cdot m_X &\leq a \cdot \hat{T}a \cdot m_X^\circ \cdot m_X = a \cdot m_X \cdot m_X^\circ \cdot m_X\\ &\leq a \cdot m_X = a \cdot \hat{T}a = a \cdot \hat{T}a_0 \cdot T\alpha = a \cdot T\alpha\,. \qquad \square\end{aligned}$$

III.5.5 Dualizing $(\mathbb{T},\mathcal{V})$-categories

In Section III.1.3 we introduced the dual X^{op} of a $\mathcal{V}$-category $X = (X,a)$ as $X^{\mathrm{op}} = (X,a^\circ)$, for a commutative quantale $\mathcal{V}$. This definition cannot be used directly for $(\mathbb{T},\mathcal{V})$-categories in general since $a^\circ : X \nrightarrow TX$ does not have the correct type, and in this section we will discuss one possible way of dealing with this problem. Roughly speaking, we consider only those $(\mathbb{T},\mathcal{V})$-categories $X = (X,a)$ where $a = a_0 \cdot \alpha$, for some $\alpha : TX \to X$, and where $a_0 = a \cdot e_X$ is the underlying $\mathcal{V}$-category structure, then dualize just (X,a_0) and combine the result with α; hence the structure of X^{op} is given by $a_0^\circ \cdot \alpha$. This defines, however, in general only a $(\mathbb{T},\mathcal{V})$-graph (see Proposition III.5.5.3). We therefore consider the concept in this context.

Throughout this section we assume that

- *$\mathcal{V}$ is commutative.*

A $(\mathbb{T}, \mathcal{V})$-graph $X = (X, a)$ is called *dualizable* whenever $a_0 = a \cdot e_X$ is transitive and $a = a_0 \cdot \alpha$, for some map $\alpha : TX \to X$. For a dualizable $(\mathbb{T}, \mathcal{V})$-graph $X = (X, a)$, we write X_0 to denote its underlying $\mathcal{V}$-category $X_0 = (X, a_0)$. We consider TX as a discrete $\mathcal{V}$-category, so that $\alpha : TX \to X_0$ is a $\mathcal{V}$-functor. With this notation, $a_0 \cdot \alpha = \alpha_*$ (see Section III.1.3) and, if $\alpha_* = a = \beta_*$, also $\alpha^* = \beta^*$, and therefore

$$a_0^\circ \cdot \alpha = (\alpha^*)^\circ = (\beta^*)^\circ = a_0^\circ \cdot \beta \,.$$

III.5.5.1 Lemma *Let $X = (X, a)$ be a dualizable $(\mathbb{T}, \mathcal{V})$-graph. Then $(X, a_0^\circ \cdot \alpha)$ is a dualizable $(\mathbb{T}, \mathcal{V})$-graph as well, where the underlying $\mathcal{V}$-category of $(X, a_0^\circ \cdot \alpha)$ is $(X_0)^{\mathrm{op}}$.*

Proof It suffices to show $a_0^\circ = a_0^\circ \cdot \alpha \cdot e_X$. From $a = a_0 \cdot \alpha$ we infer $a_0 = a_0 \cdot \alpha \cdot e_X = (\alpha \cdot e_X)_*$, hence $a_0 = (\alpha \cdot e_X)^*$, and therefore $a_0^\circ = a_0^\circ \cdot \alpha \cdot e_X$. □

The *dual $(\mathbb{T}, \mathcal{V})$-graph* of a dualizable $(\mathbb{T}, \mathcal{V})$-graph $X = (X, a)$ is then defined as $X^{\mathrm{op}} = (X, a_0^\circ \cdot \alpha)$. This definition is independent of the choice of $\alpha : TX \to X$ by the calculation given before Lemma III.5.5.1.

Every $\mathbb{T}$-cocomplete $(\mathbb{T}, \mathcal{V})$-category, seen as a $(\mathbb{T}, \mathcal{V})$-graph, is dualizable. In particular, every $\mathcal{V}$-category is dualizable, and its dual in the sense above is just the usual dual. In Section III.5.8 we will see another important example of a dualizable $(\mathbb{T}, \mathcal{V})$-graph.

III.5.5.2 Lemma *For a $(\mathbb{T}, \mathcal{V})$-functor $f : (X, a) \to (Y, b)$ between dualizable $(\mathbb{T}, \mathcal{V})$-graphs with $a = a_0 \cdot \alpha$ and $b = b_0 \cdot \beta$, the map $f : X \to Y$ defines also a $(\mathbb{T}, \mathcal{V})$-functor $f^{\mathrm{op}} : X^{\mathrm{op}} \to Y^{\mathrm{op}}$ if and only if $f \cdot \alpha \simeq \beta \cdot Tf$.*

Proof As for Lemma III.5.3.1. □

III.5.5.3 Proposition *Assume that the lax extension $\hat{\mathbb{T}}$ of $\mathbb{T}$ is associative and let $X = (X, a)$ be a $\mathbb{T}$-cocomplete $(\mathbb{T}, \mathcal{V})$-category. Then the following assertions are equivalent:*

(i) *the $(\mathbb{T}, \mathcal{V})$-graph X^{op} is a $(\mathbb{T}, \mathcal{V})$-category;*
(ii) *X satisfies $a \cdot \hat{T} a = a \cdot m_X$;*
(iii) *X is representable.*

Proof By Proposition III.5.4.4, (iii) $\Longrightarrow$ (ii); and the implication (iii) $\Longrightarrow$ (i) can be shown as in Section III.5.3. Assume now that X^{op} is a $(\mathbb{T}, \mathcal{V})$-category. Since X is a $(\mathbb{T}, \mathcal{V})$-category,

$$(\alpha \cdot T\alpha)_* = a_0 \cdot \alpha \cdot T\alpha \le a_0 \cdot \alpha \cdot \hat{T} a_0 \cdot T\alpha \le a_0 \cdot \alpha \cdot m_X = (\alpha \cdot m_X)_* \,;$$

similarly, since X^{op} is a $(\mathbb{T}, \mathcal{V})$-category,

$$a_0^\circ \cdot \alpha \cdot T\alpha \le a_0^\circ \cdot \alpha \cdot m_X \,,$$

and therefore $(\alpha \cdot T\alpha)^* \leq (\alpha \cdot m_X)^*$. Consequently, $(\alpha \cdot T\alpha)_* = (\alpha \cdot m_X)_*$ by Exercise II.4.E, hence $a \cdot \hat{T}a \geq a \cdot m_X$. The inequality $a \cdot \hat{T}a \leq a \cdot m_X$ we get from X being a $(\mathbb{T}, \mathcal{V})$-category, therefore $a \cdot \hat{T}a = a \cdot m_X$. $\square$

In conclusion, $(f : X \to Y) \mapsto (f^{\mathrm{op}} : X^{\mathrm{op}} \to Y^{\mathrm{op}})$ defines a functor

$$(-)^{\mathrm{op}} : (\mathbb{T}, \mathcal{V})\text{-}\mathsf{RepCat} \to (\mathbb{T}, \mathcal{V})\text{-}\mathsf{RepCat}\,,$$

which makes the diagram

$$\begin{array}{ccc} (\mathbb{T}, \mathcal{V})\text{-}\mathsf{RepCat} & \xrightarrow{(-)^{\mathrm{op}}} & (\mathbb{T}, \mathcal{V})\text{-}\mathsf{RepCat} \\ {\scriptstyle (-)_0}\downarrow & & \downarrow{\scriptstyle (-)_0} \\ \mathcal{V}\text{-}\mathsf{Cat} & \xrightarrow[(-)^{\mathrm{op}}]{} & \mathcal{V}\text{-}\mathsf{Cat} \end{array}$$

commutative.

III.5.6 The ultrafilter monad on Top

We consider the ultrafilter monad $\mathbb{\beta}$ with its Barr extension $\overline{\mathbb{\beta}}$ to Rel, so that $(\mathbb{\beta}, 2)\text{-}\mathsf{Cat} \cong \mathsf{Top}$. By Section III.5.4, $\mathbb{\beta}$ extends to a monad on Top, again denoted by $\mathbb{\beta}$. For a topological space X, βX is the space of all ultrafilters of the set X where, for $\mathfrak{X} \in \beta\beta X$ and $\mathfrak{x} \in \beta X$, one has $\mathfrak{X} \longrightarrow \mathfrak{x}$ precisely when $m_X(\mathfrak{X}) \leq \mathfrak{x}$ (see Examples III.5.3.7), i.e. precisely when for every open set $A \in \mathfrak{x}$, one has $A^{\mathbb{\beta}} = \{\mathfrak{a} \in \beta X \mid A \in \mathfrak{a}\} \in \mathfrak{X}$. Therefore:

© **III.5.6.1 Lemma** *The sets $A^{\mathbb{\beta}}$ (with $A \subseteq X$ open) form a base of the topology of βX.*

Important note. For a topological space X (usually assumed to be completely regular), the space βX of ultrafilters on X should not be confused with the Čech–Stone compactification of X. However, in the discussion following Proposition III.5.6.2, we show how the Čech–Stone compactification of X may be obtained from the space βX defined here.

Diagram (III.5.4.ii) specializes to

$$\begin{array}{ccccc} \mathsf{Top}^{\mathbb{\beta}} & \xrightarrow{\simeq} & \mathsf{OrdCompHaus} & \longrightarrow & \mathsf{CompHaus} \\ \downarrow & & \downarrow & & \downarrow \\ \mathsf{Top} & \longrightarrow & \mathsf{Ord} & \longrightarrow & \mathsf{Set}\,. \end{array}$$

Here the category $\mathsf{OrdCompHaus}$ of ordered compact Hausdorff spaces and their morphisms (see Examples III.5.2.1) appears as the category of Eilenberg–
© Moore algebras for the ultrafilter monad $\mathbb{\beta}$ on both Ord and Top.

The inclusion functor $\mathsf{CompHaus} \longrightarrow \mathsf{Top}$ factors as

$$\mathsf{CompHaus} \xrightarrow{(-)_d} \mathsf{Ord}^{\mathbb{B}} \xrightarrow{K} \mathsf{Top}\,,$$

so its left adjoint, usually called *Čech–Stone compactification*, can be taken as $\pi_0 \cdot M$, where $M \dashv K$ (see Theorem III.5.3.5) and $\pi_0 \dashv (-)_d$. Recall that, for an ordered compact Hausdorff space $X = (X, \leq, \alpha)$, the graph $R \subseteq X \times X$ of the order relation $\leq$ is closed and therefore compact Hausdorff. The reflection $q : X \longrightarrow \pi_0(X)$ is actually the coequalizer

$$R \underset{p_2}{\overset{p_1}{\rightrightarrows}} X \xrightarrow{q} \pi_0(X) \tag{III.5.6.i}$$

in $\mathsf{CompHaus}$, which is usually different from the coequalizer in Set (see Exercise III.5.A). However, if the graph $E \subseteq X \times X$ of the equivalence relation induced by $\leq$ is closed, then E is compact Hausdorff, and the coequalizer (III.5.6.i) in $\mathsf{CompHaus}$ can be constructed as in Set. This is certainly the case when $\leq$ is *confluent*: see Exercise III.5.A.

III.5.6.2 Proposition *For a topological space $X = (X, a)$, the order relation $\hat{a} = \overline{\beta} a \cdot m_X^\circ$ on βX is confluent if and only if for all disjoint closed subsets $A, B \subseteq X$ there exist open subsets $U, V \subseteq X$ with $A \subseteq U$, $B \subseteq V$, and $U \cap V = \varnothing$.*

Proof Assume first that $\hat{a} = \overline{\beta} a \cdot m_X^\circ$ is confluent. Let $A, B \subseteq X$ be closed subsets of X with the property that every open neighborhood of A intersects every open neighborhood of B. Hence, by Corollary II.1.13.3, there is an ultrafilter $\mathfrak{x} \in \beta X$ with

$$\{U \cap V \mid U, V \in \mathcal{O}X, A \subseteq U, B \subseteq V\} \subseteq \mathfrak{x}\,.$$

By definition, the filter generated by the filter base $\{A\}$ is disjoint from the ideal generated by $\{U \in \mathcal{O}X \mid U \notin \mathfrak{x}\}$, so Corollary II.1.13.5 guarantees the existence of an ultrafilter $\mathfrak{a} \in \beta X$ with $A \in \mathfrak{a}$ and $\mathfrak{x} \leq \mathfrak{a}$. A similar argument yields an ultrafilter $\mathfrak{b} \in \beta X$ with $B \in \mathfrak{b}$ and $\mathfrak{x} \leq \mathfrak{b}$. By hypothesis, there is some $\mathfrak{y} \in \beta X$ with $\mathfrak{a} \leq \mathfrak{y}$ and $\mathfrak{b} \leq \mathfrak{y}$. Hence, $A \in \mathfrak{y}$ and $B \in \mathfrak{y}$, so that $A \cap B \neq \varnothing$.

Assume now that the condition on disjoint closed subsets is satisfied. Let $\mathfrak{x}, \mathfrak{a}, \mathfrak{b} \in \beta X$ with $\mathfrak{x} \leq \mathfrak{a}$ and $\mathfrak{x} \leq \mathfrak{b}$. Let $A \in \mathfrak{a}$ and $B \in \mathfrak{b}$ be closed. Then $U \in \mathfrak{x}$ and $V \in \mathfrak{x}$, and therefore $U \cap V \neq \varnothing$, for all open subsets $U, V \subseteq X$ with $A \subseteq U$ and $B \subseteq V$. Consequently, $A \cap B \neq \varnothing$. Hence, by Corollary II.1.13.3, there is an an ultrafilter $\mathfrak{y} \in \beta X$ with

$$\{A \cap B \mid A, B \subseteq X \text{ closed}, A \in \mathfrak{a}, B \in \mathfrak{b}\} \subseteq \mathfrak{y}\,,$$

so that $\mathfrak{a} \leq \mathfrak{y}$ and $\mathfrak{b} \leq \mathfrak{y}$. □

A topological space with the property described in Proposition III.5.6.2 is called *normal*. By the discussion preceding this proposition, we have:

III.5.6.3 Corollary *The Čech–Stone compactification of a normal space X is isomorphic to the space of connected components of βX with respect to the order relation of* Example III.5.3.7(1).

A topological space X is β-cocomplete if every ultrafilter $\mathfrak{x}$ has a generic convergence point $x_0 \in X$, so that

$$\mathfrak{x} \longrightarrow x \iff x_0 \leq x$$

for all $x \in X$, or equivalently $\lim \mathfrak{x} = \overline{\{x_0\}}$, where $\lim \mathfrak{x}$ denotes the set of limit points of $\mathfrak{x}$. A subset of the form $\overline{\{x\}}$, for $x \in X$, is a trivial example of an *irreducible* closed subset, i.e. of a non-empty closed subset $A \subseteq X$ with the property that, whenever $A \subseteq A_1 \cup A_2$ for closed subsets $A_1, A_2 \subseteq X$, then $A \subseteq A_1$ or $A \subseteq A_2$. A topological space X is called *sober* if every irreducible closed subset $A \subseteq X$ is of the form $A = \overline{\{x\}}$, for a unique $x \in X$; without this uniqueness requirement, X is *weakly sober*. By Exercise III.5.D, every irreducible closed subset $A \subseteq X$ is the set $A = \lim \mathfrak{x}$ of limit points of some ultrafilter $\mathfrak{x} \in \beta X$. Hence, we obtain the following result.

III.5.6.4 Proposition *A topological space X is β-cocomplete if and only if*

(1) *for every $\mathfrak{x} \in \beta X$, $\lim \mathfrak{x}$ is irreducible, and*
(2) *X is weakly sober.*

Proof The result follows immediately from the definitions. □

Note that (1) implies in particular that X is compact.

To understand better the condition $a \cdot \overline{\beta} a = a \cdot m_X$, we introduce the way-below relation on the lattice of open subsets of X: for $A \subseteq X$ open, A is *way-below* B, written as $A \lll B$, if every open cover of B has a finite sub-cover of A; i.e. Whenever $B \subseteq \bigcup_{i \in I} B_i$ with open subsets B_i, there exists a finite subset $K \subseteq I$ such that $A \subseteq \bigcup_{i \in K} B_i$. This relation can be equivalently expressed in the language of ultrafilter convergence: $A \lll B$ precisely when every ultrafilter $\mathfrak{x}$ on A has a limit point in B (Exercise III.5.E). Furthermore, X is called *core-compact* if, for every point $x \in X$ and every open neighborhood B of x, there exists an open neighborhood A of x with $A \lll B$.

III.5.6.5 Remark We will see in Theorem III.5.8.5 that the core-compact spaces are precisely the exponentiable objects (see II.4.4) in the category Top.

III.5.6.6 Proposition *A topological space X with convergence relation $a : \beta X \rightarrow\!\!\!\!\!\rightarrow X$ is core-compact if and only if $a \cdot \overline{\beta} a = a \cdot m_X$.*

Proof Assume first that X is core-compact. Since $a \cdot \overline{\beta}a \leq a \cdot m_X$ for every topological space, we only have to show $a \cdot \overline{\beta}a \geq a \cdot m_X$. Let $\mathcal{X} \in \beta\beta X$ and $x \in X$, where x is a limit point of $m_X(\mathcal{X})$. Hence, $B^{\mathbb{\beta}} \in \mathcal{X}$ for every open neighborhood B of x. Let now A be an open neighborhood of x, and choose any open neighborhood B of x with $B \lll A$. Then

$$\lim{}^{-1} A = \{\chi \in \beta X \mid \lim \chi \cap A \neq \varnothing\} \supseteq B^{\mathbb{\beta}} \in \mathcal{X}\,,$$

where $\lim \chi$ denotes the set of all limit points of χ. Hence, the neighborhood filter f of x is disjoint from the ideal $j = \{A \subseteq X \mid \lim^{-1} A \notin \mathcal{X}\}$; therefore, by Corollary II.1.13.5, there exists some $\chi \in \beta X$ disjoint from j (so that $\mathcal{X}\ \overline{\beta}a\ \chi$, see Example III.1.10.3(3)), and χ contains every open neighborhood of x, i.e. $\chi\ a\ x$. Assume now that X is not core-compact, i.e. there is some $x \in X$ and some open neighborhood B of x so that for every open neighborhood A of x there is some ultrafilter $y \in \beta X$ with $A \in y$ and $\lim y \cap B = \varnothing$. Consequently, there is some $\mathcal{X} \in \beta\beta X$ containing $\{y \in \beta X \mid \lim y \cap B = \varnothing\}$ and $A^{\mathbb{\beta}}$ for every open neighborhood A of x. Then $m_X(\mathcal{X})$ converges to x, but $\mathcal{X}\ \overline{\beta}a\ \chi$ implies that $B \notin \chi$, so χ cannot converge to x. □

III.5.6.7 Remark It is often easier to check core-compactness of a topological space X by just looking at the elements of a *subbase* for $\mathcal{O}X$ (i.e. a subset of $\mathcal{O}X$ whose set of finite intersections forms a base for $\mathcal{O}X$, see II.1.9). Given a set X and a subset $\mathcal{B} \subseteq PX$ of the powerset of X (with no further axioms), one defines core-compactness and convergence $a : \beta X \nrightarrow X$ for $(X, \mathcal{B})$ in the same way as one does for topological spaces, and we note that the topology $\langle\mathcal{B}\rangle$ generated by $\mathcal{B}$ has the same convergence as $\mathcal{B}$. We wish to conclude that $a \cdot \overline{\beta}a = a \cdot m_X$ implies that $(X, \mathcal{B})$ is core-compact; however, the preceding proof uses that the collection of all neighborhoods of $x \in X$ forms a filter, a fact that is not necessarily true if $\mathcal{B}$ is just any subset of PX. For the filter f generated by all $\mathcal{B}$-neighborhoods of x, we still have $f \cap j = \varnothing$ if the convergence a satisfies the following condition: every ultrafilter has a smallest convergence point with respect to the order relation $a \cdot e_X$. Hence, under this condition, a topological space X is core-compact if X is core-compact with respect to a subbase since

$$\begin{aligned}
(X, \mathcal{B}) \text{ is core-compact } &\Longrightarrow\ a \cdot \overline{\beta}a = a \cdot m_X \text{ for the convergence } a \text{ of } \mathcal{B}\\
&\Longleftrightarrow\ a \cdot \overline{\beta}a = a \cdot m_X \text{ for the convergence } a \text{ of } \langle\mathcal{B}\rangle\\
&\Longleftrightarrow\ (X, \langle\mathcal{B}\rangle) \text{ is core-compact.}
\end{aligned}$$

In fact, this argument works for any other property of a topological space which can be equivalently expressed in terms of opens and in terms of ultrafilter convergence, without using the axioms of a topology. Another important example is compactness: a topological space X is compact if X is compact with respect to a subbase. This result is known as *Alexander's subbase lemma.*

Combining Propositions III.5.4.4, III.5.6.4, and III.5.6.6 gives

© **III.5.6.8 Proposition** *A topological space X is representable if and only if*

(1) *X is core-compact,*
(2) *for every* $\chi \in \beta X$, $\lim \chi$ *is irreducible, and*
(3) *X is weakly sober.*

III.5.7 Representable topological spaces

In this section we present a more detailed analysis of representable topological spaces.

First, we will see that a representable topological space is not only core-compact, but is also even locally compact; a topological space is *locally compact* if the neighborhood filter of every point $x \in X$ has a base formed by compact neighborhoods of x.

© **III.5.7.1 Lemma** *Every representable topological space is locally compact.*

Proof By Lemma III.5.6.1, the topology on βX is generated by all sets of the form

$$A^{\beta} = \{a \in \beta X \mid A \in a\},$$

where $A \subseteq X$ is open. Furthermore, for any ultrafilter $\mathcal{X} \in \beta\beta X$ with $A^{\beta} \in \mathcal{X}$, we have $m_X(\mathcal{X}) \in A^{\beta}$, and therefore A^{β} is compact. For any $\chi \in \beta X$,

$$\{A^{\beta} \mid A \in \chi\}$$

is a base of the neighborhood filter of χ; and we conclude that βX is locally compact. If X is representable, then X is a split subobject of βX (since $\alpha : \beta X \to X$ can be chosen such that $\alpha(e_X(x)) = x$) and hence also locally compact (Exercise III.5.F). □

© For a core-compact space X, condition (1) of Proposition III.5.6.4 is equivalent to the following stability property of the way-below relation $\ll$: for open subsets $U_1, \dots, U_n$ and $V_1, \dots, V_n$ ($n \in \mathbb{N}$) of X, with $U_i \ll V_i$ for each $1 \le i \le n$, also $\bigcap_i U_i \ll \bigcap_i V_i$ (Exercise III.5.H). Note that it is enough to consider the cases $n = 0$ and $n = 2$, and for $n = 0$ this condition reads as $X \ll X$, which just means that X is compact. Saying that a topological space X with this stability property is *stable*, we obtain the following result.

© **III.5.7.2 Theorem** *A topological space X is representable if and only if X is locally compact, weakly sober, and stable.*

We remark that representable T0-spaces are also called *stably compact* in the literature, where, however, the stability condition on the way-below relation is usually replaced by the requirement that the compact down-closed subsets of

X are closed under finite intersection. To see that these conditions are indeed equivalent, first note that a representable space X is compact, and that the binary intersections of pairs of compact down-closed subsets are compact: if $A, B \subseteq X$ are compact and down-closed, and $A \cap B \in \chi \in \beta X$, then any smallest convergence point of χ belongs to both A and B and therefore also to $A \cap B$. Consequently, in a representable space, the finite intersection of compact down-closed subsets is again compact. For the converse implication, we use the following description of the way-below relation on the lattice of opens of a locally compact space.

III.5.7.3 Lemma *Let X be locally compact and let $U, V \subseteq X$ be open. Then $U \ll V$ if and only if $U \subseteq K \subseteq V$ for some compact and down-closed $K \subseteq X$.*

Proof Assume first that $U \ll V$. For every $x \in V$ there is a compact neighborhood K of x with $K \subseteq V$. Since $V = \bigcup\{K \subseteq V \mid K \text{ is a compact neighborhood of some } x \in V\}$ and $U \ll V$, there is some compact K with $U \subseteq K \subseteq V$. Furthermore, if $A \subseteq W$ for some $A \subseteq X$ and some open subset $W \subseteq X$, then also $\downarrow A \subseteq W$, since open subsets are down-closed. In particular, the down-closure of a compact subset is compact, and therefore K above can be chosen down-closed. Conversely, if $U \subseteq K \subseteq V$ for some compact $K \subseteq X$, then every open cover of V contains a finite sub-cover of K and hence also of U. □

III.5.7.4 Remark If the compact open subsets of a locally compact space X form a base of the topology of X, then K in the proof of Lemma III.5.7.3 can be chosen as a compact open subset of X.

From Lemma III.5.7.3 we deduce at once that, for a locally compact space, stability of the way-below relation under finite intersection follows from stability of compact down-sets under finite intersection.

III.5.7.5 Theorem *A topological space is representable if and only if it is locally compact, weakly sober, and if the finite intersection of compact down-closed subsets is compact.* ©

We now turn our attention to pseudo-homomorphisms. Recall from Section III.5.4 that a pseudo-homomorphism between representable topological spaces is a continuous map $f : X \to Y$ that preserves the smallest convergence points of ultrafilters.

III.5.7.6 Proposition *Let $f : X \to Y$ be a continuous map between representable topological spaces. Then the following assertions are equivalent:* ©

(i) *f is a pseudo-homomorphism;*
(ii) *for every compact down-closed subset $K \subseteq Y$, $f^{-1}(K)$ is compact;*
(iii) *for all open subsets $U, V \subseteq Y$, $U \ll V$ implies $f^{-1}(U) \ll f^{-1}(V)$.*

Proof Assume first (i) and let $K \subseteq Y$ be compact, $\chi \in \beta X$ with $f^{-1}(K) \in \chi$, and let x be a smallest convergence point of χ. Then $f(x)$ is a smallest convergence point of $\beta f(\chi)$ and, since K is compact and $K \in \beta f(\chi)$, we have $f(x) \in K$. Therefore, $x \in f^{-1}(K)$, and we have shown that $f^{-1}(K)$ is compact, i.e. (i) $\Longrightarrow$ (ii). The implication (ii) $\Longrightarrow$ (iii) follows from Lemma III.5.7.3. Assume now (iii) and let $x \in X$ be a smallest convergence point of $\chi \in \beta X$. Assume that $\beta f(\chi) \to y \in Y$. Let $V \subseteq Y$ be any open neighborhood of y and choose some open $U \subseteq X$ with $y \in U \lll V$. Then $f^{-1}(U) \lll f^{-1}(V)$ and $f^{-1}(U) \in \chi$, hence $x \in f^{-1}(V)$ and therefore $f(x) \in V$. We conclude that $f(x) \leq y$, so (iii) $\Longrightarrow$ (i). □

© **III.5.7.7 Corollary** *Let X be a representable topological space, and $\mathsf{2}$ as in* Example III.5.3.2. *A continuous map $\varphi : X \to \mathsf{2}$ is a pseudo-homomorphism if and only if the open set $\varphi^{-1}(0) \subseteq X$ is compact.*

© **III.5.7.8 Proposition**

(1) *Let $(X, \leq, \alpha)$ be an ordered compact Hausdorff space and let $a = (\leq) \cdot \alpha$ be its induced topology. A subset $A \subseteq X$ is open in (X, a) if and only if A is down-closed and open in the compact Hausdorff space (X, α).*

(2) *Let X be a representable space, let $\chi \in \beta X$, and let $x_0 \in X$ be a smallest convergence point of χ. For any $x \in X$, $x \leq x_0$ if and only if χ contains all complements of compact down-sets B with $x \notin B$.*

Proof To see (1), let $(X, \leq, \alpha)$ be an ordered compact Hausdorff space and $A \subseteq X$. Let $\varphi : X \to \mathsf{2}$ be the characteristic map of the complement $X \setminus A$ of A. Then

$$\begin{aligned} A \text{ is open} &\iff \varphi : (X, (\leq) \cdot \alpha) \to \mathsf{2} \text{ is continuous} \\ &\iff \varphi : (X, \leq) \to \mathsf{2} \text{ is monotone and} \\ &\qquad \varphi : (X, \alpha) \to \mathsf{2} \text{ is continuous (by Lemma III.5.3.1)} \\ &\iff A \text{ is down-closed in } (X, \leq) \text{ and open in } (X, \alpha). \end{aligned}$$

To see (2), first let $x \leq x_0$. Then χ cannot contain any compact down-sets B with $x \notin B$. Assume now that χ contains these subsets. Take a neighborhood B of x_0, where B is a compact down-set. Then $x \in B$, since otherwise $B \in \chi$ and $X \setminus B \in \chi$. □

© **III.5.7.9 Corollary** *Let X be a representable space. Then the topology of X^{op} is generated by the complements of compact down-sets B of X. Furthermore, taking smallest convergence points of an ultrafilter on X with respect to its original topology describes the ultrafilter convergence with respect to the topology generated by all open sets of X and all open sets of X^{op}.*

© **III.5.7.10 Corollary** *Let $(X, \leq, \alpha)$ be an ordered compact Hausdorff space with separated order. The topology of (X, α) is generated by the open subsets* *and the complements of compact down-closed subsets of the representable space* $(X, (\leq) \cdot \alpha)$.

III.5.8 Exponentiable topological spaces

In Section III.4 we saw that the category Top can be fully embedded into the (locally) Cartesian closed category PsTop (see Examples III.4.1.3 and Corollary III.4.5.2), and that Top itself is not Cartesian closed (see Exercise III.4.G). In this section, we provide a characterization of exponentiable topological spaces, i.e. of those spaces X where $(-) \times X : \mathsf{Top} \to \mathsf{Top}$ has a right adjoint.

Let X be a topological space with convergence $a : \beta X \nrightarrow X$ (but we write more intuitively $\chi \to x$ instead of $\chi\ a\ x$) and let $\mathsf{2}$ be the Sierpiński space (see Examples III.5.3.2). We form the exponential $\mathsf{2}^X$ in PsTop, and write $\pi_1 : \mathsf{2}^X \times X \to \mathsf{2}^X$ and $\pi_2 : \mathsf{2}^X \times X \to X$ for the projection maps and $\varepsilon : \mathsf{2}^X \times X \to \mathsf{2}$ denotes the evaluation map. In the following we will think of the elements of $\mathsf{2}^X$ as closed subsets of X. For any subset $V \subseteq X$, we put

$$V^{\Diamond} = \{A \subseteq X \mid A \text{ closed}, A \cap V \neq \varnothing\}.$$

The topology of X we denote as $\mathcal{O}X$, and $\mathcal{O}(x)$ stands for the collection of open neighborhoods of $x \in X$.

III.5.8.1 Proposition *For every topological space* X*, the pseudotopological space* $\mathsf{2}^X$ *is dualizable. The underlying order of* $\mathsf{2}^X$ *is subset inclusion;* $p \longrightarrow A \iff \mu(p) \subseteq A$ *for* $p \in \beta(\mathsf{2}^X)$ *and* $A \subseteq X$ *closed, where* $\mu(p) = \bigcap_{\mathcal{A} \in p} \overline{\bigcup \mathcal{A}}$.

Proof Recall from Examples III.4.5.3 that the convergence structure of $\mathsf{2}^X$ is given by

$$p \longrightarrow A \iff \begin{cases} \text{for all } x \in X,\ w \in \beta(\mathsf{2}^X \times X) \text{ with } \beta\pi_1(w) = p,\ \chi := \beta\pi_2(w) : \\ \chi \longrightarrow x \implies (\beta\varepsilon(w) = 1 \implies x \in A), \end{cases}$$
$$\iff \begin{cases} \text{for all } x \in X,\ w \in \beta(\mathsf{2}^X \times X) \text{ with } \beta\pi_1(w) = p,\ \chi := \beta\pi_2(w) : \\ (\chi \longrightarrow x \ \&\ \beta\varepsilon(w) = 1) \implies x \in A, \end{cases}$$
$$\iff \begin{cases} \text{for all } x \in X : \\ p\ (a \cdot (\overline{\beta}\varepsilon))\ x \implies x \in A\,; \end{cases}$$

for $p \in \beta(\mathsf{2}^X)$ and $A \subseteq X$ closed, where in the last line we interpret $\varepsilon : \mathsf{2}^X \times X \longrightarrow \mathsf{2}$ as the membership relation $\mathsf{2}^X \nrightarrow X$. We put

$$\mu(p) = \{x \in X \mid p\ (a \cdot (\overline{\beta}\varepsilon))\ x\},$$

for $p \in \mathsf{2}^X$, and then

$$p \longrightarrow A \iff \mu(p) \subseteq A\,.$$

Furthermore,

$$\begin{aligned}\mu(p) &= \{x \in X \mid \exists \chi \in \beta X : (p(\overline{\beta}\varepsilon)\,\chi \;\&\; \chi \longrightarrow x)\}\\ &= \{x \in X \mid \forall V \in \mathcal{O}(x), \mathcal{A} \in p\, \exists A \in \mathcal{A}, y \in V : y \in A\}\\ &= \{x \in X \mid \forall V \in \mathcal{O}(x), \mathcal{A} \in p\ V^{\Diamond} \cap \mathcal{A} \neq \varnothing\}\\ &= \{x \in X \mid \forall V \in \mathcal{O}(x)\ V^{\Diamond} \in p\}\,.\end{aligned}$$

We also note that $V^{\Diamond} \cap \mathcal{A} \neq \varnothing$ is equivalent to $V \cap \bigcup \mathcal{A} \neq \varnothing$, and therefore

$$\begin{aligned}\mu(p) &= \{x \in X \mid \forall \mathcal{A} \in p, V \in \mathcal{O}(x)\ V \cap \textstyle\bigcup \mathcal{A} \neq \varnothing\}\\ &= \{x \in X \mid \forall \mathcal{A} \in p\ x \in \overline{\textstyle\bigcup \mathcal{A}}\}\\ &= \textstyle\bigcap_{\mathcal{A} \in p} \overline{\bigcup \mathcal{A}}\,.\end{aligned}$$

Consequently, $\mu(p)$ is a closed subset of X, for every $p \in \beta(2^X)$; $\mu(p) = A$ for $p = \dot{A}$ the principal ultrafilter generated by $A \subseteq X$ closed. Therefore $\dot{A} \longrightarrow B \iff A \subseteq B$. □

III.5.8.2 Proposition *For every topological space X, the pseudotopological space $(2^X)^{\mathrm{op}}$ is topological, where the topology of $(2^X)^{\mathrm{op}}$ is generated by the sets $V^{\Diamond}$ with $V \subseteq X$ open.*

Proof By definition, the convergence of $(2^X)^{\mathrm{op}}$ is given by

$$\begin{aligned}p \longrightarrow A &\iff A \subseteq \textstyle\bigcap_{\mathcal{A} \in p} \overline{\bigcup \mathcal{A}}\\ &\iff \forall x \in A\, \forall \mathcal{A} \in p\, (x \in \overline{\textstyle\bigcup \mathcal{A}})\\ &\iff \forall x \in A\, \forall \mathcal{A} \in p\, \forall V \in \mathcal{O}(x)\, (V \cap \textstyle\bigcup \mathcal{A} \neq \varnothing)\\ &\iff \forall x \in A\, \forall V \in \mathcal{O}(x)\, \forall \mathcal{A} \in p\, (V^{\Diamond} \cap \mathcal{A} \neq \varnothing)\\ &\iff \forall V \in \mathcal{O}X\, (A \in V^{\Diamond} \implies V^{\Diamond} \in p)\,;\end{aligned}$$

i.e. it is generated by the sets

$$V^{\Diamond} = \{A \subseteq X \mid A \text{ closed}, A \cap V \neq \varnothing\} \qquad (V \subseteq X \text{ open})\,,$$

and therefore it is the convergence of the *topology* generated by these sets. □

We find it remarkable that, although 2^X is topological if and only if X is exponentiable (see Proposition III.5.8.4 below), its dual belongs always to Top. The topological space $VX := (2^X)^{\mathrm{op}}$ is usually referred to as the *lower-Vietoris space*.

III.5.8.3 Lemma *Let X be a pseudotopological space. Then $(-)^X$: PsTop $\to$ PsTop preserves initial sources (with respect to the canonical forgetful functor to* Set*).*

Proof Let $(f_i : Y \to Y_i)_{i\in I}$ be initial in PsTop. Let $p \in \beta(Y^X)$ and $h \in Y^X$ so that, for all $i \in I$, $\beta(f_i^X)(p) \longrightarrow f_i^X(h)$. Let $w \in \beta(Y^X \times X)$ and $x \in X$ with $\beta\pi_1(w) = p$ and $\beta\pi_2(w) \longrightarrow x$. Then, for all $i \in I$,

$$\beta\pi_1(\beta(f_i^X \times 1_X)(w)) = \beta\pi_1(w) \longrightarrow x \quad \text{and}$$
$$\beta\pi_2(\beta(f_i^X \times 1_X)(w)) = \beta(f_i^X)(p) \longrightarrow f_i^X(h)\,,$$

therefore $\beta(f_i^X \times 1_X)(w) \longrightarrow (f_i \cdot h, x)$ and

$$\beta f_i(\beta\varepsilon_Y(w)) = \beta\varepsilon_{Y_i}(\beta(f_i^X \times 1_X)(w)) \longrightarrow f_i(h(x))\,.$$

Hence, by hypothesis, $\beta\varepsilon_Y(w) \longrightarrow h(x)$. This proves $p \longrightarrow h$. □

III.5.8.4 Proposition *Let X be a topological space. Then the following assertions are equivalent:*

(i) *X is exponentiable in* Top*;*
(ii) *the pseudotopological space $\mathsf{2}^X$ is topological;*
(iii) *for every topological space Y, the pseudotopological space Y^X is topological.*

Proof Assume first that X is exponentiable in Top. We write temporarily $[X, -]$ for the right adjoint of $(-) \times X : \mathsf{Top} \to \mathsf{Top}$, and ε' denotes the counit of $(-) \times X \dashv [X, -]$; in particular, we consider $\varepsilon'_{\mathsf{2}} : [X, \mathsf{2}] \times X \to \mathsf{2}$. By the universal property of $\varepsilon : \mathsf{2}^X \times X \to \mathsf{2}$ in PsTop, there is a unique map $t : [X, \mathsf{2}] \to \mathsf{2}^X$ in PsTop making

$$\begin{array}{ccc} [X,\mathsf{2}] \times X & \xrightarrow{\;t\times 1_X\;} & \mathsf{2}^X \times X \\ & {\scriptstyle \varepsilon'_{\mathsf{2}}} \searrow \quad \swarrow {\scriptstyle \varepsilon_{\mathsf{2}}} & \\ & \mathsf{2} & \end{array}$$

commute. For any $p \longrightarrow A$ in $\mathsf{2}^X$, we let $(\mathsf{2}^X)_{p,A}$ be the topological space of all closed subsets of X where, besides the principal convergence, only $p \longrightarrow A$ (see Exercise III.5.C). Then there exists a continuous map $s_{p,A} : (\mathsf{2}^X)_{p,A} \to [X, \mathsf{2}]$ making

$$\begin{array}{ccc} (\mathsf{2}^X)_{p,A} \times X & \xrightarrow{\;s_{p,A}\times 1_X\;} & [X,\mathsf{2}] \times X \\ & {\scriptstyle \varepsilon_{\mathsf{2}}} \searrow \quad \swarrow {\scriptstyle \varepsilon_{\mathsf{2}}} & \\ & \mathsf{2} & \end{array}$$

commute. Hence, for all $p \longrightarrow A$ in $\mathsf{2}^X$, $t \cdot s_{p,A}$ is the identity map; and therefore $s : [X, \mathsf{2}] \to \mathsf{2}^X$ is an isomorphism. In particular, $\mathsf{2}^X$ is topological. Assume now that $\mathsf{2}^X$ is topological. Let Y be a topological space. Then the source $\mathsf{PsTop}(Y, \mathsf{2})$ is initial. Hence, also $\mathsf{PsTop}(Y^X, \mathsf{2}^X)$ is initial by Lemma III.5.8.3, and therefore Y^X is topological. Finally, the implication (iii) $\Longrightarrow$ (i) is clear. □

We can now derive a characterization of exponentiable topological spaces.

III.5.8.5 Theorem *Let X be a topological space. Then X is exponentiable if and only if X is core-compact (see* Proposition III.5.6.6).

Proof The space X is exponentiable if and only if $(VX)^{\mathrm{op}}$ is topological, which by Proposition III.5.5.3 is equivalent to VX being core-compact. Finally, VX is core-compact if and only if X is core-compact (see Exercise III.5.I). □

We also note that VX is β-cocomplete, for any topological space X, but VX is only representable if X is core-compact. For X core-compact and $K \subseteq X$ compact, $K^{\Diamond}$ is a compact down-set in VX, and therefore its complement is open in VX^{op} (see Corollary III.5.7.9). Assume now that X is even locally compact. Then one easily verifies that the sets

$$(K^{\Diamond})^{\complement} = \{A \in VX \mid A \cap K = \varnothing\} \qquad (K \subseteq X \text{ compact})$$

generate the convergence of $2^X = (VX)^{\mathrm{op}}$. Hence, if we interpret the elements of 2^X as open subsets of X, the topology of 2^X is generated by the sets

$$\{V \subseteq X \text{ open} \mid K \subseteq V\} \qquad (K \subseteq X \text{ compact}) .$$

This topology is known as the compact-open topology.

III.5.9 Representable approach spaces

The situation for approach spaces is quite similar to the one for topological spaces when one considers the Barr extension $\overline{\beta}$ of β to P_+-Rel. This extension then yields a lifting of the ultrafilter monad β to $\mathsf{App} \simeq (\beta, \mathsf{P}_+)$-$\mathsf{Cat}$. In this case, diagram (III.5.4.ii) becomes

$$\begin{array}{ccccc}
\mathsf{App}^{\beta} & \xrightarrow{\simeq} & \mathsf{MetCompHaus} & \longrightarrow & \mathsf{CompHaus} \\
\downarrow & & \downarrow & & \downarrow \\
\mathsf{App} & \longrightarrow & \mathsf{Met} & \longrightarrow & \mathsf{Set} .
\end{array}$$

Here $\mathsf{MetCompHaus}$ denotes the category of metric compact Hausdorff spaces and morphisms (see Example III.5.2.1(4)). An approach space $X = (X, a)$ is representable if and only if $a \cdot \overline{\beta} a = a \cdot m_X$, and, moreover, every ultrafilter χ has a generic convergence point x_0, with $a_0 = a \cdot e_X$ the underlying metric; this latter condition means

$$a(\chi, x) = a_0(x_0, x)$$

for all $x \in X$. Similar to the situation in Top, one calls a non-expansive map $\varphi : X \to [0, \infty]$ *irreducible* if $\inf_{x \in X} \varphi(x) = 0$ and, if for any non-expansive maps $\varphi_1, \varphi_2 : X \to [0, \infty]$ with $\varphi(x) \geq \min\{\varphi_1(x), \varphi_2(x)\}$ (for all $x \in X$), one has $\varphi \geq \varphi_1$ or $\varphi \geq \varphi_2$. Recall from Examples III.5.3.2 that we consider $[0, \infty]$

as an approach space with convergence $\lambda(u, u) = u \ominus (\sup_{A \in u} \inf_{v \in A} v)$, for $u \in \beta[0, \infty]$ and $u \in [0, \infty]$. A typical example of an irreducible non-expansive map is $\varphi = a_0(x, -)$, for $x \in X$ (see Exercise III.5.J). An approach space X is called *sober* whenever every irreducible non-expansive map $\varphi : X \to [0, \infty]$ is of the form $\varphi = a_0(x, -)$, for a unique $x \in X$; X is called *weakly sober* if such $x \in X$ is not necessarily unique.

III.5.9.1 Lemma *Let $X = (X, a)$ be an approach space and let $\varphi : X \to [0, \infty]$ be an irreducible non-expansive map. Then there exists some $\chi \in \beta X$ with $\varphi = a(\chi, -)$.* ©

Proof We freely make use of Exercise III.5.J and the notation introduced there. Let $\varphi : X \to [0, \infty]$ be irreducible. For every $u \in [0, \infty]$, $u > 0$, put $A_u = \{x \in X \mid \varphi(x) \leq u\}$; by hypothesis, $A_u \neq \varnothing$. Then $\varphi_{A_u} \leq \varphi$ since, with $A := \{x \in X \mid \varphi_{A_u}(x) \leq \varphi(x)\}$, one has $A_u \subseteq A$ and $0 < \inf\{\varphi(x) \mid x \in X,\ x \notin A\} =: v$. Then $\varphi(x) \geq \min\{\varphi_{A_u}(x), v\}$, but $v \leq \varphi$ is not possible since $\inf_{x \in X} \varphi(x) = 0$, therefore $\varphi_{A_u} \leq \varphi$. The down-directed set

$$f = \{A_u \mid u \in [0, \infty],\ u > 0\}$$

is disjoint from

$$j = \{B \subseteq X \mid \varphi_B \not\leq \varphi\},$$

and j is an ideal since φ is irreducible. Hence, by Corollary II.1.13.5 there is some ultrafilter $\chi \in \beta X$ with $f \subset \chi$ and $\chi \cap j = \varnothing$. Then

$$a(\chi, -) = \sup_{A \in \chi} \varphi_A \leq \varphi,$$

and $\varphi \leq a(\chi, -)$ since $\sup_{A \in \chi} \inf_{x \in A} \varphi(x) = 0$. □

Following the example of topological spaces, we call an approach space $X = (X, a)$ *core-compact* whenever $u \cdot \overline{\beta} a = a \cdot m_X$, and we call X *stable* whenever $a(\chi, -)$ is irreducible, for every $\chi \in \beta X$. With this terminology we have the following result.

III.5.9.2 Theorem *An approach space $X = (X, a)$ is representable if and only if X is weakly sober, stable, and core-compact.* ©

Proof If X is representable, then X is core-compact (see Proposition III.5.4.4) and, for every $\chi \in \beta X$, $a(\chi, -)$ is irreducible since $a(\chi, -) = a_0(x_0, -)$ for some $x_0 \in X$. We also conclude that X is weakly sober since every irreducible non-expansive map $\varphi : X \to [0, \infty]$ is of the form $\varphi = a(\chi, -)$ for some $\chi \in \beta X$ (by Lemma III.5.9.1). Assume now that X is weakly sober, stable, and core-compact. Then every ultrafilter $\chi \in \beta$ has a generic convergence point since $a(\chi, -)$ is irreducible and X is weakly sober. Since X is core-compact, the assertion follows from Proposition III.5.4.4. □

Exercises

III.5.A *Connected components of an ordered set.* Show that the left adjoint $\pi_0 : \mathsf{Ord} \to \mathsf{Set}$ of $(-)_d : \mathsf{Set} \to \mathsf{Ord}$, $X \mapsto (X, =)$ sends an ordered set $X = (X, \longrightarrow)$ to $X/\!\sim$, where $x \sim y$ precisely when there exists a path $x \longrightarrow \bullet \longleftarrow \cdots y$. The reflection map $q : X \to X/\!\sim$ is the coequalizer of $p_1, p_2 : R \to X$, where $R \subseteq X \times X$ is the graph of the order relation $\longrightarrow$ of X. Furthermore, if $\longrightarrow$ is *confluent*, so that for all $x \longrightarrow y$, $x \longrightarrow y'$ there exists $z \in X$ with $y \longrightarrow z$ and $y \longrightarrow z$

$$\begin{array}{ccccc} & & x & & \\ & \swarrow & & \searrow & \\ y & & & & y' \\ & \searrow & & \swarrow & \\ & & z\,, & & \end{array}$$

then $x \sim y$ if and only if $x \longrightarrow z \longleftarrow y$ for some $z \in X$.

III.5.B *Connected components of a $\mathcal{V}$-category.* Let $\mathcal{V}$ be a quantale with $k = \top$ the top element of $\mathcal{V}$. Furthermore, assume that $u \otimes v = \bot$ implies $u = \bot$ or $v = \bot$, for all $u, v \in \mathcal{V}$, so that $o : \mathcal{V} \to 2$ defined by ($o(x) = 1 \iff x > \bot$) is a lax homomorphism of quantales (see also Exercise II.1.I). Show that the left adjoint $\pi_0 : \mathcal{V}\text{-}\mathsf{Cat} \to \mathsf{Set}$ of $(-)_d : \mathsf{Set} \to \mathcal{V}\text{-}\mathsf{Cat}$, $X \mapsto (X, 1_X)$ is the composite

$$\mathcal{V}\text{-}\mathsf{Cat} \xrightarrow{B_o} \mathsf{Ord} \xrightarrow{\pi_0} \mathsf{Set}\,,$$

where B_o is defined as in Section III.3.5.

III.5.C *A topology with chosen convergence.* Let X be a set, let $\chi_0 \in \beta X$ be an ultrafilter on X, and $x_0 \in X$. Define a relation $a : \beta X \nrightarrow X$ by

$$\chi \; a \; x \text{ whenever } \begin{cases} \chi = \chi_0 \text{ and } x = x_0\,, & \text{or} \\ \chi = \dot{x}\,. \end{cases}$$

Show that a is the convergence of a topology on X, i.e. $1_X \leq a \cdot e_X$ and $a \cdot \overline{\beta} a \leq a \cdot m_X$.

III.5.D *Irreducible closed sets.* Let X be a topological space and let $A \subseteq X$ be a non-empty closed subset of X. Show that A is irreducible if and only if, for all open subsets $U, V \subseteq X$, if $U \cap V \cap A = \varnothing$, then $U \cap A = \varnothing$ or $V \cap A = \varnothing$.
© Conclude that A is the set of limit points of some filter χ with $A \in \chi$. Give an example of a compact topological space X with an ultrafilter χ where $\lim \chi$ is not irreducible.

© **III.5.E** *The way-below relation via ultrafilter convergence.* Let X be a topological space and let $A, B \subseteq X$ be open subsets of X. Show that $A \ll B$ if and only if every ultrafilter $\chi \in \beta X$ with $A \in \chi$ has a limit point in B. In particular, X is compact if and only if $X \lll X$ if and only if every ultrafilter of X converges.

III.5.F *Split subobjects of locally compact spaces.* Let $f : X \to Y$ and let $g : Y \to X$ be continuous maps between topological spaces with $g \cdot f = 1_X$, and assume that Y is locally compact. Show that X is locally compact.

III.5.G *Local compactness versus core-compactness.* Let X be a topological space. For $B \subseteq X$ open, consider ©

$$\lim{}^{-1}(B) = \{\chi \in \beta X \mid \lim \chi \cap B \neq \varnothing\},$$

where $\lim \chi$ denotes the set of convergence points of χ. Show that $\lim^{-1}(B)$ is open in βX if and only if, for every $x \in B$, there is some open neighborhood U of x with $U \lll B$. Hence, the following statements hold.

(1) For a core-compact space X, the subspace

$$\lim{}^{-1}(X) = \{\chi \in \beta X \mid \chi \text{ converges to some } x \in X\}$$

of βX is locally compact.

(2) If X is core-compact and every convergent ultrafilter $\chi \in \beta X$ has a smallest convergence point, then the map $\lim^{-1}(X) \to X$ that associates to every convergent ultrafilter a (tacitly chosen) smallest convergence point is continuous; therefore, X is locally compact.

(3) If X is Hausdorff, then

$$\begin{aligned} X \text{ is core-compact} &\iff X \text{ is locally compact} \\ &\iff \text{every point of } X \text{ has a compact neighborhood.} \end{aligned}$$

III.5.H *Stable spaces.* Let X be a topological space. Show that ©

(1) if $\lim \chi$ is irreducible for every $\chi \in \beta X$, then X is stable, and
(2) if X is stable and core compact, then $\lim \chi$ is irreducible for every $\chi \in \beta X$.

III.5.I *Local compactness of VX.* Consider the lower-Vietoris space VX of Section III.5.8, for a topological space X. Let $x \in X$ and let $U, U_i \subseteq X$ $(i \in I)$ be open. Then the following hold:

(1) $\overline{\{x\}} \in U^{\Diamond}$ if and only if $x \in U$;
(2) $\left(\bigcup_{i \in I} U_i\right)^{\Diamond} = \bigcup_{i \in I} U_i^{\Diamond}$;
(3) X is core-compact if and only if VX is core-compact.

Hint. For (3), use Remark III.5.6.7.

III.5.J *The approach space* $[0, \infty]$. Consider the Barr extension $\overline{\mathbb{\beta}}$ of the ultrafilter monad $\mathbb{\beta}$ to P_+-Rel. ©

(1) Show that $([0, \infty], \mu, \xi)$ is a metric compact Hausdorff space, where $\mu(u, v) = v \ominus u$ and $\xi : \beta[0, \infty] \to [0, \infty]$, $u \mapsto \sup_{A \in u} \inf_{u \in A} u$.

(2) Let $X = (X, a)$ be an approach space ($= (\beta, \mathsf{P}_+)$-category). Show:

(a) $a(\mathfrak{x}, -) : X \to [0, \infty]$ is non-expansive, for every $\mathfrak{x} \in \beta X$;

(b) $\varphi_A : X \to [0, \infty]$, $x \mapsto \inf\{a(\mathfrak{x}, x) \mid \mathfrak{x} \in \beta X, A \in \mathfrak{x}\}$, is non-expansive, for every $A \subseteq X$. Furthermore, $\varphi_{A\cup B}(x) = \min\{\varphi_A(x), \varphi_B(x)\}$, for all $A, B \subseteq X$ and $x \in X$.

(c) $a(\dot{x}, -) : X \to [0, \infty]$ is irreducible, for every $x \in X$.

(3) Let $X = (X, a)$ be an approach space, $\mathfrak{x} \in \beta X$ and $x \in X$. Show

$$a(\mathfrak{x}, x) = \sup_{A\in\mathfrak{x}} \varphi_A(x) .$$

Conclude that

$$\begin{aligned} a(\mathfrak{x}, -) = \sup\{\varphi \mid \varphi : X \to [0, \infty] \text{ non-expansive},\\ \sup_{A\in\mathfrak{x}} \inf_{x\in X} \varphi(x) = 0\} . \end{aligned}$$

III.5.K *Lifting* $\tilde{T}$. Show that the functor $\tilde{T} : \mathsf{Set} \to (\mathbb{T}, \mathcal{V})\text{-}\mathsf{Cat}$ of Proposition III.3.3.6 factors through the functor $T : (\mathbb{T}, \mathcal{V})\text{-}\mathsf{Cat} \to (\mathbb{T}, \mathcal{V})\text{-}\mathsf{Cat}$ of Section III.5.4.

III.5.L *Fundamental adjunctions revisited.* With the notations of Sections III.3.6 and III.5.4, show that the diagrams

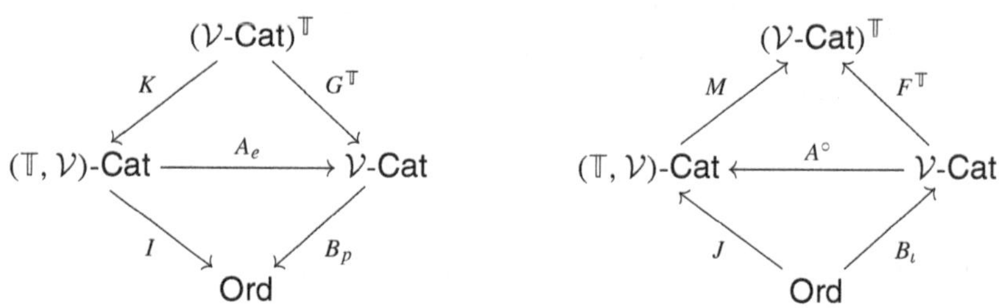

commute (where I denotes the induced-order functor, see Section III.3.3). Describe $J := A^\circ B_\iota$ and show

$$G^{\mathbb{T}} M J(X, =) = (TX, \widehat{1_X^\sharp}) = (TX, \hat{T}1_X) \quad \text{(see the proof of Theorem III.5.3.5)},$$

$$K F^{\mathbb{T}} B_\iota(X, =) = \tilde{T}X \quad \text{(see Proposition III.3.3.6)}.$$

Notes on Chapter III

The Notes on Chapter IV give some information on the history of the axiomatization of convergence in topology, which culminated in the Manes–Barr characterization of a topological space in terms of an abstract ultrafilter convergence relation (see [Manes, 1969, 1974], [Barr, 1970]). With the motivation taken from these papers and from Lawvere's landmark contribution [Lawvere, 1973], the theory of $(\mathbb{T}, \mathcal{V})$-categories as presented in this chapter started with [Clementino and Hofmann, 2003] and [Clementino and Tholen, 2003]. Whereas the term lax extension of a monad $\mathbb{T}$ to $\mathcal{V}$-Rel was first understood in a more restrictive sense than the one used here (dealing only with flat extensions), the set of axioms presented in Sections III.1.4 and III.1.5 was shaped in [Seal, 2005]. The construction of the lax extension of a functor and monad to Rel of, respectively, Section III.1.10 and Section III.1.12 stems from [Barr, 1970]. The deep connection between extensions of functors to categories of internal relations and the preservation of weak pullbacks was first exposed in [Trnková, 1977]. The Beck–Chevalley condition belongs to the folklore domain of higher-level category theory and is credited independently to both name givers. The weaker notion of taut monad appears first (under the name Alexandrov monad) in Möbus' thesis [Möbus, 1981] and was re-introduced in [Manes, 2002]. The double-categorical presentation of lax extensions presented in Section III.1.13 appears in [Cruttwell and Shulman, 2010] following a suggestion by Paré.

The quantaloid $\mathcal{V}$-Rel described in Section III.1.1 was introduced in a more general form in [Betti *et al.*, 1983] and extensively used in a particular case by Rosebrugh and Wood [2002]. In these papers, the quantale $\mathcal{V}$ is allowed to be a monoidal category or even a bicategory, and the term $\mathcal{V}$-matrix is used instead of $\mathcal{V}$-relation. This term was adopted also in the first studies of lax algebras for a Set-monad laxly extended to $\mathcal{V}$-Rel, as given in [Clementino and Tholen, 2003] (for $\mathcal{V}$ a symmetric monoidal closed category), as well as in the subsequent paper [Clementino, Hofmann, and Tholen, 2004b] (for $\mathcal{V}$ a commutative and unital quantale). The latter paper also introduced the Kleisli convolution of Definition III.1.7.1 (under the name co-Kleisli composition). The associativity criterion of Proposition III.1.9.4 for this operation is original, while the identification of maps (in Lawvere's sense) in the 2-category $\mathcal{V}$-Rel as given in Proposition III.1.2.1 goes back to [Freyd and Scedrov, 1990] and [Clementino and Hofmann, 2009].

Of course, for a symmetric monoidal closed category $\mathcal{V}$, the notion and theory of $(\mathbb{T}, \mathcal{V})$-categories as initiated in [Clementino and Tholen, 2003] builds on the theory of $\mathcal{V}$-categories, as introduced in [Eilenberg and Kelly, 1966] and developed further in [Kelly, 1982], after their significance in the context of the subject of this book had been emphasized in [Lawvere, 1973]. The concept of module (called bimodule by Lawvere) was originally introduced (under the name distributor, but often also called profunctor) by Bénabou; see [Bénabou, 2000].

The presentation of metric spaces as enriched categories (Example III.1.3.1(2)) is due to Lawvere [1973]. This description motivated numerous works on the reconciliation of order, metric, and category theory; see in particular the work of Flagg and his coauthors on continuity spaces [Flagg, 1992, 1997; Flagg and Kopperman, 1997; Flagg, Sünderhauf, and Wagner, 1996] and (metric) generalizations of domain theory as in [Bonsangue, van Breugel, and Rutten, 1998] and [Wagner, 1994]. The probabilistic metric spaces presented in Section III.2.1 were introduced in [Menger, 1942]; for more information, see [Schweizer and Sklar, 1983]. They were recognized as enriched categories in [Flagg, 1992] and as such were further investigated in [Chai, 2009] and [Hofmann and Reis, 2013]. The monadicity of compact Hausdorff spaces exposed in Theorem III.2.3.3 is due to Manes [1969], and the ensuing presentation of topological spaces as relational algebras (Theorem III.2.2.5) was established in [Barr, 1970]. Approach spaces were introduced in [Lowen, 1989], and a comprehensive presentation of their theory can be found in [Lowen, 1997]. The description of approach spaces as lax algebras (Theorem III.2.4.5) was established in [Clementino and Hofmann, 2003]. The lax-algebraic description of closure spaces of Section III.2.5 together with the introduction of their metric version of Exercise III.2.G appeared first in [Seal, 2005].

The easily established but important property of topologicity of categories of lax algebras over **Set** along with the investigation of algebraic functors, change-of-base functors, and induced orders was already present in the initial papers on the subject (see [Clementino and Hofmann, 2003], [Clementino and Tholen, 2003], [Clementino *et al.*, 2004b] and [Seal, 2009]). Of course, these types of functors were previously studied in the monad and enriched-category contexts. Universality of coproducts in these categories was recognized in [Mahmoudi, Schubert, and Tholen, 2006]; for the study of quotient structures, see [Hofmann, 2005].

The notions of pseudotopological and pretopological spaces were introduced in [Choquet, 1948] (Example III.4.1.3(2)). Partial products of topological spaces first appeared in [Pasynkov, 1965] and were studied in [Dyckhoff, 1984]. The categorical notion and its linkage with exponentiable morphisms (studied in the general categorical as well as the topological realm by Niefield [1982]) was established in [Dyckhoff and Tholen, 1987]. The notion of quasitopos (see Section III.4.8) was given by Penon [1973]. Machado proved [1973] that **PsTop** is Cartesian closed, and Wyler showed [1976] that **PsTop** is a quasitopos (in fact, the quasitopos hull of **Top**), see Example III.4.8.5(1). The corresponding facts about **PsApp** can be found in [Colebunders and Lowen, 1988, 1989]. Under the conditions of Corollary III.4.8.2, but in the broader context of a monoidal category $\mathcal{V}$, the quasitopos property of $(\mathbb{T}, \mathcal{V})$-**Gph** was established in [Clementino, Hofmann, and Tholen, 2003a]. Final density of $(\mathbb{T}, \mathcal{V})$-**Cat** in this category as stated in Theorem III.4.9.2 originates with [Clementino and Hofmann, 2012].

The theme of Section III.5 is motivated by the equivalence between Nachbin's ordered compact Hausdorff spaces (introduced in [Nachbin, 1950]) and stably compact spaces, which were first described in [Gierz *et al.*, 1980]. Another source of inspiration is Hermida [2000]; see [Hermida, 2000] on representable multicategories which established a similar correspondence in the context of multicategories. The $(\mathbb{T}, \mathcal{V})$-framework presented in the first subsections stems largely from [Tholen, 2009] and is augmented by crucial ingredients from [Clementino and Hofmann, 2009], such as the functor M of Theorem III.5.3.5 that in essence facilitates the notion of representability of a $(\mathbb{T}, \mathcal{V})$-category. Representable T0-spaces are known as stably compact spaces; for more information, we refer to [Gierz *et al.*, 2003]; [Jung, 2004]; [Lawson, 2011] (see also the comment after Theorem III.5.7.2). The notion of dual stably compact spaces goes back to [de Groot, 1967]; [de Groot, Strecker, and Wattel, 1967]; [Hochster, 1969] (see Corollary III.5.7.9). The Čech–Stone compactification of a completely regular topological space, briefly referred to in Section III.5.6, was introduced in [Čech, 1937] and [Stone, 1937] building on [Tychonoff, 1930]. The characterization of normal topological spaces in terms of convergence (Proposition III.5.6.2) was first obtained in [Möbus, 1981]. The characterization of exponentiable topological spaces as precisely the core-compact ones (Theorem III.5.8.5) is due to [Day and Kelly, 1970]; for more information, see [Isbell, 1986]. The characterization of core-compactness via convergence (Proposition III.5.6.6) stems from [Möbus, 1981, 1983] and [Pisani, 1999]. However, the approach taken in this book is quite distinct from the one in these sources as it makes essential use of the Vietoris construction, which has its roots in [Vietoris, 1922]. The notion of sobriety for approach spaces featured in the characterization of representable approach spaces in Section III.5.9 was introduced in [Banaschewski, Lowen, and Van Olmen, 2006] and further studied in [Van Olmen, 2005], which uses the term approach prime map for what is called (in resemblance to the topological counterpart) irreducible non-expansive map in this book.

Suggestions for further reading: [Bentley, Herrlich, and Lowen, 1991]; [Birkedal, Støvring, and Thamsborg, 2010]; [Bunge, 1974]; [Clementino and Hofmann, 2009]; [Gerlo, Vandersmissen, and Van Olmen, 2006]; [Herrlich, Colebunders, and Schwarz, 1991]; [Hofmann and Waszkiewicz, 2011]; [Kopperman, 1988]; [Kostanek and Waszkiewicz, 2011]; [Lowen, 2013]; [Rutten, 1998]; [van Breugel, 2001]; [Van Olmen and Verwulgen, 2010]; [Waszkiewicz, 2009].

IV

Kleisli monoids

Dirk Hofmann, Robert Lowen, Rory Lucyshyn-Wright, and Gavin J. Seal

This chapter revolves around an alternative presentation of $(\mathbb{T}, \mathcal{V})$-$\mathsf{Cat}$ as the category $\mathbb{T}$-Mon of monoids in the hom-set of a Kleisli category that has the advantage of avoiding explicit use of relations or lax extensions. Our role model is given by the filter monad $\mathbb{F}$ for which $\mathbb{F}$-$\mathsf{Mon} \cong \mathsf{Top}$. After obtaining an isomorphism

$$\mathbb{T}\text{-}\mathsf{Mon} \cong (\mathbb{T}, 2)\text{-}\mathsf{Cat}$$

in Section IV.1, in Section IV.2 we use the isomorphisms

$$(\beta, 2)\text{-}\mathsf{Cat} \cong \mathsf{Top} \cong (\mathbb{F}, 2)\text{-}\mathsf{Cat}$$

as role models to compare $(\mathbb{S}, \mathcal{V})$-categories with $(\mathbb{T}, \mathcal{V})$-categories for a monad morphism $\alpha : \mathbb{S} \to \mathbb{T}$ and a general $\mathcal{V}$ in lieu of the embedding $\beta \to \mathbb{F}$ and $\mathcal{V} = 2$. In Section IV.3, we prove that any $(\mathbb{T}, \mathcal{V})$-category obtained from an associative lax extension can also be presented as a $(\Pi, 2)$-category, in effect passing all needed information provided by $\mathcal{V}$, $\mathbb{T}$ and its lax extension to $\mathcal{V}$-Rel into a new monad $\Pi = \Pi(\mathbb{T}, \mathcal{V})$. In Section IV.4, we identify the injective $(\mathbb{T}, 2)$-categories as precisely the $\mathbb{T}$-algebras by exploiting the fact that the forgetful functor $\mathsf{Set}^{\mathbb{T}} \to (\mathbb{T}, 2)$-$\mathsf{Cat}$ is monadic of Kock–Zöberlein type. Finally, in Section IV.5 we focus on the filter monad and investigate the interplay between $(\mathbb{F}, 2)$-categories and $\mathbb{F}$-algebras in the context of ordered sets.

IV.1 Kleisli monoids and lax algebras

In Section II.1.9, a topological space is defined as a set equipped with a collection of subsets closed under finite intersections and arbitrary unions, and

Monoidal Topology: A Categorical Approach to Order, Metric, and Topology, eds. Dirk Hofmann, Gavin J. Seal, and Walter Tholen. Published by Cambridge University Press.

in Exercise II.1.G an equivalent description in terms of a set with a finitely additive closure operation is given. In Section III.2.2, convergence of ultrafilters is used as a defining structure. In this section, we present two filter-based counterparts that avoid the Axiom of Choice: the first focuses on neighborhood filters (Proposition IV.1.1.1) and serves as a model for the *Kleisli monoids* introduced in Section IV.1.3; the second concentrates on filter convergence (Corollary IV.1.5.4) and is facilitated by a new general construction of a lax extension in Section IV.1.4, namely the *Kleisli extension* of a monad.

IV.1.1 Topological spaces via neighborhood filters

A topological space can be entirely defined in terms of its neighborhood systems by way of the filter monad $\mathbb{F} = (F, m, e)$ described in Example II.3.1.1(5). From a categorical viewpoint, it is convenient for the map $\tau_X : PX \longrightarrow FX$ that sends a set A to the principal filter $\dot{A}$ (see II.1.12) to be monotone; hence, the set FX of filters on X is ordered by the *refinement order*:

$$\mathfrak{x} \leq \mathfrak{y} \iff \mathfrak{x} \supseteq \mathfrak{y} ,$$

for all $\mathfrak{x}, \mathfrak{y} \in FX$. A filter $\mathfrak{x}$ is *finer* than $\mathfrak{y}$, or $\mathfrak{y}$ is *coarser* than $\mathfrak{x}$, if $\mathfrak{x} \supseteq \mathfrak{y}$.

Given a topology $\mathcal{O}X$ and $x \in X$, the collection of all open sets that contain x spans the neighborhood filter $\nu(x)$ of x:

$$A \in \nu(x) \iff \exists U \in \mathcal{O}X \ (x \in U \subseteq A) ,$$

for all $A \subseteq X$. This defines a map $\nu : X \longrightarrow FX$ that sends a point of a topological space to its neighborhood filter and is such that $e_X(x) = \{A \subseteq X \mid x \in A\}$ contains $\nu(x)$ for all $x \in X$, i.e.

$$e_X \leq \nu \tag{IV.1.1.i}$$

in the pointwise refinement order. To relate ν with the filter monad multiplication, we define for $A \subseteq X$ the set $A^{\mathbb{F}}$ of filters that contain A,

$$A^{\mathbb{F}} := \{\mathfrak{a} \in FX \mid A \in \mathfrak{a}\} ,$$

and recall that an open set is a neighborhood of each of its points (see Exercise II.1.Q); thus, in particular, for all $x \in X$ and $A \subseteq X$, one has

$$\begin{aligned} A \in \nu(x) &\iff \exists B \in \nu(x) \, \forall y \in B \ (A \in \nu(y)) \\ &\iff \exists B \in \nu(x) \, \forall y \in B \ (\nu(y) \in A^{\mathbb{F}}) \\ &\iff \exists B \in \nu(x) \ (B \subseteq \nu^{-1}(A^{\mathbb{F}})) \\ &\iff \nu^{-1}(A^{\mathbb{F}}) \in \nu(x) \\ &\iff A^{\mathbb{F}} \in F\nu \cdot \nu(x) \\ &\iff A \in m_X \cdot F\nu \cdot \nu(x) . \end{aligned}$$

This last expression is simply the Kleisli composition of $\nu : X \to FX$ with itself, so the previous equivalences show that

$$\nu \circ \nu \leq \nu . \qquad \text{(IV.1.1.ii)}$$

By (IV.1.1.i) and (IV.1.1.ii), a topology on X determines a monoid in the ordered hom-set $\mathsf{Set}_{\mathbb{F}}(X, X)$ of the Kleisli category $\mathsf{Set}_{\mathbb{F}}$. Consider now a continuous map $f : X \to Y$ between topological spaces, with $\nu : X \to FX$ and $\mu : Y \to FY$ the corresponding neighborhood filter maps. If $B \subseteq Y$ is a neighborhood of $f(x)$, there exists an open set $U \subseteq B$ containing $f(x)$; thus, $f^{-1}(U)$ is an open set with $x \in f^{-1}(U) \subseteq f^{-1}(B)$, and $f^{-1}(B)$ is an element of $\nu(x)$. Thanks to the equivalence

$$f^{-1}(B) \in \nu(x) \iff B \in Ff \cdot \nu(x)$$

(for all $B \subseteq Y$ and $x \in X$), we deduce $\mu \cdot f(x) \subseteq Ff \cdot \nu(x)$ for all $x \in X$, i.e.

$$Ff \cdot \nu \leq \mu \cdot f .$$

Instead of considering a map $f : X \to Y$, one can look at its image $f_\natural = e_Y \cdot f : X \to FY$ under the left adjoint $\mathsf{Set} \to \mathsf{Set}_{\mathbb{F}}$ (see Section II.3.6), and this last condition becomes

$$f_\natural \circ \nu \leq \mu \circ f_\natural \qquad \text{(IV.1.1.iii)}$$

by naturality of e. Not only do the neighborhood filters of topological spaces have properties nicely expressible in the language of the Kleisli category of $\mathbb{F}$, but also the conditions (IV.1.1.i), (IV.1.1.ii), and (IV.1.1.iii) are sufficient to describe topological spaces and continuous maps.

IV.1.1.1 Proposition *The category* Top *of topological spaces and continuous maps is isomorphic to the category* $\mathbb{F}$-Mon *whose objects are pairs* (X, ν), *with* $\nu : X \to FX$ *a monoid in* $\mathsf{Set}_{\mathbb{F}}(X, X)$:

$$\nu \circ \nu \leq \nu , \quad e_X \leq \nu ,$$

and whose morphisms $f : (X, \nu) \to (Y, \mu)$ *are maps* $f : X \to Y$ *such that*

$$f_\natural \circ \nu \leq \mu \circ f_\natural .$$

Proof The previous discussion shows that the neighborhood filters of a topological space define a map $\nu : X \to FX$ satisfying the required properties. Conversely, given a monoid $\nu : X \to FX$ in $\mathsf{Set}_{\mathbb{F}}(X, X)$, we define open sets as those $U \subseteq X$ that are neighborhoods of each of their points:

$$U \in \mathcal{O}X \iff \forall x \in X \; (x \in U \implies U \in \nu(x)) .$$

Straightforward verifications show that the set $\mathcal{O}X$, ordered by inclusion, is closed under arbitrary suprema as well as under finite infima, and that $U \in \mathcal{O}Y$

implies $f^{-1}(U) \in \mathcal{O}X$ when $f : X \to Y$ satisfies $Ff \cdot \nu \leq \mu \cdot f$. One therefore has two functors

$$\mathsf{Top} \to \mathbb{F}\text{-}\mathsf{Mon} \qquad \text{and} \qquad \mathbb{F}\text{-}\mathsf{Mon} \to \mathsf{Top}$$

whose composites are routinely verified to be the identities on Top and $\mathbb{F}\text{-}\mathsf{Mon}$. $\square$

IV.1.2 Power-enriched monads

The ultrafilter-convergence presentation of topological spaces of Section III.2.2 uses the algebra of relations in an essential way. In turn, relations are precisely the morphisms of the Kleisli category associated to the powerset monad: $\mathsf{Set}_{\mathbb{P}} = \mathsf{Rel}$. The passage from neighborhood to convergence structure presented further on in Section IV.1.5 exploits the interaction of filters and relations via the *principal filter monad morphism* $\tau : \mathbb{P} \to \mathbb{F}$ whose components $\tau_X : PX \to FX$ send a set $A \in PX$ to the principal filter $\dot{A} \in FX$. This monad morphism allows us to place the study of neighborhood systems, appearing in Proposition IV.1.1.1 as morphisms of the Kleisli category $\mathsf{Set}_{\mathbb{F}}$, in a more general context. The following proposition recalls that a monad morphism $\tau : \mathbb{P} \to \mathbb{T}$ relates $\mathsf{Set}_{\mathbb{T}}$ with both Rel and Sup via functors

$$\mathsf{Rel} \to \mathsf{Set}_{\mathbb{T}} \to \mathsf{Sup}$$

(Exercises II.3.H and II.3.I).

IV.1.2.1 Proposition *For a monad $\mathbb{T} = (T, m, e)$ on Set, one has a one-to-one correspondence between*

(i) *monad morphisms $\tau : \mathbb{P} \to \mathbb{T}$;*

(ii) *extensions E of the functor $F_{\mathbb{T}} : \mathsf{Set} \to \mathsf{Set}_{\mathbb{T}}$ along the functor $(-)_{\circ} : \mathsf{Set} \to \mathsf{Rel}$ of III.1.2:*

$$\begin{array}{ccc} \mathsf{Rel} & \xrightarrow{E} & \mathsf{Set}_{\mathbb{T}} \\ {\scriptstyle (-)_{\circ}}\uparrow & \nearrow_{F_{\mathbb{T}}} & \\ \mathsf{Set} & & \end{array}$$

(iii) *liftings L of the functor $G_{\mathbb{T}} : \mathsf{Set}_{\mathbb{T}} \to \mathsf{Set}$ along the forgetful functor $\mathsf{Sup} \to \mathsf{Set}$:*

$$\begin{array}{ccc} \mathsf{Set}_{\mathbb{T}} & \xrightarrow{L} & \mathsf{Sup} \\ & {}_{G_{\mathbb{T}}}\searrow & \downarrow \\ & & \mathsf{Set} \end{array}$$

(iv) *complete lattice structures on TX such that $Tf : TX \to TY$ and $m_X : TTX \to TX$ are sup-maps for all maps $f : X \to Y$ and sets X.*

Proof To simplify the proof, we identify Rel with $\mathsf{Set}_{\mathbb{P}}$ (Example II.3.6.2) and Sup with $\mathsf{Set}^{\mathbb{P}}$ via the isomorphism $\mathsf{Set}^{\mathbb{P}} \cong \mathsf{Sup}$ of Example II.3.2.2(2).

(i) $\iff$ (ii): This is a direct consequence of Exercise II.3.I. Here, the functor E sends a $\mathsf{Set}_{\mathbb{P}}$-morphism $r : X \to PY$ to the map $Er = \tau_Y \cdot r : X \to TY$.

(i) $\iff$ (iii): The equivalence follows from Exercise II.3.H. Note that L sends a map $f : X \to TY$ to the $\mathbb{P}$-homomorphism $m_Y \cdot Tf : (TX, m_X \cdot \tau_{TX}) \to (TY, m_Y \cdot \tau_{TY})$.

(iii) $\iff$ (iv): The functor $G_{\mathbb{T}}$ of (iii) sends a map $g : X \to TY$ to $m_Y \cdot Tg : TX \to TY$, so that (with $g = 1_{TY}$ or $g = e_Y \cdot f$) condition (iv) is just an element-wise restatement of (iii). □

For a morphism $\tau : \mathbb{P} \to \mathbb{T}$ of monads on Set, condition (iii) equips the underlying set TX of a free $\mathbb{T}$-algebra with the separated order given by

$$\mathfrak{x} \leq \mathfrak{y} \iff m_X \cdot \tau_{TX}(\{\mathfrak{x}, \mathfrak{y}\}) = \mathfrak{y} \tag{IV.1.2.i}$$

for all $\mathfrak{x}, \mathfrak{y} \in TX$. The hom-sets $\mathsf{Set}(X, TY)$ become separated ordered sets via the induced pointwise order:

$$f \leq g \iff \forall x \in X\ (f(x) \leq g(x))$$

for all $f, g : X \to TY$. Composition on the right is always monotone, but composition on the left $(-)^{\mathbb{T}} \cdot f : \mathsf{Set}_{\mathbb{T}}(Y, Z) \to \mathsf{Set}_{\mathbb{T}}(X, Z)$ may fail to be so; see Exercise IV.1.C (here, $(-)^{\mathbb{T}} = m_Z \cdot T(-)$ denotes the monad extension operation of II.3.7). To remedy this, and therefore make $\mathsf{Set}_{\mathbb{T}}$ into a separated ordered category, it suffices that $(-)^{\mathbb{T}}$ be monotone:

$$f \leq g \implies f^{\mathbb{T}} \leq g^{\mathbb{T}},$$

for all $f, g : X \to TY$. If this condition is satisfied, then the functors $E : \mathsf{Rel} \to \mathsf{Set}_{\mathbb{T}}$ and $L : \mathsf{Set}_{\mathbb{T}} \to \mathsf{Sup}$ of Proposition IV.1.2.1 become 2-functors between ordered categories.

IV.1.2.2 Definition A *power-enriched monad* is a pair $(\mathbb{T}, \tau)$ composed of a monad $\mathbb{T}$ on Set and a monad morphism $\tau : \mathbb{P} \to \mathbb{T}$ such that

$$f \leq g \implies f^{\mathbb{T}} \leq g^{\mathbb{T}}, \tag{IV.1.2.ii}$$

for all $f, g : X \to TY$. A *morphism* $\alpha : (\mathbb{S}, \sigma) \to (\mathbb{T}, \tau)$ of power-enriched monads is a monad morphism $\alpha : \mathbb{S} \to \mathbb{T}$ such that $\tau = \alpha \cdot \sigma$:

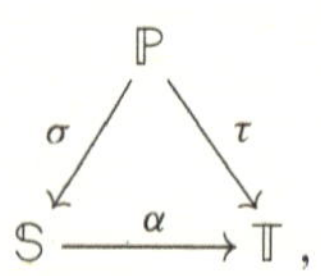

so the category of power-enriched monads is a full subcategory of the comma category $\mathbb{P}/\,\mathsf{MND}_{\mathsf{Set}}$ (see also Exercise IV.1.A). When working with power-enriched monads $(\mathbb{T}, \tau)$, we will often assume a fixed choice of τ, and speak of "the power-enriched monad $\mathbb{T}$."

IV.1.2.3 Examples

(1) There are two trivial monads on Set (Exercise II.3.A), but only one is power-enriched, namely the terminal monad $\mathbb{1}$ whose functor sends all sets to a singleton $\{\star\}$; the components of its structure morphism $\mathbb{P} \to \mathbb{1}$ are the unique maps $!_X : PX \to \{\star\}$. The other monad does not even have a structure morphism, as there is no map from $P\varnothing = \{\star\}$ to $\varnothing$.

(2) The powerset monad $\mathbb{P}$ with the identity structure $1_{\mathbb{P}} : \mathbb{P} \to \mathbb{P}$ is power-enriched. Hence, $(\mathbb{P}, 1_{\mathbb{P}})$ is an initial object in the category of power-enriched monads and their morphisms. The order on the sets PX coming from (IV.1.2.i) is simply subset inclusion because the supremum operation is given by arbitrary union.

(3) The filter monad $\mathbb{F}$ is power-enriched via the principal filter natural transformation $\tau : P \to F$ which yields a monad morphism $\tau : \mathbb{P} \to \mathbb{F}$. The order on FX defined by (IV.1.2.i) is the refinement order introduced in Section IV.1.1, and suprema in FX are given by intersections.

(4) The ultrafilter monad β is not power-enriched: for the set $X = \varnothing$, one observes that $\beta X = \varnothing$ cannot be a complete lattice.

(5) The up-set monad has at least two different structure morphisms $\sigma, \tau : \mathbb{P} \to \mathbb{U}$, defined componentwise for $A \in PX$ by

$$\sigma_X(A) = \{B \subseteq X \mid A \cap B \neq \varnothing\} \qquad \text{and} \qquad \tau_X(A) = \{B \subseteq X \mid A \subseteq B\}$$

(τ is just the extension of the principal filter natural transformation). The order induced on UX by σ is given by subset inclusion, while the one induced by τ is opposite, i.e. τ induces the *refinement order* on up-sets: for all $\mathfrak{x}, \mathfrak{y} \in UX$,

$$\mathfrak{x} \leq \mathfrak{y} \iff \mathfrak{x} \supseteq \mathfrak{y} .$$

These morphisms demonstrate that the morphism $\mathbb{P} \to \mathbb{T}$ given with a power-enriched monad is indeed a structure and not a property of the monad.

(6) Both monad morphisms of the previous example can be extended to the double-powerset monad to give $\sigma, \tau : \mathbb{P} \to \mathbb{P}^2$. However, neither of these satisfy the condition (IV.1.2.ii) (see Exercise IV.1.C).

IV.1.3 $\mathbb{T}$-monoids

Motivated by Proposition IV.1.1.1, we introduce the category of monoids in the hom-sets of a Kleisli category.

IV.1.3.1 Definition Let $\mathbb{T} = (T, m, e)$ be a monad on a category X whose Kleisli category $\mathsf{X}_{\mathbb{T}}$ is a separated ordered category. The category $\mathbb{T}$-Mon of $\mathbb{T}$-*monoids* (or *Kleisli monoids*) has as objects pairs (X, ν), where X is an X-object, and its structure $\nu : X \to TX$ is a *transitive* and *reflexive* $\mathsf{X}_{\mathbb{T}}$-morphism:

$$\nu \circ \nu \leq \nu\,, \qquad e_X \leq \nu$$

(where $\circ$ is composition of the Kleisli category $\mathsf{X}_{\mathbb{T}}$); a morphism $f : (X, \nu) \to (Y, \mu)$ is an X-morphism $f : X \to Y$ satisfying:

$$Tf \cdot \nu \leq \mu \cdot f \qquad \text{or equivalently} \qquad f_\natural \circ \nu \leq \mu \circ f_\natural\,,$$

where $f_\natural := e_Y \cdot f$. In the case where $\mathbb{T} = (\mathbb{T}, \tau)$ is a power-enriched monad, the order on the hom-sets of $\mathsf{Set}_{\mathbb{T}}$ depends on τ; however, we will often assume that τ is given implictly, and denote a category of Kleisli monoids by $\mathbb{T}$-Mon rather than by $(\mathbb{T}, \tau)$-Mon.

We hasten to remark that, in the presence of the reflexivity condition, transitivity can be expressed as an equality $\nu \circ \nu = \nu$, since

$$\nu = \nu \circ e_X \leq \nu \circ \nu \leq \nu\,.$$

Idempotent structures are also preserved by the functor $G_{\mathbb{T}} = (-)^{\mathbb{T}} : \mathsf{X}_{\mathbb{T}} \to \mathsf{X}$:

$$\nu^{\mathbb{T}} \cdot \nu^{\mathbb{T}} = (\nu \circ \nu)^{\mathbb{T}} = \nu^{\mathbb{T}}\,. \tag{IV.1.3.i}$$

IV.1.3.2 Examples

(1) For $\mathbb{T} = \mathbb{1}$ the terminal monad, Kleisli monoids are simply pairs $(X, !_X : X \to \{\star\})$, and morphisms are maps $f : X \to Y$. In other words, the category of Kleisli monoids is isomorphic to Set:

$$\mathbb{1}\text{-}\mathsf{Mon} \cong \mathsf{Set}\,.$$

(2) In the case of the powerset monad (together with its identity structure $1_{\mathbb{P}}$), $\mathbb{P}$-Mon is the category of ordered sets. Indeed, a map $\nu : X \to PX$ is precisely a relation on X, and the transitivity and reflexivity conditions translate as reflexivity and transitivity of ν; because the set PX is ordered by set-inclusion, ν is the down-set map $\downarrow_X : X \to PX$ (see Section II.1.7). A map $f : X \to Y$ is a morphism of $\mathbb{P}$-Mon if and only if it preserves the relations, i.e. if and only if f is a monotone map. Hence,

$$\mathbb{P}\text{-}\mathsf{Mon} \cong \mathsf{Ord}\,.$$

(3) Proposition IV.1.1.1 and Example IV.1.2.3(3) show that when $\mathbb{F}$ is equipped with the principal filter morphism $\tau : \mathbb{P} \to \mathbb{F}$, $\mathbb{F}$-Mon is the category of topological spaces and continuous maps:

$$\mathbb{F}\text{-}\mathsf{Mon} \cong \mathsf{Top}\,.$$

(4) With the principal filter morphism $\tau : \mathbb{P} \to \mathbb{U}$ of Example IV.1.2.3(5), the category of $\mathbb{U}$-monoids is isomorphic to the category of interior spaces:

$$\mathbb{U}\text{-}\mathsf{Mon} \cong \mathsf{Int}$$

(one can proceed as in the proof of Proposition IV.1.1.1 or more syntactically as in Exercise IV.1.D). In fact, the monad morphism $\sigma : \mathbb{P} \to \mathbb{U}$ of Example IV.1.2.3(5) yields

$$\mathbb{U}\text{-}\mathsf{Mon} \cong \mathsf{Cls}\ ,$$

so that the structures τ and σ return isomorphic categories of $\mathbb{U}$-monoids.

IV.1.3.3 Proposition *A morphism of power-enriched monads $\alpha : (\mathbb{S}, \sigma) \to (\mathbb{T}, \tau)$ induces a functor*

$$\mathbb{S}\text{-}\mathsf{Mon} \to \mathbb{T}\text{-}\mathsf{Mon}$$

that sends (X, ν) to $(X, \alpha_X \cdot \nu)$ and commutes with the underlying-set functors.

Proof Thanks to the functor $\mathsf{Set}_\alpha : \mathsf{Set}_\mathbb{S} \to \mathsf{Set}_\mathbb{T}$ that sends ν to $\alpha_X \cdot \nu$ (Exercise II.3.I), the claim easily follows from the fact that α_X is monotone (Exercise IV.1.A). □

IV.1.4 The Kleisli extension

Categories of lax algebras depend upon the lax extension of a monad $\mathbb{T}$ on Set to $\mathcal{V}$-Rel. The Barr extension (Section III.1.10) provides a construction of a lax extension by viewing a relation $r : X \nrightarrow Y$ as a composite $r = q \cdot p^\circ$ (where p, q are projection maps); the Kleisli extension introduced in the following exploits relations as morphisms of the Kleisli category $\mathsf{Set}_\mathbb{P} = \mathsf{Rel}$ (Example II.3.6.2). Hence, we will often be working with maps $r : X \to PY$ representing relations $r : X \nrightarrow Y$, and will indifferently use the notations $\mathbb{P} = (P, \bigcup, \{-\})$ or $(P, (-)^\mathbb{P}, \{-\})$ for the powerset monad, and $\mathbb{T} = (T, m, e)$ or $(T, (-)^\mathbb{T}, e)$ for an arbitrary monad on Set, together with the corresponding expressions for monad morphisms $\tau : \mathbb{P} \to \mathbb{T}$ (see Section II.3.7).

Let us denote by $(-)^\flat : \mathsf{Rel}^{\mathrm{op}} \to \mathsf{Set}_\mathbb{P}$ the functor that is identical on sets and sends a relation $r : X \nrightarrow Y$ to the map $r^\flat : Y \to PX$ representing the opposite relation $r^\circ : Y \nrightarrow X$:

$$x\ r\ y \iff x \in r^\flat(y)\ .$$

By composition with the functors $E : \mathsf{Set}_\mathbb{P} \to \mathsf{Set}_\mathbb{T}$ and $L : \mathsf{Set}_\mathbb{T} \to \mathsf{Sup}$ of Proposition IV.1.2.1, one obtains a functor

$$(-)^\tau : \mathsf{Rel}^{\mathrm{op}} \xrightarrow{(-)^\flat} \mathsf{Set}_\mathbb{P} \xrightarrow{E} \mathsf{Set}_\mathbb{T} \xrightarrow{L} \mathsf{Sup}$$

that sends a set X to TX, and a relation $r : X \nrightarrow Y$ to the map $r^\tau : TY \to TX$ defined by

$$r^\tau := m_X \cdot T(\tau_X \cdot r^\flat) = (\tau_X \cdot r^\flat)^\mathbb{T}\ .$$

IV.1.4.1 Definition Given a power-enriched monad $(\mathbb{T}, \tau)$, the *Kleisli extension* $\check{T}$ of $\mathbb{T}$ to Rel (with respect to τ) is described by the functions $\check{T} = \check{T}_{X,Y}$: $\mathsf{Rel}(X, Y) \to \mathsf{Rel}(TX, TY)$ (indexed by sets X and Y), with

$$\mathfrak{x}\ (\check{T}r)\ \mathfrak{y} \iff \mathfrak{x} \le r^{\tau}(\mathfrak{y}) \qquad \text{(IV.1.4.i)}$$

for all relations $r : X \nrightarrow Y$, and $\mathfrak{x} \in TX$, $\mathfrak{y} \in TY$, or, equivalently,

$$(\check{T}r)^{\flat} = \downarrow_{TX} \cdot r^{\tau} : TY \to PTX\ .$$

IV.1.4.2 Examples

(1) In the case of the terminal power-enriched monad $(\mathbb{1}, !)$, the Kleisli extension of a relation $r : X \nrightarrow Y$ is $\{\star\} \nrightarrow \{\star\}$ with constant value $\top$.

(2) To obtain an explicit description of the Kleisli extension of the powerset monad $(\mathbb{P}, 1_{\mathbb{P}})$, observe that

$$\begin{aligned} A \subseteq r^{1_{\mathbb{P}}}(B) &\iff A \subseteq \textstyle\bigcup_X \cdot Pr^{\flat}(B) \iff \forall x \in A\ \exists y \in B\ (x \in r^{\flat}(y)) \\ &\iff A \subseteq r^{\circ}(B)\ , \end{aligned}$$

for a relation $r : X \nrightarrow Y$, and $A \in PX$, $B \in PY$, where $r^{\circ}(B) = \{x \in X \mid \exists y \in B\ (x \in r^{\flat}(y))\}$, as in Example III.1.10.3(2). Hence,

$$A\ (\check{P}r)\ B \iff A \subseteq r^{\circ}(B)\ .$$

Here, we obtain the lax extension $\check{P}$ introduced in Example III.1.4.2(2).

(3) Let $\mathbb{T} = \mathbb{F}$ be the filter monad and let $\tau : \mathbb{P} \to \mathbb{T}$ be the principal filter natural transformation. For a relation $r : X \nrightarrow Y$, $A \subseteq X$, and $\mathfrak{y} \in FY$, we have

$$\begin{aligned} A \in m_X \cdot F(\tau_X \cdot r^{\flat})(\mathfrak{y}) &\iff (\tau_X \cdot r^{\flat})^{-1}(A^{\mathbb{F}}) \in \mathfrak{y} \\ &\iff \{y \in Y \mid A \in \tau_X \cdot r^{\flat}(y)\} \in \mathfrak{y} \\ &\iff \{y \in Y \mid r^{\flat}(y) \subseteq A\} \in \mathfrak{y} \\ &\iff \exists B \in \mathfrak{y}\ (r^{\circ}(B) \subseteq A)\ . \end{aligned}$$

This shows precisely that

$$r^{\tau}(\mathfrak{y}) = \uparrow_{PX}\{r^{\circ}(B) \mid B \in \mathfrak{y}\}\ ,$$

so $\mathfrak{x}\ (\check{F}r)\ \mathfrak{y} \iff \mathfrak{x} \supseteq r^{\tau}(\mathfrak{y})$ or, if we use the notation $r^{\circ}[\mathfrak{y}] = \{r^{\circ}(B) \mid B \in \mathfrak{y}\}$,

$$\mathfrak{x}\ (\check{F}r)\ \mathfrak{y} \iff \mathfrak{x} \supseteq r^{\circ}[\mathfrak{y}]\ . \qquad \text{(IV.1.4.ii)}$$

The Kleisli extension of the filter monad returns the lax extension $\check{F}$ of Example III.1.10.3(4).

(4) The Kleisli extension of the up-set monad $\mathbb{U}$, equipped with the principal filter natural transformation $\tau : \mathbb{P} \to \mathbb{U}$, is obtained as for the filter monad in the previous example, so that

$$\mathfrak{x}\ (\check{U}r)\ \mathfrak{y} \iff \mathfrak{x} \supseteq r^{\circ}[\mathfrak{y}]$$

for all maps $r : X \longrightarrow PY$ and up-sets $\mathfrak{x} \in UX$, $\mathfrak{y} \in UY$.

To prove that $\check{T} : \mathsf{Rel} \longrightarrow \mathsf{Rel}$ is indeed a lax extension of the Set-functor T, it is convenient to express the former as a composite of lax functors. In view of this, we remark that $\check{T}r$ (for a relation $r : X \nrightarrow Y$) can be written as

$$\check{T}r = (r^\tau)^* : TX \nrightarrow TY\,,$$

where $(-)^* : \mathsf{Ord} \longrightarrow \mathsf{Mod}^{\mathrm{op}}$ is the functor that sends a monotone map $f : X \to Y$ to the module $f^* = f^\circ \cdot (\leq_Y) : Y \nrightarrow X$ (see II.1.4 and II.2.2). The Kleisli extension is therefore a functor

$$\check{T} : \mathsf{Rel}^{\mathrm{op}} \xrightarrow{(-)^\tau} \mathsf{Sup} \longrightarrow \mathsf{Ord} \xrightarrow{(-)^*} \mathsf{Mod}^{\mathrm{op}}$$

(here, $\mathsf{Sup} \longrightarrow \mathsf{Ord}$ is the forgetful functor). There is moreover a lax functor $\mathsf{Mod} \longrightarrow \mathsf{Rel}$ that assigns to a module its underlying relation: composition of modules is composition of relations, identity modules are order relations, and $1_X \leq (\leq_X)$ for any ordered set X. Hence, with $E : \mathsf{Set}_{\mathbb{P}} \longrightarrow \mathsf{Set}_{\mathbb{T}}$ and $L : \mathsf{Set}_{\mathbb{T}} \longrightarrow \mathsf{Sup}$ denoting the functors from (ii) and (iii) of Proposition IV.1.2.1, the Kleisli extension $\check{T}^{\mathrm{op}}$ can be decomposed as the top line of the commutative diagram

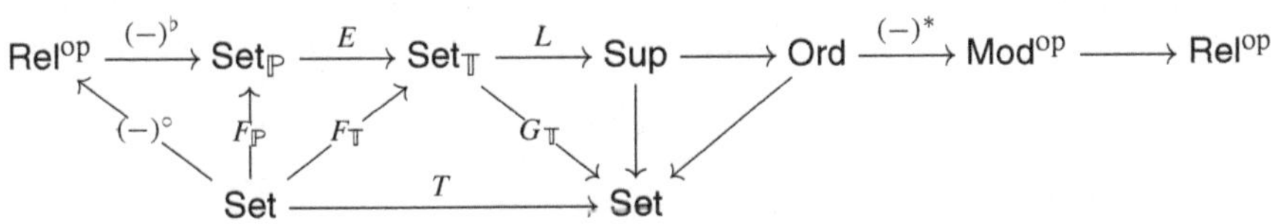

(IV.1.4.iii)

in which all arrows except $\mathsf{Mod}^{\mathrm{op}} \longrightarrow \mathsf{Rel}^{\mathrm{op}}$ are functors, and $\mathsf{Mod}^{\mathrm{op}} \longrightarrow \mathsf{Rel}^{\mathrm{op}}$ is a lax functor that fails only to preserve identities (the unnamed arrows are all forgetful).

IV.1.4.3 Proposition *Given a power-enriched monad $(\mathbb{T}, \tau)$, the Kleisli extension $\check{T}$ of T to* Rel *yields a lax extension $\check{\mathbb{T}} = (\check{T}, m, e)$ of $\mathbb{T} = (T, m, e)$ to* Rel.

Proof The fact that $\check{T} : \mathsf{Rel} \longrightarrow \mathsf{Rel}$ is a lax functor follows from its decomposition as lax functors preserving composition in the first line of (IV.1.4.iii). The lax extension condition $(Tf)^\circ \leq \check{T}(f^\circ)$ can be deduced from the diagram

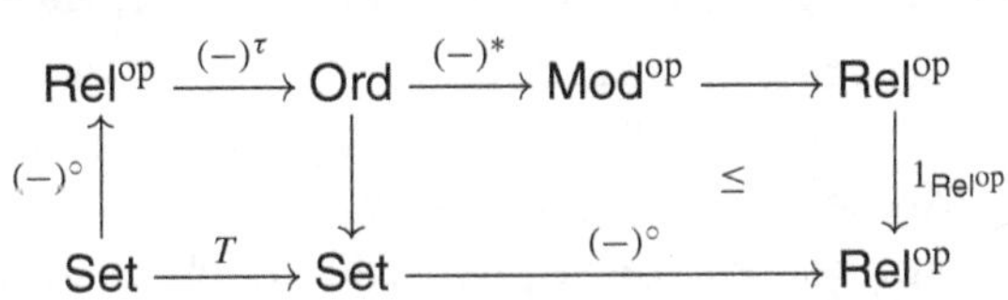

in which the first line is $\check{T}^{\mathrm{op}}$. The second lax extension condition $\check{T}(h^\circ \cdot r) = (Th)^\circ \cdot \hat{T}r$ for all relations $r : X \nrightarrow Y$ and maps $h : Z \longrightarrow Y$ (see Proposition III.1.4.3(3)) comes from the equivalences

$$\mathfrak{x}\ (\check{T}(h^\circ \cdot r))\ \mathfrak{z} \iff \mathfrak{x} \le r^\tau \cdot (h^\circ)^\tau(\mathfrak{z}) \iff \mathfrak{x} \le r^\tau \cdot Th(\mathfrak{z})$$
$$\iff \mathfrak{x}\ ((Th)^\circ \cdot \check{T}r)\ \mathfrak{z}$$

for all $\mathfrak{x} \in TX$, $\mathfrak{z} \in TZ$. To verify oplaxness of $e : 1_{\mathsf{Rel}} \to \check{T}$, we use that $\tau_X = m_X \cdot \tau_{TX} \cdot Pe_X = \bigvee_{TX} \cdot Pe_X$. Given a relation $r : X \nrightarrow Y$, and $x \in X$, $y \in Y$ with $x\ r\ y$, one then has

$$e_X(x) \le \bigvee_{x' \in r^\flat(y)} e_X(x') = \tau_X \cdot r^\flat(y) = (\tau_X \cdot r^\flat)^{\mathbb{T}} \cdot e_Y(y) = r^\tau \cdot e_Y(y)\,,$$

as required. For proving oplaxness of $m : \check{T}\check{T} \to \check{T}$, recall that $m_X \cdot \tau_{TX} \cdot {\downarrow_{TX}} = \bigvee_{TX} \cdot {\downarrow_{TX}} = 1_{TX}$, and note that

$$(r^\tau)^{\mathbb{T}} = (r^\tau \cdot 1_{TY})^{\mathbb{T}} = ((\tau_X \cdot r^\flat)^{\mathbb{T}} \cdot 1_{TY})^{\mathbb{T}} = (\tau_X \cdot r^\flat)^{\mathbb{T}} \cdot 1_{TY}^{\mathbb{T}} = r^\tau \cdot m_Y\,.$$

Thus, if $\mathcal{X} \in TTX$ and $\mathcal{Y} \in TTY$ are such that $\mathcal{X}\ (\check{T}\check{T}r)\ \mathcal{Y}$, or equivalently $\mathcal{X} \le (\check{T}r)^\tau(\mathcal{Y})$, then

$$\begin{aligned} m_X(\mathcal{X}) &\le m_X \cdot (\check{T}r)^\tau(\mathcal{Y}) \\ &= 1_{TX}^{\mathbb{T}} \cdot (\tau_{TX} \cdot {\downarrow_{TX}} \cdot r^\tau)^{\mathbb{T}}(\mathcal{Y}) \\ &= (1_{TX}^{\mathbb{T}} \cdot \tau_{TX} \cdot {\downarrow_{TX}} \cdot r^\tau)^{\mathbb{T}}(\mathcal{Y}) \\ &= (r^\tau)^{\mathbb{T}}(\mathcal{Y}) \\ &= r^\tau \cdot m_Y(\mathcal{Y})\,, \end{aligned}$$

which concludes the proof. □

IV.1.4.4 Remark Since the Kleisli extension provides the monad $\mathbb{T}$ with a lax extension, there is a natural order on TX associated with $\check{T}$ (see III.3.3); on TX there is also the order (IV.1.2.i) induced by the monad morphism $\tau : \mathbb{P} \to \mathbb{T}$. Since the first order $\check{T}1_X$ is defined via the second:

$$\mathfrak{x}\ (\check{T}1_X)\ \mathfrak{y} \iff \mathfrak{x} \le \mathfrak{y}$$

(Definition (IV.1.4.i)), the orders are equivalent. Let us emphasize that $\check{T}$ fails to preserve identity relations unless $\mathbb{T} = \mathbb{1}$ is the terminal power-enriched monad.

IV.1.5 Topological spaces via filter convergence

In this section, we show that $(\mathbb{F}, 2)$-Cat is isomorphic to $\mathbb{F}$-Mon $\cong$ Top (Proposition IV.1.1.1), i.e. we present topological spaces as sets equipped with a transitive and reflexive convergence relation $a : FX \nrightarrow X$.

The correspondence between convergence and neighborhoods can be formalized as in III.2.2 via maps

$$\mathrm{conv} : \mathsf{Set}(X, FX) \to \mathsf{Rel}(FX, X) \quad\text{and}\quad \mathrm{nbhd} : \mathsf{Rel}(FX, X) \to \mathsf{Set}(X, FX)\,.$$

In fact, one can without further thought replace the filter monad $\mathbb{F}$ with a power-enriched monad $(\mathbb{T}, \tau)$. By identifying $\mathsf{Rel}(TX, X)$ with $\mathsf{Set}(X, PTX)$, isomorphic as ordered sets, we define

$$\mathrm{conv}(\nu) = \downarrow_{TX} \cdot \nu \qquad \text{and} \qquad \mathrm{nbhd}(r) = \bigvee_{TX} \cdot r^{\flat}$$

for all maps $\nu : X \to TX$ and relations $r : TX \nrightarrow X$. In pointwise notation, these maps may be written as

$$\chi \ \mathrm{conv}(\nu)\ y \iff \chi \le \nu(y) \quad \text{and} \quad \mathrm{nbhd}(r)(y) = \bigvee\{\chi \in TX \mid \chi \in r^{\flat}(y)\},$$

for all $y \in X$ and $\chi \in TX$, as a direct generalization of the fact that, in a topological space X, a filter χ converges to a point y precisely when χ is finer than the neighborhood filter of y.

IV.1.5.1 Proposition *With* $\mathsf{Set}(X, TX)$ *and* $\mathsf{Rel}(TX, X)$ *ordered pointwise, the monotone maps defined above form an adjunction* $\mathrm{nbhd} \dashv \mathrm{conv} : \mathsf{Set}(X, TX) \to \mathsf{Rel}(TX, X)$ *for all sets* X.

Moreover, the fixpoints of $(\mathrm{conv} \cdot \mathrm{nbhd})$ *are precisely the unitary relations, and* conv *is fully faithful, so that the fixpoints of* $(\mathrm{nbhd} \cdot \mathrm{conv})$ *are the maps* $\nu : X \to TX$.

Proof The equivalence

$$\mathrm{nbhd}(r) \le \nu \iff r^{\flat} \le \mathrm{conv}(\nu)$$

(for all maps $\nu : X \to TX$ and relations $r : TX \nrightarrow X$) follows directly from the adjunction $\bigvee_{TX} \dashv \downarrow_{TX}$. Similarly, from $\bigvee_{TX} \cdot \downarrow_{TX} = 1_{TX}$ follows that $\mathrm{nbhd} \cdot \mathrm{conv} = 1$, i.e. conv is fully faithful (see Corollary II.1.5.2).

Remark IV.1.4.4 shows that $(\check{T} 1_X)^{\flat} = \downarrow_{TX}$. Hence, if $r : TX \nrightarrow X$ is unitary, then

$$\begin{aligned}
r^{\flat} &= (\downarrow_{TX})^{\mathbb{P}} \cdot r^{\flat} && (r \text{ right unitary in III.1.7.3})\\
&= (\downarrow_{TX})^{\mathbb{P}} \cdot (e_X^{\circ} \cdot \check{T} r \cdot m_X^{\circ})^{\flat} && (r \text{ left unitary})\\
&= (\downarrow_{TX})^{\mathbb{P}} \cdot ((m_X^{\circ})^{\flat})^{\mathbb{P}} \cdot (\check{T} r)^{\flat} \cdot e_X && ((e_X^{\circ})^{\flat} = \{-\}_{TX} \cdot e_X)\\
&= (\downarrow_{TX} \cdot m_X)^{\mathbb{P}} \cdot (\check{T} r)^{\flat} \cdot e_X && (g^{\mathbb{P}} \cdot f^{\mathbb{P}} = (g^{\mathbb{P}} \cdot f)^{\mathbb{P}})\\
&= (\downarrow_{TX} \cdot m_X)^{\mathbb{P}} \cdot \downarrow_{TTX} \cdot (\tau_{TX} \cdot r^{\flat})^{\mathbb{T}} \cdot e_X && (\text{definition of } \check{T})\\
&= (\downarrow_{TX} \cdot m_X)^{\mathbb{P}} \cdot \downarrow_{TTX} \cdot \tau_{TX} \cdot r^{\flat} && (f^{\mathbb{T}} \cdot e_X = f)\\
&\ge (\downarrow_{TX} \cdot m_X)^{\mathbb{P}} \cdot \{-\}_{TTX} \cdot \tau_{TX} \cdot r^{\flat} && (\downarrow_{TTX} \ge \{-\}_{TTX})\\
&= \downarrow_{TX} \cdot m_X \cdot \tau_{TX} \cdot r^{\flat} && (f^{\mathbb{P}} \cdot \{-\}_X = f)\\
&= \mathrm{conv} \cdot \mathrm{nbhd}(r) && (m_X \cdot \tau_{TX} = \textstyle\bigvee_{TX})\\
&\ge r^{\flat} && (\mathrm{nbhd} \dashv \mathrm{conv}),
\end{aligned}$$

and we may conclude that $\mathrm{conv} \cdot \mathrm{nbhd}(r) = r$ (via the understood identification of $\mathsf{Rel}(TX, X)$ with $\mathsf{Set}(X, PTX)$). Conversely, if r is a fixpoint of $\mathrm{conv} \cdot \mathrm{nbhd}$, then r is of the form $\mathrm{conv}(\nu)$ for some $\nu : X \to TX$, and one sees that $r \cdot \check{T}1_X \leq r$, so r is right unitary. To prove that r is left unitary, we must verify that $e_X^\circ \cdot \check{T}r \leq r \cdot m_X$. Thus, suppose that $\mathfrak{X}\ (\check{T}r)\ e_X(y)$ holds. By definition of $\check{T}$, we have $\mathfrak{X} \leq r^\tau \cdot e_X(y)$, or equivalently $\mathfrak{X} \leq \tau_{TX} \cdot r^\flat(y)$. Applying m_X to each side of this inequality, we obtain $m_X(\mathfrak{X}) \leq m_X \cdot \tau_{TX} \cdot r^\flat(y)$. This means precisely that $m_X(\mathfrak{X}) \leq \bigvee_{TX} \cdot r^\flat(y) = \mathrm{nbhd}(r)(y)$, or $m_X(\mathfrak{X})\ (\mathrm{conv} \cdot \mathrm{nbhd}(r))\ y$, which is $m_X(\mathfrak{X})\ r\ y$ by the fixpoint condition. □

IV.1.5.2 Proposition *The adjoint maps* nbhd *and* conv *defined above are monoid homomorphisms between* $\mathsf{Set}_{\mathbb{T}}(X, X)$ *and* $(\mathbb{T}, \mathcal{V})$-$\mathsf{URel}^{\mathrm{op}}(X, X)$*, i.e. they satisfy*

$$\mathrm{nbhd}(s \circ r) = \mathrm{nbhd}(r) \circ \mathrm{nbhd}(s)\,, \qquad \mathrm{conv}(\mu) \circ \mathrm{conv}(\nu) = \mathrm{conv}(\nu \circ \mu)\,,$$
$$\mathrm{nbhd}(1_X^\sharp) = e_X\,, \qquad \mathrm{conv}(e_X) = 1_X^\sharp\,,$$

for all unitary relations $r, s : TX \nrightarrow X$*, and maps* $\mu, \nu : X \to TX$*.*

Proof The equality $\mathrm{nbhd}(1_X^\sharp) = e_X$ follows immediately from the definition of $1_X^\sharp$, as

$$\mathfrak{X}\ 1_X^\sharp\ y \iff \mathfrak{X}\ (e_X^\circ \cdot \check{T}1_X)\ y \iff \mathfrak{X} \leq e_X(y)$$

for all $\mathfrak{X} \in TX$ and $y \in X$. The multiplication $m_X = 1_{TX}^{\mathbb{T}}$ of the monad $\mathbb{T}$ is a sup-map and $1_{TX}^{\mathbb{T}} \cdot \tau_{TX} = \bigvee_{TX}$ (Proposition IV.1.2.1), so

$$1_{TX}^{\mathbb{T}} \cdot \tau_{TX} \cdot Pm_X \cdot {\downarrow_{TTX}} = \bigvee_{TX} \cdot Pm_X \cdot {\downarrow_{TTX}} = m_X \cdot \bigvee_{TTX} \cdot {\downarrow_{TTX}} = 1_{TX}^{\mathbb{T}}\,.$$

By definition of $\check{T}$, we then obtain

$$1_{TX}^{\mathbb{T}} \cdot (\tau_{TX} \cdot r^\flat)^{\mathbb{T}} = 1_{TX}^{\mathbb{T}} \cdot \tau_{TX} \cdot Pm_X \cdot {\downarrow_{TTX}} \cdot (\tau_{TX} \cdot r^\flat)^{\mathbb{T}} = 1_{TX}^{\mathbb{T}} \cdot \tau_{TX} \cdot Pm_X \cdot (\check{T}r)^\flat\,. \tag{IV.1.5.i}$$

Therefore,

$$\begin{aligned}
&\mathrm{nbhd}(r) \circ \mathrm{nbhd}(s) \\
&= (1_{TX}^{\mathbb{T}} \cdot \tau_{TX} \cdot r^\flat)^{\mathbb{T}} \cdot 1_{TX}^{\mathbb{T}} \cdot \tau_{TX} \cdot s^\flat && (1_{TX}^{\mathbb{T}} \cdot \tau_{TX} = \textstyle\bigvee_{TX}) \\
&= ((1_{TX}^{\mathbb{T}} \cdot \tau_{TX} \cdot r^\flat)^{\mathbb{T}})^{\mathbb{T}} \cdot \tau_{TX} \cdot s^\flat && (g^{\mathbb{T}} \cdot f^{\mathbb{T}} = (g^{\mathbb{T}} \cdot f)^{\mathbb{T}}) \\
&= (1_{TX}^{\mathbb{T}} \cdot (\tau_{TX} \cdot r^\flat)^{\mathbb{T}})^{\mathbb{T}} \cdot \tau_{TX} \cdot s^\flat && ((g^{\mathbb{T}} \cdot f)^{\mathbb{T}} = g^{\mathbb{T}} \cdot f^{\mathbb{T}}) \\
&= (1_{TX}^{\mathbb{T}} \cdot \tau_{TX} \cdot Pm_X \cdot (\check{T}r)^\flat)^{\mathbb{T}} \cdot \tau_{TX} \cdot s^\flat && (\text{by (IV.1.5.i) above}) \\
&= 1_{TX}^{\mathbb{T}} \cdot (\tau_{TX} \cdot Pm_X \cdot (\check{T}r)^\flat)^{\mathbb{T}} \cdot \tau_{TX} \cdot s^\flat && ((g^{\mathbb{T}} \cdot f)^{\mathbb{T}} = g^{\mathbb{T}} \cdot f^{\mathbb{T}}) \\
&= 1_{TX}^{\mathbb{T}} \cdot \tau_{TX} \cdot (Pm_X \cdot (\check{T}r)^\flat)^{\mathbb{P}} \cdot s^\flat && (\text{naturality of } \tau) \\
&= \textstyle\bigvee_{TX} \cdot ((m_X^\circ)^\flat)^{\mathbb{P}} \cdot ((\check{T}r)^\flat)^{\mathbb{P}} \cdot s^\flat && (((m_X^\circ)^\flat)^{\mathbb{P}} = Pm_X) \\
&= \textstyle\bigvee_{TX} \cdot (s \cdot \check{T}r \cdot m_X^\circ)^\flat && (\mathsf{Rel} = \mathsf{Set}_{\mathbb{P}}) \\
&= \mathrm{nbhd}(s \circ r)\,.
\end{aligned}$$

The equalities for conv then follow directly from the fact that conv and nbhd are inverse of each other on fixpoints (Proposition IV.1.5.1). □

IV.1.5.3 Theorem *Given a power-enriched monad $(\mathbb{T},\tau)$ equipped with its Kleisli extension $\check{T}$, there is an isomorphism*

$$(\mathbb{T},2)\text{-}\mathsf{Cat} \cong \mathbb{T}\text{-}\mathsf{Mon}$$

that commutes with the underlying-set functors.

Proof For a $(\mathbb{T},2)$-algebra (X,r), Proposition IV.1.5.2 implies that $(X,\mathrm{nbhd}(r))$ is a $\mathbb{T}$-monoid, and conversely if (X,ν) is a $\mathbb{T}$-monoid, then $(X,\mathrm{conv}(\nu))$ is a $(\mathbb{T},2)$-algebra (one also uses the fact that nbhd and conv are monotone). Moreover, Proposition IV.1.5.1 entails that these objects are in bijective correspondence.

We are therefore left to show that this correspondence is functorial. Consider first a $(\mathbb{T},2)$-functor $f:(X,r)\to(X,s)$, so that $r\cdot(Tf)^\circ\le f^\circ\cdot s$. We have

$$\begin{aligned}
Tf\cdot\mathrm{nbhd}(r) &= Tf\cdot\textstyle\bigvee_{TX}\cdot r^\flat \\
&= \textstyle\bigvee_{TX}\cdot PTf\cdot r^\flat && (Tf \text{ sup-map}) \\
&= \textstyle\bigvee_{TX}\cdot(r\cdot(Tf)^\circ)^\flat && (PTf=(((Tf)^\circ)^\flat)^{\mathbb{P}}) \\
&\le \textstyle\bigvee_{TX}\cdot(f^\circ\cdot s)^\flat && (f \text{ a } (\mathbb{T},2)\text{-functor}) \\
&= \textstyle\bigvee_{TX}\cdot(s^\flat)^{\mathbb{P}}\cdot d_X\cdot f && ((f^\circ)^\flat=\{-\}_X\cdot f) \\
&= \textstyle\bigvee_{TX}\cdot s^\flat\cdot f && (g^{\mathbb{P}}\cdot\{-\}_X=g) \\
&= \mathrm{nbhd}(s)\cdot f\,;
\end{aligned}$$

hence, f is a morphism of $\mathbb{T}$-monoids. Consider now $f:(X,\nu)\to(Y,\mu)$ satisfying $Tf\cdot\nu\le\mu\cdot f$. Then $\chi\ \mathrm{conv}(\nu)\ y$ means $\chi\le\nu(y)$, so we have $Tf(\chi)\le Tf\cdot\nu(y)\le\mu\cdot f(y)$, and can therefore conclude that $Tf(\chi)\ \mathrm{conv}(\mu)\ y$; i.e. f is a $(\mathbb{T},2)$-functor between the corresponding $(\mathbb{T},2)$-categories. □

IV.1.5.4 Corollary *The category* Top *of topological spaces is isomorphic to the category* $(\mathbb{F},2)$-Cat *whose objects are pairs (X,a), with $a:FX\nrightarrow X$ a relation representing convergence and, when a and $\hat{F}a$ are denoted by $\longrightarrow$, satisfying*

$$\mathfrak{X}\longrightarrow\mathfrak{y}\ \&\ \mathfrak{y}\longrightarrow z\implies\textstyle\sum\mathfrak{X}\longrightarrow z\qquad\textit{and}\qquad\dot{x}\longrightarrow x\,,$$

for all $x,z\in X$, $\mathfrak{y}\in FX$, and $\mathfrak{X}\in FFX$; here $\mathfrak{X}\longrightarrow\mathfrak{y}\iff\mathfrak{X}\supseteq a^\circ[\mathfrak{y}]$. The morphisms are the convergence-preserving maps $f:X\to Y$:

$$\chi\longrightarrow y\implies f[\chi]\longrightarrow f(y)$$

for all $y\in X$, $\chi\in FX$.

Proof Proposition IV.1.1.1 together with Theorem IV.1.5.3 yield an isomorphism between Top and $(\mathbb{F},2)$-Cat, and the statement is just an explicit description of the latter category using the Kleisli extension of the filter monad (IV.1.4.ii). □

Naturally, the same statement holds for the category Cls of closure spaces, for which we now give a more compact form.

IV.1.5.5 Corollary *For the up-set monad* $\mathbb{U}$ *equipped with the Kleisli extension associated with the principal filter natural transformation, there is an isomorphism*

$$\mathsf{Cls} \cong (\mathbb{U}, 2)\text{-}\mathsf{Cat}$$

that commutes with the underlying-set functors.

Proof This is another application of Theorem IV.1.5.3 with Exercise IV.1.D in the case of the up-set monad. □

Exercises

IV.1.A *The slice under* $\mathbb{P}$. Let $\mathbb{S} = (S, n, d)$, $\mathbb{T} = (T, m, e)$ be monads on Set, and $\alpha : \mathbb{S} \longrightarrow \mathbb{T}$ a monad morphism. The following statements are equivalent for objects $\sigma : \mathbb{P} \longrightarrow \mathbb{S}$ and $\tau : \mathbb{P} \longrightarrow \mathbb{T}$ of the comma category $\mathbb{P}/\mathsf{MND}_{\mathsf{Set}}$:

(i) α is a morphism $\sigma \longrightarrow \tau$;
(ii) $\alpha_X : SX \longrightarrow TX$ is a sup-map for all sets X.

IV.1.B *Constructing the power-enrichment.* Let $\mathbb{T} = (T, m, e)$ be a monad on Set. Suppose that the sets TX are equipped with an order that make them complete lattices, and is such that $Tf : TX \longrightarrow TY$ and $m_X : TTX \longrightarrow TX$ are sup-maps for all maps $f : X \longrightarrow Y$ and sets X. The monad morphism $\tau : \mathbb{P} \longrightarrow \mathbb{T}$ of Proposition IV.1.2.1 is then given componentwise by

$$\tau_X(A) = \textstyle\bigvee_{x \in A} e_X(x) .$$

IV.1.C *The double-powerset monad is not power-enriched.* Consider the maps $f, g : \{a, b\} \longrightarrow P^2\{\star\}$ given by

$$f(a) = \{\varnothing\} , \quad f(b) = \{\{\star\}\},$$
$$g(a) = \{\varnothing\} , \quad g(b) = \{\varnothing, \{\star\}\} .$$

One has $f(x) \subseteq g(x)$ for all $x \in \{a, b\}$, but if $\chi = \{\{a\}, \{b\}\} \in P^2\{a, b\}$, then $f^{\mathbb{P}^2}(\chi) \not\subseteq g^{\mathbb{P}^2}(\chi)$, since

$$f^{\mathbb{P}^2}(\chi) = \{\varnothing, \{\star\}\} , \quad g^{\mathbb{P}^2}(\chi) = \{\{\star\}\} ,$$

where $(-)^{\mathbb{P}^2} = m_X \cdot P^2(-)$ comes from the double-powerset monad $\mathbb{P}^2$. Hence, neither the principal filter natural transformation $\tau : \mathbb{P} \longrightarrow \mathbb{P}^2$ nor the natural transformation $\sigma : \mathbb{P} \longrightarrow \mathbb{P}^2$ of Example IV.1.2.3(6) make $\mathbb{P}^2$ into a power-enriched monad.

IV.1.D *Kleisli monoids from the double-powerset monad.* The adjunction $P^\bullet \dashv (P^\bullet)^{\mathrm{op}} : \mathsf{Set}^{\mathrm{op}} \to \mathsf{Set}$ of Example II.2.5.1(6) yields a one-to-one correspondence

$$\frac{X \xrightarrow{f} P^2Y}{PX \xleftarrow{P^\bullet f} PY\,.} \tag{IV.1.5.ii}$$

Both hom-sets $\mathsf{Set}(X, P^2Y)$ and $\mathsf{Set}(PY, PX)$ inherit a pointwise order, the first from the refinement order on P^2Y and the second from the inclusion order on PX. If $f, g \in \mathsf{Set}(X, P^2Y)$ and $P^\bullet f, P^\bullet g \in \mathsf{Set}(PY, PX)$ are maps obtained via (IV.1.5.ii), then

$$f \le g \iff P^\bullet g \le P^\bullet f\,.$$

If we denote by PSet the category whose objects are sets and morphisms from X to Y are maps $f : PX \to PY$, then the correspondence (IV.1.5.ii) describes an isomorphism of ordered categories:

$$\mathsf{Set}_{\mathbb{P}^2} \cong \mathsf{PSet}^{\mathrm{co\,op}}\,.$$

Moreover, a map $\nu : X \to P^2X$ factors through $UX \hookrightarrow P^2X$ (where UX is the set of up-sets on X) precisely when $P^\bullet\nu$ is monotone, and $\nu : X \to UX$ factors through $FX \hookrightarrow UX$ (where FX is the set of filters on X) if and only if $P^\bullet\nu$ preserves finite intersections. There are therefore isomorphisms (see II.2.1)

$$\mathbb{U}\text{-}\mathsf{Mon} \cong \mathsf{Int} \qquad \text{and} \qquad \mathbb{F}\text{-}\mathsf{Mon} \cong \mathsf{Top}\,.$$

IV.1.E *Finitary up-sets.* An up-set $a \in UX$ is *finitary* if, for all $A \subseteq X$,

$$A \in a \implies \exists F \in a\,(F \text{ finite } \& \ F \subseteq A)\,;$$

the set of all finitary up-sets on a set X is denoted by $U_{\mathrm{fin}}X$. Show that the components of the up-set monad $\mathbb{U}$ restrict to such elements to yield the *finitary-up-set monad* $\mathbb{U}_{\mathrm{fin}}$ on Set. With every set $U_{\mathrm{fin}}X$ ordered by subset inclusion, $\mathbb{U}_{\mathrm{fin}}$ is power-enriched.

The category of $\mathbb{U}_{\mathrm{fin}}$-monoids is isomorphic to the full subcategory $\mathsf{Cls}_{\mathrm{fin}}$ of closure spaces Cls (see III.2.5):

$$\mathbb{U}_{\mathrm{fin}}\text{-}\mathsf{Mon} \cong \mathsf{Cls}_{\mathrm{fin}}\,.$$

IV.1.F *The clique monad and closure spaces.* A *clique* a on a set X is a subset of PX such that

(1) $A, B \in a,\ A \cap B \neq \varnothing$
(2) $A \in a,\ A \subseteq B \implies B \in a$,

for all $A, B \in PX$. The set of all cliques on X is denoted by CX, and the double-powerset monad $\mathbb{P}^2$ restricts to such sets to yield the *clique monad* $\mathbb{C}$. The Kleisli category of the clique monad is a separated ordered category, and there is an isomorphism

$$\mathbb{U}\text{-}\mathsf{Mon} \cong \mathbb{C}\text{-}\mathsf{Mon}\ .$$

Since $\mathbb{U}$-Mon is isomorphic to the category Cls of closure spaces, cliques provide an alternative structure for the study of closure spaces via the isomorphism Cls $\cong$ $\mathbb{C}$-Mon.

IV.1.G *The Kleisli extension to* Rel *is associative.* For a power-enriched monad $(\mathbb{T}, \tau)$, the monoid homomorphisms nbhd and conv of Section IV.1.5 can be extended to yield monotone maps

$$\text{nbhd} = \text{nbhd}_{X,Y} : \mathsf{Rel}(TX, Y) \longrightarrow \mathsf{Set}(Y, TX)\ ,$$
$$\text{conv} = \text{conv}_{Y,X} : \mathsf{Set}(Y, TX) \longrightarrow \mathsf{Rel}(TX, Y)$$

that form an adjunction nbhd $\dashv$ conv for all sets X, Y. When $\mathbb{T}$ is equipped with its Kleisli extension, one has

$$\text{nbhd}(s \circ r) = \text{nbhd}(r) \circ \text{nbhd}(s)\ , \qquad \text{conv}(\mu) \circ \text{conv}(\nu) = \text{conv}(\nu \circ \mu)\ ,$$
$$\text{nbhd}(1_X^\sharp) = e_X\ , \qquad \text{conv}(e_X) = 1_X^\sharp\ ,$$

for all unitary relations $r, s : TX \nrightarrow Y$, and maps $\mu, \nu : Y \longrightarrow TX$. As a consequence, the Kleisli extension $\check{\mathbb{T}}$ is associative, and one can form the category $(\mathbb{T}, 2)$-URel.

The maps nbhd and conv define functors nbhd : $\mathsf{Set}_{\mathbb{T}} \longrightarrow (\mathbb{T}, 2)\text{-}\mathsf{URel}^{\mathrm{op}}$ and conv : $(\mathbb{T}, 2)\text{-}\mathsf{URel}^{\mathrm{op}} \longrightarrow \mathsf{Set}_{\mathbb{T}}$ that determine a 2-isomorphism

$$\mathsf{Set}_{\mathbb{T}} \cong (\mathbb{T}, 2)\text{-}\mathsf{URel}^{\mathrm{op}}\ .$$

IV.1.H *Unitary relations as sup-maps.* For a power-enriched monad $(\mathbb{T}, \tau)$ and a relation $r : TX \nrightarrow Y$, the following statements are equivalent:

(i) r is unitary;
(ii) $r^\flat = \downarrow_{TX} \bigvee r^\flat$;
(iii) $r(-, y) : TX \longrightarrow \mathsf{2}^{\mathrm{op}}$ is a sup-map for all $y \in Y$.

IV.2 Lax extensions of monads

In Corollary IV.1.5.4 we effectively established an isomorphism $(\mathbb{F}, 2)$-Cat $\longrightarrow$ $(\beta, 2)$-Cat, both categories being isomorphic to Top. It turns out that this isomorphism may be thought of as induced by the monad morphism $\beta \longrightarrow \mathbb{F}$. More generally, in this section we seek sufficient conditions for a monad morphism $\alpha : \mathbb{S} \longrightarrow \mathbb{T}$ into a power-enriched monad $\mathbb{T}$ to induce an isomorphism

$$A_\alpha : (\mathbb{T}, \mathcal{V})\text{-}\mathsf{Cat} \longrightarrow (\mathbb{S}, \mathcal{V})\text{-}\mathsf{Cat}$$

when $\mathbb{S}$ and $\mathbb{T}$ are equipped with adequate lax extensions; here, A_α is the algebraic functor of α; see Section III.3.4. For this, we proceed in two steps: we first obtain isomorphisms

$$(\mathbb{S}, 2)\text{-}\mathsf{Cat} \cong \mathbb{T}\text{-}\mathsf{Mon} \cong (\mathbb{T}, 2)\text{-}\mathsf{Cat}\,,$$

and then consider the case where $\mathsf{2}$ is replaced by an arbitrary quantale $\mathcal{V}$ (Theorems IV.2.3.3 and IV.2.5.3). Each of these steps is facilitated by the construction of lax extensions induced by a lax extension $\hat{\mathbb{T}}$ of $\mathbb{T}$ to Rel. Specifically, in Section IV.2.1 we "transfer" the lax extension from $\mathbb{T}$ to $\mathbb{S}$ along α, and in Section IV.2.4 we describe a process of generating a lax extension of $\mathbb{T}$ to $\mathcal{V}$-Rel from a lax extension to Rel.

IV.2.1 Initial extensions

In III.1.10 and Section IV.1.4, two constructions of a lax extension of a monad to Rel are given: the Barr extension and the Kleisli extension. In practice, the Barr extension can be extracted from the Kleisli extension of a larger monad. For example, the Barr extension of the ultrafilter functor $\beta : \mathsf{Set} \to \mathsf{Set}$:

$$\mathfrak{x}\ (\overline{\beta}r)\ \mathfrak{y} \iff \mathfrak{x} \supseteq r^\circ[\mathfrak{y}]$$

(for all relations $r : X \nrightarrow Y$, $\mathfrak{x} \in \beta X$, $\mathfrak{y} \in \beta Y$) is the restriction to ultrafilters of the Kleisli extension of the filter functor

$$\mathfrak{x}\ (\check{F}r)\ \mathfrak{y} \iff \mathfrak{x} \supseteq r^\circ[\mathfrak{y}]$$

(for all $\mathfrak{x} \in FX$, $\mathfrak{y} \in FY$).

More generally, if $\alpha : S \to T$ is a natural transformation of Set-functors, and $\hat{T}$ is a lax extension of T to $\mathcal{V}$-Rel, the *initial extension* of S induced by α is the lax extension $\hat{S}$ given by

$$\hat{S}r := \alpha_Y^\circ \cdot \hat{T}r \cdot \alpha_X\,,$$

for any $\mathcal{V}$-relation $r : X \nrightarrow Y$. In pointwise notation, the definition becomes

$$\hat{S}r(\mathfrak{x}, \mathfrak{y}) = \hat{T}r(\alpha_X(\mathfrak{x}), \alpha_Y(\mathfrak{y}))\,,$$

for all $\mathfrak{x} \in SX$, $\mathfrak{y} \in SY$. Before showing that $\hat{S}$ is indeed a lax extension of S if $\hat{T}$ is one of T (Proposition IV.2.1.1), we briefly discuss the "initial" terminology.

Recall from Section III.3.4 that a morphism of lax extensions $\alpha : (S, \hat{S}) \to (T, \hat{T})$ is a natural transformation $\alpha : S \to T$ that extends to an oplax transformation $\hat{S} \to \hat{T}$:

$$\alpha_Y \cdot \hat{S}r \leq \hat{T}r \cdot \alpha_X\,,$$

for all $\mathcal{V}$-relations $r : X \nrightarrow Y$. If $U : \mathcal{V}\text{-}\mathsf{LXT} \to \mathsf{Set}^{\mathsf{Set}}$ denotes the forgetful functor from the category of lax extensions to $\mathcal{V}$-Rel (see Exercise III.3.A) that sends $\hat{T}$ to T, then the initial extension is an U-initial morphism in the sense of Section II.5.6. Indeed, consider a natural transformation $\lambda : R \to S$ with a lax extension $\hat{R}$ of R; then λ is a morphism of lax extensions if and only if $\alpha \cdot \lambda : R \to T$ is one:

$$\alpha_Y \cdot \lambda_Y \cdot \hat{R}r \leq \hat{T}r \cdot \alpha_X \cdot \lambda_X \iff \lambda_Y \cdot \hat{R}r \leq \alpha_Y^\circ \cdot \hat{T}r \cdot \alpha_X \cdot \lambda_X \iff \lambda_Y \cdot \hat{R}r \leq \hat{S}r \cdot \lambda_X$$

for all relations $r : X \nrightarrow Y$.

IV.2.1.1 Proposition *For a lax extension $\hat{T}$ to $\mathcal{V}$-*Rel *of a* Set*-functor T, and a natural transformation $\alpha : S \to T$, the initial extension $\hat{S}$ of S induced by α is a lax extension of S.*

*Furthermore, if $\hat{T}$ belongs to a lax extension to $\mathcal{V}$-*Rel *of a monad $\mathbb{T} = (T, m, e)$ and $\alpha : \mathbb{S} \to \mathbb{T}$ is a monad morphism, then $\hat{S}$ also belongs to a lax extension of $\mathbb{S} = (S, n, d)$.*

Proof Since $\hat{T}$ preserves the order on the hom-sets $\mathcal{V}$-Rel(X, Y), it is immediate that $\hat{S}$ does too. If $r : X \nrightarrow Y$ and $s : Y \nrightarrow Z$ are $\mathcal{V}$-relations, then

$$\begin{aligned}\hat{S}s \cdot \hat{S}r &= \alpha_Z^\circ \cdot \hat{T}s \cdot \alpha_Y \cdot \alpha_Y^\circ \cdot \hat{T}r \cdot \alpha_X \\ &\leq \alpha_Z^\circ \cdot \hat{T}s \cdot \hat{T}r \cdot \alpha_X \\ &\leq \alpha_Z^\circ \cdot \hat{T}(s \cdot r) \cdot \alpha_X = \hat{S}(s \cdot r)\,,\end{aligned}$$

because $\hat{T}$ is a lax functor. As α is a natural transformation, we have $\alpha_Y \cdot Sf = Tf \cdot \alpha_X$, which can be written $Sf \leq \alpha_Y^\circ \cdot Tf \cdot \alpha_X$ or equivalently $(Sf)^\circ \leq \alpha_X^\circ \cdot (Tf)^\circ \cdot \alpha_Y$, in $\mathcal{V}$-Rel; the extension conditions for $\hat{S}$ then follow because they are satisfied for $\hat{T}$.

Finally, suppose that $\hat{T}$ yields a lax extension of the monad $\mathbb{T}$. Since $e : 1_{\mathcal{V}\text{-Rel}} \to \hat{T}$ is oplax, then so is $d : 1_{\mathcal{V}\text{-Rel}} \to \hat{S}$: indeed, α is a monad morphism, so we have $\alpha \cdot d = e$, and

$$r \leq e_Y^\circ \cdot \hat{T}r \cdot e_X = d_Y^\circ \cdot \hat{S}r \cdot d_X\,,$$

as expected. To verify oplaxness of n, we use that $m \cdot T\alpha \cdot \alpha S = \alpha \cdot n$:

$$\begin{aligned}\hat{S}\hat{S}r &= \alpha_{SY}^\circ \cdot \hat{T}(\alpha_Y^\circ \cdot \hat{T}r \cdot \alpha_X) \cdot \alpha_{SX} \\ &= \alpha_{SY}^\circ \cdot (T\alpha_Y)^\circ \cdot \hat{T}\hat{T}r \cdot T\alpha_X \cdot \alpha_{SX} \\ &\leq \alpha_{SY}^\circ \cdot (T\alpha_Y)^\circ \cdot m_Y^\circ \cdot \hat{T}r \cdot m_X \cdot T\alpha_X \cdot \alpha_{SX} = n_Y^\circ \cdot \hat{S}r \cdot n_X\,,\end{aligned}$$

as required. □

From this last result one infers that, in the presence of the initial extension $\hat{S}$ of S, the maps $\alpha_X : SX \to TX$ become order-embeddings with respect to the orders Section III.3.3 induced by the lax extensions:

$$\mathfrak{x} \leq \mathfrak{y} \iff \alpha_X(\mathfrak{x}) \leq \alpha_X(\mathfrak{y})$$

for all $\mathfrak{x}, \mathfrak{y} \in SX$. In fact, this condition witnesses the smooth interaction of the initial and Kleisli extensions, as we will see next.

IV.2.1.2 Proposition *A morphism $\alpha : (\mathbb{S},\sigma) \to (\mathbb{T},\tau)$ of power-enriched monads becomes a morphism $\alpha : \check{S} \to \check{T}$ of the Kleisli extensions to* Rel.

When the sets SX and TX are equipped with the orders (IV.1.2.i) *induced by σ and τ, respectively, the components α_X are order-embeddings if and only if the initial extension of S induced by α is the Kleisli extension of S.*

Proof Observe that $\alpha_X \cdot r^\sigma = r^\tau \cdot \alpha_Y$ for any relation $r : X \nrightarrow Y$. Therefore,

$$\mathfrak{x} \le r^\sigma(\mathfrak{y}) \implies \alpha_X(\mathfrak{x}) \le \alpha_X \cdot r^\sigma(\mathfrak{y}) = r^\tau \cdot \alpha_Y(\mathfrak{y})$$

for all $\mathfrak{x} \in SX$, $\mathfrak{y} \in SY$ (α_X is monotone by Exercise IV.1.A). This implies that α is a morphism between the respective Kleisli extensions. If α_X is an order-embedding, the implication above is an equivalence, so that α is initial. Conversely, if the initial extension of S is the Kleisli extension, the equivalence also holds, and we can conclude that α_X is an order-embedding by choosing $r = 1_X$. □

IV.2.1.3 Examples

(1) For every monad $\mathbb{S}$ on Set, there is a unique monad morphism $! : \mathbb{S} \to \mathbb{1}$ into the terminal monad. When the latter is equipped with its largest lax extension $\star^\top : \mathcal{V}\text{-Rel} \to \mathcal{V}\text{-Rel}$ (such that $\star^\top r(\star,\star) = \top$ for all $\mathcal{V}$-relation $r : X \nrightarrow Y$ as in Example III.1.4.2(3)), the initial extension $\hat{S}$ of $\star^\top$ induced by ! is given by

$$\hat{S}r(\mathfrak{x},\mathfrak{y}) = \top\,,$$

for all $\mathfrak{x} \in SX$, $\mathfrak{y} \in SY$; hence, $\hat{\mathbb{S}} = \mathbb{S}^\top$.

(2) It is obvious from Example IV.1.4.2(4) that the Kleisli extension $\check{F}$ of F can be obtained as the restriction to filters of the Kleisli extension $\check{U}$ of the up-set functor U; in this case, the components $\alpha_X : FX \to UX$ of α are the embeddings, and $\check{F}$ is the initial extension of $\check{U}$ induced by α. Similarly, the Barr extension of the ultrafilter functor can be obtained by restriction of the Kleisli extension $\check{F}$ of the filter functor, and the lax extension of the identity monad can also be seen as a restriction of the ultrafilter functor (via the principal ultrafilter natural transformation). Thus, the chain of natural transformations, whose respective components are all embeddings, as described in Example II.3.1.1(5),

$$1_{\mathsf{Set}} \to \beta \to F \to U,$$

yields the following chain of initial extensions:

$$1_{\mathsf{Rel}} \to \overline{\beta} \to \check{F} \to \check{U}\,.$$

IV.2.2 Sup-dense and interpolating monad morphisms

The isomorphism $(\mathbb{S},2)$-Cat $\cong (\mathbb{T},2)$-Cat that we are aiming for requires that the monads $\mathbb{S}$ and $\mathbb{T}$ be sufficiently compatible. The conditions we present continue to be guided by the case where $\mathbb{S}=\beta$, $\mathbb{T}=\mathbb{F}$, and $\alpha:\beta\hookrightarrow\mathbb{F}$ is the inclusion of the set of ultrafilters into the set of filters.

Consider a monad morphism $\alpha:\mathbb{S}\to\mathbb{T}$, where $\mathbb{T}=(T,m,e)$ is a power-enriched monad with structure $\tau:\mathbb{P}\to\mathbb{T}$ and equipped with its Kleisli extension $\check{T}$, and $\mathbb{S}=(S,n,d)$ is a monad equipped with its initial extension $\hat{S}$ induced by α. To be able to exploit the adjunction $\bigvee_{TX}\dashv\downarrow_{TX}$, as in Proposition IV.1.5.2, we introduce the transformation $\alpha^{\bigvee}:PS\to T$ via

$$\alpha_X^{\bigvee}:=\bigvee P\alpha_X=m_X\cdot\tau_{TX}\cdot P\alpha_X=\alpha_X^{\mathbb{T}}\cdot\tau_{SX}\,,$$

or equivalently $\alpha_X^{\bigvee}(\mathcal{A})=\bigvee\alpha_X(\mathcal{A})$ for all $\mathcal{A}\subseteq SX$. Each $\alpha_X^{\bigvee}$ preserves suprema, and therefore has a right adjoint, denoted by $\alpha_X^{\downarrow}:TX\to PSX$, so that

$$\alpha_X^{\downarrow}(f)=\{\mathfrak{x}\in SX\mid\alpha_X(\mathfrak{x})\le f\}=\alpha_X^{-1}\cdot\downarrow_{TX}(f)$$

for all $f\in TX$. The maps $\alpha_X^{\downarrow}$ allow for a convenient description of the initial extension $\hat{S}$ of S. Indeed, for a relation $r:X\to\!\!\!\!\!\!+\;Y$, we have

$$(\hat{S}r)^{\flat}=\alpha_X^{\downarrow}\cdot r^{\tau}\cdot\alpha_Y\,,$$

so the order relation $\hat{S}1_X$ on SX is given by $(\hat{S}1_X)^{\flat}=\alpha_X^{\downarrow}\cdot\alpha_X$.

The monad morphism $\alpha:\mathbb{S}\to\mathbb{T}$ is *sup-dense* if one has

$$\alpha^{\bigvee}\cdot\alpha^{\downarrow}=1_{TX}\,; \tag{IV.2.2.i}$$

in pointwise notation this says that every element of TX can be expressed as a supremum of α_X-images of elements of SX:

$$\forall f\in TX\;\exists\mathcal{A}\subseteq SX\;(f=\bigvee\alpha_X(\mathcal{A}))\,.$$

When $\mathbb{S}$ is a submonad of $\mathbb{T}$ and the embedding is sup-dense, we simply say that $\mathbb{S}$ is *sup-dense* in $\mathbb{T}$.

The morphism $\alpha:\mathbb{S}\to\mathbb{T}$ is *interpolating* for a relation $r:SX\to\!\!\!\!\!\!+\;X$ if

$$\alpha_X^{\downarrow}\cdot\alpha_X^{\bigvee}\cdot r^{\flat}\le(\downarrow_{SX}\cdot n_X)^{\mathbb{P}}\cdot(\hat{S}r)^{\flat}\cdot d_X \tag{IV.2.2.ii}$$

holds. This condition expands to $\alpha_X^{\downarrow}\cdot\alpha_X^{\bigvee}\cdot r^{\flat}\le(\alpha_X^{\downarrow}\cdot\alpha_X\cdot n_X)^{\mathbb{P}}\cdot\alpha_{SX}^{\downarrow}\cdot\tau_{SX}\cdot r^{\flat}$ and can be written pointwise as

$$\begin{aligned}&\alpha_X(\mathfrak{x})\le\bigvee\{\alpha_X(\mathfrak{y})\mid\mathfrak{y}\;r\;y\}\\&\qquad\Longrightarrow\exists\mathfrak{X}\in SSX\;(\mathfrak{x}\le n_X(\mathfrak{X})\;\&\;\alpha_{SX}(\mathfrak{X})\le\tau_{SX}\cdot r^{\flat}(y))\end{aligned}$$

for all $\mathfrak{x} \in SX$, $y \in X$. If $\mathbb{S}$ is a submonad of $\mathbb{T}$, the preceding condition naturally has a simpler expression, and may be represented graphically by

$$\mathfrak{x} \leq m_X \cdot \tau_{SX} \cdot r^\flat(y) \quad \Longrightarrow \quad \exists \mathcal{X} : \quad \begin{array}{c} \mathcal{X} \leq \tau_{SX} \cdot r^\flat(y) \\ \big\downarrow \\ \mathfrak{x} \leq m_X(\mathcal{X}). \end{array}$$

A monad morphism $\alpha : \mathbb{S} \to \mathbb{T}$ is *interpolating* if it is interpolating for all relations $r : SX \nrightarrow X$. If $\mathbb{S}$ is a submonad of $\mathbb{T}$ and the embedding is interpolating, we may simply say that $\mathbb{S}$ is interpolating *in* $\mathbb{T}$.

Note that α is interpolating whenever it is a morphism of power-enriched monads $\alpha : (\mathbb{S}, \sigma) \to (\mathbb{T}, \tau)$. Indeed, since $\{-\}_{SSX} \leq \alpha^{\downarrow}_{SX} \cdot \alpha_{SX}$, we have

$$\begin{aligned} \alpha^{\downarrow}_X \cdot \alpha^{\vee}_X \cdot r^\flat &= \alpha^{\downarrow}_X \cdot \alpha^{\mathbb{T}}_X \cdot \alpha_{SX} \cdot \sigma_{SX} \cdot r^\flat \\ &= (\alpha^{\downarrow}_X \cdot \alpha^{\mathbb{T}}_X \cdot \alpha_{SX})^{\mathbb{P}} \cdot \{-\}_{SSX} \cdot \sigma_{SX} \cdot r^\flat \\ &\leq (\alpha^{\downarrow}_X \cdot \alpha^{\mathbb{T}}_X \cdot \alpha_{SX})^{\mathbb{P}} \cdot \alpha^{\downarrow}_{SX} \cdot \alpha_{SX} \cdot \sigma_{SX} \cdot r^\flat \\ &= (\alpha^{\downarrow}_X \cdot \alpha_X \cdot n_X)^{\mathbb{P}} \cdot \alpha^{\downarrow}_{SX} \cdot \tau_{SX} \cdot r^\flat . \end{aligned}$$

IV.2.2.1 Examples

(1) Any power-enriched monad $\mathbb{T} = (T, m, e)$ comes with the interpolating monad morphism $\alpha = e : \mathbb{I} \to \mathbb{T}$. Indeed, using $\alpha^{\mathbb{T}}_X = 1_{TX}$ and $\{-\}_X \leq \alpha^{\downarrow}_X \cdot \alpha_X$, we have

$$\alpha^{\downarrow}_X \cdot \alpha^{\vee}_X \cdot r^\flat = \alpha^{\downarrow}_X \cdot \tau_X \cdot r^\flat = (\{-\}_X)^{\mathbb{P}} \cdot \alpha^{\downarrow}_X \cdot \tau_X \cdot r^\flat \leq (\alpha^{\downarrow}_X \cdot \alpha_X)^{\mathbb{P}} \cdot \alpha^{\downarrow}_X \cdot \tau_X \cdot r^\flat$$

for all relations $r : X \nrightarrow X$.

(2) If $\mathbb{S} = \mathbb{P}$ is the powerset monad embedded in $\mathbb{T} = \mathbb{F}$ via the principal filter morphism $\tau : \mathbb{P} \to \mathbb{F}$, then the interpolation condition is immediate since τ is a morphism of power-enriched monads.

(3) Consider the filter monad $\mathbb{F}$ with the principal monad morphism $\tau : \mathbb{P} \to \mathbb{F}$. © Every filter is the supremum (i.e. the intersection) of all ultrafilters finer than it (Corollary II.1.13.4), so the ultrafilter monad $\mathbb{\beta}$ is sup-dense in $\mathbb{F}$.

Let us verify that $\mathbb{\beta}$ is interpolating in $\mathbb{F}$. For ultrafilters $\mathfrak{x}, \mathfrak{y}$ on X and a relation $r : \beta X \nrightarrow X$, suppose $\mathfrak{x} \leq \sum r^\tau(\mathfrak{y})$ (with $\sum$ denoting the monad multiplication of $\mathbb{\beta}$), i.e.

$$\forall B \in \mathfrak{y}\ (r^\flat(B) \subseteq A^{\mathbb{\beta}} \implies A \in \mathfrak{x}),$$

for all $A \subseteq X$ (where $A^{\mathbb{\beta}} = \{z \in \beta X \mid A \in z\}$, see Section III.2.2). If there existed $A \in \mathfrak{x}$ and $B \in \mathfrak{y}$ with $A^{\mathbb{\beta}} \cap r^\flat(B) = \varnothing$, we would have $r^\flat(B) \subseteq (A^{\mathbb{\beta}})^{\complement} = (A^{\complement})^{\mathbb{\beta}}$, so that $A^{\complement} \in \mathfrak{x}$, a contradiction. Therefore, $A^{\mathbb{\beta}} \cap r^\flat(B) \neq \varnothing$ for all $A \in \mathfrak{x}$ and $B \in \mathfrak{y}$, and there exists an ultrafilter $\mathcal{X}$ on βX that refines both $\{A^{\mathbb{\beta}} \mid A \in \mathfrak{x}\}$ and $r^\tau(\mathfrak{y})$. In particular, $\Sigma_X(\mathcal{X}) = \mathfrak{x}$.

By setting $y = \dot{y}$, we observe that $r^{\tau}(y) = \tau_{\beta X} \cdot r^{\flat}(y)$, so the interpolation condition is verified.

IV.2.3 $(\mathbb{S}, 2)$-categories as Kleisli monoids

Given a lax extension $\hat{\mathbb{S}}$ to $\mathcal{V}$-Rel of a monad $\mathbb{S}$ on Set, and a non-trivial quantale $\mathcal{V}$ (see III.1.2), recall that (see III.4.1)

$$(\mathbb{S}, \mathcal{V})\text{-UGph}$$

is the category whose object are pairs (X, r) with $r : SX \nrightarrow X$ a unitary $\mathcal{V}$-relation, so

$$e_X^{\circ} \cdot \hat{S}r \cdot m_X^{\circ} \leq r \qquad \text{and} \qquad r \cdot \hat{S}1_X \leq r\,,$$

and whose morphisms $f : (X, r) \to (Y, s)$ are maps $f : X \to Y$ satisfying

$$f \cdot r \leq s \cdot Sf\,.$$

In addition, given a functor T : Set $\to$ Set that lifts tacitly along the forgetful functor Ord $\to$ Set, we can consider the *lax comma category*

$$(1_{\mathsf{Set}} \downarrow T)_{\leq}$$

whose objects are pairs (X, ν) with a map $\nu : X \to TX$, and whose morphisms are maps $f : X \to Y$ with $Tf \cdot \nu \leq \mu \cdot f$:

$$\begin{array}{ccc} X & \xrightarrow{\;f\;} & Y \\ {\scriptstyle \nu}\downarrow & \leq & \downarrow{\scriptstyle \mu} \\ TX & \xrightarrow[\;Tf\;]{} & TY\,. \end{array}$$

We now present conditions which will give us an isomorphism between the two categories $(\mathbb{S}, 2)$-UGph and $(1_{\mathsf{Set}} \downarrow T)_{\leq}$ that restricts to an isomorphism

$$(\mathbb{S}, 2)\text{-Cat} \cong \mathbb{T}\text{-Mon}\,.$$

Hypotheses

We consider a monad morphism $\alpha : \mathbb{S} \to \mathbb{T}$, where $\mathbb{T} = (T, m, e)$ is a monad power-enriched by $\tau : \mathbb{P} \to \mathbb{T}$ and equipped with its Kleisli extension $\check{T}$, and $\mathbb{S} = (S, n, d)$ is a monad equipped with the initial extension $\hat{S}$ induced by α. The sets SX and TX are equipped with the orders III.3.3 induced by the respective lax extensions, and hom-sets of Set are ordered pointwise.

Following Section IV.1.5, we define for all sets X an adjunction

$$\mathsf{Set}(X, TX) \underset{\text{nbhd}}{\overset{\text{conv}}{\underset{\longleftarrow}{\overset{\longrightarrow}{\top}}}} \mathsf{Rel}(SX, X)$$

by exploiting the Ord-isomorphism $\mathsf{Rel}(SX, X) \cong \mathsf{Set}(X, PSX)$. More concretely, we set

$$\mathrm{conv} : \mathsf{Set}(X, TX) \to \mathsf{Set}(X, PSX)\,, \qquad \nu \mapsto \alpha_X^{\downarrow} \cdot \nu\,,$$
$$\mathrm{nbhd} : \mathsf{Set}(X, PSX) \to \mathsf{Set}(X, TX)\,, \qquad r^\flat \mapsto \alpha_X^{\bigvee} \cdot r^\flat\,.$$

The adjunction nbhd $\dashv$ conv follows from $\alpha_X^{\bigvee} \dashv \alpha_X^{\downarrow}$; to verify this, one can use the pointwise notation of conv and nbhd:

$$\chi \;\mathrm{conv}(\nu)\; y \iff \alpha_X(\chi) \le \nu(y) \qquad \text{and} \qquad \mathrm{nbhd}(r)(y) = \textstyle\bigvee_{TX} \alpha_X(r^\flat(y))\,,$$

for all relations $r : SX \nrightarrow X$ and maps $\nu : X \to TX$. Naturally, conv and nbhd restrict to mutually inverse isomorphisms between the sets of fixpoints of (conv $\cdot$ nbhd) and of (nbhd $\cdot$ conv).

The fixpoints of (nbhd $\cdot$ conv) are exactly the maps $\nu : X \to TX$ such that

$$\forall y \in X\, \exists \mathcal{A} \subseteq SX\; (\nu(y) = \textstyle\bigvee \alpha_X(\mathcal{A}))\,.$$

Hence, (nbhd $\cdot$ conv) $= 1_{\mathsf{Set}(X,TX)}$ precisely when α is sup-dense. In turn, a relation $r : SX \nrightarrow X$ is a fixpoint of (conv $\cdot$ nbhd) precisely when

$$\forall y \in X\; (\alpha_X(\chi) \le \textstyle\bigvee \alpha_X(r^\flat(y)) \implies \chi\; r\; y)\,.$$

One obtains the following generalizations of Proposition IV.1.5.1, Proposition IV.1.5.2, and Theorem IV.1.5.3.

IV.2.3.1 Lemma *For the adjunction* nbhd $\dashv$ conv : $\mathsf{Set}(X, TX) \to \mathsf{Rel}(SX, X)$ *defined above, the following hold:*

(1) Fix(nbhd $\cdot$ conv) $= \mathsf{Set}(X, TX)$ *if and only if α is sup-dense;*
(2) *a relation $r : SX \nrightarrow X$ is a fixpoint of* (conv $\cdot$ nbhd) *if and only if it is unitary and α is interpolating for r.*

Proof The first point is immediate from the preceding discussion.

For a unitary relation $r : SX \nrightarrow X$, we obtain, as in the proof of Proposition IV.1.5.1,

$$r^\flat = (\downarrow_{SX} \cdot n_X)^{\mathbb{P}} \cdot (\hat{S}r)^\flat \cdot d_X\,. \tag{IV.2.3.i}$$

If moreover α is interpolating for r, then

$$\alpha_X^{\downarrow} \cdot \alpha_X^{\bigvee} \cdot r^\flat \le (\downarrow_{SX} \cdot n_X)^{\mathbb{P}} \cdot (\hat{S}r)^\flat \cdot d_X = r^\flat\,,$$

and r is indeed a fixpoint of (conv $\cdot$ nbhd).

Conversely, if $r : SX \nrightarrow X$ is a fixpoint of (conv $\cdot$ nbhd), it is of the form conv(ν) for a map $\nu : X \to TX$, so $r \cdot \hat{S}1_X \le r$, and r is right unitary. Moreover, if $\mathcal{X}\; \hat{S}r\; d_X(y)$ holds, we can apply $m_X \cdot T\alpha_X$ to each side of the inequality $\alpha_{SX}(\mathcal{X}) \leq r^{\tau} \cdot e_X(y)$ to conclude that $n_X(\mathcal{X})\ r\ y$ by the fixpoint condition, i.e. r is left unitary. As r is unitary, (IV.2.3.i) holds, and implies that α is interpolating for r. □

IV.2.3.2 Proposition *The adjoint maps* nbhd *and* conv *defined above satisfy*

$$\mathrm{nbhd}(s \circ r) \leq \mathrm{nbhd}(r) \circ \mathrm{nbhd}(s) , \qquad \mathrm{conv}(\mu) \circ \mathrm{conv}(\nu) \leq \mathrm{conv}(\nu \circ \mu) ,$$
$$\mathrm{nbhd}(1_X^{\sharp}) = e_X , \qquad \mathrm{conv}(e_X) = 1_X^{\sharp} ,$$

for all relations $r, s : SX \nrightarrow X$ *and maps* $\mu, \nu : X \to TX$ *(where* $1_X^{\sharp} = d_X^{\circ} \cdot \hat{S}1_X$*).*

Moreover, if α *is sup-dense, then* $\mathrm{nbhd}(s \circ r) = \mathrm{nbhd}(r) \circ \mathrm{nbhd}(s)$ *for all relations* $r, s : TX \nrightarrow X$*. If in addition* α *is interpolating, then* $\mathrm{conv}(\mu) \circ \mathrm{conv}(\nu) = \mathrm{conv}(\nu \circ \mu)$ *also holds.*

Proof The displayed equalities follow from the fact that

$$\mathfrak{x}\ (1_X^{\sharp})\ y \iff \mathfrak{x}\ (d_X^{\circ} \cdot \alpha_X^{\circ} \cdot \check{T}1_X \cdot \alpha_X)\ y \iff \alpha_X(\mathfrak{x}) \leq e_X(y)$$

for all $\mathfrak{x} \in SX$ and $y \in X$.

To show that $\mathrm{nbhd}(s \circ r) \leq \mathrm{nbhd}(r) \circ \mathrm{nbhd}(s)$, we first note that

$$\alpha_X^{\vee} \cdot Pn_X = (\alpha_X \cdot 1_{SX}^{\mathbb{S}})^{\mathbb{T}} \cdot \tau_{SSX} = (\alpha_X^{\mathbb{T}} \cdot \alpha_{SX})^{\mathbb{T}} \cdot \tau_{SSX} = \alpha_X^{\mathbb{T}} \cdot \alpha_{SX}^{\vee} ;$$

by composing these equalities with $\alpha_{SX}^{\downarrow} \cdot r^{\tau} \cdot \alpha_X$ on the right, we obtain

$$\alpha_X^{\vee} \cdot Pn_X \cdot \alpha_{SX}^{\downarrow} \cdot r^{\tau} \cdot \alpha_X \leq \alpha_X^{\mathbb{T}} \cdot r^{\tau} \cdot \alpha_Y . \qquad \text{(IV.2.3.ii)}$$

We can now proceed as in Proposition IV.1.5.2:

$$\begin{aligned}
&\mathrm{nbhd}(r) \circ \mathrm{nbhd}(s) \\
&= (\alpha_X^{\vee} \cdot r^{\flat})^{\mathbb{T}} \cdot \alpha_X^{\vee} \cdot s^{\flat} \\
&= \alpha_X^{\mathbb{T}} \cdot (\tau_{SX} \cdot r^{\flat})^{\mathbb{T}} \cdot \alpha_X^{\mathbb{T}} \cdot \tau_{SX} \cdot s^{\flat} && ((g^{\mathbb{T}} \cdot f)^{\mathbb{T}} = g^{\mathbb{T}} \cdot f^{\mathbb{T}}) \\
&= (\alpha_X^{\mathbb{T}} \cdot r^{\tau} \cdot \alpha_X)^{\mathbb{T}} \cdot \tau_{SX} \cdot s^{\flat} && (g^{\mathbb{T}} \cdot f^{\mathbb{T}} = (g^{\mathbb{T}} \cdot f)^{\mathbb{T}}) \\
&\geq (\alpha_X^{\vee} \cdot Pn_X \cdot (\hat{S}r)^{\flat})^{\mathbb{T}} \cdot \tau_{SX} \cdot s^{\flat} && (\text{by (IV.2.3.ii)}) \\
&= \alpha_X^{\mathbb{T}} \cdot (\tau_{SX} \cdot Pn_X \cdot (\hat{S}r)^{\flat})^{\mathbb{T}} \cdot \tau_{SX} \cdot s^{\flat} && ((g^{\mathbb{T}} \cdot f)^{\mathbb{T}} = g^{\mathbb{T}} \cdot f^{\mathbb{T}}) \\
&= \alpha_X^{\mathbb{T}} \cdot \tau_{SX} \cdot (Pn_X \cdot (\hat{S}r)^{\flat})^{\mathbb{P}} \cdot s^{\flat} && (\tau \text{ natural transformation}) \\
&= \alpha_X^{\vee} \cdot (((n_X^{\circ})^{\flat})^{\mathbb{P}} \cdot (\hat{S}r)^{\flat})^{\mathbb{P}} \cdot s^{\flat} && (Pn_X = ((n_X^{\circ})^{\flat})^{\mathbb{P}}) \\
&= \alpha_X^{\vee} \cdot (s \cdot \hat{S}r \cdot n_X^{\circ})^{\flat} && (\mathsf{Rel} = \mathsf{Set}_{\mathbb{P}}) \\
&= \mathrm{nbhd}(s \circ r) .
\end{aligned}$$

The inequality for conv follows from the adjunction $\mathrm{nbhd} \dashv \mathrm{conv}$.

If α is sup-dense, the inequality in (IV.2.3.ii) becomes an equality, so that $\mathrm{nbhd}(r) \circ \mathrm{nbhd}(s) = \mathrm{nbhd}(s \circ r)$. For maps $\mu, \nu : X \to TX$, the $(\mathbb{S}, 2)$-relations $\mathrm{conv}(\mu)$ and $\mathrm{conv}(\nu)$ are unitary, and therefore so is $\mathrm{conv}(\mu) \circ \mathrm{conv}(\nu)$ (Exercise III.1.N). The claim for conv then follows from $\mathrm{nbhd}(r) \circ \mathrm{nbhd}(s) = \mathrm{nbhd}(s \circ r)$ and the fact that nbhd and conv form a bijection between $\mathsf{Set}(X, TX)$ and the set of all unitary $(\mathbb{S}, 2)$-relations $r : X \nrightarrow X$ (see Lemma IV.2.3.1). □

IV.2.3.3 Theorem *Let $(\mathbb{T}, \tau)$ be a power-enriched monad together with a monad morphism $\alpha : \mathbb{S} \to \mathbb{T}$, and suppose that $\mathbb{T}$ is equipped with its Kleisli extension $\check{T}$, and $\mathbb{S}$ with the initial extension of $\check{T}$ induced by α. If α is sup-dense, there is a full reflective embedding $(1_{\mathsf{Set}} \downarrow T)_{\leq} \hookrightarrow (\mathbb{S}, 2)\text{-}\mathsf{UGph}$ that commutes with the underlying-set functors and restricts to a full reflective embedding*

$$\mathbb{T}\text{-}\mathsf{Mon} \hookrightarrow (\mathbb{S}, 2)\text{-}\mathsf{Cat}\,.$$

If α is also interpolating, this functor is an isomorphism.

Proof For every map $\nu : X \to TX$, the relation $\mathrm{conv}(\nu)$ is a fixpoint of $(\mathrm{conv} \cdot \mathrm{nbhd})$, and is therefore unitary by Lemma IV.2.3.1. Similarly, a unitary relation $r : SX \nrightarrow X$ yields a map $\mathrm{nbhd}(r) : X \to TX$. Thus, we can consider the functors

$$C : (1_{\mathsf{Set}} \downarrow T)_{\leq} \to (\mathbb{S}, 2)\text{-}\mathsf{UGph},$$
$$N : (\mathbb{S}, 2)\text{-}\mathsf{UGph} \to (1_{\mathsf{Set}} \downarrow T)_{\leq}$$

defined on objects by $C(X, \nu) = (X, \mathrm{conv}(\nu))$ and $N(X, r) = (X, \mathrm{nbhd}(r))$, and leaving maps untouched (the fact that C and N send morphisms to morphisms follows easily from the definitions; see also Exercise IV.1.G). The adjunction $\mathrm{nbhd} \dashv \mathrm{conv}$ yields an adjunction $N \dashv C$. Lemma IV.2.3.1 shows that if α is sup-dense, then C is a full reflective embedding, and Proposition IV.2.3.2 yields that C restricts to a functor $\mathbb{T}\text{-}\mathsf{Mon} \hookrightarrow (\mathbb{S}, 2)\text{-}\mathsf{Cat}$. Finally, if α is also interpolating, then C is an isomorphism by Lemma IV.2.3.1 again. □

Theorem IV.1.5.3 now appears as a direct consequence of this more general result, since $\alpha = 1_{\mathbb{T}}$ is both sup-dense and interpolating. Moreover, the category of Kleisli monoids provides a link between presentations of lax algebras.

IV.2.3.4 Proposition *If $\alpha : \mathbb{S} \to \mathbb{T}$ is sup-dense as in* Theorem IV.2.3.3, *the algebraic functor*

$$A_{\alpha} : (\mathbb{T}, 2)\text{-}\mathsf{Cat} \to (\mathbb{S}, 2)\text{-}\mathsf{Cat}$$

of Section III.3.4 *is a full reflective embedding. If α is also interpolating, A_{α} is an isomorphism.*

Proof The isomorphism $(\mathbb{T}, 2)\text{-}\mathsf{Cat} \cong \mathbb{T}\text{-}\mathsf{Mon}$ of Theorem IV.1.5.3 composed with the full reflective embedding $\mathbb{T}\text{-}\mathsf{Mon} \hookrightarrow (\mathbb{S}, 2)\text{-}\mathsf{Cat}$ of Theorem IV.2.3.3

sends $(X, a : TX \nrightarrow X)$ to $(X, a \cdot \alpha_X : SX \nrightarrow X)$. Hence, this composition is precisely the algebraic functor A_α. When α is interpolating, A_α is an isomorphism by Theorem IV.2.3.3. □

IV.2.3.5 Examples

(1) Depending on whether a relation r on a set X is seen as a map

$$r : X \times X \to 2\,, \quad r : X \to PX\,, \quad \text{or} \quad r : PX \times X \to 2\,,$$

the category Ord of ordered sets is described, respectively, as any of the three categories

$$2\text{-Cat} \cong \mathbb{P}\text{-Mon} \cong (\mathbb{P}, 2)\text{-Cat}\,.$$

© (2) Whether ultrafilter convergence, neighborhood systems, or filter convergence is chosen as the defining structure, the category Top of topological spaces appears as

$$(\beta, 2)\text{-Cat} \cong \mathbb{F}\text{-Mon} \cong (\mathbb{F}, 2)\text{-Cat}\,.$$

IV.2.4 Strata extensions

To pass from $(\mathbb{T}, 2)$-categories to $(\mathbb{T}, \mathcal{V})$-categories, one views a $\mathcal{V}$-relation $r : X \nrightarrow Y$ as a family of 2-relations $(r_v : X \nrightarrow Y)_{v\in\mathcal{V}}$ indexed by $\mathcal{V}$, via

$$x\, r_v\, y \iff v \le r(x, y) \qquad \text{(IV.2.4.i)}$$

for all $x \in X$, $y \in Y$, and $v \in \mathcal{V}$. The relation r_v is referred to as the *v-stratum* of r. Conversely, given a family $(r_v : X \nrightarrow Y)_{v\in\mathcal{V}}$ of relations, one can define a $\mathcal{V}$-relation $r : X \nrightarrow Y$ by setting

$$r(x, y) := \bigvee\{v \in \mathcal{V} \mid x\, r_v\, y\}\,.$$

Starting with a $\mathcal{V}$-relation $r : X \nrightarrow Y$, the $\mathcal{V}$-relation obtained from the family $(r_v : X \nrightarrow Y)_{v\in\mathcal{V}}$ is r again. The family $(r_v)_{v\in\mathcal{V}}$ obtained from r is not arbitrary, as one has, for example, if $u, v \in \mathcal{V}$

$$u \le v \implies r_v \le r_u\,.$$

In fact, a $\mathcal{V}$-indexed family comes from a $\mathcal{V}$-relation precisely when the family can be identified with an inf-map $r_{(-)} : \mathcal{V}^{\mathrm{op}} \to \mathsf{Rel}(X, Y)$.

IV.2.4.1 Lemma *For any sets X and Y, if the set $\mathsf{Inf}(\mathcal{V}^{\mathrm{op}}, \mathsf{Rel}(X, Y))$ is ordered pointwise, then the preceding correspondence between $\mathcal{V}$-relations and $\mathcal{V}$-indexed families of relations describes an* Ord*-isomorphism*

$$\mathcal{V}\text{-}\mathsf{Rel}(X, Y) \cong \mathsf{Inf}(\mathcal{V}^{\mathrm{op}}, \mathsf{Rel}(X, Y))\,.$$

Proof If $(r_v)_{v\in\mathcal{V}}$ is obtained from a $\mathcal{V}$-relation r, one has

$$x\left(\bigwedge_{v\in A} r_v\right) y \iff \forall v\in A\ (x\, r_v\, y) \iff \forall v\in A\ (v\le r(x,y)) \\ \iff \bigvee A \le r(x,y)$$

for all $A\subseteq\mathcal{V}$, $x\in X$, $y\in Y$, so $r_{(-)}$ is an indeed an inf-map $r_{(-)}:\mathcal{V}^{\mathrm{op}}\to \mathsf{Rel}(X,Y)$. A straightforward verification shows that if r was originally the image of a family $(r'_v)_{v\in\mathcal{V}}$ then $(r_v)_{v\in\mathcal{V}} = (r'_v)_{v\in\mathcal{V}}$, and the discussion preceding the statement allows us to conclude. □

Suppose that a lax extension $\hat{T}:\mathsf{Rel}\to\mathsf{Rel}$ of a functor $T:\mathsf{Set}\to\mathsf{Set}$ is given. For every $\mathcal{V}$-relation $r: X \nrightarrow Y$, we set

$$\hat{T}_{\mathcal{V}} r(\mathfrak{x},\mathfrak{y}) := \bigvee\{v\in\mathcal{V} \mid \mathfrak{x}\ (\hat{T} r_v)\ \mathfrak{y}\}\,. \tag{IV.2.4.ii}$$

The assignment

$$\hat{T}_{\mathcal{V}}:\mathcal{V}\text{-}\mathsf{Rel}\to\mathcal{V}\text{-}\mathsf{Rel}\,,\quad (r:X\nrightarrow Y)\mapsto(\hat{T}_{\mathcal{V}} r: TX\to TY)\,,$$

is called the *strata extension* of $\hat{T}$ along $\mathsf{Rel}\to\mathcal{V}\text{-}\mathsf{Rel}$.

IV.2.4.2 Proposition *Let T be a* Set*-functor. If $\hat{T}$ is a lax extension of T to* Rel*, then the strata extension $\hat{T}_{\mathcal{V}}$ of $\hat{T}$ is a lax extension of T to $\mathcal{V}$-*Rel.

Proof We first note that in $\mathcal{V}$-Rel

$$Tf\le\hat{T}f\le\hat{T}_{\mathcal{V}}f\,,\quad (Tf)^\circ\le\hat{T}(f^\circ)\le\hat{T}_{\mathcal{V}}(f^\circ)$$

because both f and f° take values in $\{\bot,k\}$. For $\mathcal{V}$-relations $r,r': X\nrightarrow Y$ with $r\le r'$, one has $r_v\le r'_v$, and then $\hat{T}r_v\le\hat{T}r'_v$ for all $v\in\mathcal{V}$, which implies $\hat{T}_{\mathcal{V}}r\le\hat{T}_{\mathcal{V}}r'$. To verify $\hat{T}_{\mathcal{V}}s\cdot\hat{T}_{\mathcal{V}}r\le\hat{T}_{\mathcal{V}}(s\cdot r)$ for all $\mathcal{V}$-relations $r:X\nrightarrow Y$ and $s:Y\nrightarrow Z$, consider $\mathfrak{x}\in TX$, $\mathfrak{y}\in TY$, and $\mathfrak{z}\in TZ$. If there are $u,v\in\mathcal{V}$ with $\mathfrak{x}\ \hat{T}r_u\ \mathfrak{y}$ and $\mathfrak{y}\ \hat{T}s_v\ \mathfrak{z}$, then $\mathfrak{x}\ \hat{T}(s_v\cdot r_u)\ \mathfrak{z}$ holds by the hypothesis on $\hat{T}$; but $s_v\cdot r_u\le(r\cdot s)_{u\otimes v}$, so $\mathfrak{x}\ (\hat{T}(r\cdot s)_{u\otimes v})\ \mathfrak{z}$ holds. Hence,

$$\hat{T}_{\mathcal{V}}r(\mathfrak{x},\mathfrak{y})\otimes\hat{T}_{\mathcal{V}}s(\mathfrak{y},\mathfrak{z}) = \bigvee\{u\in\mathcal{V}\mid\mathfrak{x}\ (\hat{T}r_u)\ \mathfrak{y}\}\otimes\bigvee\{v\in\mathcal{V}\mid\mathfrak{y}\ (\hat{T}s_v)\ \mathfrak{z}\} \\ \le\bigvee\{u\otimes v\in\mathcal{V}\mid\mathfrak{x}\ (\hat{T}(r\cdot s)_{u\otimes v})\ \mathfrak{z}\}\le\hat{T}_{\mathcal{V}}(r\cdot s)(\mathfrak{x},\mathfrak{z})$$

concludes the proof. □

IV.2.4.3 Proposition *For $\mathcal{V}$ completely distributive, the strata extension $\hat{T}_{\mathcal{V}}$ of a lax extension $\hat{\mathbb{T}}$ of $\mathbb{T}$ to* Rel *yields a lax extension of $\mathbb{T}$ to $\mathcal{V}$-*Rel.

Proof For any $\mathcal{V}$-relation $r: X\nrightarrow Y$, one has $r_v\le e_Y^\circ\cdot\hat{T}r_v\cdot e_X$ for all $v\in\mathcal{V}$, so that $r\le e_Y^\circ\cdot\hat{T}_{\mathcal{V}}r\cdot e_X$. To see that the monad multiplication is oplax, let us first show that, for each $v\in\mathcal{V}$ and $u\ll v$, we have $(\hat{T}_{\mathcal{V}}r)_v\le\hat{T}r_u$. Indeed, consider $\mathfrak{x}\in TX$, $\mathfrak{y}\in TY$, $v\in\mathcal{V}$ such that $\mathfrak{x}\ (\hat{T}_{\mathcal{V}}r)_v\ \mathfrak{y}$, i.e. $v\le\hat{T}_{\mathcal{V}}r(\mathfrak{x},\mathfrak{y})$; by definition of $\ll$ in a completely distributive lattice, there is for each $u\ll v$ a

$u' \in \mathcal{V}$ with $u \le u'$ and $\mathfrak{x}\ (\hat{T} r_{u'})\ \mathfrak{y}$, so that $\mathfrak{x}\ (\hat{T} r_u)\ \mathfrak{y}$. As $\hat{T}$ is monotone, and $\mathcal{V}$ is completely distributive, we can therefore write for $\mathcal{X} \in TTX$ and $\mathcal{Y} \in TTY$:

$$\begin{aligned}\bigvee\{v \in \mathcal{V} \mid \mathcal{X}\ \hat{T}(\hat{T}_{\mathcal{V}} r)_v\ \mathcal{Y}\} &\le \bigvee\{u \in \mathcal{V} \mid \mathcal{X}\ (\hat{T}\hat{T} r_u)\ \mathcal{Y}\} \\ &\le \bigvee\{u \in \mathcal{V} \mid m_X(\mathcal{X})\ (\hat{T} r_u)\ m_Y(\mathcal{Y})\}\,.\end{aligned}$$

These inequalities then yield $\hat{T}_{\mathcal{V}}\hat{T}_{\mathcal{V}} r(\mathcal{X}, \mathcal{Y}) \le \hat{T}_{\mathcal{V}} r(m_X(\mathcal{X}), m_Y(\mathcal{Y}))$, as required. □

IV.2.4.4 Remarks

(1) For a lax extension $\hat{T}$ of T to Rel, one has, for $\mathcal{V} = 2$,

$$\hat{T}_2 r(\mathfrak{x}, \mathfrak{y}) = \top \iff \mathfrak{x}\ r\ \mathfrak{y}$$

for all relations $r : X \rightarrow\!\!\!\!\!\shortmid\ Y$, and $\mathfrak{x} \in TX$, $\mathfrak{y} \in TY$, i.e. the strata extension of $\hat{T}$ to 2-Rel = Rel returns $\hat{T}$:

$$\hat{T}_2 = \hat{T}\,.$$

(2) The strata extension to $\mathcal{V}$-Rel of the identity functor 1_{Rel}, which is a lax extension of 1_{Set} to Rel (see Example III.1.5.2), returns the identity functor $1_{\mathcal{V}\text{-}\mathsf{Rel}}$.

The Barr extension of the ultrafilter monad $\mathbb{B}$ to P_+-Rel given in III.2.4 is simply the strata extension of the Barr extension $\overline{\mathbb{B}}$ to Rel along Rel $\to$ P_+-Rel. More generally, for a completely distributive $\mathcal{V}$, the *Barr extension* $\overline{\mathbb{B}}$ of β to $\mathcal{V}$-Rel is given by the strata extension to $\mathcal{V}$-Rel of its Barr extension to Rel:

$$\overline{\beta} r(\mathfrak{x}, \mathfrak{y}) := \overline{\beta}_{\mathcal{V}} r(\mathfrak{x}, \mathfrak{y}) = \bigvee\{v \in \mathcal{V} \mid \mathfrak{x}\ (\overline{\beta} r_v)\ \mathfrak{y}\}\,,$$

and one obtains the corresponding equivalent expression,

$$\overline{\beta} r(\mathfrak{x}, \mathfrak{y}) = \bigwedge_{A \in \mathfrak{x}, B \in \mathfrak{y}} \bigvee_{x \in A, y \in B} r(x, y)\,,$$

for all $\mathcal{V}$-relations $r : X \rightarrow\!\!\!\!\!\shortmid\ Y$, and $\mathfrak{x} \in \beta X$, $\mathfrak{y} \in \beta Y$. Proposition IV.2.4.3 then offers a generalization of Proposition III.2.4.3, as follows.

IV.2.4.5 Corollary *For $\mathcal{V}$ completely distributive, the Barr extension $\overline{\mathbb{B}} = (\overline{\beta}, m, e)$ is a flat lax extension to $\mathcal{V}$-Rel of the ultrafilter monad $\mathbb{B} = (\beta, m, e)$.*

Proof By Proposition IV.2.4.3, we need to verify only that the Barr extension is flat. For $v \in \mathcal{V}$ with $\bot < v$, one has

$$(1_X)_v(x, y) = \begin{cases} 1_X(x, y) & \text{if } v \le k, \\ \bot & \text{otherwise,} \end{cases}$$

for all $x, y \in X$. Moreover, if $r = \bot_X : X \nrightarrow X$ is the constant relation to $\bot$, i.e. the empty relation, then $\overline{\beta}r = \bot_X$. By definition, one therefore has $\overline{\beta}_{\mathcal{V}}1_X(\mathfrak{x}, \mathfrak{y}) = \beta 1_X(\mathfrak{x}, \mathfrak{y})$ for all $\mathfrak{x}, \mathfrak{y} \in \beta X$, so $\overline{\beta}_{\mathcal{V}}$ is flat. □

For a power-enriched monad $\mathbb{T}$ and completely distributive $\mathcal{V}$, the *Kleisli extension* $\check{\mathbb{T}}$ of $\mathbb{T}$ to $\mathcal{V}$-Rel is given by the strata extension to $\mathcal{V}$-Rel of its Kleisli extension to Rel:

$$\check{T}r(\mathfrak{x}, \mathfrak{y}) := \check{T}_{\mathcal{V}}r(\mathfrak{x}, \mathfrak{y}) = \bigvee\{v \in \mathcal{V} \mid \mathfrak{x} \leq (r_v)^{\tau}(\mathfrak{y})\},$$

for all $\mathcal{V}$-relations $r : X \nrightarrow Y$, and $\mathfrak{x} \in TX$, $\mathfrak{y} \in TY$. As in the case of the Barr extension, there are alternative descriptions of the lax extensions given in Examples IV.1.4.2 when $\mathcal{V}$ is completely distributive; see Exercise IV.2.H.

The strata extension has a very good behavior with respect to the order induced on the sets TX and to initial extensions.

IV.2.4.6 Proposition *Let $\hat{T}$ be a lax extension to* Rel *of a* Set-*functor T. The order induced on TX by the strata extension $\hat{T}_{\mathcal{V}}$ of $\hat{T}$ is the order induced by $\hat{T}$.*

Proof Suppose first that $\mathfrak{x}, \mathfrak{y} \in TX$ are such that $\mathfrak{x}\ \hat{T}1_X\ \mathfrak{y}$. Since $1_X = (1_X)_k$ for $1_X : X \to X$ seen as a $\mathcal{V}$-relation, we have $k \leq \hat{T}_{\mathcal{V}}1_X(\mathfrak{x}, \mathfrak{y})$ by definition of the strata extension to $\mathcal{V}$-Rel. If, on the other hand, one has $k \leq \hat{T}_{\mathcal{V}}1_X(\mathfrak{x}, \mathfrak{y})$, because $\mathcal{V}$ is non-trivial (by convention, see Section IV.2.3), there exists $v \in \mathcal{V}$ such that $\bot < v$ and $\mathfrak{x}\ \hat{T}(1_X)_v\ \mathfrak{y}$; as $(1_X)_v = (1_X)_k = 1_X$, because 1_X takes values in $\{\bot, k\}$, we can conclude that $\mathfrak{x}\ \hat{T}1_X\ \mathfrak{y}$. □

IV.2.4.7 Proposition *If $\alpha : (S, \hat{S}) \to (T, \hat{T})$ is a morphism of lax extensions, then $\alpha : (S, \hat{S}_{\mathcal{V}}) \to (T, \hat{T}_{\mathcal{V}})$ is one too. Moreover, if $\hat{S}$ is the initial extension of S to* Rel *induced by $\alpha : S \to T$, then $\hat{S}_{\mathcal{V}}$ is the initial extension of S to $\mathcal{V}$-*Rel *induced by α (where T is equipped with its respective lax extensions $\hat{T}$ and $\hat{T}_{\mathcal{V}}$).*

Proof By definition of a morphism of lax extensions and of the strata extension,

$$\begin{aligned}\hat{S}_{\mathcal{V}}r(\mathfrak{x}, \mathfrak{y}) &= \bigvee\{v \in \mathcal{V} \mid \mathfrak{x}\ (\hat{S}r_v)\ \mathfrak{y}\} \leq \bigvee\{v \in \mathcal{V} \mid \alpha_X(\mathfrak{x})\ (\hat{T}r_v)\ \alpha_Y(\mathfrak{y})\}\\ &= \hat{T}_{\mathcal{V}}r(\alpha_X(\mathfrak{x}), \alpha_Y(\mathfrak{y}))\end{aligned}$$

for all $\mathfrak{x} \in SX$, $\mathfrak{y} \in SY$, and $\mathcal{V}$-relation $r : X \nrightarrow Y$. Hence, α is a morphism between the respective strata extensions, and if $\hat{S}$ is initial then $\hat{S}_{\mathcal{V}}$ is initial. □

IV.2.5 $(\mathbb{S}, \mathcal{V})$-categories as Kleisli towers

The strata extension of a lax extension (Section IV.2.4) constructs a lax extension $\hat{\mathbb{T}}_{\mathcal{V}}$ to $\mathcal{V}$-Rel from a lax extension $\hat{\mathbb{T}}$ to Rel of a monad $\mathbb{T}$ on Set. Such a lax extension $\hat{\mathbb{T}}_{\mathcal{V}}$ then determines a category of $(\mathbb{T}, \mathcal{V})$-categories. In contrast, the tower construction that we describe here determines $(\mathbb{T}, \mathcal{V})$-categories directly from a category of $(\mathbb{T}, 2)$-categories by exploiting the isomorphism

$$\mathcal{V}\text{-}\mathsf{Rel}(X, Y) \cong \mathsf{Inf}(\mathcal{V}^{\mathrm{op}}, \mathsf{Rel}(X, Y))$$

of Lemma IV.2.4.1. Under this isomorphism, reflexive and transitive $(\mathbb{T}, \mathcal{V})$-relations $a : X \nrightarrow X$ correspond to certain inf-maps $\mathcal{V}^{\mathrm{op}} \to \mathsf{Rel}(TX, X)$ whose characterization depends only on the original lax extension $\hat{\mathbb{T}}$ to Rel (Proposition IV.2.5.2). In fact, these inf-maps corestrict to fixpoints of nbhd $\cdot$ conv so that Proposition IV.1.5.1 allows us to consider instead inf-maps $\mathcal{V}^{\mathrm{op}} \to \mathsf{Set}(X, TX)$, in effect relating $(\mathbb{T}, \mathcal{V})$-category structures with $\mathbb{T}$-monoid structures. In this section, we pursue this idea in the presence of a sup-dense and interpolating monad morphism $\alpha : \mathbb{S} \to \mathbb{T}$, considering $\mathsf{Rel}(SX, X)$ in lieu of $\mathsf{Rel}(TX, X)$ and using Lemma IV.2.3.1 instead of Proposition IV.1.5.1. We then obtain Theorem IV.2.5.2 as a generalization of Theorem IV.2.3.3.

The following lemma will be used throughout without explicit mention in our proofs.

IV.2.5.1 Lemma *For all maps $f : X \to Y$, $g : Y \to Z$, $\mathcal{V}$-relations $r : X \nrightarrow Y$, $s : Y \nrightarrow Z$, and $u, v \in \mathcal{V}$,*

$$s_v \cdot f = (s \cdot f)_v\,, \quad g^\circ \cdot r_v = (g^\circ \cdot r)_v\,, \qquad \text{and} \qquad s_u \cdot r_v \leq (s \cdot r)_{v \otimes u}\,.$$

Proof The expressions follow directly from the definition of the strata of a relation (see Section IV.2.4). □

IV.2.5.2 Proposition *Let $\hat{\mathbb{T}}$ be a lax extension of $\mathbb{T}$ to Rel and $\mathcal{V}$ a quantale. Via the isomorphism $\mathcal{V}\text{-}\mathsf{Rel}(TX, X) \cong \mathsf{Inf}(\mathcal{V}^{\mathrm{op}}, \mathsf{Rel}(TX, X))$ of* Lemma IV.2.4.1, *there is a one-to-one correspondence between transitive and reflexive $(\mathbb{T}, \mathcal{V})$-relations $r : X \nrightarrow X$ and inf-maps $r_{(-)} : \mathcal{V}^{\mathrm{op}} \to (\mathbb{T}, \mathcal{V})\text{-}\mathsf{URel}(X, X)$ that satisfy*

$$r_v \circ r_u \leq r_{u \otimes v} \qquad \text{and} \qquad e_X^\circ \leq r_k$$

for all $u, v \in \mathcal{V}$.

Proof For a transitive $\mathcal{V}$-relation $r : TX \to X$, we show $r_v \cdot (\hat{T}_{\mathcal{V}} r)_u \leq r_{u \otimes v} \cdot m_X$ for all $u, v \in \mathcal{V}$: since $\hat{T} r_u \leq (\hat{T}_{\mathcal{V}} r)_u$ by definition, $r_u \circ r_v \leq r_{v \otimes u}$ will follow. If $z \in X$ and $\mathfrak{X} \in TTX$ are such that $\mathfrak{X}\ (r_v \cdot (\hat{T}_{\mathcal{V}} r)_u)\ z$, then there exists $y \in TX$ with $v \leq r(y, z)$ and $u \leq \hat{T}_{\mathcal{V}} r(\mathfrak{X}, y)$. Therefore,

$$u \otimes v \leq \bigvee\nolimits_{y \in TX} \hat{T}_{\mathcal{V}} r(\mathfrak{X}, y) \otimes r(y, z) = r \cdot \hat{T}_{\mathcal{V}} r(\mathfrak{X}, z) \leq r \cdot m_X(\mathfrak{X}, z)\,,$$

and $(r \cdot m_X)_{u \otimes v} = r_{u \otimes v} \cdot m_X$ yields the required inequality. If $r : TX \nrightarrow X$ is a reflexive $\mathcal{V}$-relation, then $e_X^\circ \leq r_k$ because e_X° takes values in $\{\bot, k\}$.

Conversely, assume that $r_v \circ r_u \leq r_{u \otimes v}$ holds for all $u, v \in \mathcal{V}$. One has to show that $T_{\mathcal{V}} r(\mathfrak{X}, y) \otimes r(y, z) \leq r(m_X(\mathfrak{X}), z)$ for all $z \in X$, $y \in TX$, $\mathfrak{X} \in TTX$.

Choose elements $z, y, \mathfrak{X}$ and set $v := r(y, z)$, $A := \{u \in \mathcal{V} \mid \mathfrak{X}\ (\hat{T}r_u)\ y\}$, so that $\hat{T}_{\mathcal{V}}r(\mathfrak{X}, y) \otimes r(y, z) = \bigvee_{u \in A} u \otimes v$. Since by hypothesis $r_v \cdot \hat{T}r_u \leq r_{u \otimes v} \cdot m_X$, we have $u \otimes v \leq r(m_X(\mathfrak{X}), z)$ for all $u \in A$. Hence, r is transitive. Reflexivity of r follows from $e_X^{\circ} \leq r_k \leq r$. □

Let now $(\mathbb{T}, \tau)$ be a power-enriched monad, so that all hom-sets $\mathsf{Set}_{\mathbb{T}}(X, X)$ are complete lattices when equipped with the pointwise order (see Proposition IV.1.2.1). To extend Theorem IV.2.3.3 to $\mathcal{V}$-relations, we consider the category

$$(\mathbb{T}, \mathcal{V})\text{-}\mathsf{Mon}$$

whose objects are *Kleisli* $(\mathbb{T}, \mathcal{V})$*-towers*, i.e. pairs $(X, \nu^{(-)})$ with X a set and $\nu^{(-)} : \mathcal{V}^{\mathrm{op}} \to \mathsf{Set}_{\mathbb{T}}(X, X)$ an inf-map satisfying

$$\nu^u \circ \nu^v \leq \nu^{u \otimes v} \qquad \text{and} \qquad e_X \leq \nu^k$$

for all $u, v \in \mathcal{V}$. A morphism of Kleisli $(\mathbb{T}, \mathcal{V})$-towers $f : (X, \nu^{(-)}) \to (Y, \mu^{(-)})$ is a map $f : X \to Y$ such that

$$Tf \cdot \nu^v \leq \mu^v \cdot f$$

for all $v \in \mathcal{V}$.

IV.2.5.3 Theorem *Let* $(\mathbb{T}, \tau)$ *be a power-enriched monad, let* $\alpha : \mathbb{S} \to \mathbb{T}$ *be a sup-dense and interpolating monad morphism, and suppose that* $\mathcal{V}$ *is completely distributive. With respect to the initial extension* $\hat{S}_{\mathcal{V}}$ *of* S *induced by* α *(where* $\mathbb{T}$ *is equipped with its Kleisli extension* $\check{\mathbb{T}}_{\mathcal{V}}$*), there is an isomorphism*

$$(\mathbb{S}, \mathcal{V})\text{-}\mathsf{Cat} \cong (\mathbb{T}, \mathcal{V})\text{-}\mathsf{Mon}$$

that commutes with the underlying-set functors.

Proof By Lemma IV.2.3.1, a Kleisli $(\mathbb{T}, \mathcal{V})$-tower $(X, \nu^{(-)})$ corresponds to a pair $(X, r_{(-)})$, with $r_{(-)} : \mathcal{V}^{\mathrm{op}} \to (\mathbb{S}, 2)\text{-}\mathsf{URel}(X, X)$ an inf map (set $r_v := \mathrm{conv}(\nu^v)$ so that $\nu^v = \mathrm{nbhd}(r_v)$). By Proposition IV.2.3.2, the conditions for $\nu^{(-)}$ to be a Kleisli $(\mathbb{T}, \mathcal{V})$ tower translate as

$$r_v \circ r_u \leq r_{u \otimes v} \qquad \text{and} \qquad e_X^{\circ} \leq r_k$$

for all $u, v \in \mathcal{V}$. Proposition IV.2.5.2 then yields a bijective correspondence between Kleisli $(\mathbb{T}, \mathcal{V})$-towers $(X, \nu^{(-)})$ and $(\mathbb{T}, \mathcal{V})$-categories (X, r).

Consider now a morphism $f : (X, \nu^{(-)}) \to (Y, \mu^{(-)})$ of Kleisli $(\mathbb{T}, \mathcal{V})$-towers. By Theorem IV.2.3.3, $f \cdot \mathrm{conv}(\nu^v) \leq \mathrm{conv}(\mu^v) \cdot Tf$ for all $v \in \mathcal{V}$; hence, by definining r and s as the $(\mathbb{S}, \mathcal{V})$-relations corresponding, respectively, to the families $(\mathrm{conv}(\nu^v) : SX \nrightarrow X)_{v \in \mathcal{V}}$ and $(\mathrm{conv}(\mu^v) : SY \nrightarrow Y)_{v \in \mathcal{V}}$ via Lemma IV.2.4.1, one obtains that $f : (X, r) \to (Y, s)$ is an $(\mathbb{S}, \mathcal{V})$-functor. Conversely, if $f : (X, r) \to (Y, s)$ is an $(\mathbb{S}, \mathcal{V})$-functor, then

$$f \cdot r_v = (f \cdot r)_v \leq (s \cdot Tf)_v = s_v \cdot Tf$$

for all $v \in \mathcal{V}$. This proves that the mentioned correspondence on objects is functorial. □

IV.2.5.4 Corollary *Let $(\mathbb{T}, \tau)$ be a power-enriched monad and suppose that $\mathcal{V}$ is completely distributive. With respect to the Kleisli extension $\check{\mathbb{T}}_{\mathcal{V}}$, there is an isomorphism*

$$(\mathbb{T}, \mathcal{V})\text{-}\mathsf{Cat} \cong (\mathbb{T}, \mathcal{V})\text{-}\mathsf{Mon}\,.$$

Proof Apply Theorem IV.2.5.3 to $\alpha = 1_{\mathbb{T}}$, which is both sup-dense and interpolating. □

IV.2.5.5 Examples

(1) The category Met of metric spaces (see Example III.1.3.1(2)) is isomorphic to any of the categories

$$\mathsf{P}_{+}\text{-}\mathsf{Cat} \cong (\mathbb{P}, \mathsf{P}_{+})\text{-}\mathsf{Mon} \cong (\mathbb{P}, \mathsf{P}_{+})\text{-}\mathsf{Cat}$$

(with $\mathbb{P}$ equipped with its Kleisli extension to $\mathsf{P}_{+}\text{-}\mathsf{Rel}$).

© (2) The category App of approach spaces can equivalently be described as

$$(\beta, \mathsf{P}_{+})\text{-}\mathsf{Cat} \cong (\mathbb{F}, \mathsf{P}_{+})\text{-}\mathsf{Mon} \cong (\mathbb{F}, \mathsf{P}_{+})\text{-}\mathsf{Cat}$$

(with β equipped with its Barr extension and $\mathbb{F}$ with its Kleisli extension to $\mathsf{P}_{+}\text{-}\mathsf{Rel}$). We use here the isomorphism $(\beta, \mathsf{P}_{+})\text{-}\mathsf{Cat} \cong \mathsf{App}$ of Theorem III.2.4.5, and therefore the Axiom of Choice, to prove these results. Nevertheless, the isomorphism $(\mathbb{F}, \mathsf{P}_{+})\text{-}\mathsf{Mon} \cong \mathsf{App}$ can be established without the use of the Axiom of Choice; see Exercise IV.2.K.

© (3) The category of "many-valued neighborhood spaces" is obtained as

$$(\beta, [0,1])\text{-}\mathsf{Cat} \cong (\mathbb{F}, [0,1])\text{-}\mathsf{Mon} \cong (\mathbb{F}, [0,1])\text{-}\mathsf{Cat}\,,$$

where the frame $[0, 1]$ is considered a quantale (see II.1.10). Analogous results hold for $\mathbb{F}$ replaced by $\mathbb{U}$, giving rise to extensions of the category of interior (or closure) spaces.

© (4) $(\beta, 2^2)\text{-}\mathsf{Cat}$ is the category BiTop of bitopological spaces and bicontinuous maps (Exercise III.2.D). This category can also be described as $(\mathbb{F}, 2^2)\text{-}\mathsf{Mon}$, or as $(\mathbb{F}, 2^2)\text{-}\mathsf{Cat}$, avoiding the use of the Axiom of Choice.

(5) Tower extensions allow us effectively to describe $(\mathbb{T}, \mathcal{V})$-categories for different lattices $\mathcal{V}$. For instance, $(\mathbb{F}, \{0,1,2\})\text{-}\mathsf{Cat}$, where the frame $\{0, 1, 2\}$ is considered a quantale, is isomorphic to the category whose objects are triples $(X, \mathcal{O}_0X, \mathcal{O}_1X)$, with topologies $\mathcal{O}_0X \subseteq \mathcal{O}_1X$, and whose morphisms are maps which are continuous with respect to both topologies.

(6) For the quantale $\mathsf{3} = (\bot, k, \top)$ (see Example II.1.10.1(2)), there are isomorphisms

$$\mathsf{3}\text{-}\mathsf{Cat} \cong (\mathbb{P}, \mathsf{3})\text{-}\mathsf{Mon} \cong (\mathbb{P}, \mathsf{3})\text{-}\mathsf{Cat}\,.$$

Objects of $\mathsf{3\text{-}Cat}$ are sets X with a map $r_{(-)} : \mathsf{3} \to \mathsf{Rel}(X, X)$ such that $r_\perp$ satisfies $x \, r_\perp \, x$ for all $x \in X$, r_k is a reflexive and transitive relation, and $r_\top$ is a relation such that $r_\top \le r_k$, $r_\top \cdot r_k \le r_\top$, and $r_k \cdot r_\top \le r_\top$; morphisms of $\mathsf{3\text{-}Cat}$ are maps that preserve the relations r_k and $r_\top$. Therefore, $\mathsf{3\text{-}Cat}$ is isomorphic to the category whose objects are ordered sets $(X, \le_X)$ equipped with an *auxiliary relation*, i.e. a module $a : X \nrightarrow X$ such that $a \le (\le_X)$ (see II.1.4), and whose morphisms are monotone maps preserving the auxiliary relation.

Exercises

IV.2.A *The category of lax extensions to* $\mathcal{V}$-Rel. The forgetful functor $U : \mathcal{V}\text{-}\mathsf{LXT} \to \mathsf{Set}^{\mathsf{Set}}$ (that sends a lax extension $\hat{T} : \mathcal{V}\text{-}\mathsf{Rel} \to \mathcal{V}\text{-}\mathsf{Rel}$ to its underlying functor $T : \mathsf{Set} \to \mathsf{Set}$) is topological: if $(T_i, \hat{T}_i)_{i \in I}$ is a family of lax extensions to $\mathcal{V}$-Rel, then every family $(\alpha_i : S \to T_i)_{i \in I}$ of natural transformations admits a U-initial lifting $(\alpha_i : (S, \hat{S}) \to (T_i, \hat{T}_i))_{i \in I}$ with

$$\hat{S}r := \bigwedge_{i \in I} (\alpha_i)_Y^\circ \cdot \hat{T}_i r \cdot (\alpha_i)_X \, ,$$

or equivalently

$$\hat{S}r(\mathfrak{x}, \mathfrak{y}) = \bigwedge_{i \in I} \hat{T}r((\alpha_i)_X(\mathfrak{x}), (\alpha_i)_Y(\mathfrak{y})) \, ,$$

for all $\mathcal{V}$-relations $: X \nrightarrow Y$, $\mathfrak{x} \in TX$, and $\mathfrak{y} \in TY$. Moreover, if all α_i are morphisms of monads $\alpha_i : \mathbb{S} \to \mathbb{T}_i$, and all $\hat{T}_i$ are lax extensions of $\mathbb{T}_i = (T_i, m_i, e_i)$, then $\hat{S}$ is a lax extension of $\mathbb{S} = (S, n, d)$.

IV.2.B *The Zariski topology on a set of ultrafilters.* Consider the set βX of all ultrafilters on a set X, and the closure operation $c : P\beta X \to P\beta X$ given by

$$\mathfrak{x} \in c(\mathcal{A}) \iff \mathfrak{x} \le \bigvee \mathcal{A} \iff \mathfrak{x} \supseteq \bigcap \mathcal{A}$$

for all $\mathfrak{x} \in \beta X$, $\mathcal{A} \subseteq \beta X$. This closure operation defines the *Zariski topology* on βX whose closed sets can be identified with the set of filters $F\beta X$.

A relation $r : \beta X \nrightarrow Y$ is therefore a fixpoint of $(\mathrm{conv} \cdot \mathrm{nbhd}) : \mathsf{Rel}(\beta X, X) \to \mathsf{Rel}(\beta X, X)$ exactly when $r^\flat(y)$ is closed in this topology on βX for all $y \in X$.

IV.2.C *The neighborhood map as a monoid homomorphism.* For a power-enriched monad $\mathbb{T}$ equipped with its Kleisli extension $\check{\mathbb{T}}$ to Rel, a monad morphism $\alpha : \mathbb{S} \to \mathbb{T}$ inducing the initial extension on $\mathbb{S}$, and relations $r, s : SX \nrightarrow X$, the map

$$\mathrm{nbhd} : \mathsf{Set}(X, PSX) \to \mathsf{Set}(X, TX)$$

satisfies $\mathrm{nbhd}(s \circ r) = \mathrm{nbhd}(r) \circ \mathrm{nbhd}(s)$ if and only if one of the following equivalent conditions hold:

(i) $\alpha_X^{\mathbb{T}} \cdot r^\tau \cdot \alpha_X = \alpha_X^{\mathsf{V}} \cdot Pn_X \cdot (\hat{S}r)^\flat$;
(ii) $\alpha_X^{\mathbb{T}} \cdot r^\tau \cdot \alpha_X \leq \alpha_X^{\mathsf{V}} \cdot Pn_X \cdot (\hat{S}r)^\flat$;
(iii) $\alpha_X^{\mathbb{T}} \cdot r^\tau \cdot \alpha_X(y) \leq \bigvee\{\alpha_X \cdot n_X(\mathcal{X}) \mid \mathcal{X} \in SSX : \alpha_{SX}(\mathcal{X}) \leq r^\tau \cdot \alpha_X(y)\}$ for all $y \in SX$.

Hint. For (ii) $\Longrightarrow$ (i), use that $\alpha_X^{\mathbb{T}} \cdot r^\tau \cdot \alpha_X = \alpha_X^{\mathsf{V}} \cdot Pn_X \cdot (\hat{S}r)^\flat$.

IV.2.D *Full coreflective subcategories of* $\mathbb{T}$-Mon. Consider a monad morphism $\alpha : \mathbb{S} \to \mathbb{T}$ into a power-enriched monad $(\mathbb{T}, \tau)$ equipped with its Kleisli extension $\check{T}$, where $\mathbb{S}$ is equipped with its initial extension $\hat{S}$ induced by α. If α is interpolating and satisfies

$$\alpha_X^{\mathbb{T}} \cdot r^\tau \cdot \alpha_X \leq \alpha_X^{\mathsf{V}} \cdot Pn_X \cdot (\hat{S}r)^\flat$$

(see Exercise IV.2.C) for all unitary relations $r : SX \nrightarrow X$, there is a full coreflective embedding $(\mathbb{S}, 2)\text{-}\mathsf{UGph} \hookrightarrow (1_{\mathsf{Set}} \downarrow T)_{\leq}$ that commutes with the underlying-set functors and restricts to

$$(\mathbb{S}, 2)\text{-}\mathsf{Cat} \hookrightarrow \mathbb{T}\text{-}\mathsf{Mon}\,.$$

By composing this embedding with the isomorphism $\mathbb{T}\text{-}\mathsf{Mon} \cong (\mathbb{T}, 2)\text{-}\mathsf{Cat}$, one obtains a coreflective embedding $(\mathbb{S}, 2)\text{-}\mathsf{Cat} \to (\mathbb{T}, 2)\text{-}\mathsf{Cat}$ that sends $(X, r : SX \nrightarrow X)$ to $(X, \hat{r} : TX \nrightarrow X)$, where

$$\hat{r}(f, y) = \bigwedge\{r(\chi, y) \mid \chi \in SX : \alpha_X(\chi) \leq f\}\,,$$

for all $f \in TX$, $y \in X$. In particular, the principal filter monad morphisms $\mathbb{P} \to \mathbb{F} \to \mathbb{U}$ yield full coreflective embeddings

$$\mathsf{Ord} \hookrightarrow \mathsf{Top} \hookrightarrow \mathsf{Cls}\,.$$

IV.2.E *Ultracliques and closure spaces.* With the set of cliques CX of Exercise IV.1.F ordered by reverse inclusion, an *ultraclique* is either the empty clique or a minimal proper element of CX. Alternatively, a non-empty ultraclique χ is an up-set in PX such that

$$A \in \chi \iff A^{\complement} \notin \chi$$

for all $A \in PX$. Although the existence of "sufficiently many" ultracliques requires the Axiom of Choice (to obtain, for example, a clique version of Proposition II.1.13.2), these structures appear to be less elusive than ultrafilters. For example, if the set X has at least three distinct elements $x, y, z \in X$, then a non-principal ultraclique is given by

$$\uparrow_{PX}\{\{x, y\}, \{y, z\}, \{z, x\}\}\,.$$

The clique monad restricts to ultracliques to form the *ultraclique monad* $\mathbb{K} = (\kappa, m, e)$. With the clique and ultraclique monads both equipped with their initial extensions induced by their inclusions in $\mathbb{U}$, there are isomorphisms

$$(\mathbb{U}, 2)\text{-}\mathsf{Cat} \cong \mathbb{U}\text{-}\mathsf{Mon} \cong (\mathbb{K}, 2)\text{-}\mathsf{Cat}\,.$$

IV.2.F *A flat extension of the ultraclique monad.* The initial extension of the non-empty ultraclique functor (see Exercise IV.2.E) is flat, although the functor does not satisfy the Beck–Chevalley condition.

Hint. For the first statement, investigate the induced order on a set of non-empty ultracliques. For the second, consider the maps

$$\{x\} \overset{f}{\hookrightarrow} \{x, y\} \underset{g}{\longleftarrow} \{x, y, z\}\,,$$

where $g(y) = g(z) = y$ and $g(x) = x$.

IV.2.G *Unitary $\mathcal{V}$-relations are sup-maps.* Let $\mathcal{V}$ be a completely distributive quantale and let $(\mathbb{T}, \tau)$ be a power-enriched monad. For a unitary $(\mathbb{T}, \mathcal{V})$-relation $r : X \nrightarrow Y$ (with respect to the Kleisli extension of $\mathbb{T}$ to $\mathcal{V}$-Rel), the map $r(-, y) : TX \to \mathcal{V}^{\mathrm{op}}$ preserves suprema for all $y \in Y$ (see also Exercise IV.1.H).

IV.2.H *Alternative descriptions of lax extensions.* If the lattice $\mathcal{V}$ is completely distributive, the Kleisli extensions to $\mathcal{V}$-Rel of the powerset, filter, and up-set functors of Examples IV.1.4.2, can equivalently be expressed as

$$\begin{aligned}
\check{P}r(A, B) &= \textstyle\bigwedge_{x\in A} \bigvee_{y\in B} r(x, y)\,,\\
\check{F}r(\mathfrak{f}, \mathfrak{g}) &= \textstyle\bigwedge_{B\in \mathfrak{g}} \bigvee_{A\in \mathfrak{f}} \bigwedge_{x\in A} \bigvee_{y\in B} r(x, y)\,,\\
\check{U}r(\mathfrak{x}, \mathfrak{y}) &= \textstyle\bigwedge_{B\in \mathfrak{y}} \bigvee_{A\in \mathfrak{x}} \bigwedge_{x\in A} \bigvee_{y\in B} r(x, y)\,,
\end{aligned}$$

respectively, where $r : X \nrightarrow Y$ is a $\mathcal{V}$-relation, $A \in PX$, $B \in PY$, $\mathfrak{f} \in FX$, $\mathfrak{g} \in FY$, $\mathfrak{x} \in UX$, and $\mathfrak{y} \in UY$.

IV.2.I *Monad retractions induce equivalences of categories.* Let $\hat{\mathbb{T}}$ be a lax extension of $\mathbb{T}$ and let $\alpha : \mathbb{S} \to \mathbb{T}$ be a retraction of monads (so there exists a monad morphism $\rho : \mathbb{T} \to \mathbb{S}$ such that $\alpha \cdot \rho = 1_{\mathbb{T}}$). If $\hat{\mathbb{S}}$ is the initial extension induced by α, then $(\mathbb{S}, \mathcal{V})$-$\mathsf{Cat}$ and $(\mathbb{T}, \mathcal{V})$-$\mathsf{Cat}$ are equivalent categories.

IV.2.J *Topologicity of the functor* $(\mathbb{T}, 2)$-$\mathsf{UGph} \to \mathsf{Set}$. For a lax extension $\hat{\mathbb{T}}$ of a monad $\mathbb{T}$ on Set, the forgetful functor $U : (\mathbb{T}, \mathcal{V})$-$\mathsf{UGph} \to \mathsf{Set}$ is topological.

IV.2.K *Approach spaces as* $(\mathbb{F}, \mathsf{P}_+)$*-categories.* Consider the filter monad $\mathbb{F}$ with the principal filter natural transformation $\tau : \mathbb{P} \to \mathbb{F}$ and the extended real half-line P_+. The category of reflexive and transitive P_+-towers of $(\mathbb{F}, 2)$-UGph is isomorphic to the category of reflexive and transitive P_+-towers of PSet (see

Exercise IV.1.D). Moreover, P_+-towers of PSet $(c_v : PX \to PX)_{v\in\mathcal{V}}$ and maps $\delta : X \times PX \to \mathsf{P}_+$ determine each other via

$$\delta(x, A) := \inf\{v \in [0, \infty] \mid x \in c_v(A)\} \quad \text{and} \quad c_v(A) = \{x \in X \mid \delta(x, A) \le v\},$$

and this correspondence induces an isomorphism $(\mathbb{F}, \mathsf{P}_+)\text{-}\mathsf{Cat} \cong \mathsf{App}$.

IV.2.L *Quantales as* $(\mathbb{T}, \mathcal{V})$*-categories.* Let $(\mathbb{T}, \tau)$ be a power-enriched monad, with $\mathbb{T} = (T, m, e)$, and let $\mathcal{V}$ be a completely distributive quantale (with totally below relation $\ll$, see II.1.11). The map $\alpha : \mathcal{V} \to T\mathcal{V}$ defined by

$$\alpha(v) := \bigvee\{\chi \in T\mathcal{V} \mid \forall u \in \mathcal{V}\ (u \ll v \implies \chi \le \tau_{\mathcal{V}}(\uparrow u))\}$$

is an inf-map $\alpha : \mathcal{V}^{\mathrm{op}} \to T\mathcal{V}$. Hence, α has a left adjoint $\xi : T\mathcal{V} \to \mathcal{V}^{\mathrm{op}}$ in Ord. Verify that the pair $(\mathcal{V}, \alpha)$ is a $\mathbb{T}$-monoid, and therefore that $(\mathcal{V}, \xi_*)$ is a $(\mathbb{T}, \mathsf{2})$-category (see II.1.4).

The map α extends to $\tilde{\alpha} : \mathcal{V}^{\mathrm{op}} \to \mathsf{Set}(\mathcal{V}, T\mathcal{V})$ via

$$\tilde{\alpha}(u)(v) := \alpha(u \otimes v),$$

so that $(\mathcal{V}, \tilde{\alpha})$ is a $\mathcal{V}$-tower of $\mathbb{T}\text{-}\mathsf{Mon}$, i.e. an object of $(\mathbb{T}, \mathcal{V})\text{-}\mathsf{Mon}$. The structure $\mathcal{V}$-relation of the corresponding $(\mathbb{T}, \mathcal{V})$-category $(X, r_{\tilde{\alpha}})$ is given by

$$r_{\tilde{\alpha}}(\chi, v) = \xi(\chi) \multimapdotinv v$$

(see II.1.10).

IV.3 Lax algebras as Kleisli monoids

In this section, we show that an *associative* lax extension to $\mathcal{V}\text{-}\mathsf{Rel}$ of a monad $\mathbb{T}$ always has an associated power-enriched monad Π that, when equipped with its Kleisli extension to Rel, returns the same category of lax algebras:

$$(\mathbb{T}, \mathcal{V})\text{-}\mathsf{Cat} \cong \Pi\text{-}\mathsf{Mon}$$

(Theorem IV.3.2.2). In Section IV.3.3, an alternative description of the monad Π is given in the case of approach spaces, i.e. when $\mathbb{T} = \beta$ and $\mathcal{V} = \mathsf{P}_+$.

IV.3.1 The ordered category $(\mathbb{T}, \mathcal{V})\text{-}\mathsf{URel}$

Given an associative lax extension to $\mathcal{V}\text{-}\mathsf{Rel}$ of a monad $\mathbb{T}$ on Set, recall from III.1.9 that sets with unitary $(\mathbb{T}, \mathcal{V})$-relations form the category $(\mathbb{T}, \mathcal{V})\text{-}\mathsf{URel}$, with Kleisli convolution as composition and the identity on X given by $1_X^\sharp = e_X^\circ \cdot \hat{T}1_X$. Moreover, $(\mathbb{T}, \mathcal{V})\text{-}\mathsf{URel}$ forms an ordered category when the hom-sets $(\mathbb{T}, \mathcal{V})\text{-}\mathsf{URel}(X, Y) \subseteq \mathcal{V}\text{-}\mathsf{Rel}(TX, Y)$ are equipped with the pointwise order inherited from $\mathcal{V}\text{-}\mathsf{Rel}$, since the Kleisli convolution preserves this order on the left and right.

IV.3.1.1 Lemma *Let $\hat{\mathbb{T}}$ be a lax extension of $\mathbb{T}$ to $\mathcal{V}$-*Rel*. For a family of unitary $(\mathbb{T},\mathcal{V})$-relations $(\varphi_i : X \nrightarrow Y)_{i\in I}$, the $(\mathbb{T},\mathcal{V})$-relation $\bigwedge_{i\in I}\varphi_i$ is unitary.*

Proof The statement follows from the inequalities

$$\left(\bigwedge_{i\in I}\varphi_i\right)\cdot\hat{T}1_X \le \bigwedge_{i\in I}(\varphi_i\cdot\hat{T}1_X) = \bigwedge_{i\in I}\varphi_i\,,$$
$$e_Y^\circ\cdot\hat{T}\left(\bigwedge_{i\in I}\varphi_i\right)\cdot m_X^\circ \le \bigwedge_{i\in I}(e_Y^\circ\cdot\hat{T}\varphi_i\cdot m_X^\circ) = \bigwedge_{i\in I}\varphi_i\,,$$

since the inequalities in the other direction always hold. □

This lemma implies that the ordered category $(\mathbb{T},\mathcal{V})$-URel has complete hom-sets. However, in general $(\mathbb{T},\mathcal{V})$-URel is not a quantaloid since $\varphi\circ(-)$ typically does not preserve suprema. The situation is better for composition on the right, and when the maps $(-)\circ\varphi$ preserve suprema we say that $(\mathbb{T},\mathcal{V})$-URel is a *right-sided* quantaloid.

IV.3.1.2 Proposition *Let $\hat{\mathbb{T}}$ be an associative lax extension of $\mathbb{T}$ to $\mathcal{V}$-*Rel*. Then, for every $(\mathbb{T},\mathcal{V})$-relation $\varphi : X \nrightarrow Y$ and every set Z, the monotone map*

$$(-)\circ\varphi : (\mathbb{T},\mathcal{V})\text{-Rel}(Y,Z) \to (\mathbb{T},\mathcal{V})\text{-Rel}(X,Z)$$

*has a right adjoint $(-)\multimapdotinv\varphi : (\mathbb{T},\mathcal{V})\text{-Rel}(X,Z) \to (\mathbb{T},\mathcal{V})\text{-Rel}(Y,Z)$ that sends $\psi : X \nrightarrow Z$ to $\psi \bullet\!\!- (\hat{T}\varphi\cdot m_X^\circ)$ (see Exercise III.1.E). Moreover, if $\varphi : X \nrightarrow Y$ and $\psi : X \nrightarrow Z$ are unitary, then $\psi \multimapdotinv \varphi$ is also unitary, and consequently $(\mathbb{T},\mathcal{V})$-*URel* is a right-sided quantaloid.*

Proof Consider a $(\mathbb{T},\mathcal{V})$-relation $\varphi : X \nrightarrow Y$. For all $\gamma : Y \nrightarrow Z$ and $\psi : X \nrightarrow Z$,

$$\gamma\circ\varphi\le\psi \iff \gamma\cdot\hat{T}\varphi\cdot m_X^\circ\le\psi \iff \gamma\le\psi \bullet\!\!- (\hat{T}\varphi\cdot m_X^\circ)\,,$$

hence the map

$$(-)\multimapdotinv\varphi : (\mathbb{T},\mathcal{V})\text{-Rel}(X,Z) \to (\mathbb{T},\mathcal{V})\text{-Rel}(Y,Z)\,,$$
$$\psi \mapsto (\psi\multimapdotinv\varphi) := \psi \bullet\!\!- (\hat{T}\varphi\cdot m_X^\circ)$$

is right adjoint to $(-)\circ\varphi$. Suppose now that $\varphi : X \nrightarrow Y$ and $\psi : X \nrightarrow Z$ are unitary. By associativity of the Kleisli convolution and Proposition III.1.9.4,

$$(1_Z^\sharp\circ(\psi\multimapdotinv\varphi))\circ\varphi \le 1_Z^\sharp\circ\psi = \psi$$

and

$$((\psi\multimapdotinv\varphi)\circ 1_Y^\sharp)\circ\varphi \le (\psi\multimapdotinv\varphi)\circ\varphi \le \psi\,;$$

therefore, $1_Z^\sharp\circ(\psi\multimapdotinv\varphi) \le \psi\multimapdotinv\varphi$ and $(\psi\multimapdotinv\varphi)\circ 1_Y^\sharp \le \psi\multimapdotinv\varphi$. □

Recall from Exercise III.1.M that a map $f : X \to Y$ gives rise to a unitary $(\mathbb{T}, \mathcal{V})$-relation $f^\sharp : Y \nrightarrow X$ via

$$f^\sharp = f^\circ \cdot e_Y^\circ \cdot \hat{T}1_Y = e_X^\circ \cdot (Tf)^\circ \cdot \hat{T}1_Y = e_X^\circ \cdot \hat{T}(f^\circ)\,,$$

where $1_X^\sharp = (1_X)_\sharp$ is the identity morphism on X in $(\mathbb{T}, \mathcal{V})$-URel.

IV.3.1.3 Lemma *Let $\hat{\mathbb{T}}$ be a lax extension of $\mathbb{T}$ to $\mathcal{V}$-Rel. For a unitary $(\mathbb{T}, \mathcal{V})$-relation $\varphi : X \nrightarrow Y$, one has*

$$f^\sharp \circ \varphi = f^\circ \cdot \varphi$$

for all maps $f : Z \to Y$.

Proof One computes $f^\sharp \circ \varphi = f^\circ \cdot e_Y^\circ \cdot \hat{T}1_Y \cdot \hat{T}\varphi \cdot m_X^\circ = f^\circ \cdot (1_X^\sharp \circ \varphi) = f^\circ \cdot \varphi$.□

If $\hat{\mathbb{T}}$ is an associative lax extension of $\mathbb{T}$ to $\mathcal{V}$-Rel, then $f^\sharp \circ g^\sharp = (g \cdot f)^\sharp$ for all maps $f : X \to Y$ and $g : Y \to Z$ in Set:

$$f^\sharp \circ g^\sharp = f^\circ \cdot g^\sharp = f^\circ \cdot g^\circ \cdot e_Z^\circ \cdot \hat{T}1_Z = (g \cdot f)^\sharp\,.$$

Hence, there is a functor

$$(-)^\sharp : \mathsf{Set} \to (\mathbb{T}, \mathcal{V})\text{-}\mathsf{URel}^{\mathrm{op}}$$

that maps objects identically. We now proceed to show that this functor is left adjoint to the contravariant hom-functor

$$(\mathbb{T}, \mathcal{V})\text{-}\mathsf{URel}(-, 1) : (\mathbb{T}, \mathcal{V})\text{-}\mathsf{URel}^{\mathrm{op}} \to \mathsf{Set}\,,$$

where $1 = \{\star\}$ denotes a singleton. We identify elements $x \in X$ with maps $x : 1 \to X$, and with a unitary $(\mathbb{T}, \mathcal{V})$-relation $\psi : X \nrightarrow Y$ we associate the map $\psi^\flat : Y \to (\mathbb{T}, \mathcal{V})\text{-}\mathsf{URel}(X, 1)$ defined by

$$\psi^\flat(y) := y^\sharp \circ \psi = y^\circ \cdot \psi = \psi(-, y)$$

for all $y \in Y$ (the second equality follows from Lemma IV.3.1.3, and the third by definition of composition in $\mathcal{V}$-Rel); here, $\psi(-, y)(\mathfrak{x}, \star) := \psi(\mathfrak{x}, y)$.

Lemma IV.3.1.4 shows that unitariness of a $(\mathbb{T}, \mathcal{V})$-relation $\varphi : X \nrightarrow Y$ can be tested just by using elements of Y.

IV.3.1.4 Lemma *Let $\hat{\mathbb{T}}$ be an associative lax extension of $\mathbb{T}$ to $\mathcal{V}$-Rel, and let $\varphi : X \nrightarrow Y$ be a $(\mathbb{T}, \mathcal{V})$-relation. Then φ is unitary if and only if $y^\circ \cdot \varphi$ is unitary for all $y \in Y$.*

Proof If φ is unitary, then $y^\circ \cdot \varphi = y^\sharp \circ \varphi$ is unitary as well (see III.1.9). To verify the other implication, suppose that $y^\circ \cdot \varphi$ is unitary for all $y \in Y$. Then one has

$$y^\circ \cdot (\varphi \circ e_X^\circ) = y^\circ \cdot \varphi \cdot \hat{T}1_X = y^\circ \cdot \varphi$$

and

$$y^\circ\cdot(e_Y^\circ\circ\varphi) = y^\circ\cdot e_Y^\circ\cdot\hat{T}\varphi\cdot m_X^\circ = e_1^\circ\cdot(Ty)^\circ\cdot\hat{T}\varphi\cdot m_X^\circ = e_1^\circ\cdot\hat{T}(y^\circ\cdot\varphi)\cdot m_X^\circ = y^\circ\cdot\varphi$$

for all $y\in Y$, so $\varphi\circ e_X^\circ = \varphi$ and $e_Y^\circ\circ\varphi = \varphi$. □

IV.3.1.5 Proposition *Let $\hat{\mathbb{T}}$ be an associative lax extension of $\mathbb{T}$ to $\mathcal{V}$-*Rel.

(1) *For a set X, the product $1^X = \prod_{x\in X} 1_x$ in $(\mathbb{T},\mathcal{V})$-*URel *(with $1_x = 1$ for all $x\in X$) exists, and can be chosen as $1^X = X$ with projections $\pi_x = x^\sharp : X \nrightarrow 1$ ($x\in X$).*
(2) *The contravariant hom-functor*

$$(\mathbb{T},\mathcal{V})\text{-URel}(-,1) : (\mathbb{T},\mathcal{V})\text{-URel}^{\mathrm{op}} \to \mathsf{Set}$$

has $(-)^\sharp : \mathsf{Set}\to(\mathbb{T},\mathcal{V})\text{-URel}^{\mathrm{op}}$ as left adjoint. The unit and counit of the associated adjunction are given by the Yoneda maps

$$\mathbf{y}_X = (1_X^\sharp)^\flat : X \to (\mathbb{T},\mathcal{V})\text{-URel}(X,1)\ , \quad x\mapsto x^\sharp$$

and the evaluation relations

$$\varepsilon_X : X \nrightarrow (\mathbb{T},\mathcal{V})\text{-URel}(X,1)\ , \quad \varepsilon_X(\mathfrak{x},\psi) = \psi(\mathfrak{x},\star)\ ,$$

respectively.

Proof (1): For a family $(\phi_x : Y\nrightarrow 1)_{x\in X}$ of unitary $(\mathbb{T},\mathcal{V})$-relations, one can define a $(\mathbb{T},\mathcal{V})$-relation $\phi : Y\nrightarrow X$ by setting $\phi(\mathfrak{y},x) = \phi_x(\mathfrak{y},\star)$ for all $\mathfrak{y}\in TY$. Since $x^\circ\cdot\phi = \phi_x$ is unitary for all $x\in X$, then so is ϕ by Lemma IV.3.1.4. Unicity of the connecting morphism $\phi : Y\nrightarrow X$ follows from its definition.

(2): By Exercise II.2.L, the functor $H = (\mathbb{T},\mathcal{V})$-URel$(-,1) : (\mathbb{T},\mathcal{V})$-URel^{op} $\to$ Set has a left adjoint $F = 1^{(-)}$ that sends a set X to its product $1^X = X$, and a map $f : X\to Y$ to the unitary $(\mathbb{T},\mathcal{V})$-relation $f^\sharp : Y\nrightarrow X$. The same exercise yields that $\mathbf{y}$ is indeed the unit of the adjunction and ε its counit. □

IV.3.2 The discrete presheaf monad

For an associative lax extension $\hat{\mathbb{T}}$ of $\mathbb{T}$ to $\mathcal{V}$-Rel, the adjunction described in Proposition IV.3.1.5 induces a monad

$$\mathbb{\Pi} = \mathbb{\Pi}(\mathbb{T},\mathcal{V}) = \mathbb{\Pi}(\mathbb{T},\mathcal{V},\hat{\mathbb{T}}) = (\Pi,\mathbf{m},\mathbf{y})$$

on Set, where

$$\Pi X = (\mathbb{T},\mathcal{V})\text{-URel}(X,1)\ , \qquad \Pi f(\psi) = \psi\circ f^\sharp\ ,$$
$$\mathbf{m}_X(\Psi) = \Psi\circ\varepsilon_X\ , \qquad \mathbf{y}_X(x) = x^\sharp\ ;$$

for all $x\in X$, $f : X\to Y$, and unitary $(\mathbb{T},\mathcal{V})$-relations $\psi : X\nrightarrow 1$, $\Psi : \Pi X\nrightarrow 1$. We call $\mathbb{\Pi}$ the *discrete presheaf monad* associated to $\hat{\mathbb{T}}$.

The fully faithful comparison functor $L : \mathsf{Set}_{\sqcap} \to (\mathbb{T}, \mathcal{V})\text{-}\mathsf{URel}^{\mathrm{op}}$ (Proposition II.3.6.1) is bijective on objects since the left adjoint $(-)^{\sharp} : \mathsf{Set} \to (\mathbb{T}, \mathcal{V})\text{-}\mathsf{URel}^{\mathrm{op}}$ is so. That is, the Kleisli category $\mathsf{Set}_{\sqcap}$ of $\sqcap$ is isomorphic to $(\mathbb{T}, \mathcal{V})\text{-}\mathsf{URel}^{\mathrm{op}}$. Explicitly, $LX = X$ for each set X, and L sends a morphism $r : X \to \Pi Y$ in $\mathsf{Set}_{\sqcap}$ to the unitary $(\mathbb{T}, \mathcal{V})$-relation $r^{\sharp} \circ \varepsilon_Y : Y \nrightarrow X$. By Lemma IV.3.1.3, $r^{\sharp} \circ \varepsilon_Y(y, x) = r(x)(y, \star)$ for all $x \in X$ and $y \in TY$, so the inverse of L sends a unitary $(\mathbb{T}, \mathcal{V})$-relation $\varphi : Y \nrightarrow X$ to $\varphi^{\flat} : X \to \Pi Y$.

Since $(\mathbb{T}, \mathcal{V})$-$\mathsf{URel}$ is a right-sided quantaloid (Proposition IV.3.1.2), $\Pi X = (\mathbb{T}, \mathcal{V})\text{-}\mathsf{URel}(X, 1)$ is a complete lattice when equipped with its pointwise order:

$$\psi_1 \leq \psi_2 \iff \forall \mathfrak{x} \in TX \, (\psi_1(\mathfrak{x}) \leq \psi_2(\mathfrak{x})) \, ,$$

for all unitary $(\mathbb{T}, \mathcal{V})$-relations $\psi_1, \psi_2 : X \nrightarrow 1$. By Proposition IV.3.1.2, the map $(-) \circ \varphi : \Pi Y \to \Pi X$ preserves suprema for every unitary $(\mathbb{T}, \mathcal{V})$-relation $\varphi : X \nrightarrow Y$; in particular, $\Pi f = (-) \circ f^{\sharp}$ and $\mathbf{m}_X = (-) \circ \varepsilon_X$ preserve suprema for every set X and every map $f : X \to Y$. The order on the sets ΠX therefore corresponds to a monad morphism $\tau : \mathbb{P} \to \sqcap$ from the powerset monad $\mathbb{P}$ (Proposition IV.1.2.1). As a consequence, the hom-sets of $\mathsf{Set}_{\sqcap}$ are complete ordered sets when ordered pointwise:

$$f_1 \leq f_2 \iff \forall x \in X \, (f_1(x) \leq f_2(x))$$

for all maps $f_1, f_2 : X \to \Pi Y$. One then has for unitary $(\mathbb{T}, \mathcal{V})$-relations $\varphi_1, \varphi_2 : Y \nrightarrow X$:

$$\varphi_1 \leq \varphi_2 \iff \forall x \in X \, \forall y \in TY \, (\varphi_1(y, x) \leq \varphi_2(y, x)) \iff \varphi_1^{\flat} \leq \varphi_2^{\flat} \, .$$

We therefore obtain an isomorphism (see also the discussion after Lemma IV.3.1.3)

$$(\mathbb{T}, \mathcal{V})\text{-}\mathsf{URel}(Y, X) \to \mathsf{Set}(X, (\mathbb{T}, \mathcal{V})\text{-}\mathsf{URel}(Y, 1)) \, , \quad \varphi \mapsto (\varphi^{\flat} : x \mapsto \varphi(-, x))$$

in Ord. Since $(\mathbb{T}, \mathcal{V})$-$\mathsf{URel}$ is an ordered category, this shows that $\mathsf{Set}_{\sqcap}$ is also an ordered category.

IV.3.2.1 Proposition *Let $\hat{\mathbb{T}}$ be an associative lax extension to $\mathcal{V}$-Rel of a monad $\mathbb{T}$ on Set. The discrete presheaf monad $\sqcap = \sqcap(\mathbb{T}, \mathcal{V})$ is power-enriched, and the comparison functor*

$$L : \mathsf{Set}_{\sqcap} \to (\mathbb{T}, \mathcal{V})\text{-}\mathsf{URel}^{\mathrm{op}}$$

is a 2-isomorphism. Moreover, for every $f : X \to Y$ in Set, one has $(f^{\sharp})^{\flat} = \mathbf{y}_Y \cdot f$.

Proof The functor L and its inverse have already been described. Since the Kleisli comparison functor L makes the diagram

$$\begin{array}{ccc} \mathsf{Set}_{\mathbb{\Pi}} & \xrightarrow{\quad L \quad} & (\mathbb{T},\mathcal{V})\text{-}\mathsf{URel}^{\mathrm{op}} \\ & {\scriptstyle F_{\mathbb{\Pi}}}\nwarrow \qquad \nearrow{\scriptstyle (-)^{\sharp}} & \\ & \mathsf{Set} & \end{array}$$

commute, $(f^{\sharp})^{\flat} = L^{-1}(f^{\sharp}) = F_{\mathbb{\Pi}} f = \mathbf{y}_Y \cdot f$. □

IV.3.2.2 Theorem *Given an associative lax extension* $\hat{\mathbb{T}}$ *of* $\mathbb{T}$ *to* $\mathcal{V}$-Rel, *there is an isomorphism*

$$(\mathbb{T},\mathcal{V})\text{-}\mathsf{Cat} \cong \mathbb{\Pi}\text{-}\mathsf{Mon}$$

that commutes with the underlying-set functors.

Proof Recall that every reflexive and transitive $(\mathbb{T},\mathcal{V})$-relation $a : X \nrightarrow X$ is also unitary. By Proposition IV.3.2.1, sending a to $a^{\flat}$ defines a bijection between reflexive and transitive $(\mathbb{T},\mathcal{V})$-relations $a : X \nrightarrow X$ and maps $\nu : X \to \Pi X$ satisfying $\mathbf{y}_X \le \nu$ and $\nu \circ \nu \le \nu$. Furthermore, a map $f : X \to Y$ is a $(\mathbb{T},\mathcal{V})$-functor $f : (X,a) \to (Y,b)$ if and only if $a \circ f^{\sharp} \le f^{\sharp} \circ b$; this condition is equivalent to $(\mathbf{y}_Y \cdot f) \circ a^{\flat} \le b^{\flat} \circ (\mathbf{y}_Y \cdot f)$ in $\mathsf{Set}_{\mathbb{\Pi}}$, i.e. $f : (X, a^{\flat}) \to (Y, b^{\flat})$ is a morphism of Kleisli monoids. □

Since $\mathbb{\Pi}$ is power-enriched, we can consider the corresponding Kleisli extension $\check{\mathbb{\Pi}}$ of $\mathbb{\Pi}$ to Rel. Explicitly, the extension $\check{\Pi}$ of the Set-functor Π to Rel sends a relation $r : X \nrightarrow Y$ to the relation $\check{\Pi} r : \Pi X \nrightarrow \Pi Y$ defined by

$$\psi_1 \, \check{\Pi} r \, \psi_2 \iff \psi_1 \le \psi_2 \cdot \hat{T} r$$

for all unitary $(\mathbb{T},\mathcal{V})$-relations $\psi_1 : X \nrightarrow 1$, $\psi_2 : Y \nrightarrow 1$ (Exercise IV.3.B). With respect to this extension of $\mathbb{\Pi}$, we consider the category $(\mathbb{\Pi}, 2)$-Cat.

IV.3.2.3 Corollary *Given an associative lax extension* $\hat{\mathbb{T}}$ *of* $\mathbb{T}$ *to* $\mathcal{V}$-Rel, *there is an isomorphism*

$$(\mathbb{T},\mathcal{V})\text{-}\mathsf{Cat} \cong (\mathbb{\Pi}, 2)\text{-}\mathsf{Cat}$$

that commutes with the underlying-set functors, where $\mathbb{\Pi} = \mathbb{\Pi}(\mathbb{T},\mathcal{V})$ *is equipped with its Kleisli extension* $\check{\mathbb{\Pi}}$.

Proof Theorem IV.3.2.2 yields the isomorphism $(\mathbb{T},\mathcal{V})$-$\mathsf{Cat} \cong \mathbb{\Pi}$-$\mathsf{Mon}$, and Theorem IV.1.5.3 yields the isomorphism $\mathbb{\Pi}$-$\mathsf{Mon} \cong (\mathbb{\Pi}, 2)$-$\mathsf{Cat}$. □

IV.3.2.4 Remark The Kleisli extension of a power-enriched monad $\check{\mathbb{\Pi}} = \check{\mathbb{\Pi}}(\mathbb{T},\mathcal{V},\hat{\mathbb{T}})$ is in particular associative (see Exercise IV.1.G). Hence, the transition from $\hat{\mathbb{T}}$ to $\check{\mathbb{\Pi}}$ preserves associativity while maintaining the same categories of lax algebras (up to isomorphism).

IV.3.2.5 Proposition *Let $\hat{\mathbb{T}}$ be an associative lax extension of $\mathbb{T}$ to $\mathcal{V}$-Rel. Then the natural transformation* $\mathbf{Y}$ *defined componentwise by*

$$\mathbf{Y}_X : TX \to \Pi X\,, \quad \mathfrak{x} \mapsto \hat{T}1_X(-,\mathfrak{x})$$

is a monad morphism $\mathbf{Y} : \mathbb{T} \to \Pi(\mathbb{T},\mathcal{V})$.

Proof The left adjoint functor $(-)^\sharp : \mathsf{Set} \to (\mathbb{T},\mathcal{V})\text{-}\mathsf{URel}^{\mathrm{op}}$ extends to a functor

$$(-)^\sharp : \mathsf{Set}_{\mathbb{T}} \to (\mathbb{T},\mathcal{V})\text{-}\mathsf{URel}^{\mathrm{op}}$$

sending $r : X \rightharpoonup Y$ to $r^\sharp := r^\circ \cdot \hat{T}1_Y : Y \nrightarrow X$. Indeed, the identity $1_X : X \rightharpoonup X$ in $\mathsf{Set}_{\mathbb{T}}$ (which is the map $e_X : X \to TX$) goes to the identity $1_X^\sharp = e_X^\circ \cdot \hat{T}1_X$ in $(\mathbb{T},\mathcal{V})$-URel; and for $r : X \rightharpoonup Y$ and $s : Y \rightharpoonup Z$ one has

$$\begin{aligned} r^\sharp \circ s^\sharp &= r^\circ \cdot \hat{T}1_Y \cdot \hat{T}(s^\circ \cdot \hat{T}1_Z) \cdot m_Z^\circ = r^\circ \cdot \hat{T}(s^\circ \cdot \hat{T}1_Z) \cdot m_Z^\circ \\ &= r^\circ \cdot (Ts)^\circ \cdot \hat{T}\hat{T}1_Z \cdot m_Z^\circ = (m_Z \cdot Ts \cdot r)^\circ \cdot \hat{T}1_Z\,, \end{aligned}$$

hence $(-)^\sharp$ preserves composition. By definition, the diagram

$$\begin{array}{ccc} \mathsf{Set}_{\mathbb{T}} & \xrightarrow{\ (-)^\sharp\ } & (\mathbb{T},\mathcal{V})\text{-}\mathsf{URel}^{\mathrm{op}} \\ & {\scriptstyle F_{\mathbb{T}}}\nwarrow \quad \nearrow{\scriptstyle (-)^\sharp} & \\ & \mathsf{Set} & \end{array}$$

commutes, and therefore $(-)^\sharp : \mathsf{Set}_{\mathbb{T}} \to (\mathbb{T},\mathcal{V})\text{-}\mathsf{URel}^{\mathrm{op}}$ induces a monad morphism $\mathbf{Y} : \mathbb{T} \to \Pi(\mathbb{T},\mathcal{V})$. Its component $\mathbf{Y}_X$ is the composite (see II.3.I)

$$TX \xrightarrow{\ \mathrm{y}_{TX}\ } (\mathbb{T},\mathcal{V})\text{-}\mathsf{URel}(TX,1) \xrightarrow{\ (-)\circ 1_{TX}{}^\sharp\ } (\mathbb{T},\mathcal{V})\text{-}\mathsf{URel}(X,1)\,,$$

hence $\mathbf{Y}_X(\mathfrak{x}) = \mathfrak{x}^\circ \cdot e_{TX}^\circ \cdot \hat{T}1_{TX} \cdot \hat{T}\hat{T}1_X \cdot m_X^\circ = \mathfrak{x}^\circ \cdot \hat{T}1_X = \hat{T}1_X(-,\mathfrak{x})$. □

IV.3.2.6 Examples

(1) For the identity monad $\mathbb{I}$ on Set extended to the identity monad on Rel, one has

$$(\mathbb{I},2)\text{-}\mathsf{URel} \cong \mathsf{Rel}\,,$$

and the monad $\Pi = \Pi(\mathbb{I},2)$ is isomorphic to the powerset monad $\mathbb{P}$. Hence, $\mathsf{Ord} \cong (\mathbb{P},2)\text{-}\mathsf{Cat}$ by Corollary IV.3.2.3. The monad morphism $\mathbf{Y} : \mathbb{I} \to \mathbb{P}$ is necessarily given by the unit of $\mathbb{P}$.

(2) More generally, for the identity monad $\mathbb{I}$ on Set extended to the identity monad on $\mathcal{V}$-Rel, one has

$$(\mathbb{I},\mathcal{V})\text{-}\mathsf{URel} \cong \mathcal{V}\text{-}\mathsf{Rel}\,,$$

and the monad $\Pi = \Pi(\mathbb{I},\mathcal{V})$ is isomorphic to the $\mathcal{V}$-powerset monad $\mathbb{P}_{\mathcal{V}}$ (see Exercise III.1.D). Hence, $\mathcal{V}\text{-}\mathsf{Cat} \cong (\mathbb{P}_{\mathcal{V}},2)\text{-}\mathsf{Cat}$ by Corollary IV.3.2.3.

As above, the monad morphism $\mathbf{Y} : \mathbb{I} \to \mathbb{P}_{\mathcal{V}}$ is necessarily given by the unit of $\mathbb{P}_{\mathcal{V}}$.

The monoid $(\mathcal{V}, \otimes, k)$ induces a monad $\mathbb{V}$ on Set with functor $\mathcal{V} \times (-)$ (see Exercise II.3.B), and for each set X there is a map $\alpha_X : \mathcal{V} \times X \to \mathcal{V}^X$ defined by

$$\alpha_X(u, x)(y) = \begin{cases} u & \text{if } x = y, \\ \bot & \text{else.} \end{cases}$$

One easily verifies that these maps yield a monad morphism $\alpha : \mathbb{V} \to \mathbb{P}_{\mathcal{V}}$ (Exercise IV.3.D). It is also clear that α is sup-dense; therefore, when considering the $\mathcal{V}$-powerset monad $\mathbb{P}_{\mathcal{V}}$ with its Kleisli extension and $\mathbb{V}$ with the initial extension induced by α, we obtain (see Proposition IV.2.3.4) a full reflective embedding

$$A_\alpha : \mathcal{V}\text{-}\mathsf{Cat} \hookrightarrow (\mathbb{V}, 2)\text{-}\mathsf{Cat}\,.$$

However, A_α is not an equivalence in general. Indeed, for $\mathcal{V} = \mathsf{P}_+$ and a metric space $X = (X, a)$, one has

$$(u, x)\ (a \cdot \alpha_X)\ y \iff a(x, y) \le u\,,$$

for all $x, y \in X$ and $u \in [0, \infty]$. Consider $X = \{a, b\}$ with the relation $\longrightarrow : [0, \infty] \times X \nrightarrow X$ defined by

$$(u, a) \longrightarrow b \iff 0 < u$$

and $(u, a) \longrightarrow a$, $(u, b) \longrightarrow b$ for all $u \in [0, \infty]$. Then X is indeed a $(\mathbb{V}, 2)$-category but $\longrightarrow$ is not induced by a metric on X.

(3) For the powerset monad $\mathbb{P}$ on Set with its Kleisli extension $\check{\mathbb{P}}$ to Rel, one has an isomorphism

$$(-)^\sharp : \mathsf{Set}_{\mathbb{P}} \to (\mathbb{\Gamma}, 2)\text{-}\mathsf{URel}^{\mathrm{op}}$$

commuting with the left adjoints from Set, and therefore $\mathbf{Y} : \mathbb{P} \to \mathbb{\Pi}(\mathbb{P}, 2)$ is an isomorphism.

(4) Consider the ultrafilter monad β on Set with its Barr extension $\overline{\beta}$ to Rel (Example III.1.10.3(3)). In this case, the monad $\mathbb{\Pi} = \mathbb{\Pi}(\beta, 2)$ is isomorphic to the filter monad $\mathbb{F}$ on Set. Indeed, a unitary $(\beta, 2)$-relation $\psi : X \nrightarrow 1$ may be identified with a set $\mathcal{A} \subseteq \beta X$ of ultrafilters on X with the property that $\chi \supseteq \bigcap \mathcal{A}$ implies $\chi \in \mathcal{A}$. Therefore, the map

$$\delta_X : (\beta, 2)\text{-}\mathsf{URel}(X, 1) \to FX\,, \quad \mathcal{A} \mapsto \bigcap \mathcal{A}$$

is a bijection. Let us show that $\delta = (\delta_X)$ is indeed a monad morphism $\delta : \mathbb{\Pi} \to \mathbb{F}$. Recall from Example II.3.1.1(4) that the filter monad on Set is induced by the adjunction

$$\mathsf{SLat}^{\mathrm{op}} \underset{(P^\bullet)^{\mathrm{op}}}{\overset{\mathsf{SLat}(-,2)}{\underset{\longleftarrow}{\overset{\longrightarrow}{\top}}}} \mathsf{Set}\,.$$

Since SLat is equivalent to the category $\mathsf{SLat}^{\mathrm{co}}$ of join-semilattices and their homomorphisms, $\mathbb{F}$ is also induced by the adjunction

$$\mathsf{SLat}^{\mathrm{coop}} \underset{(P^\bullet)^{\mathrm{op}}}{\overset{\mathsf{SLat}(-,2)}{\underset{\longleftarrow}{\overset{\longrightarrow}{\top}}}} \mathsf{Set}\,.$$

In the former case, a filter $f \in FX$ corresponds to the characteristic map $\chi_f : PX \to \mathsf{2}$ with

$$\chi_f(A) = 1 \iff A \in f\,,$$

whereas in the latter case f corresponds to the map $\chi'_f : PX \to \mathsf{2}$

$$\chi'_f(A) = 1 \iff A^{\complement} \notin f\,.$$

The covariant hom-functor $(\beta, \mathsf{2})\text{-}\mathsf{URel}(1, -) : (\beta, \mathsf{2})\text{-}\mathsf{URel} \to \mathsf{Set}$ lifts to a functor $L : (\beta, \mathsf{2})\text{-}\mathsf{URel} \to \mathsf{SLat}^{\mathrm{co}}$, since, for every $\psi : X \nrightarrow Y$ in $(\beta, \mathsf{2})\text{-}\mathsf{URel}$, the map

$$\psi \circ (-) : PX \to PY\,, \quad A \mapsto \{y \in Y \mid \exists \mathfrak{x} \in \beta X\ (A \in \mathfrak{x}\ \&\ \mathfrak{x}\ \psi\ y)\}$$

preserves finite suprema (we use here the identification $(\beta, \mathsf{2})\text{-}\mathsf{URel}(1, X) \cong PX$). It follows from Lemma IV.3.1.3 that the diagram

$$\begin{array}{ccc} (\beta, \mathsf{2})\text{-}\mathsf{URel}^{\mathrm{op}} & \xrightarrow{\;L^{\mathrm{op}}\;} & \mathsf{SLat}^{\mathrm{coop}} \\ & {\scriptstyle (-)^\sharp}\nwarrow \quad \nearrow{\scriptstyle (P^\bullet)^{\mathrm{op}}} & \\ & \mathsf{Set} & \end{array}$$

commutes up to natural isomorphism, and therefore L^{op} induces an isomorphism of monads $\Pi \to \mathbb{F}$ whose component at X is precisely δ_X. The composite $\delta \cdot \mathbf{Y} : \beta \to \Pi \to \mathbb{F}$ is the canonical monad morphism $\beta \to \mathbb{F}$.

(5) By Corollary IV.3.2.3 and Theorem III.2.4.5,

$$\mathsf{App} \cong (\beta, \mathsf{P}_+)\text{-}\mathsf{Cat} \cong (\Pi, \mathsf{2})\text{-}\mathsf{Cat}\,.$$

In Section 3.3 we provide an alternative description of $\Pi = \Pi(\beta, \mathsf{P}_+)$.

IV.3.3 Approach spaces

For an alternative description of $\Pi(\beta, \mathsf{P}_+)$ (where the ultrafilter monad β is provided with the extension $\overline{\beta}$ of β to P_+-Rel of Section III.2.4), we consider the category

$$\mathsf{Met}_{(\vee, \perp, +)}$$

of separated metric spaces $X = (X, d)$, the underlying order of which given by

$$x \leq y \iff 0 = d(x, y) ,$$

has finite suprema, and which admit an action $+ : X \times [0, \infty] \to X$ (denoted here as a right action) satisfying

$$d(x + u, y) = d(x, y) \ominus u , \tag{IV.3.3.i}$$

for all $x, y \in X$ and $u \in [0, \infty]$ (recall from II.1.10.1(3) that "$\ominus$" denotes truncated subtraction); a morphism of $\mathsf{Met}_{(\vee,\perp,+)}$ is a non-expansive map that preserves finite suprema and the action of $[0, \infty]$. Equation (IV.3.3.i) implies immediately that the monotone map $d(x, -) : X \to \mathsf{P}_+$ has $x + (-) : \mathsf{P}_+ \to X$ as a left adjoint, therefore a separated metric space admits at most one such action.

The metric space $[0, \infty] = ([0, \infty], \mu)$ with $\mu(u, v) = v \ominus u$ $(u, v \in [0, \infty])$ belongs to $\mathsf{Met}_{(\vee,\perp,+)}$ since its underlying order is the natural $\geq$, which has finite suprema, and the action of $[0, \infty]$ is given by the usual addition $+$. Similarly, for every set X, the set $[0, \infty]^X$ with the metric

$$[\varphi, \varphi'] = \sup\{\mu(\varphi(x), \varphi'(x)) \mid x \in X\}$$

(for all $\varphi, \varphi' \in [0, \infty]^X$) and the action given by pointwise addition belongs to $\mathsf{Met}_{(\vee,\perp,+)}$. In fact, in $\mathsf{Met}_{(\vee,\perp,+)}$ one has

$$[0, \infty]^X \cong \textstyle\prod_{x\in X}[0, \infty]_x ,$$

where $[0, \infty]_x = ([0, \infty], \mu)$ for all $x \in X$. Therefore, as in Proposition IV.3.1.5, the contravariant hom-functor

$$\mathsf{Met}_{(\vee,\perp,+)}(-, [0, \infty]) : \mathsf{Met}^{\mathrm{op}}_{(\vee,\perp,+)} \to \mathsf{Set}$$

has a left adjoint

$$[0, \infty]^{(-)} : \mathsf{Set} \to \mathsf{Met}^{\mathrm{op}}_{(\vee,\perp,+)} .$$

The functor J of the monad $\mathbb{J}$ induced on Set by this adjunction is given by

$$JX = \mathsf{Met}_{(\vee,\perp,+)}([0, \infty]^X, [0, \infty]) ,$$

with the unit $X \to JX$ defined by evaluation $x \mapsto (\varphi \mapsto \varphi(x))$, and the multiplication $JJX \to JX$ defined by $\Psi \mapsto (\varphi \mapsto \Psi(\mathrm{ev}_\varphi))$, with $\mathrm{ev}_\varphi(\Phi) = \Phi(\varphi)$ for all $\Phi \in JX$, $\varphi \in [0, \infty]^X$.

We show now that $\mathbb{J}$ is isomorphic to the monad $\mathbb{\Pi}$. To this end, we first observe that a map $\varphi : X \to [0, \infty]$ can be interpreted as a unitary $(\mathbb{\beta}, \mathsf{P}_+)$-relation $\varphi : 1 \nrightarrow X$; in particular, every element $u \in [0, \infty]$ can be seen as a unitary $(\mathbb{\beta}, \mathsf{P}_+)$-relation $u : 1 \nrightarrow 1$. From this perspective, the distance $[\varphi, \varphi'] \in [0, \infty]$ (where $\varphi, \varphi' \in [0, \infty]^X$) is precisely the lifting $\varphi \multimap \varphi'$ of φ' along φ in $(\mathbb{\beta}, \mathsf{P}_+)$-$\mathsf{URel}$

(see Example II.4.8.3), and the action $\varphi + u$ is the composite $\varphi \circ u$. Every unitary (β, P_+)-relation $\psi : X \nrightarrow Y$ defines a mapping

$$\psi \circ (-) : (\beta, \mathsf{P}_+)\text{-}\mathsf{URel}(1, X) \longrightarrow (\beta, \mathsf{P}_+)\text{-}\mathsf{URel}(1, Y)$$

that clearly preserves the action of $[0, \infty]$, and from

$$\psi \circ \varphi \circ (\varphi \multimap \varphi') \geq \psi \circ \varphi'$$

follows that $\varphi \multimap \varphi' \geq (\psi \circ \varphi) \multimap (\psi \circ \varphi')$, for all $\varphi, \varphi' : 1 \nrightarrow X$. To see that $\psi \circ (-)$ preserves finite suprema, note that

$$\psi \circ \varphi = \inf_{\chi \in \beta X} \psi(\chi, -) + \xi(\beta\varphi(\chi)) = \psi \cdot \hat{\varphi}\,,$$

where $\hat{\varphi} : 1 \nrightarrow \beta X$ is defined as $\hat{\varphi}(\chi) = \xi(\beta\varphi(\chi))$ with

$$\xi(a) := \sup_{A \in a} \inf_{u \in A} u = \inf_{A \in a} \sup_{u \in A} u\,,$$

for all $a \in \beta[0, \infty]$. Being left adjoint, $\psi \cdot (-)$ preserves all suprema, and $\varphi \mapsto \hat{\varphi}$ preserves the bottom element; $\varphi \mapsto \hat{\varphi}$ also preserves binary suprema since $\min : [0, \infty] \times [0, \infty] \longrightarrow [0, \infty]$ is continuous. All said, the covariant hom-functor $K := (\beta, \mathsf{P}_+)\text{-}\mathsf{URel}(1, -)$ takes values in $\mathsf{Met}_{(\vee,\bot,+)}$. Moreover, it follows from Lemma IV.3.1.3 that the diagram

$$\begin{array}{ccc} (\beta, \mathsf{P}_+)\text{-}\mathsf{URel}^{\mathrm{op}} & \xrightarrow{\quad K^{\mathrm{op}} \quad} & \mathsf{Met}^{\mathrm{op}}_{(\vee,\bot,+)} \\ & {\scriptstyle (-)^\sharp}\nwarrow \qquad \nearrow{\scriptstyle [0,\infty]^{(-)}} & \\ & \mathsf{Set} & \end{array}$$

commutes up to natural isomorphism, and therefore K^{op} induces a monad morphism $\delta : \mathbb{\Pi} \longrightarrow \mathbb{J}$. A small computation shows that, for every set X, the map

$$\delta_X : \Pi X \longrightarrow \mathsf{Met}_{(\vee,\bot,+)}([0, \infty]^X, [0, \infty])$$

sends $\psi : X \nrightarrow 1$ to $\psi \circ (-) : [0, \infty]^X \longrightarrow [0, \infty]$.

IV.3.3.1 Theorem $\delta : \mathbb{\Pi} \longrightarrow \mathbb{J}$ *is an isomorphism.*

Proof Every subset $A \subseteq X$ can be seen as an element of $[0, \infty]^X$, namely as the function $A = \theta_A : X \longrightarrow [0, \infty]$ sending $x \in A$ to 0 and everything else to ∞. With this interpretation, for every $\Phi : [0, \infty]^X \longrightarrow [0, \infty]$ in $\mathsf{Met}_{(\vee,\bot,+)}$ we set

$$\Gamma(\Phi)(\chi) = \sup_{A \in \chi} \Phi(A)\,.$$

Then $\psi = \Gamma(\Phi) : X \nrightarrow 1$ is indeed unitary since one obtains

$$\xi(\beta\psi(\mathcal{X})) \geq \psi(m_X(\mathcal{X}))$$

from the inequality $\Phi(A) \leq \Gamma(\Phi)(\chi)$, for all $A \in \chi$. If Φ is of the form $\Phi = \psi \circ (-)$ for some unitary $\psi : X \nrightarrow 1$, then

$$\Gamma(\Phi)(\chi) = \sup_{A\in\chi} \psi \circ A = \sup_{A\in\chi} \psi \cdot \hat{A} \leq \psi \cdot \sup_{A\in\chi} \hat{A} \, .$$

Since $\hat{A}(y) = 0$ if $A \in y$ and $\hat{A}(y) = \infty$ if $A \notin y$, one obtains $\sup_{A\in\chi} \hat{A}(\chi) = 0$ and $\sup_{A\in\chi} \hat{A}(y) = \infty$ for $y \neq \chi$, and consequently $\Gamma(\Phi) \leq \psi$. To see that

$$\psi(\chi) \leq \Gamma(\Phi)(\chi) = \sup_{A\in\chi} \inf_{y\in\beta A} \psi(y) \, ,$$

we use Lemma III.2.4.2, which guarantees the existence of some $\mathcal{X} \in \beta\beta X$ with

$$\{\beta A \mid A \in \chi\} \subseteq \mathcal{X} \qquad \text{and} \qquad \sup_{A\in\chi} \inf_{y\in\beta A} \psi(y) \geq \xi(\beta\psi(\mathcal{X})) \, .$$

Since $\psi : X \nrightarrow 1$ is unitary,

$$\xi(\beta\psi(\mathcal{X})) \geq \psi(m_X(\mathcal{X})) = \psi(\chi) \, .$$

Now let $\Phi : [0,\infty]^X \to [0,\infty]$ in $\mathsf{Met}_{(\vee,\perp,+)}$. We show first that $\Gamma(\Phi) \circ (-)$ coincides with Φ on subsets $B \subseteq X$ of X. Indeed,

$$\Gamma(\Phi) \circ B = \inf_{\chi\in\beta X} \sup_{A\in\chi} \Phi(A) + \hat{B}(\chi) = \inf_{\chi\in\beta B} \sup_{A\in\chi} \Phi(A) \geq \Phi(B) \, .$$

To see that $\inf_{\chi\in\beta B} \sup_{A\in\chi} \Phi(A) \leq \Phi(B)$, we apply Corollary II.1.13.5 (if $\Phi(B) < \infty$) to the filter base $\{B\}$ and the ideal $j = \{A \subseteq X \mid \Phi(A) > \Phi(B)\}$. Hence, there is some ultrafilter $\chi \in \beta X$ with $B \in \chi$ and $\chi \cap j = \varnothing$, and therefore

$$\sup_{A\in\chi} \Phi(A) \leq \Phi(B) \, .$$

To finish the proof, we show that any $\Phi : [0,\infty]^X \to [0,\infty]$ is completely determined by its restriction to subsets $B \subseteq X$ of X. Since

$$\begin{aligned}\Phi(\varphi) &= \sup_{u<\infty} \min\{\Phi(\varphi), (\Phi(0) + u)\} = \sup_{u<\infty} \min\{\Phi(\varphi), \Phi(u)\} \\ &= \sup_{u<\infty} \Phi(\min\{\varphi, u\}) \, ,\end{aligned}$$

Φ is determined by its restriction to bounded maps $\varphi : X \to [0,\infty]$. Now let $\varphi : X \to [0,\infty]$ be a bounded map, let $\varepsilon > 0$, and let N be any natural number with $\varphi(x) < N \cdot \varepsilon$, for all $x \in X$. For every natural number n with $0 \leq n < N$, we set $A_n = \{x \in X \mid n \cdot \varepsilon \leq \varphi(x) < (n+1) \cdot \varepsilon\}$ and $u_n = n \cdot \varepsilon$, and define

$$\varphi_\varepsilon = \min\{A_n + u_n \mid 0 \leq n < N\}$$

(note that finite suprema in $[0,\infty]^X$ are given by pointwise minima). Then $\Phi(\varphi_\varepsilon) \leq \Phi(\varphi) \leq \Phi(\varphi_\varepsilon) + \varepsilon$ and $\Phi(\varphi_\varepsilon) = \min\{\Phi(A_n) + u_n \mid 0 \leq n < N\}$, which proves that Φ is determined by its effect on subsets of X. □

IV.3.4 Revisiting change of base

In Section IV.3.2 we devised a construction that incorporates a quantale into a monad. In this section we extend this construction to quantale morphisms and show that certain change-of-base functors correspond to algebraic functors.

Recall from III.3.5 that every lax homomorphism $\varphi : \mathcal{V} \to \mathcal{W}$ of quantales induces a lax functor

$$\varphi : \mathcal{V}\text{-}\mathsf{Rel} \to \mathcal{W}\text{-}\mathsf{Rel}$$

that sends the $\mathcal{V}$-relation $r : X \times Y \to \mathcal{V}$ to $\varphi r : X \times Y \to \mathcal{W}$. Moreover, $\varphi : \mathcal{V}\text{-}\mathsf{Rel} \to \mathcal{W}\text{-}\mathsf{Rel}$ is actually a functor if $\varphi : \mathcal{V} \to \mathcal{W}$ is a homomorphism of quantales.

Now let $\mathbb{T} = (T, m, e)$ be a monad on Set together with associative lax extensions

$$\hat{T} : \mathcal{V}\text{-}\mathsf{Rel} \to \mathcal{V}\text{-}\mathsf{Rel} \qquad \text{and} \qquad \check{T} : \mathcal{W}\text{-}\mathsf{Rel} \to \mathcal{W}\text{-}\mathsf{Rel}\,.$$

A lax homomorphism of quantales $\varphi : \mathcal{V} \to \mathcal{W}$ is compatible with these lax extensions if $\check{T}(\varphi r) \leq \varphi(\hat{T} r)$ for all $r : X \nrightarrow Y$ in $\mathcal{V}$-Rel, and we say that $\varphi : \mathcal{V} \to \mathcal{W}$ is *strictly compatible* with $\hat{T}$ and $\check{T}$ if $\check{T}(\varphi r) = \varphi(\hat{T} r)$ for all $r : X \nrightarrow Y$ in $\mathcal{V}$-Rel.

IV.3.4.1 Proposition *If $\varphi : \mathcal{V} \to \mathcal{W}$ is a lax homomorphism of quantales compatible with associative lax extensions $\hat{T}$ and $\check{T}$ of a monad $\mathbb{T} = (T, m, e)$ to $\mathcal{V}$-Rel and $\mathcal{W}$-Rel, respectively, then $r \mapsto \varphi r$ defines a lax functor*

$$\varphi : (\mathbb{T}, \mathcal{V})\text{-}\mathsf{URel} \to (\mathbb{T}, \mathcal{W})\text{-}\mathsf{URel}$$

such that $(-)^\sharp \leq \varphi^{\mathrm{op}} \cdot (-)^\sharp$. Moreover, $\varphi : (\mathbb{T}, \mathcal{V})\text{-}\mathsf{URel} \to (\mathbb{T}, \mathcal{W})\text{-}\mathsf{URel}$ is even a functor making

$$\begin{array}{ccc} (\mathbb{T}, \mathcal{V})\text{-}\mathsf{URel}^{\mathrm{op}} & \xrightarrow{\ \varphi^{\mathrm{op}}\ } & (\mathbb{T}, \mathcal{W})\text{-}\mathsf{URel}^{\mathrm{op}} \\ & {\scriptstyle (-)^\sharp} \nwarrow \quad \nearrow {\scriptstyle (-)^\sharp} & \\ & \mathsf{Set} & \end{array}$$

commute provided that $\varphi : \mathcal{V} \to \mathcal{W}$ is a homomorphism of quantales strictly compatible with the lax extensions $\hat{T}$ and $\check{T}$ of $\mathbb{T}$.

Proof Let $r : X \rightharpoonup Y$ and $s : Y \rightharpoonup Z$ be morphisms in $(\mathbb{T}, \mathcal{V})$-$\mathsf{URel}$. Then

$$\begin{aligned} (\varphi s) \circ (\varphi r) &= (\varphi s) \cdot \check{T}(\varphi r) \cdot m_X^\circ \\ &\leq (\varphi s) \cdot (\varphi \hat{T} r) \cdot (\varphi m_X^\circ) \\ &\leq \varphi(s \circ r) \end{aligned}$$

with equality if $\varphi : \mathcal{V} \to \mathcal{W}$ is a homomorphism of quantales strictly compatible with the lax extensions $\hat{T}$ and $\check{T}$ of $\mathbb{T}$. If $f : X \to Y$ is map, then

$$f^\sharp = e_Y^\circ \cdot \check{T}(f^\circ) \le \varphi(e_Y^\circ \cdot \hat{T}(f^\circ)) = \varphi f^\sharp ,$$

and, as above, we have equality if $\varphi : \mathcal{V} \to \mathcal{W}$ is a homomorphism of quantales strictly compatible with the lax extensions $\hat{T}$ and $\check{T}$ of $\mathbb{T}$. Note that this applies in particular to the identity $1_X^\sharp$ in $(\mathbb{T}, \mathcal{V})$-URel. □

IV.3.4.2 Lemma *Let $\varphi : \mathcal{V} \to \mathcal{W}$ be a homomorphism of quantales strictly compatible with associative lax extensions $\hat{T}$ and $\check{T}$ of $\mathbb{T} = (T, m, e)$ to $\mathcal{V}$-*Rel *and $\mathcal{W}$-*Rel*, respectively. Then the right adjoint $\psi : \mathcal{W} \to \mathcal{V}$ of φ is a lax homomorphism of quantales compatible with these lax extensions of $\mathbb{T}$.*

Proof For $u, v \in \mathcal{W}$,

$$\psi(u) \otimes \psi(v) \le \psi\varphi(\psi(u) \otimes \psi(v)) = \psi(\varphi\psi(u) \otimes \varphi\psi(v)) \le \psi(u \otimes v) \quad \text{and}$$
$$k \le \psi\varphi(k) = \psi(l) .$$

Similarly, for a $\mathcal{W}$-relation $r : X \nrightarrow Y$, we obtain $\hat{T}(\psi r) \le \psi\varphi\hat{T}(\psi r) = \psi\check{T}(\varphi\psi r) \le \psi(\check{T}r)$. □

IV.3.4.3 Proposition *Every homomorphism $\varphi : \mathcal{V} \to \mathcal{W}$ of quantales strictly compatible with lax extensions $\hat{T}$ and $\check{T}$ of $\mathbb{T} = (T, m, e)$ to $\mathcal{V}$-*Rel *and $\mathcal{W}$-*Rel *induces a morphism*

$$\Pi(\mathbb{T}, \varphi) : \Pi(\mathbb{T}, \mathcal{V}) \to \Pi(\mathbb{T}, \mathcal{W})$$

of power-enriched monads. The component of $\Pi(\mathbb{T}, \varphi)$ at a set X is given by

$$(\mathbb{T}, \mathcal{V})\text{-URel}(X, 1) \to (\mathbb{T}, \mathcal{W})\text{-URel}(X, 1) , \quad r \mapsto \varphi r .$$

Proof First, $\varphi : \mathcal{V} \to \mathcal{W}$ induces a functor $\varphi : (\mathbb{T}, \mathcal{V})$-URel $\to (\mathbb{T}, \mathcal{W})$-URel so that

$$\begin{array}{ccc} (\mathbb{T}, \mathcal{V})\text{-URel}^{\mathrm{op}} & \xrightarrow{\varphi^{\mathrm{op}}} & (\mathbb{T}, \mathcal{W})\text{-URel}^{\mathrm{op}} \\ & {\scriptstyle (-)^\sharp} \nwarrow \quad \nearrow {\scriptstyle (-)^\sharp} & \\ & \mathsf{Set} & \end{array}$$

commutes. A quick computation shows that the X-component of the monad morphism $\Pi(\mathbb{T}, \varphi)$ induced by φ^{op} has the form described above. Secondly, the right adjoint ψ of φ induces a lax functor $\psi : (\mathbb{T}, \mathcal{W})$-URel $\to (\mathbb{T}, \mathcal{V})$-URel, therefore the map $s \mapsto \psi s$ defines a right adjoint to

$$(\mathbb{T}, \mathcal{V})\text{-URel}(X, 1) \to (\mathbb{T}, \mathcal{W})\text{-URel}(X, 1) , \quad r \mapsto \varphi r .$$

Hence, each component of $\Pi(\mathbb{T}, \varphi)$ preserves suprema. □

IV.3.4.4 Theorem *Let $\varphi : \mathcal{V} \to \mathcal{W}$ be a homomorphism of quantales strictly compatible with the lax extensions $\hat{T}$ and $\check{T}$ of $\mathbb{T} = (T, m, e)$ to $\mathcal{V}$-Rel and $\mathcal{W}$-Rel, and let $\psi : \mathcal{W} \to \mathcal{V}$ be the right adjoint of φ. Then the change-of-base functor B_ψ corresponds to the algebraic functor $A_{\Pi(\mathbb{T},\varphi)}$, i.e. the diagram*

$$\begin{array}{ccc} (\mathbb{T}, \mathcal{W})\text{-Cat} & \longrightarrow & (\Pi(\mathbb{T}, \mathcal{W}), 2)\text{-Cat} \\ {\scriptstyle B_\psi}\downarrow & & \downarrow{\scriptstyle A_{\Pi(\mathbb{T},\varphi)}} \\ (\mathbb{T}, \mathcal{V})\text{-Cat} & \longrightarrow & (\Pi(\mathbb{T}, \mathcal{V}), 2)\text{-Cat} \end{array}$$

commutes. Here the horizontal arrows represent the isomorphism of Corollary IV.3.2.3.

Proof Let (X, a) be a $(\mathbb{T}, \mathcal{W})$-category, $\rho \in \Pi(\mathbb{T}, \mathcal{V})(X)$, and $x \in X$. Then $\rho \longrightarrow x$ in the $(\Pi(\mathbb{T}, \mathcal{V}), 2)$-category obtained via the upper-right path of the diagram if and only if

$$\varphi\rho \leq a(-, x) \, ,$$

which is equivalent to

$$\rho \leq \psi a(-, x) \, ;$$

and this means precisely $\rho \longrightarrow x$ in the $(\Pi(\mathbb{T}, \mathcal{V}), 2)$-category obtained via the lower-left path of the diagram. □

Exercises

IV.3.A *The functor* $(-)_\sharp : \mathsf{Set} \to (\hat{\mathbb{T}}, \mathcal{V})\text{-}\mathsf{URel}$. Let $\hat{\mathbb{T}}$ be a lax extension to $\mathcal{V}$-Rel of a monad $\mathbb{T} = (T, m, e)$ on Set. There is a functor $(-)_\sharp : \mathsf{Set} \to (\hat{\mathbb{T}}, \mathcal{V})\text{-}\mathsf{URel}$ that sends a map $f : X \to Y$ to the unitary $(\mathbb{T}, \mathcal{V})$-relation $f_\sharp : X \rightharpoonup Y$ given by

$$f_\sharp := e_Y^\circ \cdot \hat{T}1_X \cdot Tf = e_Y^\circ \cdot \hat{T}f \, .$$

One has in particular

$$\varphi \circ f_\sharp = \varphi \cdot Tf$$

for all unitary $(\mathbb{T}, \mathcal{V})$-relations $\varphi : Y \rightharpoonup Z$. If $\hat{\mathbb{T}}$ is associative, $f_\sharp$ and $f^\sharp$ form an adjunction $f_\sharp \dashv f^\sharp$ in the ordered category $(\mathbb{T}, \mathcal{V})$-URel.

IV.3.B *The Kleisli extension of* Π. For an associative lax extension $\hat{\mathbb{T}}$ of $\mathbb{T}$ to $\mathcal{V}$-Rel and its associated power-enriched monad $\Pi = (\Pi, \mathbf{m}, \mathbf{y})$, the power-enrichment $\tau : \mathbb{P} \to \Pi$ is given componentwise by

$$\tau_X(A) = \bigvee_{x \in A} x^\sharp \, .$$

For a relation $r : X \nrightarrow Y$, one has

$$\tau_X \cdot r^\flat = (r \cdot 1_X^\sharp)^\flat \, ,$$

where the $(-)^\flat$ operation on the left comes from the construction of the Kleisli extension (Section IV.1.4), and the $(-)^\flat$ on the right is the one defined in Section IV.3.1. Hence, $r^\tau(\psi_2) = \psi_2 \circ (r \cdot 1_X^\sharp) = \psi_2 \cdot \hat{T}r$, so that

$$\psi_1 \, (\check{\Pi} r) \, \psi_2 \iff \psi_1 \leq \psi_2 \cdot \hat{T}r$$

for all unitary $(\mathbb{T}, \mathcal{V})$-relations $\psi_1 : X \to\!\!\!\!\!+ \, 1$, $\psi_2 : Y \to\!\!\!\!\!+ \, 1$.

IV.3.C *Discrete presheaf monad for the list monad.* Since the list monad $\mathbb{L} = (L, m, e)$ on Set is Cartesian (see Exercise III.1.Q), the Barr extension $\overline{\mathbb{L}}$ of $\mathbb{L}$ to Rel is associative, and, moreover, every $(\mathbb{L}, 2)$-relation $\varphi : X \to\!\!\!\!\!+ \, Y$ is unitary. With $\Pi = \Pi(\mathbb{L}, 2)$ denoting the discrete presheaf monad associated to $\overline{\mathbb{L}}$, the functor Π can be identified with the composite PL of the powerset functor P with L. Via this identification, $\mathbf{y}_X$ sends $x \in X$ to $\{(x)\}$ and Πf sends $A \subseteq LX$ to $\{Lf(x_1, \ldots, x_k) \mid (x_1, \ldots, x_k) \in A\}$, and the multiplication sends $\mathcal{A} \in PLPLX$ to

$$\{(x_{1,1}, \ldots, x_{1,i_1}, \ldots, x_{k,1}, \ldots, x_{k,i_k}) \mid (A_1, \ldots, A_k) \in \mathcal{A},$$
$$(x_{j,1}, \ldots, x_{j,i_j}) \in A_j\} \in PLX \, .$$

IV.3.D *A monad morphism* $\alpha : \mathbb{V} \to \mathbb{P}_{\mathcal{V}}$. The maps α_X of Example IV.3.2.6(2) form the components of a monad morphism $\alpha : \mathbb{V} \to \mathbb{P}_{\mathcal{V}}$ that is sup-dense but not interpolating.

IV.3.E *Continuous* P_+*-actions on compact Hausdorff spaces.* The convergence map

$$\xi : \beta[0, \infty] \to [0, \infty] \, , \quad \mathfrak{x} \mapsto \sup_{A \in \mathfrak{x}} \inf_{x \in A} x$$

yields the standard topology on $[0, \infty]$, with respect to which addition $+ : [0, \infty] \times [0, \infty] \to [0, \infty]$ is continuous. There is a distributive law δ (see Section II.3.8) of the ultrafilter monad β over the monad $\mathbb{V}$ induced by the monoid $([0, \infty], +, 0)$ (see Example IV.3.2.6(2)) defined by

$$\delta_X : \beta([0, \infty] \times X) \to [0, \infty] \times \beta X \, , \quad w \mapsto (\xi(\beta\pi_1(w)), \beta\pi_2(w)) \, ,$$

for each set X. The induced monad $\mathbb{T} = ([0, \infty] \times \beta, \check{m}, \check{e})$ has its multiplication given by

$$\check{m}_X : [0, \infty] \times \beta([0, \infty] \times \beta X) \to [0, \infty] \times \beta X \, , \quad (v, \mathcal{W}) \mapsto (v + \xi(a), m_X(\mathcal{X}))$$

(with $a = \beta\pi_1(\mathcal{W})$, $\mathcal{X} = \beta\pi_2(\mathcal{W})$ and m the multiplication of the ultrafilter monad) and its unit by

$$\check{e}_X : X \to [0, \infty] \times \beta X \, , \quad x \mapsto (0, e_X(x))$$

(with e the unit of the ultrafilter monad). Describe the category of $\mathbb{T}$-algebras.

IV.3.F *Approach spaces as* $(\mathbb{\pi}, 2)$*-categories.* There is a monad morphism $\alpha : \mathbb{\pi} \to \mathbb{\Pi}$ from the monad $\mathbb{\pi}$ of Exercise IV.3.E to the monad $\mathbb{\Pi}$ of Example IV.3.2.6(5) whose component at a set X is given by

$$\alpha_X(u, \chi)(y) = \begin{cases} u & \text{if } \chi = y, \\ \bot & \text{otherwise.} \end{cases}$$

This monad morphism $\alpha : \mathbb{\pi} \to \mathbb{\Pi}$ is sup-dense and interpolating, so the algebraic functor

$$A_\alpha : (\mathbb{\Pi}, 2)\text{-}\mathsf{Cat} \to (\mathbb{\pi}, 2)\text{-}\mathsf{Cat}$$

is an isomorphism, and $\mathsf{App} \cong (\mathbb{\pi}, 2)\text{-}\mathsf{Cat}$.

Hint. To see that α is interpolating, observe that, for each $S \subseteq [0, \infty]$ and $u \in [0, \infty]$ with $S \neq \varnothing$ and $u = \inf S$, there is some $a \in \beta[0, \infty]$ with $S \in a$ and $\xi(a) = u$.

IV.4 Injective lax algebras as Eilenberg–Moore algebras

Our primary goal in this section is to show that certain injective objects in $\mathbb{T}$-Mon are the Eilenberg–Moore algebras of the monad $\mathbb{T}$. The motivating example is given by continuous lattices, which are the injective objects of the category Top_0 of T0-spaces; however, we do not assume any prior knowledge of these particular ordered structures and will derive the necessary concepts in Section IV.4.4. A consequence of the formal approach is a unified treatment for the description of Sup as a category of injectives for Ord, Cnt for Top, Dst for Cls, Frm for $\mathsf{Cls}_{\text{fin}}$, and $\mathsf{Sup}^{\mathsf{P}_+}$ for Met.

IV.4.1 Eilenberg–Moore algebras as Kleisli monoids

The embedding $\mathsf{Set}^{\mathbb{T}} \hookrightarrow (\mathbb{T}, \mathcal{V})\text{-}\mathsf{Cat}$ of Proposition III.1.6.5 describes $\mathbb{T}$-algebras as $(\mathbb{T}, \mathcal{V})$-algebras when the lax extension of $\mathbb{T}$ is flat. Here, we show that a similar situation can occur when the extension is not flat.

By Exercise II.3.H, a morphism $\tau : \mathbb{P} \to \mathbb{T}$ from the powerset monad $\mathbb{P}$ to a monad $\mathbb{T} = (T, m, e)$ on Set yields a functor

$$\mathsf{Set}^{\tau} : \mathsf{Set}^{\mathbb{T}} \to \mathsf{Set}^{\mathbb{P}}, \qquad (X, a) \mapsto (X, a \cdot \tau_X)$$

that commutes with the respective underlying-set functors. Via the isomorphism $\mathsf{Set}^{\mathbb{P}} \cong \mathsf{Sup}$ (Example II.3.2.2(2)), every $\mathbb{P}$-algebra structure $s : PX \to X$ represents a supremum with right adjoint $\downarrow_X := s^{\dashv} : X \to PX$ the down-set map of X, i.e. the *separated* order relation on X (Section II.1.7). Hence, every $\mathbb{T}$-algebra morphism becomes a sup-map $f : (X, a \cdot \tau_X) \to (Y, b \cdot \tau_Y)$ that has a right adjoint

$$f^{\dashv} : Y \to X$$

in Ord. In particular, a $\mathbb{T}$-algebra structure $a : TX \to X$ has a right adjoint $a^{\dashv} : X \to TX$ which satisfies

$$a^{\dashv} \circ a^{\dashv} \leq a^{\dashv} \quad \text{and} \quad e_X \leq a^{\dashv} . \tag{IV.4.1.i}$$

Indeed, the second inequality follows by adjunction from $a \cdot e_X = 1_X$; the first is a consequence of $a \cdot Ta = a \cdot m_X$, since this identity implies $m_X \leq a^{\dashv} \cdot a \cdot Ta$, and therefore

$$m_X \cdot T(a^{\dashv}) \cdot a^{\dashv} \leq a^{\dashv} \cdot a \cdot Ta \cdot T(a^{\dashv}) \cdot a^{\dashv} = a^{\dashv}$$

($a \cdot a^{\dashv} = 1_X$ because $1_X = a \cdot e_X \leq a \cdot a^{\dashv} \leq 1_X$), i.e. $(X, a^{\dashv})$ is a $\mathbb{T}$-monoid. If $f : (X, a) \to (Y, b)$ is a $\mathbb{T}$-algebra homomorphism, then $b \cdot Tf = f \cdot a$ yields

$$Tf \cdot a^{\dashv} \leq b^{\dashv} \cdot f$$

by adjunction, so $f : (X, a^{\dashv}) \to (Y, b^{\dashv})$ is a morphism of $\mathbb{T}$-monoids. Hence, when $(\mathbb{T}, \tau)$ is power-enriched, there is a functor

$$\mathsf{Set}^{\mathbb{T}} \to \mathbb{T}\text{-}\mathsf{Mon} , \quad (X, a) \mapsto (X, a^{\dashv})$$

that commutes with the respective underlying-set functors. In particular, $(TX, m_X^{\dashv})$ is a $\mathbb{T}$-monoid, and every $\mathbb{T}$-monoid structure ν on X becomes a $\mathbb{T}$-monoid morphism $\nu : (X, \nu) \to (TX, m_X^{\dashv})$, since $\nu \circ \nu \leq \nu$ equivalently means

$$T\nu \cdot \nu \leq m_X^{\dashv} \cdot \nu .$$

We note that, unlike the functor $\mathsf{Set} \hookrightarrow (\mathbb{T}, \mathcal{V})\text{-}\mathsf{Cat}$ of Proposition III.1.6.5, the functor $\mathsf{Set}^{\mathbb{T}} \to \mathbb{T}\text{-}\mathsf{Mon}$ fails to be full in general, as Example IV.4.1.1 shows.

IV.4.1.1 Example For the powerset monad $\mathbb{P}$, the functor $\mathsf{Set}^{\mathbb{P}} \to \mathbb{P}\text{-}\mathsf{Mon}$ simply describes the forgetful functor

$$\mathsf{Set}^{\mathbb{P}} \simeq \mathsf{Sup} \to \mathsf{Ord} \cong \mathbb{P}\text{-}\mathsf{Mon}$$

(see Example IV.1.3.2(2)).

IV.4.2 Monads on categories of Kleisli monoids

In order to characterize injective objects of $\mathbb{T}$-Mon, it is convenient to introduce a monad $\mathbb{T}'$ derived from the original monad $\mathbb{T}$ on Set, as follows. Consider a morphism $\alpha : \mathbb{S} \to \mathbb{T}$ of power-enriched monads $\mathbb{S} = (S, n, d)$ and $\mathbb{T} = (T, m, e)$. By composing the functor $\mathsf{Set}^{\alpha} : \mathsf{Set}^{\mathbb{T}} \to \mathsf{Set}^{\mathbb{S}}$ (Exercise II.3.H) with $G : \mathsf{Set}^{\mathbb{S}} \to \mathbb{S}\text{-}\mathsf{Mon}$ (Section IV.4.1), we obtain a functor

$$G^{\mathbb{T}'} : \mathsf{Set}^{\mathbb{T}} \to \mathbb{S}\text{-}\mathsf{Mon}$$

that sends a $\mathbb{T}$-algebra (X, a) to the $\mathbb{S}$-monoid $(X, (a \cdot \alpha_X)^{\dashv})$ and leaves maps unchanged. This functor $G^{\mathbb{T}'}$ is right adjoint. Rather than describing its left adjoint explicitly, we construct below the induced monad $\mathbb{T}'$ on $\mathbb{S}$-Mon via its Kleisli triple components T', e'_X, and $(-)^{\mathbb{T}'}$ (Proposition IV.4.2.2). Theorem IV.4.3.2 then shows that the new monad has the same Eilenberg–Moore category as $\mathbb{T}$, a fact that justifies a posteriori the notation for the functor $G^{\mathbb{T}'}$.

In Section IV.4.4, we exploit the case where $\alpha = \tau : \mathbb{P} \to \mathbb{F}$ is the principal filter monad morphism to identify the category $\mathsf{Set}^{\mathbb{F}}$ as the category Cnt of continuous lattices via the monadic functor $\mathsf{Set}^{\mathbb{F}} \to \mathsf{Ord} \cong \mathbb{P}$-$\mathsf{Mon}$. In Section IV.4.6, we show that when $\alpha = 1_{\mathbb{T}} : \mathbb{T} \to \mathbb{T}$ is the identity, the monadic functor $\mathsf{Set}^{\mathbb{T}} \to \mathbb{T}$-$\mathsf{Mon}$ is Kock–Zöberlein; this property then facilitates our study of injective objects in Section IV.4.6.

Construction

(1) Given an $\mathbb{S}$-monoid structure $\mu : X \to SX$, the map $\nu := \alpha_X \cdot \mu$ yields a $\mathbb{T}$-monoid structure on X (Proposition IV.1.3.3). We define the set $T'X$ of $\nu^{\mathbb{T}}$-*invariants* as the equalizer in Set of the pair $(\nu^{\mathbb{T}}, 1_{TX})$:

$$T'X \xrightarrow{q_X} TX \underset{1_{TX}}{\overset{\nu^{\mathbb{T}}}{\rightrightarrows}} TX \, .$$

The universal property of q_X implies the existence of a map $p_X : TX \to T'X$ such that

$$q_X \cdot p_X = \nu^{\mathbb{T}} \qquad \text{and} \qquad p_X \cdot q_X = 1_{T'X} \, .$$

Indeed, as $\nu^{\mathbb{T}} \cdot \nu^{\mathbb{T}} = \nu^{\mathbb{T}}$ (by (IV.1.3.i)), there is a map $p_X : TX \to T'X$ with $q_X \cdot p_X = \nu^{\mathbb{T}}$; therefore, we have $q_X \cdot p_X \cdot q_X = \nu^{\mathbb{T}} \cdot q_X = q_X$, so that $p_X \cdot q_X = 1_{T'X}$ by unicity of the induced map $T'X \to T'X$.

The set $T'X$ can be equipped with the $\mathbb{S}$-monoid structure $\omega_X : T'X \to ST'X$ given by

$$\omega_X := Sp_X \cdot (m_X \cdot \alpha_{TX})^{\dashv} \cdot q_X \, .$$

Lemma IV.4.2.1 shows that $q_X : (T'X, \omega_X) \to (TX, (m_X \cdot \alpha_{TX})^{\dashv})$ is also an equalizer in $\mathbb{S}$-Mon. This ensures that the maps e'_X and $f^{\mathbb{T}'}$ defined in the following points are morphisms of $\mathbb{S}$-monoids.

(2) Since $\nu^{\mathbb{T}} \cdot \nu = \nu$, there exists a map $e'_X : X \to T'X$ such that $q_X \cdot e'_X = \nu$:

$$\begin{array}{ccccc} X & & & & \\ {\scriptstyle e'_X}\downarrow & \searrow^{\nu} & & & \\ T'X & \xrightarrow[q_X]{} & TX & \underset{1_{TX}}{\overset{\nu^{\mathbb{T}}}{\rightrightarrows}} & TX \, . \end{array}$$

This yields a morphism of $\mathbb{S}$-monoids $e'_X : (X, \mu) \to (T'X, \omega_X)$. Since $q_X \cdot e'_X = \nu$ and $p_X \cdot q_X = 1_{T'X}$, one can equivalently obtain e'_X as either

$$e'_X = p_X \cdot \nu \qquad \text{or} \qquad e'_X = p_X \cdot e_X$$

because $p_X \cdot \nu = p_X \cdot \nu^{\mathbb{T}} \cdot e_X = p_X \cdot q_X \cdot p_X \cdot e_X = p_X \cdot e_X$.

(3) If (Y, μ_Y) is another $\mathbb{S}$-monoid, and $f : (Y, \mu_Y) \to (T'X, \omega_X)$ is an $\mathbb{S}$-monoid morphism, then

$$\nu^{\mathbb{T}} \cdot (q_X \cdot f)^{\mathbb{T}} = (\nu^{\mathbb{T}} \cdot q_X \cdot f)^{\mathbb{T}} = (q_X \cdot f)^{\mathbb{T}} .$$

Hence, there exists a unique map $f^{\mathbb{T}'} : T'Y \to T'X$ making the following diagram commute:

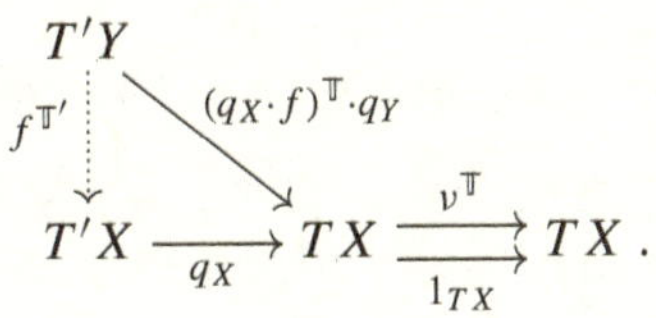

This yields a morphism of $\mathbb{S}$-monoids $f^{\mathbb{T}'} : (T'Y, \omega_Y) \to (T'X, \omega_X)$ that can also be obtained directly as

$$f^{\mathbb{T}'} = p_X \cdot (q_X \cdot f)^{\mathbb{T}} \cdot q_Y .$$

IV.4.2.1 Lemma *For a morphism $\alpha : \mathbb{S} \to \mathbb{T}$ of power-enriched monads $\mathbb{S} = (S, n, d)$ and $\mathbb{T} = (T, m, e)$, the map*

$$q_X : (T'X, \omega_X) \to (TX, (m_X \cdot \alpha_{TX})^{\dashv})$$

(defined in the previous construction) is an equalizer in $\mathbb{S}$-Mon. *As a consequence,*

$$e'_X : (X, \mu) \to (T'X, \omega_X) \qquad \textit{and} \qquad f^{\mathbb{T}'} : (T'Y, \omega_Y) \to (T'X, \omega_X)$$

are morphisms of $\mathbb{S}$-monoids.

Proof To verify that $\omega_X : T'X \to ST'X$ is an $\mathbb{S}$-monoid structure, observe that

$$d_{T'X} = d_{T'X} \cdot p_X \cdot q_X = Sp_X \cdot d_{TX} \cdot q_X \le Sp_X \cdot (m_X \cdot \alpha_{TX})^{\dashv} \cdot q_X = \omega_X$$

because $(m_X \cdot \alpha_{TX}) \cdot d_{TX} = 1_{TX}$, and

$$\begin{aligned}
&\omega_X^{\mathbb{S}} \cdot \omega_X \\
&= Sp_X \cdot n_{TX} \cdot S(m_X \cdot \alpha_{TX})^{\dashv} \cdot S\nu^{\mathbb{T}} \cdot (m_X \cdot \alpha_{TX})^{\dashv} \cdot q_X && (q_X \cdot p_X = \nu^{\mathbb{T}}) \\
&\le Sp_X \cdot n_{TX} \cdot (S(m_X \cdot \alpha_{TX}))^{\dashv} \cdot S\nu^{\mathbb{T}} \cdot (m_X \cdot \alpha_{TX})^{\dashv} \cdot q_X && (m_X \cdot \alpha_{TX} \text{ a retraction}) \\
&\le Sp_X \cdot n_{TX} \cdot (Sm_X \cdot S\alpha_{TX})^{\dashv} \cdot (m_X \cdot \alpha_{TX})^{\dashv} \cdot \nu^{\mathbb{T}} \cdot q_X && (\nu^{\mathbb{T}} \text{ morphism in } \mathbb{S}\text{-Mon}) \\
&= Sp_X \cdot n_{TX} \cdot (m_X \cdot \alpha_{TX} \cdot n_{TX})^{\dashv} \cdot \nu^{\mathbb{T}} \cdot q_X && (\alpha \text{ monad morphism}) \\
&= Sp_X \cdot (m_X \cdot \alpha_{TX})^{\dashv} \cdot q_X = \omega_X && (n_{TX} \cdot n_{TX}^{\dashv} = 1_{STX}).
\end{aligned}$$

One can reason similarly to obtain

$$Sq_X\cdot\omega_X = S\nu^{\mathbb{T}}\cdot(m_X\cdot\alpha_{TX})^{\dashv}\cdot q_X \le (m_X\cdot\alpha_{TX})^{\dashv}\cdot\nu^{\mathbb{T}}\cdot q_X = (m_X\cdot\alpha_{TX})^{\dashv}\cdot q_X\,,$$

so that $q_X : (T'X, \omega_X) \to (TX, (m_X\cdot\alpha_{TX})^{\dashv})$ is a morphism in $\mathbb{S}$-Mon. Suppose now that $g : (Y, \mu_Y) \to (TX, (m_X\cdot\alpha_{TX})^{\dashv})$ is a morphism in $\mathbb{S}$-Mon satisfying $\nu^{\mathbb{T}}\cdot g = g$. Since $q_X : T'X \to TX$ is an equalizer of $(\nu^{\mathbb{T}}, 1_{TX})$ in Set, there exists a unique map $h : Y \to T'X$ with $g = q_X\cdot h$; moreover,

$$Sh\cdot\mu_Y = Sp_X\cdot Sg\cdot\mu_Y \le Sp_X\cdot(m_X\cdot\alpha_{TX})^{\dashv}\cdot g = \omega_X\cdot h\,,$$

which shows that $h : (Y, \mu_Y) \to (T'X, \omega_X)$ is a morphism in $\mathbb{S}$-Mon. As a consequence, q_X is an equalizer in $\mathbb{S}$-Mon, and e'_X, $f^{\mathbb{T}'}$ are the underlying maps of the corresponding unique $\mathbb{S}$-monoid morphisms into $(T'X, \omega_X)$. □

IV.4.2.2 Proposition *If $\alpha : \mathbb{S} \to \mathbb{T}$ is a morphism of power-enriched monads, then the construction described in* (1)–(3) *above defines a Kleisli triple $(T', (-)^{\mathbb{T}'}, e')$ on $\mathbb{S}$-Mon. Hence, $\mathbb{T}' = (T', m', e')$ is a monad on $\mathbb{S}$-Mon with multiplication $m'_X = 1^{\mathbb{T}'}_{T'X} = p_X\cdot q^{\mathbb{T}}_X\cdot q_{T'Y}$.*

Proof Lemma IV.4.2.1 ensures that the components T', $(-)^{\mathbb{T}'}$, and e' are of the appropriate type to yield a Kleisli triple on $\mathbb{S}$-Mon. For an $\mathbb{S}$-monoid morphism $h : (X, \mu) \to (TY, (\alpha_{TY}\cdot m_Y)^{\dashv})$, we observe that if $\nu = \alpha_X\cdot\mu$ then

$$h^{\mathbb{T}}\cdot\nu = h$$

since

$$m_Y\cdot Th\cdot\alpha_X\cdot\mu \le m_Y\cdot\alpha_{TY}\cdot(m_Y\cdot\alpha_{TY})^{\dashv}\cdot h = h = m_Y\cdot Th\cdot e_X \le m_Y\cdot Th\cdot\alpha_X\cdot\mu\,.$$

The conditions (II.3.7.i) are now easily verified by using this observation. The monad $\mathbb{T}'$ is then obtained via the Kleisli triple (see Section II.3.7). □

IV.4.2.3 Examples

(1) If $\mathbb{S} = \mathbb{T} = \mathbb{P}$, then a $\mathbb{P}$-monoid is an ordered set $(X, \downarrow_X)$ (Example IV.1.3.2(2)). In this case, an element $A \in PX$ is a $\downarrow^{\mathbb{T}}_X$-invariant precisely when $\downarrow^{\mathbb{P}}_X(A) = A$, i.e. when A is down-closed:

$$\bigcup\{\downarrow_X x \mid x \in A\} = A\,.$$

The construction in Section IV.4.2 therefore describes the down-set monad $\mathbb{D}\mathrm{n} = (\mathrm{Dn}, \bigvee_{\mathrm{Dn}}, \downarrow)$ on Ord (Example II.4.9.3).

(2) If $\mathbb{S} = \mathbb{T} = \mathbb{F}$ is the filter monad, then an $\mathbb{F}$-monoid is a topological space (X, ν), where $\nu : X \to FX$ is the neighborhood filter map (Section IV.1.1). A filter $a \in FX$ is $\nu^{\mathbb{F}}$-invariant if and only if a is spanned by open sets of X:

$$A \in \nu^{\mathbb{F}}(a) \iff \nu^{-1}(A^{\mathbb{F}}) \in a \iff \{x \in X \mid A \in \nu(x)\} \in a\,,$$

so $\nu^{\mathbb{F}}(a) = a$ means that for every $A \in a$ the interior must also be in a. The *open-filter monad* on $\mathsf{Top} \cong \mathbb{F}\text{-}\mathsf{Mon}$, i.e. the monad $\mathbb{F}' = (F', m', e')$ obtained via Proposition IV.4.2.2, can therefore be described as follows. The functor $F' : \mathsf{Top} \to \mathsf{Top}$ sends a topological space X to the set $F'X$ of filters *in* $\mathcal{O}X$ equipped with its Scott topology; F' also sends a continuous map $f : X \to Y$ to the continuous map $F'f : F'X \to F'Y$ given by $O \in F'f(a) \iff f^{-1}(O) \in a$ for all $a \in F'Y$. The unit $\nu = e'_X : X \to F'X$ sends a point x to its set of open neighborhoods, and the multiplication $m'_X : F'F'X \to F'X$ is the restriction of the Kowalsky sum.

IV.4.3 Eilenberg–Moore algebras over $\mathbb{S}$-Mon

For power-enriched monads $\mathbb{S}$ and $\mathbb{T}$, the monad $\mathbb{T}'$ on $\mathbb{S}$-Mon is derived from $\mathbb{T}$ via the morphism $\alpha : \mathbb{S} \to \mathbb{T}$. Using the notations of Section IV.4.2, we now show that the category of $\mathbb{T}'$-algebras is isomorphic to $\mathsf{Set}^{\mathbb{T}}$.

IV.4.3.1 Lemma *Let $U : \mathbb{S}\text{-}\mathsf{Mon} \to \mathsf{Set}$ denote the forgetful functor. The maps p_X of* Section IV.4.2 *form the components of a natural transformation $p : TU \to UT'$ that defines a lifting of U through $(G^{\mathbb{T}'}, G^{\mathbb{T}})$* (Exercise II.3.H).

Proof An $\mathbb{S}$-monoid morphism $f : (Y, \mu_Y) \to (X, \mu)$ yields a $\mathbb{T}$-monoid morphism $f : (Y, \nu_Y) \to (TX, \nu)$ (where $\nu_Y = \alpha_Y \cdot \mu_Y$ and $\nu = \alpha_X \cdot \mu$), so that

$$\nu \cdot f = (\nu \cdot f)^{\mathbb{T}} \cdot e_Y \leq (\nu \cdot f)^{\mathbb{T}} \cdot \nu_Y = \nu^{\mathbb{T}} \cdot Tf \cdot \nu_Y \leq \nu^{\mathbb{T}} \cdot \nu \cdot f = \nu \cdot f\,.$$

Therefore, $(\nu \cdot f)^{\mathbb{T}} \cdot \nu_Y = \nu \cdot f$, and using $T'f = (e'_X \cdot f)^{\mathbb{T}'}$ we obtain

$$\begin{aligned} T'f \cdot p_Y &= p_X \cdot (\nu \cdot f)^{\mathbb{T}} \cdot \nu_Y^{\mathbb{T}} = p_X \cdot ((\nu \cdot f)^{\mathbb{T}} \cdot \nu_Y)^{\mathbb{T}} = p_X \cdot (\nu \cdot f)^{\mathbb{T}} \\ &= p_X \cdot \nu^{\mathbb{T}} \cdot Tf = p_X \cdot Tf\,, \end{aligned}$$

so that $p : TU \to UT'$ is a natural transformation. Moreover, for $m'_X = (1_{T'X})^{\mathbb{T}'} = p_X \cdot (q_X)^{\mathbb{T}} \cdot q_{T'X}$,

$$\begin{aligned} m'_X \cdot p_{T'X} \cdot Tp_X &= p_X \cdot (q_X)^{\mathbb{T}} \cdot (\alpha_{T'X} \cdot \omega_X)^{\mathbb{T}} \cdot Tp_X \\ &= p_X \cdot (m_X \cdot T\nu^{\mathbb{T}} \cdot \alpha_{TX} \cdot (m_X \cdot \alpha_{TX})^{\dashv} \cdot q_X)^{\mathbb{T}} \cdot Tp_X \\ &= p_X \cdot (\nu^{\mathbb{T}} \cdot m_X \cdot \alpha_{TX} \cdot (m_X \cdot \alpha_{TX})^{\dashv} \cdot q_X)^{\mathbb{T}} \cdot Tp_X \\ &= p_X \cdot (\nu^{\mathbb{T}} \cdot q_X)^{\mathbb{T}} \cdot Tp_X \\ &= p_X \cdot (q_X)^{\mathbb{T}} \cdot Tp_X \\ &= p_X \cdot \nu^{\mathbb{T}} \cdot m_X \\ &= p_X \cdot m_X\,. \end{aligned}$$

Since $e'_X = p_X \cdot e_X$, we conclude that r does indeed define a lifting of U through $(G^{\mathbb{T}'}, G^{\mathbb{T}})$. □

IV.4.3.2 Theorem *If $\alpha : \mathbb{S} \longrightarrow \mathbb{T}$ is a morphism of power-enriched monads, there is an isomorphism*

$$\mathsf{Set}^{\mathbb{T}} \cong \mathbb{S}\text{-}\mathsf{Mon}^{\mathbb{T}'}$$

of Eilenberg–Moore categories that commutes with the underlying-set functors.

Proof Suppose first that (X, a) is a $\mathbb{T}$-algebra. One obtains an $\mathbb{S}$-monoid (X, μ), with $\mu = (a \cdot \alpha_X)^{\dashv}$, that can be equipped with the structure $a' : (T'X, \omega_X) \longrightarrow (X, \mu)$ defined by

$$a' := a \cdot q_X \ .$$

Since $a : (TX, m_X) \longrightarrow (X, a)$ is a $\mathbb{T}$-algebra morphism, its $G^{\mathbb{T}'}$-image is an $\mathbb{S}$-monoid morphism, and therefore so is a'. To see that a' satisfies the algebra conditions for the monad $\mathbb{T}'$, we use the definition of e'_X:

$$a' \cdot e'_X = a \cdot q_X \cdot e'_X = a \cdot \alpha_X \cdot (a \cdot \alpha_X)^{\dashv} = 1_X \ .$$

Suppose now that $f, g : (Y, \mu_Y) \longrightarrow (T'X, \omega_X)$ are morphisms in $\mathbb{S}$-Mon satisfying $a' \cdot f = a' \cdot g$, or equivalently $a \cdot q_X \cdot f = a \cdot q_X \cdot g$; since a is a $\mathbb{T}$-algebra structure, one has $a \cdot (q_X \cdot f)^{\mathbb{T}} = a \cdot (q_X \cdot g)^{\mathbb{T}}$ (Exercise II.3.G), so that

$$a' \cdot f^{\mathbb{T}'} = a \cdot q_X \cdot f^{\mathbb{T}'} = a \cdot (q_X \cdot f)^{\mathbb{T}} \cdot q_Y = a \cdot (q_X \cdot g)^{\mathbb{T}} \cdot q_Y = a \cdot q_X \cdot g^{\mathbb{T}'} = a' \cdot g^{\mathbb{T}'} \ .$$

Therefore, $((X, \mu), a')$ is a $\mathbb{T}'$-algebra. A $\mathbb{T}$-algebra morphism $f : (X, a) \longrightarrow (Y, a_Y)$ yields a morphism $f : (X, (a \cdot \alpha_X)^{\dashv}) \longrightarrow (Y, (a_Y \cdot \alpha_Y)^{\dashv})$ in $\mathbb{S}$-Mon. To verify that $a'_Y \cdot (e'_Y \cdot f)^{\mathbb{T}'} = f \cdot a'$, we first observe

$$\begin{aligned} a_Y \cdot (\alpha_Y \cdot (a_Y \cdot \alpha_Y)^{\dashv} \cdot f)^{\mathbb{T}} &= a_Y \cdot m_X \cdot T(\alpha_Y \cdot (a_Y \cdot \alpha_Y)^{\dashv}) \cdot Tf \\ &= a_Y \cdot Ta_Y \cdot T(\alpha_Y \cdot (a_Y \cdot \alpha_Y)^{\dashv}) \cdot Tf \\ &= a_Y \cdot Tf \ . \end{aligned}$$

Hence,

$$\begin{aligned} a'_Y \cdot (e'_Y \cdot f)^{\mathbb{T}'} &= a_Y \cdot (q_Y \cdot e'_Y \cdot f)^{\mathbb{T}} \cdot q_X \\ &= a_Y \cdot (\alpha_Y \cdot (a_Y \cdot \alpha_Y)^{\dashv} \cdot f)^{\mathbb{T}} \cdot q_X \\ &= f \cdot a \cdot q_X \\ &= f \cdot a' \ , \end{aligned}$$

which proves that $f : ((X, \mu), a') \longrightarrow ((Y, \mu_Y), a'_Y)$ is a morphism of $\mathbb{T}'$-algebras. Thus, the assignment of $((X, (a \cdot \alpha_X)^{\dashv}), a \cdot q_X)$ to a $\mathbb{T}$-algebra (X, a) yields a functor $K : \mathsf{Set}^{\mathbb{T}} \longrightarrow \mathbb{S}\text{-}\mathsf{Mon}^{\mathbb{T}'}$ that commutes with the underlying-set functors.

The lifting $p : TU \longrightarrow UT'$ of U (see Lemma IV.4.3.1) yields a functor $\tilde{U} : \mathbb{S}\text{-}\mathsf{Mon}^{\mathbb{T}'} \longrightarrow \mathsf{Set}^{\mathbb{T}}$ that sends a $\mathbb{T}'$-algebra $((X, \mu), a')$ to (X, a), where $a : TX \longrightarrow X$ is defined by

$$a := a' \cdot p_X \ ,$$

and is invariant on maps.

Given a $\mathbb{T}$-algebra (X, a), the structure of $\tilde{U}K(X, a)$ is described by

$$a \cdot q_X \cdot p_X = a \cdot m_X \cdot T(\alpha_X \cdot (a \cdot \alpha_X)^{\dashv}) = a \cdot T(a \cdot \alpha_X) \cdot T(a \cdot \alpha_X)^{\dashv} = a\,.$$

To study the image of a $\mathbb{T}'$-algebra $((X, \mu), a')$ via $K\tilde{U}$, note first that $a' : (T'X, \omega_X) \to (X, \mu)$ is a morphism in $\mathbb{S}$-Mon. Thus, after setting $\nu = \alpha_X \cdot \mu$ and observing that $(m_X \cdot \alpha_{TX}) \cdot S(\alpha_X \cdot \mu) = \nu^{\mathbb{T}} \cdot \alpha_X$, one obtains

$$\begin{aligned} 1_{SX} &= S(a' \cdot p_X) \cdot S(\alpha_X \cdot \mu) \\ &\leq S(a' \cdot p_X) \cdot (m_X \cdot \alpha_{TX})^{\dashv} \cdot \nu^{\mathbb{T}} \cdot \alpha_X \\ &= Sa' \cdot \omega_X \cdot p_X \cdot \alpha_X \\ &\leq \mu \cdot (a' \cdot p_X \cdot \alpha_X)\,. \end{aligned}$$

This inequality, combined with $(a' \cdot p_X \cdot \alpha_X) \cdot \mu = 1_X$, yields

$$\mu = (a' \cdot p_X \cdot \alpha_X)^{\dashv}\,.$$

Hence, the image under $K\tilde{U}$ of the $\mathbb{T}'$-algebra $((X, \mu), a')$ is the $\mathbb{T}'$-algebra with underlying $\mathbb{S}$-monoid $(X, (a' \cdot p_X \cdot \alpha_X)^{\dashv}) = (X, \mu)$ and structure

$$a' \cdot p_X \cdot q_X = a'\,.$$

One concludes that K and $\tilde{U}$ are inverse of each other, and $\mathsf{Set}^{\mathbb{T}} \cong \mathbb{S}\text{-}\mathsf{Mon}^{\mathbb{T}'}$. □

In general, a $\mathbb{T}$-monoid (X, ν) on a category X is *separated* whenever ν is a monomorphism. For a power-enriched monad $\mathbb{T}$, this condition amounts to the requirement that the initial order induced on X by $\nu : X \to TX$,

$$x \leq y \iff \nu(x) \leq \nu(y) \qquad \text{(IV.4.3.i)}$$

for all $x, y \in X$, is separated. The full subcategory of $\mathbb{T}$-Mon whose objects are the separated Kleisli monoids is denoted by $\mathbb{T}\text{-}\mathsf{Mon}_{\mathrm{sep}}$.

IV.4.3.3 Corollary *Given a morphism $\alpha : \mathbb{S} \to \mathbb{T}$ of power-enriched monads, the monad $\mathbb{T}'$ of* Proposition IV.4.2.2 *restricts to a monad on $\mathbb{S}\text{-}\mathsf{Mon}_{\mathrm{sep}}$, and there is an isomorphism*

$$\mathsf{Set}^{\mathbb{T}} \cong (\mathbb{S}\text{-}\mathsf{Mon}_{\mathrm{sep}})^{\mathbb{T}'}$$

that commutes with the underlying-set functors.

Proof The results of Section IV.4.2 leading up to Theorem IV.4.3.2 can be reproduced by replacing $\mathbb{S}$-Mon by $\mathbb{S}\text{-}\mathsf{Mon}_{\mathrm{sep}}$. Indeed, the functor $G^{\mathbb{T}'} : \mathsf{Set}^{\mathbb{T}} \to \mathbb{S}\text{-}\mathsf{Mon}$ factors through $\mathbb{S}\text{-}\mathsf{Mon}_{\mathrm{sep}} \hookrightarrow \mathbb{S}\text{-}\mathsf{Mon}$, and $(T'X, \omega_X)$ is separated for all $\mathbb{S}$-monoids (X, μ): setting $\nu = \alpha_X \cdot \mu$, one has

$$\begin{aligned} p_X \cdot (m_X \cdot \alpha_{TX}) \cdot Sq_X \cdot \omega_X &= p_X \cdot (m_X \cdot \alpha_{TX}) \cdot S\nu^{\mathbb{T}} \cdot (m_X \cdot \alpha_{TX})^{\dashv} \cdot q_X \\ &= p_X \cdot \nu^{\mathbb{T}} \cdot (m_X \cdot \alpha_{TX}) \cdot (m_X \cdot \alpha_{TX})^{\dashv} \cdot q_X = 1_{T'X}, \end{aligned}$$

i.e. ω_X is a section. □

IV.4.3.4 Examples

(1) The initial order on a $\mathbb{P}$-monoid $(X, \downarrow)$ is given by

$$x \leq y \iff \downarrow x \leq \downarrow y$$

for all $x, y \in X$. Hence, if $\downarrow : X \to PX$ is a monomorphism, then $\downarrow x = \downarrow y$ implies $x = y$, so X is a separated ordered set, and we have $\mathbb{P}\text{-}\mathsf{Mon}_{\mathrm{sep}} \cong \mathsf{Ord}_{\mathrm{sep}}$. Hence, with $\mathbb{T} = \mathbb{S} = \mathbb{P}$, Theorem IV.4.3.2 and Corollary IV.4.3.3 state that the forgetful functors

$$\mathsf{Sup} \to \mathsf{Ord} \qquad \text{and} \qquad \mathsf{Sup} \to \mathsf{Ord}_{\mathrm{sep}}$$

are strictly monadic (see Example IV.4.1.1).

(2) The initial order induced on a topological space, seen as an $\mathbb{F}$-monoid, by its neighborhood map $\nu : X \to FX$ is its underlying order (see Section II.1.9). Hence, the isomorphism $\mathbb{F}\text{-}\mathsf{Mon} \cong \mathsf{Top}$ identifies separated $\mathbb{F}$-monoids with T0-spaces (Exercise IV.4.B). With Top_0 the corresponding full category of Top, we therefore have $\mathbb{F}\text{-}\mathsf{Mon}_{\mathrm{sep}} \cong \mathsf{Top}_0$. With $\mathbb{S} = \mathbb{F}$ or $\mathbb{S} = \mathbb{P}$, we obtain strictly monadic functors

$$\mathsf{Set}^{\mathbb{F}} \to \mathsf{Top}\,, \quad \mathsf{Set}^{\mathbb{F}} \to \mathsf{Top}_0\,, \quad \mathsf{Set}^{\mathbb{F}} \to \mathsf{Ord}\,, \quad \text{and} \quad \mathsf{Set}^{\mathbb{F}} \to \mathsf{Ord}_{\mathrm{sep}}\,,$$

thanks to Theorem IV.4.3.2 and Corollary IV.4.3.3. The identification of $\mathsf{Set}^{\mathbb{F}}$ as a category of lattices requires more work and is considered in Section IV.4.4.

IV.4.4 Continuous lattices

For a monad morphism $\alpha : \mathbb{P} \to \mathbb{T}$, the study of $\mathbb{T}$-algebras can be facilitated by the study of the monad $\mathbb{T}'$ on $\mathsf{Ord} \cong \mathbb{P}\text{-}\mathsf{Mon}$, as a consequence of Theorem IV.4.3.2. This approach leads us to the identification of continuous lattices as the Ini-injective objects of Top, for Ini the class of U-initial morphisms (with $U : \mathsf{Top} \to \mathsf{Set}$ forgetful).

A complete lattice X is *continuous* if the restriction of the supremum map $\bigvee_X : \mathrm{Dn}\, X \to X$ to the set $\mathrm{Idl}\, X$ of ideals in X has a left adjoint $\downarrow\!\!\!\downarrow_X$:

$$\downarrow\!\!\!\downarrow_X \dashv \bigvee_X : \mathrm{Idl}\, X \to X$$

(compare with complete distributivity, Section II.1.11). The category of continuous lattices and inf-maps which preserve up-directed suprema is denoted by

$$\mathsf{Cnt}\,.$$

Since our focus in this section is on convergence of filters rather than that of ideals, from now on we will concentrate on the dual notion of *cocontinuous* lattices.

A complete lattice X is *cocontinuous* if the restriction of the infimum map $\bigwedge_X : \mathrm{Up}\, X \to X$ to the set $\mathrm{Fil}\, X$ of filters in X has a right adjoint $\Uparrow_X$:

$$\bigwedge_X \dashv \Uparrow_X : X \to \mathrm{Fil}\, X .$$

The order on $\mathrm{Fil}\, X$ is induced by the order on $\mathrm{Up}\, X$, and is therefore given by reverse inclusion (see Section II.1.7); hence, one has

$$\bigwedge S \le a \iff S \supseteq \Uparrow a$$

for all $S \in \mathrm{Fil}\, X$, $a \in X$, and

$$\Uparrow a = \bigcap \{S \in \mathrm{Fil}\, X \mid \bigwedge S \le a\}$$

by Proposition II.1.8.2. Note also that $\bigwedge \Uparrow a = a$ for all $a \in X$, i.e. $\Uparrow_X : X \to FX$ is a fully faithful embedding. Cocontinuous lattices with sup-maps that preserve down-directed infima form a 2-category which is 2-isomorphic to $\mathsf{Cnt}^{\mathrm{co}}$ (the isomorphism sends $f : X \to Y$ to $f^{\mathrm{op}} : X^{\mathrm{op}} \to Y^{\mathrm{op}}$). Hence, in the following, we will write $\mathsf{Cnt}^{\mathrm{co}}$ to designate the category of cocontinuous lattices.

The function Fil defined on ordered sets is the object part of a 2-functor $\mathrm{Fil} : \mathsf{Ord} \to \mathsf{Ord}$ that is the restriction of the up-set 2-functor Up to filters: for a map $f : X \to Y$, one has

$$\mathrm{Fil}\, f(A) = \textstyle\bigcup_{x \in A} \uparrow f(x)$$

for all filters $A \subseteq X$. The up-set map of an ordered set X corestricts to a monotone map $\uparrow_X : X \to \mathrm{Fil}\, X$, and union yields the infimum map $\bigwedge_{\mathrm{Fil}\, X} : \mathrm{Fil}\,\mathrm{Fil}\, X \to \mathrm{Fil}\, X$ to form the *ordered-filter monad*

$$\mathbb{F}\mathrm{il} = (\mathrm{Fil}, \textstyle\bigwedge_{\mathrm{Fil}}, \uparrow)$$

on Ord. There is a distributive law $\lambda : \mathrm{Dn}\,\mathrm{Fil} \to \mathrm{Fil}\,\mathrm{Dn}$ of the down-set monad

$$\mathbb{D}\mathrm{n} = (\mathrm{Dn}, \textstyle\bigvee_{\mathrm{Dn}}, \downarrow)$$

(Example II.4.9.3) over $\mathbb{F}\mathrm{il}$, whose components are given by

$$\lambda_X(\mathfrak{x}) = \{A \in \mathrm{Dn}\, X \mid \forall B \in \mathfrak{x}\ (A \cap B \neq \varnothing)\}$$

for all ordered sets X and $\mathfrak{x} \in \mathrm{Dn}\,\mathrm{Fil}\, X$ (Exercise IV.4.A). The resulting monad is the *down-set-filter monad*

$$\mathbb{F}\mathrm{il}\mathbb{D}\mathrm{n} = (\mathrm{Fil}\,\mathrm{Dn}, \textstyle\bigwedge_{\mathrm{Fil}\,\mathrm{Dn}} \cdot \mathrm{Fil} \bigvee_{\mathrm{Fil}\,\mathrm{Dn}}, \uparrow \circ \downarrow) .$$

IV.4.4.1 Proposition *The category* $\mathsf{Cnt}^{\mathrm{co}}$ *is strictly monadic over* Ord. *More precisely, there is a 2-isomorphism*

$$\mathsf{Cnt}^{\mathrm{co}} \cong \mathsf{Ord}^{\mathbb{F}\mathrm{il}\mathbb{D}\mathrm{n}}$$

which commutes with the forgetful functors to Ord.

Proof For a cocontinuous lattice X, the diagram

$$X \underset{\uparrow_X}{\overset{\bigwedge_X}{\underset{\longrightarrow}{\overset{\longleftarrow}{\bot}}}} \mathrm{Fil}\, X \underset{\mathrm{Fil}\downarrow_X}{\overset{\mathrm{Fil}\bigvee_X}{\underset{\longrightarrow}{\overset{\longleftarrow}{\bot}}}} \mathrm{Fil}\,\mathrm{Dn}\, X \qquad \text{(IV.4.4.i)}$$

motivates us to define a map $a := \bigwedge_X \cdot \mathrm{Fil}\bigvee_X$. Let us verify that a yields the structure morphism of an $\mathbb{Fil}\mathbb{Dn}$-algebra. The unit condition $a \cdot \mathrm{Fil}\downarrow_X \cdot \uparrow_X = 1_X$ is immediate, so we need to verify only the multiplication condition $a \cdot \mathrm{Fil}\,\mathrm{Dn} a = a \cdot \bigwedge_{\mathrm{Fil}\,\mathrm{Dn}\, X} \cdot \mathrm{Fil}\bigvee_{\mathrm{Fil}\,\mathrm{Dn}\, X}$:

$$
\begin{aligned}
&a \cdot \mathrm{Fil}\,\mathrm{Dn} a \\
&\quad = \textstyle\bigwedge_X \cdot \mathrm{Fil}(\bigvee_X \cdot \mathrm{Dn} a) \\
&\quad = \textstyle\bigwedge_X \cdot \mathrm{Fil}(a \cdot \bigvee_{\mathrm{Fil}\,\mathrm{Dn}\, X}) && (a \text{ left adjoint by (IV.4.4.i)}) \\
&\quad = \textstyle\bigwedge_X \cdot \bigwedge_{\mathrm{Fil}\, X} \cdot \mathrm{Fil}\mathrm{Fil}\bigvee_X \cdot \mathrm{Fil}\bigvee_{\mathrm{Fil}\,\mathrm{Dn}\, X} && (\textstyle\bigwedge_X \text{ preserves infima}) \\
&\quad = \textstyle\bigwedge_X \cdot \mathrm{Fil}\bigvee_X \cdot \bigwedge_{\mathrm{Fil}\,\mathrm{Dn}\, X} \cdot \mathrm{Fil}\bigvee_{\mathrm{Fil}\,\mathrm{Dn}\, X} && (\textstyle\bigwedge_{\mathrm{Fil}} \text{ natural transformation}).
\end{aligned}
$$

Consider now an $\mathbb{Fil}\mathbb{Dn}$-algebra $(X, a : \mathrm{Fil}\,\mathrm{Dn}\, X \to X)$. Since there is a monad morphism $\uparrow \mathrm{Dn} : \mathbb{Dn} \to \mathbb{Fil}\mathbb{Dn}$, the ordered set X is a $\mathbb{Dn}$-algebra, i.e. a complete lattice with supremum given by $\bigvee_X = a \cdot \uparrow_{\mathrm{Dn}\, X}$. The $\mathbb{Fil}\mathbb{Dn}$-algebra morphism a is a sup-map and a retraction, so it has a fully faithful right adjoint $a^{\dashv} : X \to \mathrm{Fil}\,\mathrm{Dn}\, X$, and we have the adjunctions

$$X \underset{a^{\dashv}}{\overset{a}{\underset{\longrightarrow}{\overset{\longleftarrow}{\bot}}}} \mathrm{Fil}\,\mathrm{Dn}\, X \underset{\mathrm{Fil}(a\cdot\uparrow_{\mathrm{Dn}\, X})}{\overset{\mathrm{Fil}\downarrow_X}{\underset{\longrightarrow}{\overset{\longleftarrow}{\top}}}} \mathrm{Fil}\, X\,.$$

Since the components of the ordered-filter monad multiplication are $\bigwedge_{\mathrm{Fil}\, X}$, we have in particular $\bigwedge_{\mathrm{Fil}\,\mathrm{Dn}\, X} \cdot \mathrm{Fil}\uparrow_{\mathrm{Dn}\, X} = 1_{\mathrm{Fil}\,\mathrm{Dn}\, X}$. By naturality of $\downarrow$ and the algebra multiplication condition,

$$a \cdot \mathrm{Fil}\downarrow_X \cdot \mathrm{Fil}(a \cdot \uparrow_{\mathrm{Dn}\, X}) = a \cdot \textstyle\bigwedge_{\mathrm{Fil}\,\mathrm{Dn}\, X} \cdot \mathrm{Fil}\bigvee_{\mathrm{Fil}\,\mathrm{Dn}\, X} \cdot \mathrm{Fil}\downarrow_{\mathrm{Fil}\,\mathrm{Dn}\, X} \cdot \mathrm{Fil}\uparrow_{\mathrm{Dn}\, X} = a\,.$$

One deduces that $a \cdot \mathrm{Fil}\downarrow_X$ admits $\mathrm{Fil}(a \cdot \uparrow_{\mathrm{Dn}\, X}) \cdot a^{\dashv}$ as a right adjoint. As the infimum operation on X (obtained via the monad morphism $\mathrm{Fil}\downarrow : \mathbb{Fil} \to \mathbb{Fil}\mathbb{Dn}$) is precisely $\bigwedge_X = a \cdot \mathrm{Fil}\downarrow_X$, the ordered set X is a cocontinuous lattice (and $a = \bigwedge_X \cdot \mathrm{Fil}\bigvee_X$ by the previous displayed identity).

Finally, a cocontinuous lattice morphism $f : X \to Y$ is also an $\mathbb{Fil}\mathbb{Dn}$-algebra morphism, and a morphism $f : (X, a) \to (Y, b)$ of $\mathbb{Fil}\mathbb{Dn}$-algebras naturally preserves both suprema and down-directed infima because it is both a $\mathbb{Dn}$-algebra and an $\mathbb{Fil}$-algebra morphism. □

The monad $\mathbb{Fil}\mathbb{Dn}$ on Ord is the monad $\mathbb{F}'$ on $\mathbb{P}$-Mon obtained in Section IV.4.2 from the filter monad $\mathbb{F}$ on Set. Indeed, consider the principal filter natural transformation $\tau : \mathbb{P} \to \mathbb{F}$ and a $\mathbb{P}$-monoid $(X, \downarrow_X)$ (Example IV.1.3.2(2)). The construction of Section IV.4.2 associates to this ordered set the topological space

(X, ν) whose neighborhood map is given at each $x \in X$ by the principal filter of $\downarrow_X x \in PX$:

$$\nu(x) = \uparrow_{PX}(\downarrow_X x) .$$

Hence, (X, ν) is the Alexandroff space associated with the ordered set X (Example II.5.10.5), and open sets are the down-closed sets. Thanks to Example IV.4.2.3(2), the set of $\nu^{\mathbb{F}}$-invariant filters can be identified with the set of filters on $\mathrm{Dn}\,X$, so that $\mathbb{F}' = \mathbb{F}\mathbb{il}\mathbb{D}\mathbb{n}$ is the down-set-filter monad on Ord. Proposition IV.4.4.1 and Theorem IV.4.3.2 now yield the isomorphism

$$\mathsf{Cnt}^{\mathrm{co}} \cong \mathsf{Set}^{\mathbb{F}} .$$

The functor $G^{\mathbb{F}'} : \mathsf{Set}^{\mathbb{F}} \to \mathbb{F}\text{-}\mathsf{Mon}$ of Section IV.4.2 sending a $\mathbb{T}$-algebra (X, a) to the $\mathbb{T}$-monoid $(X, a^{\dashv})$ therefore describes the functor $\mathsf{Cnt}^{\mathrm{co}} \to \mathsf{Top}$ that equips a cocontinuous lattice with its *Scott topology* and sends a filtered-inf-preserving sup-map to a continuous map (see also Exercise IV.4.D).

Moreover, separated $\mathbb{F}$-monoids are precisely T0-spaces, and separated $\mathbb{P}$-monoids are separated ordered sets (Examples IV.4.3.4), so Corollary IV.4.3.3 tells us that the forgetful functor

$$\mathsf{Cnt}^{\mathrm{co}} \to \mathsf{Ord}_{\mathrm{sep}}$$

is strictly monadic.

IV.4.5 Kock–Zöberlein monads on $\mathbb{T}$-Mon

The powerset monad $\mathbb{P}$ induces the Kock–Zöberlein monad $\mathbb{D}\mathbb{n}$ on Ord (Example IV.4.2.3(1)), and we will see that $\mathbb{F}$ yields such a monad on Top (Example IV.4.2.3(2)). Before proving in Theorem IV.4.5.3 that an arbitrary power-enriched monad $\mathbb{T}$ induces a Kock–Zöberlein monad $\mathbb{T}'$ on $\mathbb{T}$-Mon, as in the cases $\mathbb{T} = \mathbb{P}$ or $\mathbb{T} = \mathbb{F}$, we must first show that $\mathbb{T}$-Mon is an ordered category.

IV.4.5.1 Lemma *If* $\mathbb{T} = (T, m, e)$ *is a power-enriched monad, the initial order* (IV.4.3.i) *induced by the Kleisli monoid structures yields functors*

$$\mathbb{T}\text{-}\mathsf{Mon} \to \mathsf{Ord} \qquad \textit{and} \qquad \mathbb{T}\text{-}\mathsf{Mon}_{\mathrm{sep}} \to \mathsf{Ord}_{\mathrm{sep}}$$

that commute with the underlying-set functors. As a consequence, $\mathbb{T}$-Mon *is an ordered and* $\mathbb{T}$-$\mathsf{Mon}_{\mathrm{sep}}$ *a separated ordered category.*

Proof To see that morphisms in $\mathbb{T}$-Mon are monotone, consider $f : (X, \nu) \to (Y, \mu)$ and suppose that $x, y \in X$ are such that $x \le y$, or equivalently $\nu(x) \le \nu(y)$. Then

$$e_Y \cdot f(x) = Tf \cdot e_X(x) \le Tf \cdot \nu(x) \le Tf \cdot \nu(y) \le \mu \cdot f(y) ;$$

by composing with $\mu^{\mathbb{T}}$ on the left, one obtains $\mu \cdot f(x) \le \mu \cdot f(y)$, so that $f(x) \le f(y)$ by definition of the order on Y. Hence, $\mathbb{T}$-monoid morphisms are monotone, and we have the claimed functors into Ord and $\mathsf{Ord}_{\mathrm{sep}}$. Since hom-sets of $\mathbb{T}$-Mon are ordered pointwise, we have monotone maps

$$f \cdot (-) : \mathbb{T}\text{-}\mathsf{Mon}(Z, X) \to \mathbb{T}\text{-}\mathsf{Mon}(Z, Y) \qquad \text{and}$$
$$(-) \cdot g : \mathbb{T}\text{-}\mathsf{Mon}(X, Y) \to \mathbb{T}\text{-}\mathsf{Mon}(Z, Y)$$

for all $f \in \mathbb{T}\text{-}\mathsf{Mon}(X, Y)$ and $g \in \mathbb{T}\text{-}\mathsf{Mon}(Z, X)$, and $\mathbb{T}$-Mon is a separated ordered category. □

The initial order allows us to refine our analysis of the maps q_X and p_X defined in Section IV.4.2.

IV.4.5.2 Lemma *Let $\mathbb{T} = (T, m, e)$ be a power-enriched monad, let $\mathbb{T}'$ be its induced monad on $\mathbb{T}$-Mon (with $\mathbb{S} = \mathbb{T}$ and $\alpha = 1_{\mathbb{T}}$ in* IV.4.2*), and let (X, ν) be a $\mathbb{T}$-monoid. Then, with respect to the initial order on $T'X$ induced by ω_X, the section $q_X : (T'X, \omega_X) \to (TX, m_X^{\dashv})$ is an order-embedding and the retraction $p_X : (TX, m_X^{\dashv}) \to (T'X, \omega_X)$ is a $\mathbb{T}$-monoid morphism. In fact, $p_X \dashv q_X$.*

Proof By Lemmata IV.4.2.1 and IV.4.5.1, the maps q_X are monotone. Suppose that for a $\mathbb{T}$-monoid (X, ν) there are $\mathfrak{x}, \mathfrak{y} \in T'X$ with $q_X(\mathfrak{x}) \leq q_X(\mathfrak{y})$, or equivalently, such that $m_X^{\dashv} \cdot q_X(\mathfrak{x}) \leq m_X^{\dashv} \cdot q_X(\mathfrak{y})$. One can compose each side of this last inequality with the monotone map Tp_X on the left to obtain $\omega_X(\mathfrak{x}) \leq \omega_X(\mathfrak{y})$, i.e. $\mathfrak{x} \leq \mathfrak{y}$ by definition of the order on $T'X$.

Since $\mathbb{T}$ is power-enriched, $e_X \leq \nu$ so $1_{TX} = e_X^{\mathbb{T}} \leq \nu^{\mathbb{T}} = q_X \cdot p_X$ and

$$Tp_X \cdot m_X^{\dashv} \leq Tp_X \cdot m_X^{\dashv} \cdot q_X \cdot p_X = \omega_X \cdot p_X \, .$$

Hence, p_X is a $\mathbb{T}$-monoid morphism.

The last statement then follows from $1_{TX} \leq \nu^{\mathbb{T}} = q_X \cdot p_X$ and $p_X \cdot q_X = 1_{T'X}$. □

IV.4.5.3 Theorem *For a power-enriched monad $\mathbb{T}$, the derived monad $\mathbb{T}'$ on $\mathbb{T}$-Mon (or $\mathbb{T}$-$\mathsf{Mon}_{\mathrm{sep}}$) is of Kock–Zöberlein type.*

Proof Since $\mathbb{T}$ is power-enriched, one has for $\mathbb{T}$-monoid morphisms $f, g : (X, \nu) \to (Y, \mu)$

$$f \leq g \implies (e'_Y \cdot f)^{\mathbb{T}'} = p_Y \cdot (q_Y \cdot e'_Y \cdot f)^{\mathbb{T}} \cdot q_X \leq p_Y \cdot (q_Y \cdot e'_Y \cdot g)^{\mathbb{T}} \cdot q_X = (e'_Y \cdot g)^{\mathbb{T}'} ,$$

so T' is a 2-functor (since $(e'_Y \cdot f)^{\mathbb{T}'} = T'f$).

To verify that $m'_X \dashv e'_{T'X}$ for all $\mathbb{T}$-monoids (X, ν), it suffices to verify $1_{T'X} \leq e'_{T'X} \cdot m'_X$ (because $m'_X \cdot e'_{T'X} = 1_{T'X}$ always holds). In view of this, we write

$$\begin{aligned} q_{T'X} &\leq Tp_X \cdot m_X^{\dashv} \cdot m_X \cdot Tq_X \cdot q_{T'X} \\ &= Tp_X \cdot m_X^{\dashv} \cdot (q_X)^{\mathbb{T}} \cdot q_{T'X} \\ &= Tp_X \cdot m_X^{\dashv} \cdot (\nu^{\mathbb{T}} \cdot q_X)^{\mathbb{T}} \cdot q_{T'X} \\ &= Tp_X \cdot m_X^{\dashv} \cdot q_X \cdot p_X \cdot (q_X)^{\mathbb{T}} \cdot q_{T'X} \\ &= \omega_X \cdot m'_X = q_{T'X} \cdot e'_{T'X} \cdot m'_X \, . \end{aligned}$$

Hence, $1_{T'X} \le e'_{T'X} \cdot m'_X$ because $q_{T'X}$ is an order-embedding (Lemma IV.4.5.2). The same proof obviously holds if $\mathbb{T}$-Mon is replaced by $\mathbb{T}$-Mon$_{\text{sep}}$. □

IV.4.5.4 Examples

(1) If $\mathbb{T} = \mathbb{P}$, Example IV.4.2.3(1) shows that $\mathbb{P}' = \mathbb{D}\mathbb{n}$ is the down-set monad on Ord. Theorem IV.4.5.3 states that this monad is Kock–Zöberlein, and we recover Example II.4.9.3.

(2) The open-filter monad of Example IV.4.2.3(2) is the monad $\mathbb{F}'$ on $\mathbb{F}$-Mon $\cong$ Top derived from the filter monad $\mathbb{F}$, and is therefore Kock–Zöberlein.

IV.4.6 Eilenberg–Moore algebras and injective Kleisli monoids

We are now ready to look into the identification of injective objects that motivated this Section IV.4.

Let $U : \mathbb{T}\text{-Mon} \to \mathsf{Set}$ be the forgetful functor associated to a power-enriched monad $\mathbb{T}$. We denote by Ini the class Ini U of all U-initial $\mathbb{T}$-monoid morphisms, and by RegMono the class RegMono($\mathbb{T}$-Mon) of all U-initial $\mathbb{T}$-monoid morphisms whose underlying maps are monomorphisms (Exercise II.5.D).

In Theorem IV.4.6.3, we identify $\mathcal{M}$-injective Kleisli monoids when $\mathcal{M}$ = Ini or $\mathcal{M}$ = RegMono as $\mathbb{T}$-algebras (see Exercise II.5.M). If A is an ordered category, we denote by $\mathcal{M}$-Inj(A) the category of $\mathcal{M}$-injective A-objects with left adjoint A-morphisms.

When $\mathbb{T}$ is power-enriched, every map Tf is a morphism of $\mathbb{T}$-monoids $Tf : (TX, m_X^{\dashv}) \to (TY, m_Y^{\dashv})$. In fact, its right adjoint $(Tf)^{\dashv}$ is a right adjoint $(Tf)^{\dashv} : (TY, m_Y^{\dashv}) \to (TX, m_X^{\dashv})$ in $\mathbb{T}$-Mon. Indeed, $(-)^{\mathbb{T}}$ applied to $Tf \cdot (Tf)^{\dashv} \le 1_{TX}$ yields the inequality in

$$Tf \cdot m_X \cdot T(Tf)^{\dashv} = m_Y \cdot TTf \cdot T(Tf)^{\dashv} \le m_Y ,$$

so that $T(Tf)^{\dashv} \cdot m_Y^{\dashv} \le m_X^{\dashv} \cdot (Tf)^{\dashv}$ by adjunction.

For the associated monad $\mathbb{T}' = (T', m', e')$ on $\mathbb{T}$-Mon (induced as in Section IV.4.2 by the identity monad morphism $\alpha = 1_{\mathbb{T}} : \mathbb{T} \to \mathbb{T}$), if $f : (X, \nu) \to (Y, \mu)$ is a $\mathbb{T}$-monoid morphism, one can define the right adjoint $\mathbb{T}$-monoid morphism $(T'f)^{\dashv} : (T'Y, \omega_Y) \to (T'X, \omega_X)$ by

$$(T'f)^{\dashv} := p_X \cdot (Tf)^{\dashv} \cdot q_Y .$$

Indeed, $(T'f)^{\dashv}$ is a composite of $\mathbb{T}$-monoid morphisms (Lemma IV.4.5.2), and one verifies that

$$1_{T'X} \le (T'f)^{\dashv} \cdot T'f \qquad \text{and} \qquad T'f \cdot (T'f)^{\dashv} \le 1_{T'Y}$$

by using the identity $T'f = p_Y \cdot Tf \cdot q_X$.

IV.4.6.1 Lemma *If $\mathbb{T}$ is a power-enriched monad, the U-initial $\mathbb{T}$-monoid structure on X induced by $f : X \to U(Y, \mu)$ is*

$$\nu := (Tf)^{\dashv} \cdot \mu \cdot f .$$

As a consequence, $e'_X : (X, \nu) \to (T'X, \omega_X)$ is U-initial.

Proof The fact that $(Tf)^{\dashv} \cdot \mu \cdot f$ defines an initial $\mathbb{T}$-monoid morphism $f : (X, \nu) \to (Y, \mu)$ follows from straightforward verifications. The initial structure on X induced by $e'_X : X \to U(T'X, \omega_X)$ is therefore

$$(Te'_X)^{\dashv} \cdot \omega_X \cdot e'_X = (Tp_X \cdot T\nu)^{\dashv} \cdot Tp_X \cdot m_X^{\dashv} \cdot q_X \cdot p_X \cdot e_X = (T\nu)^{\dashv} \cdot m_X^{\dashv} \cdot \nu .$$

Furthermore, $(\nu^{\mathbb{T}})^{\dashv} = e_X^{\mathbb{T}} \cdot (\nu^{\mathbb{T}})^{\dashv} \leq \nu^{\mathbb{T}} \cdot (\nu^{\mathbb{T}})^{\dashv} \leq 1_{TX}$, so

$$\nu \leq (\nu^{\mathbb{T}})^{\dashv} \cdot \nu^{\mathbb{T}} \cdot \nu = (\nu^{\mathbb{T}})^{\dashv} \cdot \nu \leq \nu ,$$

i.e. $(T\nu)^{\dashv} \cdot m_X^{\dashv} \cdot \nu = \nu$, so the last claim is verified. □

IV.4.6.2 Lemma *Let $\mathbb{T}$ be a power-enriched monad. In the notations of Section IV.4.2 (with $\mathbb{S} = \mathbb{T}$ and $\alpha = 1_{\mathbb{T}}$), a morphism of $\mathbb{T}$-monoids $f : (X, \nu) \to (Y, \mu)$ satisfies*

$$\nu = (Tf)^{\dashv} \cdot \mu \cdot f \iff e'_X = (T'f)^{\dashv} \cdot e'_Y \cdot f .$$

Proof Suppose that $\nu = (Tf)^{\dashv} \cdot \mu \cdot f$ holds. Then

$$(T'f)^{\dashv} \cdot e'_Y \cdot f = p_X \cdot (Tf)^{\dashv} \cdot q_Y \cdot e'_Y \cdot f = p_X \cdot (Tf)^{\dashv} \cdot \mu \cdot f = p_X \cdot \nu = e'_X .$$

Conversely, if $e'_X = (T'f)^{\dashv} \cdot e'_Y \cdot f$, then, recalling that $T'f \cdot p_X = p_Y \cdot Tf$ by Lemma IV.4.3.1 and that $q_X = p_X^{\dashv}$ by Lemma IV.4.5.2, we can write

$$\begin{aligned} \nu = q_X \cdot e'_X &= p_X^{\dashv} \cdot (T'f)^{\dashv} \cdot e'_Y \cdot f = (p_Y \cdot Tf)^{\dashv} \cdot e'_Y \cdot f \\ &= (Tf)^{\dashv} \cdot q_Y \cdot e'_Y \cdot f = (Tf)^{\dashv} \cdot \mu \cdot f . \end{aligned}$$ □

IV.4.6.3 Theorem *If $\mathbb{T}$ is power-enriched, there is an isomorphism of categories*

$$\mathsf{Ini\text{-}Inj}(\mathbb{T}\text{-}\mathsf{Mon}) \cong (\mathbb{T}\text{-}\mathsf{Mon})^{\mathbb{T}'}$$

which commutes with the forgetful functors to $\mathbb{T}$-Mon. In particular, the Ini-*injective $\mathbb{T}$-monoids are precisely the $\mathbb{T}'$-algebras.*

Proof Consider a $\mathbb{T}$-monoid (X, ν) that is an Ini-injective object. By Lemma IV.4.6.1, the $\mathbb{T}$-monoid morphism $e'_X : (X, \nu) \to (T'X, \omega_X)$ is U-initial, so there is a $\mathbb{T}$-monoid morphism $a' : (T'X, \omega_X) \to (X, \nu)$ that extends $1_X : (X, \nu) \to (X, \nu)$ along e'_X:

$$\begin{array}{ccc} (X,\nu) & \xrightarrow{e'_X} & (T'X,\omega_X) \\ & {}_{1_X}\searrow & \vdots\, a' \\ & & (X,\nu)\,, \end{array} \qquad \text{(IV.4.6.i)}$$

i.e. such that $a' \cdot e'_X = 1_X$. Moreover, by using that $\mathbb{T}'$ is Kock–Zöberlein and e' a natural transformation, we get

$$1_{T'X} = T'a' \cdot T'e'_X \leq T'a' \cdot e'_{T'X} = e'_X \cdot a'\,,$$

so $a' \dashv e'_X$. Also, since $\mathbb{T}'$ is a Kock–Zöberlein monad, $a' \cdot T'a' \simeq a' \cdot m'_X$ (i.e. $a' \cdot T'a' \leq a' \cdot m'_X$ and $a' \cdot m'_X \leq a' \cdot T'a'$, see Section II.4.9); this equivalence is induced by the *separated* order on $T'X$, so it is actually an equality $a' \cdot T'a' = a' \cdot m'_X$. Hence, (X, a') is a $\mathbb{T}'$-algebra. Finally, let us return to the definition of a': by the adjunction $a' \dashv e'_X$, the morphism a' is determined up to equivalence in X; since $a' \cdot p_X \cdot \nu = 1_X$ implies that the order on X is separated, a' is really *uniquely* determined.

Suppose that $f : (X, \nu) \to (Y, \mu)$ is a morphism between Ini-injective $\mathbb{T}$-monoids that has a right adjoint $f^{\dashv} : (Y, \mu) \to (X, \nu)$ in $\mathbb{T}$-Mon. The previous construction yields $\mathbb{T}$-algebras (X, a') and (Y, b'), respectively. One then has

$$\begin{aligned} b' \cdot T'f &= b' \cdot p_Y \cdot Tf \cdot q_X && \\ &\leq b' \cdot p_Y \cdot Tf \cdot q_X \cdot e'_X \cdot a' && (a' \dashv e'_X) \\ &\leq b' \cdot p_Y \cdot \mu \cdot f \cdot a' && (Tf \cdot \nu \leq \mu \cdot f) \\ &= f \cdot a' && (b' \cdot p_Y \cdot \mu = b' \cdot e'_Y = 1_Y)\,, \end{aligned}$$

i.e. $b' \cdot T'f \leq f \cdot a'$. In the same way, one obtains $a' \cdot T'(f^{\dashv}) \leq f^{\dashv} \cdot b'$, which is equivalent to $f \cdot a' \leq b' \cdot T'f$, and we can conclude that $f : (X, a') \to (Y, b')$ is a morphism of $\mathbb{T}$-algebras.

Conversely, any $\mathbb{T}'$-algebra $((X, \nu), a')$ makes the diagram (IV.4.6.i) commute. Thus, given a U-initial morphism $j : (Y, \mu) \to (Z, \zeta)$, and a morphism $f : (Y, \mu) \to (X, \nu)$, one can define $\overline{f} := a' \cdot T'f \cdot (T'j)^{\dashv} \cdot e'_Z$:

$$\begin{array}{ccccc} (T'Y,\omega_X) & \xleftarrow{(T'j)^{\dashv}} & (T'Z,\omega_Z) & \xleftarrow{e'_Z} & (Z,\zeta) \\ & {}_{T'f}\searrow & & & \vdots\,\overline{f} \\ & & (T'X,\omega_X) & \xrightarrow[a']{} & (X,\nu)\,. \end{array}$$

The morphism $\overline{f} : (Z, \zeta) \to (X, \nu)$ does indeed extend f along j, since by Lemmata IV.4.6.2 and IV.4.6.1 one has

$$\overline{f} \cdot j = a' \cdot T'f \cdot (T'j)^{\dashv} \cdot e'_Z \cdot j = a' \cdot T'f \cdot e'_Y = a' \cdot e'_X \cdot f = f\,.$$

This proves that (X, ν) is an Ini-injective object of $\mathbb{T}$-Mon.

If $f : ((X, \nu), a') \to ((Y, \mu), b')$ is a morphism of $\mathbb{T}'$-algebras, then the $\mathbb{T}$-monoid morphism $f^{\dashv} : (Y, \mu) \to (X, \nu)$ defined by

$$f^{\dashv} = a' \cdot p_X \cdot (Tf)^{\dashv} \cdot \mu$$

is right adjoint to f. Indeed, one readily verifies the inequalities

$$1_X \leq f^{\dashv} \cdot f \qquad \text{and} \qquad f \cdot f^{\dashv} \leq 1_Y \,.$$

The passages from Ini-injective Kleisli monoids to Eilenberg–Moore algebras and back described above obviously define functors that commute with the forgetful functors to $\mathbb{T}$-Mon and that are inverse to each other. □

IV.4.6.4 Corollary *If $\mathbb{T}$ is power-enriched, there is an isomorphism between the category of* RegMono*-injective separated $\mathbb{T}$-monoids (with left adjoint morphisms) and the category of $\mathbb{T}'$-algebras (with $\mathbb{T}'$ seen as a monad on $\mathbb{T}$-$\mathsf{Mon}_{\mathrm{sep}}$):*

$$\text{RegMono-}\mathsf{Inj}(\mathbb{T}\text{-}\mathsf{Mon}_{\mathrm{sep}}) \cong (\mathbb{T}\text{-}\mathsf{Mon}_{\mathrm{sep}})^{\mathbb{T}'} \,.$$

Moreover, the functors forming the isomorphism commute with the forgetful functors to $\mathbb{T}$-Mon.

Proof The proof that an Ini-injective object is an Eilenberg–Moore algebra in Theorem IV.4.6.3 relies on the fact that the structure morphism $\nu : X \to TX$ of a Kleisli monoid (X, ν) belongs to the class of U-initial morphisms. By definition, the structure morphisms of *separated* Kleisli monoids have monic underlying maps, so the same proof yields the stated isomorphism. □

IV.4.6.5 Corollary *Given an associative lax extension $\hat{\mathbb{T}}$ of $\mathbb{T}$ to $\mathcal{V}$-Rel, there is an isomorphism*

$$\mathsf{Ini}\text{-}\mathsf{Inj}((\mathbb{T}, \mathcal{V})\text{-}\mathsf{Cat}) \cong (\mathbb{\Pi}\text{-}\mathsf{Mon})^{\mathbb{\Pi}'} \,.$$

In particular, the Ini*-injective $(\mathbb{T}, \mathcal{V})$-categories are precisely the $\mathbb{\Pi}'$-algebras (where $\mathbb{\Pi}$ is the discrete presheaf monad associated to $\hat{\mathbb{T}}$).*

Proof By Proposition IV.3.2.1, the discrete presheaf monad associated to $\hat{\mathbb{T}}$ is power-enriched. The result then follows directly from the isomorphisms of Theorems IV.4.6.3 and IV.3.2.2. □

The following direct consequences of Theorem IV.4.6.3 and its corollaries allow us in particular to identify the injective objects of $\mathsf{Ord}_{\mathrm{sep}}$ and Top_0.

IV.4.6.6 Examples

(1) Since $\mathsf{Set}^{\mathbb{P}} \cong \mathsf{Sup}$ and $\mathbb{P}$-$\mathsf{Mon} \cong \mathsf{Ord}$, complete lattices are both the Ini-injective objects in Ord and the RegMono-injective objects in $\mathsf{Ord}_{\mathrm{sep}}$ (in this case, the regular monomorphisms are the order-embeddings).

(2) Cocontinuous lattices can equivalently be seen as the Ini-injective objects in Top or the RegMono-injective objects in Top_0 via the isomorphisms $\mathsf{Set}^{\mathbb{F}} \cong \mathsf{Cnt}^{\mathrm{co}}$, and $\mathbb{F}$-Mon $\cong$ Top or $\mathbb{F}\text{-}\mathsf{Mon}_{\mathrm{sep}} \cong \mathsf{Top}_0$, see Exercise IV.4.B (here, regular monomorphisms are the topological embeddings).
(3) Completely distributive lattices are the Ini-injective objects of the category Cls of closure spaces (Exercise IV.4.E), and frames are the Ini-injective objects of the category $\mathsf{Cls}_{\mathrm{fin}}$ of finitary closure spaces (Exercise IV.4.F).
(4) For a given quantale $\mathcal{V}$, the Ini-injective objects of $\mathcal{V}$-Cat are the $\mathcal{V}$-actions in Sup (Exercise IV.4.G).
(5) Quantales are the Ini-injective objects of the category MultiOrd $\cong$ $(\mathbb{L}, 2)$-Cat of *multi-ordered sets* and their morphisms (see Section V.1.4 and Exercise IV.4.H).

Exercises

IV.4.A *A distributive law for the down-set-filter monad.* The down-set-filter monad $\mathbb{F}\mathrm{il}\mathbb{D}\mathrm{n}$ is a composite monad obtained from a distributive law of $\mathbb{D}\mathrm{n}$ over $\mathbb{F}\mathrm{il}$ (see Section IV.4.4). The monad $\mathbb{D}\mathrm{n}$ is of Kock–Zöberlein type and $\mathbb{F}\mathrm{il}$ is of dual Kock–Zöberlein type, but $\mathbb{F}\mathrm{il}\mathbb{D}\mathrm{n}$ is neither. Moreover, if (X, a) is a $\mathbb{D}\mathrm{n}$-algebra, then the order on X is necessarily separated, so the 2-isomorphisms of Example II.4.9.3 become

$$\mathsf{Ord}^{\mathbb{D}\mathrm{n}} \cong \mathsf{Sup} \qquad \text{and} \qquad \mathsf{Ord}^{\mathbb{U}\mathrm{p}} \cong \mathsf{Inf}\,.$$

Hence, $\mathsf{Ord}^{\mathbb{F}\mathrm{il}}$ is 2-isomorphic to the category of ordered sets in which every down-directed set has an infima, with maps preserving down-directed infima.

IV.4.B *T0-spaces.* Via the isomorphism $\mathbb{F}$-Mon $\cong$ Top of Proposition IV.1.1.1, the category $\mathbb{F}\text{-}\mathsf{Mon}_{\mathrm{sep}}$ of separated $\mathbb{F}$-monoids is isomorphic to the category Top_0 of T0-spaces.

IV.4.C *Separated representable Kleisli monoids.* Let $\mathbb{T} = (T, m, e)$ be a power-enriched monad with its Kleisli extension $\check{T}$ to Rel. It follows from Section III.5.1 that $\check{T}$ induces a monad $\mathbb{T} = (T, m, e)$ on Ord. Show that:

(1) the functor $T : \mathsf{Set} \to \mathsf{Set}$ sends surjections to split epimorphisms;
(2) the full subcategory $(\mathsf{Ord}^{\mathbb{T}})_{\mathrm{sep}}$ of $\mathsf{Ord}^{\mathbb{T}}$ spanned by all separated ordered sets is reflective;
(3) the composite functor $(\mathsf{Ord}^{\mathbb{T}})_{\mathrm{sep}} \hookrightarrow \mathsf{Ord}^{\mathbb{T}} \xrightarrow{K} (\mathbb{T}, 2)\text{-}\mathsf{Cat}$ (see Section III.5.3) is strictly monadic;
(4) there is an isomorphism $\mathsf{Set}^{\mathbb{T}} \cong (\mathsf{Ord}^{\mathbb{T}})_{\mathrm{sep}}$ that commutes with the underlying-set functors;
(5) the functor $\mathsf{Set}^{\mathbb{T}} \to \mathbb{T}$-Mon of Section IV.4.2 factors as

$$\mathsf{Set}^{\mathbb{T}} \cong (\mathsf{Ord}^{\mathbb{T}})_{\mathrm{sep}} \to \mathsf{Ord}^{\mathbb{T}} \to (\mathbb{T}, 2)\text{-}\mathsf{Cat} \cong \mathbb{T}\text{-}\mathsf{Mon}\,.$$

IV.4.D *Open sets and continuity in the Scott topology.* The open sets of the Scott topology on a cocontinuous lattice X are those down-sets $O \in \mathrm{Dn}\, X$ such that

$$\forall A \in \mathrm{Fil}\, X \; \left(\bigwedge A \in O \implies A \cap O \neq \varnothing\right) .$$

If Y is another cocontinuous lattice, then a map $f : X \to Y$ is continuous if and only if it preserves down-directed infima.

IV.4.E *Completely distributive lattices as Eilenberg–Moore algebras.* The Eilenberg–Moore category of the up-set monad $\mathbb{U}$ on Set is the category Dst of completely distributive lattices and maps that preserve all infima and suprema:

$$\mathsf{Set}^{\mathbb{U}} \cong \mathsf{Dst} .$$

The Ini-injective objects of the category Cls of closure spaces are precisely the completely distributive lattices (use Example IV.1.3.2(4)).

IV.4.F *Frames as Eilenberg–Moore algebras.* The Eilenberg–Moore category of the finitary-up-set monad of Exercise IV.1.E is the category frames and frame homomorphisms:

$$\mathsf{Set}^{\mathbb{U}_{\mathrm{fin}}} \cong \mathsf{Frm} .$$

The Ini-injective objects of the category $\mathsf{Cls}_{\mathrm{fin}}$ of finitary closure spaces are precisely frames (use Example IV.1.E).

IV.4.G *Categories associated to the $\mathcal{V}$-powerset monad.* The $\mathcal{V}$-powerset monad (Exercise III.1.D) is power-enriched, and

$$\mathbb{P}_{\mathcal{V}}\text{-}\mathsf{Mon} \cong \mathcal{V}\text{-}\mathsf{Cat} \qquad \text{and} \qquad \mathsf{Set}^{\mathbb{P}_{\mathcal{V}}} \cong \mathsf{Sup}^{\mathcal{V}}$$

(see Example II.4.3.1(3)). Hence, the Ini-injective objects of $\mathcal{V}\text{-}\mathsf{Cat}$ are the $\mathcal{V}$-actions in Sup.

IV.4.H *Quantales* There is a distributive law $\delta : LP \to PL$ of the list monad $\mathbb{L}$ over the powerset monad $\mathbb{P}$ given by

$$\delta(A_1, \ldots, A_n) = \{(a_1, \ldots, a_n) \mid a_i \in A_i\} .$$

The resulting composite monad $\mathbb{P}\mathbb{L}$ is power-enriched, and is induced by the monadic composition of the monadic forgetful functors $\mathsf{Qnt} \to \mathsf{Mon}$ and $\mathsf{Mon} \to \mathsf{Set}$ (see Exercise II.2.T and Example II.3.2.2(1)). Show that $\mathbb{P}\mathbb{L}$ is isomorphic to the discrete presheaf monad of the list monad, so that

$$\mathbb{P}\mathbb{L}\text{-}\mathsf{Mon} \cong (\mathbb{L}, 2)\text{-}\mathsf{Cat}$$

(see Exercise IV.3.C).

IV.5 Domains as lax algebras and Kleisli monoids

It is shown in Section IV.4.4 that the cocontinuous lattices are the Eilenberg–Moore algebras of the filter monad $\mathbb{F}$ on Set, whereas in Corollary IV.1.5.4 we saw that the category $(\mathbb{F}, 2)\text{-}\mathsf{Cat}$ of *lax* algebras of the Kleisli extension of the filter monad is isomorphic to the category Top of all topological spaces. We saw in Section IV.4.4 that there is a monadic functor

$$\mathsf{Cnt}^{\mathrm{co}} \cong \mathsf{Set}^{\mathbb{F}} \longrightarrow \mathbb{F}\text{-}\mathsf{Mon} \cong (\mathbb{F}, 2)\text{-}\mathsf{Cat} \cong \mathsf{Top}$$

which associates to each cocontinuous lattice its Scott topology (see Exercise IV.4.D). We shall see that the lax algebra thus associated with each cocontinuous lattice is *strict*, meaning that it satisfies the lax multiplicative Eilenberg–Moore law *strictly*. Further, we show in Theorem IV.5.4.1 that the strict lax algebras of $\mathbb{F}$ constitute a class of topological spaces that not only generalize the continuous lattices, but in fact also provide a close topological generalization of the broader class of *continuous dcpos* (see Section IV.5.2). The spaces in question – which we call the *observable realization spaces* – capture an essential domain-theoretic approximation property in topological terms, and the continuous dcpos occur among these spaces as precisely the *sober* observable realization spaces. In Section IV.5.9 we will characterize those strict lax algebras that are associated to continuous lattices.

In Theorem IV.5.7.2 we also show that the observable realization spaces are Kleisli monoids of the *ordered-filter monad* $\mathbb{F}\mathrm{il}$ on Ord.

IV.5.1 Modules and adjunctions

Recall from Section II.2.2 that the category Ord of ordered sets and monotone maps is manifested within the category Mod of ordered sets and modules as their morphisms via functors

$$(-)_* : \mathsf{Ord} \longrightarrow \mathsf{Mod} \quad \text{and} \quad (-)^* : \mathsf{Ord}^{\mathrm{op}} \longrightarrow \mathsf{Mod}$$

which send a monotone map $f : X \to Y$ to the modules

$$f_* = (\leq_Y) \cdot f : X \nrightarrow Y \qquad \text{and} \qquad f^* = f^\circ \cdot (\leq_Y) : Y \nrightarrow X\,,$$

respectively, where $f^\circ : Y \nrightarrow X$ is the converse relation of f (given by $y\ f^\circ\ x \iff y = f(x)$). Endowing the hom-sets $\mathsf{Ord}(X, Y)$ and $\mathsf{Mod}(X, Y)$ with the pointwise and inclusion orders, respectively, we find that, for monotone maps $f, g : X \to Y$,

$$f_* \geq g_* \iff f \leq g \iff f^* \leq g^*\,,$$

so that we have monotone maps

$$(-)_* : \mathsf{Ord}(X, Y)^{\mathrm{op}} \to \mathsf{Mod}(X, Y) \quad \text{and} \quad (-)^* : \mathsf{Ord}(X, Y) \to \mathsf{Mod}(Y, X)$$

that are fully faithful (when viewed as functors).

We recall (see Section II.1.4) that, for monotone maps $f : X \to Y, g : Y \to X$, one has

$$f \dashv g \iff \forall x \in X, y \in Y\ (f(x) \le y \iff x \le g(y)) \iff f_* = g^* .$$

As a quantaloid, Mod is in particular an ordered category and hence a 2-category. For any monotone map $f : X \to Y$ we have

$$1_X^* \le f^* \cdot f_* \quad \text{and} \quad f_* \cdot f^* \le 1_Y^* ,$$

so that f_* is left adjoint to f^* in the 2-category Mod; in symbols, $f_* \dashv f^*$.

IV.5.2 Cocontinuous ordered sets

Section IV.4.4 defines a cocontinuous lattice X as a complete lattice for which the infimum map $\bigwedge_X : \mathrm{Fil}\,X \to X$ has a right adjoint $\Uparrow_X$; if such a right adjoint exists, it is given by

$$x \in \Uparrow a \iff \forall S \in \mathrm{Fil}\,X\ \left(\bigwedge S \le a \implies x \in S\right) . \qquad \text{(IV.5.2.i)}$$

Considering an arbitrary ordered set X instead of a complete lattice, let us define a relation $(\lll) : X \nrightarrow X$ by

$$a \lll x \iff \forall S \in \mathrm{Fil}_\wedge X\ \left(\bigwedge S \le a \implies x \in S\right) ,$$

where $\mathrm{Fil}_\wedge X$ is the set of all filters S in X for which a meet $\bigwedge S$ exists. Letting $\Uparrow x := \{y \in X \mid x \lll y\}$ for each $x \in X$, we say that the ordered set X is *cocontinuous* if

$$\Uparrow x \in \mathrm{Fil}_\wedge X \qquad \text{and} \qquad \bigwedge \Uparrow x \cong x \qquad \text{(IV.5.2.ii)}$$

for each $x \in X$. For example, if X is a *co-dcpo*, i.e. if X is separated and $\mathrm{Fil}_\wedge X = \mathrm{Fil}\,X$, then the infimum map $\bigwedge_X : \mathrm{Fil}\,X \to X$ has a right adjoint $\Uparrow_X$ if and only if X is cocontinuous; if this is the case, the right adjoint is given by $\Uparrow_X x = \Uparrow x$ and we say that X is a *cocontinuous co-dcpo*. The dual notion, that of a *continuous dcpo*, is more often employed. The term *dcpo* was originally introduced as an acronym for *directed-complete partial order*.

For any ordered set X, we may define the *Scott topology* on X as the collection of all *Scott-open* sets, i.e. of all down-sets $U \in \mathrm{Dn}\,X$ such that

$$\forall S \in \mathrm{Fil}_\wedge X\ \left(\bigwedge S \in U \implies \exists x \in S \cap U\right) .$$

As in Exercise IV.4.D with regard to cocontinuous lattices, one may readily verify that a function $f : X \to Y$ between ordered sets is continuous with respect to the Scott topologies on X and Y if and only if f preserves meets of down-directed subsets. Furthermore, the Scott topology on an ordered set X is *order-compatible*, in the sense that the underlying order associated to the Scott topology of an ordered set X coincides with the order relation possessed by X. In a topological space X endowed with its underlying order, we say that a point x *specializes* a point y if $x \leq y$.

The nature of the Scott topology is related to the preservation of down-directed meets by the neighborhood-filter map $\nu : X \to FX$ of a topological space X endowed with its underlying order. For any down-directed subset $S \subseteq X$ with a meet in X, the image $\nu(S)$ is down-directed in FX by monotonicity, and directed meets in FX are given by union, so to say that ν preserves the meet $\bigwedge S$ means that

$$\nu\left(\bigwedge S\right) = \bigcup_{z \in S} \nu(z) .$$

IV.5.2.1 Definition Let X be a topological space endowed with its underlying order, and let $S \subseteq X$ be a down-directed subset of X with a meet $\bigwedge S$ in X. We say that $\bigwedge S$ is a *topological meet* of S in X if the neighborhood-filter map $\nu : X \to FX$ preserves this meet.

Given an ordered set X, the Scott topology on X is the largest order-compatible topology whose neighborhood filter map ν preserves all directed meets, as follows:

IV.5.2.2 Proposition *Let X be a topological space endowed with its underlying order. The following are equivalent:*

(i) *$\mathcal{O}X$ is coarser than the Scott topology on X; i.e. every open set is Scott-open;*
(ii) *every directed meet in X is a topological meet.*

Proof This is Exercise IV.5.A. □

We shall also require the following lemma.

IV.5.2.3 Lemma *Let X be a cocontinuous ordered set. Then for any $y \in X$ the set*

$$\downarrow\!\!\downarrow y := \{x \in X \mid x \lll y\}$$

is Scott-open.

Proof Suppose that $S \in \mathrm{Fil}_{\bigwedge} X$ and $\bigwedge S \in \downarrow\!\!\downarrow y$. Observe that

$$\bigwedge S = \bigwedge\left\{\bigwedge \uparrow\!\!\uparrow x \mid x \in S\right\} = \bigwedge\left(\bigcup\left\{\uparrow\!\!\uparrow x \mid x \in S\right\}\right) .$$

Since $\bigcup\{\uparrow\!\!\uparrow x \mid x \in S\}$ is the union of a down-directed subset of $\mathrm{Fil}\, X$, it is an element of $\mathrm{Fil}\, X$. The result follows. □

One can provide an alternative characterization of the property of cocontinuity of an ordered set with reference to the Scott topology. For any topological space X with its underlying order and with neighborhood-filter map $\nu : X \to FX$, we may define a relation $(\prec) : X \nrightarrow X$ by

$$x \prec y \iff \downarrow y \in \nu(x)\,,$$

so in particular the set

$$\rotatebox{0}{$\updownarrow$}\, y := \{x \in X \mid x \prec y\}$$

is open for all $y \in X$. Since $\downarrow y$ is the set of points that specialize y, the statement that $x \prec y$ for points $x, y \in X$ is in general stronger than the statement that x specializes y, and we say that x *observably specializes* y if $x \prec y$. We refer to $(\prec)$ as the *observable specialization relation* associated to X.

In the case that X is a cocontinuous ordered set under the Scott topology, we find that $(\prec) = (\lll)$ as follows. For an ordered set X and a relation $r : X \nrightarrow X$, we say that r is *approximating* if, for each $x \in X$, $r(x)$ is a filter in X and x is a meet of $r(x)$. For example, X is cocontinuous if and only if $(\lll)$ is approximating.

IV.5.2.4 Lemma *For an ordered set X endowed with the Scott topology, if either* $(\lll)$ *or* $(\prec)$ *is approximating, then* $(\lll) = (\prec)$.

Proof See Exercise IV.5.B. □

IV.5.2.5 Proposition *For an ordered set X endowed with the Scott topology, the following are equivalent:*

(i) *X is cocontinuous;*
(ii) *the relation* $(\prec)$ *is approximating;*
(iii) *each* $x \in X$ *is a topological meet of* $\updownarrow x := \{y \in X \mid x \prec y\}$.

Proof This is Exercise IV.5.C. □

IV.5.3 Observable realization spaces

The third characterization of continuity in ordered sets in Proposition IV.5.2.5 affords an interesting generalization to arbitrary topological spaces:

IV.5.3.1 Proposition *Let X be a topological space endowed with its underlying order. There is an embedding of ordered sets* $\iota : \mathrm{Fil}\,X \to FX$ *given by* $S \mapsto {\uparrow_{PX}}\{\downarrow y \mid y \in S\}$, *and the following are equivalent:*

(i) *each* $x \in X$ *is a topological meet of* $\updownarrow x$;
(ii) *there is a map* $n : X \to \mathrm{Fil}\,X$ *such that the neighborhood-filter map* $\nu : X \to FX$ *factors as*

$$X \xrightarrow{\ n\ } \mathrm{Fil}\,X \overset{\iota}{\hookrightarrow} FX\,;$$

(iii) *there is a module* $c : \mathrm{Fil}\,X \nrightarrow X$ *such that the convergence relation* $a : FX \nrightarrow X$ *of X factors as*

$$FX \xrightarrow{\ \iota^*\ } \mathrm{Fil}\,X \xrightarrow{\ c\ } X\,;$$

(iv) *for all* $U \in \mathcal{O}X$ *and* $x \in U$, *there exists* $y \in U$ *with* $x \prec y$.

If these conditions hold, the map n of (ii) and the relation c of (iii) are uniquely determined as

$$n = \Uparrow : X \to \mathrm{Fil}X \quad \textit{and} \quad c = a \cdot \iota = \Uparrow^* : \mathrm{Fil}X \nrightarrow X$$

with $S \xrightarrow{c} x \iff \iota(S) \supseteq \nu(x) \iff S \supseteq \Uparrow x$, *and* $\Uparrow : X \to \mathrm{Fil}X$ *is a fully faithful monotone map.*

Proof It is readily verified that ι as given is well defined, injective, monotone, and fully faithful.

Suppose (iv) holds. Then for any $x \in X$ we have that

$$\bigcup_{y \in \Uparrow x} \nu(y) \supseteq \nu(x) ,$$

and since $\Uparrow x \subseteq \uparrow x$ this inclusion is in fact an equality, from which it follows that x is a meet of $\Uparrow x$. Further, $\Uparrow x$ is down-directed, since if $y, z \in \Uparrow x$ then the sets

$$\Downarrow y = \{t \in X \mid t \prec y\} \qquad \text{and} \qquad \Downarrow z = \{t \in X \mid t \prec z\}$$

are open neighborhoods of x, and hence by (iv) there is $u \in \Uparrow x$ with $u \in \Downarrow y \cap \Downarrow z \subseteq \downarrow y \cap \downarrow z$ as needed. Thus, x is a topological meet of $\Uparrow x$, showing that (i) holds.

Suppose (i) holds. Then for any $x \in X$

$$\nu(x) = \bigcup_{y \in \Uparrow x} \nu(y) ,$$

and hence (iv) holds. We find also that (ii) holds, as follows. The preceding equation implies that $\nu(x) \subseteq \iota(\Uparrow x)$ since open sets are down-closed. This inclusion may be replaced by an equality, since for any $y \in \Uparrow x$ we have that $\nu(x) \ni \downarrow y$, so we find that ν factors through the map $\Uparrow : X \to \mathrm{Fil}X \ x \mapsto \Uparrow x$ as $\nu = \iota \cdot \Uparrow$.

If there is a map n as in (ii), then for any $x \in X$ we find that

$$n(x) \ni y \iff \iota(n(x)) \ni \downarrow y \iff \nu(x) \ni \downarrow y \iff x \prec y$$

for each $y \in X$, whence $n(x) = \Uparrow x$, and we find also that (iv) holds. Further, since ν and ι are monotone and fully faithful and $\iota \cdot n = \nu$, it follows that n is monotone and fully faithful. Letting $c := n^*$, we have that

$$a = \nu^* = (\iota \cdot n)^* = n^* \cdot \iota^* = c \cdot \iota^* ,$$

so (iii) holds.

Suppose we have a relation c as in (iii). Since ι is fully faithful, we have that $\iota^* \cdot \iota_* = 1^*_{\mathrm{Fil}X}$, and hence

$$a \cdot \iota = a \cdot \iota_* = c \cdot \iota^* \cdot \iota_* = c .$$

Further,

$$1_X \leq 1^*_X \leq \nu^* \cdot \nu_* = a \cdot \nu_* = c \cdot \iota^* \cdot \nu_* = a \cdot \iota_* \cdot \iota^* \cdot \nu_* = \nu^* \cdot \iota_* \cdot \iota^* \cdot \nu_* = \nu^* \cdot \iota \cdot \iota^\circ \cdot \nu_* ,$$

so for each $x \in X$ there is $S \in \mathrm{Fil}\,X$ with $\nu(x) \supseteq \iota(S) \supseteq \nu(x)$, whence $\nu(x) = \iota(S)$, and one verifies that $S = \updownarrow x$ and hence $\nu(x) = \iota(\updownarrow x)$. Therefore, (ii) holds with $n := \updownarrow$. □

IV.5.3.2 Definition For a topological space with convergence relation $a : FX \nrightarrow X$, we find that the associated relation $a \cdot \iota : \mathrm{Fil}\,X \nrightarrow X$ of Proposition IV.5.3.1 is such that $S \xrightarrow{a \cdot \iota} x$ if and only if S enters every open neighborhood of x, and if this is the case then we say that S *observably realizes* x. Note that S observably realizes x if and only if x is in the closure of S.

If the equivalent conditions of Proposition IV.5.3.1 hold for a topological space X, we say that X is an *observable realization space*.

IV.5.3.3 Example Every cocontinuous ordered set is an observable realization space when endowed with the Scott topology. In fact, we show in Section IV.5.8 that the cocontinuous co-dcpos under the Scott topology are precisely the *sober* observable realization spaces. For a cocontinuous co-dcpo, the map $\updownarrow : X \to \mathrm{Fil}\,X$ is given by $x \mapsto \Uparrow x$, and the *observable realization relation* $c = a \cdot \iota : \mathrm{Fil}\,X \nrightarrow X$ is given by

$$S \xrightarrow{a \cdot \iota} x \iff \bigwedge S \leq x\,;$$

the associated filter convergence relation $a = c \cdot \iota^*$ is called the relation of *Scott convergence*.

Thus, when we express the domain-theoretic *approximation property* of continuity in topological rather than order-theoretic terms, relative to an arbitrary topology, we obtain a characterization of the observable realization spaces (Proposition IV.5.3.1(i)). Moreover, we find that these spaces can also be characterized via condition IV.5.3.1(iii) as those in which filter convergence reduces to a relation of convergence of directed sets that captures the essential topological character of the notion of *convergence by directed meets* supported by cocontinuous co-dcpos. Thus, these spaces provide our general notion of a *domain*, such that in Section IV.5.8 we shall see that the continuous dcpos are those domains that are *sober*.

The following lemma entails in particular that every observable realization space carries a topology finer than the Scott topology on its specialization order (i.e. a topology in which every Scott-open subset is open). The proof is straightforward.

IV.5.3.4 Lemma *Suppose X is a topological space with its underlying order such that $(\prec)$ is approximating. Then $\mathcal{O}X$ is finer than the Scott topology on X; i.e. every Scott-open subset is open.*

IV.5.4 Observable realization spaces as lax algebras

We now observe that the observable realization spaces may be characterized among arbitrary topological spaces as those whose corresponding $(\mathbb{F}, 2)$-category (with respect to the Kleisli extension $\check{\mathbb{F}}$, see Section IV.1.4) satisfies the lax Eilenberg–Moore multiplicative law up to equality rather than simply an inequality, as follows:

IV.5.4.1 Theorem *Let X be a topological space with associated $(\mathbb{F}, 2)$-category (X, a) (where $a : FX \nrightarrow X$ is the filter convergence relation associated to X). Then*

$$a \cdot m_X = a \cdot \check{F}a \iff X \text{ is an observable realization space.}$$

Thus, the isomorphism $(\mathbb{F}, 2)\text{-}\mathsf{Cat} \cong \mathsf{Top}$ of Corollary IV.1.5.4 restricts to an isomorphism

$$(\mathbb{F}, 2)\text{-}\mathsf{Cat}_{=} \cong \mathsf{ObsReSp}$$

between the full subcategory $(\mathbb{F}, 2)\text{-}\mathsf{Cat}_{=}$ of $(\mathbb{F}, 2)\text{-}\mathsf{Cat}$, consisting of those lax algebras of $\mathbb{F}$ that satisfy the multiplicative law up to equality, and the full subcategory $\mathsf{ObsReSp}$ of Top with objects all observable realization spaces.

Proof Since $\mathbb{F}$ is a power-enriched monad, we know by Proposition IV.1.2.1 that $m_X : FFX \to FX$ is a sup-map and hence has a right adjoint $m_X^{\dashv} : FX \to FFX$ given by

$$m_X^{\dashv}(\mathfrak{x}) = \uparrow_{PFX}\{A^{\mathbb{F}} \mid A \in \mathfrak{x}\},$$

for all $\mathfrak{x} \in FX$ (where $A^{\mathbb{F}} = \{\mathfrak{y} \in FX \mid A \in \mathfrak{y}\}$). Also, with reference to (IV.1.4.i), there is a monotone map $a^{\tau} : FX \to FFX$, given by $a^{\tau}(\mathfrak{y}) = \uparrow_{PFX}\{a^{\circ}(A) \mid A \in \mathfrak{y}\}$, such that $\check{F}a = (a^{\tau})^*$. Note also that $a = \nu^*$, where $\nu : X \to FX$ is the neighborhood filter map of X. Letting $\mathcal{O}_X(x)$ denote the set of open neighborhoods of a point $x \in X$, we may reason as follows:

$$\begin{aligned}
& a \cdot m_X = a \cdot \check{F}a \\
\iff\ & a \cdot m_X \le a \cdot \check{F}a && ((X, a) \in (\mathbb{F}, 2)\text{-}\mathsf{Cat}) \\
\iff\ & a \cdot (m_X)_* \le a \cdot \check{F}a \\
\iff\ & \nu^* \cdot (m_X)_* \le \nu^* \cdot \check{F}a \\
\iff\ & \nu^* \cdot (m_X^{\dashv})^* \le \nu^* \cdot \check{F}a && (m_X \dashv m_X^{\dashv}) \\
\iff\ & \nu^* \cdot (m_X^{\dashv})^* \le \nu^* \cdot (a^{\tau})^* \\
\iff\ & (m_X^{\dashv} \cdot \nu)^* \le (a^{\tau} \cdot \nu)^* \\
\iff\ & m_X^{\dashv} \cdot \nu \le a^{\tau} \cdot \nu \\
\iff\ & \forall x \in X \left(\uparrow\{A^{\mathbb{F}} \mid A \in \nu(x)\} \supseteq \uparrow\{a^{\circ}(A) \mid A \in \nu(x)\}\right) \\
\iff\ & \forall x \in X \left(\uparrow\{V^{\mathbb{F}} \mid V \in \mathcal{O}_X(x)\} \supseteq \uparrow\{a^{\circ}(U) \mid U \in \mathcal{O}_X(x)\}\right) \\
\iff\ & \forall x \in X, U \in \mathcal{O}_X(x) \left(\exists V \in \mathcal{O}_X(x)\ (V^{\mathbb{F}} \subseteq a^{\circ}(U))\right).
\end{aligned}$$

This final condition is equivalent to condition (iv) of Proposition IV.5.3.1, as follows. Consider any $x \in X$ and $U \in \mathcal{O}_X(x)$. For any $V \in \mathcal{O}_X(x)$, we have that

$$V^{\mathbb{F}} \subseteq a^\circ(U) \iff \uparrow\{V\} \in a^\circ(U)\,,$$

and moreover

$$\begin{aligned} \exists V \in \mathcal{O}_X(x)\ (V^{\mathbb{F}} \subseteq a^\circ(U)) &\iff \exists V \in \mathcal{O}_X(x), y \in U\ (\uparrow\{V\} \xrightarrow{a} y) \\ &\iff \exists y \in U\ (x \prec y)\,, \end{aligned}$$

with the last equivalence holding as follows. First, if $V \in \mathcal{O}_X(x)$, $y \in U$, and $\uparrow\{V\} \xrightarrow{a} y$, then any $z \in V$ specializes y since $\nu(z) \supseteq \uparrow\{V\} \supseteq \nu(y)$, so that $x \in V \subseteq \downarrow y$ and hence $x \prec y$. Conversely, if $x \prec y$ then there is $V \in \mathcal{O}X$ with $x \in V \subseteq \downarrow y$, so for any open neighborhood U of y, $V \subseteq \downarrow y \subseteq U$, whence we have that $\uparrow\{V\} \supseteq \nu(y)$; equivalently, $\uparrow\{V\} \xrightarrow{a} y$. □

IV.5.4.2 Remark Related to the characterization of the observable realization spaces given in Theorem IV.5.4.1, we already mentioned in Remark III.5.6.5 that $(\beta, 2)$-$\mathsf{Cat}_=$, taken with respect to the Barr extension, is isomorphic to the full subcategory of Top consisting of exponentiable spaces (see Proposition III.5.6.6 and Theorem III.5.8.5). In fact, if the last condition in the first chain of equivalences in the proof of Theorem IV.5.4.1 is modified by substituting the set $V^{\beta} = \{\chi \in V^{\mathbb{F}} \mid \chi \text{ is an ultrafilter}\}$ in place of $V^{\mathbb{F}}$, we obtain the following result.

© **IV.5.4.3 Corollary** *Every observable realization space is exponentiable in* Top.

IV.5.5 Observable specialization systems

We now show that observable realization spaces may be axiomatized quite directly as a set equipped with a binary relation $(\prec)$. We then find that this axiomatization in turn leads to a description of these spaces as Kleisli monoids.

IV.5.5.1 Definition Let X be a set and $(\prec) : X \nrightarrow X$ a relation. We say that $(X, (\prec))$ is an *observable specialization system* if the following statements hold for all $x, y, z \in X$:

(1) $x \prec y \prec z \implies x \prec z$;
(2) $\exists u \in X\ (x \prec u)$;
(3) $x \prec y, z \implies (\exists u : x \prec u \prec y, z)$;
(4) $x \prec y\ \&\ (\forall u\ (z \prec u \implies y \prec u)) \implies x \prec z$.

If (1)–(3) hold, we say that $(X, \prec)$ is an *abstract basis*.

The relation $(\prec)$ induces an order on X, given by

$$x \leq y \iff \rotatebox{90}{$\rightarrowtail$}\, x \supseteq \rotatebox{90}{$\rightarrowtail$}\, y \tag{IV.5.5.i}$$

for $x, y \in X$, where $\updownarrow x := \{y \in X \mid x \prec y\}$. With respect to this order, note that (4) requires exactly that $(\prec) : X \nrightarrow X$ is a module, since one always has $(\prec) \cdot (\leq) \subseteq (\prec)$.

The category **ObsSys** has objects all observable specialization systems and morphisms all maps $f : (X, \prec) \to (Y, \prec)$ such that

$$\forall x \in X\ y \in Y\ (f(x) \prec y \implies \exists u \in X\ (x \prec u\ \&\ f(\updownarrow u) \subseteq \updownarrow y))\ ,$$

where $\updownarrow y := \{z \in Y \mid z \prec y\}$.

IV.5.5.2 Theorem *There is an isomorphism*

$$\mathsf{ObsReSp} \cong \mathsf{ObsSys}$$

which commutes with the underlying-set functors and sends an observable realization space X to the pair $(X, \prec)$ consisting of the underlying set and observable specialization relation $(\prec)$ of X.

Quite generally, if $(X, \prec)$ is an abstract basis, then $\mathcal{B} := \{\updownarrow y \mid y \in X\}$ is a base for a topology on X, under which X is an observable realization space, and if $(X, \prec)$ is moreover an observable specialization system, then $(\prec)$ coincides with the observable specialization relation of this observable realization space.

Proof If X is an observable realization space, then, by Proposition IV.5.3.1, $\updownarrow_X : X \to \mathrm{Fil}\, X$ is a fully faithful monotone map, so the underlying order $(\leq_X)$ is the order induced by $(\prec_X)$ (IV.5.5.i), and it is an easy exercise to verify that (1)–(4) hold for $(\prec_X)$.

Next suppose $(X, \prec)$ is an abstract basis. It is readily verified that $\mathcal{B}$ as given is a base for a topology on X (meaning that for each $x \in X$ the set $\{B \in \mathcal{B} \mid x \in B\}$ is either empty or down-directed) and that the order induced by $(\prec)$ (IV.5.5.i) coincides with the associated underlying order $(\leq_X)$ of the resulting topological space X. It then follows by (1) that $(\prec) \leq (\leq_X)$, and in turn that $(\prec) \leq (\prec_X)$ using (3). We may use this and (3) to deduce, by means of Proposition IV.5.3.1(iv), that X is an observable realization space.

Supposing moreover that $(X, \prec)$ also satisfies (4), we now show that $(\prec_X) \leq (\prec)$ as well. If $x \prec_X y$, there is some u with $x \in \updownarrow u \subseteq \downarrow_X y$, and by (3) there is some v with $x \prec v \prec u$, whence $x \prec v \leq_X y$, so by (4) $x \prec y$.

Noting also that for an observable realization space X the base $\mathcal{B}$ associated to $(X, \prec_X)$ is a base for X, the result follows. □

Thus, while the topological concept of a domain embodied by the observable realization spaces is not the usual order-theoretic one, in which the specialization order is primitive and the topology is derivative, the observable realization spaces are nevertheless determined by a transitive relation, but one that is not necessarily reflexive (nor irreflexive). For these spaces, *observable specialization*, rather than *specialization*, provides the primitive notion.

IV.5.6 Ordered abstract bases and round filters

For an ordered set X, recall that a module $(\prec) : X \nrightarrow X$ is an *auxiliary relation* on X if $(\prec) \leq (\leq_X)$ (see also Example IV.2.5.5(6)). Observe that any auxiliary relation $(\prec)$ on X is necessarily transitive, since $(\prec) \cdot (\prec) \leq (\prec) \cdot (\leq_X) = (\prec)$.

IV.5.6.1 Definition For an auxiliary relation $(\prec)$ on an ordered set X, we say that $(X, \prec)$ is an *ordered abstract basis* if the underlying set X and the relation $(\prec)$ constitute an abstract basis.

Ordered abstract bases provide a mild generalization of observable specialization systems, as follows:

IV.5.6.2 Proposition *Let X be a set and $(\prec) : X \nrightarrow X$ a relation. A pair $(X, \prec)$ is an observable specialization system if and only if $(X, \prec)$ is an ordered abstract basis when X is endowed with the order induced by $(\prec)$(see* (IV.5.5.i)*).*

The following alternative axioms lead us to a description of ordered abstract bases and observable specialization systems as Kleisli monoids in Section IV.5.7:

IV.5.6.3 Proposition *For an ordered set X and a relation $(\prec) : X \nrightarrow X$, the pair $(X, (\prec))$ is an ordered abstract basis if and only if the following statements hold, where $\Uparrow x := \{y \in X \mid x \prec y\}$ for each $x \in X$:*

(1) *$(\prec)$ is an auxiliary relation on X;*
(2) *for each $x \in X$, $\Uparrow x$ is down-directed;*
(3) *for any $x, z \in X$, if $x \prec z$ then there is some $y \in X$ with $x \prec y \prec z$.*

IV.5.6.4 Definition Given an ordered set X equipped with an auxiliary relation $(\prec)$, we say that a filter S in X is a *round filter* in $(X, (\prec))$ if for any $y \in S$ there is some $x \in S$ with $x \prec y$. Observe that $(X, (\prec))$ is an ordered abstract basis if and only if $\Uparrow x$ is a round filter for each $x \in X$.

IV.5.7 Domains as Kleisli monoids of the ordered-filter monad

For an ordered set X, modules $(\prec) : X \nrightarrow X$ correspond bijectively to monotone maps $n : X \longrightarrow \mathrm{Up}\, X$ via

$$x \prec y \iff n(x) \ni y ,$$

and we have $n(x) = \Uparrow x$. In the case that $(X, \prec)$ is an ordered abstract basis, we have that each $n(x)$ is a round filter, so we obtain a map $n : X \longrightarrow \mathrm{Fil}\, X$. In fact, we may employ this correspondence to obtain an axiomatization of observable realization spaces and ordered abstract bases as Kleisli monoids of the ordered-filter monad $\mathbb{F}\mathrm{il} = (\mathrm{Fil}, \uparrow, \bigcup)$ (see Section IV.4.4).

IV.5.7.1 Lemma *For a monotone map $n : X \longrightarrow \mathrm{Up}\, X$ with corresponding module $(\prec) : X \nrightarrow X$, the following are equivalent:*

(i) $n : X \to \operatorname{Up} X$ *restricts to an* $\mathbb{F}\text{il}$*-monoid structure* $n' : X \to \operatorname{Fil} X$*;*
(ii) $(X, \prec)$ *is an ordered abstract basis.*

Proof (i) is equivalent to the statement that n restricts to a (monotone) map $n' : X \to \operatorname{Fil} X$ with

$$\uparrow_X x \leq n' \qquad \text{and} \qquad \bigcup \cdot \operatorname{Fil} n' \cdot n' \leq n' .$$

The first inequality requires that $\uparrow_X \supseteq \upharpoonleft x$ for all $x \in X$, or equivalently that $(\prec) \leq (\leq_X)$; since $(\prec)$ is a module, this is equivalent to the requirement that $(\prec)$ be an auxiliary relation on X. Next observe that

$$\begin{aligned} y \in (\bigcup \cdot \operatorname{Fil} n' \cdot n')(x) &\iff \uparrow_X y \in (\operatorname{Fil} n' \cdot n')(x) \\ &\iff \exists u \in n(x)\ (\uparrow_X y \subseteq n(u)) \\ &\iff \exists u\ (u \in n(x)\ \&\ y \in n(u)) \end{aligned}$$

for all $x, y \in X$, so that the second inequation requires that

$$\forall x, y\ (x \prec y \implies \exists u\ (x \prec u \prec y)) . \qquad \square$$

IV.5.7.2 Theorem *Let X be an ordered set. There is a bijection between* $\mathbb{F}\text{il}$*-monoids $(X, n : X \to \operatorname{Fil} X)$ and ordered abstract bases $(X, \prec)$, wherein we associate to each Kleisli monoid (X, n) the ordered abstract basis whose auxiliary relation $(\prec)$ is the module associated to the composite $X \xrightarrow{n} \operatorname{Fil} X \hookrightarrow \operatorname{Up} X$ under the bijection* $\mathsf{Ord}(X, \operatorname{Up} X) \cong \mathsf{Mod}(X, X)$ *given above.*

Consequently, observable specialization systems $(X, \prec)$ correspond bijectively to Kleisli monoids (X, n) in which $n : X \to \operatorname{Fil} X$ is fully faithful, *and hence there are isomorphisms*

$$\mathsf{ObsReSp} \cong \mathsf{ObsSys} \cong \mathbb{F}\text{il}\text{-}\mathsf{Mon}_{\mathrm{ff}}$$

which commute with the underlying-set functors, where $\mathbb{F}\text{il}\text{-}\mathsf{Mon}_{\mathrm{ff}}$ *is the full subcategory of* $\mathbb{F}\text{il}\text{-}\mathsf{Mon}$ *consisting of those Kleisli monoids (X, n) with n fully faithful.*

Proof Lemma IV.5.7.1 yields the bijection between Kleisli monoids and ordered abstract bases, and an ordered abstract basis $(X, \prec)$ carries the induced order if and only if its mate $n : X \to \operatorname{Fil} X$ is fully faithful.

Consider observable realization spaces X,Y with associated Kleisli monoid structures n_X, n_Y and a function $f : X \to Y$. If f is continuous, then f is monotone, so we may assume that f is monotone. Then one may employ the characterization of morphisms given in Theorem IV.5.5.2 to verify that f is continuous if and only if

$$\forall x \in X, y \in Y\ (f(x) \prec y \implies \exists u \in X\ (x \prec u\ \&\ f(u) \prec y)) ,$$

using the fact that $(\prec_X)$ is interpolating and $(\prec_X) \leq (\leq_X)$. On the other hand, $f : (X, n_X) \to (Y, n_Y)$ is a morphism of Kleisli monoids if and only if $\mathrm{Fil}\, f \cdot n_X \leq n_Y \cdot f$, equivalently if and only if

$$\forall x \in X, y \in Y\ (f(x) \prec y \implies \exists v\ (x \prec v\ \&\ f(v) \leq y))\,.$$

This is entailed by continuity, since $(\prec_Y) \leq (\leq_Y)$, and entails continuity, since if $f(x) \prec y$ then we may interpolate to obtain y' with $f(x) \prec y' \prec y$. □

IV.5.8 Continuous dcpos as sober domains

Herein we show that the continuous dcpos under the Scott topology are precisely the *sober* observable realization spaces. Moreover, we find that the observable realization spaces are closed under *sobrification*, and hence that the continuous dcpos occur as a reflective subcategory of ObsReSp. The sobrification of an observable realization space X may be formed by topologizing the set of round filters in X.

For each point x in a topological space X, the *open neighborhood filter* $\mathcal{O}_X(x) = \mathcal{O}X \cap \dot{x}$ of x is a filter in the frame $\mathcal{O}X$. The filter $p = \mathcal{O}_X(x)$ is *completely prime*, meaning that, for any family $(U_i \in \mathcal{O}X)_{i \in I}$,

$$p \ni \bigcup_{i \in I} U_i \iff \exists i \in I\ (p \ni U_i)\,. \qquad \text{(IV.5.8.i)}$$

Quite generally, given any frame O, we refer to the completely prime filters in O as the *points* of O, and we denote the set of all such points by $\mathrm{pt}\, O$. The idea is that we may liken the elements $u \in O$ to open sets in a topological space and the points $p \in \mathrm{pt}\, O$ to points in a topological space, interpreting statements of the form $p \ni u$ as statements of membership of a point in an open set. We can make this concrete by defining $u^{\square} := \{p \in \mathrm{pt}\, O \mid p \ni u\}$ for each $u \in O$ and observing that there is a resulting frame homomorphism

$$O \to P\, \mathrm{pt}\, O\,, \quad u \mapsto u^{\square}$$

whose image is thus a topology on $\mathrm{pt}\, O$, so that the points of O are in fact the points of a topological space $\mathrm{pt}\, O$, which we call the *spatialization* of O. However, the frame $\mathcal{O}\, \mathrm{pt}\, O$ need not be isomorphic to O, and in fact the discipline of *locale theory* pursues a formulation of general topology in which frames, rather than the usual topological spaces, are taken as the primitive objects of study (see e.g. [Johnstone, 1982]).

In the case that O is the frame $\mathcal{O}X$ of open sets of a topological space X, the given frame homomorphism $\mathcal{O}X \to \mathcal{O}\, \mathrm{pt}\, \mathcal{O}X$ has a retraction given by $\mathcal{U} \mapsto \{x \in X \mid \mathcal{O}_X(x) \in \mathcal{U}\}$, and hence is actually an isomorphism of frames $\mathcal{O}X \cong \mathcal{O}\, \mathrm{pt}\, \mathcal{O}X$. The original space X is manifested within the spatialization $\mathrm{pt}\, \mathcal{O}X$ via an *initial* continuous map $\mathcal{O}_X : X \to \mathrm{pt}\, \mathcal{O}X$ – the *open neighborhood filter map* – and this map is an embedding if and only if X is T0. If this map is a bijection, then, since it is initial, it is a *homeomorphism* (i.e. an isomorphism in Top) $X \cong \mathrm{pt}\,\mathcal{O}X$, and we say that X is *sober*, in which case the points of X may be viewed as order-theoretic features inherent in the frame $\mathcal{O}X$ of opens. Regardless, the isomorphism of frames $\mathcal{O}X \cong \mathcal{O}\,\mathrm{pt}\,\mathcal{O}X$ induces an evident bijection $\mathrm{pt}\,\mathcal{O}X \cong \mathrm{pt}\,\mathcal{O}\,\mathrm{pt}\,\mathcal{O}X$, and in fact this bijection coincides with the open neighborhood filter map of the space $\mathrm{pt}\,\mathcal{O}X$, which is thus always sober and is called the *sobrification* of X.

For any frame O, there is a bijection $\mathrm{pt}\,O \cong \mathsf{Frm}(O, 2)$ given by associating to a completely prime filter $p \in \mathrm{pt}\,O$ its characteristic function. Thus, we may alternatively construct the spatialization of O by taking $\mathrm{pt}\,O := \mathsf{Frm}(O, 2)$ under the topology given by analogy with the above. In fact, it is trivially verified that the contravariant hom-functor $\mathsf{Frm}(-, 2) : \mathsf{Frm}^{\mathrm{op}} \longrightarrow \mathsf{Set}$ sends each frame homomorphism $h : O \longrightarrow N$ to a *continuous* map $\mathrm{pt}\,h : \mathrm{pt}\,N \longrightarrow \mathrm{pt}\,O$ and hence factors through a functor $\mathrm{pt} : \mathsf{Frm}^{\mathrm{op}} \longrightarrow \mathsf{Top}$. Composing this functor with the functor $\mathcal{O} : \mathsf{Top} \longrightarrow \mathsf{Frm}^{\mathrm{op}}$ that sends a continuous map $f : X \longrightarrow Y$ to the frame homomorphism $\mathcal{O}Y \longrightarrow \mathcal{O}X$ given by inverse image, we obtain a covariant functor

$$\mathsf{Top} \xrightarrow{\ \mathcal{O}\ } \mathsf{Frm}^{\mathrm{op}} \xrightarrow{\ \mathrm{pt}\ } \mathsf{Top}$$

which sends a space X to its sobrification. This functor factors through the full subcategory Sob of Top consisting of sober spaces as $\mathsf{Top} \xrightarrow{R} \mathsf{Sob} \overset{J}{\hookrightarrow} \mathsf{Top}$, and the maps $\mathcal{O}_X : X \longrightarrow \mathrm{pt}\,\mathcal{O}X$, which are isomorphisms for precisely the objects of Sob, serve as the components of a natural transformation $\rho : 1_{\mathsf{Top}} \longrightarrow JR$. One may verify immediately that $\rho JR = JR\rho$ and then use this fact and the naturality of ρ to verify that ρ is the unit of an adjunction $R \dashv J$, so that Sob is a reflective subcategory of Top.

For observable realization spaces, we have the following:

IV.5.8.1 Proposition *Let X be an observable realization space. Then the sobrification $\mathrm{pt}\,\mathcal{O}X$ of X is an observable realization space and is homeomorphic to the space $\mathrm{Fil}_\circ X$ of all round filters in X under a topology consisting of the sets $U^\Diamond = \{R \in \mathrm{Fil}_\circ X \mid \exists y\ (R \ni y \in U)\}$ with $U \in \mathcal{O}X$, and there is a commutative diagram*

$$\begin{array}{ccc} & X & \\ {}^{\uparrow}\swarrow & & \searrow{}^{\mathcal{O}_X} \\ \mathrm{Fil}_\circ X & \xrightarrow{\ \cong\ } & \mathrm{pt}\,\mathcal{O}X \end{array}$$

in Top whose bottom row is an isomorphism.

Proof It is straightforward to verify that the map $\mathcal{O}X \longrightarrow P\,\mathrm{Fil}_\circ X$ given by $U \mapsto U^\Diamond$ is a frame homomorphism, so the image of this map is a topology

$\mathcal{O}\,\mathrm{Fil}_\circ X$ on $\mathrm{Fil}_\circ X$. Moreover, since X is an observable realization space, we have that for each $U \in \mathcal{O}X$, $x \in X$

$$x \in U \iff \exists y\,(x \prec y \in U) \iff \Uparrow x \in U^{\Diamond}\,, \tag{IV.5.8.ii}$$

so the given frame homomorphism $\mathcal{O}X \to \mathcal{O}\,\mathrm{Fil}_\circ X$ has a retraction $\mathcal{U} \mapsto \{x \in X \mid \Uparrow x \in \mathcal{U}\}$ and hence is an isomorphism of frames $\mathcal{O}X \cong \mathcal{O}\,\mathrm{Fil}_\circ X$. Thus, there is a homeomorphism $\mathrm{pt}\,\mathcal{O}X \cong \mathrm{pt}\,\mathcal{O}\,\mathrm{Fil}_\circ X$, wherein each completely prime filter $q \in \mathrm{pt}\,\mathcal{O}\,\mathrm{Fil}_\circ X$ is the image $q = \{U^{\Diamond} \mid U \in p\}$ of a unique $p \in \mathrm{pt}\,\mathcal{O}X$. It is readily verified that $R = \{y \in X \mid p \ni \Downarrow y\}$ is a round filter, and then for any $U \in \mathcal{O}X$ we find that

$$p \ni U \iff p \ni \bigcup_{y \in U} \Downarrow y \iff \exists y \in U\,(p \ni \Downarrow y) \iff R \in U^{\Diamond}\,,$$

whence $q = \mathcal{O}_{\mathrm{Fil}_\circ X}(R)$. This shows that the initial continuous map $\mathcal{O}_{\mathrm{Fil}_\circ X} : \mathrm{Fil}_\circ X \to \mathrm{pt}\,\mathcal{O}\,\mathrm{Fil}_\circ X$ is surjective, and we show below that this map is injective and hence a homeomorphism, so we obtain a composite homeomorphism $\mathrm{Fil}_\circ X \cong \mathrm{pt}\,\mathcal{O}\,\mathrm{Fil}_\circ X \cong \mathrm{pt}\,\mathcal{O}X$, which, by (IV.5.8.ii), commutes with the maps $\Uparrow$, $\mathcal{O}_X$ as in the above diagram.

Indeed, $\mathrm{Fil}_\circ X$ is T0, since its underlying order $(\leq)$ is the reverse inclusion order, as follows. We have that

$$R \in (\Downarrow y)^{\Diamond} \iff \exists u \in X\,(R \ni u \prec y) \iff R \ni y$$

for any $R \in \mathrm{Fil}_\circ X$, $y \in X$. Thus, for $R, S \in \mathrm{Fil}_\circ X$, one readily verifies the following implications:

$$R \leq S \implies \forall y \in X\,(R \in (\Downarrow y)^{\Diamond} \Longleftarrow S \in (\Downarrow y)^{\Diamond}) \iff R \supseteq S \implies R \leq S\,,$$

thus obtaining the required conclusion.

Lastly, we show that $\mathrm{Fil}_\circ X$ is an observable realization space via Proposition IV.5.3.1(iv). Suppose $R \in U^{\Diamond}$. Then

$$R \in U^{\Diamond} = \left(\bigcup_{y \in U} \Downarrow y\right)^{\Diamond} = \bigcup_{y \in U} \left(\Downarrow y\right)^{\Diamond}$$

and hence there is $y \in U$ with $R \in (\Downarrow y)^{\Diamond}$. We know that $\Uparrow y \in \mathrm{Fil}_\circ X$, and each $S \in (\Downarrow y)^{\Diamond}$ specializes $\Uparrow y$ since $S \ni y$ and hence $S \supseteq \Uparrow y$. Thus, we have that $R \prec \Uparrow y \in U^{\Diamond}$. □

Before characterizing the sober observable realization spaces as precisely the continuous dcpos under the Scott topology, we note the following well-known result concerning sober spaces:

IV.5.8.2 Proposition *Let X be a sober space. Then, under the underlying order, X is a co-dcpo, and $\mathcal{O}X$ is coarser than the Scott topology on X (i.e. every open set is Scott-open).*

Proof It is straightforward to verify that the union of a directed subset of $\mathrm{pt}\,\mathcal{O}X$ is a completely prime filter on $\mathcal{O}X$ and hence that directed meets in $\mathrm{pt}\,\mathcal{O}X$ are given by union (as is the case in TX), so that these directed meets are moreover preserved by the monotone map $\uparrow_{PX} : \mathrm{pt}\,\mathcal{O}X \to TX$. Further, the top row of the following commutative diagram is an Ord-isomorphism:

$$\begin{array}{ccc} X & \xrightarrow[\mathcal{O}_X]{\cong} & \mathrm{pt}\,\mathcal{O}X \\ & {}_{\nu}\searrow \quad \swarrow_{\uparrow_{PX}} & \\ & TX & \end{array}$$

and hence X is a separated ordered set with all directed meets such that these meets are preserved by ν. Therefore, X is a co-dcpo, and by Proposition IV.5.2.2 $\mathcal{O}X$ is coarser than the Scott topology. □

In observable realization spaces, these conditions are also sufficient for sobriety, and we obtain the following result:

IV.5.8.3 Theorem *For any topological space X endowed with the underlying order, the following are equivalent:*

(i) *X is a sober observable realization space;*
(ii) *X is a cocontinuous co-dcpo and carries the Scott topology.*

Proof First suppose (i). Then, by Proposition IV.5.8.2, X is a co-dcpo and $\mathcal{O}X$ is coarser than the Scott topology, so by Lemma IV.5.3.4 $\mathcal{O}X$ coincides with the Scott topology. Thus, since X is an observable realization space, the co-dcpo X is cocontinuous, by Corollary IV.5.2.5 and Proposition IV.5.3.1.

Next, suppose (ii). Since X is a cocontinuous co-dcpo carrying the Scott topology, X is an observable realization space. X is T0, since it carries a separated underlying order, so to show that X is sober it suffices by Proposition IV.5.8.1 to show that the map $\Uparrow : X \to \mathrm{Fil}_\circ X$ is surjective, so let $R \in \mathrm{Fil}_\circ X$. Since X is a co-dcpo and carries the Scott topology, by Proposition IV.5.2.2 there is a topological meet $\bigwedge R$ of R in X, so that $\mathcal{O}_X\left(\bigwedge R\right) = \bigcup_{y\in R} \mathcal{O}_X(y)$. Thus, for any $z \in X$,

$$\bigwedge R \prec z \iff \mathcal{O}_X\left(\bigwedge R\right) \ni \Downarrow z \iff \exists y\,(R \ni y \in \Downarrow z) \iff R \ni z\,,$$

since R is a round filter, so $\Uparrow \bigwedge R = R$. □

Letting $\mathsf{CntDcpo}^{\mathrm{co}}$ be the category of cocontinuous co-dcpos and directed-meet-preserving maps, we obtain the following:

IV.5.8.4 Corollary *There is a reflective full embedding*

$$\mathsf{CntDcpo}^{\mathrm{co}} \hookrightarrow \mathsf{ObsReSp}$$

which commutes with the underlying-set functors and endows a cocontinuous co-dcpo with its Scott topology and whose image is the full subcategory of ObsReSp *consisting of sober observable realization spaces. A left adjoint to this embedding may be taken as a restriction of the composite*

$$\mathsf{Top} \xrightarrow{\mathrm{pt}\,\mathcal{O}} \mathsf{Top} \xrightarrow{S} \mathsf{Ord}$$

of the sobrification functor pt $\mathcal{O}$ *and the concrete functor* S*, which sends a topological space to its underlying order.*

Proof By Theorem IV.5.8.3 and the remarks following the definition of the Scott topology, there is an embedding $\mathsf{CntDcpo}^{\mathrm{co}} \hookrightarrow \mathsf{ObsReSp}$, as described, and this embedding restricts to an isomorphism $\mathsf{CntDcpo}^{\mathrm{co}} \cong \mathsf{Sob} \cap \mathsf{ObsReSp}$ with inverse a restriction of S. As discussed above, the embedding $\mathsf{Sob} \hookrightarrow \mathsf{Top}$ has a left adjoint $R : \mathsf{Top} \to \mathsf{Sob}$ which is a restriction of pt $\mathcal{O}$, and since ObsReSp is closed under sobrification by Proposition IV.5.8.1, R restricts to a left adjoint of the inclusion $\mathsf{Sob} \cap \mathsf{ObsReSp} \hookrightarrow \mathsf{ObsReSp}$, thus yielding the required reflector. □

IV.5.9 Cocontinuous lattices among lax algebras

A cocontinuous lattice is a complete lattice that is cocontinuous (see Section IV.5.2). In Section IV.4.4 it was shown that the category of $\mathsf{Cnt}^{\mathrm{co}}$ of cocontinuous lattices and directed-meet-preserving sup-maps is isomorphic to the category $\mathsf{Set}^{\mathbb{F}}$ of Eilenberg–Moore algebras of the filter monad $\mathbb{F}$. By Theorem IV.5.8.3, when we endow a cocontinuous lattice with its Scott topology, we obtain a (sober) observable realization space. The resulting functor $\mathsf{Cnt}^{\mathrm{co}} \to \mathsf{ObsReSp}$ is faithful, but not full, and, given the isomorphism $\mathsf{ObsReSp} \cong (\mathbb{F}, 2)\text{-}\mathsf{Cat}_=$ of Theorem IV.5.4.1, we may form an evident commutative diagram

$$\begin{array}{ccccc} \mathsf{Cnt}^{\mathrm{co}} & \longrightarrow & \mathsf{ObsReSp} & \hookrightarrow & \mathsf{Top} \\ \downarrow \cong & & \downarrow \cong & & \downarrow \cong \\ \mathsf{Set}^{\mathbb{F}} & \longrightarrow & (\mathbb{F}, 2)\text{-}\mathsf{Cat}_= & \hookrightarrow & (\mathbb{F}, 2)\text{-}\mathsf{Cat}\,. \end{array}$$

The functors in the leftmost square allow us to consider cocontinuous lattices as spaces, $\mathbb{F}$-algebras, and strict lax algebras of $\mathbb{F}$, as follows.

IV.5.9.1 Proposition *Let X be a cocontinuous lattice.*

(1) *The relation $a : FX \nrightarrow X$ of Scott convergence of filters (see* Example IV.5.3.3*) is determined by a map*

$$l : FX \to X\,, \quad \mathfrak{x} \mapsto \bigwedge\{z \in X \mid \downarrow z \in \mathfrak{x}\} = \bigwedge_{A \in \mathfrak{x}} \bigvee A\,,$$

in the sense that

$$a = l_* = (\leq_X) \cdot l\,.$$

(2) *By* (1), *the neighborhood filter map* $\nu : X \to FX$ *is related to* l *via* $l_* = a = \nu^*$, *and hence* $l \dashv \nu$.
(3) *The isomorphism* $\mathsf{Cnt}^{\mathrm{co}} \cong \mathsf{Set}^{\mathbb{F}}$ *of* Section IV.4.4 *associates to* X *the* $\mathbb{F}$-*algebra* (X, l).
(4) *The given functor* $\mathsf{Cnt}^{\mathrm{co}} \to \mathsf{ObsReSp}$ *endows the cocontinuous lattice* X *with its Scott topology, and the isomorphism* $\mathsf{ObsReSp} \cong (\mathbb{F}, 2)\text{-}\mathsf{Cat}_=$ *of* Theorem IV.5.4.1 *associates to this space the strict lax algebra* $(X, l_* : FX \nrightarrow X)$.

Proof The verification is straightforward. □

We now characterize the image of the given composite $\mathsf{Cnt}^{\mathrm{co}} \cong \mathsf{Set}^{\mathbb{F}} \to (\mathbb{F}, 2)\text{-}\mathsf{Cat}_=$, thus characterizing the cocontinuous lattices among strict lax algebras.

IV.5.9.2 Theorem *Among lax algebras* $(X, a : FX \nrightarrow X)$ *of* $\mathbb{F}$, *those associated to cocontinuous lattices in the preceding commutative diagram are precisely those that are*

(1) *strict,*
(2) *map-determined, in the sense that* $a = l_*$ *for some monotone map* $l : FX \to X$, *and*
(3) *T0, in the sense that* $(\leq_X)$ *is separated,*

when we endow X *with the underlying order* $(\leq_X) = a \cdot e_X$.

Thus, the category of cocontinuous lattices and directed-meet-preserving maps is evidently isomorphic to the full subcategory of $(\mathbb{F}, 2)\text{-}\mathsf{Cat}_=$ *whose objects satisfy* (2) *and* (3).

Proof By the above, it is clear that if X is a cocontinuous lattice, then the associated lax algebra satisfies all three conditions. Conversely, consider an arbitrary $(X, a) \in \mathrm{ob}(\mathbb{F}, 2)\text{-}\mathsf{Cat}$ satisfying (1)–(3). Since (X, a) is strict, the associated topological space is an observable realization space by Theorem IV.5.4.1. Therefore, by Proposition IV.5.2.2, Lemma IV.5.3.4, and Proposition IV.5.2.5, it suffices to show that X is a complete lattice and that the neighborhood-filter map $\nu : X \to FX$ preserves directed meets.

Since a is the convergence relation of the associated space, we have that $l_* = a = \nu^*$. Thus,

$$FX \underset{l}{\overset{\nu}{\underset{\longrightarrow}{\overset{\longleftarrow}{\top}}}} X\,,$$

so, since FX is a complete lattice and $\nu : X \to FX$ is a fully faithful functor, it follows that X is a complete lattice and that ν preserves all meets. □

IV.5.9.3 Remark Condition IV.5.9.2(2) means

$$\chi \; a \; x \iff l(\chi) \leq x$$

for all $\chi \in FX$, $x \in X$. In other words, l chooses for every $\chi \in FX$ a smallest convergence point which, in conjunction with (3), is uniquely determined. Conversely, if every filter χ has a least convergence point $l(\chi)$, we obtain a map $l : FX \to X$ which is automatically monotone and satisfies $a = l_*$.

Exercises

IV.5.A *Topological meets and the Scott topology.* For a topological space X endowed with its underlying order, the statement that every open subset of X is Scott-open is equivalent to the statement that every directed meet in X is a topological meet (Proposition IV.5.2.2).

IV.5.B *Comparing $\lll$ and $\prec$.* Let X be an ordered set endowed with the Scott topology. If either $(\lll)$ or $(\prec)$ is approximating, then $(\lll) = (\prec)$ (see Lemma IV.5.2.4).

Hint. Note that we have $(\prec) \leq (\lll)$ in general. Employ Lemma IV.5.2.3.

IV.5.C *Characterizations of cocontinuity.* For an ordered set X endowed with the Scott topology, the following are equivalent:

(i) X is cocontinuous;
(ii) the relation $(\prec)$ is approximating;
(iii) each $x \in X$ is a topological meet of $\Uparrow x := \{y \in X \mid x \prec y\}$.

(See Proposition IV.5.2.5.)

IV.5.D *Closure properties of the class of observable realization spaces.* Show that $\mathsf{ObsReSp}$ is closed in Top under finite products, arbitrary coproducts, and retracts.

Notes on Chapter IV

The provision of a satisfactory notion of convergence has been one of the major concerns of topology from its very beginnings. In his pioneering work [Fréchet, 1905, 1906], Fréchet not only introduced metric spaces (the designation is due to Hausdorff), but also defined sequential convergence in an abstract way. However, his concept of convergence turned out to be insufficient as it lacks in particular the ability to deal with iterated limits. A more general type of convergence based on directed sets rather than sequences was introduced by Moore and his student Smith [Moore, 1915; Moore and Smith, 1922], and Birkhoff [1937] characterized T1-topological spaces in terms of this *net* convergence (the term net was coined by Kelley [1950]). At first glance, the notion of net resembles closely the notion of sequence; however, already the definition of subnet turns out to be more sophisticated than the corresponding notion of subsequence. Birkhoff also showed that net convergence can be equivalently substituted by a notion of convergence based on certain systems of sets that he had introduced earlier [Birkhoff, 1935], called filter bases in today's language. At approximately the same time, Cartan [1937a,b] introduced the notion of *filter* convergence, and this idea became central in Bourbaki's treatment of topology [Bourbaki, 1942]. The notions of filter and ultrafilter and their predecessors *per se* had emerged earlier in various mathematical contexts; in fact, the *ideale Verdichtungsstelle* mentioned by Riesz [1908] may be considered a precursor of the notion of ultrafilter, and the Polish School of measure theorists and functional analysts had studied the dual concept of ideal, with Tarski [1930] in effect proving that every filter is contained in an ultrafilter.

Grimeisen [1960, 1961] characterized topological spaces among pretopological spaces as precisely those which have the "iterated limit propery." Later on, Cook and Fischer [1967] presented four axioms that characterize topological spaces in terms of filter convergence. Barr [1970] extended Manes' equational presentation of compact Hausdorff spaces (see [Manes, 1969]) to characterize topological spaces in terms of two simple inequalities for an ultrafilter convergence relation that are amenable to algebraic manipulation. The fact that Barr's two ultrafilter convergence axioms may be taken *verbatim* as filter convergence axioms already to fully characterize topological spaces appeared only in [Seal, 2005]; a slightly weaker version of this result had been established earlier in [Pisani, 1999].

The original definition of a topological space given by Hausdorff [1914] used neighborhood systems as the primary structure. (His axioms included the "Hausdorff" separation axiom, which, however, is independent of his other axioms.) The fact that this definition is equivalent to that of a monoid in the Kleisli category of the filter monad on Set (Proposition IV.1.1.1) was first observed in [Gähler *et al.*, 1992]. Therein, the authors introduced the notion of a preordered monad that is at the origin of our power-enriched monads (Section IV.1.2), and their associated category of monadic topologies that, similarly to Kleisli monoids (Section IV.1.3), are defined as monoids in the Kleisli category of the given monad. Höhle's notion of a topological space object (see [Höhle, 2001]) uses the same idea in a framework different from the one used in this book, but his work targeted many of the categories also of interest here. The presentation of closure spaces and the up-set monad as in Example IV.1.3.2(4) appeared in [Seal, 2009], which also defined the Kleisli and strata extensions of a power-enriched monad as described in Sections IV.1.4 and IV.2.4. The strata extensions of a general lax monad first appeared in [Clementino and Hofmann, 2004b]. Initial extensions of a functor and of a monad (Section IV.2.1) were introduced in [Schubert and Seal, 2008] and [Colebunders, Lowen, and Rosiers, 2011], respectively. The Kleisli towers in Section IV.2.5 are a particular case of the tower extension construction over a topological functor from [Zhang, 2000], where Zhang generalized the presentation of approach spaces as towers [Lowen, 1989].

The construction of the discrete presheaf monad (as a power-enriched monad) from an arbitrary monad with an associative lax extension to $\mathcal{V}$-Rel as given in Section IV.3 – which enables the surprising result that $(\mathbb{T}, \mathcal{V})$-categories may always be considered as relational algebras (see Corollary IV.3.2.3) – was motivated by the presentation of approach spaces as relational algebras in [Lowen and Vroegrijk, 2008]. The alternative description of the discrete presheaf monad $\mathbb{\Pi}$ in Section IV.3.3 is similar to that one of the monad π in [Colebunders *et al.*, 2011].

The identification of Eilenberg–Moore algebras for the filter monad on Set or Top_0 as continuous lattices (Section IV.4.4) is due to Day [1975]. Wyler [1981] noted that Cnt is also monadic over Sup and $\mathsf{CompHaus}$. The fact that completely distributive lattices are the Eilenberg–Moore algebras for the up-set monad on Set (Exercise IV.4.E) was noted in [Pedicchio and Wood, 1999]; the corresponding observation for frames (Exercise IV.4.F) over Set can be found in [Johnstone, 1982]. The description of $\mathbb{P}_{\mathcal{V}}$-algebras as $\mathcal{V}$-actions in Sup (Exercise IV.4.G) comes from Pedicchio and Tholen [1989] (following the more technical description of Machner [1985]); Pedicchio and Tholen [1989] also remarked that $\mathsf{Set}^{\mathcal{V}}$ is monadic over Ord and $\mathcal{V}$-Cat. Further developments in the context of categories enriched in a quantaloid can be found in [Stubbe, 2007]. The description of the distributive law of the list monad over the powerset monad and its category of Eilenberg–Moore algebras (Exercise IV.4.H) appeared in [Manes and Mulry, 2007]. Finally, the distributive law of Exercise IV.4.A can be found in [Schubert, 2006] and uses the distributive law of Dn over Up described in [Marmolejo, Rosebrugh, and Wood, 2002] that, in turn, finds its source in a distributive law for the frame monad over Ord mentioned by Linton at a meeting in 1984. The proof that the RegMono-injective T0-spaces are the continuous lattices endowed with their Scott topology (Example IV.4.6.6(2)) goes back to Scott [1972], but the treatment we give of the subject in Section IV.4.6 – and in particular the proof of Theorem IV.4.6.3 – is more directly inspired by Escardó [1998]. The power-enriched treatment was developed in [Seal, 2010, 2011].

Section IV.5 is based largely on [Lucyshyn-Wright, 2009, 2011], in which Theorems IV.5.4.1 and IV.5.9.2 were first proved. The proof of Lemma IV.5.2.3 is derived from the proof (for dcpos) that was contributed by Escardó to [Gierz *et al.*, 2003]. The given relation of Scott convergence (Example IV.5.3.3) was studied (with respect to the dual order) by Erné [1981] (under the name of s_3-convergence), considered there among two other generalizations to arbitrary preorders of Scott's original notion of convergence in continuous lattices [Scott, 1972].

The class of spaces that we call the observable realization spaces appeared in papers by Banaschewski [1977]; Erné [1980, 1981, 1991, 1999]; Erné and Wilke [1983]; Ershov [1993]; Hoffmann [1979, 1981]; Lawson [1979, 1997], but were often identified via quite different and less elementary characterizations. These spaces were called C-spaces by Erné. The basic role of these spaces in domain theory was noted by Ershov [1993], who had earlier developed domain theory in topological terms. The domain-theoretic relevance of C-spaces was also underscored in [Lucyshyn-Wright, 2009, 2011]. Lemma IV.5.3.4 is part of Exercise II.34 of [Gierz *et al.*, 2003], contributed by Escardó and Heckmann.

The axioms for an observable specialization system given in Definition IV.5.5.1 and the main result of Theorem IV.5.5.2 were given by Hoffmann [1981] (building upon work in [Hoffmann, 1979]), whereas the axioms (1)–(3) defining an abstract basis were given by Smyth [1977/78] as a way of generating continuous dcpos. Each abstract basis gives rise to an associated continuous dcpo, the set of round ideals of the basis (see, for example, [Abramsky and Jung, 1994]). Herein, following [Lucyshyn-Wright, 2009], we regard those abstract bases satisfying (4) as domains in their own right. These domains – the observable realization spaces – need not be sober, and the space of round ideals (or, rather, round filters) associated to such a space is in fact the sobrification of that space and is a continuous dcpo under the Scott topology (see Section IV.5.8). With regard to ordered abstract bases (see Definition IV.5.6.1), note that auxiliary relations with the given properties were studied by Hoffmann [1981].

Suggestions for further reading: [Adámek *et al.*, 2003]; [Gutierres and Hofmann, 2007, 2013]; [Hofmann and Tholen, 2006]; [Johnstone, 1979]; [Kowalsky, 1954a,b]; [McShane, 1952]; [Tukey, 1940]; [Wyler, 1995].

V

Lax algebras as spaces

Maria Manuel Clementino, Eva Colebunders, and Walter Tholen

Looking at $(\mathbb{T}, \mathcal{V})$-categories as geometric objects, in this chapter we explore fundamental topological properties like compactness and Hausdorff separation, along with low-separation properties, regularity, normality, and extremal disconnectedness. We do so first by taking a more traditional object-oriented view, before highlighting the central role of proper and open maps in $(\mathbb{T}, \mathcal{V})$-Cat. Each of these classes gives a "topology" on $(\mathbb{T}, \mathcal{V})$-Cat and, along with other classes, allows us to investigate relativized topological properties in an axiomatic categorical setting. We also explore the categorical notion of connected object in $(\mathbb{T}, \mathcal{V})$-Cat.

V.1 Hausdorff separation and compactness

Topological spaces have been described as lax algebras via ultrafilter convergence in III.2.2, i.e. as sets X equipped with a relation $a : \beta X \nrightarrow X$ satisfying the two requirements of a $(\beta, 2)$-category. The relation a will actually be a map if, when we write $\mathfrak{x} \longrightarrow x$ instead of $\mathfrak{x}\, a\, x$,

$$\forall x, y \in X, \mathfrak{z} \in TX\ (\mathfrak{z} \longrightarrow x \;\&\; \mathfrak{z} \longrightarrow y \implies x = y), \text{ i.e. } a \cdot a^\circ \leq 1_X\,,$$

and

$$\forall \mathfrak{x} \in TX\ \exists x \in X\ (\mathfrak{x} \longrightarrow x), \text{ i.e. } 1_{TX} \leq a^\circ \cdot a\,.$$

The first property identifies X as a Hausdorff space, and the second one as a compact space (see III.2.3). This observation leads us to consider the following notions in the general context of a *monad* $\mathbb{T} = (T, m, e)$ *with a lax extension* $\hat{T}$ *to* $\mathcal{V}$-Rel, *where* $\mathcal{V} = (\mathcal{V}, \otimes, k)$ *is a quantale.*

Monoidal Topology: A Categorical Approach to Order, Metric, and Topology, eds. Dirk Hofmann, Gavin J. Seal, and Walter Tholen. Published by Cambridge University Press.

V.1.1 Basic definitions and properties

As we are considering topologically inspired properties of $(\mathbb{T}, \mathcal{V})$-categories, we will most often refer to the objects of $(\mathbb{T}, \mathcal{V})$-Cat as $(\mathbb{T}, \mathcal{V})$-*spaces*, and to its morphisms as $(\mathbb{T}, \mathcal{V})$-*continuous maps*. Objects and morphisms of $\mathcal{V}$-Cat are simply called $\mathcal{V}$-*spaces* and $\mathcal{V}$-*continuous maps*, respectively.

V.1.1.1 Definition A $(\mathbb{T}, \mathcal{V})$-space (X, a) is *Hausdorff* if

$$a \cdot a^\circ \leq 1_X ,$$

and it is *compact* if

$$1_{TX} \leq a^\circ \cdot a .$$

The resulting full subcategories of $(\mathbb{T}, \mathcal{V})$-Cat are denoted by

$$(\mathbb{T}, \mathcal{V})\text{-}\mathsf{Cat}_{\mathsf{Haus}} \qquad \text{and} \qquad (\mathbb{T}, \mathcal{V})\text{-}\mathsf{Cat}_{\mathsf{Comp}} ,$$

respectively, and their intersection is denoted by $(\mathbb{T}, \mathcal{V})\text{-}\mathsf{Cat}_{\mathsf{CompHaus}}$.

V.1.1.2 Proposition *Let (X, a) be a $(\mathbb{T}, \mathcal{V})$-space.*

(1) *(X, a) is Hausdorff if and only if, for all $x, y \in X$ and $z \in TX$,*

$$\bot < a(z, x) \otimes a(z, y) \implies x = y \qquad \text{and} \qquad a(z, x) \otimes a(z, x) \leq k ,$$

where the latter condition holds trivially when $\mathcal{V}$ is integral.

(2) *(X, a) is compact if and only if, for all $\mathfrak{x} \in TX$,*

$$k \leq \textstyle\bigvee_{z \in X} a(\mathfrak{x}, z) \otimes a(\mathfrak{x}, z) . \tag{V.1.1.i}$$

Proof The assertions follow immediately from

$$(a \cdot a^\circ)(x, y) = \bigvee\nolimits_{z \in TX} a(z, x) \otimes a(z, y)$$

and

$$(a^\circ \cdot a)(\mathfrak{x}, \mathfrak{y}) = \bigvee\nolimits_{z \in X} a(\mathfrak{x}, z) \otimes a(\mathfrak{y}, z)$$

for all $x, y \in X$, $\mathfrak{x}, \mathfrak{y} \in TX$. □

V.1.1.3 Remark If the quantale $\mathcal{V}$ is *superior*, i.e. if

$$k \leq \textstyle\bigvee_{i \in I} u_i \otimes u_i \iff k \leq \bigvee_{i \in I} u_i \tag{V.1.1.ii}$$

for all families $u_i \in \mathcal{V}$, $i \in I$, then the compactness condition (V.1.1.i) simplifies to

$$k \leq \bigvee_{z \in X} a(\chi, z)\,, \qquad \text{(V.1.1.iii)}$$

for all $\chi \in TX$. Note that for $\mathcal{V}$ integral (so that $k = \top$) one has $u \otimes u \leq u \otimes k = u$, so that the implication $\Longrightarrow$ of (V.1.1.ii) holds trivially. If $\mathcal{V}$ is a frame with $\otimes = \wedge$, or when $\mathcal{V} = \mathsf{P}_+$, the implication $\Longleftarrow$ of (V.1.1.ii) also holds, so that in this case the simplified compactness condition (V.1.1.iii) applies.

An example of a non-superior quantale is given in Remark V.1.4.3(1).

V.1.1.4 Examples

(1) Let $\mathbb{T} = \mathbb{I}$ be the identity monad (identically extended to $\mathcal{V}$-Rel). Then, for $x \neq y$ in a Hausdorff $(\mathbb{T}, \mathcal{V})$-space (X, a), we must have $a(x, y) = \bot$. Consequently, if $k = \top$ in $\mathcal{V}$, necessarily $a = 1_X$, i.e. (X, a) is discrete. However, quite trivially, (X, a) is always compact. Briefly, if $\mathcal{V}$ is integral, then $\mathcal{V}\text{-Cat}_{\text{Comp}} = \mathcal{V}\text{-Cat}$, *while* $\mathcal{V}\text{-Cat}_{\text{Haus}}$ *is the full coreflective subcategory of discrete* $\mathcal{V}$*-categories in* $\mathcal{V}$-Cat.

(2) In Top $\cong (\beta, 2)$-Cat (see Section III.2.2), the Hausdorff separation property and the compactness property of Definition V.1.1.1 are equivalent to the usual notions that are expressed in terms of open sets, provided that the Axiom of Choice is granted: a topological space X is Hausdorff if and only if for any distinct points x, y one finds disjoint open subsets $U \ni x$, $V \ni y$, and X is compact if and only if any open cover of X (i.e. any set $\mathcal{U}$ of open subsets with $\bigcup \mathcal{U} = X$) contains a finite subcover (see Propositions III.2.3.1 and III.2.3.2). Briefly,

$$(\beta, 2)\text{-Cat}_{\text{Haus}} \cong \text{Haus}\,, \quad (\beta, 2)\text{-Cat}_{\text{Comp}} \cong \text{Comp}\,.$$

(3) In Top $\cong (\mathbb{F}, 2)$-Cat (see Corollary IV.1.5.4), since the filter PX on X converges to every point in X, a Hausdorff $(\mathbb{F}, 2)$-space can have at most one point: $(\mathbb{F}, 2)\text{-Cat}_{\text{Haus}}$ is equivalent to $\{\varnothing, 1\}$. Considering the filter $\{X\}$, one sees that a compact $(\mathbb{F}, 2)$ space X must contain a point whose only neighborhood is X. Since every other filter on X converges to that point as well, this property characterizes compactness in $(\mathbb{F}, 2)$-Cat. Spaces with this property are called *supercompact* since they may equivalently be described by the property that every open cover of X contains at most one open subset of X (see Exercise V.1.A).

(4) If we replace the filter monad by the submonad $\mathbb{F}_{\text{p}} = (F_{\text{p}}, m, e)$, where $F_{\text{p}}X = FX \setminus \{PX\}$ is the set of *proper* filters on X, equipped with its Kleisli extension $\hat{\mathbb{F}}_{\text{p}}$, then the Hausdorff property assumes its usual meaning: $(\mathbb{F}_{\text{p}}, 2)\text{-Cat}_{\text{Haus}} \cong \text{Haus}$. However, $(\mathbb{F}_{\text{p}}, 2)\text{-Cat}_{\text{Comp}} = (\mathbb{F}, 2)\text{-Cat}_{\text{Comp}} \setminus \{\varnothing\}$. Consequently, compact Hausdorff objects in $(\mathbb{F}_{\text{p}}, 2)$-Cat can have at most one point (see Exercise V.1.A).

(5) An object (X, a) in (β, P_+)-Cat $\cong$ App (see Theorem III.2.4.5) is Hausdorff precisely when $a(z, x) < \infty$ & $a(z, y) < \infty$ implies $x = y$, for

all $x, y \in X$, $z \in \beta X$. Obviously, (X, a) is Hausdorff if and only if its *pseudotopological modification* (X, oa) of (X, a) is Hausdorff. (Here $o : [0, \infty]^{\mathrm{op}} \to 2 = \{\bot, \top\}$ is the "optimist's map" with $(o(v) = \top \iff v < \infty)$, which induces a functor

$$(\mathbb{B}, \mathsf{P}_+)\text{-}\mathsf{Cat} \cong \mathsf{App} \to (\mathbb{B}, 2)\text{-}\mathsf{Gph} \cong \mathsf{PsTop}\,,$$

see Exercise II.1.I and Examples III.4.1.3; a pseudotopological space $(X, \longrightarrow)$ is called Hausdorff if $z \longrightarrow x$ & $z \longrightarrow y$ implies $x = y$, for all $x, y \in X$, $\chi \in \beta X$.) The approach space (X, a) is compact if and only if

$$\inf_{x \in X} a(\chi, x) = 0$$

for all $\chi \in \beta X$, a property called 0-*compact* in approach theory (see [Lowen, 1988]).

V.1.1.5 Proposition

(1) $(\mathbb{T}, \mathcal{V})\text{-}\mathsf{Cat}_{\mathrm{Haus}}$ *is closed under non-empty mono-sources in* $(\mathbb{T}, \mathcal{V})\text{-}\mathsf{Cat}$*; it is closed under all mono-sources, and therefore it is strongly epireflective in* $(\mathbb{T}, \mathcal{V})\text{-}\mathsf{Cat}$ *if* $\mathcal{V}$ *is integral or* $T1 \cong 1$.

(2) $(\mathbb{T}, \mathcal{V})\text{-}\mathsf{Cat}_{\mathrm{Comp}}$ *is closed under those sinks* $g_i : (X_i, a_i) \to (Y, b)$ *in* $(\mathbb{T}, \mathcal{V})\text{-}\mathsf{Cat}$ *for which* $(Tg_i)_{i \in I}$ *is epic in* Set.

Proof (1): For a mono-source $(f_i : (X, a) \to (Y_i, b_i))_{i \in I}$ with $I \neq \varnothing$ in $(\mathbb{T}, \mathcal{V})\text{-}\mathsf{Cat}$, one has $1_X = \bigwedge_{i \in I} f_i^\circ \cdot f_i$ (see Proposition III.1.2.2). Hence, if all (Y_i, b_i) are Hausdorff, from

$$a \cdot a^\circ \leq (f_i^\circ \cdot b_i \cdot Tf_i) \cdot ((Tf_i)^\circ \cdot b_i^\circ \cdot f_i) \leq f_i^\circ \cdot b_i \cdot b_i^\circ \cdot f_i \leq f_i^\circ \cdot f_i$$

one obtains $a \cdot a^\circ \leq 1_X$, as desired. If $I = \varnothing$, then $|X| \leq 1$, and the Hausdorff criterion of Proposition V.1.1.2(1) is trivially satisfied when $k = \top$ or $T1 \cong 1$. The additional claim follows with Proposition II.5.10.1.

(2): Similarly, the hypothesis on the sink $(g_i)_{i \in I}$ guarantees $1_{TY} = \bigvee_{i \in I} Tg_i \cdot (Tg_i)^\circ$. Consequently, with all (X_i, a_i) compact, one obtains

$$b^\circ \cdot b \geq (Tg_i \cdot a_i^\circ \cdot g_i^\circ) \cdot (g_i \cdot a_i \cdot (Tg_i)^\circ) \geq Tg_i \cdot a_i^\circ \cdot a_i \cdot (Tg_i)^\circ \geq Tg_i \cdot (Tg_i)^\circ$$

and, hence, $b^\circ \cdot b \geq 1_{TY}$. □

V.1.1.6 Corollary

© (1) *For a surjective* $(\mathbb{T}, \mathcal{V})$*-continuous map* $g : (X, a) \to (Y, b)$ *between* $(\mathbb{T}, \mathcal{V})$*-spaces, if* (X, a) *is compact, then* (Y, b) *is compact.*

(2) *If the* Set*-functor* T *preserves small coproducts,* $(\mathbb{T}, \mathcal{V})\text{-}\mathsf{Cat}_{\mathrm{Comp}}$ *is closed under small epi-sinks in* $(\mathbb{T}, \mathcal{V})\text{-}\mathsf{Cat}$*. The same statement holds for finite coproducts with closure under finite epi-sinks.*

Proof (1): g is a split epimorphism in Set and is preserved by T.

(2): Decompose $(Tg_i)_{i\in I}$ as

$$TX_i \longrightarrow \coprod_i TX_i \xrightarrow{\ d\ } T(\coprod_i X_i) \xrightarrow{\ Tg\ } TY\,.$$

Here the induced morphism g is epic and therefore preserved by T, and the canonical comparison morphism d is bijective by hypothesis. It actually suffices to know that d is surjective to be able to conclude that $(Tg_i)_{i\in I}$ is epic and to apply Proposition V.1.1.5. □

V.1.1.7 Example Compact spaces are closed under finite epi-sinks in $\mathsf{Top} \cong (\mathbb{\beta}, \mathsf{2})\text{-}\mathsf{Cat}$ (since $\beta : \mathsf{Set} \to \mathsf{Set}$ preserves finite coproducts), but not under countable epi-sinks (consider $(n : 1 \to \mathbb{N})_{n\in\mathbb{N}}$ with $\mathbb{N}$ discrete). Likewise, 0-compact spaces are closed under finite epi-sinks in $\mathsf{App} \cong (\mathbb{\beta}, \mathsf{P}_+)\text{-}\mathsf{Cat}$, but not under countable epi-sinks.

Finally, Hausdorffness and compactness are preserved by algebraic functors (see Section III.3.4).

V.1.1.8 Proposition *Let $\alpha : (\mathbb{S}, \hat{S}) \to (\mathbb{T}, \hat{T})$ be a morphism of lax extensions. Then the associated algebraic functor $A_\alpha : (\mathbb{T}, \mathcal{V})\text{-}\mathsf{Cat} \to (\mathbb{S}, \mathcal{V})\text{-}\mathsf{Cat}$ preserves Hausdorffness and compactness.*

Proof Preservation of Hausdorffness follows from $(a \cdot \alpha_X) \cdot (a \cdot \alpha_X)^\circ = a \cdot \alpha_X \cdot \alpha_X^\circ \cdot a^\circ \leq a \cdot a^\circ \leq 1_X$, and preservation of compactness follows from $(a \cdot \alpha_X)^\circ \cdot (a \cdot \alpha_X) = \alpha_X^\circ \cdot a^\circ \cdot a \cdot \alpha_X \geq \alpha_X^\circ \cdot \alpha_X \geq 1_{TX}$. □

V.1.2 Tychonoff Theorem, Čech–Stone compactification

For $\mathbb{T}$ and $\mathcal{V}$ as in Section V.1.1, a $(\mathbb{T}, \mathcal{V})$-space (X, a) is by definition compact Hausdorff when $a : TX \nrightarrow X$ is a map in the ordered category $\mathcal{V}$-Rel. By Proposition III.1.2.1, a is then actually a Set-map $TX \to X$, provided that $\mathcal{V}$ is integral and lean. In that case, the $\mathcal{V}$-relational inequalities

$$a \cdot Ta \leq a \cdot \hat{T}a \leq a \cdot m_X \qquad \text{and} \qquad 1_X \leq a \cdot e_X$$

between Set-maps must actually be equalities; likewise, the $(\mathbb{T}, \mathcal{V})$-continuity condition $f \cdot a \leq b \cdot Tf$ for $f : (X, a) \to (Y, b)$ must actually be an equality if (X, a), (Y, b) are both compact Hausdorff. This proves the following result.

V.1.2.1 Proposition *Let $\mathcal{V}$ be integral and lean. Then $(\mathbb{T}, \mathcal{V})\text{-}\mathsf{Cat}_{\mathrm{CompHaus}}$ is precisely the full subcategory of $\mathsf{Set}^{\mathbb{T}}$ containing those $\mathbb{T}$-algebras (X, a) with $a \cdot \hat{T}a = a \cdot m_X$. In particular, if $\hat{\mathbb{T}}$ is flat, then*

$$(\mathbb{T}, \mathcal{V})\text{-}\mathsf{Cat}_{\mathrm{CompHaus}} = \mathsf{Set}^{\mathbb{T}}\,.$$

V.1.2.2 Examples

(1) The category of 0-compact approach spaces with Hausdorff pseudotopological modification (see Example V.1.1.4(5)) is isomorphic to the category of compact Hausdorff spaces:

$$(\beta, \mathsf{P}_+)\text{-}\mathsf{Cat}_{\mathsf{CompHaus}} \cong (\beta, 2)\text{-}\mathsf{Cat}_{\mathsf{CompHaus}} \cong \mathsf{Set}^{\beta} \cong \mathsf{CompHaus}\,.$$

(2) Example V.1.1.4.(3) shows that flatness of $\hat{T}$ is essential for $(\mathbb{T}, \mathcal{V})\text{-}\mathsf{Cat}_{\mathsf{CompHaus}} = \mathsf{Set}^{\mathbb{T}}$: while $\mathsf{Set}^{\mathbb{F}}$ is the category of cocontinuous lattices (see IV.4.4), $(\mathbb{F}, 2)\text{-}\mathsf{Cat}_{\mathsf{CompHaus}}$ contains just singletons.

Proposition V.1.2.1 entails a *Tychonoff Theorem* for compact Hausdorff objects in $(\mathbb{T}, \mathcal{V})\text{-}\mathsf{Cat}$, with the needed hypotheses on $\mathbb{T}$ and $\mathcal{V}$ granted as follows. With (X_i, a_i) in $\mathsf{Set}^{\mathbb{T}}$ for all $i \in I$, the product $X = \prod_{i\in I} X_i$ in $\mathsf{Set}^{\mathbb{T}}$ carries the structure $a : TX \to X$ with $\pi_i \cdot a = a_i \cdot T\pi_i$ (where π_i are the product projections for all $i \in I$), and using Proposition III.1.2.2 one obtains that a coincides with the product structure formed in $(\mathbb{T}, \mathcal{V})\text{-}\mathsf{Cat}$ (see Proposition III.3.1.1):

$$a = (\textstyle\bigwedge_{i\in I} \pi_i^\circ \cdot \pi_i) \cdot a = \textstyle\bigwedge_{i\in I} \pi_i^\circ \cdot \pi_i \cdot a = \textstyle\bigwedge_{i\in I} \pi_i^\circ \cdot a_i \cdot T\pi_i\,.$$

With the help of the Adjoint Functor Theorem II.2.12.1 we can now go one step further and guarantee the existence of a *Čech–Stone compactification* for $(\mathbb{T}, \mathcal{V})$-spaces.

V.1.2.3 Theorem *Let $\mathcal{V}$ be integral and lean, and let $\hat{T}$ be flat. Then $(\mathbb{T}, \mathcal{V})\text{-}\mathsf{Cat}_{\mathsf{CompHaus}}$ is reflective in $(\mathbb{T}, \mathcal{V})\text{-}\mathsf{Cat}_{\mathsf{Haus}}$, which, in turn, is strongly epireflective in $(\mathbb{T}, \mathcal{V})\text{-}\mathsf{Cat}$.*

Proof The embedding $\mathsf{Set}^{\mathbb{T}} \hookrightarrow (\mathbb{T}, \mathcal{V})\text{-}\mathsf{Cat}$ preserves not only small products, but also equalizers. Indeed, the $\mathbb{T}$-algebra structure of the equalizer $j : E \hookrightarrow X$ of $f, g : (X, a) \to (Y, b)$ in $\mathsf{Set}^{\mathbb{T}}$ is the map $a_0 : TE \to E$ with $j \cdot a_0 = a \cdot Tj$; hence, $a_0 = j^\circ \cdot a \cdot Tj$, which is the structure of the equalizer formed in $(\mathbb{T}, \mathcal{V})\text{-}\mathsf{Cat}$. Consequently, $\mathsf{Set}^{\mathbb{T}}$ is closed under small limits in $(\mathbb{T}, \mathcal{V})\text{-}\mathsf{Cat}$ and therefore in $(\mathbb{T}, \mathcal{V})\text{-}\mathsf{Cat}_{\mathsf{Haus}}$ (which is strongly epireflective in $(\mathbb{T}, \mathcal{V})\text{-}\mathsf{Cat}$, see Proposition V.1.1.5).

In order to build a solution set in $\mathsf{Set}^{\mathbb{T}} = (\mathbb{T}, \mathcal{V})\text{-}\mathsf{Cat}_{\mathsf{CompHaus}}$ for (X, a) in $(\mathbb{T}, \mathcal{V})\text{-}\mathsf{Cat}_{\mathsf{Haus}}$, consider any $(\mathbb{T}, \mathcal{V})$-continuous $f : (X, a) \to (Y, b)$ with (Y, b) in $\mathsf{Set}^{\mathbb{T}}$ and form the least $\mathbb{T}$-subalgebra $\langle M\rangle$ of (Y, b) containing $M = f(X)$, which may be constructed as the image $h(TM)$ of the $\mathbb{T}$-homomorphism $h : (TM, m_M) \to (Y, b)$ with $h \cdot e_M = (M \hookrightarrow Y)$. Hence, f factors as $(X \to \langle M\rangle \hookrightarrow Y)$, where the cardinality of $\langle M\rangle$ cannot exceed the cardinality of TX. Hence a solution set for (X, a) can be given by a representative system of non-isomorphic $\mathbb{T}$-algebras (Y, b) whose cardinalities do not exceed that of TX. □

Theorem V.1.2.3 relies on the algebraic realization of $(\mathbb{T}, \mathcal{V})\text{-}\mathsf{Cat}_{\mathsf{CompHaus}}$ as $\mathsf{Set}^{\mathbb{T}}$. The theme of Hausdorff-separation and compactness will be taken up again in Sections V.3 and V.4, through an algebraic approach to proper maps.

V.1.3 Compactness for Kleisli-extended monads

The example $\mathsf{Top} \cong (\beta, 2)\text{-}\mathsf{Cat} \cong (\mathbb{F}, 2)\text{-}\mathsf{Cat}$ shows that the notion of compactness may change dramatically when changing the parameters $\mathbb{T}$ or $\mathcal{V}$; see Examples V.1.1.4(2) and (3). Here, we wish to explore this change when $(\mathbb{T}, \mathcal{V})\text{-}\mathsf{Cat}$ is presented relationally as $(\Pi(\mathbb{T}, \mathcal{V}), 2)\text{-}\mathsf{Cat}$, as in Corollary IV.3.2.3.

With that goal in mind, let us first look at relational $\mathbb{T}$-algebras in general. Independently of the lax extension of $\mathbb{T}$ to Rel, for (X, a) in $(\mathbb{T}, 2)\text{-}\mathsf{Cat}$ one has (writing $\chi \longrightarrow x$ instead of $a(\chi, x) = \top$):

$$(X, \longrightarrow) \text{ is Hausdorff} \iff \forall x, y \in X,\ z \in TX\ (z \longrightarrow x \ \&\ z \longrightarrow y \implies x = y)\,,$$
$$(X, \longrightarrow) \text{ is compact} \iff \forall \chi \in TX\ \exists z \in X\ (\chi \longrightarrow z)\,.$$

In the case where $\mathbb{T} = (T, m, e)$ is power-enriched via $\tau : \mathbb{P} \to \mathbb{T}$, so that $\tau : (\mathbb{P}, \check{P}) \to (\mathbb{T}, \check{T})$ is a morphism of the respective Kleisli extensions, the algebraic functor

$$A_\tau : (\mathbb{T}, 2)\text{-}\mathsf{Cat} \longrightarrow (\mathbb{P}, 2)\text{-}\mathsf{Cat} \cong \mathsf{Ord}$$

simply equips $(X, \longrightarrow)$ with the induced order considered in Section III.3.3:

$$x \le y \iff \tau(\{x\}) \longrightarrow y \iff e_X(x) \longrightarrow y\,.$$

V.1.3.1 Proposition *For a monad $\mathbb{T}$ power-enriched by $\tau : \mathbb{P} \to \mathbb{T}$ and $(X, \longrightarrow)$ in $(\mathbb{T}, 2, \check{\mathbb{T}})\text{-}\mathsf{Cat}$, one has:*

1. *$(X, \longrightarrow)$ is Hausdorff if and only if $|X| \le 1$.*
2. *If $(X, \longrightarrow)$ is compact, X has a largest element, with the converse statement holding when $\tau(X)$ is the largest element in TX.*

Proof (1): The complete lattice TX has a least element p. By Proposition III.3.3.6 and Remark IV.1.4.4 one has

$$(p \le \chi \ \&\ \chi \longrightarrow x) \implies (p \longrightarrow x)$$

for all $x \in X$, $\chi \in TX$. Since $e_X(x) \longrightarrow x$, one has $p \longrightarrow x$ for all $x \in X$, and (1) follows.

(2): For $(X, \longrightarrow)$ compact there is x_0 with $\tau(X) \longrightarrow x_0$. Since $\tau(X) = \bigvee_{x \in X} e_X(x)$ (see Exercise IV.1.B), one obtains $e_X(x) \longrightarrow x_0$ and, hence, $x \le x_0$ for all $x \in X$. Conversely, assuming x_0 a largest element in X, with Exercise IV.1.H one concludes $\tau(X) \longrightarrow x_0$, which, with $\tau(X)$ the largest element in TX, gives $\chi \longrightarrow x_0$ for all $\chi \in TX$. □

For $\mathbb{T} = \mathbb{F}$ the filter monad, the least element p of FX is the filter PX, while the largest element $\tau(X)$ is the filter $\{X\}$. Compactness in $(\mathbb{F}, 2)$-Cat (i.e. supercompactness, see Example V.1.1.4(3)) may seem like a rare property at first sight. However, the following Proposition shows that in general there is a rich supply of such spaces.

Recall from IV.4.1 and Theorem IV.1.5.3 that for a power-enriched monad $\mathbb{T}$ one has the functors

$$\begin{array}{ccccc} \mathsf{Set}^{\mathbb{T}} & \longrightarrow & \mathbb{T}\text{-}\mathsf{Mon} & \longrightarrow & (\mathbb{T}, 2)\text{-}\mathsf{Cat}, \\ (X, \alpha) & \longmapsto & (X, \alpha^{\dashv}) = (X, \nu) & \longmapsto & (X, \longrightarrow)\,, \end{array}$$

where $\alpha^{\dashv}$ is right adjoint to the sup-map $\alpha : TX \to X$, and

$$\chi \longrightarrow x \iff \chi \le \nu(x) \iff \chi \le \alpha^{\dashv}(x) \iff \alpha(\chi) \le x \qquad \text{(V.1.3.i)}$$

for all $x \in X$, $\chi \in TX$. In particular, setting $x = \alpha(\chi)$, one obtains the following result.

V.1.3.2 Proposition *For a power-enriched monad $\mathbb{T}$, every $\mathbb{T}$-algebra is compact when considered as a $(\mathbb{T}, 2)$-space via* (V.1.3.i). *In fact, for $(X, \alpha) \in \mathsf{Set}^{\mathbb{T}}$ and every $\chi \in TX$ there is a least point $x \in X$ with $\chi \longrightarrow x$.*

For $\mathbb{T} = \mathbb{F}$, one obtains from Section IV.4.4:

V.1.3.3 Corollary *Every continuous lattice endowed with its Scott topology is supercompact.*

Let us now consider the discrete presheaf monad

$$\mathbb{\Pi} = \mathbb{\Pi}(\mathbb{T}, \mathcal{V}, \hat{\mathbb{T}})$$

induced by a monad $\mathbb{T}$ and an associative lax extension $\hat{\mathbb{T}}$ to $\mathcal{V}$-Rel and apply Proposition V.1.3.2 to $\mathbb{\Pi}$ with its Kleisli extension (see Corollary IV.3.2.3):

$$\begin{array}{ccccc} \mathsf{Set}^{\mathbb{\Pi}} & \longrightarrow & \mathbb{\Pi}\text{-}\mathsf{Mon} & \longrightarrow & (\mathbb{\Pi}, 2)\text{-}\mathsf{Cat} \cong (\mathbb{T}, \mathcal{V})\text{-}\mathsf{Cat} \\ (X, \alpha) & \longmapsto & (X, \alpha^{\dashv}) & \longmapsto & (X, \longrightarrow)\,. \end{array}$$

V.1.3.4 Corollary *For a monad $\mathbb{T}$ with an associative lax extension $\hat{\mathbb{T}}$ to $\mathcal{V}$-Rel, every $\mathbb{\Pi}(\mathbb{T}, \mathcal{V})$-algebra is compact as an object of* $(\mathbb{\Pi}, 2)$-Cat.

V.1.3.5 Example Consider $\mathbb{T} = \beta$ with its Barr extension to $\mathcal{V} = \mathsf{P}_{+} = ([0, \infty]^{\mathrm{op}}, +, 0)$, which carries a $\mathbb{\Pi}(\mathbb{T}, \mathcal{V})$-algebra structure

$$\begin{aligned} \alpha : \Pi[0, \infty] = (\mathbb{T}, \mathcal{V})\text{-}\mathsf{URel}(1, V) &\to (\mathbb{T}, \mathcal{V})\text{-}\mathsf{URel}(1, 1) \cong [0, \infty] \\ \varphi &\mapsto \varphi \circ \iota = \inf\nolimits_{\chi \in [0,\infty]} \varphi(\chi) + \xi(\chi)\,, \end{aligned}$$

where $\iota : 1 \nrightarrow V$ is given by the identity map on $[0,\infty]$ and $\xi(\chi) = \sup_{A\in\chi}\inf_{u\in A} u$ for all $\chi \in \beta[0,\infty]$; see Section IV.3.3. (Note that the convergence map $\xi : \beta[0,\infty] \to [0,\infty]$ provides $[0,\infty]$ with the standard topology of $[0,\infty]$.) By Corollary V.1.3.4, the space $([0,\infty], \longrightarrow)$ with

$$\varphi \longrightarrow v \iff \inf_{\chi\in\beta[0,\infty]}(\varphi(\chi) + \xi(\chi)) \geq v$$

becomes compact in $(\mathbb{\Pi}, 2)$-Cat $\cong$ App.

Finally, let us examine compactness in $(\mathbb{\Pi}(\mathbb{T}, \mathcal{V}), 2)$-Cat beyond spaces that arise from $\mathbb{\Pi}$-algebras. Recall from Section IV.3.2 the isomorphisms

$$\begin{array}{ccccc} (\mathbb{T},\mathcal{V})\text{-Cat} & \xrightarrow{\cong} & \mathbb{\Pi}\text{-Mon} & \xrightarrow{\cong} & (\mathbb{\Pi}, 2)\text{-Cat} \\ (X,a) & \longmapsto & (X,a^\flat) & \longmapsto & (X,\longrightarrow)\,, \end{array}$$

where $a^\flat : X \to \Pi X$ is given by $a^\flat(x) = x^\circ \cdot a = a(-,x)$, and

$$\varphi \longrightarrow x \iff \varphi \leq a(-,x)$$

for all $\varphi \in \Pi X$.

V.1.3.6 Proposition *For a monad $\mathbb{T}$ with a flat associative lax extension to $\mathcal{V}$-Rel and a $(\mathbb{T},\mathcal{V})$-space (X,a), each of the following statements implies the next:*

(i) *$(X,\longrightarrow)$ is compact in $(\mathbb{\Pi}(\mathbb{T},\mathcal{V}), 2)$-Cat;*
(ii) *$\forall \chi \in TX\, \exists x \in X\, (a(\chi, x) \geq k)$;*
(iii) *(X,a) is compact in $(\mathbb{T},\mathcal{V})$-Cat.*

Proof For (i) $\Longrightarrow$ (ii), we observe that compactness of $(X,\longrightarrow)$ means

$$\forall \varphi \in \Pi X\; \exists x \in X\; (\varphi \longrightarrow x)\,.$$

Hence, given $\chi \in TX$ we may exploit this property for $\varphi = \mathbf{Y}_X(\chi)$, with $\mathbf{Y} : \mathbb{T} \to \mathbb{\Pi}$ as in Proposition IV.3.2.5. With $\varphi \longrightarrow x \in X$ we then have

$$k = \varphi(\chi) \leq a(\chi, x)\,,$$

as desired. (ii) $\Longrightarrow$ (iii) is immediate. □

V.1.3.7 Remark For $\mathbb{T} = \mathbb{I}$ the identity monad identically extended to $\mathcal{V}$-Rel, $\mathbb{\Pi}(\mathbb{I},\mathcal{V}) = \mathbb{P}_{\mathcal{V}}$ is the $\mathcal{V}$-powerset monad and $(\mathbb{\Pi}, 2)$-Cat $\cong$ $\mathcal{V}$-Cat; see Corollary IV.3.2.3.

For $\mathcal{V} = 2$, so that $\mathcal{V}$-Cat $\cong$ Ord and $\varphi \in \mathbb{P}_2 X$ corresponds to $A \subseteq X$, one has

$$A \longrightarrow x \iff A \subseteq {\downarrow} x\,,$$

so that compactness of $(X,\leq)$ in $(\mathbb{\Pi}, 2)$-Cat means existence of a largest element in X (as already confirmed in Proposition V.1.3.1(2)). When $\mathcal{V} = \mathsf{P}_+$, so that $\mathcal{V}$-Cat $\cong$ Met, for $\varphi \in \Pi X$ one has

$$\varphi \longrightarrow x \iff \varphi \geq a(-,x)$$

for a metric space (X, a) considered as a $(\Pi, 2)$-space. With φ constantly 0 one sees:

$$\begin{aligned}(X, \longrightarrow) \text{ compact} &\iff \exists x \in X\, \forall y \in X\ (a(y, x) = 0) \\ &\iff (X, \leq) \text{ compact},\end{aligned}$$

where $\leq$ is the underlying order of (X, a), making $(X, \leq)$ an object of $(\Pi(\mathbb{I}, 2), 2)$-Cat.

This invariance property of compactness is now shown to hold in full generality, not just for $\mathbb{T} = \mathbb{I}$.

As in Theorem IV.3.4.4, *let* $\varphi : (\mathcal{V}, \otimes, k) \to (\mathcal{W}, \otimes, l)$ *be a homomorphism of integral quantales with a right adjoint* ψ *(in* Ord*); furthermore, for a monad* $\mathbb{T}$ *we assume* φ *to be strictly compatible with the associative lax extensions* $\hat{\mathbb{T}}$, $\tilde{\mathbb{T}}$ *to* $\mathcal{V}$-Rel, $\mathcal{W}$-Rel, *respectively, so that*

$$\tilde{T}(\varphi r) = \varphi(\hat{T} r)$$

for all $\mathcal{V}$*-relations* r. With $\Pi_{\mathcal{V}} = \Pi(\mathbb{T}, \mathcal{V})$, $\Pi_{\mathcal{W}} = \Pi(\mathbb{T}, \mathcal{W})$ there is then a morphism

$$\varphi^{\mathbb{T}} = \Pi(\mathbb{T}, \varphi) : \Pi_{\mathcal{V}} \to \Pi_{\mathcal{W}}$$

of power-enriched monads whose algebraic functor commutes with the change-of-base functor induced by ψ:

$$\begin{array}{ccc} (\mathbb{T}, \mathcal{W})\text{-Cat} & \longrightarrow & (\Pi_{\mathcal{W}}, 2)\text{-Cat} \\ \downarrow B_\psi & & \downarrow A_{\varphi^{\mathbb{T}}} \\ (\mathbb{T}, \mathcal{V})\text{-Cat} & \longrightarrow & (\Pi_{\mathcal{V}}, 2)\text{-Cat} \end{array} \qquad \begin{array}{ccc} (X, a) & \longmapsto & (X, \longrightarrow) \\ \downarrow & & \downarrow \\ (X, \psi a) & \longmapsto & (X, \rightsquigarrow), \end{array}$$

where

$$\begin{aligned} \rho \longrightarrow x &\iff \rho \leq a(-, x), \\ \theta \rightsquigarrow x &\iff \theta \leq \psi a(-, x) \iff \varphi\theta \leq a(-, x), \end{aligned}$$

for all $\rho \in \Pi_{\mathcal{W}} X, \theta \in \Pi_{\mathcal{V}} X, x \in X$.

V.1.3.8 Theorem *For* $(X, a) \in (\mathbb{T}, \mathcal{W})$-Cat,

$$(X, \longrightarrow) \textit{ compact in } (\Pi_{\mathcal{W}}, 2)\text{-Cat} \iff (X, \rightsquigarrow) \textit{ compact in } (\Pi_{\mathcal{V}}, 2)\text{-Cat}.$$

Proof " $\Longrightarrow$ " follows from Proposition V.1.1.8. For " $\Longleftarrow$ ", consider $\theta : TX \nrightarrow 1$ with $\theta(\chi) = \top$ for all $\chi \in TX$. Since

$$\theta \cdot \hat{T} 1_X \geq \theta \text{ and } e_1^\circ \cdot \hat{T}\theta \cdot m_X^\circ \geq e_1^\circ \cdot \hat{T}\theta \cdot e_{TX} \geq \theta,$$

θ is unitary when $k = \top$. Compactness of $(X, \rightsquigarrow)$ gives $x_0 \in X$ with $\theta \longrightarrow x_0$, so that $\theta \leq \psi a(-, x_0)$ implies $\varphi\theta \leq a(-, x_0)$, with $\varphi\theta$ the top element in $\Pi_{\mathcal{W}} X$. Consequently, $\rho \longrightarrow x_0$ for all $\rho \in \Pi_{\mathcal{W}} X$. ☐

V.1.3.9 Corollary *An approach space is compact as an object of* $(\mathbb{\Pi}(\beta, \mathsf{P}_+), \mathsf{2})$-$\mathsf{Cat}$ *if and only if its topological coreflection is supercompact.*

Proof Apply Theorem V.1.3.8 with $\varphi = \iota : \mathsf{2} \to \mathsf{P}_+$ and $\psi = p$ the "pessimist's map" (see Exercise II.1.I). □

V.1.4 Examples involving monoids

For a monoid (H, μ, η) we consider the monad

$$\mathbb{H} = (H \times (-), m, e)$$

on Set, with $m_X = \mu \times 1_X : H \times H \times X \to H \times X$ and $e_X = \langle \eta, 1_X \rangle : X \to H \times X$ (see Exercise II.3.B). The Barr extension $\overline{H} : \mathsf{Rel} \to \mathsf{Rel}$ is a lax extension of the monad $\mathbb{H}$, and it is given explicitly by

$$(\alpha, x)\ \overline{H}r\ (\beta, y) \iff \alpha = \beta \;\&\; x\ r\ y\,,$$

for $r : X \nrightarrow Y$, $x \in X$, $y \in Y$, and $\alpha, \beta \in H$. For a relation $a : H \times X \nrightarrow X$, when we write $x \xrightarrow{\alpha} y$ instead of $(\alpha, x)\ a\ y$ (i.e. "x is related to y with weight α"), an $(\mathbb{H}, \mathsf{2})$-category (X, a) is characterized by the conditions

$$x \xrightarrow{\eta} x \quad \text{and} \quad (x \xrightarrow{\alpha} y \;\&\; y \xrightarrow{\beta} z \implies x \xrightarrow{\beta\cdot\alpha} z)\,,$$

where $\beta \cdot \alpha = \mu(\beta, \alpha)$ in H; a morphism $f : (X, a) \to (Y, b)$ must satisfy

$$x \xrightarrow{\alpha} y \implies f(x) \xrightarrow{\alpha} f(y)$$

for all $x, y \in X$, $\alpha \in H$. For an $(\mathbb{H}, \mathsf{2})$-category (X, a), one has:

$$(X, a) \text{ is compact} \iff \forall x \in X, \alpha \in H\ \exists y \in Y\ (x \xrightarrow{\alpha} y)\,,$$

$$(X, a) \text{ is Hausdorff} \iff \forall x, y, z \in X, \alpha \in H\ (x \xrightarrow{\alpha} y \;\&\; x \xrightarrow{\alpha} z \implies y = z)\,,$$

and the conjunction of both properties makes $a : H \times X \to X$ an action of H on X: $x \xrightarrow{\alpha} y$ means $y = \alpha \cdot x$ when we write the action multiplicatively. These observations illustrate the assertions of Proposition V.1.2.1 and Theorem V.1.2.3.

V.1.4.1 Corollary *The category of compact Hausdorff* $(\mathbb{H}, \mathsf{2})$*-spaces and* $(\mathbb{H}, \mathsf{2})$*-continuous maps is isomorphic to the category of* H*-actions and equivariant maps:*

$$(\mathbb{H}, \mathsf{2})\text{-}\mathsf{Cat}_{\mathrm{CompHaus}} \cong \mathsf{Set}^{\mathbb{H}}\,.$$

It is reflective in $(\mathbb{H}, \mathsf{2})$-$\mathsf{Cat}_{\mathrm{Haus}}$*, which, in turn, is strongly epireflective in* $(\mathbb{H}, \mathsf{2})$-Cat.

Our arrow notation for the structure of an $(\mathbb{H}, \mathsf{2})$-category (X, a) emphasizes that X is actually the object set of a small category, denoted again by X, with hom-sets

$$X(x, y) = \{(x, \alpha, y) \mid \alpha \in H, x \xrightarrow{\alpha} y \text{ in } (X, a)\}\,;$$

moreover, this category comes with a faithful functor

$$p : X \to H , \quad (x, \alpha, y) \mapsto \alpha ,$$

with H considered as a one-object category. (Hence, p identifies the objects while leaving the morphisms intact.) Considering now more generally small categories X which come equipped with an H*-valued norm*, i.e. a functor $n : \mathsf{X} \to H$, we see that there is a full embedding

$$E : (\mathbb{H}, 2)\text{-}\mathsf{Cat} \hookrightarrow \mathsf{Cat}/H , \quad (X, a) \mapsto (X, p) .$$

V.1.4.2 Proposition *The functor E is reflective and identifies $(\mathbb{H}, 2)$-categories as those small categories over H whose norm is faithful.*

Proof The reflection of an object (X, n) in Cat/H into $(\mathbb{H}, 2)\text{-}\mathsf{Cat}$ may be formed by $X := \mathrm{ob}\,\mathsf{X}$ with $(\mathbb{H}, 2)$-structure

$$x \xrightarrow{\alpha} y \iff \exists \varphi \in \mathsf{X}(x, y)\ (n(\varphi) = \alpha)$$

which makes $n : (\mathsf{X}, n) \to (X, p)$ a morphism in Cat/H and, in fact, the reflection morphism, as one easily verifies. Furthermore, n as a morphism in Cat/H becomes an isomorphism if and only if the functor $n : \mathsf{X} \to H$ is faithful. Consequently, $(\mathbb{H}, 2)\text{-}\mathsf{Cat}$ is equivalent to the full subcategory of Cat/H whose objects have a faithful norm. □

V.1.4.3 Remarks

(1) The category $(\mathbb{H}, 2)\text{-}\mathsf{Cat}$ may also be presented as $\mathcal{V}\text{-}\mathsf{Cat}$ with $\mathcal{V}$ the powerset of H, ordered by inclusion, and equipped with the tensor product $\otimes$ defined by

$$A \otimes B = \{\alpha \cdot \beta \mid \alpha \in A,\ \beta \in B\} ,$$

which preserves unions (see Exercise II.1.M). Then $\{\eta\}$ is the tensor unit, but not the top element of $\mathcal{V}$, hence $\mathcal{V}$ is not integral unless $H = 1$. Moreover, if, for instance, H is a non-trivial group, $\mathcal{V}$ is not superior: let $A = \{\alpha, \alpha^{-1}\}$, with $\alpha \neq \eta$; then $\{\eta\} \not\subseteq A$, although $\{\eta\} \subseteq A \otimes A$.

It is easy to check that if, to each $\mathcal{V}$-space (X, a), we assign the $(\mathbb{H}, 2)$-space $(X, \tilde{a})$, with $\tilde{a}((\alpha, x), y) = \top$ if and only if $\alpha \in a(x, y)$, every $\mathcal{V}$-functor $f : (X, a) \to (Y, b)$ is an $(\mathbb{H}, 2)$-functor $f : (X, \tilde{a}) \to (Y, \tilde{b})$, and in fact this correspondence defines an isomorphism $\mathcal{V}\text{-}\mathsf{Cat} \to (\mathbb{H}, 2)\text{-}\mathsf{Cat}$ (see also Example III.3.5.2(2)). We note, however, that Hausdorff and compact $\mathcal{V}$-spaces do not coincide with Hausdorff and compact $(\mathbb{H}, 2)$-spaces.

(2) In the trivial case $H = 1$, we have $\mathbb{H} = \mathbb{I}$, so that $(\mathbb{H}, 2)\text{-}\mathsf{Cat}$ reproduces $2\text{-}\mathsf{Cat} \cong \mathsf{Ord}$, also identified as the full subcategory of Cat given by small categories X for which $\mathsf{X} \to \mathbf{1}$ is faithful.

We now turn to the list monad $\mathbb{L} = (L, m, e)$ induced by the right adjoint functor $\mathsf{Mon} \to \mathsf{Set}$ (see Example II.3.1.1(2)). The Barr extension $\overline{L} : \mathsf{Rel} \to \mathsf{Rel}$ of the list monad is given by

$$(x_1, \dots, x_n)\ \overline{L}r\ (y_1, \dots, y_m) \iff n = m\ \&\ x_1\ r\ y_1\ \&\dots\ \&\ x_n\ r\ y_n$$

for all $(x_1, \dots, x_n) \in LX$, $(y_1, \dots, y_m) \in LY$, and relations $r : X \nrightarrow Y$; it is obviously an associative flat lax extension of $\mathbb{L}$ (by Corollary III.1.12.2 and Exercise III.1.Q). An $(\mathbb{L}, 2)$-category is a *multi-ordered set*, i.e. a set X with a relation $a : LX \nrightarrow X$ that, when we write

$$(x_1, \dots, x_n) \vdash y \quad \text{for} \quad (x_1, \dots, x_n)\ a\ y\,,$$

must satisfy $(x) \vdash x$ and the transitivity condition

$$\begin{aligned}&(x_1^1, \dots, x_{n_1}^1) \vdash y_1, \dots, (x_1^m, \dots, x_{n_m}^m) \vdash y_m\ \&\ (y_1, \dots, y_m) \vdash z\\ &\implies (x_1^1, \dots, x_{n_m}^m) \vdash z\,.\end{aligned}$$

For an $(\mathbb{L}, 2)$-continuous map $f : (X, a) \to (Y, b)$ one must have the monotonicity condition

$$(x_1, \dots, x_n) \vdash y \implies (f(x_1), \dots, f(x_n)) \vdash f(y)\,.$$

An $(\mathbb{L}, 2)$-space (X, a) is compact precisely when, for all integer $n \geq 0$,

$$\forall x_1, \dots, x_n \in X\ \exists z \in X\ ((x_1, \dots, x_n) \vdash z)\,,$$

and (X, a) is Hausdorff if and only if, for all $n \geq 0$,

$$\forall x_1, \dots, x_n, y, z \in X\ ((x_1, \dots, x_n) \vdash y\ \&\ (x_1, \dots, x_n) \vdash z \implies y = z)\,.$$

In particular, compactness forces $X \neq \varnothing$ (from the case $n = 0$). The conjunction of both properties makes $(x_1, \dots, x_n) \vdash y$ the n-ary extension of a binary monoid operation on X.

V.1.4.4 Corollary *The category of compact Hausdorff* $(\mathbb{L}, 2)$*-spaces and* $(\mathbb{L}, 2)$*-continuous maps is isomorphic to the category of monoids and their homomorphisms:*

$$(\mathbb{L}, 2)\text{-}\mathsf{Cat}_{\mathrm{CompHaus}} \cong \mathsf{Mon}\,.$$

It is reflective in $(\mathbb{L}, 2)\text{-}\mathsf{Cat}_{\mathrm{Haus}}$*, which, in turn, is strongly epireflective in* $(\mathbb{L}, 2)\text{-}\mathsf{Cat}$.

Proof The result is a direct consequence of Proposition V.1.2.1 and Theorem V.1.2.3. □

Exercises

V.1.A *Supercompact topological spaces.* Show that the compact objects in $(\mathbb{F}_p, 2)\text{-}\mathsf{Cat}$ are precisely the supercompact topological spaces (see Example V.1.1.4(4)). Furthermore,

$$(\mathbb{F}_p, 2)\text{-}\mathsf{Cat}_{\mathrm{Haus}} \cong \mathsf{Haus} \quad \text{and} \quad (\mathbb{F}, 2)\text{-}\mathsf{Cat}_{\mathrm{CompHaus}} \cong \{X \in \mathsf{Top} \mid |X| \leq 1\}\,.$$

V.1.B *Algebraic functors induced by monoid homomorphisms*

(1) For a homomorphism $\alpha : H \to K$ of monoids one obtains a morphism $\alpha : \mathbb{H} \to \mathbb{K}$ of monads on Set, which, in turn, induces an algebraic functor $A_\alpha : (\mathbb{K}, 2)\text{-}\mathsf{Cat} \to (\mathbb{H}, 2)\text{-}\mathsf{Cat}$ (see Exercise II.3.B and Section V.1.4).
(2) Considering that the monad associated to the trivial monoid 1 is the identity monad $\mathbb{I}$ on Set, show that for every monoid H there are algebraic functors $A : (\mathbb{H}, 2)\text{-}\mathsf{Cat} \to \mathsf{Ord}$ and $P : \mathsf{Ord} \to (\mathbb{H}, 2)\text{-}\mathsf{Cat}$ with $AP = 1_{\mathsf{Ord}}$.

V.1.C *Induced algebraic functors that are not adjoint.* Let $H = (\{0, 1\}, +, 0)$ and $K = (\{0, 1\}, \cdot, 1)$ be the monoids that make up the ring $\mathbb{Z}/2\mathbb{Z}$.

(1) Show that $(\mathbb{H}, 2)\text{-}\mathsf{Cat}$ is isomorphic to the category of ordered sets $(X, \leq)$ that come with an additional relation $\prec$ satisfying

$$(u \leq x \prec y \leq z \implies u \prec z) \qquad \text{and} \qquad (x \prec y \prec z \implies x \leq z)\,.$$

(2) Show that $(\mathbb{K}, 2)\text{-}\mathsf{Cat}$ may be described as the category of ordered sets $(X, \leq)$ that come with an additional relation $\prec$ satisfying

$$u \leq x \prec y \prec z \leq w \implies u \prec w\,.$$

(3) Identify the compact and Hausdorff objects in $(\mathbb{H}, 2)\text{-}\mathsf{Cat}$ and $(\mathbb{K}, 2)\text{-}\mathsf{Cat}$.
(4) Describe the functors A and P of Exercise V.1.B for both H and K, and show that in both cases A and P fail to be adjoint to each other.

V.1.D *Superior but not integral.* The powerset of the multiplicative monoid $\{0, 1\}$ yields a four-element commutative and superior, but non-integral, quantale.

V.1.E *Characterizing superior completely distributive quantales.* If the $\otimes$-neutral element k of $\mathcal{V}$ satisfies $k = \bigvee_{u \ll k} u$, then the following conditions are equivalent:

(i) $\forall u_i \in \mathcal{V}\, (i \in I)\ (k \leq \bigvee_{i \in I} u_i \implies k \leq \bigvee_{i \in I} u_i \otimes u_i)$;
(ii) $k \leq \bigvee_{u \ll k} u \otimes u$.

Conclude that an integral completely distributive quantale is superior if and only if (ii) holds.

V.2 Low separation, regularity, and normality

We continue to work with a monad $\mathbb{T} = (T, m, e)$ on Set, laxly extended by $\hat{\mathbb{T}}$ to $\mathcal{V}$-Rel, for a quantale $\mathcal{V} = (\mathcal{V}, \otimes, k)$.

V.2.1 Order separation

Every $(\mathbb{T}, \mathcal{V})$-category (X, a) comes with an underlying order, defined by

$$x \leq y \iff k \leq a(e_X(x), y)$$

for all $x, y \in X$ (see Proposition III.3.3.1). We recall that (X, a) is *separated* if its underlying order is separated; i.e. if for all $x, y \in X$

$$x \leq y \;\&\; y \leq x \implies x = y\,.$$

Hence, for $\mathbb{T} = \mathbb{I}$ and $\mathcal{V} = 2$ or P_+, this notion returns the terminology used for ordered sets and metric spaces (see Section II.1.3 and Examples III.1.3.1). For topological spaces, whether considered as objects of $(\mathbb{\beta}, 2)$-Cat or $(\mathbb{F}, 2)$-Cat, separation means *T0-separation*:

$$\dot{x} \longrightarrow y \;\&\; \dot{y} \longrightarrow x \implies x = y\,;$$

equivalently, the map $\nu : X \to FX$ that assigns to every point its neighborhood filter is injective. An approach space (X, a), as an object of $(\mathbb{\beta}, \mathsf{P}_+)$-Cat, is separated if and only if for all $x, y \in X$

$$a(\dot{x}, y) = 0 = a(\dot{y}, x) \implies x = y\,;$$

equivalently, in terms of its approach distance δ, if

$$\delta(x, \{y\}) = 0 = \delta(y, \{x\}) \implies x = y\,.$$

V.2.1.1 Proposition *Let (X, a) be a $(\mathbb{T}, \mathcal{V})$-space.*

(1) *If (X, a) is Hausdorff, then (X, a) is separated.*
(2) *If (X, a) is separated, any $(\mathbb{T}, \mathcal{V})$-continuous map $f : (2, \top) \to (X, a)$ from a two-element indiscrete $(\mathbb{T}, \mathcal{V})$-space is constant, and this property is equivalent to (X, a) being separated if $a(\mathfrak{a}, i) = \top$ for all $\mathfrak{a} \in T2 \setminus \{e_2(i)\}$, $i \in 2 = \{0, 1\}$.*
(3) *The full subcategory $(\mathbb{T}, \mathcal{V})$-$\mathsf{Cat}_{\mathrm{sep}}$ of separated $(\mathbb{T}, \mathcal{V})$-spaces is closed under mono-sources in $(\mathbb{T}, \mathcal{V})$-Cat.*

Proof (1): If (X, a) is Hausdorff, already $k \leq a(e_X(x), y)$ implies

$$\bot < k = k \otimes k \leq a(e_X(x), x) \otimes a(e_X(x), y)$$

and then $x = y$.

(2): For a $(\mathbb{T}, \mathcal{V})$-functor $f : (2, \top) \to (X, a)$ one has

$$k \leq \top \leq a(Tf(e_2(0)), f(1)) = a(e_X(f(0)), f(1))$$

and, likewise, $k \leq a(e_X(f(1)), f(0))$; hence, $f(0) = f(1)$ if (X, a) is separated. Conversely, if $x \leq y$ and $y \leq x$ in (X, a), the additional hypothesis makes $f : (0 \mapsto x, 1 \mapsto y)$ $(\mathbb{T}, \mathcal{V})$-continuous, hence constant.

(3): This is a straightforward exercise. □

Closure under mono-sources makes $(\mathbb{T}, \mathcal{V})$-$\mathsf{Cat}_{\mathrm{sep}}$ strongly epireflective in $(\mathbb{T}, \mathcal{V})$-$\mathsf{Cat}$. We give an easy ad hoc description of the reflector, modeled after the situation in Top. For a topological space X, the quotient topology on the T0-reflection $X/\!\sim$ with $(x \sim y \iff \overline{\{x\}} = \overline{\{y\}})$ has the special property that it makes the projection $p : X \to X/\!\sim$ not only O-final, but also O-initial, with respect to $O : \mathsf{Top} \to \mathsf{Set}$. This property is crucial for establishing the reflection for $(\mathbb{T}, \mathcal{V})$-spaces.

© **V.2.1.2 Theorem** *For a $(\mathbb{T}, \mathcal{V})$-space (X, a), the projection $p : X \to X/\!\sim$ of the separated-order reflection, given by*

$$x \sim y \iff x \leq y \;\&\; y \leq x \,,$$

serves also as a reflection into $(\mathbb{T}, \mathcal{V})$-$\mathsf{Cat}_{\mathrm{sep}}$ when $X/\!\sim$ is endowed with the $(\mathbb{T}, \mathcal{V})$-category structure $\tilde{a} = p \cdot a \cdot (Tp)^\circ$, making p both O-final and O-initial with respect to the functor $O : (\mathbb{T}, \mathcal{V})$-$\mathsf{Cat} \to \mathsf{Set}$.

Proof By definition of $\sim$, one has $p^\circ \cdot p \leq a \cdot e_X$, which implies

$$\begin{aligned} \tilde{a} \cdot e_{X/\sim} &= p \cdot a \cdot (Tp)^\circ \cdot e_{X/\sim} \\ &\geq p \cdot a \cdot e_X \cdot p^\circ \\ &\geq p \cdot p^\circ = 1_{X/\sim} \end{aligned}$$

as well as

$$\begin{aligned} p^\circ \cdot \tilde{a} \cdot Tp &= p^\circ \cdot p \cdot a \cdot (Tp)^\circ \cdot Tp \\ &\leq a \cdot e_X \cdot a \cdot \hat{T}(a \cdot e_X) \\ &\leq a \cdot e_X \cdot a \\ &\leq a \cdot \hat{T}a \cdot e_{TX} \\ &= a \,. \end{aligned}$$

We can now conclude that

© $$\begin{aligned} \tilde{a} \cdot \hat{T}\tilde{a} \cdot m^\circ_{X/\sim} &= p \cdot p^\circ \cdot \tilde{a} \cdot Tp \cdot (Tp)^\circ \cdot \hat{T}\tilde{a} \cdot TTp \cdot (TTp)^\circ \cdot m^\circ_{X/\sim} \\ &\leq p \cdot a \cdot \hat{T}(p^\circ \cdot \tilde{a} \cdot Tp) \cdot m^\circ_X \cdot (Tp)^\circ \\ &\leq p \cdot a \cdot \hat{T}a \cdot m^\circ_X \cdot (Tp)^\circ \\ &= p \cdot a \cdot (Tp)^\circ = \tilde{a} \,. \end{aligned}$$

(Here we used the fact that Tp and TTp are surjective since p is surjective, the Axiom of Choice granted.) Consequently, $(X/\!\sim, \tilde{a})$ is a $(\mathbb{T}, \mathcal{V})$-space, and p is both O-initial and O-final. Quite trivially, $(X/\sim, \tilde{a})$ is separated, since $p(x) \leq p(y)$ implies

$$\begin{aligned} k &\leq \tilde{a}(e_{X/\sim}(p(x), p(y)) \\ &\leq (p^\circ \cdot \tilde{a} \cdot e_{X/\sim} \cdot p)(x, y) \\ &\leq (p^\circ \cdot \tilde{a} \cdot Tp \cdot e_X)(x, y) \\ &= a(e_X(x), y) , \end{aligned}$$

so $x \leq y$.

It now suffices to show that for any $(\mathbb{T}, \mathcal{V})$-continuous map $f : (X, a) \to (Y, b)$ with (Y, b) separated one has $(x \sim y \implies f(x) = f(y))$, since the induced map $\tilde{f} : X/\sim \to Y$ with $\tilde{f} \cdot p = f$ is $(\mathbb{T}, \mathcal{V})$-continuous by O-finality of p. Moreover, $p(x) = p(y)$ implies

$$k \leq a(e_X(x), y) \leq b(Tf \cdot e_X(x), f(y)) = b(e_Y(f(x)), f(y))$$

and, likewise, $k \leq b(e_Y(f(x)), f(y))$, so $f(x) = f(y)$ by separatedness of (Y, b). □

V.2.1.3 Corollary *The separated-order reflection of a $\mathcal{V}$-category carries its reflection into $\mathcal{V}$-$\mathsf{Cat}_{\mathrm{sep}}$; i.e. the solid-arrow pullback diagram*

$$\begin{array}{ccc} \mathcal{V}\text{-}\mathsf{Cat}_{\mathrm{sep}} & \longrightarrow & \mathsf{Ord}_{\mathrm{sep}} \\ \uparrow \dashv \downarrow & & \uparrow \dashv \downarrow \\ \mathcal{V}\text{-}\mathsf{Cat} & \longrightarrow & \mathsf{Ord} \end{array}$$

in CAT *also commutes with its dotted-arrow left adjoints.*

V.2.2 Between order separation and Hausdorff separation

For a $(\mathbb{T}, \mathcal{V})$-space (X, a), we now consider an array of low separation and symmetry conditions, with the terminology borrowed from the role model $\mathsf{Top} \cong (\beta, 2)$-$\mathsf{Cat}$, as follows:

(T0) $(a \cdot e_X) \wedge (a \cdot e_X)^\circ \leq 1_X$;

(T1) $a \cdot e_X \leq 1_X$;

(R0) $(a \cdot e_X)^\circ \leq a \cdot e_X$;

(R1) $a \cdot a^\circ \leq a \cdot e_X$.

The following scheme shows how these conditions are related with the Hausdorff separation condition $a \cdot a^\circ \leq 1_X$ and with separatedness of (X, a).

V.2.2.1 Proposition *The following implication holds for a $(\mathbb{T}, \mathcal{V})$-space (X, a):*

$$\begin{array}{ccccc} \textit{Hausdorff} & \iff & \textit{T1} & \& & \textit{R1} \\ \Downarrow & & \Downarrow & & \Downarrow \\ \textit{T1} & \iff & \textit{T0} & \& & \textit{R0} \\ & & \Downarrow & & \\ & & \textit{separated.} & & \end{array}$$

Proof

Hausdorff $\Longrightarrow$ T1 & R1: Since $e_X \leq a^\circ$, one has $a \cdot e_X \leq a \cdot a^\circ \leq 1_X \leq a \cdot e_X$.
T1 & R1 $\Longrightarrow$ Hausdorff: One has $a \cdot a^\circ \leq a \cdot e_X \leq 1_X$.
T1 $\Longrightarrow$ T0 & R0: One has $(a \cdot e_X) \wedge (a \cdot e_X)^\circ \leq a \cdot e_X \leq 1_X$ and $(a \cdot e_X)^\circ \leq 1_X \leq a \cdot e_X$.
T0 & R0 $\Longrightarrow$ T1: One has $a \cdot e_X = (a \cdot e_X)^{\circ\circ} \leq (a \cdot e_X)^\circ$, so $a \cdot e_X = (a \cdot e_X) \wedge (a \cdot e_X)^\circ \leq 1_X$.
R1 $\Longrightarrow$ R0: One has $(a \cdot e_X)^\circ = e_X^\circ \cdot a^\circ \leq a \cdot a^\circ \leq a \cdot e_X$.
T0 $\Longrightarrow$ separated: If $k \leq a(e_X(x), y) \wedge a(e_X(y), x)$, then $\bot < k \leq ((a \cdot e_X) \wedge (a \cdot e_X)^\circ)(x, y) \leq 1_X(x, y)$, hence $x = y$. □

V.2.2.2 Examples

(1) In Ord $\cong$ 2-Cat, T0-separation coincides with separation (see II.1.3), while R1 = R0 means symmetry, i.e. the order is an equivalence relation. Also in Met $\cong$ P_+-Cat, R1 = R0 assumes the usual meaning of symmetry: $a(x, y) = a(y, x)$ for all points in the metric space (X, a). Like Hausdorffness, T1 means being discrete. When (X, a) is symmetric, even T0 means being discrete and is therefore considerably stronger than being order separated. But a two-point space $X = \{u, v\}$ with $a(u, v) = \infty$ and all other distances 0 is T0 but not T1.

(2) In Top $\cong$ $(\beta, 2)$-Cat, or in Top $\cong$ $(\mathbb{F}, 2)$-Cat, T0 means order separated and T1 means that singleton sets are closed, i.e. these conditions assume their usual meanings. Likewise for R0 (if $y \in \overline{\{x\}}$, then $x \in \overline{\{y\}}$) and R1 (if $U \cap W \neq \varnothing$ for all neighborhoods U of x and W of y, then $\overline{\{x\}} = \overline{\{y\}}$). In App $\cong$ (β, P_+)-Cat one has the following straightforward characterizations:

$$(X, a) \text{ is T0} \iff \forall x, y \in X\ (a(\dot{x}, y) < \infty\ \&\ a(\dot{y}, x) < \infty \implies x = y)\ ;$$
$$(X, a) \text{ is T1} \iff \forall x, y \in X\ (a(\dot{x}, y) < \infty \implies x = y)\ ;$$
$$(X, a) \text{ is R0} \iff \forall x, y \in X\ (a(\dot{y}, x) = a(\dot{x}, y))\ ;$$
$$(X, a) \text{ is R1} \iff \forall x, y \in X\ \forall z \in \beta X\ (\delta(y, \{x\}) \leq a(z, x) + a(z, y))\ ,$$

where δ is the approach distance of (X, a).

V.2.2.3 Proposition

(1) *Let $\mathcal{V}$ be an integral quantale. The T0 and T1 properties are closed under mono-sources in* $(\mathbb{T}, \mathcal{V})$-Cat*. Hence, the corresponding full subcategories are strongly epireflective in* $(\mathbb{T}, \mathcal{V})$-Cat.

(2) *The R0 and R1 properties are closed under O-initial sources in* $(\mathbb{T}, \mathcal{V})$-Cat*, where* $O : (\mathbb{T}, \mathcal{V})\text{-Cat} \to \text{Set}$*. Hence, the corresponding full subcategories are both mono- and epireflective in* $(\mathbb{T}, \mathcal{V})$-Cat.

Proof Let $f_i : (X, a) \to (Y_i, b_i)$ $(i \in I)$ be a source in $(\mathbb{T}, \mathcal{V})$-**Cat** that is assumed to be monic when all (Y_i, b_i) are assumed to be T0 or T1, and O-initial when all (Y_i, b_i) are assumed to be R0 or R1.

Suppose that all (Y_i, b_i) are T0; then

$$\begin{aligned} a \cdot e_X \wedge (a \cdot e_X)^\circ &\leq \textstyle\bigwedge_{i,j\in I}(f_i^\circ \cdot b_i \cdot Tf_i \cdot e_X) \wedge (f_j^\circ \cdot b_j \cdot Tf_j \cdot e_X)^\circ \\ &\leq \textstyle\bigwedge_{i\in I}(f_i^\circ \cdot b_i \cdot e_{Y_i} \cdot f_i) \wedge (f_i^\circ \cdot e_{Y_i}^\circ \cdot b_i^\circ \cdot f_i) \\ &\leq \textstyle\bigwedge_{i\in I} f_i^\circ \cdot (b_i \cdot e_{Y_i} \wedge (b_i \cdot e_{Y_i})^\circ) \cdot f_i \\ &\leq \textstyle\bigwedge_{i\in I} f_i^\circ \cdot f_i = 1_X \qquad ((Y_i, b_i) \text{ are T0}). \end{aligned}$$

The proof for T1 is similar.

If all (Y_i, b_i) are R0, then

$$\begin{aligned} (a \cdot e_X)^\circ &\leq \textstyle\bigwedge_{i\in I}(f_i^\circ \cdot b_i \cdot Tf_i \cdot e_X)^\circ \\ &= \textstyle\bigwedge_{i\in I} f_i^\circ \cdot (b_i \cdot e_{Y_i})^\circ \cdot f_i \\ &\leq \textstyle\bigwedge_{i\in I} f_i^\circ \cdot b_i \cdot e_{Y_i} \cdot f_i \qquad ((Y_i, b_i) \text{ are R0}) \\ &= (\textstyle\bigwedge_{i\in I} f_i^\circ \cdot b_i \cdot Tf_i) \cdot e_X = a \cdot e_X \,. \end{aligned}$$

Finally, if all (Y_i, b_i) are R1, then

$$\begin{aligned} a \cdot a^\circ &\leq \textstyle\bigwedge_{i\in I}(f_i^\circ \cdot b_i \cdot Tf_i) \wedge \bigwedge_{j\in I}(f_j^\circ \cdot b_j \cdot Tf_j)^\circ \\ &\leq \textstyle\bigwedge_{i\in I} f_i^\circ \cdot b_i \cdot Tf_i \cdot (Tf_i)^\circ \cdot b_i^\circ \cdot f_i \\ &\leq \textstyle\bigwedge_{i\in I} f_i^\circ \cdot b_i \cdot b_i^\circ \cdot f_i \\ &\leq \textstyle\bigwedge_{i\in I} f_i^\circ \cdot b_i \cdot e_{Y_i} \cdot f_i \qquad ((Y_i, b_i) \text{ are R1}) \\ &= (\textstyle\bigwedge_{i\in I} f_i^\circ \cdot b_i \cdot Tf_i) \cdot e_X = a \cdot e_X \,. \end{aligned}$$

□

V.2.3 Regular spaces

Throughout this section we assume that

- *$\mathcal{V}$ is commutative.*

In order to formulate regularity and normality for $(\mathbb{T}, \mathcal{V})$-spaces, we will make essential use of the $\mathcal{V}$-relation (see Theorem III.5.3.5)

$$(TX \xrightarrow{\hat{a}} TX) := (TX \xrightarrow{m_X^\circ} TTX \xrightarrow{\hat{T}a} TX)\,.$$

For every $(\mathbb{T}, \mathcal{V})$-space (X, a), the $\mathcal{V}$-relation $\hat{a} = \hat{T}a \cdot m_X^\circ$ is a $\mathcal{V}$-graph structure on TX:

$$1_{TX} \leq \hat{T}1_X = \hat{T}(e_X^\circ) \cdot m_X^\circ \leq \hat{T}a \cdot m_X^\circ\,.$$

Therefore, both $(X, a \cdot \hat{a})$ and $(X, a \cdot \hat{a}^\circ)$ are $(\mathbb{T}, \mathcal{V})$-graphs. While the inequality $a \cdot \hat{a} \leq a$ is exactly the transitivity condition for a, the condition $a \cdot \hat{a}^\circ \leq a$ encodes an interesting separation property of a.

V.2.3.1 Definition A $(\mathbb{T}, \mathcal{V})$-space (X, a) is *regular* if $a \cdot \hat{a}^\circ \leq a$, i.e. $a \cdot m_X \cdot (\hat{T}a)^\circ \leq a$, or, in pointwise form,

$$\hat{T}a(\mathfrak{Y}, \mathfrak{x}) \otimes a(m_X(\mathfrak{Y}), z) \leq a(\mathfrak{x}, z) ,$$

for all $\mathfrak{Y} \in TTX$, $\mathfrak{x} \in TX$, and $z \in X$.

We denote the resulting full subcategory of $(\mathbb{T}, \mathcal{V})$-**Cat** by $(\mathbb{T}, \mathcal{V})$-$\mathsf{Cat}_{\mathrm{reg}}$.

V.2.3.2 Examples

(1) If $\mathbb{T} = \mathbb{I}$ is the identity monad (identically extended to $\mathcal{V}$-**Rel**), then $a \cdot \hat{a}^\circ = a \cdot a^\circ$, so that $a \cdot a^\circ \leq a$ if and only if $a = a^\circ$. Indeed, if $a = a^\circ$ then $a \cdot a^\circ \leq a$ follows from transitivity. Conversely, if $a \cdot a^\circ \leq a$, then $a^\circ \leq a \cdot a^\circ \leq a$. Hence, *for $\mathcal{V}$-spaces regularity means symmetry.*

(2) A topological space X considered as a $(\beta, 2)$-space (X, a) is regular if and only if it is *regular* in the usual sense, i.e. if, for $x \in X$ and $A \subseteq X$ closed with $x \notin A$, there exist open sets $U, W \subseteq X$ with $x \in U$, $A \subseteq W$, and $U \cap W = \varnothing$. To prove this claim we first recall that, as observed in Examples III.5.3.7, for $\mathfrak{x}, \mathfrak{y} \in \beta X$, with $(\preceq) = \hat{a}$,

$$\begin{aligned} \mathfrak{y} \preceq \mathfrak{x} &\iff \forall A \subseteq X,\ A \text{ closed}(A \in \mathfrak{y} \implies A \in \mathfrak{x}) \\ &\iff \forall A \subseteq X,\ A \text{ open}(A \in \mathfrak{x} \implies A \in \mathfrak{y}) . \end{aligned}$$

Furthermore, the condition $a \cdot \hat{a}^\circ \leq a$ means that

$$\mathfrak{y} \preceq \mathfrak{x} \;\&\; \mathfrak{y} \longrightarrow x \implies \mathfrak{x} \longrightarrow x ,$$

for all $\mathfrak{x}, \mathfrak{y} \in \beta X$ and $x \in X$. Therefore, if we write $\mathfrak{y} \dashrightarrow \mathfrak{x}$ whenever $\mathfrak{y} \preceq \mathfrak{x}$, we may depict this property as follows:

$$\begin{array}{ccc} & \mathfrak{y} & \\ \swarrow\!\!\cdot\cdot & & \searrow \\ \mathfrak{x} & & x \end{array} \quad \implies \quad \mathfrak{x} \longrightarrow x .$$

If X is regular in the usual topological sense, the closed neighborhoods of $x \in X$ form a neighborhood base for x. Therefore, if $\mathfrak{y} \longrightarrow x$, i.e. if $\mathfrak{y}$ contains the neighborhoods of x, and every closed subset A of X belonging to $\mathfrak{y}$ also belongs to $\mathfrak{x}$, then clearly $\mathfrak{x} \longrightarrow x$. If X is not regular, i.e. if there is $x \in X$ with a neighborhood W that does not contain any closed neighborhood of x, consider an ultrafilter $\mathfrak{x}$ containing all the closed neighborhoods of x and $X \setminus W$, and an ultrafilter $\mathfrak{y}$ containing all the neighborhoods of x. Then $\mathfrak{y} \preceq \mathfrak{x}$ and $\mathfrak{y} \longrightarrow y$, and therefore $a \cdot \hat{a}^\circ(\mathfrak{x}, x) = \top$. However, by construction, one does not have $\mathfrak{x} \longrightarrow x$, so $a \cdot \hat{a} \leq a$ fails.

(3) Let $\mathbb{T} = \mathbb{H}$ as in Section V.1.4. For an $(\mathbb{H}, 2)$-space (X, a), the order $(\preceq) = \hat{a}$ is defined by

$$\begin{aligned} (\beta, y) \preceq (\alpha, x) &\iff \overline{H}a \cdot m_X^\circ((\beta, y), (\alpha, x)) = \top \\ &\iff \exists \gamma \in H\ (\beta = \alpha \cdot \gamma \;\&\; y \xrightarrow{\ \gamma\ } x) , \end{aligned}$$

for $(\alpha, x), (\beta, y) \in H \times X$. Hence (X, a) is regular if and only if

$$\begin{array}{ccc} & y & \\ {}^{\beta}\swarrow & & \searrow^{\alpha\cdot\beta} \\ x & & z \end{array} \quad \Longrightarrow \quad x \xrightarrow{\alpha} z\ .$$

for all $\alpha, \beta \in H$ and all $x, y, z \in X$.

(4) Let $\mathbb{T}$ be the list monad $\mathbb{L}$ as in Section V.1.4. If (X, a) is a multi-ordered set, then, for $(\preceq) = \hat{a}$ and $(x_1, \ldots, x_n), (y_1, \ldots, y_m) \in LX$, one has $(y_1, \ldots, y_m) \preceq (x_1, \ldots, x_n)$ if and only if there is an n-partition of m, i.e. there exist $1 \leq m_1 < m_2 < \ldots < m_n = m$, such that

$$(y_1, \ldots, y_{m_1}) \vdash x_1, \ldots, (y_{m_{n-1}+1}, \ldots, y_m) \vdash x_n\ .$$

Hence, (X, a) is regular if and only if for all $(x_1, \ldots, x_n), (y_1, \ldots, y_m) \in LX$, $z \in X$, one has

$$(y_1, \ldots, y_m) \preceq (x_1, \ldots, x_n)\ \&\ (y_1, \ldots, y_m) \vdash z \Longrightarrow (x_1, \ldots, x_n) \vdash z\ .$$

(5) Let $\mathbb{T} = \beta$ be the ultrafilter monad and let $\mathcal{V} = \mathsf{P}_+$. An approach space, considered as a (β, P_+)-space (X, a), is regular if and only if, for any $\mathfrak{x}, \mathfrak{y} \in \beta X$ and $x \in X$,

$$a(\mathfrak{x}, x) \leq \hat{a}(\mathfrak{y}, \mathfrak{x}) + a(\mathfrak{y}, x)\ ,$$

where $\hat{a}(\mathfrak{y}, \mathfrak{x}) = \inf\{u \in [0, \infty] \mid \forall A \in \mathfrak{y}\ (A^{(u)} \in \mathfrak{x})\}$ (see Examples III.5.3.7). Analogously to the characterization for topological spaces, one can write

$$\begin{array}{ccc} & \mathfrak{y} & \\ {}^{u}\swarrow & & \searrow^{v} \\ \mathfrak{x} & & x \end{array} \quad \Longrightarrow \quad \mathfrak{x} \xrightarrow{\leq u+v} x\ .$$

V.2.3.3 Proposition *For a (β, P_+)-space (X, a) with approach distance δ, the following conditions are equivalent:*

(i) *(X, a) is regular;*

(ii) *for every filter f on X, $u \in [0, \infty]$ and every $B \subseteq X$ with $B \cap F^{(u)} \neq \varnothing$ for all $F \in f$,*

$$\delta(x, B) \leq \sup\nolimits_{A \in \mathcal{A}} \delta(x, A) + u\ ,$$

with $\mathcal{A} = \{A \subseteq X \mid \forall F \in f\ (A \cap F \neq \varnothing)\}$;

(iii) *for every ultrafilter $\mathfrak{y}$ on X, $u \in [0, \infty]$ and every $B \subseteq X$ with $B \cap A^{(u)} \neq \varnothing$ for all $A \in \mathfrak{y}$,*

$$\delta(x, B) \leq a(\mathfrak{y}, x) + u\ .$$

Proof (i) $\Longrightarrow$ (ii): Let f be a filter on X, $u \in [0, \infty]$ and let $B \subseteq X$ with $B \cap F^{(u)} \neq \varnothing$ for every $F \in f$. Let $\mathfrak{x}$ be an ultrafilter on X such that $B \in \mathfrak{x}$

and $F^{(u)} \in \chi$ whenever $F \in f$. Since $\{W \subseteq X \mid W^{(u)} \notin \chi\}$ is an ideal disjoint from f, by Corollary II.1.13.5 there exists an ultrafilter y such that $f \subseteq y$ and $W^{(u)} \in y$ whenever $W \in \chi$. Then, for every $x \in X$,

$$\delta(x, B) \leq a(\chi, x) \leq \hat{a}(y, \chi) + a(y, x) \leq u + a(y, x) \leq u + \sup_{A \in \mathcal{A}} \delta(x, A) .$$

(ii) $\Longrightarrow$ (iii) is straightforward. For (iii) $\Longrightarrow$ (i), consider $\chi, y \in \beta X$, $x \in X$, and $u \in [0, \infty]$ such that $A^{(u)} \in \chi$ whenever $A \in y$; for every $B \in \chi$, one then has $\delta(x, B) \leq a(y, x) + u$, and therefore

$$\begin{aligned} a(\chi, x) = \sup_{B \in \chi} \delta(x, B) &\leq a(y, x) + \inf\{u \mid \forall A \in y\ (A^{(u)} \in \chi)\} \\ &= a(y, x) + \hat{a}(y, \chi) . \end{aligned}$$

□

Here are some expected general assertions about regularity.

V.2.3.4 Proposition

(1) *If $\mathcal{V}$ is lean and integral and $\hat{T}$ is flat, then every compact Hausdorff $(\mathbb{T}, \mathcal{V})$-space is regular.*
(2) $(\mathbb{T}, \mathcal{V})$-$\mathsf{Cat}_{\mathrm{reg}}$ *is closed under O-initial sources (where $O : (\mathbb{T}, \mathcal{V})$-$\mathsf{Cat} \to \mathsf{Set}$ is the forgetful functor), and hence is both epi- and mono-reflective in* $(\mathbb{T}, \mathcal{V})$-$\mathsf{Cat}$.

Proof (1): Let (X, a) be a compact Hausdorff $(\mathbb{T}, \mathcal{V})$-space. If $\mathcal{V}$ is lean and integral, a is a map and $a \cdot Ta = a \cdot m_X$. Since $\hat{T}$ is flat,

$$a \cdot \hat{a}^\circ = a \cdot m_X \cdot (Ta)^\circ = a \cdot Ta \cdot (Ta)^\circ \leq a \cdot T(a \cdot a^\circ) \leq a .$$

(2): Let $(f_i : (X, a) \to (Y_i, b_i))_{i \in I}$ be an O-initial source, i.e. $a = \bigwedge_{i \in I} f_i^\circ \cdot b_i \cdot Tf_i$, with (Y_i, b_i) regular for every $i \in I$. Then, for every $i \in I$,

$$\begin{aligned} a \cdot m_X \cdot (\hat{T}a)^\circ &\leq f_i^\circ \cdot b_i \cdot Tf_i \cdot m_X \cdot (\hat{T}(f_i^\circ \cdot b_i \cdot Tf_i))^\circ \\ &= f_i^\circ \cdot b_i \cdot Tf_i \cdot m_X \cdot ((Tf_i)^\circ \cdot \hat{T}b_i \cdot TTf_i)^\circ \\ &= f_i^\circ \cdot b_i \cdot Tf_i \cdot m_X \cdot (TTf_i)^\circ \cdot (\hat{T}b_i)^\circ \cdot Tf_i \\ &= f_i^\circ \cdot b_i \cdot m_{Y_i} \cdot TTf_i \cdot (TTf_i)^\circ \cdot (\hat{T}b_i)^\circ \cdot Tf_i \\ &\leq f_i^\circ \cdot b_i \cdot m_{Y_i} \cdot (\hat{T}b_i)^\circ \cdot Tf_i \qquad ((Y_i, b_i) \text{ are regular}) \\ &\leq f_i^\circ \cdot b_i \cdot Tf_i , \end{aligned}$$

and therefore $a \cdot m_X \cdot (\hat{T}a)^\circ \leq \bigwedge_{i \in I} f_i^\circ \cdot b_i \cdot Tf_i = a$. □

V.2.3.5 Remarks

(1) That a regular $(\mathbb{T}, \mathcal{V})$-space does not need to be Hausdorff (or even separated) can be seen already at the level of $\mathcal{V}$-spaces: while Hausdorffness means discreteness, regularity means symmetry.

(2) As shown in Theorem III.5.3.5, if $\hat{\mathbb{T}}$ is associative, for a $(\mathbb{T}, \mathcal{V})$-space (X, a), $\hat{a}$ is not only reflexive, but also transitive, i.e. $(TX, \hat{a})$ is a $\mathcal{V}$-space. Moreover, the $(\mathbb{T}, \mathcal{V})$-space (X, a) is regular whenever the $\mathcal{V}$-space $(TX, \hat{a})$ is regular. Indeed, if $\hat{a} = \hat{a}^\circ$, then using the equality $a = a \cdot \hat{a}$ leads to

$$a \cdot \hat{a}^\circ = a \cdot \hat{a} \cdot \hat{a}^\circ \leq a \cdot \hat{a} = a .$$

The converse statement is not true in general (see Exercise V.2.F).

(3) A necessary condition for the $(\mathbb{T}, \mathcal{V})$-space (X, a) to be regular is the regularity of the $\mathcal{V}$-space $(X, a \cdot e_X)$, provided that the lax extension $\hat{T}$ of T is *symmetric*, i.e. $\hat{T}(r^\circ) = (\hat{T}r)^\circ$ for all r in $\mathcal{V}$-Rel. The following result generalizes this remark.

V.2.3.6 Proposition *Let $\alpha : (\mathbb{S}, \hat{S}) \to (\mathbb{T}, \hat{T})$ be a morphism of symmetric lax extensions. Then the algebraic functor $A_\alpha : (\mathbb{T}, \mathcal{V})$-$\mathsf{Cat} \to (\mathbb{S}, \mathcal{V})$-$\mathsf{Cat}$ preserves regularity.*

Proof For (X, a) regular in $(\mathbb{T}, \mathcal{V})$-$\mathsf{Cat}$ one has (with $\mathbb{S} = (S, n, d)$):

$$\begin{aligned}
a \cdot \alpha_X \cdot \widehat{a \cdot \alpha_X}^\circ &= a \cdot \alpha_X \cdot n_X \cdot (S\alpha_X)^\circ \cdot (\hat{S}a)^\circ \\
&= a \cdot m_X \cdot T\alpha_X \cdot \alpha_{SX} \cdot (S\alpha_X)^\circ \cdot \hat{S}(a^\circ) && (\alpha \text{ monad morphism}) \\
&\leq a \cdot m_X \cdot T\alpha_X \cdot (T\alpha_X)^\circ \cdot \alpha_{TX} \cdot \hat{S}(a^\circ) && (T\alpha \cdot \alpha S = \alpha T \cdot S\alpha) \\
&\leq a \cdot m_X \cdot \hat{T}(a^\circ) \cdot \alpha_X && (\alpha \text{ lax extension morphism}) \\
&\leq a \cdot \hat{a}^\circ \cdot \alpha_X \\
&\leq a \cdot \alpha_X .
\end{aligned}$$

□

For $(\mathbb{T}, \mathcal{V}) = (\beta, \mathsf{P}_+)$ one concludes in particular that the underlying metric of a regular approach space is symmetric: see Example V.2.3.2(1).

V.2.4 Normal and extremally disconnected spaces

Throughout this section we assume that

- *$\mathcal{V}$ is commutative;*
- *$\hat{\mathbb{T}}$ is associative* (see Remark V.2.3.5(2)).

Recall that a topological space is normal if, for all A, B closed with $A \cap B = \varnothing$, there are open sets U and W such that $A \subseteq U$, $B \subseteq W$, and $U \cap W = \varnothing$. It was shown in Proposition III.5.6.2 that normality for a $(\beta, 2)$-space (X, a) can be expressed by using the ordered set $(\beta X, \hat{a})$:

V.2.4.1 Proposition *For a topological space X presented as a $(\beta, 2)$-space (X, a), the following conditions are equivalent:*

(i) *X is a normal topological space;*

(ii) $\hat{a} \cdot \hat{a}^\circ \leq \hat{a}^\circ \cdot \hat{a}$.

Proposition V.2.4.1 leads us to the following definition.

V.2.4.2 Definition A $(\mathbb{T}, \mathcal{V})$-space (X, a) is *normal* if

$$\hat{a} \cdot \hat{a}^\circ \leq \hat{a}^\circ \cdot \hat{a} \,, \tag{V.2.4.i}$$

i.e.

$$\hat{a}(\mathfrak{z}, \mathfrak{x}) \otimes \hat{a}(\mathfrak{z}, \mathfrak{y}) \leq \bigvee\nolimits_{\mathfrak{w} \in TX} (\hat{a}(\mathfrak{x}, \mathfrak{w}) \otimes \hat{a}(\mathfrak{y}, \mathfrak{w})) \,,$$

for all $\mathfrak{x}, \mathfrak{y}, \mathfrak{z} \in TX$.

V.2.4.3 Proposition *For a $(\mathbb{T}, \mathcal{V})$-space (X, a), the following conditions are equivalent:*

(i) *(X, a) is normal;*
(ii) *$(TX, \hat{a})$ is a normal $\mathcal{V}$-space;*
(iii) *$(TX, \hat{a}^\circ \cdot \hat{a})$ is a $\mathcal{V}$-space.*

Proof For (i) $\iff$ (ii), we just note that normality for $(TX, \hat{a})$ means $\hat{a} \cdot \hat{a}^\circ \leq \hat{a}^\circ \cdot \hat{a}$.

(ii) $\Longrightarrow$ (iii): The structure $\hat{a}^\circ \cdot \hat{a}$ is obviously reflexive. If $\hat{a} \cdot \hat{a}^\circ \leq \hat{a}^\circ \cdot \hat{a}$, then

$$\hat{a}^\circ \cdot \hat{a} \cdot \hat{a}^\circ \cdot \hat{a} \leq \hat{a}^\circ \cdot \hat{a}^\circ \cdot \hat{a} \cdot \hat{a} \leq \hat{a}^\circ \cdot \hat{a} \,.$$

(iii) $\Longrightarrow$ (i): Transitivity of $\hat{a}^\circ \cdot \hat{a}$ and reflexivity of $\hat{a}$ and $\hat{a}^\circ$ give

$$\hat{a} \cdot \hat{a}^\circ \leq \hat{a}^\circ \cdot \hat{a} \cdot \hat{a}^\circ \cdot \hat{a} \leq \hat{a}^\circ \cdot \hat{a} \,,$$

i.e. the $(\mathbb{T}, \mathcal{V})$-space (X, a) is normal. □

Reversing the inequality (V.2.4.i) has an interesting topological meaning. It leads us to consider extremally disconnected objects. Recall that a topological space X is *extremally disconnected* if the closure of every open set in X is open.

V.2.4.4 Proposition *For a topological space X presented as a $(\beta, 2)$-space (X, a), the following conditions are equivalent:*

(i) *X is extremally disconnected;*
(ii) *for all open subsets U, W of X, if $U \cap W = \varnothing$ then $\overline{U} \cap \overline{W} = \varnothing$;*
(iii) *$\hat{a}^\circ \cdot \hat{a} \leq \hat{a} \cdot \hat{a}^\circ$.*

Proof (i) $\Longrightarrow$ (ii): Let U, W be open subsets of X. If $U \cap W = \varnothing$, then $\overline{U} \cap W = \varnothing$, and, since $\overline{U}$ is open, $\overline{U} \cap \overline{W} = \varnothing$.

(ii) $\Longrightarrow$ (i): Let U be an open subset of X and $W = X \setminus \overline{U}$. Then $U \cap W = \varnothing$, and therefore $\overline{U} \cap \overline{W} = \varnothing$. This implies that $\overline{U}$ is open since

$$\overline{U} \subseteq X \setminus \overline{W} \subseteq X \setminus W = \overline{U} \,.$$

To show (ii) $\iff$ (iii), first we point out that, for any ultrafilters $\mathfrak{x}, \mathfrak{y}$ on X and with $\hat{a} = (\preceq)$,

$$\begin{aligned} \hat{a}^\circ \cdot \hat{a}(\mathfrak{x}, \mathfrak{y}) = \top &\iff \exists \mathfrak{z} \in \beta X \ (\mathfrak{x} \preceq \mathfrak{z} \text{ and } \mathfrak{y} \preceq \mathfrak{z}) \\ &\iff \exists \mathfrak{z} \in \beta X \ (\forall B \subseteq X, B \text{ closed}, B \in \mathfrak{x} \cup \mathfrak{y} \implies B \in \mathfrak{z}), \\ \hat{a} \cdot \hat{a}^\circ(\mathfrak{x}, \mathfrak{y}) = \top &\iff \exists \mathfrak{w} \in \beta X \ (\mathfrak{w} \preceq \mathfrak{x} \text{ and } \mathfrak{w} \preceq \mathfrak{y}) \\ &\iff \exists \mathfrak{w} \in \beta X \ (\forall A \subseteq X, A \text{ open}, A \in \mathfrak{x} \cup \mathfrak{y} \implies A \in \mathfrak{w}). \end{aligned}$$

Assuming (ii), let $\mathfrak{x}, \mathfrak{y}, \mathfrak{z}$ be ultrafilters on X with $\mathfrak{x} \preceq \mathfrak{z}$ and $\mathfrak{y} \preceq \mathfrak{z}$. For any open subsets U, W of X, if $U \in \mathfrak{x}$ and $W \in \mathfrak{y}$, then $\overline{U}, \overline{W} \in \mathfrak{z}$, and therefore $\overline{U} \cap \overline{W} \neq \varnothing$, which implies with (ii) that $U \cap W \neq \varnothing$. The filter base

$$\{U \cap W \mid U, W \text{ open subsets of } X, U \in \mathfrak{x}, W \in \mathfrak{y}\}$$

is contained in an ultrafilter $\mathfrak{w}$. By construction, $\mathfrak{w} \preceq \mathfrak{x}$ and $\mathfrak{w} \preceq \mathfrak{y}$, and therefore $\hat{a} \cdot \hat{a}^\circ(\mathfrak{x}, \mathfrak{y}) = \top$.

Conversely, assume that (ii) does not hold. Thus, there are disjoint open subsets U and W of X with $\overline{U} \cap \overline{W} \neq \varnothing$. Let $\mathfrak{z}$ be an ultrafilter containing $\overline{U} \cap \overline{W}$. Consider the ultrafilters $\mathfrak{x}$ and $\mathfrak{y}$ with filter bases

$$\begin{aligned} \mathcal{B}_{\mathfrak{x}} &= \{A \mid A \text{ open and } (A \in \mathfrak{z} \text{ or } A = U)\}, \\ \mathcal{B}_{\mathfrak{y}} &= \{A \mid A \text{ open and } (A \in \mathfrak{z} \text{ or } A = W)\}. \end{aligned}$$

Then $\mathfrak{x} \preceq \mathfrak{z}$ and $\mathfrak{y} \preceq \mathfrak{z}$, but there is no $\mathfrak{w}$ with $\mathfrak{w} \preceq \mathfrak{x}$ and $\mathfrak{w} \preceq \mathfrak{y}$, i.e. (iii) fails. □

V.2.4.5 Definition A $(\mathbb{T}, \mathcal{V})$-space (X, a) is *extremally disconnected* if

$$\hat{a}^\circ \cdot \hat{a} \leq \hat{a} \cdot \hat{a}^\circ ;$$

i.e.

$$\hat{a}(\mathfrak{x}, \mathfrak{z}) \otimes \hat{a}(\mathfrak{y}, \mathfrak{z}) \leq \bigvee_{\mathfrak{w} \in TX} (\hat{a}(\mathfrak{w}, \mathfrak{x}) \otimes \hat{a}(\mathfrak{w}, \mathfrak{y}))$$

for all $\mathfrak{x}, \mathfrak{y}, \mathfrak{z} \in TX$.

V.2.4.6 Remark A $\mathcal{V}$-space (X, a) is normal if and only if (X, a°) is extremally disconnected.

Using Proposition V.2.4.3 we obtain:

V.2.4.7 Corollary *For a $(\mathbb{T}, \mathcal{V})$-space (X, a), the following conditions are equivalent:*

(i) (X, a) *is extremally disconnected;*
(ii) $(TX, \hat{a})$ *is an extremally disconnected* $\mathcal{V}$*-space;*
(iii) $(TX, \hat{a}^\circ)$ *is a normal* $\mathcal{V}$*-space;*
(iv) $(TX, \hat{a} \cdot \hat{a}^\circ)$ *is a* $\mathcal{V}$*-space.*

V.2.4.8 Examples

(1) Let $\mathbb{T} = \mathbb{I}$ be the identity monad (identically extended to $\mathcal{V}$-Rel). A $\mathcal{V}$-space (X, a) is normal if and only if

$$\forall x, y, z \in X \; (a(x, y) \otimes a(x, z) \le \bigvee_{s \in X} a(y, s) \otimes a(z, s)) \,. \qquad \text{(V.2.4.ii)}$$

If we write $x \xrightarrow{\beta} y$ whenever $a(x, y) = \beta$, diagrammatically this condition can be represented as follows:

$\forall$ 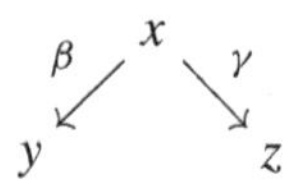 $\beta \otimes \gamma \le \bigvee_{s \in X} \beta_s \otimes \gamma_s :$ 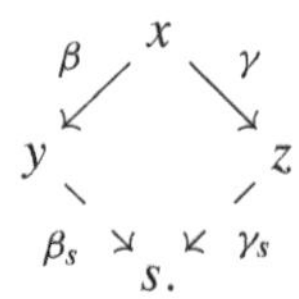

Extremally disconnected $\mathcal{V}$-spaces are described in a similar way, with the arrows reversed.

For $\mathcal{V} = \mathsf{2}$, an ordered set $(X, \le)$, considered as a $\mathsf{2}$-space, is normal if and only if the order $\le$ is confluent (see Exercise III.5.A). In particular, this shows that a normal $(\mathbb{T}, \mathcal{V})$-space does not need to be regular (see also Exercise V.2.H).

A regular $\mathcal{V}$-space, i.e. a symmetric $\mathcal{V}$-space, is trivially normal and extremally disconnected.

(2) Let β be the ultrafilter monad. By Proposition V.2.4.3, in $(\beta, \mathsf{2})$-Cat a topological space (X, a) is normal if and only if the order $\preceq$ on βX is confluent: for $\mathfrak{x}, \mathfrak{y}, \mathfrak{z} \in \beta X$, with $\mathfrak{x} \preceq \mathfrak{y}$, $\mathfrak{x} \preceq \mathfrak{z}$, there exists $\mathfrak{w} \in \beta X$ with $\mathfrak{y} \preceq \mathfrak{w}$ and $\mathfrak{z} \preceq \mathfrak{w}$:

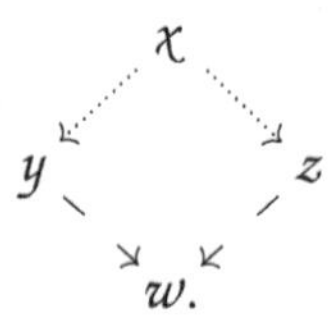

(3) Let $\mathbb{T} = \mathbb{H}$ as in Example V.2.3.2(3). Since $(\preceq) = \hat{a}$ is transitive, normality of an $(\mathbb{H}, \mathsf{2})$-space is equivalent to the condition $\hat{a} \cdot \hat{a}^\circ \le \hat{a}^\circ \cdot \hat{a}$. We recall that

$$(\beta, y) \preceq (\alpha, x) \iff \exists \gamma \in H \; (\beta = \alpha \cdot \gamma \;\&\; y \xrightarrow{\gamma} x) \,.$$

According to this description, when $(\beta, y) \preceq (\alpha, x)$, we say that (β, y) is a multiple of (α, x) and that (α, x) is a divisor of (β, y). For any $(\alpha, x), (\beta, y) \in H \times X$,

$$(\alpha, x) \; (\hat{a} \cdot \hat{a}^\circ) \; (\beta, y) \iff \exists (\gamma, z) \in H \times X \; ((\alpha, x) \succeq (\gamma, z) \preceq (\beta, y)) \,,$$

i.e. (α, x) and (β, y) have a common multiple, and

$$(\alpha, x)\ (\hat{a}^\circ \cdot \hat{a})\ (\beta, y) \iff \exists(\delta, w) \in H \times X\ ((\alpha, x) \preceq (\delta, w) \succeq (\beta, y))\ ,$$

i.e. (α, x) and (β, y) have a common divisor. Hence (X, a) is normal if and only if any pair of elements (α, x), (β, y) of $H \times X$ with a common multiple has a common divisor. That is, for all $x, y, z \in X$, $\alpha, \beta, \alpha_1, \beta_1 \in H$ such that $\alpha \cdot \alpha_1 = \beta \cdot \beta_1$ and $x \xleftarrow{\alpha_1} z \xrightarrow{\beta_1} y$, there exists $w \in X$, $\delta, \alpha_2, \beta_2 \in H$ with $\alpha = \delta \cdot \alpha_2$, $\beta = \delta \cdot \beta_2$ and $x \xrightarrow{\alpha_2} w \xleftarrow{\beta_2} y$:

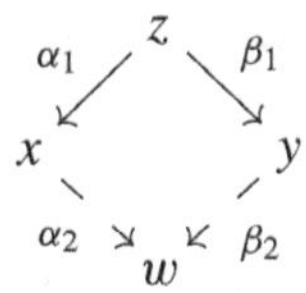

with $\delta \cdot \alpha_2 \cdot \alpha_1 = \alpha \cdot \alpha_1 = \beta \cdot \beta_1 = \delta \cdot \beta_2 \cdot \beta_1$. On reversing the arrows, one obtains a description of extremally disconnected $(\mathbb{H}, 2)$-spaces.

(4) If $\mathbb{T} = \mathbb{L}$ is the list monad, then again we can use the condition $\hat{a} \cdot \hat{a}^\circ \le \hat{a}^\circ \cdot \hat{a}$ to study normality of a multi-ordered set (X, a). It is easy to check that (X, a) is normal if and only if for all $(x_1, \ldots, x_n), (y_1, \ldots, y_m), (z_1, \ldots, z_l) \in LX$ the following condition holds: if there exist partitions $1 \le i_1 < \cdots < i_n = l$ and $1 \le j_1 < \cdots < j_m = l$ of l such that

$$((z_1, \ldots, z_{i_1}), \ldots, (z_{i_{n-1}+1}, \ldots, z_l)) \vdash (x_1, \ldots, x_n)\ \&$$
$$((z_1, \ldots, z_{j_1}), \ldots, (z_{j_{m-1}+1}, \ldots, z_l)) \vdash (y_1, \ldots, y_m)$$

then there exist $(w_1, \ldots, w_k) \in LX$ and k-partitions $1 \le n_1 < \cdots < n_k = n$ and $1 \le m_1 < \cdots < m_k = m$, of n and m, such that

$$((x_1, \ldots, x_{n_1}), \ldots, (x_{n_{k-1}+1}, \ldots, x_n)) \vdash (w_1, \ldots, w_k)\ \&$$
$$((y_1, \ldots, y_{m_1}), \ldots, (y_{m_{k-1}+1}, \ldots, y_m)) \vdash (w_1, \ldots, w_k)$$

(where $\vdash$ abbreviates $\overline{L}a$).

(5) In $\mathsf{Met} \cong (\mathbb{I}, \mathsf{P}_+)\text{-}\mathsf{Cat}$, a metric space (X, a) is normal if and only if, for every $x, y, z \in X$,

$$a(z, x) + a(z, y) \ge \inf_{w \in X} a(x, w) + a(y, w)\ .$$

Likewise in $\mathsf{App} \cong (\mathbb{B}, \mathsf{P}_+)\text{-}\mathsf{Cat}$, an approach space (X, a) is normal if, and only if, for any $\mathfrak{x}, \mathfrak{y}, \mathfrak{z} \in \beta X$,

$$\hat{a}(\mathfrak{z}, \mathfrak{x}) + \hat{a}(\mathfrak{z}, \mathfrak{y}) \ge \inf_{\mathfrak{w} \in \beta X} \hat{a}(\mathfrak{x}, \mathfrak{w}) + \hat{a}(\mathfrak{y}, \mathfrak{w})\ ,$$

where $\hat{a}(\mathfrak{x}, \mathfrak{y}) = \inf\{u \in [0, \infty] \mid \forall A \in \mathfrak{x}\ (A^{(u)} \in \mathfrak{y})\}$.

Normal approach spaces will be investigated in Section V.2.5. Here we give conditions on $\mathbb{T}$ and $\mathcal{V}$ in general for compact Hausdorff $(\mathbb{T}, \mathcal{V})$-spaces to be normal.

V.2.4.9 Proposition *If $\hat{T}$ is flat, every $\mathbb{T}$-algebra is a normal $(\mathbb{T}, \mathcal{V})$-space.*

Proof First we remark that, when $a : TX \to X$ is a map, the $\mathcal{V}$-space $(TX, \hat{a} = Ta \cdot m_X^\circ)$ is completely determined by its underlying order (given by $B_p : \mathcal{V}\text{-}\mathsf{Cat} \to 2\text{-}\mathsf{Cat} = \mathsf{Ord}$). Therefore, to show normality of the $\mathbb{T}$-algebra (X, a), we have to check that $(\preceq) = \hat{a}$ is confluent. Let $\mathfrak{x}, \mathfrak{y}, \mathfrak{z} \in TX$ with $\mathfrak{x} \preceq \mathfrak{y}$ and $\mathfrak{x} \preceq \mathfrak{z}$; i.e. there exist $\mathfrak{Y}, \mathfrak{Z} \in TTX$ such that $m_X(\mathfrak{Y}) = m_X(\mathfrak{Z}) = \mathfrak{x}$ and $Ta(\mathfrak{Y}) = \mathfrak{y}$, $Ta(\mathfrak{Z}) = \mathfrak{z}$. For $y = a(\mathfrak{y})$ and $z = a(\mathfrak{z})$, using the equality $a \cdot Ta = a \cdot m_X$ we conclude that

$$y = a \cdot Ta(\mathfrak{Y}) = a \cdot m_X(\mathfrak{Y}) = a(\mathfrak{x}) = a \cdot m_X(\mathfrak{Z}) = a \cdot Ta(\mathfrak{Z}) = z\ .$$

Now it is easy to conclude $\mathfrak{y} \preceq e_X(y)$, since, for $\mathfrak{W} = e_{TX}(\mathfrak{y})$, $m_X(\mathfrak{W}) = \mathfrak{y}$, and $Ta(\mathfrak{W}) = Ta(e_{TX}(\mathfrak{y})) = e_X(a(\mathfrak{y})) = e_X(y)$; an analogous argument shows that $\mathfrak{z} \preceq e_X(y)$. □

Using Proposition V.1.2.1 we reach the following conclusion:

V.2.4.10 Corollary *If $\mathcal{V}$ is integral and lean and $\hat{T}$ is flat, then every compact Hausdorff $(\mathbb{T}, \mathcal{V})$-space is normal.*

V.2.5 Normal approach spaces

Normality of an approach space (X, a) implies a strong separation property that can be expressed in terms of its approach distance. In preparation for that we first show the following result.

V.2.5.1 Lemma *For subsets A, B of an approach space X and any real $u > 0$, the following are equivalent:*

(i) $\forall C \subseteq X\ (A \cap C^{(u)} \neq \varnothing$ *or* $B \cap (X \setminus C)^{(u)} \neq \varnothing)$;
(ii) $\exists \mathfrak{x}, \mathfrak{y}, \mathfrak{z} \in \beta X\ \forall C \in \mathfrak{z}\ (A \cap C^{(u)} \in \mathfrak{x}$ *and* $B \cap C^{(u)} \in \mathfrak{y})$.

Proof (i) $\Longrightarrow$ (ii): We must show the existence of an ultrafilter $\mathfrak{z}$ such that both $\{A\} \cup \{C^{(u)} \mid C \in \mathfrak{z}\}$ and $\{B\} \cup \{C^{(u)} \mid C \in \mathfrak{z}\}$ generate proper filters on X. Failing that, for every $\mathfrak{z} \in \beta X$ we could choose $D \in \mathfrak{z}$ with $A \cap D^{(u)} = \varnothing$ or $B \cap D^{(u)} = \varnothing$; moreover, by compactness of βX, finitely many such sets $D_1, \dots, D_n$ could be found with $\bigcup_i D_i = X$. Then, with

$$C := \bigcup\{D_i \mid A \cap D_i^{(u)} = \varnothing\}$$

one would trivially have $A \cap C^{(u)} = \varnothing$, and also $B \cap (X \setminus C)^{(u)} = \varnothing$ since $X \setminus C \subseteq \bigcup\{D_i \mid B \cap D_i^{(u)} = \varnothing\}$, in contradiction to (i).

(ii) $\Longrightarrow$ (i): For $C \subseteq X$ one either has $C \in \mathfrak{z}$ and then $A \cap C^{(u)} \in \mathfrak{x}$, or $X \setminus C \in \mathfrak{z}$ and then $B \cap (X \setminus C)^{(u)} \in \mathfrak{z}$. □

V.2.5.2 Theorem *For an approach space (X, a), each of the following statements implies the next:*

(i) *(X, a) is normal in (β, P_+)-*Cat;

(ii) *for all ultrafilters x, y, z on X and any real $w > 0$,*

$$\hat{a}(z, x) < w \;\&\; \hat{a}(z, y) < w \implies \exists w \in \beta X \; (\hat{a}(x, w) < 2w \;\&\; \hat{a}(y, w) < 2w)$$

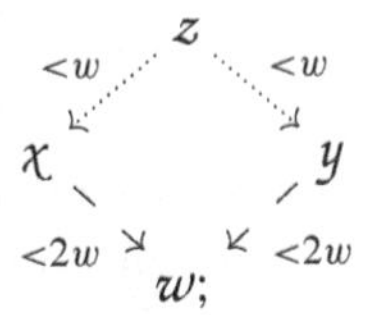

(iii) *for all $A, B \subseteq X$ and any real $v > 0$,*

$$A^{(v)} \cap B^{(v)} = \varnothing \implies \exists u > 0 \; \exists C \subseteq X \; (A^{(u)} \cap C^{(u)} = \varnothing = B^{(u)} \cap (X \setminus C)^{(u)}) \, ;$$

(iv) *for all ultrafilters x, y, z on X,*

$$\hat{a}(z, x) = 0 = \hat{a}(z, y) \implies \exists w \in \beta X \; (\hat{a}(x, w) = 0 = \hat{a}(y, w)) \, .$$

Proof (i) $\implies$ (ii) is an immediate consequence of the defining property of normality:

$$\hat{a}(z, x) + \hat{a}(z, y) \geq \inf_{w \in \beta X} (\hat{a}(x, w) + \hat{a}(y, w)) \, .$$

(ii) $\implies$ (iii): Given $v > 0$, we claim that any $u \leq v/4$ has the property described by (iii). Indeed, assuming failure of (iii), by Lemma V.2.5.1 we would have $A, B \subseteq X$ with $A^{(v)} \cap B^{(v)} = \varnothing$ and ultrafilters x, y, z on X such that

$$\forall C \in z \; (A^{(u)} \cap C^{(u)} \in x \text{ and } B^{(u)} \cap C^{(u)} \in y) \, .$$

Since $\hat{a}(z, x) = \inf\{w \mid \forall C \in z \; (C^{(w)} \in x)\}$, one has $u \geq \hat{a}(z, x)$ and, likewise, $u \geq \hat{a}(z, y)$. From (ii) (with $w = \frac{3}{2}u$) one obtains $w \in \beta X$ with $\hat{a}(x, w) < 3u$ and $\hat{a}(y, w) < 3u$; therefore, since $A^{(u)} \in x$ and $B^{(u)} \in y$,

$$A^{(v)} \cap B^{(v)} \supseteq (A^{(u)})^{(3u)} \cap (B^{(u)})^{(3u)} \in w \, ,$$

which contradicts $A^{(v)} \cap B^{(v)} = \varnothing$.

(iii) $\implies$ (iv): Consider $x, y, z \in \beta X$ with $\hat{a}(z, x) = 0 = \hat{a}(z, y)$. For any $A \in x$, $B \in y$, and $u > 0$ one has

$$\forall C \in z \; (A \cap C^{(u)} \in x \text{ and } B \cap C^{(u)} \in y) \, ,$$

which, by Lemma V.2.5.1, means equivalently

$$\forall C \in z \; (A \cap C^{(u)} \neq \varnothing \text{ or } B \cap (X \setminus C)^{(u)} \neq \varnothing) \, ;$$

in particular,

$$\forall C \in z\,(A^{(u)} \cap C^{(u)} \neq \varnothing \text{ or } B^{(u)} \cap (X \setminus C)^{(u)} \neq \varnothing)\,.$$

By hypothesis (iii), $A^{(v)} \cap B^{(v)} \neq \varnothing$ for all $A \in \mathfrak{x}$, $B \in \mathfrak{y}$, $v > 0$. With an ultrafilter $\mathfrak{w}$ containing the filter base given by all these non-empty sets, one obtains

$$\forall v > 0\,\forall A \in \mathfrak{x}\,\forall B \in \mathfrak{y}\,(A^{(v)} \in \mathfrak{w} \text{ and } B^{(v)} \in \mathfrak{w})\,,$$

i.e. $\hat{a}(\mathfrak{x}, \mathfrak{w}) = 0 = \hat{a}(\mathfrak{y}, \mathfrak{w})$. □

V.2.5.3 Remark None of the three implications of Theorem V.2.5.2 is reversible, as the following three examples show. We point out that all three examples are in particular metric spaces, showing that the four conditions are distinct already at the level of metric spaces.

(1) Let $X = \{x, y, z, w\}$ with $a(z, x) = a(z, y) = a(z, w) = 1$, $a(x, w) = a(y, w) = 2$, $a(x', x') = 0$ for any $x' \in X$, and $a(x', y') = \infty$ elsewhere. Then it is straightforward to check that (X, a) satisfies (ii), although $a(z, y) + a(z, x) < \inf_{w' \in X}(a(x, w') + a(y, w')) = 4$.
(2) Let $X = \{x, y, z, w\}$ with $a'(x', y') = a(x', y')$ except for $a'(x, w) = a'(y, w) = 3$. Then (ii) clearly fails. Since distinct points are at distance at least 1, for any $A \subseteq X$ and any $u < 1$ one has $A^{(u)} = A$, so (iii) holds.
(3) Let $Y = \{x, y\} \cup \{x_n \,;\, n \in \mathbb{N}\}$, with $b(x_n, x) = 1/n = b(x_n, y)$, $b(x', x') = 0$ for all $x' \in X$, and $b(x', y') = \infty$ elsewhere. Then $\{x\}^{(1)} \cap \{y\}^{(1)} = \{x\} \cap \{y\} = \varnothing$ although, for any $C \subseteq X$, either C or $X \setminus C$ has an infinite number of x_n, and so, for any $u > 0$, either $x, y \in C^{(u)}$ or $x, y \in (X \setminus C)^{(u)}$, i.e. (iii) does not hold. Moreover, $b(z, x) = 0$ only if $z = x$, and so (iv) holds.

Exercises

V.2.A *Preservation of low separation by algebraic functors.* Let $\alpha : (\mathbb{S}, \hat{S}) \to (\mathbb{T}, \hat{T})$ be a morphism of lax extensions and let $A_\alpha : (\mathbb{T}, \mathcal{V})\text{-}\mathsf{Cat} \to (\mathbb{S}, \mathcal{V})\text{-}\mathsf{Cat}$ be the induced algebraic functor. Prove that (X, a) is T0, T1, or R0 if and only if $A_\alpha(X, a)$ has the respective property. Furthermore, if (X, a) is R1 or Hausdorff, then $A_\alpha(X, a)$ has the respective property, with the converse statement holding when α_X is surjective. Exploit these statements for $\alpha = e$, i.e. when $A_\alpha = A$: $(\mathbb{T}, \mathcal{V})\text{-}\mathsf{Cat} \to \mathcal{V}\text{-}\mathsf{Cat}$.

V.2.B *Low separation in* $(\mathbb{H}, 2)\text{-}\mathsf{Cat}$. Present $(\mathbb{H}, 2)\text{-}\mathsf{Cat}$ as $\mathcal{V}\text{-}\mathsf{Cat}$ with $\mathcal{V}$ the powerset of H as in Remark V.1.4.3(1) and show that each of T0, T1, R0, and R1 changes its meaning under the change of presentation.

V.2.C *Compact T1-spaces for power-enriched monads.* When $\mathbb{T}$ is power-enriched, prove that every compact T1-space in $(\mathbb{T}, 2)$-Cat has precisely one point.

V.2.D *Strengthening R0.* We say that a $(\mathbb{T}, \mathcal{V})$-space is R0+ if $\hat{a} \cdot e_X \leq a^\circ$. Show that

(1) every R0+ space is R0;
(2) every regular space is R0+.

V.2.E *Separated reflections and Kleisli monoids.* For a monad $\mathbb{T}$ on Set © with an associative lax extension $\hat{\mathbb{T}}$ to $\mathcal{V}$-Rel, show that the separated reflection of Theorem V.2.1.2 defines also a left adjoint to the inclusion functor $((\mathbb{T}, \mathcal{V})\text{-}\mathsf{Cat}^{\mathbb{T}})_{\mathrm{sep}} \to (\mathbb{T}, \mathcal{V})\text{-}\mathsf{Cat}^{\mathbb{T}}$, where $\mathbb{T}$ is being considered as a monad on $(\mathbb{T}, \mathcal{V})$-$\mathsf{Cat}$ (see Section III.5.4). Conclude that if $\mathbb{T}$ is power-enriched, then $T'X$ as defined in Section IV.4.2 is the separated reflection of TX.

Hint. See Exercise IV.4.C.

V.2.F *Regularity of (X, a) does not imply regularity of $(TX, \hat{a})$.* Let $\mathbb{H}$ be as in Example V.2.3.2(3).

(1) Show that when $\mathbb{H} = (\mathbb{N}, \cdot, 1)$ is the multiplicative monoid of natural numbers, the $(\mathbb{H}, 2)$-space $(\mathbb{N}, a)$, with $a((\alpha, x), y) = \top$ only when $\alpha \cdot x = y$, is regular, although the ordered set $(H\mathbb{N}, \hat{a})$ is not regular (i.e. symmetric).
(2) Show that if H is a group then an $(\mathbb{H}, 2)$-space (X, a) is regular if and only if $(HX, \hat{a})$ is regular.

V.2.G *Discrete $(\mathbb{T}, \mathcal{V})$-spaces.* Show that, when $\hat{T}$ is flat, discrete $(\mathbb{T}, \mathcal{V})$-spaces are both normal and extremally disconnected (see Section III.3.2).

V.2.H *The extended real half-line.* Show that, as a P_+-space, the extended real half-line $([0, \infty], \ominus)$ is normal and extremally disconnected, but not regular.

V.2.I *Normal Hausdorff spaces in $(\mathbb{L}, 2)$-Cat.* Let $(X, \vdash)$ be a normal Hausdorff space in $(\mathbb{L}, 2)$-Cat. Then

$$x \cdot y = z \iff (x, y) \vdash z$$

establishes a partially defined binary operation $\cdot$ on X such that, whenever $x \cdot y$, $y \cdot z$ are defined, then $(x \cdot y) \cdot z$, $x \cdot (y \cdot z)$ are defined and equal, or $x \cdot y = x$ and $y \cdot z = z$.

V.3 Proper and open maps

The power of the notion of compact Hausdorff $(\mathbb{T}, \mathcal{V})$-space arises from its equational description as an Eilenberg–Moore algebra (see Proposition V.1.2.1). In this section we consider the equationally defined classes of proper and of open maps in $(\mathbb{T}, \mathcal{V})$-Cat, which in Section V.4, inter alia, lead us to equationally defined modifications of the notions of compact and Hausdorff $(\mathbb{T}, \mathcal{V})$-space, as defined in Section V.1.1.

Throughout the section, $\mathcal{V} = (\mathcal{V}, \otimes, k)$ is a quantale and $\mathbb{T} = (T, m, e)$ is a Set-monad with a lax extension $\hat{\mathbb{T}}$ to $\mathcal{V}$-Rel.

V.3.1 Finitary stability properties

For $(\mathbb{T}, \mathcal{V})$-spaces (X, a), (Y, b), $(\mathbb{T}, \mathcal{V})$-continuity of a map $f : X \to Y$ is equivalently expressed by the inequalities

$$f \cdot a \leq b \cdot Tf \qquad \text{and} \qquad a \cdot (Tf)^\circ \leq f^\circ \cdot b\ .$$

Considering equalities in either case leads us to the two key notions of this subsection.

V.3.1.1 Definition A $(\mathbb{T}, \mathcal{V})$-continuous map $f : (X, a) \to (Y, b)$ is *proper* if

$$b \cdot Tf \leq f \cdot a\ ,$$

and $f : (X, a) \to (Y, b)$ is *open* if

$$f^\circ \cdot b \leq a \cdot (Tf)^\circ\ ,$$

as in Definition III.4.3.1.

Hence, f is proper if and only if

$$\forall \chi \in TX, y \in Y\ (b(Tf(\chi), y) \leq \bigvee\nolimits_{z \in f^{-1}y} a(\chi, z))\ , \tag{V.3.1.i}$$

in which case the inequality is actually an equality; and f is open if and only if

$$\forall x \in X\ \forall \mathfrak{y} \in TY\ (b(\mathfrak{y}, f(x)) \leq \bigvee\nolimits_{\mathfrak{z} \in (Tf)^{-1}\mathfrak{y}} a(\mathfrak{z}, x))\ ,$$

in which case the inequality is again an equality.

In order to emphasize their dependency on $\mathbb{T}$ and $\mathcal{V}$, whenever needed we speak more precisely of $(\mathbb{T}, \mathcal{V})$-*proper* maps and $(\mathbb{T}, \mathcal{V})$-*open* maps.

V.3.1.2 Remarks

(1) For $\mathcal{V}$ commutative, one has the dualization functor

$$(-)^{\mathrm{op}} : \mathcal{V}\text{-Cat} \to \mathcal{V}\text{-Cat}\ , \quad X = (X, a) \mapsto X^{\mathrm{op}} = (X, a^\circ)\ ,$$

which maps morphisms identically. For $\mathbb{T} = \mathbb{I}$ (identically extended to $\mathcal{V}$-Rel) and $f : (X, a) \to (Y, b)$ one has

$$b \cdot f \leq f \cdot a \iff f^\circ \cdot b^\circ \leq a^\circ \cdot f^\circ\ ,$$

i.e.

$$f \text{ proper} \iff f^{\mathrm{op}} \text{ open}.$$

Hence, open is dual to proper.

(2) *When $\mathcal{V}$ is integral and superior (see* Remark V.1.1.3*), the $(\mathbb{T}, \mathcal{V})$-functor $(X, a) \to (1, \top)$ is proper if and only if (X, a) is compact.*

(3) *When $\mathcal{V}$ is integral and $T1 \cong 1$, then the $(\mathbb{T}, \mathcal{V})$-functor $(X, a) \to (1, \top)$ is open.* (See also Exercise V.3.C.)

V.3.1.3 Examples

(1) In $\mathsf{Ord} \cong \mathsf{2}\text{-}\mathsf{Cat}$, a monotone map $f : (X, \leq) \to (Y, \leq)$ is proper if and only if

$$f(x) \leq y \implies \exists z \in f^{-1}y\ (x \leq z), \qquad \begin{array}{ccc} x & \leq & z \\ | & & \vdots \\ f(x) & \leq & y, \end{array}$$

and it is open if and only if

$$y \leq f(x) \implies \exists z \in f^{-1}y\ (z \leq x), \qquad \begin{array}{ccc} z & \leq & x \\ \vdots & & | \\ y & \leq & f(x). \end{array}$$

In terms of the down- and up-closure operations (see II.1.7), one has

$$\begin{aligned} f \text{ proper} &\iff \forall x \in X\ (\uparrow_Y f(x) \subseteq f(\uparrow_X x)) \\ &\iff \forall A \subset X\ (\uparrow_Y f(A) \subseteq f(\uparrow_X A)) \\ &\iff \forall y \in Y\ (f^{-1}(\downarrow_Y y) \subseteq \downarrow_X (f^{-1}y)) \\ &\iff \forall B \subseteq Y\ (f^{-1}(\downarrow_Y B) \subseteq \downarrow_X f^{-1}(B)), \end{aligned}$$

and all inclusions may equivalently be replaced by equalities. Openness of f is characterized by the order-dual conditions.

(2) If Ord is presented as $(\mathbb{P}, \mathsf{2})$-Cat (with the powerset monad laxly extended by $\check{P}$; see Example III.1.6.2(1)), the meaning of proper map changes from the previous example. Indeed, in this presentation, an ordered set $(X, \leq)$ is considered as a $(\mathbb{P}, \mathsf{2})$-space $(X, \lll)$ via

$$A \lll y \iff \forall x \in A\ (x \leq y)$$

for all $A \subseteq X$, $y \in X$, and with $\Uparrow_X A := \{x \in X \mid A \lll x\}$ one obtains for a monotone map f:

$$f \text{ is } (\mathbb{P}, \mathsf{2})\text{-proper} \iff \forall A \subseteq X\ (\Uparrow_Y f(A) \subseteq f(\Uparrow_X A)).$$

Such maps must necessarily be surjective ($A = \varnothing$) and $\mathsf{2}$-proper ($A = \{x\}$), but not conversely. However, the meaning of openness stays the same as in the preceding example:

$$f \text{ is } (\mathbb{P}, 2)\text{-open} \iff f \text{ is } 2\text{-open} .$$

(3) In $\mathsf{Met} \cong \mathsf{P}_+\text{-}\mathsf{Cat}$, a non-expansive map $f : (X, a) \to (Y, b)$ is proper if and only if

$$b(f(x), y) = \inf\{a(x, z) \mid z \in X,\, f(z) = y\}$$

for all $x \in X$, $y \in Y$. Openness is characterized dually.

(4) For a monoid H and $\mathbb{H}$ as in Section V.1.4, an $(\mathbb{H}, 2)$-map $f : X \to Y$ is proper if

$$f(x) \xrightarrow{\alpha} y \implies \exists z \in f^{-1}y\, (x \xrightarrow{\alpha} z)$$

and open if

$$y \xrightarrow{\alpha} f(x) \implies \exists z \in f^{-1}y\, (z \xrightarrow{\alpha} x)$$

(for all $x \in X$, $y \in Y$, $\alpha \in H$).

(5) For the list monad $\mathbb{L}$ as in Section V.1.4, an $(\mathbb{L}, 2)$-map $f : X \to Y$ is proper if

$$(f(x_1), \ldots, f(x_n)) \vdash y \implies \exists z \in f^{-1}y\, ((x_1, \ldots, x_n) \vdash z)$$

and open if

$$(y_1, \ldots, y_n) \vdash f(x) \implies \exists z_i \in f^{-1}(y_i)\, (i = 1, \ldots, n)\, ((z_1, \ldots, z_n) \vdash x)$$

(for all $x, x_i \in X$, $y, y_i \in Y$).

The more important examples $\mathsf{Top} \cong (\beta, 2)\text{-}\mathsf{Cat}$ and $\mathsf{App} \cong (\beta, \mathsf{P}_+)\text{-}\mathsf{Cat}$ will be discussed in Section V.3.4, after we have established the following stability properties.

V.3.1.4 Proposition

(1) *The classes of proper maps and of open maps in* $(\mathbb{T}, \mathcal{V})\text{-}\mathsf{Cat}$ *are both closed under composition and contain all isomorphisms.*

(2) *For* $(\mathbb{T}, \mathcal{V})$*-continuous maps* $f : (X, a) \to (Y, b)$, $m : (Y, b) \to (Z, c)$, *if* m *is injective and* $m \cdot f$ *proper or open,* f *is also proper or open, respectively.*

(3) *For* $(\mathbb{T}, \mathcal{V})$*-continuous maps* $e : (X, a) \to (Y, b)$, $g : (Y, b) \to (Z, c)$, *if* e *is surjective and* $g \cdot e$ *proper or open,* g *is also proper or open, respectively.*

(4) *If* $\mathcal{V}$ *is Cartesian closed, every pullback of a proper map is proper; if, in addition,* T *satisfies BC, every pullback of an open map is open.*

Proof (1): is immediate.

(2): If m is injective, then $m^\circ \cdot m = 1_Y$ and $(Tm)^\circ \cdot Tm = 1_{TY}$ (see Proposition III.1.2.2 and Exercise III.1.P). Hence, if $m \cdot f$ is proper, then

$$b \cdot Tf = b \cdot (Tm)^\circ \cdot Tm \cdot Tf \le m^\circ \cdot c \cdot T(m \cdot f) = m^\circ \cdot m \cdot f \cdot a = f \cdot a$$

follows, and if $m \cdot f$ is open one obtains

$$f^\circ \cdot b = f^\circ \cdot m^\circ \cdot m \cdot b \le f^\circ \cdot m^\circ \cdot c \cdot Tm = c \cdot (Tf)^\circ \cdot (Tm)^\circ \cdot Tm = c \cdot (Tf)^\circ .$$

(3): If e is surjective, then $(Te) \cdot (Te)^\circ = 1_{TY}$, so that $g \cdot e$ proper implies

$$c \cdot Tg = c \cdot Tg \cdot Te \cdot (Te)^\circ \le g \cdot e \cdot a \cdot (Te)^\circ \le g \cdot e \cdot e^\circ \cdot b \le g \cdot b .$$

The case where $g \cdot e$ is open is treated similarly.
(4): Pullback stability of openness was shown in Proposition III.4.3.8. Keeping the notation used there, one shows the same property for proper maps, but without requiring BC for T, as follows:

$$\begin{aligned} b \cdot Tq &= (b \wedge b) \cdot Tq \\ &\le ((g^\circ \cdot c \cdot Tg) \wedge b) \cdot Tq \\ &= (g^\circ \cdot c \cdot Tg \cdot Tq) \wedge b \cdot Tq \\ &= (g^\circ \cdot c \cdot Tf \cdot Tp) \wedge b \cdot Tq \\ &= (g^\circ \cdot f \cdot a \cdot Tp) \wedge b \cdot Tq && (f \text{ proper}) \\ &\le (q \cdot p^\circ \cdot a \cdot Tp) \wedge b \cdot Tq \\ &= q \cdot ((p^\circ \cdot a \cdot Tp) \wedge (q^\circ \cdot b \cdot Tq)) && (\mathcal{V} \text{ Cartesian closed by Lemma III.4.3.7}) \\ &= q \cdot d . \end{aligned}$$

□

V.3.2 First characterization theorems

In this section we show that openness of $(\mathbb{T}, \mathcal{V})$-Cat is often described as openness in $\mathcal{V}$-Cat, and then we explore to what extent this reduction is possible also for proper maps. We then prove a first generalization of the characterization of proper maps in Top as the closed maps with compact fibers.

V.3.2.1 Remark By Corollary III.1.4.4, every lax extension $\hat{T}$ of T satisfies

$$\hat{T}(r \cdot f) = \hat{T}r \cdot Tf \text{ and } \hat{T}(g^\circ \cdot r) = (Tg)^\circ \cdot \hat{T}r$$

for all $f : X \to Y, r : Y \nrightarrow Z, g : W \to Z$. We say that $\hat{T}$ is *left-whiskering* if

$$\hat{T}(h \cdot r) = Th \cdot \hat{T}r$$

for all $r : Y \nrightarrow Z, h : Z \to W$. This condition implies in particular $\hat{T}h = Th \cdot \hat{T}1_Z$, which is also sufficient for the general case when $\hat{T}$ preserves composition and, a fortiori, when $\hat{\mathbb{T}}$ is associative (see Proposition III.1.9.4). Similarly one says that $\hat{T}$ is *right-whiskering* if

$$\hat{T}(s \cdot f^\circ) = \hat{T}s \cdot (Tf)^\circ$$

for all $f : X \to Y, s : X \nrightarrow Z$, which implies $\hat{T}(f^\circ) = \hat{T}1_X \cdot (Tf)^\circ$ without loss of information when $\hat{T}$ preserves composition. Note that *a flat associative lax extension is always left- and right-whiskering*. Hence, the Barr extensions of $\mathbb{\beta}$ to Rel and to $\mathcal{V}$-Rel are left- and right-whiskering. The Kleisli extension $\check{\mathbb{F}}$ to Rel of the filter monad is right- but not left-whiskering: see Exercise V.3.K. In Examples III.1.9.6 we gave an example of a non-associative flat lax extension which is left- and right-whiskering.

If $m^\circ : \hat{T} \to \hat{T}\hat{T}$ is a natural transformation, in particular if $\hat{\mathbb{T}}$ is associative, the problem of characterizing the proper and open maps may often be reduced to the $\mathcal{V}$-Cat case, via the composite functor

$$\begin{array}{ccccc} (\mathbb{T},\mathcal{V})\text{-}\mathsf{Cat} & \xrightarrow{\ M\ } & (\mathcal{V}\text{-}\mathsf{Cat})^{\mathbb{T}} & \xrightarrow{\ G^{\mathbb{T}}\ } & \mathcal{V}\text{-}\mathsf{Cat} \\ (X,a) & \longmapsto & (TX, \hat{T}a\cdot m_X^\circ, m_X) & \longmapsto & (TX, \hat{T}a\cdot m_X^\circ) \end{array}$$

of Section V.2.3 and Theorem III.5.3.5, which, by abuse of notation, we denote by T again. Indeed, with $\hat{T}$ left-whiskering or right-whiskering one obtains the following criteria.

V.3.2.2 Proposition *Assume $m^\circ : \hat{T} \to \hat{T}\hat{T}$ to be a natural transformation and $f : (X,a) \to (Y,b)$ to be $(\mathbb{T},\mathcal{V})$-continuous.*

(1) *If $\hat{T}$ is left-whiskering,*

$$f \text{ is } (\mathbb{T},\mathcal{V})\text{-proper} \implies Tf \text{ is } \mathcal{V}\text{-proper}\,.$$

(2) *If $\hat{T}$ right-whiskering,*

$$f \text{ is } (\mathbb{T},\mathcal{V})\text{-open} \iff Tf \text{ is } \mathcal{V}\text{-open}\,.$$

Proof (1): From $b\cdot Tf \le f\cdot a$, one obtains

$$\begin{aligned} \hat{T}b\cdot m_Y^\circ\cdot Tf &\le \hat{T}b\cdot \hat{T}\hat{T}f\cdot m_X^\circ && (m^\circ \text{ natural})\\ &\le \hat{T}(b\cdot \hat{T}f)\cdot m_X^\circ && (\hat{T} \text{ lax functor})\\ &= \hat{T}(b\cdot Tf)\cdot m_X^\circ && (b \text{ right unitary})\\ &= \hat{T}(f\cdot a)\cdot m_X^\circ && (f \text{ is } (\mathbb{T},\mathcal{V})\text{-proper})\\ &= Tf\cdot \hat{T}a\cdot m_X^\circ && (\hat{T} \text{ is left-whiskering}). \end{aligned}$$

(2): From $f^\circ\cdot b \le a\cdot (Tf)^\circ$, one derives

$$\begin{aligned} (Tf)^\circ\cdot \hat{T}b\cdot m_Y^\circ &= \hat{T}(f^\circ\cdot b)\cdot m_Y^\circ \\ &\le \hat{T}(a\cdot (Tf)^\circ)\cdot m_Y^\circ && (f \text{ is } (\mathbb{T},\mathcal{V})\text{-open})\\ &= \hat{T}a\cdot (TTf)^\circ\cdot m_Y^\circ && (\hat{T} \text{ is right-whiskering})\\ &= \hat{T}a\cdot m_X^\circ\cdot (Tf)^\circ\,. \end{aligned}$$

Conversely, from $(Tf)^\circ \cdot \hat{T}b \cdot m_Y^\circ \leq \hat{T}a \cdot m_X^\circ \cdot (Tf)^\circ$ we can derive

$$\begin{aligned} f^\circ \cdot b &\leq f^\circ \cdot b \cdot e_{TY}^\circ \cdot e_{TY} \\ &\leq e_X^\circ \cdot \hat{T}(f^\circ \cdot b) \cdot e_{TY} \\ &\leq e_X^\circ \cdot (Tf)^\circ \cdot \hat{T}b \cdot m_Y^\circ \\ &\leq e_X^\circ \cdot \hat{T}a \cdot m_X^\circ \cdot (Tf)^\circ \\ &= (e_X^\circ \circ a) \cdot (Tf)^\circ = a \cdot (Tf)^\circ , \end{aligned}$$

since a is left unitary. □

Lax extensions for which the implication of Proposition V.3.2.2(1) becomes a logical equivalence are, from a topological perspective, rare, but can be fully characterized, as follows.

V.3.2.3 Proposition *Assume $m^\circ : \hat{T} \to \hat{T}\hat{T}$ to be a natural transformation and $\hat{T}$ to be left-whiskering. Consider the following assertions:*

(i) *every $(\mathbb{T}, \mathcal{V})$-continuous map $f : (X, a) \to (Y, b)$ such that Tf is $\mathcal{V}$-proper is $(\mathbb{T}, \mathcal{V})$-proper;*
(ii) *every function $f : X \to Y$ gives a $(\mathbb{T}, \mathcal{V})$-proper map $f : (X, 1_X^\sharp) \to (Y, 1_Y^\sharp)$;*
(iii) *$e : 1 \to T$ satisfies the Beck–Chevalley condition.*

Then (i) $\iff$ (ii) $\Longleftarrow$ (iii)*, and all are equivalent when $\hat{T}$ is flat.*

Proof (i) $\Longrightarrow$ (ii): We first determine how $T : (\mathbb{T}, \mathcal{V})\text{-}\mathsf{Cat} \to \mathcal{V}\text{-}\mathsf{Cat}$ maps discrete $(\mathbb{T}, \mathcal{V})$-spaces and prove $T(X, 1_X^\sharp) = (TX, \hat{T}1_X)$. Indeed, since m° is a natural transformation,

$$\widehat{1_X^\sharp} = \hat{T}(e_X^\circ \cdot \hat{T}1_X) \cdot m_X^\circ = (Te_X)^\circ \cdot \hat{T}\hat{T}1_X \cdot m_X^\circ = (Te_X)^\circ \cdot m_X^\circ \cdot \hat{T}1_X = \hat{T}1_X .$$

Since $\hat{T}$ is left-whiskering, $Tf \cdot \hat{T}1_X = \hat{T}f = \hat{T}1_Y \cdot Tf$, so that

$$Tf : (TX, \hat{T}1_X) \to (TY, \hat{T}1_Y)$$

is $\mathcal{V}$-proper, for every function $f : X \to Y$. By hypothesis (i), the map $f : (X, 1_X^\sharp) \to (Y, 1_Y^\sharp)$ is $(\mathbb{T}, \mathcal{V})$-proper.

(ii) $\Longrightarrow$ (i): For $f : (X, a) \to (Y, b)$, let $Tf : (TX, \hat{a}) \to (TY, \hat{b})$ be $\mathcal{V}$-proper, and let $f : (X, 1_X^\sharp) \to (Y, 1_Y^\sharp)$ be $(\mathbb{T}, \mathcal{V})$-proper. Since $\hat{T}$ is left-whiskering, we obtain

$$
\begin{aligned}
b \cdot Tf &= (e_Y^\circ \circ b) \cdot Tf \\
&= e_Y^\circ \cdot \hat{T}b \cdot m_Y^\circ \cdot Tf \\
&= e_Y^\circ \cdot Tf \cdot \hat{T}a \cdot m_X^\circ && (Tf \text{ is } \mathcal{V}\text{-proper}) \\
&= e_Y^\circ \cdot Tf \cdot \hat{T}1_X \cdot \hat{T}a \cdot m_X^\circ \\
&= e_Y^\circ \cdot \hat{T}1_X \cdot Tf \cdot \hat{T}a \cdot m_X^\circ && (\hat{T} \text{ is left-whiskering}) \\
&= 1_Y^\sharp \cdot Tf \cdot \hat{T}a \cdot m_X^\circ \\
&= f \cdot 1_X^\sharp \cdot \hat{T}a \cdot m_X^\circ && (f : (X, 1_X^\sharp) \to (Y, 1_Y^\sharp) \text{ is } (\mathbb{T}, \mathcal{V})\text{-proper}) \\
&= f \cdot (e_X^\circ \circ a) \\
&= f \cdot a .
\end{aligned}
$$

(iii) $\Longleftrightarrow$ (ii): $f \cdot e_X^\circ = e_Y^\circ \cdot Tf$ implies $f \cdot e_X^\circ \cdot \hat{T}1_X = e_Y^\circ \cdot Tf \cdot \hat{T}1_X = e_Y^\circ \cdot \hat{T}1_Y \cdot Tf$ since $\hat{T}$ is left-whiskering, and the reverse implication holds if $\hat{T}$ is flat. □

V.3.2.4 Remarks

(1) For a monoid H, consider the flat extension $\overline{H}$ of the associated monad $\mathbb{H}$ (see Section V.1.4). An $(\mathbb{H}, 2)$-functor $f : (X, a) \to (Y, b)$ is proper if it satisfies

$$f(x) \xrightarrow{\alpha} y \implies \exists z \in f^{-1}y \ (x \xrightarrow{\alpha} z)$$

for all $x \in X$, $y \in Y$, $\alpha \in H$. The lax extension satisfies the hypothesis of Proposition V.3.2.3, and the unit morphism of the monad satisfies BC. Proper $(\mathbb{H}, 2)$-functors f are therefore equivalently described as those $(\mathbb{H}, 2)$-functors for which $1_H \times f$ is proper.

(2) For the ultrafilter monad β, the unit fails to satisfy BC: see Proposition III.1.12.4; indeed, the naturality square for the map $X \to 1$ with X infinite fails to be a BC-square. Consequently, by Proposition V.3.2.3, there are maps $f : X \to Y$ in $\mathsf{Top} \cong (\beta, 2)\text{-}\mathsf{Cat}$ for which βf is proper, but f is not. We will see in Proposition V.3.4.5 that Proposition V.3.2.3 gives the $(\mathbb{T}, \mathcal{V})$-categorical reason for the existence of closed maps that are not stable under pullback.

We can now give a complete characterization of $(\mathbb{T}, \mathcal{V})$-proper maps in terms of the condition that Tf be $\mathcal{V}$-proper which, as we will show in Proposition V.3.4.5, generalizes the characterization of proper maps in $\mathsf{Top} = (\beta, 2)\text{-}\mathsf{Cat}$ as the closed maps with compact fibers, and similarly in $\mathsf{App} \cong (\beta, \mathsf{P}_+)\text{-}\mathsf{Cat}$. But in order to be able to talk about fibers of f, we should first clarify that very term. For each $y \in Y$, the assignment $* \mapsto y$ defines a $(\mathbb{T}, \mathcal{V})$-functor $y : (1, 1^\sharp) \to (Y, b)$, where $1^\sharp = e_1^\circ \cdot \hat{T}1_1$ is the discrete structure on $1 = \{*\}$ (see Section III.3.2); explicitly, for $w \in T1$,

$$1^\sharp(w,*) = \hat{T}1_1(w, e_1(*))\ .$$

By a *fiber of f on y* we mean the pullback $(f^{-1}y, \tilde{a}) \to (1, 1^\sharp)$ of f along the $(\mathbb{T}, \mathcal{V})$-functor $y : (1, 1^\sharp) \to (Y, b)$. We note that $(f^{-1}y, \tilde{a}) \to (X, a)$ is a monomorphism, but in general not a regular monomorphism, i.e. $\tilde{a}$ does not need to be the restriction of $a : TX \times X \to V$ to $T(f^{-1}y) \times f^{-1}y$:

$$\begin{aligned} \tilde{a}(\chi, x) &= a(\chi, x) \wedge 1^\sharp(T!(\chi), *) \qquad\qquad \text{(with } ! : f^{-1}y \to 1) \\ &= a(\chi, x) \wedge \hat{T}1_X(T!(\chi), e_1(*))\ , \end{aligned}$$

for every $\chi \in T(f^{-1}y)$ and $x \in f^{-1}y$. However, when $T1 \cong 1$ and $\mathcal{V}$ is integral, from $e_1^{-1} = e_1^\circ \le e_1^\circ \cdot \hat{T}1$ one obtains $1^\sharp = \top$, and every $(f^{-1}y, \tilde{a})$ becomes a subspace of (X, a).

Cartesian closedness of $\mathcal{V}$ suffices to insure that proper $(\mathbb{T}, \mathcal{V})$-functors have proper fibers, see Proposition V.3.1.4.

We can now prove a first characterization theorem.

V.3.2.5 Theorem *Let $\mathcal{V}$ be Cartesian closed and T be taut, with $\hat{T}$ left-whiskering and m° a natural transformation. Then a $(\mathbb{T}, \mathcal{V})$-functor $f : (X, a) \to (Y, b)$ is proper if and only if all of its fibers are proper and the $\mathcal{V}$-functor $Tf : (TX, \hat{a}) \to (TY, \hat{b})$ is proper.*

Proof If f is proper, the fibers of f are proper by Proposition V.3.1.4 and Tf is proper by Proposition V.3.2.2.

Conversely, assume that all fibers of f are proper in $(\mathbb{T}, \mathcal{V})$-Cat and Tf is proper in $\mathcal{V}$-Cat. Since

$$b = b \cdot e_{TY}^\circ \cdot m_Y^\circ \le e_Y^\circ \cdot \hat{T}b \cdot m_Y^\circ = e_Y^\circ \cdot \hat{b}\ ,$$

for all $\chi \in TX$, $y \in Y$ one obtains

$$\begin{aligned} b \cdot Tf(\chi, y) &= b(Tf(\chi), y) \\ &\le \hat{b}(Tf(\chi), e_Y(y)) \\ &= \textstyle\bigvee_{z \in (Tf)^{-1}(e_Y(y))} \hat{a}(\chi, z) \qquad\qquad (Tf \text{ proper}) \\ &= \textstyle\bigvee_{z \in (Tf)^{-1}(e_Y(y))} (\hat{T}a \cdot m_X^\circ)(\chi, z) \\ &= \textstyle\bigvee_{z \in (Tf)^{-1}(e_Y(y))} \bigvee_{\mathfrak{X} \in m_X^{-1}\chi} \hat{T}a(\mathfrak{X}, z) \otimes k\ . \end{aligned}$$

Since tautness of T guarantees that the following diagram is a pullback

$$\begin{array}{ccc} T(f^{-1}y) & \xrightarrow{T!} & T1 \\ \downarrow & & \downarrow{\scriptstyle Ty} \\ TX & \xrightarrow{Tf} & TY\ , \end{array}$$

every $z \in (Tf)^{-1}(e_Y(y)) = (Tf)^{-1}(Ty(e_1(*)))$ satisfies $z \in T(f^{-1}y)$ and $T!(z) = e_1(*)$. Using propriety of $(f^{-1}y, \tilde{a}) \to (1, 1^\sharp)$, one obtains

$$\begin{aligned}
\bigvee_{z\in(Tf)^{-1}(e_Y(y))} \bigvee_{\mathcal{X}\in m_X^{-1}\chi} \hat{T}a(\mathcal{X}, z) \otimes k &\leq \bigvee_{z\in(T!)^{-1}(e_1(*))} \bigvee_{\mathcal{X}\in m_X^{-1}\chi} \hat{T}a(\mathcal{X}, z) \\
&\qquad \otimes \bigvee_{x\in f^{-1}y} \tilde{a}(z, x) \\
&\leq \bigvee_{z\in(T!)^{-1}(e_1(*))} \bigvee_{\mathcal{X}\in m_X^{-1}\chi} \\
&\qquad \bigvee_{x\in f^{-1}y} \hat{T}a(\mathcal{X}, z) \otimes a(z, x) \\
&\leq \bigvee_{\mathcal{X}\in m_X^{-1}\chi} \bigvee_{x\in f^{-1}y} a(m_X(\mathcal{X}), x) \\
&\leq \bigvee_{x\in f^{-1}y} a(\chi, x) \\
&= (f \cdot a)(\chi, y)\,.
\end{aligned}$$

Hence, f is proper. □

Next we show that propriety of fibers trivializes whenever the unit e of the monad $\mathbb{T}$ satisfies BC – a rather restrictive condition, as we have seen in Section III.1.12 and Exercise III.1.Q.

V.3.2.6 Proposition *If $\mathcal{V}$ is integral and $e^\circ : \hat{T} \to 1$ is a natural transformation, then any $(\mathbb{T}, \mathcal{V})$-functor has proper fibers.*

Proof For a $(\mathbb{T}, \mathcal{V})$-functor $f : (X, a) \to (Y, b)$ and $y \in Y$, we must show that the diagram

$$\begin{array}{ccc}
T(f^{-1}y) & \xrightarrow{T!} & T1 \\
{\scriptstyle \tilde{a}}\big\downarrow & & \big\downarrow{\scriptstyle 1^\sharp} \\
f^{-1}y & \xrightarrow{!} & 1
\end{array}$$

commutes; for that, it suffices to consider $\chi \in T(f^{-1}y)$ with $1^\sharp(T!(\chi), *) = \hat{T}1(T!(\chi), e_1(*)) > \bot$ and show $\tilde{a}(\chi, *) = \top$. From the commutativity of the diagram

$$\begin{array}{ccccc}
T(f^{-1}y) & \xrightarrow{T!} & T1 & \xrightarrow{\hat{T}1} & T1 \\
{\scriptstyle e^\circ}\big\downarrow & & \big\downarrow{\scriptstyle e_1^\circ} & & \big\downarrow{\scriptstyle e_1^\circ} \\
f^{-1}y & \xrightarrow{!} & 1 & \xrightarrow{1} & 1
\end{array}$$

we first obtain

$$\bot < e_1^\circ \cdot \hat{T}1 \cdot T!(\chi, *) = e_1^\circ \cdot T!(\chi, *) = ! \cdot e^\circ(\chi, *) = \bigvee_{x\in f^{-1}y} e^\circ(\chi, x) = k\,,$$

and then

$$! \cdot \tilde{a}(\chi, *) \geq ! \cdot e^\circ(\chi, x) = k = \top\,.$$ □

V.3.2.7 Corollary *Under the hypotheses of* Theorem V.3.2.5 *and* Proposition V.3.2.6, *a* $(\mathbb{T}, \mathcal{V})$*-functor* $f : (X, a) \to (Y, b)$ *is proper if and only if the* $\mathcal{V}$*-functor* Tf *is proper.*

V.3.3 Notions of closure

Our next aim is to characterize $(\mathbb{T}, \mathcal{V})$-proper and $(\mathbb{T}, \mathcal{V})$-open maps in terms of suitable notions of closure. Since the functor T of the monad $\mathbb{T}$ preserves injections (see Exercise III.1.P), for a subset $A \subseteq X$ we may assume $TA \subseteq TX$; with $\mathcal{A} \subseteq TX$, $v \in V$ we define

$$\mathcal{A}^{[v]} := \{x \in X \mid \textstyle\bigvee_{\mathfrak{x} \in \mathcal{A}} a(\mathfrak{x}, x) \geq v\}\,,$$
$$A^{(v)} := TA^{[v]}\,,$$
$$\overline{A} := \textstyle\bigcup_{v > \bot} A^{(v)} = \{x \in X \mid \exists \mathfrak{x} \in TA\ (a(\mathfrak{x}, x) > \bot)\}\,,$$

and call $A^{(v)}$ and $\overline{A}$ the *v-closure* and *grand closure* of A, respectively.

V.3.3.1 Proposition *The grand closure defines a hereditary* $\mathcal{M}$*-closure operator on* $(\mathbb{T}, \mathcal{V})$-$\mathsf{Cat}$, *with* $\mathcal{M}$ *the class of embeddings, and so does the* v*-closure, for any fixed* $v \leq k$ *in* V. *In general, neither operator is idempotent.*

Proof See Exercise V.3.A. □

We can now analyse to what extent these closures help us characterize propriety and openness of maps.

V.3.3.2 Proposition *The following statements on a* $(\mathbb{T}, \mathcal{V})$*-continuous map* $f : (X, a) \to (Y, b)$ *satisfy the following implications:*

$$\begin{array}{ccccc} \text{(i)} & \Longrightarrow & \text{(iii)} & \Longrightarrow & \text{(v)} \\ \Downarrow & & \Downarrow & & \\ \text{(ii)} & \Longrightarrow & \text{(iv)}; & & \end{array}$$

furthermore, the vertical implications are equivalences if $\mathcal{V}$ *is completely distributive, and one has* (ii) $\iff$ (iv) *if* $\mathbb{T} = \mathbb{I}$ *is the identity monad, and* (iii) $\iff$ (v) *if* $\mathcal{V} = \mathsf{2}$.

(i) $b \cdot Tf \leq f \cdot a$, *i.e.* f *is proper;*
(ii) $\forall \mathcal{A} \subseteq TX\ \forall u \ll v$ *in* $\mathcal{V}$ $(Tf(\mathcal{A})^{[v]} \subseteq f(\mathcal{A}^{[u]}))$;
(iii) $\forall A \subseteq X\ (b \cdot Tf \cdot Ti_A \cdot !^{\circ}_{TA} \leq f \cdot a \cdot Ti_A \cdot !^{\circ}_{TA})$;
(iv) $\forall A \subseteq X\ \forall u \ll v$ *in* V $(f(A)^{(v)} \subseteq f(A^{(u)}))$;
(v) $\forall A \subseteq X\ (\overline{f(A)} \subseteq f(\overline{A}))$;

$i_A : A \hookrightarrow X$ *is the inclusion and* $!_{TA} : TA \to 1$.

Proof The implications (i) $\Longrightarrow$ (iii) and (ii) $\Longrightarrow$ (iv) are trivial, and the remaining generally valid implications become obvious once transcribed in elementwise terms, as follows:

(i) $\forall \mathfrak{x} \in TX, y \in Y\ (b(Tf(\mathfrak{x}), y) \leq \bigvee_{x \in f^{-1}y} a(\mathfrak{x}, x))$;
(ii) $\forall \mathcal{A} \subseteq TX, u \ll v \text{ in } \mathcal{V}, y \in Y\ (v \leq \bigvee_{\mathfrak{x} \in \mathcal{A}} b(Tf(\mathfrak{x}), y) \implies \exists x \in f^{-1}y\ (u \leq \bigvee_{\mathfrak{z} \in \mathcal{A}} a(\mathfrak{z}, x)))$;
(iii) $\forall A \subseteq X, y \in Y\ (\bigvee_{\mathfrak{x} \in TA} b(Tf(\mathfrak{x}), y) \leq \bigvee_{x \in f^{-1}y} \bigvee_{\mathfrak{z} \in TA} a(\mathfrak{z}, x))$;
(iv) $\forall A \subseteq X, u \ll v \text{ in } \mathcal{V}, y \in Y\ (v \leq \bigvee_{\mathfrak{x} \in TA} b(Tf(\mathfrak{x}), y) \implies \exists x \in f^{-1}y\ (u \leq \bigvee_{\mathfrak{z} \in TA} a(\mathfrak{z}, x)))$;
(v) $\forall A \subseteq X, y \in Y\ (\exists \mathfrak{x} \in TA\ (b(Tf(\mathfrak{x}), y) > \bot) \implies \exists x \in f^{-1}y, \mathfrak{z} \in TA\ (a(\mathfrak{z}, x) > \bot))$.

Here, for the reformulation of (iii) and (iv), note that $T(f(A)) = Tf(TA)$ since T preserves the surjection $A \to f(A)$.

To see (ii) $\Longrightarrow$ (i), if $\mathcal{V}$ is completely distributive, put $\mathcal{A} := \{\mathfrak{x}\}$ and $v := b(Tf(\mathfrak{x}), y)$, and for (iv) $\Longrightarrow$ (iii) put $v := \bigvee_{\mathfrak{x} \in TA} b(Tf(\mathfrak{x}), y)$. One trivially has (iv) $\Longrightarrow$ (ii) when T is the identity functor, and the elementwise formulations also show (v) $\Longrightarrow$ (iii) for $\mathcal{V} = \mathsf{2}$, where $\bigvee_i v_i = \top$ equivalently means $v_i > \bot$ for some i. □

V.3.3.3 Remarks

(1) With the reverse inequalities in Proposition V.3.3.2(i) and (iii) holding by $(\mathbb{T}, \mathcal{V})$-continuity of f, these inequalities may also be replaced by equalities. The same is true for the inclusion in (v), and when $\mathcal{V}$ is completely distributive, statements (ii) and (iv) may, respectively, be replaced by

(ii′) $\forall \mathcal{A} \subseteq TX, v \in \mathcal{V}\ (Tf(\mathcal{A})^{[v]} = \bigcap_{u \ll v} f(\mathcal{A}^{[u]}))$;
(iv′) $\forall A \subseteq X, v \in \mathcal{V}\ (f(A)^{(v)} = \bigcap_{u \ll v} f(A^{(u)}))$.

Briefly, all five conditions considered in Proposition V.3.3.2 are *equationally defined.*

(2) Whereas conditions (i)–(iv) are equivalent in $\mathcal{V}$-Cat for $\mathcal{V}$ completely distributive (and $\mathbb{T} = \mathbb{I}$), condition (v) is in general considerably weaker: in $\mathsf{Met} = \mathsf{P}_+$-$\mathsf{Cat}$, every surjective map $(X, a) \to (Y, b)$ with a, b finite satisfies condition (v)! By contrast, *in* $\mathsf{Top} = (\mathbb{\beta}, \mathsf{2})$-$\mathsf{Cat}$ *the grand closure describes the ordinary Kuratowski closure, so that each of* (iii) $\iff$ (iv) $\iff$ (v) *describes closed continuous maps* (continuous maps which preserve closed subsets), while each of (i) $\iff$ (ii) describes stably closed maps, as we will show in Proposition V.3.4.5 and Theorem V.3.4.6.

(3) For an embedding $f : (X, a) \hookrightarrow (Y, b)$ in $(\mathbb{T}, \mathcal{V})$-$\mathsf{Cat}$, one has (i) $\iff$ (iii) $\iff$ (v) in Proposition V.3.3.2; these equivalent conditions are satisfied precisely when the set X is closed in (Y, b) with respect to the grand closure, i.e. when $\overline{X}^Y = X$: see Exercise V.3.A.

Proposition V.3.3.2 compares propriety with the behavior of closures under taking images; we now similarly compare openness with the behavior under taking inverse images. In order to do so, we need that

$$T(f^{-1}B) = (Tf)^{-1}(TB)$$

for all $B \subseteq Y$ and $f : X \to Y$, i.e. that T is taut (see Definition III.4.3.5).

V.3.3.4 Proposition *Let the functor T of the monad $\mathbb{T}$ be taut. The following statements on a $(\mathbb{T}, \mathcal{V})$-continuous map $f : (X, a) \to (Y, b)$ satisfy the following implications:*

$$\begin{array}{ccccc} \text{(i)} & \Longrightarrow & \text{(iii)} & \Longrightarrow & \text{(v)} \\ \Uparrow & & \Uparrow & & \\ \text{(ii)} & \Longrightarrow & \text{(iv)}; & & \end{array}$$

furthermore, the vertical implications are equivalences if $\mathcal{V}$ is completely distributive, and one has (ii) $\iff$ (iv) if $\mathbb{T} = \mathbb{I}$ is the identity monad, and (iii) $\iff$ (v) if $\mathcal{V} = 2$.

(i) $f^\circ \cdot b \leq a \cdot (Tf)^\circ$, *i.e. f is open;*
(ii) $\forall \mathcal{B} \subseteq TY, v \in \mathcal{V}\ (\bigcap_{u \ll v} f^{-1}(\mathcal{B}^{[u]}) \subseteq (Tf)^{-1}(\mathcal{B})^{[v]})$;
(iii) $\forall B \subseteq Y\ (f^\circ \cdot b \cdot Ti_B \cdot !^\circ_{TB} \leq a \cdot (Tf)^\circ \cdot Ti_B \cdot !^\circ_{TB})$;
(iv) $\forall B \subseteq Y, v \in \mathcal{V}\ (\bigcap_{u \ll v} f^{-1}(B^{(u)}) \subseteq f^{-1}(B)^{(v)})$;
(v) $\forall B \subseteq Y\ (f^{-1}(\overline{B}) \subseteq \overline{f^{-1}(B)})$.

Proof One proceeds schematically as in the proof for Proposition V.3.3.2, by transcribing the five statements in elementwise terms:

(i) $\forall x \in X, y \in TY\ (b(y, f(x)) \leq \bigvee_{z \in (Tf)^{-1}y} a(z, x))$;
(ii) $\forall \mathcal{B} \subseteq TY, v \in \mathcal{V}, x \in X\ ((\forall u \ll v\ (u \leq \bigvee_{y \in \mathcal{B}} b(y, f(x))) \implies v \leq \bigvee_{z \in (Tf)^{-1}(\mathcal{B})} a(z, x))$;
(iii) $\forall B \subseteq Y, x \in X\ (\bigvee_{y \in TB} b(y, f(x)) \leq \bigvee_{z \in (Tf)^{-1}(TB)} a(z, x))$;
(iv) $\forall B \subseteq Y, v \in \mathcal{V}, x \in X\ ((\forall u \ll v\ (u \leq \bigvee_{y \in TB} b(y, f(x)))) \implies v \leq \bigvee_{z \in (Tf)^{-1}(TB)} a(z, x))$;
(v) $\forall B \subseteq Y, x \in X\ (\exists y \in TB\ (b(y, f(x)) > \bot) \implies \exists z \in (Tf)^{-1}(TB)\ (a(z, x) > \bot))$. □

V.3.3.5 Remarks

(1) Because of the $(\mathbb{T}, \mathcal{V})$-continuity of f, the inequalities in (i) and (iii) and the inclusion in Proposition V.3.3.4(v) may equivalently be replaced by equalities; likewise in (ii) and (iv) if $\mathcal{V}$ is completely distributive.
(2) As in Proposition V.3.3.2, conditions (i)–(iv) of Proposition V.3.3.4 are equivalent in $\mathcal{V}$-Cat for $\mathcal{V}$ completely distributive (and $\mathbb{T} = \mathbb{I}$), but condition (v) is generally weaker: again, any surjective map $(X, a) \to (Y, b)$ in Met $=$ P_+-Cat with a, b finite satisfies condition (v). However, *in*

Top $=$ (β, 2)-Cat *all five conditions are equivalent and describe open maps in the usual sense, i.e. continuous maps which preserve open subsets.* Indeed, such a map $f : X \to Y$ satisfies condition (i), which reads as follows:

$$\forall x \in X, y \in \beta Y \ (y \longrightarrow f(x) \implies \exists z \in \beta X \ (f[z] = y \ \& \ z \longrightarrow x)) ,$$

$$\begin{array}{ccc} z & \longrightarrow & x \\ \vdots & & | \\ y & \longrightarrow & f(x) . \end{array}$$

(One simply takes for z an ultrafilter on X containing the filter base $\{f^{-1}B \mid B \in y\}$.)
Since (i) implies (v), it suffices to show that (v) makes f open in the usual sense. But for $A \subseteq X$ open one obtains

$$f^{-1}\overline{(Y \setminus f(A))} = \overline{f^{-1}(Y \setminus f(A))} \subseteq \overline{X \setminus A} = X \setminus A ,$$

and then $\overline{Y \setminus f(A)} \subseteq Y \setminus f(A)$, so that $f(A)$ is open.

(3) As in Proposition V.3.3.2, one has (i) $\iff$ (iii) $\iff$ (v) in Proposition V.3.3.4 for an embedding $f : (X, a) \hookrightarrow (Y, b)$, where now (v) reads as follows:

$$\forall B \subseteq Y \ (\overline{B} \cap X = \overline{B \cap X}) .$$

Letting Top $=$ (β, 2)-Cat again guide our terminology in the general context, we consider properties V.3.3.2(iii) and V.3.3.4(iii) to yield the following definition.

V.3.3.6 Definition A $(\mathbb{T}, \mathcal{V})$-continuous map $f : (X, a) \longrightarrow (Y, b)$ is *closed* if

$$b \cdot Tf \cdot Ti_A \cdot !_{TA}^\circ \le f \cdot a \cdot Ti_A \cdot !_{TA}^\circ$$

for all $A \subseteq X$, and $f : (X, a) \longrightarrow (Y, b)$ is *inversely closed* if

$$f^\circ \cdot b \cdot Ti_B \cdot !_{TB}^\circ \le a \cdot (Tf)^\circ \cdot Ti_B \cdot !_{TB}^\circ$$

for all $B \subseteq Y$.

Emphasizing the dependency on the parameters, we often add the prefixes $(\mathbb{T}, \mathcal{V})$ or $\mathcal{V}$ (when $\mathbb{T} = \mathbb{I}$). One trivially has

$$f \text{ proper} \implies f \text{ closed}, \qquad f \text{ open} \implies f \text{ inversely closed},$$

with the reversed implications holding for $\mathbb{T} = \mathbb{I}$ (see Propositions V.3.3.2 and V.3.3.4). With Propositions V.3.2.2 and V.3.2.3, we obtain:

V.3.3.7 Corollary *Assume $m^\circ : \hat{T}\hat{T} \to \hat{T}$ to be a natural transformation, and let $f : (X, a) \to (Y, b)$ be $(\mathbb{T}, \mathcal{V})$-continuous.*

(1) *If $\hat{T}$ is left-whiskering,*

$$f \text{ is proper} \implies Tf : (TX, \hat{a}) \to (TY, \hat{b}) \text{ is closed,}$$

with the reverse implication holding when $e : 1 \to T$ satisfies BC.

(2) *If $\hat{T}$ is right-whiskering,*

$$f \text{ is open } \iff Tf : (TX, \hat{a}) \to (TY, \hat{b}) \text{ is inversely closed.}$$

It is essential to consider Tf on the right-hand sides, not just f, as the following example shows.

V.3.3.8 Example For a monoid H and the flat extension on the associated monad $\mathbb{H} = (H\times(-), m, e)$, m° is natural and e satisfies BC. However, it is easy to check that the identity map $f : (\{0,1\}, a) \to (\{0,1\}, b)$, where $a((\alpha, 0), i) = \top$ and $a((\alpha, 1), 1) = \top$ only if $\alpha = 1$, and $b((\alpha, i), j) = \top$ for every $i, j \in \{0, 1\}$, $\alpha \in H$, is closed and inversely closed, but neither proper nor open.

Nevertheless, in Proposition V.3.4.5 we will show that in $\mathsf{Top} = (\mathbb{B}, 2)$-$\mathsf{Cat}$ and $\mathsf{App} = (\mathbb{B}, \mathsf{P}_+)$-$\mathsf{Cat}$ the condition that Tf be (inversely) closed may be replaced by the condition that f be (inversely) closed.

V.3.4 Kuratowski–Mrówka Theorem

We are now ready to characterize proper maps in terms of closure properties. The Kuratowski–Mrówka Theorem is the object version of that characterization and facilitates the proof in the general case. For its proof we rely in turn on Proposition III.4.9.1, which asserts that, under the hypotheses that T *preserve disjointness, $\hat{T}$ be flat, and e° is finitely strict*, for any set X and $\chi \in TX$ one can define a $(\mathbb{T}, \mathcal{V})$-space structure $c = c_\chi$ on $Z = X + 1$ by

$$c(\mathfrak{z}, z) = \begin{cases} k & \text{if } \mathfrak{z} = e_Z(z), \\ \top & \text{if } \mathfrak{z} = \chi \text{ and } z \in 1, \\ \bot & \text{otherwise,} \end{cases}$$

for all $z \in Z$, $\mathfrak{z} \in TZ$. Whenever all $(X + 1, c)$ are $(\mathbb{T}, \mathcal{V})$-spaces, in particular under the above hypotheses, we say that $(\mathbb{T}, \mathcal{V})$-$\mathsf{Cat}$ *has enough KM-test spaces.*

V.3.4.1 Theorem *Let $\mathcal{V}$ be Cartesian closed and let $(\mathbb{T}, \mathcal{V})$-$\mathsf{Cat}$ have enough KM-test spaces. Then the following assertions for a $(\mathbb{T}, \mathcal{V})$-space (X, a) are equivalent:*

(i) *$(X, a) \to (1, \top)$ is proper;*
(ii) *the product projection $X \times Y \to Y$ is closed for any $(\mathbb{T}, \mathcal{V})$-space (Y, b).*

When, in addition, $\mathcal{V}$ is integral and superior, we may extend the list by adding

(iii) *(X, a) is compact.*

Proof (i) $\Longrightarrow$ (ii) follows from Proposition V.3.1.4(3) since $X \times Y \to Y$ is a pullback of $X \to 1$. (i) $\iff$ (iii) re-states Remark V.3.1.2(2). For (ii) $\Longrightarrow$ (i) we must show

$$\bigvee\nolimits_{x\in X} a(\chi, x) = \top$$

for all $\chi \in TX$. Hence we exploit the hypothesis for $(Y, b) = (Z, c_\chi)$ and obtain, with $\Delta = \{(x, x) \mid x \in X\} \subseteq X \times Z$ and $q : (X \times Z, d) \to (Z, c)$ denoting the second product projection,

$$c \cdot Tq \cdot Ti_\Delta \cdot !^\circ_{T\Delta} \leq q \cdot d \cdot Ti_\Delta \cdot !^\circ_{T\Delta} \,. \tag{V.3.4.i}$$

With the natural bijection $\delta : X \to \Delta$, every $z \in T\Delta$ is of the form $z = T\delta(y)$ with $y \in TX$, and $Tp(z) = y = Tq(z) \in TX \subseteq TZ$ with $p : X \times Z \to X$ the first projection. Hence, an exploitation of the elementwise form of (V.3.4.i) gives (with $* \in 1 \subseteq Z$)

$$\begin{aligned} \top = c(\chi, *) &\leq \textstyle\bigvee_{x\in X} \bigvee_{y\in TX} d(T\delta(y), (x, *)) \\ &= \textstyle\bigvee_{x\in X} \bigvee_{y\in TX} a(y, x) \wedge c(y, *) \\ &= \textstyle\bigvee_{x\in X} a(\chi, x) \,, \end{aligned}$$

since for every $y \neq \chi$ one has $c(y, *) = \bot$. □

With Theorem V.3.2.5 we conclude from Theorem V.3.4.1 the desired characterization of proper maps, calling a map *stably closed* if all of its pullbacks are closed.

V.3.4.2 Corollary *Let $\mathcal{V}$ be Cartesian closed, T be taut with $T\varnothing = \varnothing$, $\hat{T}$ be flat and left-whiskering, e° be finitely strict, and m° be a natural transformation. Then the following assertions on a $(\mathbb{T}, \mathcal{V})$-continuous map $f : (X, a) \to (Y, b)$ are equivalent:*

(i) *f is proper;*
(ii) *f is stably closed, and $Tf : (TX, \hat{a}) \to (TY, \hat{b})$ is closed;*
(iii) *$f^{-1}y \to 1$ is proper for all $y \in Y$, and Tf is closed.*

When, in addition, $\mathcal{V}$ is integral and superior, we may extend the list by adding

(iv) *$f^{-1}y$ is compact for all $y \in Y$, and Tf is closed.*

Proof (i) $\Longrightarrow$ (ii): Follows from Propositions V.3.1.4(4) and V.3.2.2(1).

(ii) $\Longrightarrow$ (iii): As a pullback of f, each fiber of f is stably closed and therefore proper, by Theorem V.3.4.1.

(iii) $\Longrightarrow$ (i): Follows from Theorem V.3.2.5.

(iii) $\Longleftrightarrow$ (iv): Follows from Remark V.3.1.2(2). □

V.3.4.3 Remark The condition that Tf be closed can neither be removed from V.3.4.2(ii) nor be substituted in (iii) by the requirement that f be closed. Indeed, going back to Example V.3.3.8, one easily checks that f is stably closed but not proper, and Exercise V.3.D gives an example of a closed $(\mathbb{T}, \mathcal{V})$-continuous map with compact fibers which is not proper.

When $\mathbb{T} = \mathbb{\beta}$ is the ultrafilter monad, with its Barr extension to $\mathcal{V}$-**Rel** (see Corollary IV.2.4.5), for a $(\mathbb{\beta}, \mathcal{V})$-space (X, a) and $\mathfrak{x}, \mathfrak{y} \in \beta X$ one has, by definition,

$$\hat{a}(\mathfrak{x}, \mathfrak{y}) = \bigvee_{\mathcal{X} \in m_X^{-1}\mathfrak{x}} \overline{\beta}a(\mathcal{X}, \mathfrak{y}) = \bigvee_{\mathcal{X} \in m_X^{-1}\mathfrak{x}} \bigwedge_{\mathcal{A} \in \mathcal{X}, B \in \mathfrak{y}} \bigvee_{z \in \mathcal{A}, y \in B} a(z, y) .$$

V.3.4.4 Lemma *If $\mathcal{V}$ is a chain (and thus completely distributive), then*

$$\hat{a}(\mathfrak{x}, \mathfrak{y}) = \bigvee\{u \in V \mid \forall A \in \mathfrak{x}\ (A^{(u)} \in \mathfrak{y})\} . \tag{V.3.4.ii}$$

Proof For "$\leq$" consider any $\mathcal{X} \in \beta\beta X$ with $m_X(\mathcal{X}) = \mathfrak{x}$. It suffices to show that every $u \ll \bigwedge_{\mathcal{A} \in \mathcal{X}, B \in \mathfrak{y}} \bigvee_{z \in \mathcal{A}, y \in B} a(z, y)$ has the property that $A^{(u)} \in \mathfrak{y}$ for all $A \in \mathfrak{x}$. But if for $A \in \mathfrak{x}$ we assume $A^{(u)} \notin \mathfrak{y}$, so that $B := X \setminus A^{(u)} \in \mathfrak{y}$, considering

$$\mathcal{A} := A^{\mathbb{\beta}} = \{z \in \beta X \mid A \in z\} \in \mathcal{X} \quad (\text{since } A \in \mathfrak{x}) ,$$

we would conclude

$$u \ll \bigvee_{z \in \mathcal{A}, y \in B} a(z, y) ,$$

and therefore $A^{(u)} \cap B \neq \varnothing$, a contradiction.

For "$\geq$" consider $v \ll \bigvee\{u \in V \mid \forall A \in \mathfrak{x}\ (A^{(u)} \in \mathfrak{y})\}$ in $\mathcal{V}$. For all $A \in \mathfrak{x}$, $B \in \mathfrak{y}$, the ultrafilter $\mathfrak{y}$ contains $A^{(v)} \cap B \neq \varnothing$, so that $v \leq \bigvee_{z \in A^{\mathbb{\beta}}} a(z, y)$ for some $y \in B$, and

$$v \leq \bigwedge_{B \in \mathfrak{y}} \bigvee_{z \in A^{\mathbb{\beta}}, y \in B} a(z, y)$$

follows for every $A \in \mathfrak{x}$. Now

$$\mathcal{F} = \{\mathcal{A} \subseteq \beta X \mid A^{\mathbb{\beta}} \subseteq \mathcal{A} \text{ for some } A \in \mathfrak{x}\}$$

is a filter on βX, and

$$\mathcal{J} := \{\mathcal{B} \subseteq \beta X \mid v > \bigwedge_{B \in \mathfrak{y}} \bigvee_{z \in \mathcal{B}, y \in B} a(z, y)\}$$

is an ideal on βX that is disjoint from $\mathcal{F}$. In order to establish closure of $\mathcal{J}$ under binary union, we use the fact that the order of $\mathcal{V}$ is total, as follows: if $\mathcal{B}$ and $\mathcal{C}$ belong to $\mathcal{J}$, then $v > \bigvee_{z \in \mathcal{B}, y \in B} a(z, y)$ and $v > \bigvee_{z \in \mathcal{C}, y \in C} a(z, y)$ for some $B, C \in \mathfrak{y}$, hence

$$\begin{aligned} v &> \left(\bigvee_{z \in \mathcal{B}, y \in B} a(z, y)\right) \vee \left(\bigvee_{z \in \mathcal{C}, y \in C} a(z, y)\right) \\ &\geq \bigvee_{z \in \mathcal{B} \cup \mathcal{C}, y \in B \cap C} a(z, y) , \end{aligned}$$

and then

$$v > \bigwedge_{D \in \mathfrak{y}} \bigvee_{z \in \mathcal{B} \cup \mathcal{C}, y \in D} a(z, y)$$

since $B \cap C$ belongs to y. The filter $\mathcal{F}$ must be contained in an ultrafilter $\mathcal{X}$ which does not meet the ideal $\mathcal{J}$; by definition of $\mathcal{F}$, one has $\chi = m_X(\mathcal{X})$, and by definition of $\mathcal{J}$

$$v \leq \bigwedge\nolimits_{\mathcal{A} \in \mathcal{X}, B \in y} \bigvee\nolimits_{z \in \mathcal{A}, y \in B} a(z, y) \leq \hat{a}(\chi, y)$$

follows. □

Since the structure a can be recovered from $\hat{a}$ as

$$a(\chi, x) = \hat{a}(\chi, e_X(x)) ,$$

the equality (V.3.4.ii) shows that v-closures on subsets of X encode completely the structure a.

V.3.4.5 Proposition *Let $\mathcal{V}$ be a chain and let $f : (X, a) \to (Y, b)$ be a $(\beta, \mathcal{V})$-continuous map. Then the following hold:*

(1) *f is closed $\iff$ βf is closed $\iff$ βf is proper;*
(2) *f is inversely closed $\iff$ βf is inversely closed $\iff$ f is open.*

Proof To see (1), we observe that with Corollary V.3.3.7(1) it suffices to show that f is closed if and only if βf is proper. First, assuming f to be closed, we must verify

$$\hat{b}(f[\chi], y) \leq \bigvee\{\hat{a}(\chi, z) \mid z \in \beta X,\, f[z] = y\}$$

for all $\chi \in \beta X$, $y \in \beta Y$, where

$$\hat{b}(f[\chi], y) = \bigvee\{v \in \mathcal{V} \mid \forall B \in f[\chi]\ (B^{(v)} \in y)\} .$$

Since f is closed, for any $v \in \mathcal{V}$ contributing to the join on the right, one has $f(A^{(u)}) \in y$ for all $A \in \chi$ and every $u \ll v$. Consequently, with an ultrafilter z_u on X containing the filter base

$$\{A^{(u)} \cap f^{-1}(B) \mid A \in \chi,\, B \in y\} ,$$

one has $f[z_u] = y$ and $A^{(u)} \in z_u$ for all $A \in \chi$, i.e. $u \leq \hat{a}(\chi, z_u)$. Therefore $v = \bigvee_{u \ll v} u \leq (\beta f \cdot \hat{a})(\chi, y)$, as desired.

Conversely, assuming βf to be proper we consider $y \in f(A)^{(v)}$ for $A \subseteq X$, $v \in \mathcal{V}$, so that

$$v \leq \bigvee\nolimits_{y \ni f(A)} b(y, y) = \bigvee\nolimits_{\chi \ni A} b(f[\chi], y) .$$

Since

$$b(f[\chi], y) = \hat{b}(f[\chi], e_Y(y)) = \bigvee\{\hat{a}(\chi, z) \mid z \in \beta X,\, f[z] = e_Y(y)\}$$

with $\hat{a}(\chi, z)$ as in (V.3.4.ii), for all $u \ll v$ one obtains $\chi_u \in \beta X$ with $f[\chi_u] = e_Y(y)$ and $A^{(u)} \in \chi_u$, hence $y \in f(A^{(u)})$.

The proof of (2) is similar. □

Finally, we get a characterization theorem for $(\beta, \mathcal{V})$-spaces that has as particular instances the well-known characterizations of proper maps in $\mathsf{Top} \cong (\beta, \mathsf{2})\text{-}\mathsf{Cat}$ and $\mathsf{App} \cong (\beta, \mathsf{P}_{+})\text{-}\mathsf{Cat}$.

V.3.4.6 Theorem *Let $\mathcal{V}$ be totally ordered, integral, and superior. The following assertions are equivalent, for $f : (X, a) \to (Y, b)$ $(\beta, \mathcal{V})$-continuous:*

(i) *f is proper;*
(ii) *f is stably closed;*
(iii) *f is closed and, for every $y \in Y$, the $(\beta, \mathcal{V})$-functor $f^{-1}(y) \to 1$ is proper;*
(iv) *f is closed with compact fibers.*

V.3.4.7 Examples

(1) Proper maps in Top, introduced as the stably closed maps by Bourbaki [1989] and characterized as the closed maps with compact fibers, have a description dual to open maps (see Remark V.3.3.5(2)) in $(\beta, \mathsf{2})\text{-}\mathsf{Cat}$:

$$\forall \chi \in \beta X, y \in Y \ (f[\chi] \longrightarrow y \implies \exists \in X \ (f(z) = y \ \& \ \chi \longrightarrow z)) \, ,$$

$$\begin{array}{ccc} \chi & \longrightarrow & z \\ | & & \vdots \\ f[\chi] & \longrightarrow & y \, . \end{array}$$

(2) When Top is presented as $(\mathbb{F}_{\mathrm{p}}, \mathsf{2})\text{-}\mathsf{Cat}$ as in Example V.1.1.4(4), $(\mathbb{F}_{\mathrm{p}}, \mathsf{2})$-proper maps are called *superproper* and are characterized by the same property as in (1), except that χ is now allowed to be any proper filter. As compact spaces need not be supercompact, superproper is considerably stronger than proper.

(3) Closed maps $f : (X, a) \to (Y, b)$ in $(\beta, \mathsf{P}_{+})\text{-}\mathsf{Cat} \cong \mathsf{App}$ as introduced in Definition V.3.3.6 are characterized in terms of approach distances by

$$\forall A \subseteq X, y \in Y \ (\inf_{x \in f^{-1}y} \delta^X(x, A) \leq \delta^Y(y, f(A)))$$

(see Proposition V.3.3.2), and this is the description used in approach space theory. Hence, proper maps are also here characterized as the "classically" stably closed maps, or the closed maps with 0-compact fibers.

(4) By Proposition V.3.4.5, open maps $f : (X, a) \to (Y, b)$ in $(\beta, \mathsf{P}_{+})\text{-}\mathsf{Cat} \cong \mathsf{App}$ are equivalently described as the inversely closed maps, which, in terms of approach distances, leads to the classical description of open maps in App:

$$\forall B \subseteq Y, x \in X \ (\delta^X(x, f^{-1}(B)) \leq \delta^Y(f(x), B)) \, .$$

V.3.5 Products of proper maps

Our goal is to prove that the product $\prod_{i \in I} g_i : \prod_{i \in I} X_i \to \prod_{i \in I} Y_i$ of proper maps $g_i : (X_i, a_i) \to (Y_i, b_i)$ is proper, subject to a condition on $\mathcal{V}$ that entails its Cartesian closedness and therefore makes proper maps stable under pullback in $(\mathbb{T}, \mathcal{V})$-**Cat** (see Proposition V.3.1.4). Throughout this section, I is assumed to be small, a set.

Hence, we first point out that *for any class $\mathcal{P}$ of morphisms in a small-complete category* X *that contains all isomorphisms and is stable under pullback*, one has the following result.

V.3.5.1 Proposition *$\mathcal{P}$ is closed under small products if and only if $\mathcal{P}$ is stable under small multiple pullback in* X*; i.e., for any small multiple pullback diagram*

(V.3.5.i)

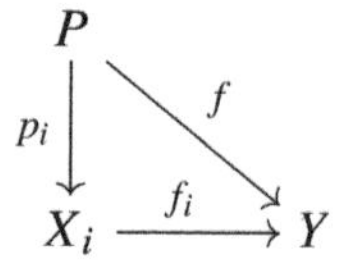

$(i \in I)$

in X*, if $f_i \in \mathcal{P}$ for all $i \in I$, then $f \in \mathcal{P}$.*

Proof The multiple pullback f in (V.3.5.i) is a pullback of $\prod_i f_i$ along $\delta_Y : Y \to Y^I$, as in

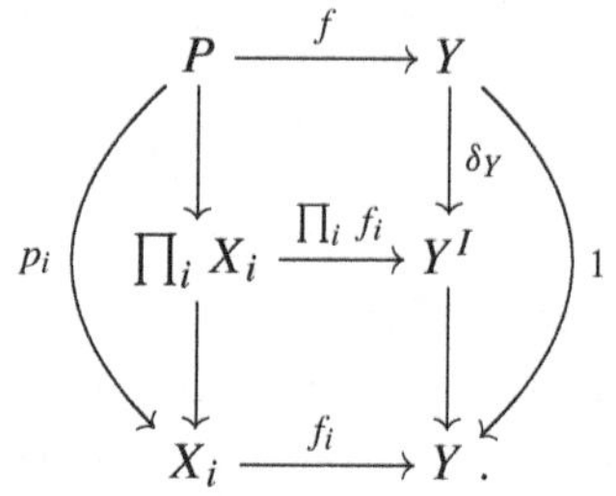

Conversely, the product $\prod_i g_i$ of morphisms $g_i : X_i \to Y_i$ is a multiple pullback of the family of pullbacks $h_i : P_i \to \prod_i Y_i = Y$ of g_i along the projection $Y \to Y_i$, as in

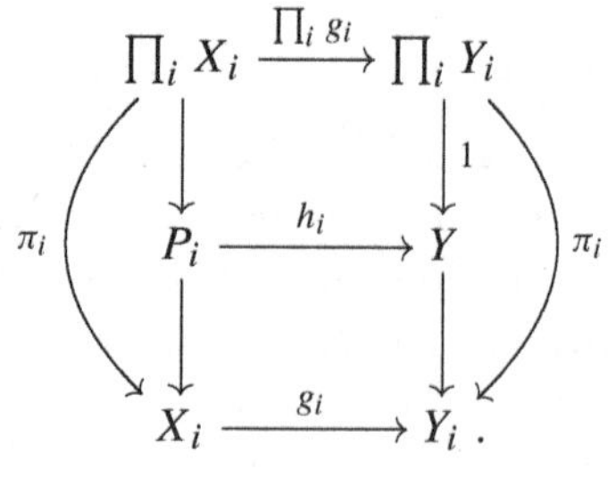

□

Hence, to verify closure of $\mathcal{P}$ under products, it suffices to prove stability under multiple pullbacks. Let us call $\mathcal{V}$-**Rel** *widely modular* if, for every non-empty family $r_i : Z \nrightarrow X_i$ $(i \in I)$ of $\mathcal{V}$-relations and every multiple pullback diagram (V.3.5.i) in **Set**, one has

$$f \cdot \bigwedge_{i \in I} p_i^{\circ} \cdot r_i = \bigwedge_{i \in I} f_i \cdot r_i \tag{V.3.5.ii}$$

in $\mathcal{V}$-**Rel**.

V.3.5.2 Proposition *If $\mathcal{V}$-Rel is widely modular, the class of proper maps is stable under multiple pullback in* $(\mathbb{T}, \mathcal{V})$-Cat.

Proof The multiple pullback of $f_i : (X_i, a_i) \to (Y, b)$ in $(\mathbb{T}, \mathcal{V})$-Cat is formed by providing the limit P of the multiple-pullback diagram (V.3.5.i) in Set with the structure $d = \bigwedge_{i\in I} p_i^\circ \cdot a_i \cdot Tp_i$. For $I = \varnothing$, the map $f = 1_Y$ is trivially proper. Otherwise, we may invoke the wide modularity of $\mathcal{V}$ and obtain

$$f \cdot d = f \cdot \bigwedge_i p_i^\circ \cdot a_i \cdot Tp_i = \bigwedge_i f_i \cdot a_i \cdot Tp_i = \bigwedge_i b \cdot Tf_i \cdot Tp_i = b \cdot Tf$$

when all f_i are proper. Hence, f is proper. □

We must now analyze the status of the hypothesis on $\mathcal{V}$-Rel in Proposition V.3.5.2. We first note that this hypothesis constitutes no restriction if all f_i are injective.

V.3.5.3 Remark If all maps $f_i : X_i \to Y$ are injective, then (V.3.5.ii) holds for all $\mathcal{V}$-relations $r_i : Z \nrightarrow X_i$ $(i \in I \neq \varnothing)$. Indeed, in that case we may assume that all maps of (V.3.5.i) are inclusion maps, with $P = \bigcap_{i\in I} X_i \subseteq Y$, and for all $z \in Z$, $y \in Y$ one has

$$\begin{aligned}(f \cdot \textstyle\bigwedge_i p_i^\circ \cdot r_i)(z, y) &= \begin{cases} \bigwedge_i r_i(z, y) & \text{if } y \in P = \bigcap_i X_i, \\ \bot & \text{otherwise} \end{cases} \\ &= (\textstyle\bigwedge_i f_i \cdot r_i)(z, y)\,.\end{aligned}$$

Here is what unrestricted wide modularity of $\mathcal{V}$-Rel means for $\mathcal{V}$.

V.3.5.4 Proposition $\mathcal{V}$-Rel *is widely modular if and only if $\mathcal{V}$ is completely distributive.*

Proof Given $f_i : X_i \to Y$ and $r_i : Z \nrightarrow X_i$ $(i \in I \neq \varnothing)$ as in (V.3.5.ii), for all $z \in Z$, $y \in Y$ one has

$$\begin{aligned}(f \cdot \textstyle\bigwedge_{i\in I} p_i^\circ \cdot r_i)(z, y) &= \textstyle\bigvee_{w\in f^{-1}y} \bigwedge_{i\in I} r_i(z, p_i(w)) \\ &= \textstyle\bigvee_{(w_i)\in\prod_i f_i^{-1}y} \bigwedge_{i\in i} r_i(z, w_i)\end{aligned}$$

and

$$(\bigwedge_{i\in I} f_i \cdot r_i)(z, y) = \bigwedge_{i\in i} \bigvee_{w_i\in f_i^{-1}y} r_i(z, w_i)\,.$$

Complete distributivity of $\mathcal{V}$ makes the right-hand sides equal (see Section II.1.11).

Conversely, let us assume $\mathcal{V}$-Rel to be widely modular and consider subsets $A_i \subseteq V$, $i \in I$. For $I = \varnothing$ one trivially has

$$\bigvee_{(a_i)\in\prod_i A_i} \bigwedge_{i\in I} a_i = \bigwedge_{i\in I} \bigvee A_i\,, \tag{V.3.5.iii}$$

and for $I \neq \varnothing$ one obtains this identity by considering the unique maps $f_i : A_i \to 1$ and the $\mathcal{V}$-relations $r_i : 1 \nrightarrow A_i$ with $r_i(*, a_i) = a_i$ for $a_i \in A_i$, $* \in 1$. □

V.3.5.5 Remarks

(1) If $\mathcal{V}$ satisfies (V.3.5.iii) (for all $A_i \subseteq V$, $i \in I$), so does its complete sublattice $\mathsf{2} = \{\bot, \top\}$. But validity of (V.3.5.iii) for $\mathsf{2}$ is, in Zermelo–Fraenkel (ZF) set theory, equivalent to the Axiom of Choice [Herrlich, 2006]. Hence, in ZF one can state Proposition V.3.5.4 more precisely as:

$$\mathcal{V}\text{-}\mathsf{Rel} \text{ is widely modular} \iff \text{AC holds and } \mathcal{V} \text{ is completely distributive}.$$

(2) An injective proper $(\mathbb{T}, \mathcal{V})$-continuous map $f : (X, a) \to (Y, b)$ is necessarily O-initial (for $O : (\mathbb{T}, \mathcal{V})\text{-}\mathsf{Cat} \to \mathsf{Set}$, see Exercises V.3.A and V.3.E), i.e. an embedding. Proper embeddings are more commonly characterized as the closed embeddings (see Remark V.3.3.5(3)), and a multiple pullback of embeddings is better known as an intersection (see Theorem II.5.3.2).

V.3.5.6 Theorem

(1) *Closed embeddings in* $(\mathbb{T}, \mathcal{V})$-Cat *are stable under intersections.*
(2) *If $\mathcal{V}$ is Cartesian closed, every product of closed embeddings is a closed embedding in* $(\mathbb{T}, \mathcal{V})$-Cat.
(3) *If $\mathcal{V}$ is completely distributive, every product of proper $(\mathbb{T}, \mathcal{V})$-continuous maps is proper.*

Proof Combine Remark V.3.3.3(3) with V.3.5.1–V.3.5.5, observing also that complete distributivity implies that $\mathcal{V}$, as a lattice, is a frame that is Cartesian closed. □

© **V.3.5.7 Corollary** *Products of proper maps in* Top *and in* App *are proper.*

Proof Apply Theorem V.3.5.6(2), noting that every chain is completely distributive; in particular, $\mathsf{2}$ and P_+ are so. □

V.3.6 Coproducts of open maps

Throughout this section we assume that

- *$m^\circ : \hat{T} \to \hat{T}\hat{T}$ is a natural transformation;*
- *$\hat{T}$ is right-whiskering.*

By Remarks III.4.3.4, these hypotheses guarantee that for a coproduct

$$t_i : (X_i, a_i) \to \coprod_{i \in I} (X_i, a_i) = (X, a) \quad (i \in I)$$

in $(\mathbb{T}, \mathcal{V})$-Cat one has

$$a = \bigvee_{i \in I} t_i \cdot a_i \cdot (Tt_i)^\circ ,$$

with each t_i open. These facts give us stability of openness under coproducts, as follows.

V.3.6.1 Proposition

(1) *If all* $f_i : (X_i, a_i) \to (Y, b)$ *in* $(\mathbb{T}, \mathcal{V})$-Cat *are open, the induced* $f : \coprod_{i \in I}(X_i, a_i) \to (Y, b)$ *is open.*
(2) *If all* $f_i : (X_i, a_i) \to (Y_i, b_i)$ *in* $(\mathbb{T}, \mathcal{V})$-Cat *are open,*

$$\coprod_{i \in I} f_i : \coprod_{i \in I}(X_i, a_i) \to \coprod_{i \in I}(Y_i, b_i)$$

is open.
(3) *Open embeddings are closed under set-theoretic union.*

Proof (1): Since $\mathcal{V}$-Rel is a quantaloid, one has

$$\begin{aligned} f^\circ \cdot b &= (\textstyle\bigvee_{i \in I} t_i \cdot t_i^\circ) \cdot f^\circ \cdot b \\ &= \textstyle\bigvee_{i \in I} t_i \cdot t_i^\circ \cdot f^\circ \cdot b \\ &= \textstyle\bigvee_{i \in I} t_i \cdot f_i^\circ \cdot b \\ &= \textstyle\bigvee_{i \in I} t_i \cdot a_i \cdot (Tf_i)^\circ && (f_i \text{ open}) \\ &= (\textstyle\bigvee_{i \in I} t_i \cdot a_i \cdot (Tt_i)^\circ) \cdot (Tf)^\circ . \end{aligned}$$

(2): One applies (1) to the open maps

$$(X_i, a_i) \xrightarrow{f_i} (Y_i, b_i) \longrightarrow \coprod_{i \in I}(Y_i, b_i) .$$

(3): If in (1) all f_i are inclusion maps, then the image of the open map f is precisely the union $\bigcup_{i \in I} X_i$ which (when provided with the O-initial structure of (Y, b)) is open in Y. □

The inclusion map $A \hookrightarrow X$ of a subset of a $(\mathbb{T}, \mathcal{V})$-space (X, a) provided with the O-initial structure is open if and only if

$$\forall x \in X, \mathfrak{x} \in TX \ (a(\mathfrak{x}, x) > \bot \ \& \ x \in A \implies \mathfrak{x} \in TA) , \qquad \text{(V.3.6.i)}$$

where we assume $T(A \hookrightarrow X) = (TA \hookrightarrow TX)$ (see Exercise III.1.P). The open subsets of X defined in this way form a topology on X if T is taut. Denoting by $\Omega(X, a)$ the resulting topological space, we obtain:

V.3.6.2 Corollary *If T is taut, there is a functor*

$$\Omega : (\mathbb{T}, \mathcal{V})\text{-Cat} \to \text{Top}$$

that preserves the underlying sets and sends $(\mathbb{T}, \mathcal{V})$-open maps to open maps. Ω also preserves and reflects coproducts.

Proof Since T is taut, open embeddings in $(\mathbb{T}, \mathcal{V})$-Cat are stable under pullbacks by Proposition III.4.3.8, so they are stable under finite intersection. Since open embeddings are also closed under unions (see Proposition V.3.6.1), the set $\Omega(X, a)$ is in fact a topological space with its open sets A defined by (V.3.6.i). Stability of open embeddings under pullback also shows that a $(\mathbb{T}, \mathcal{V})$-continuous map $f : (X, a) \to (Y, b)$ gives a continuous map $\Omega f = f : \Omega(X, a) \to \Omega(Y, b)$. As for the preservation of open subsets by Ωf when f is open in $(\mathbb{T}, \mathcal{V})$-Cat, we note that for $A \subseteq X$ open the composite morphism

$$(A \longrightarrow f(A) \hookrightarrow (Y, b)) = (A \hookrightarrow (X, a) \xrightarrow{f} (Y, b))$$

is open, so that $f(A)$ is open in (Y, b), by Proposition V.3.1.4(2).

Since Ω preserves open maps, it preserves in particular open embeddings (see Exercise V.3.E(4)) and therefore the open injections of a coproduct in $(\mathbb{T}, \mathcal{V})$-Cat. Preservation of coproducts follows, and reflection thereof likewise. □

V.3.6.3 Examples

(1) For $(\mathbb{T}, \mathcal{V}) = (\mathbb{I}, 2)$, the functor Ω describes the full coreflective embedding Ord $\hookrightarrow$ Top which provides an ordered set with its Alexandroff topology, i.e. its open sets are the down-closed sets (see Examples III.3.5.2). For $(\mathbb{T}, \mathcal{V}) = (\beta, 2)$, since (V.3.6.i) describes open sets in terms of ultrafilter convergence, Ω is the isomorphism $(\mathbb{T}, \mathcal{V})$-Cat $\cong$ Top of Theorem III.2.2.5 expressed in terms of open sets.

(2) For $(\mathbb{T}, \mathcal{V}) = (\mathbb{I}, \mathsf{P}_+)$, Ω provides a metric space (X, a) with a rather crude topology: $A \subseteq X$ is open if and only if

$$\forall x, y \in X \; (a(x, y) < \infty \;\&\; y \in A \implies x \in A) \,.$$

In particular, this choice of $(\mathbb{T}, \mathcal{V})$ shows that openness of Ωf does not imply openness of f in general. In fact, if the metric a is finite, then ΩX is indiscrete. Hence, openness of Ωf for $f : (X, a) \to (Y, b)$ in Met with a, b finite and $X \neq \varnothing$ just means that f is surjective which, in general, does not guarantee openness in Met (see Proposition V.2.3.3(3)).

(3) A set $C \subseteq X$ is open in $(X, a) \in$ App $\cong (\beta, \mathsf{P}_+)$-Cat precisely when

$$\forall B \subseteq X, x \in C \; (\delta(x, C \cap B) \leq \delta(x, B)) \,,$$

where δ is the associated approach distance of (X, a) (see Example V.3.4.7(4)). Choosing $B = \{y\}$ for $y \in X$, we see that openness of C in X in App implies openness of C in X in Met, where X carries the metric $a \cdot e_X$ (see Example III.3.5.2(2)). Consequently, for an approach space X sent by the algebraic functor A_e : App $\to$ Met (and coreflector of Met $\hookrightarrow$ App) to a space with a finite metric, we obtain from (2) that the topological space ΩX is indiscrete.

While Examples V.3.6.3 (2) and (3) indicate the limitations of the functor Ω, we will use Ω as an essential tool in Section V.5.3, especially when $\mathcal{V} = 2$.

V.3.6.4 Remark Taking as its closed sets those subsets $A \subseteq X$ for which $A \hookrightarrow (X, a)$ is proper, one obtains a functor

$$\Gamma : (\mathbb{T}, \mathcal{V})\text{-}\mathsf{Cat} \to \mathsf{Top}$$

that sends $(\mathbb{T}, \mathcal{V})$-proper maps to closed maps in Top, provided that $T : \mathsf{Set} \to \mathsf{Set}$ preserves finite coproducts; see Exercise V.3.F.

V.3.7 Preservation of space properties

For a $(\mathbb{T}, \mathcal{V})$-continuous map $f : (X, a) \to (Y, b)$, we briefly discuss cases when the codomain inherits a special property from the domain, and vice versa. We remind the reader that injectivity of f is always inherited by Tf, and the same is true for surjectivity if the Axiom of Choice is assumed. ©

V.3.7.1 Proposition

(1) *If f is injective and proper and (Y, b) is compact, (X, a) is compact.*
(2) *If Tf is surjective and (X, a) is compact, (Y, b) is compact.*

Proof (1) follows from Proposition V.1.1.5. For (2), note simply that

$$1_{TY} = Tf \cdot (Tf)^\circ \leq Tf \cdot a^\circ \cdot a \cdot (Tf)^\circ \leq b^\circ \cdot f \cdot f^\circ \cdot b \leq b^\circ \cdot b\,. \qquad \square$$

V.3.7.2 Proposition

(1) *If f is injective and (Y, b) Hausdorff, (X, a) is Hausdorff.*
(2) *If Tf is surjective, f is proper, and (X, a) is Hausdorff, (Y, b) is Hausdorff.*

Proof (1) follows from Proposition V.1.1.5. For (2), we have

$$1_Y \geq f \cdot f^\circ \geq f \cdot a \cdot a^\circ \cdot f^\circ = b \cdot Tf \cdot (Tf)^\circ \cdot b^\circ = b \cdot b^\circ\,. \qquad \square$$

For preservation of normality and extremal disconnectedness, we first consider the case $\mathbb{T} = \mathbb{I}$.

V.3.7.3 Proposition *Let (X, a), (Y, b) be $\mathcal{V}$-spaces and let $f : (X, a) \to (Y, b)$ be a proper $\mathcal{V}$-continuous map.*

(1) *If f is injective and (Y, b) is normal, (X, a) is normal.*
(2) *If f is surjective and (X, a) is normal, (Y, b) is normal.*

Proof (1) follows from

$$\begin{aligned} a \cdot a^\circ &\leq f^\circ \cdot b \cdot f \cdot f^\circ \cdot b^\circ \cdot f \leq f^\circ \cdot b \cdot b^\circ \cdot f \\ &\leq f^\circ \cdot b^\circ \cdot b \cdot f = a^\circ \cdot f^\circ \cdot f \cdot a = a^\circ \cdot a\,. \end{aligned}$$

(2) is proved analogously, using the equality $f \cdot f^\circ = 1_Y$. $\square$

V.3.7.4 Corollary *Let (X, a), (Y, b) be $\mathcal{V}$-spaces and let $f : (X, a) \to (Y, b)$ be an open $\mathcal{V}$-continuous map.*

(1) *If f is injective and (Y, b) is extremally disconnected, (X, a) is extremally disconnected.*
(2) *If f is surjective and (X, a) is extremally disconnected, (Y, b) is extremally disconnected.*

Proof Apply Proposition V.3.7.3 to the $\mathcal{V}$-functor $f : (X, a^\circ) \to (Y, b^\circ)$. □

V.3.7.5 Theorem *Let: $m^\circ : \hat{T} \to \hat{T}\hat{T}$ be a natural transformation, $\hat{T}$ be right-whiskering, and $f : (X, a) \to (Y, b)$ be a $(\mathbb{T}, \mathcal{V})$-continuous map.*

(1) (a) *If f is proper, Tf is injective, and (Y, b) is normal, then (X, a) is normal.*
    (b) *If f is proper, Tf is surjective, and (X, a) is normal, then (Y, b) is normal.*
(2) (a) *If f is open, Tf is injective, and (Y, b) is extremally disconnected, then (X, a) is extremally disconnected.*
    (b) *If f is open, Tf is surjective, and (X, a) is extremally disconnected, then (Y, b) is extremally disconnected.*

Proof We observe that the claims of (1) follow from Proposition V.3.7.3 applied to the $\mathcal{V}$-continuous map $Tf : (TX, \hat{a}) \to (TY, \hat{b})$, using commutativity of the following diagram:

$$\begin{array}{ccc}
TX & \xrightarrow{\ Tf\ } & TY \\
{\scriptstyle m_X^\circ}\big\downarrow & & \big\downarrow{\scriptstyle m_Y^\circ} \\
TTX & \xrightarrow{\ TTf\ } & TTY \\
{\scriptstyle \hat{T}a}\big\downarrow & & \big\downarrow{\scriptstyle \hat{T}b} \\
TX & \xrightarrow{\ Tf\ } & TY\,.
\end{array}$$

(2) is proved analogously. □

Exercises

V.3.A *Closure operators.*

(1) Show that, for $v \le k$ in $\mathcal{V}$, the v-closure and the grand closure provide hereditary but generally non-idempotent $\mathcal{M}$-closure operators on $(\mathbb{T}, \mathcal{V})$-**Cat**, with $\mathcal{M}$ the class of embeddings. (For non-idempotency of the grand closure, see Exercise V.3.B.)
(2) Show that proper and open embeddings can be characterized using the grand closure as stated in Remarks V.3.3.3(3) and V.3.3.5(3).

V.3.B *Idempotency of the grand closure.*

(1) For a monad $\mathbb{T}$ on Set with an associative lax extension to $\mathcal{V}$-Rel, a $(\mathbb{T}, \mathcal{V})$-space (X, a), and $A \subseteq X$ with

$$T\overline{A} \subseteq \overline{TA} = \{y \in TX \mid \exists \chi \in TA\ (\hat{a}(\chi, y) > \bot)\}\ ,$$

show $\overline{\overline{A}} = \overline{A}$. Conclude that the grand closure is idempotent whenever $T\overline{A} \subseteq \overline{TA}$ for all $A \subseteq X$.

(2) In App $\cong (\beta, \mathsf{P}_+)$-Cat, consider $([0, \infty], a)$ with structure

$$a(u, v) = v \ominus \xi(u)\ ,$$

where $\xi(u) = \sup_{C \in u} \inf_{u \in C} u$ for all $u \in \beta[0, \infty]$, $v \in [0, \infty]$ (see Exercise III.5.J). Show for $A = \{0\}$:

(a) $\overline{A} = [0, \infty)$,
(b) $\beta\overline{A} \not\subseteq \overline{\beta A}$,
(c) $\overline{\overline{A}} = [0, \infty]$.

Hint. Consider an ultrafilter y on $[0, \infty]$ containing $[0, \infty)$ and all $[v, \infty]$, $v < \infty$. Then $y \in \beta\overline{A}$, but $y \notin \overline{\beta A}$.

V.3.C *Openness of product projections.* Suppose that the functor T satisfies BC.

(1) Prove that, if $\mathcal{V}$ is integral and $T1 \cong 1$, product projections are open in $(\mathbb{T}, \mathcal{V})$-Cat.
(2) Show that both conditions are essential in (1).

V.3.D *Failure of the classical Kuratowski–Mrówka Theorem.* Let $H = \mathbb{N} \setminus \{0\}$ be the multiplicative monoid of positive integers, and let $\overline{H}$ be the flat lax extension to Rel of the associated monad $\mathbb{H}$ (see Section V.1.4). Consider the $(\mathbb{H}, 2)$-spaces (X, a), $(2, \top)$, with $X = \mathbb{N} \cup \{\infty\}$, $2 = \{0, 1\}$, and the only a-relations that hold are:

$$(\alpha, \infty)\ a\ \infty\ , \quad (\alpha, 0)\ a\ \infty\ , \quad (\alpha, 0)\ a\ 0\ , \quad (\alpha, 0)\ a\ \alpha\ , \quad (\alpha, n)\ a\ (\alpha \cdot n)$$

for $\alpha, n \in H \subseteq X$. Show that the $(\mathbb{H}, 2)$-continuous map $f : (X, a) \to (2, \top)$, defined by $f(0) = f(\infty) = 0$ and $f(n) = 1$ for $n \geq 1$, is closed, has compact fibers, but is not proper.

V.3.E *Injective and surjective proper and open maps.* For $f : (X, a) \to (Y, b)$ in $(\mathbb{T}, \mathcal{V})$-Cat and O the underlying-set functor. Show that

(1) f proper, f injective $\Longrightarrow$ f is O-initial;
(2) f proper, Tf surjective $\Longrightarrow$ f is O-final;
(3) f open, f surjective $\Longrightarrow$ f is O-final;
(4) f open, Tf injective $\Longrightarrow$ f is O-initial.

Furthermore, the O-final structure in (2) and (3) is described by $b = f \cdot a \cdot (Tf)^\circ$.

V.3.F *The functor* Γ. For a $(\mathbb{T}, \mathcal{V})$-space (X, a), declare $A \subseteq X$ to be closed if $A \hookrightarrow (X, a)$ is proper (where A carries the O-initial $(\mathbb{T}, \mathcal{V})$-structure). If T : $\mathsf{Set} \to \mathsf{Set}$ preserves finite products, this defines the object part of a functor

$$\Gamma : (\mathbb{T}, \mathcal{V})\text{-}\mathsf{Cat} \to \mathsf{Top}$$

that sends $(\mathbb{T}, \mathcal{V})$-proper maps to closed continuous maps. Describe this functor for $T = \mathbb{I}$ or $T = \beta$ and $\mathcal{V} = 2$ or $\mathcal{V} = \mathsf{P}_+$.

V.3.G *Closed subspaces of normal spaces.* For $\mathbb{T}$ with an associative and left-whiskering lax extension to $\mathcal{V}$-Rel, prove that closed subspaces of normal $(\mathbb{T}, \mathcal{V})$-spaces are normal.

V.3.H *Quasi-proper maps.* A $(\mathbb{T}, \mathcal{V})$-continuous map $f : (X, a) \to (Y, b)$ is *quasi-proper* if $b \cdot Tf \leq b \cdot e_Y \cdot f \cdot a$.

(1) Every quasi-proper map $f : (X, a) \to (Y, b)$ satisfies $b \cdot Tf = b \cdot e_Y \cdot f \cdot a$.
(2) Every proper map is quasi-proper.
(3) If $\hat{\mathbb{T}}$ is associative, every left adjoint map in $(\mathbb{T}, \mathcal{V})$-$\mathsf{Cat}$ (see Exercise III.3.F) is quasi-proper.
(4) In $\mathsf{Ord} \cong (\mathbb{I}, 2)$-$\mathsf{Cat}$, the embedding $\{0, 1\} \hookrightarrow \{0, 1, 2\}$ is left adjoint but not proper, and $\{1, 2\} \hookrightarrow \{0, 1, 2\}$ is proper but not left adjoint.
(5) A monotone map $f : X \to Y$ of ordered sets is quasi-proper as a morphism in $(\mathbb{P}, 2)$-Cat (see Example III.1.6.4(1)) if and only if f is left adjoint as a morphism in $(\mathbb{I}, 2)$-Cat.

V.3.I *Closure under composition.* Show that the classes of closed maps, of inversely closed maps, and of quasi-proper maps are all closed under composition.

V.3.J *Openness and near openness with respect to the filter monad.* Show that in $(\mathbb{F}, 2)$-Cat and $(\mathbb{F}_p, 2)$-Cat the notions of open map coincide with the usual topological notion. Nearly open maps in $(\mathbb{F}_p, 2)$-Cat (see Definition III.4.3.1) are also open, while in $(\mathbb{F}, 2)$-Cat every map is nearly open.

V.3.K *Left- but not right-whiskering, and conversely*

(1) Prove that the lax extension $\hat{\mathbb{P}}$ of the powerset monad to Rel (see Examples III.1.4.2) is left- but not right-whiskering, and that the lax extension $\check{\mathbb{P}}$ behaves conversely.
(2) The Kleisli extension $\check{\mathbb{F}}$ of the filter monad is right- but not left-whiskering.

V.4 Topologies on a category

In this section we give an axiomatic approach to considering objects in a category as spaces where the category comes equipped with a class of "proper maps." This class, called the topology of the category, determines notions of compactness and separation and allows us to exhibit their interrelations at both the object and morphism levels.

V.4.1 Topology, fiberwise topology, derived topology

The finitary stability properties of the classes of proper and of open maps in $(\mathbb{T}, \mathcal{V})$-$\mathsf{Cat}$ studied in Section V.3.1 lead us to considering classes of morphisms $\mathcal{P}$ in an arbitrary finitely complete category X and to call $\mathcal{P}$ a *topology* on X if

(1) $\mathcal{P}$ contains all isomorphisms,
(2) $\mathcal{P}$ is closed under composition,
(3) $\mathcal{P}$ is stable under pullback.

For another morphism class $\mathcal{E}$ in X, we call $\mathcal{P}$ an $\mathcal{E}$-*topology* if in addition

(4) $p \cdot e \in \mathcal{P}$ with $e \in \mathcal{E}$ implies $p \in \mathcal{P}$.

Throughout this section we require that the class $\mathcal{E}$ itself is an $\mathcal{E}$-topology. The presence of such a class $\mathcal{E}$ constitutes no restriction of generality since every topology is an $(\mathrm{Iso}\,\mathsf{X})$-topology. Note also that any pullback-stable class $\mathcal{E}$ that belongs to a factorization system $(\mathcal{E}, \mathcal{M})$ is an $\mathcal{E}$-topology (see Proposition II.5.1.1).

V.4.1.1 Examples Let $\mathsf{X} = (\mathbb{T}, \mathcal{V})$-$\mathsf{Cat}$ and $\mathcal{E}$ be the class of epimorphisms (i.e. surjective $(\mathbb{T}, \mathcal{V})$-continuous maps). Then $\mathcal{E}$ is an $\mathcal{E}$-topology, with respect to which we consider the following classes $\mathcal{P}$.

(1) The class $\mathcal{P} = \mathrm{Prop}(\mathbb{T}, \mathcal{V})$ of *proper maps* is an $\mathcal{E}$-topology if $\mathcal{V}$ is Cartesian closed; see Proposition V.3.1.4.
(2) The class $\mathrm{Open}(\mathbb{T}, \mathcal{V})$ of *open maps* is an $\mathcal{E}$-topology if $\mathcal{V}$ is Cartesian closed and T satisfies BC; see Proposition V.3.1.4.
(3) For the forgetful functor $O : (\mathbb{T}, \mathcal{V})$-$\mathsf{Cat} \to \mathsf{Set}$, the class $\mathrm{Ini}\, O$ of O-initial morphisms is an $\mathcal{E}$-topology, and so are the classes $\mathrm{Mono}((\mathbb{T}, \mathcal{V})$-$\mathsf{Cat})$ and $\mathrm{RegMono}((\mathbb{T}, \mathcal{V})$-$\mathsf{Cat}) = \mathrm{Ini}\, O \cap \mathrm{Mono}((\mathbb{T}, \mathcal{V})$-$\mathsf{Cat})$.

The class of closed continuous maps in Top satisfies properties (1), (2), and (4) of an $\mathcal{E}$-topology ($\mathcal{E} = \mathrm{Epi}\,\mathsf{Top}$), but not (3) (for example, $\mathbb{R} \to 1$ is closed, but its pullback $\mathbb{R} \times \mathbb{R} \to \mathbb{R}$ along $\mathbb{R} \to 1$ is not). However, it contains a greatest pullback-stable subclass, the class of all proper maps.

In general, we call a class $\mathcal{P}$ in X an $\mathcal{E}$-*pretopology* if it satisfies properties (1), (2), and (4). The class $\mathcal{P}$ is *hereditary* if every pullback of a morphism in $\mathcal{P}$ along an embedding lies in $\mathcal{P}$.

V.4.1.2 Proposition *An $\mathcal{E}$-pretopology $\mathcal{P}$ on X contains a largest $\mathcal{E}$-topology $\mathcal{P}^* \subseteq \mathcal{P}$. A morphism f lies in $\mathcal{P}^*$ precisely when every pullback of f lies in $\mathcal{P}$. If $\mathcal{P}$ is stable under pullback along split monomorphisms, in particular if $\mathcal{P}$ is hereditary, then f lies in $\mathcal{P}^*$ if and only if $f \times 1_Z$ lies in $\mathcal{P}$ for every object Z in X.*

Proof For the "if" part of the last claim, we consider a pullback diagram

$$\begin{array}{ccc} P & \xrightarrow{f'} & Z \\ {\scriptstyle g'}\big\downarrow & & \big\downarrow{\scriptstyle g} \\ X & \xrightarrow{f} & Y \end{array} \quad = \quad \begin{array}{ccc} P & \xrightarrow{f'} & Z \\ {\scriptstyle \langle g', f' \rangle}\big\downarrow & & \big\downarrow{\scriptstyle \langle g, 1_Z \rangle} \\ X \times Z & \xrightarrow{f \times 1_Z} & Y \times Z \\ \big\downarrow & & \big\downarrow \\ X & \xrightarrow{f} & Y \end{array} \tag{V.4.1.i}$$

and its decomposition into two pullback diagrams. The hypotheses yield

$$f \in \mathcal{P} \implies f \times 1_Z \in \mathcal{P} \implies f' \in \mathcal{P}$$

since $\langle g, 1_Z \rangle$ is split mono. □

V.4.1.3 Corollary *The class* $\mathrm{Clo}(\mathbb{T}, \mathcal{V})$ *of closed maps in* $(\mathbb{T}, \mathcal{V})$-$\mathsf{Cat}$ *is a hereditary $\mathcal{E}$-pretopology, with $\mathcal{E}$ the class of epimorphisms. One has*

$$f \in \mathrm{Clo}(\mathbb{T}, \mathcal{V})^* \iff \forall (Z, c) \in (\mathbb{T}, \mathcal{V})\text{-}\mathsf{Cat}\ (f \times 1_Z \textit{ is closed}),$$

and if $\mathbb{T} = \beta$ *and* $\mathcal{V}$ *is Cartesian closed, then* $\mathrm{Clo}(\beta, \mathcal{V})^* = \mathrm{Prop}(\beta, \mathcal{V})$.

Proof For closure of $\mathrm{Clo}(\mathbb{T}, \mathcal{V})$ under composition, see Exercise V.3.I. If the composite map

$$(Z, c) \xrightarrow{e} (X, a) \xrightarrow{f} (Y, b)$$

is closed with e surjective, then for all $A \subseteq X$ and $C = e^{-1}(A)$ one has

$$\begin{aligned} b \cdot Tf \cdot Ti_A \cdot !^\circ_{TA} &= b \cdot Tf \cdot Te \cdot Ti_C \cdot !^\circ_{TC} && (Te \text{ is surjective}) \\ &= f \cdot e \cdot c \cdot Ti_C \cdot !^\circ_{TC} \\ &\leq f \cdot b \cdot Te \cdot Ti_C \cdot !^\circ_{TC} \\ &= f \cdot b \cdot Ti_A \cdot !^\circ_{TC} \end{aligned}$$

and therefore f is closed. For a pullback diagram

$$\begin{array}{ccc} (W = f^{-1}(Z), d) & \xrightarrow{f'} & (Z, c) \\ \big\downarrow & & \big\downarrow \\ (X, a) & \xrightarrow{f} & (Y, b) \end{array} \tag{V.4.1.ii}$$

with f closed and $c = i_Z^\circ \cdot b \cdot Ti_Z$, one has $d = i_W^\circ \cdot a \cdot Ti_W$, and for all $A \subseteq W$ one obtains

$$\begin{aligned}
c \cdot Tf' \cdot Ti_A \cdot !_{TA}^\circ &= i_Z^\circ \cdot b \cdot Tf \cdot Ti_W \cdot Ti_A \cdot !_{TA}^\circ \\
&= i_Z^\circ \cdot f \cdot a \cdot Ti_W \cdot Ti_A \cdot !_{TA}^\circ \\
&= f' \cdot i_W^\circ \cdot a \cdot Ti_W \cdot Ti_A \cdot !_{TA}^\circ \qquad \text{((V.4.1.ii) is a BC-square)} \\
&= f' \cdot d \cdot Ti_A \cdot !_{TA}^\circ ,
\end{aligned}$$

so that f' is closed. Consequently, $\mathrm{Clo}(\mathbb{T}, \mathcal{V})$ is a hereditary $\mathcal{E}$-pretopology. The remaining assertions follow from Proposition V.4.1.2 and Theorem V.3.4.6. □

There are two ways of creating "new topologies from old" that are of particular importance to us now. First, given an object Z, we can "slice at Z" any topology $\mathcal{P}$ of X, by considering the class

$$\mathcal{P}_Z := D_Z^{-1}(\mathcal{P}) ,$$

where $D_Z : \mathsf{X}/Z \to \mathsf{X}$ is the "domain functor." The following result is easy to prove.

V.4.1.4 Proposition *If $\mathcal{P}$ is an $\mathcal{E}$-topology of X, then $\mathcal{P}_Z$ is an $\mathcal{E}_Z$-topology of X/Z for every object Z in X. The statement still holds if "$\mathcal{E}$-topology" is replaced everywhere by "$\mathcal{E}$-pretopology."*

We call $\mathcal{P}_Z$ the *fiberwise topology* of $\mathcal{P}$ at Z (or the *fiberwise pretopology* accordingly).

Less trivially, for any class $\mathcal{P}$, one may consider the *derived class*

$$\mathcal{P}' = \{f : X \to Y \mid \delta_f = \langle 1_X, 1_X \rangle : X \to X \times_Y X \text{ lies in } \mathcal{P}\} .$$

V.4.1.5 Proposition *For a topology $\mathcal{P}$ of X, the derived class $\mathcal{P}'$ is also a topology of X that moreover contains all monomorphisms and satisfies the cancellation condition*

$$g \cdot f \in \mathcal{P}' \implies f \in \mathcal{P}' .$$

If $\mathcal{P}$ is an $\mathcal{E}$-topology, $\mathcal{P}'$ is an $(\mathcal{E} \cap \mathcal{P})$-topology.

We call $\mathcal{P}'$ the *derived topology* of $\mathcal{P}$.

Proof A morphism f is a monomorphism if and only if δ_f is an isomorphism. Hence, $\mathcal{P}'$ contains all monomorphisms. For $f : X \to Y$ and $g : Y \to Z$ in X, let $h = g \cdot f$. There is a unique morphism $t : X \times_Y X \to X \times_Z X$ with

$h_1 \cdot t = f_1$, $h_2 \cdot t = f_2$, where (f_1, f_2) and (h_1, h_2) form the kernel pairs of f and h, respectively. The right square of

$$\begin{array}{ccccc} X & \xrightarrow{\delta_f} & X \times_Y X & \xrightarrow{f \cdot f_1} & Y \\ {\scriptstyle 1_X}\downarrow & & \downarrow{\scriptstyle t} & & \downarrow{\scriptstyle \delta_g} \\ X & \xrightarrow{\delta_h} & X \times_Z X & \xrightarrow{f \times f} & Y \times_Z Y \end{array} \qquad \text{(V.4.1.iii)}$$

is a pullback diagram since t is the equalizer of $f \cdot h_1$ and $f \cdot h_2$. Consequently, one has the implications

$$f, g \in \mathcal{P}' \implies \delta_f, \delta_g \in \mathcal{P} \implies \delta_f, t \in \mathcal{P} \implies \delta_h = t \cdot \delta_f \in \mathcal{P} \implies h \in \mathcal{P}' .$$

Since t is monic, the left square of (V.4.1.iii) is also a pullback diagram, and one concludes that

$$g \cdot f \in \mathcal{P}' \implies \delta_h \in \mathcal{P} \implies \delta_f \in \mathcal{P} \implies f \in \mathcal{P}' .$$

The pullback diagram of (f, g) can be factorized as the following outer pullback diagram:

$$\begin{array}{ccccccc} P & \xrightarrow{\delta_{f'}} & P \times_Z P & \rightrightarrows & P & \xrightarrow{f'} & Z \\ {\scriptstyle g'}\downarrow & & \downarrow{\scriptstyle g''} & & \downarrow{\scriptstyle g'} & & \downarrow{\scriptstyle g} \\ X & \xrightarrow[\delta_f]{} & X \times_Y X & \rightrightarrows & X & \xrightarrow[f]{} & Y . \end{array}$$

Since the two rightmost diagrams form a pullback diagram, the left square is also one, so

$$f \in \mathcal{P} \implies f' \in \mathcal{P} .$$

Finally, when $\mathcal{P}$ is an $\mathcal{E}$-topology, set $h = g \cdot f \in \mathcal{P}'$ with $f \in \mathcal{E} \cap \mathcal{P}$. From (V.4.1.iii) one has $\delta_g \cdot f = (f \times f) \cdot \delta_h$, with $\delta_h \in \mathcal{P}$ and $f \times f = (f \times 1) \cdot (1 \times f) \in \mathcal{P}$ (as the composite of two pullbacks of f). Since $f \in \mathcal{E}$, one has $\delta_g \in \mathcal{P}$, i.e. $g \in \mathcal{P}'$. □

V.4.1.6 Remark One has

$$(\mathrm{Iso}\,\mathsf{X})' = \mathrm{Mono}\,\mathsf{X} \qquad \text{and} \qquad (\mathrm{SplitMono}\,\mathsf{X})' = \mathrm{mor}\,\mathsf{X} .$$

In particular, for any class $\mathcal{P}$ containing $\mathrm{Iso}\,\mathsf{X}$,

$$(\mathcal{P}')' = \mathrm{mor}\,\mathsf{X} ,$$

and, for the forgetful functor $O : (\mathbb{T}, \mathcal{V})\text{-}\mathsf{Cat} \to \mathsf{Set}$,

$$(\mathrm{Ini}\, O)' = \mathrm{mor}(\mathbb{T}, \mathcal{V})\text{-}\mathsf{Cat} .$$

V.4.2 $\mathcal{P}$-compactness, $\mathcal{P}$-Hausdorffness

Let $\mathcal{P}$ be a topology of the finitely complete category X. An object X is *$\mathcal{P}$-compact* if

$$(X \to 1) \in \mathcal{P}$$

(with 1 a terminal object in X), and X is *$\mathcal{P}$-Hausdorff* if

$$(X \to 1) \in \mathcal{P}' ,$$

i.e. if $(\delta_X : X \to X \times X) \in \mathcal{P}$.

A morphism $f : X \to Y$ is *$\mathcal{P}$-proper* if f is $\mathcal{P}_Y$-compact in X/Y (see Proposition V.4.1.2), f is *$\mathcal{P}$-Hausdorff* if f is $\mathcal{P}_Y$-Hausdorff in X/Y, and f is *$\mathcal{P}$-perfect* if f is both $\mathcal{P}$-proper and $\mathcal{P}$-Hausdorff.

Since 1_Y is a terminal object in X/Y and f is the unique morphism $f \to 1_Y$ in X/Y, one sees immediately that

$$\begin{aligned} f \text{ is } \mathcal{P}\text{-proper} &\iff f \in \mathcal{P} ,\\ f \text{ is } \mathcal{P}\text{-Hausdorff} &\iff f \in \mathcal{P}' \iff (\delta_f : X \to X \times_Y X) \in \mathcal{P} . \end{aligned}$$

We examine these notions first in terms of the principal topologies for $(\mathbb{T}, \mathcal{V})$-$\mathsf{Cat}$ as considered in Examples V.4.1.1 and compare them with those introduced in Section V.1.1.

V.4.2.1 Proposition *Let $\mathcal{V}$ be a Cartesian closed quantale, and let (X, a) be a $(\mathbb{T}, \mathcal{V})$-space.*

(1) *(X, a) is* $\mathrm{Prop}(\mathbb{T}, \mathcal{V})$*-compact if and only if*

$$\forall \chi \in TX \ (\top \le \bigvee\nolimits_{z \in X} a(\chi, z)) .$$

(2) *Suppose that $\mathcal{V}$ is integral; if (X, a) is compact, (X, a) is* $\mathrm{Prop}(\mathbb{T}, \mathcal{V})$*-compact, with the converse statement holding when $\mathcal{V}$ is superior.*

(3) *(X, a) is* $\mathrm{Prop}(\mathbb{T}, \mathcal{V})$*-Hausdorff if and only if*

$$\forall x, y \in X \, \forall \mathfrak{z} \in TX \ (\bot < a(\mathfrak{z}, x) \wedge a(\mathfrak{z}, y) \implies x = y) .$$

(4) *Suppose that $\mathcal{V}$ is integral; if (X, a) is* $\mathrm{Prop}(\mathbb{T}, \mathcal{V})$*-Hausdorff, (X, a) is Hausdorff, with the converse statement holding if the map $o : \mathcal{V} \to \mathsf{2}$ with $(o(v) = \bot \iff v = \bot)$ is a lax homomorphism of quantales, i.e. if $\mathcal{V}$ satisfies $(u \otimes v = \bot \implies u = \bot$ or $v = \bot)$.*

(5) *A $(\mathbb{T}, \mathcal{V})$-continuous map $f : (X, a) \to (Y, b)$ is* $\mathrm{Prop}(\mathbb{T}, \mathcal{V})$*-Hausdorff if and only if*

$$\forall x, y \in X \ \forall \mathfrak{z} \in TX \ (f(x) = f(y) \ \& \ (\bot < a(\mathfrak{z}, x) \wedge a(\mathfrak{z}, y) \implies x = y)) .$$

Proof To see (1), observe that $(!_X : (X, a) \to (1, \top)) \in \mathrm{Prop}(\mathbb{T}, \mathcal{V})$ means $\top \cdot T\,!_X \le\, !_X \cdot a$; this, when stated elementwise, reads as claimed.

For (3), consider the structure $b = (p^\circ \cdot a \cdot Tp) \wedge (q^\circ \cdot a \cdot Tq)$ of $X \times X$ (with projections p, q); then $(\delta_X : (X, a) \to (X \times X, b)) \in \mathrm{Prop}(\mathbb{T}, \mathcal{V})$ means $b \cdot T\delta_X \leq \delta_X \cdot a$. Since, for all $x, y \in X$, $\mathfrak{z} \in TX$,

$$\begin{aligned}(b \cdot T\delta_X)(\mathfrak{z}, (x, y)) &= a(Tp \cdot T\delta_X(\mathfrak{z}), x) \wedge a(Tq \cdot T\delta_X(\mathfrak{z}), y)\\ &= a(\mathfrak{z}, x) \wedge a(\mathfrak{z}, y)\ ,\end{aligned}$$

$$(\delta_X \cdot a)(\mathfrak{z}, (x, y)) = \bigvee_{z \in \delta_X^{-1}(x,y)} a(\mathfrak{z}, z) = \begin{cases} a(\mathfrak{z}, x) & \text{if } x = y,\\ \bot & \text{otherwise,}\end{cases}$$

the criterion follows.

Finally, (2) and (4) follow from Proposition V.1.1.2, and (5) is proved as (3). □

V.4.2.2 Examples

(1) Since $\mathsf{2}$ and P_+ satisfy all additional hypotheses used in Proposition V.4.2.1, in all examples presented in V.1.1.4 the notions of $\mathrm{Prop}(\mathbb{T}, \mathcal{V})$-compactness and $\mathrm{Prop}(\mathbb{T}, \mathcal{V})$-Hausdorffness for objects are equivalent to the notions of compactness and Hausdorffness of Section V.1.1.

(2) For $\mathcal{V}$ integral and $\mathbb{T} = \mathbb{I}$ (identically extended to $\mathcal{V}$-Rel), a $\mathcal{V}$-functor $f : (X, a) \to (Y, b)$ is $\mathrm{Prop}(\mathbb{T}, \mathcal{V})$-Hausdorff if every fiber of f (as a subspace of (X, a)) is discrete.

(3) In $\mathsf{Top} \cong (\mathbb{\beta}, \mathsf{2})$-$\mathsf{Cat}$, $f : X \to Y$ is $\mathrm{Prop}(\mathbb{\beta}, \mathsf{2})$-Hausdorff if and only if any two distinct points in the same fiber of f may be separated by disjoint open neighborhoods in X. Such maps are usually called *separated* in the literature on fibered topology (see [James, 1989]). In $\mathsf{App} \cong (\mathbb{\beta}, \mathsf{P}_+)$-$\mathsf{Cat}$, a map f is $\mathrm{Prop}(\mathbb{\beta}, \mathsf{P}_+)$-Hausdorff if and only if, for all $x, y \in X$, $\mathfrak{z} \in \beta X$ with $f(x) = f(y)$, $a(\mathfrak{z}, x) < \infty$ and $a(\mathfrak{z}, y) < \infty$, one has $x = y$.

V.4.2.3 Proposition *Suppose that $\mathcal{V}$ is Cartesian closed and that T satisfies BC.*

(1) *A $(\mathbb{T}, \mathcal{V})$-space (X, a) is* $\mathrm{Open}(\mathbb{T}, \mathcal{V})$*-compact if and only if*

$$\forall \mathfrak{a} \in T1, x \in X\ (\top \leq \bigvee_{\mathfrak{x} \in (T!_X)^{-1}\mathfrak{a}} a(\mathfrak{x}, x))\ ,$$

where $!_X : X \to 1$.

(2) *If $\mathcal{V}$ is integral and $T1 \cong 1$, then every $(\mathbb{T}, \mathcal{V})$-space is* $\mathrm{Open}(\mathbb{T}, \mathcal{V})$*-compact.*

(3) *An* $\mathrm{Open}(\mathbb{T}, \mathcal{V})$*-continuous map $f : (X, a) \to (Y, b)$ is* $\mathrm{Open}(\mathbb{T}, \mathcal{V})$*-Hausdorff if and only if*

$$\begin{aligned}&\forall x \in X\ \forall \mathfrak{w} \in T(X \times_Y X)\\ &(\bot < a(Tp(\mathfrak{w}), x) \wedge a(Tq(\mathfrak{w}), x) \implies \mathfrak{w} \in T\Delta_X)\ ,\end{aligned}$$

where $p, q : X \times_Y X \longrightarrow X$ are the projections, and $w \in T\Delta_X$ means $w = T\delta_f(\chi)$ for some (uniquely determined) $\chi \in TX$.

Proof (1): The given criterion expresses $!_X^\circ \cdot \top \leq a \cdot (T\,!_X)^\circ$ in pointwise form.

(2): If $T1 \cong 1$, then $(T\,!_X)(e_X(x))$ is the only element in $T1$.

(3): Formulating $\delta_f^\circ \cdot b \leq a \cdot (T\delta_f)^\circ$ with $b = (p^\circ \cdot a \cdot Tp) \wedge (q^\circ \cdot a \cdot Tq)$ ($p, q : X \times_Y X \longrightarrow X$ projections) in pointwise form, one obtains

$$a(Tp(w), x) \wedge a(Tq(w), x) = \begin{cases} a(\chi, x) & \text{if } \exists \chi \in TX\ (T\delta_f(\chi) = w), \\ \bot & \text{otherwise,} \end{cases}$$

for all $w \in T(X \times_Y X)$, $x \in X$. Since in the first case (i.e. when $w \in T\Delta_X$) the equality holds trivially, one obtains the criterion as stated. □

V.4.2.4 Examples

(1) For $\mathcal{V}$ integral and $\mathbb{T} = \mathbb{I}$, a $\mathcal{V}$-functor is $\mathrm{Open}(\mathbb{I}, \mathcal{V})$-Hausdorff if and only if its fibers are discrete. (Hence, $\mathrm{Open}(\mathbb{I}, \mathcal{V})$-Hausdorffness is equivalent to $\mathrm{Prop}(\mathbb{I}, \mathcal{V})$-Hausdorffness; see Example V.4.2.2(2).)

(2) In $\mathsf{Top} \cong (\beta, 2)\text{-}\mathsf{Cat}$, a continuous map $f : X \longrightarrow Y$ is $\mathrm{Open}(\beta, 2)$-Hausdorff if and only if the diagonal $\Delta_X \subseteq X \times_Y X$ is open, and this means equivalently that for every point in X there is a neighborhood U of x such that $f|_U : U \longrightarrow X$ is injective; such maps are called *locally injective*. When applied to $X \longrightarrow 1$, this yields that X is $\mathrm{Open}(\beta, 2)$-Hausdorff if and only if X is discrete.

(3) An $\mathrm{Open}(\beta, \mathsf{P}_+)$-Hausdorff morphism $f : (X, a) \longrightarrow (Y, b)$ has the property that for every ultrafilter w on $X \times_Y X$ that converges to (x, x) in the induced pretopology (see Section III.3.6 and Example III.4.1.3), so that $p[w] \longrightarrow x$ and $q[w] \longrightarrow x$, one has $\Delta_X \in w$. Consequently, the neighborhood filter of any diagonal point (x, x) in the pseudotopological space $X \times_Y X$ contains Δ_X. Hence, $\mathrm{Open}(\beta, \mathsf{P}_+)$-Hausdorffness is characterized in App by local injectivity, with "local" referring to the induced pretopology. In particular, $\mathrm{Open}(\beta, \mathsf{P}_+)$-Hausdorff approach spaces are discrete.

V.4.2.5 Remark Only indiscrete $(\mathbb{T}, \mathcal{V})$-spaces are $(\mathrm{Ini}\,\mathcal{O})$-compact, but every $(\mathbb{T}, \mathcal{V})$-space is $(\mathrm{Ini}\,\mathcal{O})$-Hausdorff.

V.4.3 A categorical characterization theorem

The following easy-to-prove theorem collects important characteristic properties of $\mathcal{P}$-compact objects in any finitely complete category X with an $\mathcal{E}$-topology $\mathcal{P}$.

V.4.3.1 Theorem *The following assertions for an object X are equivalent:*

(i) *X is $\mathcal{P}$-compact;*

(ii) *every morphism $f : X \to Y$ with Y $\mathcal{P}$-Hausdorff is $\mathcal{P}$-proper;*
(iii) *there is a $\mathcal{P}$-proper morphism $f : X \to Y$ such that Y is $\mathcal{P}$-compact;*
(iv) *the projection $X \times Y \to Y$ is $\mathcal{P}$-proper for all objects Y;*
(v) *$X \times Y$ is $\mathcal{P}$-compact for every $\mathcal{P}$-compact object Y;*
(vi) *for every morphism $e : X \to Y$ in $\mathcal{E}$, the object Y is $\mathcal{P}$-compact.*

Proof (i) $\Longrightarrow$ (ii): In the graph factorization

$$\begin{array}{ccc} & X \times Y & \\ {\scriptstyle \langle 1_X, f\rangle} \nearrow & & \searrow {\scriptstyle p} \\ X & \xrightarrow[f]{} & Y \end{array}$$

$\langle 1_X, f\rangle$ is in $\mathcal{P}$ as a pullback of δ_Y, and p is in $\mathcal{P}$ as a pullback of $X \to 1$.
(ii) $\Longrightarrow$ (iii): Consider $Y = 1$.
(iii) $\Longrightarrow$ (i): $(X \longrightarrow 1\ ) = (X \xrightarrow{f} Y \longrightarrow 1)$.
(i) $\Longrightarrow$ (iv): p is a pullback of $X \to 1$.
(iv) $\Longrightarrow$ (v): $(X \times Y \longrightarrow 1) = (X \times Y \longrightarrow Y \longrightarrow 1)$.
(v) $\Longrightarrow$ (i): Consider $Y = 1$.
(i) $\Longrightarrow$ (vi): Apply the $\mathcal{E}$-topology property to $!_X = !_Y \cdot e$.
(vi) $\Longrightarrow$ (i): Consider $e = 1_X$. □

V.4.3.2 Corollary *Let the composite morphism $g \cdot f$ be $\mathcal{P}$-proper, with g $\mathcal{P}$-Hausdorff. Then f is $\mathcal{P}$-proper.*

Proof Using Proposition V.4.1.4, apply Theorem V.4.3.1(i) $\Longrightarrow$ (ii) to the morphism $f : g \cdot f \to g$ in $\mathsf{X}/\operatorname{cod}(g)$. □

Next we apply Theorem V.4.3.1 with $\mathcal{P}'$ in lieu of $\mathcal{P}$. With Proposition V.4.1.5 we obtain the following characterization of $\mathcal{P}$-separated objects.

V.4.3.3 Corollary *The following assertions for an object X are equivalent:*

(i) *X is $\mathcal{P}$-Hausdorff;*
(ii) *every morphism $f : X \to Y$ is $\mathcal{P}$-Hausdorff;*
(iii) *there is a $\mathcal{P}$-Hausdorff morphism $f : X \to Y$ such that Y is $\mathcal{P}$-Hausdorff;*
(iv) *the projection $X \times Y \to Y$ is $\mathcal{P}$-Hausdorff for all objects Y;*
(v) *$X \times Y$ is $\mathcal{P}$-Hausdorff for every $\mathcal{P}$-Hausdorff object Y;*
(vi) *for every $\mathcal{P}$-proper morphism $e : X \to Y$ in $\mathcal{E}$, the object Y is $\mathcal{P}$-Hausdorff;*
(vii) *for every equalizer diagram $E \xrightarrow{u} Z \rightrightarrows X$, the morphism u is $\mathcal{P}$-proper.*

Proof We observe that $\mathcal{P}'$-compact translates to $\mathcal{P}$-Hausdorff for objects, and that $\mathcal{P}'$-proper translates to $\mathcal{P}$-Hausdorff for morphisms. Furthermore, since $\mathcal{P}'$ contains all monomorphisms, all objects and morphisms are $\mathcal{P}'$-Hausdorff. Hence, the equivalence of (i)–(vi) follows from Theorem V.4.3.1. For the equivalence (i) $\iff$ (vii), we simply note that equalizers as in (vii) are precisely the pullbacks of $\delta_X : X \to X \times X$. □

V.4.3.4 Corollary *For a $\mathcal{P}$-perfect morphism $f : X \to Y$, one has*

$$Y \text{ is } \mathcal{P}\text{-compact and } \mathcal{P}\text{-Hausdorff} \implies X \text{ is } \mathcal{P}\text{-compact } \mathcal{P}\text{-Hausdorff},$$

with the converse implication holding when f lies in $\mathcal{E}$.

Proof Use (iii) $\implies$ (i) $\implies$ (vi) of both Theorem V.4.3.1 and Corollary V.4.3.3. □

V.4.3.5 Corollary *The full subcategory of $\mathcal{P}$-compact objects is closed under finite products in* X, *and the full subcategories of $\mathcal{P}$-Hausdorff objects and of $\mathcal{P}$-compact $\mathcal{P}$-Hausdorff objects are both closed under finite limits in* X.

Proof Closure under finite products for all three subcategories in question follows from (i) $\iff$ (v) of Theorem V.4.3.1 and Corollary V.4.3.3. Closure under equalizers for the two subcategories mentioned follows from (i) $\iff$ (iii) $\iff$ (vii) of Corollary V.4.3.3 and (i) $\iff$ (iii) of Theorem V.4.3.1. □

V.4.3.6 Remarks

(1) $\mathcal{P}$-Hausdorffness generally fails to be closed under infinite products, also in conjunction with $\mathcal{P}$-compactness, as the example $\mathsf{X} = \mathsf{Top}$ and $\mathcal{P} = \mathrm{Open}(\beta, 2)$ shows: a countable power of a two-point discrete space is no longer discrete.

(2) Theorem V.4.3.1 and its corollaries may, of course, be readily applied to the topologies $\mathcal{P} = \mathrm{Prop}(\mathbb{T}, \mathcal{V})$ and $\mathcal{P} = \mathrm{Open}(\mathbb{T}, \mathcal{V})$, provided that $\mathcal{V}$ is Cartesian closed and (for $\mathcal{P} = \mathrm{Open}(\mathbb{T}, \mathcal{V})$) that T satisfies BC. We do not formulate explicitly the emerging finitary stability properties in these two important cases. However, in what follows we highlight the most important infinite stability properties available for these two general choices of $\mathcal{P}$.

V.4.3.7 Proposition *For $\mathcal{V}$ Cartesian closed, the full subcategory of* $\mathrm{Prop}(\mathbb{T}, \mathcal{V})$-*Hausdorff spaces is closed under mono-sources in* $(\mathbb{T}, \mathcal{V})$-$\mathsf{Cat}$, *and hence is strongly epireflective in* $(\mathbb{T}, \mathcal{V})$-$\mathsf{Cat}$.

Proof Let $(f_i : (X, a) \to (Y_i, b_i))_{i \in I}$ be a point-separating source in $(\mathbb{T}, \mathcal{V})$-$\mathsf{Cat}$, with all (Y_i, b_i) $\mathrm{Prop}(\mathbb{T}, \mathcal{V})$-Hausdorff. Hence, for $x \neq y$ in X there is $j \in I$ with $f_j(x) \neq f_j(y)$. Consequently, for all $z \in TX$, one has

$$a(z, x) \wedge a(z, y) \leq b_j(Tf_j(z), f_j(x)) \wedge b_j(Tf_j(z), f_j(y)) = \bot .$$

Hence, by Proposition V.4.2.1(3), the space (X, a) is $\mathrm{Prop}(\mathbb{T}, \mathcal{V})$-Hausdorff. □

With Theorem V.3.5.6 we obtain, of course, a *Tychonoff Theorem* for $\mathrm{Prop}(\mathbb{T}, \mathcal{V})$-compactness, and then with Proposition V.4.2.1 also for compactness.

V.4.3.8 Corollary *Let $\mathcal{V}$ be completely distributive. Then every product of* $\mathrm{Prop}(\mathbb{T}, \mathcal{V})$*-compact* $(\mathbb{T}, \mathcal{V})$*-spaces is* $\mathrm{Prop}(\mathbb{T}, \mathcal{V})$*-compact. If $\mathcal{V}$ is also integral and superior, then* $(\mathbb{T}, \mathcal{V})\text{-}\mathsf{Cat}_{\mathrm{Comp}}$ *is closed under products in* $(\mathbb{T}, \mathcal{V})\text{-}\mathsf{Cat}$*, and likewise for* $(\mathbb{T}, \mathcal{V})\text{-}\mathsf{Cat}_{\mathrm{Haus}}$ *and* $(\mathbb{T}, \mathcal{V})\text{-}\mathsf{Cat}_{\mathrm{CompHaus}}$.

V.4.4 $\mathcal{P}$-dense maps, $\mathcal{P}$-open maps

Given an $\mathcal{E}$-topology $\mathcal{P}$ on a finitely complete category X, we wish to develop an internal notion of $\mathcal{P}$-open map in X. A crucial step towards this goal is the introduction of $\mathcal{P}$-dense maps, which, in the case of the principal example $\mathcal{P} = \mathrm{Prop}(\mathbb{T}, \mathcal{V})$, will be described via suitable closure operators.

V.4.4.1 Definition A morphism f in X is $\mathcal{P}$*-dense* if in any factorization $f = p\cdot h$ with $p \in \mathcal{P}$ one has $p \in \mathcal{E}$. The class of all $\mathcal{P}$-dense maps in X is denoted by $\mathcal{P}^{\mathrm{d}}$.

V.4.4.2 Proposition *Consider a composite morphism $g \cdot f$ in* X.

(1) *If $g \cdot f \in \mathcal{P}^{\mathrm{d}}$, then $g \in \mathcal{P}^{\mathrm{d}}$.*
(2) *If $f \in \mathcal{P}^{\mathrm{d}}$ and $g \in \mathcal{E}$, then $g \cdot f \in \mathcal{P}^{\mathrm{d}}$.*
(3) *If* X *has pushouts of morphisms in $\mathcal{E}$ and these belong to $\mathcal{E}$ again, then $f \in \mathcal{E}$ and $g \in \mathcal{P}^{\mathrm{d}}$ implies $g \cdot f \in \mathcal{P}^{\mathrm{d}}$.*
(4) *$\mathcal{E} \subseteq \mathcal{P}^{\mathrm{d}}$ if and only if $\mathcal{P} \cap \mathrm{SplitEpi}\,\mathsf{X} \subseteq \mathcal{E}$.*
(5) *If $(\mathcal{E}, \mathcal{M})$ is a factorization system, then $f \in \mathcal{P}^{\mathrm{d}}$ if and only if in any factorization $f = p \cdot h$ with $p \in \mathcal{P} \cap \mathcal{M}$, p is an isomorphism.*

Proof (1) is trivial. For (2) and (3), consider the diagrams

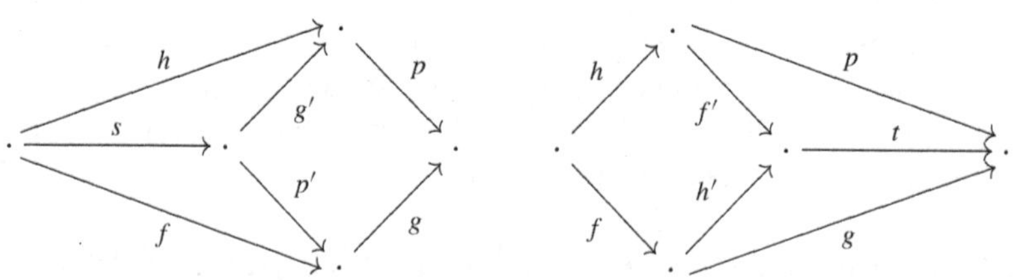

(V.4.4.i)

with $p \in \mathcal{P}$, the square being a pullback in the left diagram and a pushout in the right one, and with s, t induced morphisms. From $f \in \mathcal{P}^{\mathrm{d}}$, $p' \in \mathcal{P}$, one has $p' \in \mathcal{E}$, and from $g \in \mathcal{E}$ follows $g' \in \mathcal{E}$ and, hence, $p \cdot g' \in \mathcal{E}$, so finally $p \in \mathcal{E}$ (since $\mathcal{E}$ is an $\mathcal{E}$-topology). Similarly, from $f \in \mathcal{E}$ one obtains $f' \in \mathcal{E}$ and then $t \in \mathcal{P}$ (since $\mathcal{P}$ is an $\mathcal{E}$-topology), which implies $t \in \mathcal{E}$ when $g \in \mathcal{P}^{\mathrm{d}}$ and, hence, $p \in \mathcal{E}$.

To see (4), remark that if $\mathcal{E} \subseteq \mathcal{P}^{\mathrm{d}}$, then every identity morphism is $\mathcal{P}$-dense. Hence, if $p \cdot s = 1$ with $p \in \mathcal{P}$, we can conclude $p \in \mathcal{E}$. Conversely, $\mathcal{P} \cap$

SplitEpi $\mathsf{X} \subseteq \mathcal{E}$ implies that identity morphisms lie in $\mathcal{P}^{\mathrm{d}}$, which, with (2), implies $\mathcal{E} \subseteq \mathcal{P}^{\mathrm{d}}$.

(5) follows easily from $\mathcal{E} \cap \mathcal{M} = \mathrm{Iso}\,\mathsf{X}$ and the fact that, in an $(\mathcal{E}, \mathcal{M})$-factorization $p = m \cdot e$ of $p \in \mathcal{P}$, m is also in $\mathcal{P}$. □

By consideration of the factorization

$$f = (X \to \overline{f(X)} \hookrightarrow Y)\,,$$

the $\mathcal{P}$-dense maps in Top with $\mathcal{P}$ = {proper maps} are easily identified as the *dense* maps, i.e. as those f with $\overline{f(X)} = Y$. Using the grand closure introduced in Section V.3.3, we may proceed in the same way in $(\mathbb{T}, \mathcal{V})$-$\mathsf{Cat}$ but must take into account that this closure may fail to be idempotent (see Exercise V.3.A). Hence we consider the *idempotent hull* of the grand closure,

$$\overline{A}^{\infty} = \bigcap\{B \mid A \subseteq B \subseteq X,\, B = \overline{B}\}$$

for any $A \subseteq X$, whose inclusion map into X is $(\mathbb{T}, \mathcal{V})$-proper by Remark V.3.3.3(3). Obviously,

$$A = \overline{A} \iff A = \overline{A}^{\infty}\,.$$

V.4.4.3 Lemma *An embedding $A \hookrightarrow (X, a)$ is* $\mathrm{Prop}(\mathbb{T}, \mathcal{V})$*-dense if and only if $\overline{A}^{\infty} = X$.*

Proof The embedding $A \hookrightarrow X$ factors through $\overline{A}^{\infty} \hookrightarrow X$. If $A \hookrightarrow X$ is $\mathrm{Prop}(\mathbb{T}, \mathcal{V})$-dense, then $\overline{A}^{\infty} = X$ follows. Conversely, assuming $\overline{A}^{\infty} = X$ and considering any $(\mathbb{T}, \mathcal{V})$-proper $B \hookrightarrow X$ with $A \subseteq B$, we have $\overline{B} = B$ and, hence, $A = \overline{A}^{\infty} \subseteq \overline{B} = B$. □

V.4.4.4 Corollary *A $(\mathbb{T}, \mathcal{V})$-continuous map $f : (X, a) \to (Y, b)$ is* $\mathrm{Prop}(\mathbb{T}, \mathcal{V})$*-dense if and only if $\overline{f(X)}^{\infty} = Y$.*

Guided by Proposition V.3.3.4, we can now introduce $\mathcal{P}$-open maps as follows.

V.4.4.5 Definition A morphism $f : X \to Y$ in X is *$\mathcal{P}$-open* if, for every pullback f' of f, pulling back along f' preserves $\mathcal{P}$-density, i.e. if for all stacked pullback diagrams

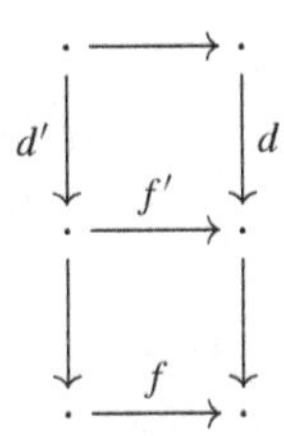

$d \in \mathcal{P}^{\mathrm{d}}$ implies $d' \in \mathcal{P}^{\mathrm{d}}$. We let $\mathcal{P}^{\mathrm{o}}$ denote the class of $\mathcal{P}$-open maps in X.

V.4.4.6 Proposition *For an $\mathcal{E}$-topology $\mathcal{P}$ on* X*, $\mathcal{P}^{\mathrm{o}}$ is also an $\mathcal{E}$-topology on* X.

Proof $\mathcal{P}^{\mathrm{o}}$ is clearly a topology. Since any pullback of $f \cdot e$ with $e \in \mathcal{E}$ is of the form $f' \cdot e'$ with $e' \in \mathcal{E}$, it suffices to show that $\mathcal{P}$-density pulls back along f if it does so along $f \cdot e$. But for the pullback diagrams

$$\begin{array}{ccccc} \cdot & \xrightarrow{e'} & \cdot & \xrightarrow{f'} & \cdot \\ {\scriptstyle d''}\downarrow & & \downarrow{\scriptstyle d'} & & \downarrow{\scriptstyle d} \\ \cdot & \xrightarrow[e]{} & \cdot & \xrightarrow[f]{} & \cdot \end{array}$$

$d \in \mathcal{P}^{\mathrm{d}}$ implies $d'' \in \mathcal{P}^{\mathrm{d}}$, which gives $e \cdot d'' = d' \cdot e' \in \mathcal{P}^{\mathrm{d}}$ by Proposition V.4.4.2(2), and therefore $d' \in \mathcal{P}^{\mathrm{d}}$. □

Our goal must now be to describe the $\mathrm{Prop}(\mathbb{T}, \mathcal{V})$-open maps, *for a Cartesian closed and integral quantale* $\mathcal{V}$ and a lax extension $\hat{\mathbb{T}}$ of the Set-monad $\mathbb{T} = (T, m, e)$. For convenience we also assume T *to satisfy the Beck–Chevalley condition*, so that not only $\mathrm{Prop}(\mathbb{T}, \mathcal{V})$, but also $\mathrm{Open}(\mathbb{T}, \mathcal{V})$ is an $\mathcal{E}$-topology on $(\mathbb{T}, \mathcal{V})$-$\mathsf{Cat}$ (see Examples V.4.1.1). Since the class $\mathcal{P}^{\mathrm{o}}$ is fully determined by the class $\mathcal{P}^{\mathrm{d}}$, which, in turn, is fully determined by the grand closure when $\mathcal{P} = \mathrm{Prop}(\mathbb{T}, \mathcal{V})$ (see Remarks V.3.3.3), $\mathrm{Prop}(\mathbb{T}, \mathcal{V})$-open maps retrieve only very little information of the quantale $\mathcal{V}$, unless $\mathcal{V} = \mathsf{2}$. To make this statement more precise, let us introduce the following auxiliary notion.

V.4.4.7 Definition A $(\mathbb{T}, \mathcal{V})$-continuous map $f : (X, a) \to (Y, b)$ is called $\mathsf{2}$-*open* if, for all $x \in X$, $\mathfrak{y} \in TY$,

$$b(\mathfrak{y}, f(x)) > \bot \implies \exists \mathfrak{x} \in TX\ (Tf(\mathfrak{x}) = \mathfrak{y}\ \&\ a(\mathfrak{x}, x) > \bot)\,.$$

The following remarks explain this terminology.

V.4.4.8 Remarks

(1) Since $\mathcal{V}$ is integral, the quantale homomorphism $\iota : \mathsf{2} \to \mathcal{V}$ has (as a monotone map) a left adjoint o (with $o(v) = \top \iff v > \bot$), which is a quantale homomorphism if and only if $\mathcal{V}$ satisfies

$$u \otimes v = \bot \implies u = \bot \text{ or } v = \bot \tag{V.4.4.ii}$$

(see Example III.3.5.2(1)).

(2) The lax extension $\hat{T} : \mathcal{V}\text{-}\mathsf{Rel} \to \mathcal{V}\text{-}\mathsf{Rel}$ may be restricted to $\mathsf{Rel} = \mathsf{2}\text{-}\mathsf{Rel}$; one simply considers the composite lax functor

$$\check{T} : (\mathsf{Rel} \xrightarrow{\ \iota\ } \mathcal{V}\text{-}\mathsf{Rel} \xrightarrow{\ \hat{T}\ } \mathcal{V}\text{-}\mathsf{Rel} \xrightarrow{\ o\ } \mathsf{Rel})$$
$$r \longmapsto \iota r \longmapsto \hat{T}(\iota r) \longmapsto o\hat{T}(\iota r)\,,$$

i.e.

$$\mathfrak{x}\ (\check{T}r)\ \mathfrak{y} \iff \hat{T}(\iota r)(\mathfrak{x}, \mathfrak{y}) > \bot\,.$$

Although $\check{T}$ is a lax extension of T (in the sense of Definition III.1.4.1) which makes $e : 1_{\mathsf{Rel}} \to \check{T}$ oplax, $\check{T}$ generally fails to make the monad multiplication m oplax. Still, with respect to $\check{T}$, we can form the category $(\mathbb{T}, \mathsf{2})$-Gph of $(\mathbb{T}, \mathsf{2})$-graphs as in III.4.1 and obtain the functor

$$(\mathbb{T}, \mathcal{V})\text{-}\mathsf{Cat} \longrightarrow (\mathbb{T}, \mathsf{2})\text{-}\mathsf{Gph}\,, \quad (X, a) \longmapsto (X, oa)\,.$$

For example, for the Barr extension of the ultrafilter monad to P_{+}-Rel, one obtains the "induced-pseudotopology" functor

$$\mathsf{App} \longrightarrow \mathsf{PsTop}\,,$$

which we encountered in III.3.6.

(3) Defining a morphism of $(\mathbb{T}, \mathsf{2})$-graphs to be open as in the case of $(\mathbb{T}, \mathcal{V})$-categories, one can now state that, for a $(\mathbb{T}, \mathcal{V})$-functor $f : (X, a) \to (Y, b)$,

$$f \text{ is 2-open} \iff f \text{ is open in } (\mathbb{T}, \mathsf{2})\text{-}\mathsf{Gph}\,.$$

(4) If $\mathcal{V} = \mathsf{2}$, then $\check{T} = \hat{T}$, and the notion of 2-openness returns precisely the notion of openness in $(\mathbb{T}, \mathsf{2})$-Cat.

V.4.4.9 Proposition *Let $\mathcal{V}$ be Cartesian closed and integral, satisfying* (V.4.4.ii), *and let T satisfy BC. Then, for the following statements on a $(\mathbb{T}, \mathcal{V})$-continuous map f of $(\mathbb{T}, \mathcal{V})$-categories, one has* (i) $\implies$ (ii) $\implies$ (iii), *with* (i) $\iff$ (ii) *holding if $\mathcal{V} = \mathsf{2}$, and* (ii) $\iff$ (iii) *if $\mathbb{T} = \mathbb{I}$:*

(i) $f \in \mathrm{Open}(\mathbb{T}, \mathcal{V})$;
(ii) *f is 2-open;*
(iii) *f is* $\mathrm{Prop}(\mathbb{T}, \mathcal{V})$*-open.*

Proof The claims about (i) and (ii) are obvious (see Remarks V.4.4.8). Now we consider a 2-open $(\mathbb{T}, \mathcal{V})$-functor $f : (X, a) \to (Y, b)$ and first show (using the grand closure of Section V.3.3)

$$f^{-1}(\overline{N}) \subseteq \overline{f^{-1}(N)}$$

for all $N \subseteq Y$. Indeed, $x \in f^{-1}(\overline{N})$ implies $b(\mathfrak{y}, f(x)) > \bot$ for some $\mathfrak{y} \in TN \subseteq TY$, and 2-openness gives some $\mathfrak{x} \in TX$ with $Tf(\mathfrak{x}) = \mathfrak{y}$ and $a(\mathfrak{x}, x) > \bot$. Moreover, since T is taut, $\mathfrak{x} \in (Tf)^{-1}(TN) = Tf^{-1}(N)$, which implies $x \in \overline{f^{-1}(N)}$.

By ordinal recursion one concludes that

$$f^{-1}(\overline{N}^{\infty}) \subseteq \overline{f^{-1}(N)}^{\infty}$$

for all $N \subseteq Y$, which implies that pulling back along f preserves $\mathrm{Prop}(\mathbb{T}, \mathcal{V})$-density (see Corollary V.4.4.4). This property holds in fact for every pullback f' of the 2-open map f, since 2-openness is stable under pullback, thanks to T

satisfying BC, as one easily shows as in Proposition V.3.1.4(4). Consequently, f is Prop($\mathbb{T}$, $\mathcal{V}$)-open.

Let now $\mathbb{T} = \mathbb{I}$, identically extended to $\mathcal{V}$-Rel. Condition (V.4.4.ii) guarantees that the grand closure is idempotent in this case. Assuming now $f:(X,a) \to (Y,b)$ in $\mathcal{V}$-Cat to be Prop($\mathbb{I}$, $\mathcal{V}$)-open, we consider $x \in X$, $y \in Y$ with $b(y, f(x)) > \bot$ and let f' be the restriction $f^{-1}Z \to Z$ of f, with the subspace $Z = \{y, f(x)\}$ of Y. Trivially, as a subspace of Z, $\{y\}$ is dense in Z, so that $f^{-1}y$ is dense in $f^{-1}Z$. Hence, $x \in f^{-1}Z = \overline{f^{-1}y}$ gives $z \in f^{-1}y$ with $a(z, x) > \bot$, which shows that f is 2-open. □

V.4.4.10 Examples

(1) A 2-open map may fail to be open, even when $\mathbb{T} = \mathbb{I}$. Indeed, for $\mathbb{R}$ with its Euclidean metric d, all surjective non-expansive maps $f : \mathbb{R} \to \mathbb{R}$ are 2-open in Met = P_+-Cat, but among them only those with

$$d(y, f(x)) = \inf_{z \in f^{-1}y} d(z, x)$$

are also open. Already, $f(x) = \frac{1}{2}x$ fails to be open in P_+-Cat.

(2) For $\mathbb{T} = \beta$ and $\mathcal{V} = 2$, so that ($\mathbb{T}$, $\mathcal{V}$)-Cat $\cong$ Top, all three conditions of Proposition V.4.4.9 are equivalent. The only critical implication left to be shown is (iii) $\Longrightarrow$ (ii). We already know that (i) (equivalently (ii)) describes open maps in Top, in the usual sense that images of open sets are open: see Remarks V.3.3.5. Hence, assuming (iii), consider an open set $A \subseteq X$ and let $N = Y \setminus f(A)$. Since the density of N in $\overline{N}$ pulls back along the restriction $f' : f^{-1}(\overline{N}) \to \overline{N}$ of f, one has that $f^{-1}(N)$ is dense in $f^{-1}(\overline{N})$, i.e. $f^{-1}(\overline{N}) = \overline{f^{-1}(N)} \subseteq \overline{X \setminus A} = X \setminus A$, and therefore $\overline{Y \setminus f(A)} \subseteq Y \setminus f(A)$, so that $f(A)$ is open in Y.

(3) Consider the monad $\mathbb{H}$ with a monoid H as in Section V.1.4. A 2-open morphism $f : X \to Y$ in ($\mathbb{H}$, 2)-Cat is (in the notation of Section V.1.4) easily characterized by

$$y \xrightarrow{\alpha} f(x) \implies \exists z \in f^{-1}y\ (z \xrightarrow{\alpha} x)\,,$$

whereas a Prop($\mathbb{H}$, 2)-open map is described by

$$y \xrightarrow{\alpha} f(x) \implies \exists z \in f^{-1}y, \beta \in H\ (z \xrightarrow{\beta} x)\,,$$

as one shows similarly to the proof of Proposition V.4.4.9(iii) $\Longrightarrow$ (ii). However, the two notions are equivalent when f is an embedding.

(4) For the list monad $\mathbb{L}$ (see Section V.1.4), a Prop($\mathbb{L}$, 2)-open map may fail to be open as well; see Exercise V.4.E. However, Prop($\mathbb{L}$, 2)-open embeddings are open in ($\mathbb{L}$, 2)-Cat.

Returning to the general setting of an $\mathcal{E}$-topology $\mathcal{P}$ in a category X, with the "new" $\mathcal{E}$-topology $\mathcal{P}^{\circ}$ at hand, one may now consider the derived topology $(\mathcal{P}^{\circ})'$ (see Proposition V.4.1.5) and explore the notions of $\mathcal{P}^{\circ}$-compactness and $\mathcal{P}^{\circ}$-Hausdorffness. While we must leave it to the reader to pursue this program in general, let us take a glimpse at $\mathsf{X} = \mathsf{Top} \cong (\beta, 2)\text{-}\mathsf{Cat}$ with $\mathcal{P} = \{\text{proper maps}\} = \mathrm{Prop}(\beta, 2)$ again, so that $\mathcal{P}^{\circ} = \{\text{open maps}\} = \mathrm{Open}(\beta, 2)$. In this case, every object is $\mathcal{P}^{\circ}$-compact, while $\mathcal{P}^{\circ}$-Hausdorffness means discreteness (see Example V.4.2.4(2)). Furthermore, for a continuous map $f : X \to Y$ one has

$$\begin{aligned} f \text{ is } \mathcal{P}^{\circ}\text{-Hausdorff} &\iff f \text{ is locally injective (see Example V.4.2.4(2))},\\ f \text{ is } \mathcal{P}^{\circ}\text{-perfect} &\iff f \text{ is } \mathcal{P}^{\circ}\text{-proper and } \mathcal{P}^{\circ}\text{-Hausdorff}\\ &\iff f \text{ is open and locally injective}\\ &\iff f \text{ is a local homeomorphism,} \end{aligned}$$

where f is a *local homeomorphism* if every point in X has an open neighborhood U such that $f(U)$ is open, and the restriction $U \to f(U)$ of f is a homeomorphism.

V.4.5 $\mathcal{P}$-Tychonoff and locally $\mathcal{P}$-compact Hausdorff objects

We continue to work *in a finitely complete category* X *equipped with an* $\mathcal{E}$*-topology*; furthermore, we now assume that $\mathcal{E}$ *belongs to a proper* $(\mathcal{E}, \mathcal{M})$*-factorization system of* X*, with* $\mathcal{E}$ *stable under pullback*. (In this case, $\mathcal{E}$ is automatically an $\mathcal{E}$-topology.)

We consider the classes

$$(\mathcal{P} \cap \mathcal{P}') \cdot \mathcal{M} = \{p \cdot m \mid m \in \mathcal{M},\ p \text{ is } \mathcal{P}\text{-perfect}\}\,,$$

$$(\mathcal{P} \cap \mathcal{P}') \cdot (\mathcal{M} \cap \mathcal{P}^{\circ}) = \{p \cdot m \mid m \in \mathcal{M} \text{ is } \mathcal{P}\text{-open},\ p \text{ is } \mathcal{P}\text{-perfect}\}\,,$$

and note that they contain all isomorphisms and are stable under pullback, but are not necessarily closed under composition.

For an object X and the unique morphism $!_X : X \to 1$, one has

$$\begin{aligned} !_X \in (\mathcal{P} \cap \mathcal{P}') \cdot \mathcal{M} &\iff \exists m : X \to K\ (m \in \mathcal{M}, K \text{ is } \mathcal{P}\text{-compact}\\ &\qquad\qquad \& \ \mathcal{P}\text{-Hausdorff})\,,\\ !_X \in (\mathcal{P} \cap \mathcal{P}') \cdot (\mathcal{M} \cap \mathcal{P}^{\circ}) &\iff \exists m : X \to K\ (m \in \mathcal{M}, K \text{ is } \mathcal{P}\text{-compact}\\ &\qquad\qquad \& \ \mathcal{P}\text{-Hausdorff})\,. \end{aligned}$$

Guided by the role model Top with $\mathcal{P} = \{\text{proper maps}\}$ and $\mathcal{E} = \{\text{surjective maps}\}$, where for a space X

$$\begin{aligned} X \text{ is Tychonoff} &\iff X \text{ subspace of a compact Hausdorff space,}\\ X \text{ is locally compact Hausdorff} &\iff X \text{ open subspace of a compact Hausdorff space,} \end{aligned}$$

in our abstract category X we say that a morphism

$$\begin{aligned} f \text{ is } \textit{$\mathcal{P}$-Tychonoff} &\iff f \in (\mathcal{P} \cap \mathcal{P}') \cdot \mathcal{M}\,,\\ f \text{ is } \textit{locally $\mathcal{P}$-perfect} &\iff f \in (\mathcal{P} \cap \mathcal{P}') \cdot (\mathcal{M} \cap \mathcal{P}^{\mathrm{o}})\,, \end{aligned}$$

and an object

$$\begin{aligned} X \text{ is } \textit{$\mathcal{P}$-Tychonoff} &\iff !_X \text{ is $\mathcal{P}$-Tychonoff,}\\ X \text{ is } \textit{locally $\mathcal{P}$-compact Hausdorff} &\iff !_X \text{ is locally $\mathcal{P}$-perfect.} \end{aligned}$$

Although the two classes of morphisms under consideration generally fail to be topologies, one may establish characteristic properties similar to those of Theorem V.4.3.1, as follows.

V.4.5.1 Proposition *The following assertions for an object X are equivalent:*

(i) *X is $\mathcal{P}$-Tychonoff;*
(ii) *every morphism $f : X \to Y$ is $\mathcal{P}$-Tychonoff;*
(iii) *there is a $\mathcal{P}$-Tychonoff morphism $f : X \to Y$ with Y $\mathcal{P}$-compact $\mathcal{P}$-Hausdorff;*
(iv) *the projection $X \times Y \to Y$ is $\mathcal{P}$-Tychonoff for all objects Y;*
(v) *$X \times Y$ is $\mathcal{P}$-Tychonoff for every $\mathcal{P}$-Tychonoff object Y.*

Proof (i) $\Longrightarrow$ (ii): With $m : X \to K$ in $\mathcal{M}$ and K $\mathcal{P}$-compact $\mathcal{P}$-Hausdorff, we consider the diagram

$$\begin{array}{ccccc} & & X & & \\ & \swarrow^{f} & \downarrow{\scriptstyle \langle f,m\rangle} & \searrow^{m} & \\ Y & \xleftarrow{\;p_1\;} & Y \times K & \xrightarrow{\;p_2\;} & K\,. \end{array}$$

Since $p_2 \cdot \langle f, m\rangle = m \in \mathcal{M}$, also $\langle f, m\rangle \in \mathcal{M}$ (see Proposition II.5.1.1(3)), and $p_1 \in \mathcal{P} \cap \mathcal{P}'$ by Theorem V.4.3.1 and Corollary V.4.3.3.

(ii) $\Longrightarrow$ (iii), (iv) $\Longrightarrow$ (i), (v) $\Longrightarrow$ (i): Choose $Y = 1$.

(iii) $\Longrightarrow$ (i): By hypothesis, $f = p \cdot m$, with $m \in \mathcal{M}$ and $p : Z \to Y$ in $\mathcal{P} \cap \mathcal{P}'$. By Theorem V.4.3.1 and Corollary V.4.3.3, p transfers the needed properties from Y to Z.

(i) $\Longrightarrow$ (iv): $X \times Y \to Y$ is a pullback of $X \to 1$.

(i) $\Longrightarrow$ (v): If $m : X \to K$, $n : Y \to L$ are in $\mathcal{M}$, and K, L are $\mathcal{P}$-compact $\mathcal{P}$-Hausdorff, then $m \times n = (m \times 1_L)\cdot(1_X \times n)$ is in $\mathcal{M}$, and $K \times L$ is $\mathcal{P}$-compact $\mathcal{P}$-Hausdorff. □

V.4.5.2 Corollary *If the composite morphism $g \cdot f$ is $\mathcal{P}$-Tychonoff, f is also $\mathcal{P}$-Tychonoff.*

Proof One argues as in the proof of Corollary V.4.3.2. □

In order to establish analogous properties for locally $\mathcal{P}$-compact Hausdorff objects, we need an additional hypothesis, as follows. We say that X *has the $\mathcal{P}$-open-closed interchange property* if every composite morphism $(X \xrightarrow{m} Y \xrightarrow{n} Z)$ with $m \in \mathcal{M} \cap \mathcal{P}$ and $n \in \mathcal{M} \cap \mathcal{P}^{\mathrm{o}}$ may be rewritten as $(X \xrightarrow{n'} W \xrightarrow{m'} Z)$ with $n' \in \mathcal{M} \cap \mathcal{P}^{\mathrm{o}}$ and $m' \in \mathcal{M} \cap \mathcal{P}$. In the role model Top, if $X \subseteq Y$ is a closed subspace and $Y \subseteq Z$ is an open subspace, one may choose $W = \overline{X}$ as the closure of X in Z.

V.4.5.3 Proposition *If X has the $\mathcal{P}$-open-closed interchange property, the following conditions are equivalent for an object X:*

(i) *X is locally $\mathcal{P}$-compact Hausdorff;*
(ii) *every morphism $f : X \to Y$ with Y $\mathcal{P}$-Hausdorff is locally $\mathcal{P}$-perfect;*
(iii) *there is a locally $\mathcal{P}$-perfect morphism $f : X \to Y$ with Y $\mathcal{P}$-compact $\mathcal{P}$-Hausdorff;*
(iv) *the projection $X \times Y \to Y$ is locally $\mathcal{P}$-perfect for all objects Y;*
(v) *$X \times Y$ is locally $\mathcal{P}$-compact Hausdorff for every locally $\mathcal{P}$-compact Hausdorff object Y.*

Proof (i) $\Longrightarrow$ (ii): Revisiting the proof of Proposition V.4.5.1(i) $\Longrightarrow$ (ii), one decomposes $\langle f, m\rangle \subset \mathcal{M}$ as

$$X \xrightarrow{\langle f, 1_X\rangle} Y \times X \xrightarrow{1_Y \times m} Y \times K\,.$$

Then $\langle f, 1_X\rangle \in \mathcal{M}$ (since $\mathcal{E} \subseteq \mathrm{Epi}\,\mathsf{X}$) and $\langle f, 1_X\rangle \in \mathcal{P}$ as a pullback of $\delta_Y : Y \to Y \times Y$ (since Y is $\mathcal{P}$-Hausdorff); furthermore, $1_Y \times m \in \mathcal{M} \cap \mathcal{P}^{\mathrm{o}}$ as a pullback of $m \in \mathcal{M} \cap \mathcal{P}^{\mathrm{o}}$. With the $\mathcal{P}$-open-closed interchange property and $\mathcal{M} \subseteq \mathrm{Mono}\,\mathsf{X}$, $\langle f, m\rangle$ is locally $\mathcal{P}$-perfect, and so is $f = p_1 \cdot \langle f, m\rangle$ because $p_1 \in \mathcal{P} \cap \mathcal{P}'$.

All other steps can be taken as in the proof of Proposition V.4.5.3. □

V.4.5.4 Corollary *Let X have the $\mathcal{P}$-open-closed interchange property. If the composite morphism $g \cdot f$ is locally $\mathcal{P}$-perfect and g is $\mathcal{P}$-perfect, then f is locally $\mathcal{P}$-perfect.*

Proof Exploit Proposition V.4.5.3(i) $\Longrightarrow$ (ii) in $\mathsf{X}/\operatorname{cod}(g)$. □

Let us now consider $\mathsf{X} = (\mathbb{T}, \mathcal{V})\text{-}\mathsf{Cat}$ *and* $\mathcal{P} = \mathrm{Prop}(\mathbb{T}, \mathcal{V})$ *with $\mathcal{V}$ Cartesian closed and integral.* Since every $\mathcal{V}$-space is $\mathcal{P}$-compact, and $\mathcal{P}$-Hausdorffness means discreteness, we note that, for $\mathbb{T} = \mathbb{I}$, being $\mathcal{P}$-Tychonoff or locally $\mathcal{P}$-compact Hausdorff also amounts to being discrete. For a general monad $\mathbb{T}$ we remark that, trivially, subspaces of $\mathcal{P}$-Tychonoff spaces and $\mathcal{P}$-open subspaces of locally $\mathcal{P}$-compact Hausdorff spaces maintain the respective properties.

V.4.5.5 Proposition *For $\mathcal{V}$ lean and superior and $\hat{\mathbb{T}}$ flat, every* $\mathrm{Prop}(\mathbb{T},\mathcal{V})$*-Tychonoff space is regular and Hausdorff in* $(\mathbb{T},\mathcal{V})$-Cat.

Proof A $\mathcal{P}$-Tychonoff space X is embeddable into a $\mathcal{P}$-compact $\mathcal{P}$-Hausdorff space K which, by Proposition V.4.2.1, is a compact Hausdorff object in $(\mathbb{T},\mathcal{V})$-Cat and therefore regular by Proposition V.2.3.4. Regularity and Hausdorffness are both inherited by subspaces. □

Let us now restrict our attention to $\mathcal{P}$-Tychonoff maps among $\mathcal{P}$-Hausdorff spaces; these are simply restrictions of proper maps in $(\mathbb{T},\mathcal{V})$-$\mathsf{Cat}_{\mathsf{Haus}}$, and one may exploit Proposition V.4.5.1 and Corollary V.4.5.2 in this case. To be able to apply Proposition V.4.5.3 to $(\mathbb{T},\mathcal{V})$-Cat, we must secure the $\mathcal{P}$-open-closed interchange property:

V.4.5.6 Proposition *If the grand closure is idempotent, then* $(\mathbb{T},\mathcal{V})$-Cat *has the* $\mathrm{Prop}(\mathbb{T},\mathcal{V})$*-open-closed interchange property.*

Proof For a closed subspace X of a $\mathcal{P}$-open subspace Y of Z, consider the closure $W=\overline{X}^Z$ of X in Z. Then W is closed in Z by the idempotency hypothesis, and since hereditariness of the grand closure makes the diagram

$$\begin{array}{ccc} X & \hookrightarrow & Y \\ \downarrow & & \downarrow \\ W & \hookrightarrow & Z \end{array}$$

a pullback, $\mathcal{P}$-openness of Y in Z gives the same property for X in W. □

V.4.5.7 Remark The idempotency hypothesis is certainly restrictive, as the case $\mathbb{T}=\beta$, $\mathcal{V}=\mathsf{P}_+$ shows (see Exercise V.3.B). Trying to strengthen Proposition V.4.5.6, one may be tempted to consider the idempotent hull of the grand closure. However, already in the general case, *the idempotent hull of the grand closure in* $(\mathbb{T},\mathcal{V})$-Cat *can be hereditary only if the grand closure is idempotent.* Indeed, assume that for $A\subseteq X$ we have $x\in\overline{\overline{A}}^X$, but $x\notin\overline{A}^X$. For the subspace $Y:=A\cup\{x\}$ one trivially has $\overline{A}^Y\subseteq\overline{A}^X\cap Y$, and therefore $x\notin\overline{A}^Y$. Consequently, $\overline{A}^Y=A$ is closed in Y. Denoting the idempotent hull by $\widetilde{A}=\overline{A}^\infty$ and assuming its hereditariness, one obtains $A=\widetilde{A}^Y=\widetilde{A}^X\cap Y\supseteq\overline{\overline{A}}^X\cap Y\ni x$, a contradiction.

V.4.5.8 Examples

(1) In $\mathsf{Top}=(\beta,2)$-Cat with $\mathcal{P}=\{\text{proper maps}\}$, $\mathcal{P}$-Tychonoff spaces are characterized as the *completely regular Hausdorff spaces* (or *Tychonoff*

spaces), i.e. as T1-spaces X with the property that for every closed set $A \subseteq X$ and every $x \in X \setminus A$ there is a continuous map $f : X \to [0, 1]$ with $f(x) = 0$ and $f(y) = 1$ for all $y \in A$. Such spaces are easily seen to be embeddable into powers of the unit interval and, hence, into a compact Hausdorff space. Locally $\mathcal{P}$-compact Hausdorff spaces are *locally compact Hausdorff spaces*, i.e. Hausdorff spaces that are locally compact (see Section III.5.7); in the presence of Hausdorff separation, these are the spaces in which every point has a compact neighborhood. Such spaces are Tychonoff spaces and are, in fact, openly embeddable into a compact Hausdorff space (see [Engelking, 1989]).

(2) For a monoid H and $\mathbb{H}$ as in Section V.1.4 and $\mathcal{P} = \{\text{proper maps}\}$ as described in Example V.3.1.3(4), an $(\mathbb{H}, 2)$-space $(X, \longrightarrow)$ is $\mathcal{P}$-Tychonoff if and only if it is embeddable into an H-action Y, i.e.

$$x \xrightarrow{\alpha} y \text{ in } X \iff \alpha \cdot x = y \text{ in } Y$$

for all $x, y \in X$, $\alpha \in H$ (with $\alpha \cdot x$ denoting the action of H on Y). Since $\mathcal{P}$-open maps are open (see Example V.4.4.10(3)), locally $\mathcal{P}$-compact Hausdorff spaces in $(\mathbb{H}, 2)$-**Cat** have the additional property that the embedding $X \hookrightarrow Y$ can be chosen to be open, i.e. if $\alpha \cdot x \in X$ for $x \in Y$, then $x \in X$. Finally, we note that $(\mathbb{H}, 2)$-**Cat** has the $\mathcal{P}$-open-closed interchange property since the grand closure is idempotent.

(3) In $(\mathbb{L}, 2)$-**Cat** with $\mathcal{P} = \mathrm{Prop}(\mathbb{L}, 2)$, $\mathcal{P}$-Tychonoff spaces are those $(X, \vdash)$ that are embeddable into monoids Y, i.e.

$$(x_1, \ldots, x_n) \vdash y \text{ in } X \iff x_1 \cdot x_2 \cdot \ldots \cdot x_n = y \text{ in } Y\,.$$

If the embedding $X \hookrightarrow Y$ is open, one has the additional property that

$$\forall x, y \in Y\ (x \cdot y \in X \implies x, y \in X)\,,$$

and this property makes $X \hookrightarrow Y$ $\mathcal{P}$-open (see Example V.4.4.10(4)) and therefore characterizes locally $\mathcal{P}$-compact Hausdorff spaces in $(\mathbb{L}, 2)$-**Cat**. In addition, since the grand closure is idempotent in $(\mathbb{L}, 2)$-**Cat**, Proposition V.4.5.3 is applicable in this category as well.

Exercises

V.4.A *$\mathcal{P}$-discrete objects.* For an $\mathcal{E}$-topology $\mathcal{P}$ on a finitely complete category X, call a morphism $f : X \to Y$ *locally $\mathcal{P}$-injective* if $\delta_f : X \to X \times_Y X$ is $\mathcal{P}$-open, and call an object X *$\mathcal{P}$-discrete* if $!_X$ is locally $\mathcal{P}$-injective. Show the equivalence of the following statements on X:

(i) X is $\mathcal{P}$-discrete;
(ii) every morphism $f : X \to Y$ is locally $\mathcal{P}$-injective;
(iii) there is a locally $\mathcal{P}$-injective morphism $f : X \to Y$ with Y $\mathcal{P}$-discrete;

(iv) the projection $X \times Y \to Y$ is locally $\mathcal{P}$-injective for all objects Y;
(v) $X \times Y$ is $\mathcal{P}$-discrete for every $\mathcal{P}$-discrete object Y;
(vi) Y is $\mathcal{P}$-discrete for every $\mathcal{P}$-open morphism $f : X \to Y$ in $\mathcal{E}$.

V.4.B *$\mathcal{P}^{\circ}$-compact objects*

(1) Let X be extensive, and let the class $\mathcal{E}$ be closed under coproducts (so that $\coprod_{i\in I} p_i : \coprod_{i\in I} X_i \to \coprod_{i\in I} Y_i$ lies in $\mathcal{E}$ whenever all $p_i \in \mathcal{E}$). Show that $\mathcal{P}^{\mathrm{d}}$ is closed under coproducts, for any $\mathcal{E}$-topology $\mathcal{P}$.
(2) For an $\mathcal{E}$-topology $\mathcal{P}$, assume $\mathcal{P}^{\mathrm{d}}$ to be closed under coproducts and let X be an object such that, for all objects U, the morphism

$$e_U : \coprod_{x:1\to X} U \to X \times U\,,$$

whose xth restriction to U is $\langle x, 1_U\rangle : U \to X \times U$, lies in $\mathcal{E}$. Show that X is $\mathcal{P}^{\circ}$-compact. Conclude that every topological space is {open map}-compact.

V.4.C *The left adjoint left-inverse topology.* Recall that $f : (X, a) \to (Y, b)$ is left adjoint to $g : (Y, b) \to (X, a)$ in $(\mathbb{T}, \mathcal{V})$-$\mathsf{Cat}$ if $g^{\circ} \cdot a = b \cdot Tf$; f is *left adjoint left-inverse* to g if, in addition, $f \cdot g = 1_Y$. Let $\mathcal{L}$ be the class of all left adjoint left-inverse maps. Show the following.

(1) $\mathcal{L}$ is a topology on $(\mathbb{T}, \mathcal{V})$-$\mathsf{Cat}$. For $\mathbb{T} = \mathbb{I}$, $\mathcal{V} = \mathsf{2}$, a monotone map $f : X \to Y$ with Y separated is left adjoint left-inverse if and only if f is proper and left adjoint.
(2) A $(\mathbb{T}, \mathcal{V})$-space (X, a) is $\mathcal{L}$-compact if and only if there is a point $x_0 \in X$ with $a(\chi, x_0) = \top$ for all $\chi \in TX$. In particular, for $\mathbb{T} = \beta$ and $\mathcal{V} = \mathsf{2}$, a topological space is $\mathcal{L}$-compact if and only if it contains a point whose only neighborhood is the space itself.
(3) If $\mathcal{V}$ is Cartesian closed and T satisfies BC, then $\mathcal{L}$ is a $\mathcal{Q}$-topology on $(\mathbb{T}, \mathcal{V})$-$\mathsf{Cat}$, where $\mathcal{Q}$ is the class of open surjections in $(\mathbb{T}, \mathcal{V})$-$\mathsf{Cat}$.

V.4.D *The exponentiable topology.* Show that, in a finitely complete category X, the class $\mathrm{Exp}\,\mathsf{X}$ of exponentiable morphisms of X forms a topology. The $\mathrm{Exp}\,\mathsf{X}$-compact objects are the exponentiable objects. For $\mathsf{X} = \mathsf{Top}$, prove that a space X is $\mathrm{Exp}\,\mathsf{X}$-Hausdorff if and only if every point in X has a Hausdorff neighborhood.

V.4.E *$\mathcal{P}$-open versus open.* Show that in $(\mathbb{H}, \mathsf{2})$-Cat and $(\mathbb{L}, \mathsf{2})$-Cat there are {proper maps}-open maps which fail to be open.

V.4.F *Nearly open maps.* Prove that the class of nearly open maps (see Definition III.4.3.1) forms an $\mathcal{E}$-pretopology in $(\mathbb{T}, \mathcal{V})$-$\mathsf{Cat}$ with $\mathcal{E}$ the class of all epimorphisms.

V.5 Connectedness

An object in a category is connected if it has no non-trivial decomposition into a coproduct. This property becomes particularly powerful when the ambient category is extensive. After a brief review of extensive categories, we explore the notion of connectedness in $(\mathbb{T}, \mathcal{V})$-Cat and exhibit the pivotal role of the category Top in this context. Stability under products is discussed at the end.

V.5.1 Extensive categories

In Corollary III.4.3.10 we gave an ad hoc definition of extensive category and provided sufficient conditions for $(\mathbb{T}, \mathcal{V})$-Cat to be extensive. Here we investigate this notion more systematically.

For every small family $(Y_i)_{i\in I}$ of objects in a category X with small-indexed coproducts and pullbacks, one has the adjunction

$$\prod_{i\in I} \mathsf{X}/Y_i \underset{\longleftarrow}{\overset{\longrightarrow}{\bot}} \mathsf{X}/\coprod_{i\in I} Y_i\ , \tag{V.5.1.i}$$

with the left adjoint given by coproduct (mapping $(f_i : X_i \to Y_i)_{i\in I}$ to $\coprod_{i\in I} f_i : \coprod_{i\in I} X_i \to \coprod_{i\in I} Y_i$) and the right adjoint by pullback along the coproduct injection t_j of $\coprod_{i\in I} Y_i$:

$$\begin{array}{ccc} X_j & \xrightarrow{s_j} & X \\ {\scriptstyle f_j}\downarrow & & \downarrow{\scriptstyle f} \\ Y_j & \xrightarrow{t_j} & \coprod_{i\in I} Y_i\ . \end{array} \tag{V.5.1.ii}$$

V.5.1.1 Definition A category X with small-indexed coproducts and pullbacks is *extensive* if the adjunction (V.5.1.i) is an equivalence of categories; equivalently, if both the counits and the units are isomorphisms, i.e. if

(1) *small coproducts are universal* in X, so that X is a coproduct of $(X_i)_{i\in I}$ with injections s_j if all diagrams (V.5.1.ii) are pullbacks, and
(2) *small coproducts are totally disjoint in* X, so that all commutative diagrams (V.5.1.ii) are pullbacks if $X \cong \coprod_{i\in I} X_i$ (and therefore $f = \coprod_{i\in i} f_i$).

X is *finitely extensive* if instead of small coproducts we consider only finite coproducts everywhere.

Note that universality of finite coproducts entails in particular that the initial object 0 must be *strict*, i.e. any morphism $f : X \to 0$ must be an isomorphism (just consider $I = \varnothing$). In fact, strictness of 0 follows already from the universality of binary coproducts, as the following result shows.

V.5.1.2 Proposition *Finite coproducts are universal in* X *if binary coproducts are.*

Proof It suffices to show that the initial object 0 in X is strict. For any morphism $f : X \to 0$,

$$\begin{array}{ccccc} X & \xrightarrow{1} & X & \xleftarrow{1} & X \\ {\scriptstyle f}\downarrow & & \downarrow{\scriptstyle f} & & \downarrow{\scriptstyle f} \\ 0 & \xrightarrow{1} & 0 & \xleftarrow{1} & 0 \end{array}$$

are pullback diagrams, with the bottom arrows representing the injections of a binary coproduct. The same is therefore true for the top arrows. To say that $X \cong X + X$ with isomorphic injections means equivalently that X is *pre-initial*, i.e. $|\mathsf{X}(X, Y)| \leq 1$ for all objects Y. In particular, the split epimorphism f must be inverse to $0 \to X$. □

V.5.1.3 Proposition *If binary coproducts are universal in* X*, the two injections of a binary coproduct are monomorphic and their pullback is pre-initial.*

Proof Let

$$\begin{array}{ccccc} P & \xrightarrow{p} & X & \xleftarrow{q} & Q \\ \downarrow & & \downarrow{\scriptstyle s} & & \downarrow \\ X & \xrightarrow{s} & X + Y & \xleftarrow{t} & Y \end{array}$$

be pullback diagrams and let s, t be coproduct injections. By hypothesis, p, q are coproduct injections as well, and since the left pullback property makes p a split epimorphism, the coproduct property makes it actually an isomorphism. Consequently, having an isomorphic projection in its kernel pair, s is a monomorphism.

This shows that coproduct injections are monic; in particular, q is monic. With the trivial pullback diagrams

$$\begin{array}{ccccc} Q & \xrightarrow{1_Q} & Q & \xleftarrow{1_Q} & Q \\ {\scriptstyle q}\downarrow & & \downarrow{\scriptstyle q} & & \downarrow{\scriptstyle 1_Q} \\ X & \xrightarrow{1_X} & X & \xleftarrow{q} & Q \end{array}$$

we see that, since the bottom arrows are the injections of $X+Q \cong X$, universality gives $Q + Q \cong Q$ with isomorphic injections, i.e. Q is pre-initial. □

V.5.1.4 Lemma *Let* X *be a category with finite coproducts and pullbacks. If binary coproducts in* X *are universal and pre-initial objects are initial, then the functor*

$$(-) + C : \mathsf{X} \to \mathsf{X}$$

reflects isomorphisms, for all objects C *in* X.

Proof For $f : A \to B$ in X, assume that $f + 1_C : A + C \to B + C$ is an isomorphism. Since coproduct injections are monic, the functor $(-) + C$ is faithful and therefore reflects mono- and epimorphisms, so that f is both monic and epic. We form the pullbacks D, E as in

$$\begin{array}{ccccc} D & \xrightarrow{\ u\ } & B & \xleftarrow{\ v\ } & E \\ & & \big\downarrow{\scriptstyle w} & & \\ {\scriptstyle a}\big\downarrow & \nearrow{\scriptstyle f} & B+C & & \big\downarrow{\scriptstyle c} \\ & & \big\downarrow{\scriptstyle (f+1_C)^{-1}} & & \\ A & \xrightarrow[\ s\ ]{} & A+C & \xleftarrow[\ t\ ]{} & C \end{array}$$

where s, t, w are coproduct injections and a, u, c, v are pullback projections. One obtains $g : A \to D$ with $a \cdot g = 1_A$ and $u \cdot g = f$, making a an isomorphism as a pullback of the monomorphism $(f + 1_C)^{-1} \cdot w$. Consequently, with u, v being coproduct injections by universality, the diagram

$$A \xrightarrow{\ f\ } B \xleftarrow{\ v\ } E$$

is also a coproduct, with f epic. With coproduct injections i, j, m, n, we have the commutative diagram

$$\begin{array}{ccccc} A & \xrightarrow{\ m\ } & A + (E + E) & \xleftarrow{\ n\ } & E + E \\ {\scriptstyle 1_A}\big\uparrow & & {\scriptstyle 1_A+i}\big\uparrow\big\uparrow{\scriptstyle 1_A+j} & & {\scriptstyle i}\big\uparrow\big\uparrow{\scriptstyle j} \\ A & \xrightarrow[\ f\ ]{} & A + E & \xleftarrow[\ v\ ]{} & E \end{array}$$

where $1_A + i = 1_A + j$ since f is epic. But then $i = j$ because n is monic, which makes E pre-initial and thus $E \cong 0$, by hypothesis. Consequently, the coproduct injection f becomes an isomorphism. $\square$

V.5.1.5 Theorem *The following conditions on a category* X *with finite coproducts and pullbacks are equivalent:*

(i) X *is finitely extensive, i.e. finite coproducts in* X *are universal and totally disjoint;*

(ii) *binary coproducts in* X *are universal and pre-initial objects are initial;*

(iii) *binary coproducts in* X *are universal and disjoint (so that the pullback of the two coproduct injections is the initial object).*

Proof (i) $\Longrightarrow$ (ii): For Q pre-initial, both rows of

$$\begin{array}{ccccc} 0 & \longrightarrow & Q & \xleftarrow{\ 1\ } & Q \\ \big\downarrow & & \big\downarrow{\scriptstyle 1} & & \big\downarrow{\scriptstyle 1} \\ Q & \xrightarrow{\ 1\ } & Q & \xleftarrow{\ 1\ } & Q \end{array} \qquad \text{(V.5.1.iii)}$$

represent binary coproducts. Total disjointness implies in particular that the left-hand side is a pullback diagram, so that we must have $Q \cong 0$.

(ii) $\Longrightarrow$ (iii) follows from Proposition V.5.1.3.

(iii) $\Longrightarrow$ (ii): If Q is pre-initial, the left-hand side of (V.5.1.iii) is a pullback diagram under the disjointness hypothesis, and $Q \cong 0$ follows.

(ii) $\Longrightarrow$ (i): By Proposition V.5.1.2, it suffices to show total disjointness of binary coproducts. Given the outer commutative diagram of

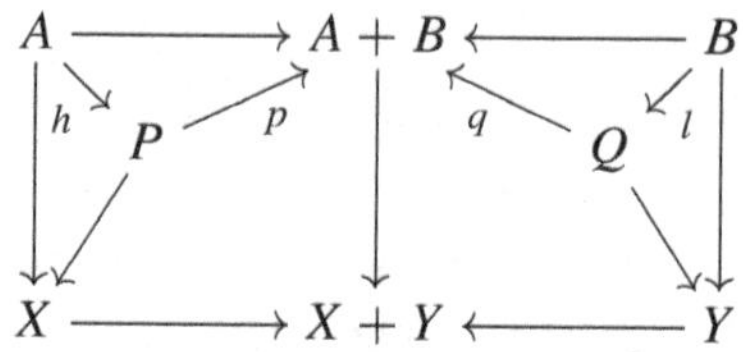

(V.5.1.iv)

(with the horizontal arrows coproduct injections), one forms the pullbacks P and Q and obtains the comparison morphisms h, l, making (V.5.1.iv) commutative. By hypothesis, $P + Q \cong A + B$ (with coproduct injections p, q), so that $h + l : A + B \to P + Q$ is an isomorphism. Since

$$h + l = (h + 1_Q) \cdot (1_A + l) = (1_P + l) \cdot (h + 1_B) ,$$

$h + 1_Q$ is split epic and $h + 1_B$ is split monic. Consequently,

$$(h + 1_Q) + 1_B \cong h + 1_{B+Q} \cong (h + 1_B) + 1_Q$$

is both split epic and split monic, i.e. an isomorphism. With Lemma V.5.1.4 we deduce that h (as well as l, by symmetry) is an isomorphism, as desired. □

V.5.1.6 Corollary *The following conditions on a category* X *with small-indexed coproducts and pullbacks are equivalent:*

(i) X *is extensive;*

(ii) *non-empty coproducts in* X *are universal and pre-initial objects are initial;*

(iii) *non-empty coproducts in* X *are universal and disjoint (so that the pullback of two coproduct injections with distinct labels is the initial object).*

Proof For any coproduct $(s_i : X_i \to X)_{i \in I}$ and $j \in I$, note that

$$X_j \xrightarrow{s_j} X \xleftarrow{\tilde{s}} \coprod_{i \neq j} X_i$$

is a binary coproduct. Therefore (ii) $\Longrightarrow$ (i) follows from V.5.1.5(ii) $\Longrightarrow$ (i), and the same is true for (ii) $\Longrightarrow$ (iii); indeed, when the pullback of s_j and $\tilde{s}$ is 0, so is the pullback of s_j and s_i for any $i \neq j$ since 0 is strict by Proposition V.5.1.2. □

We proved in Theorem III.4.3.9 that for a Cartesian closed quantale $\mathcal{V}$, a monad $\mathbb{T}$ with T taut, and an associative and right-whiskering lax extension $\hat{\mathbb{T}}$ to $\mathcal{V}$-Rel, the category $(\mathbb{T}, \mathcal{V})$-Cat is extensive. In particular, Ord, Met, Top, App, and $(\mathbb{H}, 2)$-Cat (for a monoid H) are extensive.

Frequently studied extensive categories which do not admit a topological functor to Set include Cat and $\mathsf{Rng}^{\mathrm{op}}$ (the opposite of the category of unital rings).

V.5.2 Connected objects

An object X in a category X is *connected* if $\mathsf{X}(X,-) : \mathsf{X} \to \mathsf{Set}$ preserves all small-indexed coproducts. This definition becomes especially efficient when X is extensive.

V.5.2.1 Theorem *The following assertions are equivalent for an object X in an extensive category X with a terminal object 1:*

(i) *X is connected;*
(ii) *every morphism $f : X \to \coprod_{i\in i} Y_i$ factors uniquely as $f = t_j \cdot g$ with a uniquely determined $j \in I$ (and t_j the corresponding coproduct injection);*
(iii) *every morphism $f : X \to \coprod_{i\in I} Y_i$ factors as $f = t_j \cdot g$ for some g and j;*
(iv) *$X \not\cong 0$, and every morphism $f : X \to 1+1$ factors through one of the coproduct injections of $1+1$;*
(v) *$X \not\cong 0$, and every extremal epimorphism $f : X \to Y+Z$ makes one of the coproduct injections an isomorphism;*
(vi) *$X \not\cong 0$, and $X \cong Y+Z$ implies $Y \cong 0$ or $Z \cong 0$.*

Proof The implications (i) $\iff$ (ii) $\implies$ (iii) are trivial, and for (iii) $\implies$ (iv) observe that $X \not\cong 0$ follows since there is no coproduct injection through which the empty coproduct 0 may factor.

(iv) $\implies$ (v): For any extremal epimorphism $f : X \to Y+Z$, the composite

$$X \xrightarrow{\ f\ } Y+Z \xrightarrow{\ !_Y+!_Z\ } 1+1$$

factors through one of the injections of $1+1$, and then f factors through one of the injections of $Y+Z$ since the latter is a pullback of the former, as $f = t \cdot g$ with $t : Y \to Y+Z$, say. But then the monomorphism t must be an isomorphism since the epimorphism f is extremal.

(v) $\implies$ (vi): By hypothesis, one of the coproduct injections of $X \cong Y+Z$ is an isomorphism, and since

$$\begin{array}{ccc} 0 & \longrightarrow & Z \\ \downarrow & & \downarrow \\ Y & \longrightarrow & Y+Z \end{array}$$

is a pullback diagram, one of $0 \to Y$, $0 \to Z$ is an isomorphism as well.

(vi) $\implies$ (ii): For $f : X \to \coprod_{i\in I} Y_i$, one considers the pullback diagrams (V.5.1.ii) and has $X \cong \coprod_{i\in I} X_i$ by universality of coproducts. Since $X \not\cong 0$, not all X_i may be initial. In fact, there is precisely one $j \in I$ with $X_j \not\cong 0$ since, by hypothesis, from

$$X \cong X_j + \coprod_{i \neq j} X_i$$

one obtains $\coprod_{i \neq j} X_i \cong 0$ and therefore $X_i \cong 0$ for all $i \neq j$. Consequently, the coproduct injection $X_j \to X$ is an isomorphism and thus allows for the factorization $f = t_j \cdot (f_j \cdot s_j^{-1})$, which is unique since t_j is a monomorphism. □

Let the extensive category X now come with a factorization system $(\mathcal{E}, \mathcal{M})$ and an $\mathcal{E}$-topology $\mathcal{P}$. Then one easily obtains stability of connectedness under $\mathcal{E}$-images and $\mathcal{P}$-dense extensions, as follows.

V.5.2.2 Proposition *For a morphism $h : Z \to X$ in X with Z connected, under each of the following conditions X is also connected:*

(1) *$h \in \mathcal{E}$, and coproduct injections in X lie in $\mathcal{M}$;*
(2) *$h \in \mathcal{P}^{\mathrm{d}}$, and coproduct injections in X lie in $\mathcal{M} \cap \mathcal{P}$.*

Proof Since Z is connected, every morphism $f : X \to \coprod_{i \in I} Y_i$ yields a commutative diagram

$$\begin{array}{ccc} Z & \xrightarrow{g} & Y_j \\ {\scriptstyle h}\downarrow & & \downarrow{\scriptstyle t_j} \\ X & \xrightarrow{f} & \coprod_{i \in I} Y_i \end{array}$$

with some g and coproduct injection t_j. Each of the two conditions (1), (2) produces a "diagonal" $X \to Y_j$, making f factor through t_j. □

Let now $\mathsf{X} = (\mathbb{T}, \mathcal{V})\text{-}\mathsf{Cat}$, with $\mathcal{V}$ Cartesian closed, T taut, and $\hat{\mathbb{T}}$ associative and right-whiskering, so that X is extensive (Theorem III.4.3.9), and let $\mathcal{P} = \mathrm{Prop}(\mathbb{T}, \mathcal{V})$ and $\mathcal{E}$ be the class of all epimorphisms. It is easy to see that when $\hat{T}$ *preserves bottom $\mathcal{V}$-relations* (so that $\hat{T}\bot_{X,Y} = \bot_{TX,TY}$, for the least $\mathcal{V}$-relation $\bot_{X,Y} : X \nrightarrow Y$), coproduct injections are closed embeddings, i.e. they lie in $\mathcal{M} \cap \mathcal{P}$: see Exercise V.5.B). With Corollary V.4.4.4 and Proposition V.5.2.2, one obtains:

V.5.2.3 Corollary *Let $\mathcal{V}$ be Cartesian closed, T be taut, and $\hat{\mathbb{T}}$ be associative and right-whiskering. Suppose moreover that $\hat{T}$ preserves bottom $\mathcal{V}$-relations. For a $(\mathbb{T}, \mathcal{V})$-continuous $f : X \to Y$ with $\overline{f(X)}^{\infty} = Y$, if X is connected, then Y is connected.*

It follows trivially from Theorem V.5.2.1 that $(\mathbb{T}, \mathcal{V})$-spaces X with $|X| = 1$ are connected. For $\mathbb{T} = \mathbb{I}$ the identity monad one also obtains easily the following characterization of all connected spaces:

V.5.2.4 Corollary *Let $\mathcal{V}$ be Cartesian closed. Then (X, a) is connected in $\mathcal{V}$-Cat if and only if $X \neq \varnothing$ and for all $x, y \in X$ there are $x = x_0, x_1, \ldots, x_n = y$ in X with*

$$a(x_i, x_{i+1}) \vee a(x_{i+1}, x_i) > \bot \quad (i = 0, \ldots, n-1) .$$

Proof The criterion is sufficient for connectedness of (X, a) since continuity of any map $f : X \to 1 + 1$ means equivalently

$$a(x, y) > \bot \implies f(x) = f(y)$$

for all $x, y \in X$. Conversely, considering the least equivalence relation on X that identifies all x, y with $a(x, y) > \bot$, for Z the subspace formed by the equivalence class of some $z \in X \neq \varnothing$ and $Y = X \setminus Z$, one has $X \cong Y + Z$ in $\mathcal{V}$-Cat. Connectedness of X gives $Y = \varnothing$, so that the criterion holds. □

V.5.2.5 Examples

(1) In Ord the categorical notion of connectedness retains the usual notion of a non-empty connected ordered set $(X, \leq)$: for all $x, y \in X$ one finds a "zigzag"

$$x = x_0 \leq x_1 \geq x_2 \leq x_3 \ldots x_{n-1} \leq x_n = y .$$

Connectedness in Met is less interesting: every non-empty metric space (X, d) with d finite is connected.

(2) In Top connected objects X are also characterized as expected: $X = Y \cup Z$ with $Y, Z \in \mathcal{O}X$ and $Y \cap Z = \varnothing$ only if $Y = \varnothing$ or $Z = \varnothing$; but note again that the categorical notion entails $X \neq \varnothing$.

(3) For a monoid H, a connected object $(X, \longrightarrow)$ in $(\mathbb{H}, 2)$-Cat is characterized by $X \neq \varnothing$ and the property that for all $x, y \in X$ one has

$$x = x_0 \xrightarrow{\alpha_0} x_1 \xleftarrow{\alpha_1} x_2 \xrightarrow{\alpha_2} x_3 \cdots x_{n-1} \xrightarrow{\alpha_{n-1}} x_n = y$$

for some $x_i \in X$ and $\alpha_i \in H$ $(i = 1, 2, \ldots, n-1)$.

V.5.3 Topological connectedness governs

Throughout this section we assume that

- *$\mathcal{V}$ is Cartesian closed;*
- *T is taut;*
- *$\hat{T}$ is associative and right-whiskering.*

For the extensive category $(\mathbb{T}, \mathcal{V})$-Cat we then have the functor

$$\Omega : (\mathbb{T}, \mathcal{V})\text{-Cat} \longrightarrow \mathsf{Top}$$

that provides a $(\mathbb{T}, \mathcal{V})$-space with the topology of its $(\mathbb{T}, \mathcal{V})$-open subsets; it preserves open maps and coproducts and also reflects coproducts: see

Corollary V.3.6.2. Its principal purpose arises from the following consequence of Theorem V.5.2.1:

V.5.3.1 Theorem *A $(\mathbb{T}, \mathcal{V})$-space X is connected if and only if the topological space ΩX is connected.*

Proof Let X be connected in $(\mathbb{T}, \mathcal{V})$-Cat and assume that X is the disjoint union of $(\mathbb{T}, \mathcal{V})$-open sets A, B. Then $X \cong A + B$ in $(\mathbb{T}, \mathcal{V})$-Cat (with A, B considered as subspaces of X): see Exercise III.4.B and Theorem III.4.3.3. Consequently, $A = \varnothing$ or $B = \varnothing$.

Conversely, assuming $X \cong A+B$ in $(\mathbb{T}, \mathcal{V})$-Cat and ΩX connected, coproduct preservation by Ω gives immediately $A = \varnothing$ or $B = \varnothing$, so that X must be connected in $(\mathbb{T}, \mathcal{V})$-Cat. □

A *connected component* of a $(\mathbb{T}, \mathcal{V})$-space X is a maximal (with respect to $\subseteq$) connected subspace X. By Corollary V.5.2.3, *a connected component is always $(\mathbb{T}, \mathcal{V})$-closed.* From Theorem V.5.3.1 one obtains immediately:

V.5.3.2 Corollary *The following statements on a $(\mathbb{T}, \mathcal{V})$-space X are equivalent:*

(i) *X is a coproduct of connected $(\mathbb{T}, \mathcal{V})$-spaces;*
(ii) *X is the coproduct of its connected components;*
(iii) *the connected components of X are $(\mathbb{T}, \mathcal{V})$-open;*
(iv) *ΩX is the coproduct of its connected components.*

V.5.3.3 Examples

(1) Even in Top $\cong (\beta, 2)$-Cat, a space X may fail to satisfy the equivalent conditions of Corollary V.5.3.2 (consider the subspace $\{0\} \cup \{1/n \mid n = 1, 2, \ldots\}$ of $\mathbb{R}$, for example). Every open subspace of X satisfies them precisely when X is *locally connected*, i.e. when every neighborhood of a point x contains a connected neighborhood of x.

(2) Let M be a multiplicative monoid, considered as a compact Hausdorff $(\mathbb{L}, 2)$-space; see Corollary V.1.4.4. A subset $A \subseteq M$ is closed when it is closed under the binary operation of M, and A is open in M if

$$\forall x, y \in M\ (x \cdot y \in A \implies x \in A \ \&\ y \in A)\,.$$

Hence, non-empty open sets in M must contain the neutral element, and, consequently, M is connected in $(\mathbb{L}, 2)$-Cat.

When M is commutative, for $A \subseteq M$ to be open means precisely that A is down-closed with respect to the divisibility order:

$$x|z \iff \exists y \in M\ (x \cdot y = z)\,.$$

Hence, the topology of the Alexandroff space ΩM is induced by this order.

Recall that a topological space is Alexandroff if intersections of open sets are open; see Example II.5.10.5. The connected component C_x of a point x in an Alexandroff space X is not only closed, but also open, since

$$C_x = \bigcap\nolimits_{y \in X \setminus C_x} X \setminus C_y \, .$$

We can now show that when T preserves intersections (i.e. multiple pullbacks of sinks of monomorphisms), the values of the functor Ω are always Alexandroff.

V.5.3.4 Proposition *Let $T : \mathsf{Set} \to \mathsf{Set}$ preserve intersections. Then, for every $(\mathbb{T}, \mathcal{V})$-space $X = (X, a)$, the topological space ΩX is Alexandroff and X is the coproduct of its connected components.*

Proof The second assertion follows from the first. To prove the first, consider open sets U_i $(i \in I)$ in ΩX and $x \in U$, $\mathfrak{y} \in TY$ with $a(\mathfrak{y}, x) > \bot$. Then $\mathfrak{y} \in TU_i$ by hypothesis on U_i for all $i \in I$, hence $\mathfrak{y} \in TU$ by hypothesis on T. Consequently $U \hookrightarrow X$ is $(\mathbb{T}, \mathcal{V})$-open. □

V.5.4 Products of connected spaces

Throughout this section, we assume

- *$\mathcal{V}$ is Cartesian closed and integral;*
- *T is taut and $T1 \cong 1$;*
- *$\hat{\mathbb{T}}$ is associative and right-whiskering.*

V.5.4.1 Proposition *Finite products of connected $(\mathbb{T}, \mathcal{V})$-spaces are connected.*

Proof The terminal $(\mathbb{T}, \mathcal{V})$-space $1 = (1, \top)$ is certainly connected. For X, Y connected, it suffices to show that any $(x_0, y_0), (x_1, y_1) \in X \times Y$ lie in the same connected component of $X \times Y$. The additional assumptions on T and $\mathcal{V}$ make $x_0 : 1 \to X$ a morphism, and the split monomorphism $x_0 \times 1_Y : Y \cong 1 \times Y \to X \times Y$ is preserved by Ω, making $\{x_0\} \times Y$ a connected subspace of $\Omega(X \times Y)$, by Theorem V.5.3.1; likewise for $X \times \{y_1\}$. Since $(\{x_0\} \times Y) \cap (X \times \{y_1\}) \neq \varnothing$, both $(x_0, y_0), (x_1, y_1)$ lie in the connected subspace $(\{x_0\} \times Y) \cup (X \times \{y_1\})$ of $\Omega(X \times Y)$, showing that $\Omega(X \times Y)$ is connected. By Theorem V.5.3.1 again, $X \times Y$ is connected in $(\mathbb{T}, \mathcal{V})$-$\mathsf{Cat}$. □

The hypothesis $T1 \cong 1$ is essential for Proposition V.5.4.1 to hold, and so is the restriction to finite products, as the following examples show.

V.5.4.2 Examples

(1) For the list monad $\mathbb{L}$, consider the $(\mathbb{L}, 2)$-space $(X, \vdash)$ with $X = \{x, y\}$ and $\vdash$ the least $(\mathbb{L}, 2)$-structure on X with $(x, y) \vdash x$, and $(Y, \vdash)$ with $Y = \{*\}$ discrete. Since $X \ncong \{x\} + \{y\}$ and $|Y| = 1$, the spaces X and Y are connected, but $X \times Y \cong (\{x\} \times Y) + (\{y\} \times Y)$ is not.

(2) In Ord $\cong$ ($\mathbb{I}$, 2)-Cat, the product of a family of objects in which any two elements have a lower bound is connected (by Corollary V.5.2.4). However, if we let $X_n = \{x_1 \leq x_2 \geq x_3 \leq \ldots x_n\}$ be a "zigzag" of length n, then X_n is connected, but $X = \prod_n X_n$ is not since the sequences $(x_1)_n$ and $(x_n)_n$ lie in distinct components of X. Note that Ω : Ord $\to$ Top is the coreflective embedding that provides an ordered set with its Alexandroff topology, and that the topological space $\prod_n(\Omega X_n)$ is connected (see Corollary V.5.4.4) while ΩX is not. In particular, Ω does not preserve infinite products, but it does preserve finite products.

Here is a criterion for infinite products of connected $(\mathbb{T}, \mathcal{V})$-spaces to be connected. We assume $\hat{T}$ to preserve bottom $\mathcal{V}$-relations.

V.5.4.3 Theorem *For a family $(X_i)_{i\in I}$ of connected $(\mathbb{T}, \mathcal{V})$-spaces, $X = \prod_{i\in I} X_i$ is connected if and only if there is a connected subspace A of X such that*

$$\hat{A} = \{(x_i) \in X \mid \exists (z_i) \in A \ (x_i = z_i \textit{ for all but finitely many } i \in I)\}$$

is Prop$(\mathbb{T}, \mathcal{V})$*-dense in X.*

Proof The condition is trivially necessary since we may choose $A = X$. Conversely, for $z = (z_i)_{i\in I} \in A$ and $F \subseteq I$ finite, there is a morphism $1 \to \prod_{i\in I\setminus F} X_i$ with constant value $(z_i)_{i\in I\setminus F}$ and then a split monomorphism

$$\prod_{i\in F} X_i \to \prod_{i\in F} X_i \times \prod_{i\in I\setminus F} X_i \cong X\,.$$

Since the domain is connected, as in Proposition V.5.4.1, its image

$$F_z = \{(x_i)_{i\in F} \mid \forall i \in I \setminus F \ (x_i = z_i)\}$$

is connected as well, and so is

$$\hat{A} = \bigcup_{z\in A,\, F\subseteq I \text{ finite}} F_z\,.$$

Indeed, for $x \in F_z$, $y \in G_w$ with $z, w \in A$ and $F, G \subseteq I$ finite, the connected components of x and z coincide since F_z is connected, and so do the connected components of y and w, but also of z and w since A is connected. With Corollary V.5.2.3, connectedness of X follows. □

Choosing for A any singleton subset of X, Theorem V.5.4.3 shows in particular:

© **V.5.4.4 Corollary** *The product of connected spaces in* Top $\cong$ (β, 2)-Cat *is connected.*

V.5.4.5 Remark In general, the connected subspace A in Theorem V.5.4.3 may not be chosen to be a singleton set, not even finite. Indeed, in Ord $\cong$ ($\mathbb{I}$, 2)-Cat, the product of countably many copies of the set $\mathbb{Z}$ of integers is connected (see

Example V.5.4.22). For any finite set $A \subseteq \mathbb{Z}^{\mathbb{N}}$ consider a point $x = (x_n)_{n\in\mathbb{N}}$ with $x_m < z_m$ for all $z = (z_n)_{n\in\mathbb{N}} \in A$ and $m \in \mathbb{N}$; since the idempotent grand closure is given by the up-closure, such x cannot lie in the grand closure of $\hat{A}$.

Exercises

V.5.A *Connected categories.* Confirm that connected objects in Cat are characterized as in Exercise II.2.Q.

V.5.B *Coproduct injections are closed embeddings.* Let $\hat{\mathbb{T}}$ and $\mathcal{V}$ be such that the lax extension $\hat{T} : \mathcal{V}\text{-}\mathsf{Rel} \to \mathcal{V}\text{-}\mathsf{Rel}$ preserves bottom $\mathcal{V}$-relations (see Corollary V.5.2.3). Then coproduct injections in $(\mathbb{T}, \mathcal{V})\text{-}\mathsf{RGph}$ are O-initial (for $O : (\mathbb{T}, \mathcal{V})\text{-}\mathsf{RGph} \to \mathsf{Set}$). Moreover, when $\hat{\mathbb{T}}$ is associative, the corresponding statement holds for $(\mathbb{T}, \mathcal{V})\text{-}\mathsf{Cat}$, and every coproduct injection is proper.

V.5.C *Infinite products of connected spaces.* Consider the subspaces $X_n = \{0, n\}$ $(n \in \mathbb{N})$ of $\mathbb{R}$ with their Euclidean metric. Show that the product $\prod_{n\in\mathbb{N}} X_n$ fails to be connected in Met although each X_n is connected.

V.5.D *Total disconnectedness.* A $(\mathbb{T}, \mathcal{V})$-space Y is called *totally disconnected* if each of its connected components has precisely one element. Show that the following statements hold under the hypotheses of Corollary V.5.2.3.

(1) Y is totally disconnected if and only if every $(\mathbb{T}, \mathcal{V})$-continuous map $f : X \to Y$ with X connected is constant.
(2) A $(\mathbb{U}, \mathcal{V})$-space X is connected if and only if every $(\mathbb{T}, \mathcal{V})$-continuous map $f : X \to Y$ with Y totally disconnected is constant.
(3) The full subcategory of totally disconnected $(\mathbb{T}, \mathcal{V})$-spaces is strongly epireflective in $(\mathbb{T}, \mathcal{V})\text{-}\mathsf{Cat}$.
(4) In $\mathsf{Top} \cong (\beta, 2)\text{-}\mathsf{Cat}$, an extremally disconnected T0-space must be totally disconnected, but not conversely. Indiscrete spaces are both connected and extremally disconnected.

V.5.E *Shortcomings of* Ω. While the functor $\Omega : (\mathbb{T}, \mathcal{V})\text{-}\mathsf{Cat} \to \mathsf{Top}$ of Section V.5.3 preserves open maps and open embeddings, it generally fails to preserve embeddings or finite products.

V.5.F *A right adjoint to* Ω. If $\mathcal{V} = 2$, T is taut, and $\hat{\mathbb{T}}$ is associative and right-whiskering, then the functor

$$\Omega : (\mathbb{T}, 2)\text{-}\mathsf{Cat} \to \mathsf{Top}$$

has a right adjoint which assigns to a topological space X the $(\mathbb{T}, 2)$-space $(X, \longrightarrow)$ with

$$\chi \longrightarrow x \iff \forall U \in \mathcal{O}X\ (x \in U \implies \chi \in TU)$$

for all $x \in X$, $\chi \in TX$.

Notes on Chapter V

For concrete categories endowed with a notion of closed subobject, in [Manes, 1974] Manes essentially considers stably closed maps and defines an object X to be compact and Hausdorff if, respectively, $X \longrightarrow 1$ and $X \longrightarrow X \times X$ are stably closed, just as in the axiomatic setting of Section V.4. Furthermore, for a monad $\mathbb{T}$ on **Set** endowed with its Barr extension to **Rel**, he, in essence, considers the category $(\mathbb{T}, 2)$-**Cat** and the notion of closed subobject as in Remark V.3.6.4, and gives a relational characterization of compactness and Hausdorff separation as in our Definition V.1.1.1. He also observes that the stably closed maps are equationally defined, in the same way as proper maps are defined in Section V.3. Briefly, his paper is to be considered an eminent precursor to large parts of Chapter V. There are two remarkable (but not well-known or accessible) Ph.D. theses that greatly extended Manes' ideas in an abstract relational setting, by Kamnitzer [1974a] (written under the direction of G.C.L. Brümmer; see also [Kamnitzer, 1974b]) and Möbus [1981] (written under the direction of H. Schubert; see also [Möbus, 1978]). Specifically, Kamnitzer considers T0, T1, and Hausdorff separation and compactness as used in this chapter, and our definition of R0, R1, regularity, normality, and extremal disconnectedness follows Möbus, who considers these notions in the general relational context of [Klein, 1970] and [Meisen, 1974]. Our treatment of order separation can be traced back to Marny's definition of T0-separation for topological categories [Marny, 1979].

In the particular case of approach or (β, P_+)-spaces, compactness coincides with 0-compactness as developed in [Lowen, 1988, 1997]. A study of low-separation properties in that setting goes back to [Lowen and Sioen, 2003], where order separation is called T0. The R1 property was considered by Robeys [1992], where an approach space satisfying this condition is called complemented. Regularity for approach or (β, P_+)-spaces coincides with the notion considered in [Robeys, 1992] and is further characterized in terms of the tower of the approach space in [Brock and Kent, 1998]. A notion of normality for approach spaces weaker than the one used in Section V.2 was introduced in Van Olmen's thesis [Van Olmen, 2005]; here it appears as item (iii) in Theorem V.2.5.2.

Versions of the Tychonoff Theorem and the Čech–Stone compactification appear in various contexts, including the ones already mentioned (see in particular [Clementino, Giuli, and Tholen, 1996]; [Clementino and Tholen, 1996]; [Lowen, 1997]; [Möbus, 1981]), but its treatment in the general $(\mathbb{T}, \mathcal{V})$-context as given in Proposition V.1.2.1 and Theorem V.1.2.3 draws heavily on Proposition III.1.2.1, which first appeared in [Clementino and Hofmann, 2009].

The equational definition of proper morphism as presented in $(\mathbb{T}, \mathcal{V})$-**Cat** in Section V.3 may be traced back to [Manes, 1974], called perfect by him and strongly closed in [Kamnitzer, 1974a], while we maintain Bourbaki's terminology [Bourbaki, 1989]. The corresponding definition of open morphism as used here appears first in [Möbus, 1981]. Generalizations of the Kuratowski–Mrówka Theorem, named after Kuratowski [1931] (who proved that product projections along a compact space are closed) and Mrówka [1959] (who showed that Kuratowski's property characterizes compactness) have been considered by various authors, notably in the context of closure operators by Dikranjan and Giuli [1989] and for approach spaces in [Colebunders, Lowen, and Wuyts, 2005]. There is also a fairly general version of the Kuratowski–Mrówka theorem in a different setting in Hofmann's fundamental article [Hofmann, 2007], while the construction used in Theorem V.3.4.1 relies on Proposition III.4.9.1 that draws on [Clementino and Hofmann, 2012]. The other crucial ingredient in the characterization of proper maps, Theorem V.3.2.5, appeared only recently in [Clementino and Tholen, 2013], following which Solovyov [2013] proposed the notion of closed maps as given in Definition V.3.3.6. The characterization of proper maps of approach spaces as stably closed maps or as closed maps with 0-compact fibers appears in [Colebunders *et al.*, 2005]. Open maps of convergence spaces are described in [Kent and Richardson, 1973]; for approach spaces, openness is introduced in [Lowen and Verbeeck, 1998, 2003] in terms of the associated distance, a notion coinciding with that of inversely closed maps for (β, P_+)-spaces.

Theorem V.3.5.6 on the product stability of proper maps (which entails the Tychonoff Theorem) originates with Schubert's thesis [Schubert, 2006], who also observed the crucial role of complete distributivity as in Proposition V.3.5.4. Conditions for the openness of coproduct injections are considered in [Möbus, 1981]. Our presentation of stability of openness under coproducts relies on [Mahmoudi, Schubert, and Tholen, 2006].

The first axiomatic categorical treatment of compactness and Hausdorff separation depending on a parameter $\mathcal{P}$ as in Section V.4 was given by Penon [1972]; in fact, the axioms we have imposed on a topology $\mathcal{P}$ may be considered as a finitary version of Penon's axioms, except that he does not require closure under composition. But, as emphasized in [Tholen, 1999], closure under composition is essential to exhibit fully the beautiful interplay of the two notions. A slightly different axiomatic approach, starting with a notion of closed map from which proper is derived as stably closed, is presented in [Clementino, Giuli, and Tholen, 2004a], where further topological themes like local compactness and exponentiability are being pursued in greater depth. Of course, there is a rich supply of articles recognizing and axiomatizing the key role of the class of proper or perfect maps, including [Herrlich, 1974]; [Herrlich, Salicrup, and Strecker, 1987]; [Manes, 1974], which, once a definite notion of categorical closure operator had been introduced by Dikranjan and Giuli [1987], led to many investigations of compactness and separation in that context; see in particular [Clementino *et al.*, 1996]; [Clementino and Tholen, 1996]; [Dikranjan and Giuli, 1989]. The term topology on a category, as in Section V.4, is adopted in [Schubert, 2006], and published in [Hofmann and Tholen, 2012].

Extensive categories (in the finitary sense) were studied by Carboni, Lack, and Walters [1993]. Their elegant definition as given in Section V.5 is due to Steve Schanuel, and their characterization as given in Theorem V.5.1.5 draws also on [Börger, 1994], an extended preprint of which appeared as [Börger, 1987b]. Connected objects as defined in Section V.5.2 appeared in Hoffmann's thesis [Hoffmann, 1972], and their characterization as given in Theorem V.5.2.1 draws on [Janelidze, 2004]. Theorem V.5.3.1 (for $\mathcal{V} = 2$) and its consequences are due to Clementino, Hofmann, and Montoli [2013].

Suggestions for further reading: [Höhle, 2001], [Manes, 2002, 2007, 2010]; [Jäger, 2012].

Bibliography

Abramsky, S. and A. Jung. Domain theory. In S. Abramsky, D.M. Gabbay, and T.S.E. Maibaurn, eds., *Handbook of Logic in Computer Science*. Oxford University Press, Oxford, 1994, pp. 1–168.

Adámek, J., H. Herrlich, and G.E. Strecker. *Abstract and Concrete Categories: The Joy of Cats*. Wiley, New York, 1990. Republished as *Repr. Theory Appl. Categ.*, **17** (2006).

Adámek, J., F.W. Lawvere, and J. Rosický. Continuous categories revisited. *Theory Appl. Categ.*, **11**:252–282, 2003.

Adámek, J., J. Rosický, and E.M. Vitale. *Algebraic Theories: A Categorical Introduction to General Algebra*. Cambridge University Press, Cambridge, 2011. With a foreword by F.W. Lawvere.

Akhvlediani, A. Hausdorff and Gromov distances in quantale enriched categories. MA thesis, York University, Toronto, 2008.

Akhvlediani, A., M.M. Clementino, and W. Tholen. On the categorical meaning of Hausdorff and Gromov distances, I. *Topology Appl.*, **157**(8):1275–1295, 2010.

Awodey, S. *Category Theory*. Clarendon, New York, 2006.

Banaschewski, B. Essential extensions of T_0-spaces. *Gen. Topol. Appl.*, **7**:233–246, 1977.

Banaschewski, B. and G. Bruns. Categorical characterization of the MacNeille completion. *Arch. Math. (Basel)*, **18**:369–377, 1967.

Banaschewski, B., R. Lowen, and C. Van Olmen. Sober approach spaces. *Topol. Appl.*, **153**:3059–3070, 2006.

Barr, M. Relational algebras. *Lect. Notes Math.*, **137**:39–55, 1970.

Barr, M. and C. Wells. *Toposes, Triples and Theories*. Springer, New York, 1985. Republished as *Repr. Theory Appl. Categ.*, **12** (2005).

Beck, J. Distributive laws. *Lect. Notes Math.*, **80**:119–140, 1969. Republished as *Repr. Theory Appl. Categ.*, **18** (2008).

Bénabou, J. Catégories avec multiplication. *C.R. Acad. Sci. Paris*, **256**:1887–1890, 1963.

Bénabou, J. Fibrations petites et localement petites. *C.R. Acad. Sci. Paris Sér. A*, **281**:897–900, 1975.

Bénabou, J. Fibered categories and the foundations of naive category theory. *J. Symb. Log.*, **50**:10–37, 1985.

Bénabou, J. Distributors at work. `http://www.mathematik.tu-darmstadt.de/~streicher/`, 2000. Lecture notes by T. Streicher.

Bentley, H.L., H. Herrlich, and R. Lowen. Improving constructions in topology. In H. Herrlich and H.-E. Porst, eds., *Category Theory at Work*. Heldermann, Berlin, 1991, pp. 3–20.

Betti, R., A. Carboni, R. Street, and R. Walters. Variation through enrichment. *J. Pure Appl. Algebra*, **29**:109–127, 1983.

Birkedal, L., K. Støvring, and J. Thamsborg. The category-theoretic solution of recursive metric-space equations. *Theoret. Comput. Sci.*, **411**:4102–4122, 2010.

Birkhoff, G. A new definition of limit. *Bull. Amer. Math. Soc.*, **41**:636, 1935.

Birkhoff, G. Moore–Smith convergence in general topology. *Ann. of Math. (2)*, **38**:39–56, 1937.

Birkhoff, G. *Lattice Theory*, 3rd edn. American Mathematical Society, Providence, RI, 1979.

Bonsangue, M.M., F. van Breugel, and J.J.M.M. Rutten. Generalized metric spaces: completion, topology, and powerdomains via the Yoneda embedding. *Theoret. Comput. Sci.*, **193**:1–51, 1998.

Borceux, F. *Handbook of Categorical Algebra 1. Basic Category Theory*. Cambridge University Press, Cambridge, 1994a.

Borceux, F. *Handbook of Categorical Algebra 2. Categories and Structures*. Cambridge University Press, Cambridge, 1994b.

Borceux, F. *Handbook of Categorical Algebra 3. Categories of Sheaves*. Cambridge University Press, Cambridge, 1994c.

Börger, R. Coproducts and ultrafilters. *J. Pure Appl. Algebra*, **46**:35–47, 1987a.

Börger, R. Disjoint and universal coproducts, I, II. Seminarberichte, FernUniversität, Hagen, 1987b.

Börger, R. Disjointness and related properties of coproducts. *Acta Univ. Carolin. Math. Phys.*, **35**:43–63, 1994.

Börger, R. and W. Tholen. Cantors Diagonalprinzip für Kategorien. *Math. Z.*, **160**:135–138, 1978.

Bourbaki, N. *Éléments de Mathématique. Les Structures Fondamentales de l'Analyse, III: Topologie Générale, III. Groupes Topologiques (Théorie Élémentaire) IV. Nombres Réels*. Hermann, Paris, 1942.

Bourbaki, N. *General Topology, Chapters 1–4*. Springer, Berlin, 1989.

Brock, P. and D.C. Kent. Approach spaces, limit tower spaces, and probabilistic convergence spaces. *Appl. Categ. Structures*, **5**:99–110, 1997.

Brock, P. and D.C. Kent. On convergence approach spaces. *Appl. Categ. Structures*, **6**:117–125, 1998.

Bunge, M. Coherent extensions and relational algebras. *Trans. Amer. Math. Soc.*, **197**:355–390, 1974.

Burroni, A. T-catégories (catégories dans un triple). *Cahiers Topol. Géom. Différent.*, **12**: 215–321, 1971.

Carboni, A., G.M. Kelly, and R.J. Wood. A 2-categorical approach to change of base and geometric morphisms, I. *Cahiers Topologie Géom. Différentielle*, **32**: 47–95, 1991.

Carboni, A., S. Lack, and R. Walters. Introduction to extensive and distributive categories. *J. Pure Appl. Algebra*, **84**:145–158, 1993.

Cartan, H. Filtres et ultrafiltres. *C.R. Acad. Sci. Paris*, **205**:777–779, 1937a.

Cartan, H. Théorie des filtres. *C.R. Acad. Sci. Paris*, **205**:595–598, 1937b.

Čech, E. On bicompact spaces. *Ann. of Math. (2)*, **38**:823–844, 1937.

Chai, Y.-M. A note on the probabilistic quasi-metric spaces. *J. Sichuan Univ. (Natur. Sci. Ed.)*, **46**:543–547, 2009.

Choquet, G. Convergences. *Ann. Univ. Grenoble. Sect. Sci. Math. Phys. (N.S.)*, **23**:57–112, 1948.

Clementino, M.M., E. Giuli, and W. Tholen. Topology in a category: compactness. *Portugal. Math.*, **53**:397–433, 1996.

Clementino, M.M., E. Giuli, and W. Tholen. A functional approach to general topology. In M.-C. Pedicchio and W. Tholen, eds., *Categorical Foundations*. Cambridge University Press, Cambridge, 2004a, pp. 103–164.

Clementino, M.M. and D. Hofmann. Triquotient maps via ultrafilter convergence. *Proc. Amer. Math. Soc.*, **130**:3423–3431, 2002.

Clementino, M.M. and D. Hofmann. Topological features of lax algebras. *Appl. Categ. Structures*, **11**:267–286, 2003.

Clementino, M.M. and D. Hofmann. Effective descent morphisms in categories of lax algebras. *Appl. Categ. Structures*, **12**:413–425, 2004a.

Clementino, M.M. and D. Hofmann. On extensions of lax monads. *Theory Appl. Categ.*, **13**:41–60, 2004b.

Clementino, M.M. and D. Hofmann. Exponentiation in V-categories. *Topology Appl.*, **153**:3113–3128, 2006.

Clementino, M.M. and D. Hofmann. On some special classes of continuous maps. In E. Pearl, ed., *Open Problems in Topology, II*. Elsevier, Amsterdam, 2007, pp. 367–376.

Clementino, M.M. and D. Hofmann. Relative injectivity as cocompleteness for a class of distributors. *Theory Appl. Categ.*, **21**:210–230, 2008.

Clementino, M.M. and D. Hofmann. Lawvere completeness in topology. *Appl. Categ. Structures*, **17**:175–210, 2009.

Clementino, M.M. and D. Hofmann. Descent morphisms and a van Kampen theorem in categories of lax algebras. *Topology Appl.*, **159**:2310–2319, 2012.

Clementino, M.M., D. Hofmann, and G. Janelidze. Local homeomorphisms via ultrafilter convergence. *Proc. Amer. Math. Soc.*, **133**:917–922, 2005.

Clementino, M.M., D. Hofmann, and A. Montoli. Covering morphisms in categories of relational algebras. *Appl. Categ. Structures*, 2013, in press; DOI: 10.1007/s10485-013-9349-0.

Clementino, M.M., D. Hofmann, and W. Tholen. The convergence approach to exponentiable maps. *Math. Portugal.*, **60**:139–160, 2003a.

Clementino, M.M., D. Hofmann, and W. Tholen. Exponentiability in categories of lax algebras. *Theory Appl. Categ.*, **11**:337–352, 2003b.

Clementino, M.M., D. Hofmann, and W. Tholen. One setting for all: metric, topology, uniformity, approach structure. *Appl. Categ. Structures*, **12**:127–154, 2004b.

Clementino, M.M. and W. Tholen. Tychonoff's theorem in a category. *Proc. Amer. Math. Soc.*, **124**:3311–3314, 1996.

Clementino, M.M. and W. Tholen. Metric, topology and multicategory – a common approach. *J. Pure Appl. Algebra*, **179**:13–47, 2003.

Clementino, M.M. and W. Tholen. Proper maps for lax algebras and the Kuratowski–Mrówka theorem. *Theory Appl. Categ.*, **27**:327–346, 2013.

Colebunders, E. and R. Lowen. A quasitopos containing CONV and MET as full subcategories. *Internat. J. Math. Math. Sci.*, **11**:417–438, 1988.

Colebunders, E. and R. Lowen. Topological quasitopos hulls of categories containing topological and metric objects. *Cah. Topol. Géom. Différ. Catég.*, **30**:213–227, 1989.

Colebunders, E. and R. Lowen. Metrically generated theories. *Proc. Amer. Math. Soc.*, **133**:1547–1556, 2005.

Colebunders, E., R. Lowen, and W. Rosiers. Lax algebras via initial monad morphisms: APP, TOP, MET and ORD. *Topology Appl.*, **158**:882–903, 2011.

Colebunders, E., R. Lowen, and P. Wuyts. A Kuratowski-Mrówka theorem in approach theory. *Topology Appl.*, **153**:756–766, 2005.

Colebunders, E. and G. Richter. An elementary approach to exponential spaces. *Appl. Categ. Structures*, **9**:303–310, 2001.

Cook, C.H. and H.R. Fischer. Regular convergence spaces. *Math. Ann.*, **174**:1–7, 1967.

Cruttwell, G.S.H. Normed Spaces and the Change of Base for Enriched Categories. Ph.D. thesis, Dalhousie University, Halifax, 2008.

Cruttwell, G.S.H. and M.A. Shulman. A unified framework for generalized multicategories. *Theory Appl. Categ.*, **24**:580–655, 2010.

Davey, B.A. and H.A. Priestley. *Introduction to Lattices and Order*. Cambridge University Press, Cambridge, 1990.

Day, A. Filter monads, continuous lattices and closure systems. *Canad. J. Math.*, **27**:50–59, 1975.

Day, B.J. and G.M. Kelly. On topologically quotient maps preserved by pullbacks or products. *Proc. Cambridge Philos. Soc.*, **67**:553–558, 1970.

de Groot, J. An isomorphism principle in general topology. *Bull. Amer. Math. Soc.*, **73**:465–467, 1967.

de Groot, J., G.E. Strecker, and E. Wattel. The compactness operator in general topology. In *General Topology and its Relations to Modern Analysis and Algebra, II*. Academia, Prague, 1967, pp. 161–163.

Dikranjan, D. and E. Giuli. Closure operators, I. *Topology Appl.*, **27**:129–143, 1987.

Dikranjan, D. and E. Giuli. Compactness, minimality and closedness with respect to a closure operator. In J. Adámek and S. MacLane, eds., *Categorical Topology and its Relation to Analysis, Algebra and Combinatorics*, World Scientific Publishing, Teaneck, 1989, pp. 284–296.

Dikranjan, D. and W. Tholen. *Categorical Structure of Closure Operators: With Applications to Topology, Algebra and Discrete Mathematics*. Kluwer, Dordrecht, 1995.

Duskin, J. Variations on Beck's tripleability criterion. *Lect. Notes Math.*, **106**:74–129, 1969.

Dyckhoff, R. Total reflections, partial products, and hereditary factorizations. *Topology Appl.*, **17**:101–113, 1984.

Dyckhoff, R. and W. Tholen. Exponentiable morphisms, partial products and pullback complements. *J. Pure Appl. Algebra*, **49**:103–116, 1987.

Eilenberg, S. and G.M. Kelly. Closed categories. In *Proc. Conf. Categorical Algebra*. Springer, New York, 1966, pp. 421–562.

Eilenberg, S. and J.C. Moore. Adjoint functors and triples. *Illinois J. Math.*, **9**:381–398, 1965.

Engelking, R. *General Topology*, 2nd edn. Heldermann, Berlin, 1989.

Erné, M. *Vollständig Distributive Topologien und Idempotente Relationen*. Deutsche Mathematiker-Vereinigung, Dortmund, 1980.

Erné, M. Scott convergence and Scott topology in partially ordered sets, II. In B. Bemaschewski and R.-E. Hoffmann, eds., *Continuous Lattices*, Lecture Notes in Mathematics, 871. Springer, Berlin, 1981, pp. 61–96.

Erné, M. The ABC of order and topology. In H. Herrlich and H.-E. Porst, eds., *Category Theory at Work*. Heldermann, Berlin, 1991, pp. 57–83.

Erné, M. $\mathcal{Z}$-continuous posets and their topological manifestation. *Appl. Categ. Structures*, **7**:31–70, 1999.

Erné, M. and G. Wilke. Standard completions for quasiordered sets. *Semigroup Forum*, **27**:351–376, 1983.

Ershov, Y.L. Theory of domains and nearby. In D. Bjørner, M. Broy, and I.V. Pottosin, eds., *Formal Methods in Programming and their Applications*. Lecture Notes in Computer Science 735. Springer, Berlin, 1993, pp. 1–7.

Escardó, M.H. Properly injective spaces and function spaces. *Topology Appl.*, **89**:75–120, 1998.

Fawcett, B. and R. Wood. Constructive complete distributivity, I. *Math. Proc. Camb. Phil. Soc.*, **107**:81–89, 1990.

Feferman, S. Set-theoretical foundations of category theory. In *Reports of the Midwest Category Seminar III*, Lecture Notes in Mathematics 106. Springer, Berlin, 1969, pp. 106–201.

Feferman, S. Categorical foundations and foundations of category theory. In R.E. Butts and J. Hintikka, eds., *Logic, Foundations of Mathematics, and Computability Theory*, Proc. 5th Int. Congr., London, Ontario, 1975, Part 1, 1977, pp. 149–169. *Philos. Sci.*, **9**:149–169, 1977.

Flagg, R.C. Completeness in continuity spaces. In R.A.G. Seely, ed., *Category Theory 1991*. AMS, Providence, RI, 1992, pp. 183–200.

Flagg, R.C. Quantales and continuity spaces. *Algebra Universalis*, **37**:257–276, 1997.

Flagg, R.C. and R. Kopperman. Continuity spaces: reconciling domains and metric spaces. *Theoret. Comput. Sci.*, **177**:111–138, 1997.

Flagg, R.C., P. Sünderhauf, and K. Wagner. A logical approach to quantitative domain theory. Topology Atlas, Preprint 23, http://at.yorku.ca/e/a/p/p/23.htm, 1996.

Fréchet, M. Généralisation d'un théorème de Weierstrass. *C.R. Acad. Sci. Paris*, **139**:848–850, 1905.

Fréchet, M. Sur quelques points du calcul fonctionnel. *Rend. Circ. Mat. Palermo*, **22**:1–74, 1906.

Fréchet, M. Sur divers modes de convergence d'une suite de fonctions d'une variable. *Bull. Calcutta Math. Soc.*, **11**:187–206, 1921.

Freyd, P.J. and A. Scedrov. *Categories, Allegories*. North-Holland, Amsterdam, 1990.

Gabriel, P. and F. Ulmer. *Lokal präsentierbare Kategorien*. Lecture Notes in Mathematics 221. Springer, Berlin, 1971.

Gähler, W. Monads and convergence. In *Proc. Conf. Generalized Functions, Convergence Structures and Their Applications*, Dubrovnik, Yugoslavia, 1987. Springer, New York, 1988, pp. 29–46.

Gähler, W. (ed.) *et al. Recent Developments of General Topology and its Applications. Int. Conf. in Memory of Felix Hausdorff (1868–1942)*, Berlin, Germany, March 22–28, 1992. Akademie Verlag. Math. Res. 67, Berlin, 1992, pp. 136–149.

Gerlo, A., E. Vandersmissen, and C. Van Olmen. Sober approach spaces are firmly reflective for the class of epimorphic embeddings. *Appl. Categ. Structures*, **14**:251–258, 2006.

Gierz, G., K.H. Hofmann, K. Keimel, J.D. Lawson, M.W. Mislove, and D.S. Scott. *A Compendium of Continuous Lattices*. Springer, Berlin, 1980.

Gierz, G., K.H. Hofmann, K. Keimel, J.D. Lawson, M.W. Mislove, and D.S. Scott. *Continuous Lattices and Domains*. Cambridge University Press, Cambridge, 2003.

Godement, R. *Topologie Algébrique et Théorie des Faisceaux*. Hermann, Paris, 1958.

Grimeisen, G. Gefilterte Summation von Filtern und iterierte Grenzprozesse, I. *Math. Ann.*, **141**:318–342, 1960.

Grimeisen, G. Gefilterte Summation von Filtern und iterierte Grenzprozesse, II. *Math. Ann.*, **144**:386–417, 1961.

Grothendieck, A., J.-L. Verdier, and P. Deligne. Conditions de finitude, topos et sites fibrés. Applications aux questions de passage à la limite. In *Séminaire de Géométrie Algébrique du Bois Marie 1963/64*, SGA 4, no. 6. Lecture Notes in Mathematics 270. Springer, Berlin, 1972, pp. 163–340.

Gutierres, G. and D. Hofmann. Axioms for sequential convergence. *Appl. Categ. Structures*, **15**:599–614, 2007.

Gutierres, G. and D. Hofmann. Approaching metric domains. *Appl. Categ. Structures*, **21**(6):617–650, 2013.

Hausdorff, F. *Grundzüge der Mengenlehre*. Veit, Leipzig, 1914.

Hermida, C. Representable multicategories. *Adv. Math.*, **151**:164–225, 2000.

Hermida, C. From coherent structures to universal properties. *J. Pure Appl. Algebra*, **165**:7–61, 2001.

Herrlich, H. Perfect subcategories and factorizations. In *Topics in Topol., Colloqu. Keszthely 1972*. Colloquia Math. Soc. János Bolyai 8, 1974, pp. 387–403, 1974.

Herrlich, H. *Topologie, I: Topologische Räume*. Heldermann, Berlin, 1986.

Herrlich, H. On the representability of partial morphisms in Top and in related constructs. In *Proc. 1st Conf. Categorical Algebra and its Applications*, Louvain-la-Neuve, Belgium, 1987. Lecture Notes in Mathematics 1348. Springer, Berlin, 1988a, pp. 143–153.

Herrlich, H. *Topologie, II: Uniforme Räume*. Heldermann, Berlin, 1988b.

Herrlich, H. *Axiom of Choice*. Lecture Notes in Mathematics 1876. Springer, Berlin, 2006.

Herrlich, H., E. Colebunders, and F. Schwarz. Improving Top: PrTop and PsTop. In H. Herrlich and H.-E. Porst, eds., *Category Theory at Work*. Heldermann, Berlin, 1991, pp. 21–34.

Herrlich, H. and M. Hušek. Galois connections categorically. *J. Pure Appl. Algebra*, **68**:165–180, 1990.

Herrlich, H., G. Salicrup, and G.E. Strecker. Factorizations, denseness, separation, and relatively compact objects. *Topology Appl.*, **27**:157–169, 1987.

Hochster, M. Prime ideal structure in commutative rings. *Trans. Amer. Math. Soc.*, **142**:43–60, 1969.

Höhle, U. M-valued sets and sheaves over integral commutative CL-monoids. In S.E. Rodabaugh, E.P. Klement, and U. Höhle, eds., *Applications of Category Theory to Fuzzy Subsets*. Theory and Decision Library vol. 14. Kluwer, Dordrecht, 1992, pp. 33–72.

Höhle, U. *Many-Valued Topology and its Applications*. Kluwer, Boston, 2001.

Hoffmann, R.-E. *Die kategorielle Auffassung der Initial- und Finaltopologie*. Ph.D. thesis, Ruhr-Universität, Bochum, 1972.

Hoffmann, R.-E. Projective sober spaces. In *Structure of Topological Categories*, Math.-Arbeitspapiere, vol. 18. University of Bremen, Bremen, 1979, pp. 109–153.

Hoffmann, R.-E. Continuous posets, prime spectra of completely distributive complete lattices, and Hausdorff compactifications. B. Bernaschewski and R.-E. Hoffmann, eds., *Proc. Conf. Continuous Lattices*, Bremen, 1979. Lecture Notes in Mathematics 871. Berlin, Springer, 1981, pp. 61–96.

Hofmann, D. An algebraic description of regular epimorphisms in topology. *J. Pure Appl. Algebra*, **199**:71–86, 2005.

Hofmann, D. Exponentiation for unitary structures. *Topology Appl.*, **153**:3180–3202, 2006.

Hofmann, D. Topological theories and closed objects. *Adv. Math.*, **215**:789–824, 2007.

Hofmann, D. and C.D. Reis. Probabilistic metric spaces as enriched categories. *Fuzzy Set. Syst.*, **210**:1–21, 2013.

Hofmann, D. and W. Tholen. Kleisli compositions for topological spaces. *Topology Appl.*, **153**:2952–2961, 2006.

Hofmann, D. and W. Tholen. Lax algebra meets topology. *Topology Appl.*, **159**:2434–2452, 2012.

Hofmann, D. and P. Waszkiewicz. Approximation in quantale-enriched categories. *Topology Appl.*, **158**:963–977, 2011.

Huber, P.J. Homotopy theory in general categories. *Math. Ann.*, **144**:361–385, 1961.

Isbell, J.R. Six theorems about injective metric spaces. *Comment. Math. Helv.*, **39**:65–76, 1964.

Isbell, J.R. Atomless parts of spaces. *Math. Scand.*, **31**:5–32, 1972.

Isbell, J.R. General function spaces, products and continuous lattices. *Math. Proc. Camb. Philos. Soc.*, **100**:193–205, 1986.

Jäger, G. A one-point compactification for lattice-valued convergence spaces. *Fuzzy Set. Syst.*, **190**:21–31, 2012.

James, I.M. *Fibrewise Topology*. Cambridge University Press, Cambridge, 1989.

Janelidze, G. Categorical Galois theory: revision and some recent developments. In K. Denecke *et al.*, eds., *Galois Connections and Applications*. Mathematics and its Applications (Dordrecht) 565. Kluwer, Dordrecht, 2004, pp. 139–171.

Jayewardene, R. and O. Wyler. Categories of relations and functional relations. *Appl. Categ. Structures*, **9**:279–305, 2000.

Johnstone, P.T. *Topos Theory*. Academic Press, London, 1977.

a topological topos. *Proc. London Math. Soc. (3)*, **38**:237–271, 1979.

ne Spaces. Cambridge University Press, Cambridge, 1982.

Johnstone, P.T. *Sketches of an Elephant: A Topos Theory Compendium 1*. Clarendon, New York, 2002a.

Johnstone, P.T. *Sketches of an Elephant: A Topos Theory Compendium 2*. Clarendon, New York, 2002b.

Jung, A. Stably compact spaces and the probabilistic powerspace construction. *Electron. Notes Theoret Comput. Sci.*, **87**:5–20, 2004.

Kamnitzer, S.H. *Protoreflections, Relational Algebras and Topology*. Ph.D. thesis, University of Cape Town, Cape Town, 1974a.

Kamnitzer, S.H. Topological aspects of relational algebras. *Math. Colloq. Univ. Cape Town*, **9**:23–59, 1974b.

Kelley, J.L. Convergence in topology. *Duke Math. J.*, **17**:277–283, 1950.

Kelly, G.M. *Basic Concepts of Enriched Category Theory*. Cambridge University Press, Cambridge, 1982. Republished as *Repr. Theory Appl. Categ.* **10** (2005).

Kelly, G.M. A note on relations relative to a factorization system. In *Proc. Int. Conf. Category Theory*, Como, Italy, 1990. Lecture Notes in Mathematics 1488. Springer, Berlin, 1991, pp. 249–261.

Kent, D.C. and W.K. Min. Neighborhood spaces. *Int. J. Math. Math. Sci.*, **32**:387–399, 2002.

Kent, D.C. and G.D. Richardson. Open and proper maps between convergence spaces. *Czech. Math. J.*, **23**:15–23, 1973.

Klein, A. Relations in categories. *Illinois J. Math.*, **14**:536–550, 1970.

Kock, A. Monads for which structures are adjoint to units. *J. Pure Appl. Algebra*, **104**:41–59, 1995.

Kopperman, R. All topologies come from generalized metrics. *Amer. Math. Month.*, **95**:89–97, 1988.

Kostanek, M. and P. Waszkiewicz. The formal ball model for $\mathcal{Q}$-categories. *Math. Structures Comput. Sci.*, **21**:41–64, 2011.

Kowalsky, H.-J. Beiträge zur topologischen Algebra. *Math. Nachr.*, **11**:143–185, 1954a.

Kowalsky, H.-J. Limesräume und Komplettierung. *Math. Nachr.*, **12**:301–340, 1954b.

Kreyszig, E. Interaction between general topology and functional analysis. In *Handbook of the History of General Topology*, vol. 1. Kluwer, Dordrecht, 1997, pp. 357–390.

Kruml, D. and J. Paseka. Algebraic and categorical aspects of quantales. In M. Hazewinkel, ed., *Handbook of Algebra*, vol. 5. Elsevier, Amsterdam, 2008, pp. 323–362.

Kuratowski, C. Évaluation de la classe borélienne ou projective d'un ensemble de points à l'aide des symboles logiques. *Fundam. Math.*, **17**:249–272, 1931.

Lack, S. and R. Street. The formal theory of monads, II. *J. Pure Appl. Algebra*, **175**:243–265, 2002.

Lambek, J. Deductive systems and categories, II. Standard constructions and closed categories. In *Proc. Conf. Category Theory, Homology Theory and their Applications*, Battelle Memorial Institute Seattle Research Center, 1968. Lecture Notes in Mathematics 86. Springer, Berlin, 1969, pp. 76–122.

Lawson, J. The duality of continuous posets. *Houston J. Math.*, **5**:357–386, 1979.

Lawson, J. The round ideal completion via sobrification. *Topology Proc.*, **22**:261–274, 1997.

Lawson, J. Stably compact spaces. *Math. Structures Comput. Sci.*, **21**:125–169, 2011.

Lawvere, F.W. The category of categories as a foundation for mathematics. In *Proc. Conf. Categorical Algebra*, Springer, New York, 1966, pp. 1–20.

Lawvere, F.W. Metric spaces, generalized logic, and closed categories. *Rend. Sem. Mat. Fis. Milano*, **43**:135–166, 1973. Republished as *Repr. Theory Appl. Categ.* **1** (2002).

Lawvere, F.W. and R. Rosebrugh. *Sets for Mathematics*. Cambridge University Press, Cambridge, 2003.

Lowen, R. Kuratowski's measure of noncompactness revisited. *Quart. J. Math. Oxford Ser. 2*, **39**:235–254, 1988.

Lowen, R. Approach spaces: a common supercategory of TOP and MET. *Math. Nachr.*, **141**:183–226, 1989.

Lowen, R. *Approach Spaces: The Missing Link in the Topology–Uniformity–Metric Triad.* Oxford University Press, Oxford, 1997.

Lowen, R. A survey of some categorical aspects of approach theory. In H. Herrlich and H.-E. Porst, eds., *Categorical Methods in Algebra and Topology*. University of Bremen, Bremen, 2000, pp. 267–277.

Lowen, R. *Index Calculus: Approach Theory at Work*. Springer, Berlin, 2013.

Lowen, R. and M. Sioen. A note on separation in AP. *Appl. Gen. Topol.*, **4**:475–486, 2003.

Lowen R., C. Van Olmen, and T. Vroegrijk. Functional ideals and topological theories. *Houston J. Math.*, **34**:1065–1089, 2008.

Lowen, R. and C. Verbeeck. Local compactness in approach spaces, I. *Int. J. Math. Math. Sci.*, **21**:429–438, 1998.

Lowen, R. and C. Verbeeck. Local compactness in approach spaces, II. *Int. J. Math. Math. Sci.*, **2**:109–117, 2003.

Lowen, R. and T. Vroegrijk. A new lax algebraic characterization of approach spaces. *Quad. Mat.*, **22**:199–232, 2008.

Lucyshyn-Wright, R.B.B. Monoidal domain-theoretic topology. Master's thesis, York University, Toronto, 2009.

Lucyshyn-Wright, R.B.B. Domains occur among spaces as strict algebras among lax. *Math. Structures Comput. Sci.*, **21**:647–670, 2011.

Mac Lane, S. Natural associativity and commutativity. *Rice Univ. Studies*, **49**:28–46, 1963.

Mac Lane, S. One universe as a foundation for category theory. In *Reports of the Midwest Category Seminar III*, Lecture Notes in Mathematics 106. Springer, Berlin, 1969, pp. 192–200.

Mac Lane, S. *Categories for the Working Mathematician*. Springer, New York, 1971; 2nd edn. 1998.

Mac Lane, S. and I. Moerdijk. *Sheaves in Geometry and Logic: A First Introduction to Topos Theory*. Springer, New York, 1994.

MacDonald, J. and M. Sobral. Aspects of monads. In C. Pedicchio and W. Tholen, eds., *Categorical Foundations*. Cambridge University Press, Cambridge, 2004, pp. 213–268.

Machado, A. Espaces d'Antoine et pseudo-topologies. *Cah. Topol. Géom. Différ. Catég.*, **14**:309–327, 1973.

Machner, J. T-algebras of the monad L-Fuzz. *Czech. Math. J.*, **35**(110):515–528, 1985.

McShane, E.J. Partial orderings and Moore–Smith limits. *Am. Math. Mon.*, **59**:1–11, 1952.

Mahmoudi, M., C. Schubert, and W. Tholen. Universality of coproducts in categories of lax algebras. *Appl. Categ. Structures*, **14**:243–249, 2006.

Manes, E.G. *A Triple Miscellany*. Ph.D. thesis, Wesleyan University, Middletown, 1967.

Manes, E.G. A triple theoretic construction of compact algebras. In *Sem. Triples and Categorical Homology Theory*, ETH, Zurich, 1966/67. Lecture Notes in Mathematics 80. Springer, Berlin, 1969, pp. 91–118.

Manes, E.G. Compact Hausdorff objects. *Gen. Topology Appl.*, **4**:341–360, 1974.

Manes, E.G. *Algebraic Theories*. Springer, New York, 1976.

Manes, E.G. Taut monads and $T0$-spaces. *Theor. Comp. Sci.*, **275**:79–109, 2002.

Manes, E.G. Taut monads, dynamical logic and determinism. *Electron. Notes Theoret. Comput. Sci.*, **173**:241–262, 2007.

Manes, E.G. Monads in topology. *Topology Appl.*, **157**:961–989, 2010.

Manes, E.G. and P. Mulry. Monad compositions, I: General constructions and recursive distributive laws. *Theory Appl. Categ.*, **18**:172–208, 2007.

Marmolejo, F., R. Rosebrugh, and R. Wood. A basic distributive law. *J. Pure Appl. Algebra*, **168**:209–226, 2002.

Marny, T. On epireflective subcategories of topological categories. *Gen. Topology Appl.*, **10**:175–181, 1979.

Meisen, J. Relations in regular categories. In *Symp. Localized Group Theory Homotopy Theory and Related Topics*, Battelle Memorial Institute Seattle Research Center. Lecture Notes in Mathematics 418. Springer, Berlin, 1974, pp. 196–202.

Menger, K. Statistical metrics. *Proc. Nat. Acad. Sci. USA*, **28**:535–537, 1942.

Möbus, A. (E, M)-Relationalalgebren und T_{3a}-Objekte. In *Nordwestdeutsches Kategorienseminar 1977*. Universität Bielefeld, Bielefeld, 1978, pp. 163–194.

Möbus, A. *Relational-Algebren*. Ph.D. thesis, Universität Düsseldorf, Düsseldorf, 1981.

Möbus, A. Wallman compactification of T_4-relational algebras and Mal'cev monads. In D. Pumpliin and W. Tholen, eds., *Seminarberichte*, vol. 16. Fernuniversität, Hagen, 1982, pp. 111–132.

Möbus, A. Alexandrov compactification of relational algebras. *Arch. Math. (Basel)*, **40**:526–537, 1983.

Moore, E.H. Definition of limit in general integral analysis. *Nat. Acad. Proc.*, **1**:628–632, 1915.

Moore, E.H. and H.L. Smith. A general theory of limits. *Amer. J. Math.*, **44**: 102–121, 1922.

Mrówka, S. Compactness and product spaces. *Colloq. Math.*, **7**:19–22, 1959.

Nachbin, L. Sur les espaces topologiques ordonnés. *C.R. Acad. Sci. Paris*, **226**:381–382, 1948.

Nachbin, L. *Topologia e Ordem*. University of Chicago Press, Chicago, IL, 1950. English translation: *Topology and Order*, Van Nostrand, Princeton, NJ, 1965.

Niefield, S. Cartesianness: topological spaces, uniform spaces, and affine schemes. *J. Pure Appl. Algebra*, **23**:147–167, 1982.

Pasynkov, B.A. Partial topological products. *Trans. Mosc. Math. Soc.*, **13**: 153–271, 1965.

Pedicchio, M.C. and W. Tholen. Multiplicative structures over sup-lattices. *Arch. Math. (Brno)*, **25**:107–114, 1989.

Pedicchio, M.C. and R.J. Wood. Groupoidal completely distributive lattices. *J. Pure Appl. Algebra*, **143**:339–350, 1999.

Penon, J. Objets séparés ou compacts dans une catégorie. *C.R. Acad. Sci. Paris Sér. A*, **274**:384–387, 1972.

Penon, J. Quasi-topos. *C.R. Acad. Sci. Paris Sér. A*, **276**:237–240, 1973.

Perry, R.J. Completely regular relational algebras. *Cahiers Topologie Géom. Différentielle*, **17**:125–133, 1976.

Picado, J., A. Pultr, and A. Tozzi. Locales. In M.C. Pedicchio and W. Tholen, eds., *Categorical Foundations*. Cambridge University Press, Cambridge, 2004, pp. 49–101.

Pisani, C. Convergence in exponentiable spaces. *Theory Appl. Categ.*, **5**:148–162, 1999.

Ramaley, J.F. and O. Wyler. Cauchy spaces, I, structure and uniformization theorems. *Math. Ann.*, **187**:175–186, 1970.

Reis, C.D. *Topologia via Categorias Enriquecidas*. Ph.D. thesis, University of Aveiro, Aveiro, 2013.

Reitermann, J. and W. Tholen. Effective descent maps of topological spaces. *Topology Appl.*, **57**:53–69, 1994.

Riesz, F. Stetigkeitsbegriff und abstrakte Mengenlehre. In *Atti del IV Congresso Intern. dei Matem., Bologna*, vol. 2, 1908, pp. 18–24.

Robeys, K.T. *Extensions of Products of Metric Spaces*. Ph.D. thesis, University of Antwerp, Antwerp, 1992.

Rosebrugh, R. and R. Wood. Boundedness and complete distributivity. *Appl. Categ. Structures*, **9**:437–456, 2001.

Rosebrugh, R. and R.J. Wood. Distributive laws and factorization. *J. Pure Appl. Algebra*, **175**:327–353, 2002.

Rosenthal, K.I. *Quantales and their Applications*. Addison Wesley Longman, Harlow, 1990.

Rosenthal, K.I. *The Theory of Quantaloids*. Addison Wesley Longman, Harlow, 1996.

Rutten, J.J.M.M. Weighted colimits and formal balls in generalized metric spaces. *Topology Appl.*, **89**:179–202, 1998.

Schubert, C. *Lax Algebras – A Scenic Approach*. Ph.D. thesis, Universität Bremen, Bremen, 2006.

Schubert, C. and G.J. Seal. Extensions in the theory of lax algebras. *Theory Appl. Categ.*, **21**:118–151, 2008.

Schubert, H. *Categories*. Springer, New York, 1972.

Schweizer, B. and A. Sklar. *Probabilistic Metric Spaces*. North-Holland, New York, 1983.

Scott, D. Continuous lattices. In *Conf. Toposes, Algebraic Geometry and Logic*, Dalhousie, Halifax, N.S., 1971. Lecture Notes in Mathematics 274. Springer, Berlin, 1972, pp. 97–136.

Seal, G.J. Canonical and op-canonical lax algebras. *Theory Appl. Categ.*, **14**:221–243, 2005.

Seal, G.J. A Kleisli-based approach to lax algebras. *Appl. Categ. Structures*, **17**:75–89, 2009.

Seal, G.J. Order-adjoint monads and injective objects. *J. Pure Appl. Algebra*, **214**:778–796, 2010.

Seal, G.J. On the monadic nature of categories of ordered sets. *Cah. Topol. Géom. Différ. Catég.*, **52**:163–187, 2011.

Smyth, M.B. Effectively given domains. *Theoret. Comput. Sci.*, **5**:257–274, 1977/78.

Solovyov, S.A. On the category Q-**mod**. *Algebra Universalis*, **58**:35–58, 2008.

Solovyov, S.A. On a lax-algebraic characterization of closed maps, *Appl. Categ. Structures*, 2013, in press; DOI: 10.1007/s 10485-013-9334-7.

Šostak, A.P. Fuzzy functions and an extension of the category L-Top of Chang–Goguen L-topological spaces. In *Proc. Ninth Prague Topological Symp.,* 2001, pp. 271–294. Contributed papers available at www.emis.de/procedings/TopoSym2001/00.htm.

Šostak, A.P. Fuzzy functions as morphisms in an extension of the category of Hutton L-uniform spaces. *Proc. Latv. Acad. Sci. Sect. B Nat. Exact Appl. Sci.*, **57**:121–127, 2003.

Stone, M.H. Applications of the theory of Boolean rings to general topology. *Trans. Amer. Math. Soc.*, **41**:375–481, 1937.

Street, R. The formal theory of monads. *J. Pure Appl. Algebra*, **2**:149–168, 1972.

Street, R. Fibrations and Yoneda's lemma in a 2-category. In *Proc. Category Sem.*, Sydney, 1972/73. Lecture Notes in Mathematics 420. Springer, Berlin, 1974, pp. 104–133.

Street, R. Categorical structures. In M. Hazewinkel, ed., *Handbook of Algebra*, vol. 1. Elsevier, Amsterdam, pp. 529–577. 1996.

Streicher, T. Fibred categories à la Jean Bénabou. Lecture notes. Available at `http://www.mathematik.tu-darmstadt.de/~streicher/FIBR/FibLec.pdf`, 1998–2012.

Stubbe, I. Categorical structures enriched in a quantaloid: categories, distributors and functors. *Theory Appl. Categ.*, **14**:1–45, 2005.

Stubbe, I. Categorical structures enriched in a quantaloid: tensored and cotensored categories. *Theory Appl. Categ.*, **16**:283–306, 2006.

Stubbe, I. $\mathcal{Q}$-modules are $\mathcal{Q}$-suplattices. *Theory Appl. Categ.*, **19**:50–60, 2007.

Tarski, A. Une contribution à la théorie de la mesure. *Fundam. Math.*, **15**:42–50, 1930.

Thampuran, D.V. Extended topology: filters and convergence, I. *Math. Ann.*, **158**:57–68, 1965.

Tholen, W. On Wyler's taut lift theorem. *Gen. Topology Appl.*, **8**:197–206, 1978.

Tholen, W. Semitopological functors, I. *J. Pure Appl. Algebra*, **15**:53–73, 1979.

Tholen, W. Factorizations, localizations, and the orthogonal subcategory problem. *Math. Nachr.*, **114**:63–85, 1983.

Tholen, W. A categorical guide to separation, compactness and perfectness. *Homology, Homotopy Appl.*, **1**:147–161, 1999.

Tholen, W. Ordered topological structures. *Topology Appl.*, **156**:2148–2157, 2009.

Trnková, V. Relational automata in a category and their languages. In *Proc. Int. Conf. Fundamental Computing Theory*, Poznan-Kornik, 1977. Lecture Notes in Computer Science 56. Springer, Berlin, 1977, pp. 340–355.

Tukey, J.W. *Convergence and Uniformity in Topology*. Princeton University Press, Princeton, NJ, 1940.

Tychonoff, A.N. Über die topologische Erweiterung von Räumen. *Math. Ann.*, **102**:544–561, 1930.

van Breugel, F. An introduction to metric semantics: operational and denotational models for programming and specification languages. *Theoret. Comput. Sci.*, **258**:1–98, 2001.

Van Olmen, C. A study of the interaction between frame theory and approach theory. Ph.D. thesis, University of Antwerp, 2005.

Van Olmen, C. and S. Verwulgen. A finite axiom scheme for approach frames. *Bull. Belg. Math. Soc. Simon Stevin*, **17**:899–908, 2010.

Vietoris, L. Bereiche zweiter Ordnung. *Monatsh. Math. Phys.*, **32**:258–280, 1922.

Wagner, K.R. Solving recursive domain equations with enriched categories. Ph.D. thesis, Carnegie Mellon University, Pittsburgh, 1994.

Waszkiewicz, P. On domain theory over Girard quantales. *Fund. Inform.*, **92**:169–192, 2009.

Wood, R.J. Ordered sets via adjunctions. In C. Pedicchio and W. Tholen, eds., *Categorical Foundations*. Cambridge University Press, Cambridge, 2004, pp. 5–47.

Wyler, O. On the categories of general topology and topological algebra. *Arch. Math. (Basel)*, **22**:7–17, 1971.

Wyler, O. Are there topoi in topology? In *Proc. Conf. Categorical Topology*, Mannheim, 1975. Lecture Notes in Mathematics 540. Springer, Berlin, 1976, pp. 699–719.

Wyler, O. Algebraic theories of continuous lattices. In *Continous Lattices*. Lecture Notes in Mathematics 871. Springer, Berlin, 1981, pp. 390–413.

Wyler, O. *Lectures Notes on Topoi and Quasitopoi*. World Scientific, Singapore, 1991.

Wyler, O. Convergence axioms for topology. *Ann. NY Acad. Sci.*, **806**:465–475, 1995.

Zhang, D. Tower extensions of topological constructs. *Comment. Math. Univ. Carolinae*, **41**:41–51, 2000.

Selected categories

D^{C}	category of functors from C to D and natural transformations	46
$(X \downarrow G)$	comma category of a functor G under an object X	55
$(F \downarrow A)$	comma category of a functor F over an object A	55
X/A	slice category over A	56
A/X	slice category under A	56
$\mathsf{X}^{\mathbb{T}}$	Eilenberg–Moore category of the monad $\mathbb{T}$ on X	76
$\mathsf{X}_{\mathbb{T}}$	Kleisli category of the monad $\mathbb{T}$ on X	84
AbGrp	category of Abelian groups and group homomorphisms	71
App	category of approach spaces and non-expansive maps	192
BiTop	category of bitopological spaces and bicontinuous maps	200
CAT	metacategory of categories and functors	44
Cat	category of small categories and functors	44
Cat(O)	category of small categories with object set O and functors mapping O identically	114
$\mathcal{V}$-Cat	category of $\mathcal{V}$-categories and $\mathcal{V}$-functors	150
$(\mathbb{T}, \mathcal{V})$-Cat	category of $(\mathbb{T}, \mathcal{V})$-categories and $(\mathbb{T}, \mathcal{V})$-functors	160
$(\mathbb{T}, \mathcal{V})$-Cat$_{\mathrm{CompHaus}}$	category of compact Hausdorff $(\mathbb{T}, \mathcal{V})$-categories and $(\mathbb{T}, \mathcal{V})$-functors	376

$(\mathbb{T}, \mathcal{V})$-$\mathsf{Cat}_{\mathrm{Haus}}$	category of Hausdorff $(\mathbb{T}, \mathcal{V})$-categories and $(\mathbb{T}, \mathcal{V})$-functors	376
$(\mathbb{T}, \mathcal{V}, \hat{\mathbb{T}})$-$\mathsf{Cat}$	category of $(\mathbb{T}, \mathcal{V}, \hat{\mathbb{T}})$-categories and $(\mathbb{T}, \mathcal{V})$-functors	160
Cls	category of closure spaces and continuous maps	44
$\mathsf{Cls}_{\mathrm{fin}}$	category of finitary closure spaces and continuous maps	199
Cnt	category of continuous lattices and inf-maps that preserve up-directed suprema	344
$\mathsf{CompHaus}$	category of compact Hausdorff spaces and continuous maps	140
Dst	category of completely distributive lattices and maps that preserve all infima and suprema	354
Frm	category of frames and frame homomorphisms	44
$\mathsf{Gph}(O)$	category of graphs over O and graph morphisms	113
$\mathcal{V}$-Gph	category of $\mathcal{V}$-graphs and $\mathcal{V}$-functors	233
$(\mathbb{T}, \mathcal{V})$-$\mathsf{Gph}$	category of $(\mathbb{T}, \mathcal{V})$-graphs and $(\mathbb{T}, \mathcal{V})$-functors	217
Grp	category of groups and group homomorphisms	67
Haus	category of Hausdorff topological spaces and continuous maps	130
INF	metacategory of inf-complete classes and inf-maps	142
Inf	category of complete lattices and inf-maps	43
$\mathcal{M}$-$\mathsf{Inj}(\mathsf{A})$	category of $\mathcal{M}$-injective A-objects and left adjoint A-morphisms	349
Int	category of interior spaces and continuous maps	44
Lat	category of lattices and lattice homomorphisms	44
$\mathcal{V}$-LXT	metacategory of lax extensions to $\mathcal{V}$-Rel and their morphisms	213
Met	category of metric spaces and non-expansive maps	151
MetCls	category of metric closure spaces and non-expansive maps	200
$\mathsf{MetCompHaus}$	category of metric compact Hausdorff spaces and continuous non-expansive maps	256
$\mathsf{Met}_{\mathrm{sep}}$	category of separated metric spaces and non-expansive maps	151
$\mathsf{Met}_{\mathrm{sym}}$	category of symmetric metric spaces and non-expansive maps	151

$\mathsf{MND}_{\mathsf{X}}$	category of monads on X and monad morphisms	76
Mod	category of ordered sets and modules	43
Mod_R	category of R-modules and R-linear maps	97
$\mathcal{V}$-Mod	category of $\mathcal{V}$-categories and $\mathcal{V}$-modules	153
Mon	category of monoids and monoid homomorphisms	44
$\mathbb{T}$-Mon	category of $\mathbb{T}$-monoids and their morphisms	290
$\mathbb{T}$-$\mathsf{Mon}_{\mathrm{sep}}$	category of separated $\mathbb{T}$-monoids and their morphisms	343
$(\mathbb{T}, \mathcal{V})$-Mon	category of Kleisli $\mathcal{V}$-towers and $\mathcal{V}$-tower morphisms	315
MultiOrd	category of multi-ordered sets and their morphisms	353
ObsReSp	category of observable realization spaces and continuous maps	361
ORD	metacategory of ordered classes and monotone maps	105
Ord	category of ordered sets and monotone maps	43
OrdCompHaus	category of ordered compact Hausdorff spaces and continuous monotone maps	256
$\mathsf{Ord}_{\mathrm{sep}}$	category of separated ordered sets and monotone maps	43
PrApp	category of pre-approach spaces and non-expansive maps	220
ProbMet	category of probabilistic metric spaces and probabilistically non-expansive maps	186
PrTop	category of pretopological spaces and continuous maps	219
PsApp	category of pseudo-approach spaces and non-expansive maps	219
PsTop	category of pseudotopological spaces and continuous maps	214
Qnt	category of quantales and quantale homomorphisms	44
Rel	category of sets and relations	43
$\mathcal{V}$-Rel	category of sets and $\mathcal{V}$-relations	146
$(\mathbb{T}, \mathcal{V})$-RepCat	representable $(\mathbb{T}, \mathcal{V})$-categories and pseudo-homomorphism	264

$(\mathbb{T},\mathcal{V})$-RGph	category of right-unitary $(\mathbb{T},\mathcal{V})$-graphs and $(\mathbb{T},\mathcal{V})$-functors	217
Rng	category of unital rings and unital ring homomorphisms	90
RNRel	category of sets with a reflexive numerical relation and non-expansive maps	233
RRel	category of sets with a reflexive relation and relation-preserving maps	233
SET	metacategory of classes and maps	45
Set	category of sets and maps	43
Set_*	category of pointed sets and base-point-preserving maps	243
SLat	category of semilattices and semilattice homomorphisms	44
Sob	category of sober topological spaces and continuous maps	367
SUP	metacategory of sup-complete classes and sup-maps	107
Sup	category of complete lattices and sup-maps	43
$\mathsf{Sup}^{\mathcal{V}}$	category of left $\mathcal{V}$-actions in Sup and equivariant sup-maps	100
Top	category of topological spaces and continuous maps	44
Top_0	category of T0-spaces and continuous maps	344
TopGrp	category of topological groups and continuous group homomorphisms	137
$(\mathbb{T},\mathcal{V})$-UGph	category of unitary $(\mathbb{T},\mathcal{V})$-graphs and $(\mathbb{T},\mathcal{V})$-functors	217
UltraMet	category of ultrametric spaces and non-expansive maps	199
$(\mathbb{T},\mathcal{V})$-URel	category of sets and unitary $(\mathbb{T},\mathcal{V})$-relations	165

Selected functors

$(-)^{\mathrm{op}}$	dualization functor on Ord, CAT or $\mathcal{V}$-Cat	45, 154
$(-)_*$	$\mathcal{V}$-module functor (covariant)	45, 154
$(-)^*$	$\mathcal{V}$-module functor (contravariant)	45, 154
$(-)^\circ$	graph functor (contravariant)	148
$(-)_\sharp$	unitary $(\mathbb{T}, \mathcal{V})$-relation functor (covariant)	208, 334
$(-)_d$	discrete-structure functor to a category of lax algebras	203
$(-)_i$	indiscrete-structure functor to a category of lax algebras	203
$(-)^\flat$	associated-Kleisli-morphism functor (contravariant)	291
$(-)^\tau$	Kleisli-lax-extension functor (contravariant)	291
$\mathsf{C}(A, -)$	hom-functor of the C-object A (covariant)	45
$\mathsf{C}(-, A)$	hom-functor of the C-object A (contravariant)	45
$A \multimap (-)$	right internal hom-functor of A (covariant)	101
$(-) \multimapinv A$	left internal hom-functor of A (covariant)	101
$(-) \multimap C$	right internal hom-functor of C (contravariant)	103
$C \multimapinv (-)$	left internal hom-functor of C (contravariant)	103
$C^{(-)}$	internal hom-functor of C for Cartesian structure (contravariant)	104
$(-) \circ \varphi$	left Kleisli convolution functor of φ	321
$(-) \circ\!\!- \varphi$	internal hom functor right adjoint to $(-) \circ \varphi$	321
$(-) \otimes (-)$	generic tensor functor of a monoidal category	96
$S(-)$	left-whiskering by a functor S	46
$(-)T$	right-whiskering by a functor T	46

$F^{\mathbb{T}}, G^{\mathbb{T}}$	left and right adjoint functors associated to an Eilenberg–Moore category	76
$F_{\mathbb{T}}, G_{\mathbb{T}}$	left and right adjoint functors associated to a Kleisli category	84
$!$	unique functor into the terminal category	54
$f_!$	left adjoint to the pullback functor	230
$f(-)$	image functor of f	118
$f^{-1}(-)$	inverse-image functor of f	117
f_*	co-Cartesian lift functor induced by f	129
f^*	Cartesian lift functor induced by f functor, pullback functor	129, 229
A_α	algebraic functor associated to α	207
A°	unitary $(\mathbb{T}, \mathcal{V})$-category functor	208
B_φ	change-of-base functor associated to φ	210
$\beta, \mathbb{\beta}$	ultrafilter functor and monad	75
$C, \mathbb{C}$	clique functor and monad	299
$\kappa, \mathbb{K}$	ultraclique functor and monad	319
cod	codomain functor	128
conv	convergence functor (contravariant)	300
Δ	constant diagram functor	56
$\mathrm{Dn}, \mathbb{D}\mathrm{n}$	down-set functor on Ord and monad	45, 110
dom	domain functor	141
Ev	evaluation functor	65
$F, \mathbb{F}$	filter functor and monad	75
$\mathrm{Fil}, \mathbb{F}\mathrm{il}$	filter functor on Ord and monad	345
$F_{\mathrm{p}}, \mathbb{F}_{\mathrm{p}}$	proper filter functor and monad	377
$1_{\mathsf{C}}, \mathbb{1}$	identity functor on C and monad	44, 75
I	identity functor (alternative notation)	168
$L, \mathbb{L}$	list functor and monad	75
$\mathrm{Lan}_S\, T$	left Kan extension of T along S	69
nbhd	neighborhood functor (contravariant)	300
O	underlying-set functor from a category of lax algebras	181
$\mathcal{O}$	open-set functor	45
Ω	underlying-topology functor from a category of lax algebras	427
ob	object-functor	130
$P, \mathbb{P}$	powerset functor (covariant) and monad	44, 75

P_{fin}, $\mathbb{P}_{\mathrm{fin}}$	finite-powerset functor and monad	198
Π, $\mathbb{\Pi}$	discrete presheaf functor and monad	323
$P^{\bullet}$	powerset functor (contravariant)	45
P^2, $\mathbb{P}^2$	double-powerset functor and monad	75
pt	points-of-a-frame functor	367
$P_{\mathcal{V}}$, $\mathbb{P}_{\mathcal{V}}$	$\mathcal{V}$-powerset functor and monad	180
$\mathrm{Ran}_S\, T$	right Kan extension of T along S	67
U, $\mathbb{U}$	up-set functor and monad	75
U_{fin}, $\mathbb{U}_{\mathrm{fin}}$	finitary-up-set functor and monad	299
Up, $\mathbb{U}\mathrm{p}$	up-set functor on Ord and monad	110
y	Yoneda embedding	48

Selected symbols

Symbol	Meaning	Page
©	Axiom of Choice	17
$\to$	map, morphism, functor, natural transformation, convergence, relation	18, 42, 44, 46, 190, 385
$\hookrightarrow$	inclusion map, embedding	35
⇸	relation arrow, $\mathcal{V}$-relation arrow	20
⇀⃒	$(\mathbb{T}, \mathcal{V})$-relation arrow	161
⇴	module arrow	22, 153
$\Longrightarrow$	implication, natural transformation in diagram	22, 46
$\Longleftrightarrow$	logical equivalence	20
$\dashv$	adjunction	23, 49
$\bot$	bottom element in an ordered set	28
$\top$	top element in an ordered set	28
•—	left internal hom-functor	101
—•	right internal hom-functor	101
$\multimapinv$	left internal hom-functor (alternative notation)	321
$\multimap$	right internal hom-functor (alternative notation)	329
$\models$	logical entailment	20
/	slice category (under or over), quotient	56
$\downarrow$	comma category	55, 73
$\downarrow, \leq$	down-closure in an ordered set, and associated order relation	20, 25
$\uparrow$	up-closure in an ordered set	26
$\Downarrow, \lll$	totally below operation, and associated relation	32
$\twoheaddownarrow, \ll$	way-below operation, and associated relation	268, 344
$\twoheaduparrow, \gg$	way-above operation, and associated relation	345

Symbol	Meaning	Page
$\downarrowtail$, $\prec$	observable specialization operation, and associated relation	358
$\uparrowtail$	opposite of observable specialization operation	358
$\vert$	restriction of a map, restriction of a functor	37
0	initial object	55
1	terminal object, identity morphism, identity functor, identity natural transformation	18, 42, 44, 46, 55
I	initial object (alternative notation), identity functor (alternative notation), indexing set (generic)	29, 55, 168
T	terminal object (alternative notation)	54
k	neutral element of a quantale	30
$\otimes$	tensor of a quantale or of a monoidal category	30, 96
$\times$	binary categorical product	18, 57
$+$	binary categorical coproduct	60
$\prod$	arbitrary categorical product	57
$\coprod$	arbitrary categorical coproduct	60
$\sum$	Kowalski sum of a set of subsets	35
$(\dot{-})$	principal filter	35
$\cong$	isomorphic objects	43
$\simeq$	equivalent objects	21, 43
$\wedge$	binary meet in an ordered set	28
$\vee$	binary join in an ordered set	28
$\bigwedge$	arbitrary meet in an ordered set	26
$\bigvee$	arbitrary join in an ordered set	25
inf	arbitrary meet in $[0, \infty]$ with its natural order	31
sup	arbitrary join in $[0, \infty]$ with its natural order	31
$[-, -]$	closed interval, internal hom in $\mathcal{V}$-Cat, structure of an internal hom-object in $\mathcal{V}$-Cat	31, 151, 152
$(-)$	image, inverse image of an element or a subset under a map or relation	18, 20, 25, 155
$[-]$	image, inverse image of a filter or set of subsets under a map or relation	34, 171
$(-)^\circ$	converse of a relation, converse of the graph of a map	20
$(-)^*$, $(-)_*$	modules induced by a monotone map, $\mathcal{V}$-modules induced by $\mathcal{V}$-functors	22, 153
$(-)^\sharp$, $(-)_\sharp$	unitary $(\mathbb{T}, \mathcal{V})$-relation induced by a map, discrete $(\mathbb{T}, \mathcal{V})$-structure on a set	162, 165, 166

$(-)\cdot(-)$	composition of morphisms, vertical composition of natural transformations	18, 42
$(-)\circ(-)$	horizontal composition of natural transformations, Kleisli composition, Kleisli convolution	47, 84, 161
$(\hat{-})$	lax extension of a functor or of a monad (generic)	155, 157
$\overline{(-)}$	Barr extension of a functor or of a monad, grand closure of a subset	170, 176, 415
$(\check{-})$	Kleisli extension of a functor or of a monad	292, 293
$(-)^{\flat}$	$\mathcal{V}$-powerset-valued map induced by a $\mathcal{V}$-relation	291, 322
$(-)^{\mathbb{T}}$	Eilenberg–Moore category, adjunction functors associated to an Eilenberg–Moore category, Kleisli extension operation	76, 85
$(-)^{\mathbb{F}}$	restriction of the set of filters on a set to a subset	285
$(-)^{\beta}$	restriction of the set of ultrafilters on a set to a subset	187
$(-)^{(v)}$	v-closure of a subset	415
$\overline{(-)}^{\infty}$	idempotent hull of the grand closure of a subset	443
$(-)_{\mathbb{T}}$	Kleisli category or associated adjunction functors	84
$\mathbb{T}'$	monad on $\mathbb{S}$-Mon derived from a morphism $\alpha : \mathbb{S} \to \mathbb{T}$ of power-enriched monads	340
$\mathsf{C}(-,-)$	hom-set of a category C	42
$\hom_{\mathsf{C}}(-,-)$	hom-set of a category C (alternative notation)	42
$(-)^{\mathrm{op}}$	opposite of an ordered set, monotone map, category, functor, natural transformation, $\mathcal{V}$-category or $\mathcal{V}$-functor	21, 43, 45, 47, 154
$(-)^{\mathrm{co}}$	2-cell opposite of a 2-category	105
lim	limit of a diagram, limit points of a filter	57, 268
colim	colimit of a diagram	60
sub A	subobjects of an object A	117
ob C	objects of a category C	42
mor C	morphisms of a category C	61
Iso C	class of isomorphisms of a category C	114
Epi C	class of epimorphisms of a category C	114
Mono C	class of monomorphisms of a category C	114
ExtEpi C	class of extremal epimorphisms of a category C	138

ExtMono C	class of extremal monomorphisms of a category C	115
RegEpi C	class of regular epimorphisms of a category C	138
RegMono C	class of regular monomorphisms of a category C	125
SplitMono C	class of split monomorphisms of a category C	115
StrongEpi C	class of strong epimorphisms of a category C	138
Fin U	class of final morphisms with respect to a functor U	128
Ini U	class of initial morphisms with respect to a functor U	128
Clo$(\mathbb{T}, \mathcal{V})$	class of closed maps of $(\mathbb{T}, \mathcal{V})$-$\mathsf{Cat}$	434
Open$(\mathbb{T}, \mathcal{V})$	class of open maps of $(\mathbb{T}, \mathcal{V})$-$\mathsf{Cat}$	433
Prop$(\mathbb{T}, \mathcal{V})$	class of proper maps of $(\mathbb{T}, \mathcal{V})$-$\mathsf{Cat}$	433
$\mathbf{Y}$	monad morphism $\mathbb{T} \to \mathbb{\Pi}(\mathbb{T}, \mathcal{V})$	326

Index

absolute coequalizer, 80
G-absolute pair, 80
abstract basis, 362
  ordered, 364
action, 92, 99
  left, 99
  right, 100
additive category, 111
adjoint
  left, 23, 49, 106
    left-inverse, 452
  right, 23, 49, 106
adjunction, 23, 49, 106
  descent type, 95
  functor, 45
  rules, 148
  vs. Galois correspondence, 23
admissible, 181
Alexander's subbase lemma, 269
Alexandroff
  space, 136
  topology, 136
$(\mathbb{T}, \mathcal{V})$-algebra, *see* $(\mathbb{T}, \mathcal{V})$-category
$\mathbb{T}$-algebra, 76
  free, 76
algebraic
  closure space, *see* finitary
  functor, 207
antisymmetric, *see* separated
approach distance, 192
approach space, 192
  0-compact, 378
  core-compact, 277
  pre-, 220
  pseudo-, 220
  stable, 277
arrow, *see* morphism
associated monad, 74
associative lax extension, 165
associativity law, 42
$\ll$-atomic, 33
auxiliary relation, 317

Barr extension, 194, 200, 312
base
  filter, 34
  ideal, 34
  for a topology, 29
BC condition
  for a functor, 173
  for a natural transformation, 173
BC-square, 172
Beck criterion, 80
Beck–Chevalley condition, *see* BC condition
Beck–Chevalley square, *see* BC-square
bicontinuous map, 200
bicoreflective, 117
$\mathcal{V}$-bimodule, *see* $\mathcal{V}$-module
bimorphism, 97
biproduct, 108
bireflective, 117
birelation, 147
bitopology, 200
bottom, 28
bound
  greatest lower, 26
  least upper, 26
  lower, 26
  upper, 26

Cartesian
  lifting, 129
  monad, 185
  morphism, 127
  product, 18
  square, 128
Cartesian closed
  category, 101
  locally, 227
  quantale, 225

category, 43
additive, 111
Cartesian closed, 101
cocomplete, 61
has coequalizers, 61
comma, 55
comma, lax, 306
complete, 61
concrete, 57
connected, 73
has coproducts, 61
discrete, 56
dual, 43
Eilenberg–Moore, 76
enriched, 105
has equalizers, 61
finite, 61
has finite coproducts, 61
has finite products, 61
finitely cocomplete, 61
finitely complete, 61
functor, 46
Kleisli, 84
locally small, 45
meta-, 44
monadic, 77
monadic, strictly, 77
monoidal, 96
monoidal closed, 101
monoidal, strict, 97
opposite, 43
ordered, 104
has partial products, 228
has products, 61
has pullbacks, 61
has pushouts, 61
self-dual, 45
small, 42
small-cocomplete, 61
small-complete, 61
symmetric monoidal, 96
thin, 105
$(\mathbb{T}, \mathcal{V})$-category, 158
discrete, 163
separated, 263
$\mathcal{V}$-category, 150
tensored, 215
underlying, 207
V-category, 111
2-category, 105, 111
Cayley representation, 45
Čech–Stone compactification, 267, 380
cell, 178
chain, 21
three-, 30
two-, 30
change-of-base functor, 210
chaotic structure, *see* indiscrete
characteristic morphism, 238
class, 17, 38, 43
large, 17
proper, 17
small, 17
classifier
partial-map, 241
regular, 241
$\mathcal{M}$-partial-map, 240
subobject, 239
regular, 239
$\mathcal{M}$-subobject, 238
clique, 299
monad, 299
ultra-, 318
closed, 24
Cartesian category, 101
$(\mathbb{T}, \mathcal{V})$-functor, 418
inversely, 418
under limits, 64, 115
monoidal category, 101
under $\mathcal{M}$-morphisms, 134
under products, 115
under $\mathbb{M}$-sources, 134
stably, 420
subobject, 123
closure, 123
finitary, 199
grand, 415
Kuratowski, 39
υ-closure, 415
closure operation, 24
down-, 25
finitely additive, 39
metric, 200
up-, 26
closure operator, 122
continuous, 122
extensive, 122
hereditary, 124
weakly, 123
idempotent, 123
monotone, 122
closure space, 25
algebraic, *see* finitary
finitary, 199
metric, 200
cloven fibration, 129
co-Cartesian
lifting, 129
morphism, 127
co-dcpo, 356
cocontinuous, 356
coarser, 285
cocomplete, 61
finitely, 61
$\mathbb{T}$-cocomplete, 264
cocone, 60
cocontinuous
co-dcpo, 356
functor, 63
lattice, 345

ordered set, 356
codomain, 42, 113
functor, 128
coequalizer, 60
absolute, 80
cofibration, 129
cogenerating class, 126
cogenerator, 126
coherence conditions
for Cartesian product, 19
for tensor product, 97
cokernel pair, 60
colax idempotent monad, 110
colimit, 60
creation, 78
directed, 63
comma category, 55
lax, 306
commutative
monoid, 19
quantale, 30
comonoid, 99
compact, 191
core-, 268
locally, 270
Hausdorff, 451
($\mathbb{T}$, $\mathcal{V}$)-space, 376
stably, 270
super-, 377
0-compact, 378
$\mathcal{P}$-compact, 437
compactification, Čech–Stone, 267, 380
companion
left, 116
right, 116
comparison functor, 77
compatible, 210
strictly, 332
complement, 25
complemented, 148
complete
category, 61
finitely, 61
large-, 132
lattice, 29
ordered set, 25
small-, 61
completely distributive, 32
constructively, 34
completely prime, 366
composite monad, 88
composition, 42, 111
of functors, 44
horizontal, 47, 106
Kleisli, 84
vertical, 46, 106
concrete category, 57
cone, 56
confluent, 278
conglomerate, 17
conjugate ordered category, 105
connected
category, 73
component, 73, 460
limit, 133
locally, 460
object, 457
constant
diagram, 56
sequence, 41
constructively completely distributive, *see* completely distributive
continuous
closure operator, 122
dcpo, 356
functor, 63
lattice, 344
map, 25, 29
($\mathbb{T}$, $\mathcal{V}$)-continuous, 376
$\mathcal{V}$-continuous, 376
contravariant
hom-functor, 45
powerset functor, 50
convergent sequence, 41
convolution
distance distribution maps, 186
Kleisli, 161
coproduct, 60
disjoint, 227
injection, 60
pullback stable, 226
totally disjoint, 453
universal, 453
core-compact, 268, 277
coreflection, 54
coreflective
bi-, 117
mono-, 117
$\mathcal{M}$-coreflective, 117
coreflective subcategory, 54
coreflector, 54
coseparator, *see* cogenerator
counit, 49
couniversal
arrow, 52
property, 52
covariant hom-functor, 45
cowellpowered, 122
creation
of colimits, 78
of limits, 78

dcpo, continuous, 356
dense
finally, 135
functor, 69
initially, 135
map, 443
morphism, 123
subobject, 123

$\mathcal{P}$-dense, 442
densely generating, 140
derived
  class, 435
  topology, 435
descent type, 95
diagonal morphism, 46
diagram, 56
  constant, 56
  shape, 56
diamond lattice, 40
Dirac morphism, 211
direct limit, 63
directed
  colimit, 63
  down-, 34
  graph, 113
  limit, 63
  up-, 34
disconnected
  extremally, 399
    topology, 398
  totally, 463
discrete
  category, 56
  $(\mathbb{T}, \mathcal{V})$-category, 163
  object, 127
  order, 29, 50
  presheaf monad, 323
  structure, 203
  topology, 50
$\mathcal{P}$-discrete, 451
disjoint coproduct, 227
disjointness preservation, 246
distance
  approach, 192
  distribution map, 185
distributive law, 86
  lax, 182
$\mathcal{V}$-distributor, *see* $\mathcal{V}$-module
domain, 42, 113
  functor, 141
double category, 179
double-powerset monad, 75
down
  -closed, 25
  -closure, 25
  -directed, 34
  -interior, 40
down-set, 25
  functor, 45
  monad, 110
down-set-filter monad, 345
dual
  category, 43
  $\mathcal{V}$-category, 154
  lax extension, 183
  relation, 20
  $\mathcal{V}$-relation, 146
  statement, 53
dualization, 52
  functor, 45
Duskin criterion, 82

edge, 113
Eilenberg–Moore
  algebra, *see* $\mathbb{T}$-algebra
  category, 76
embedding
  full, 44
  open, 224
  order-, 28
  Yoneda, 48
enriched category, 105
epi
  -reflective, 117
    regular, 117
  -sink, 122
    extremal, 126
    strict, 140
epic, 60
epimorphism, 60
  extremal, 116
  regular, 71
  split, 71
  strong, 138
equalizer, 58
  generalized, 66
equivalence, 23, 54
equivalent categories, 54
equivariant, 92, 100
evaluation, 50, 228
  functor, 65
exponentiable, 101
$\otimes$-exponentiable, 101
exponential, 101, 234
extension, 109
  of a functor, 93
  Kleisli, 292
  lax, *see* lax extension
  of a monad, 90
  operation, 85
extensive, 453
  category, 227
    finitely, 453
  closure operator, 122
extremal
  epi-sink, 126
  epimorphism, 116
  mono-source, 126
  monomorphism, 115

factorization class
  left, 116
  right, 116
factorization system
  for morphisms, 114
  proper, 114
  for sinks, 119
  for sources, 122

trivial, 116
faithful functor, 44
fiber, 20, 129, 413
finite, 247
fibered product, 59
fiberwise
pretopology, 435
topology, 435
fibration, 129
mono-, 130
$\mathcal{E}_0$-fibration, 130
$\mathcal{M}_0$-fibration, 130
filter
base, 34
coarser, 285
finer, 285
functor, 75
image, 34
inverse image, 34
monad, 75
neighborhood, 41
open, 366
in ordered set, 34
principal, 34, 35
proper, 34
restriction, 35
round, 364
ultra-, 35
vs. sequence, 41
on a set, 34
filtered sum, 35
finally dense, 135
finer, 285
finitary
closure, 199
closure space, 199
metric, 151
$\mathcal{V}$-relation, 151
up-set, 299
-up-set monad, 299
finite-powerset monad, 198
finitely additive, 39, 193
fixpoints, 24
flat lax extension, 156, 157
forgetful functor, 44
fork, 92
split, 79
frame, 29
homomorphism, 29
sub-, 29
free
-group functor, 67
-monoid functor, 49
-monoid monad, 75
$\mathbb{T}$-algebra, 76
Frobenius law
left, 226
right, 226
full
functor, 44
subcategory, 44
fully faithful
functor, 44
monotone map, 21
functor, 44
adjunction, 45
algebraic, 207
category, 46
change-of-base, 210
cocontinuous, 63
codomain, 128
comparison, 77
composition, 44
continuous, 63
contravariant powerset, 50
dense, 69
domain, 141
double category, 179
down-set, 45
dualization, 45
evaluation, 65
extension, 93
faithful, 44
filter, 75
forgetful, 44
free-group, 67
free-monoid, 49
full, 44
full embedding, 44
fully faithful, 44
hom-, contravariant, 45
hom-, covariant, 45
identity, 44, 75
inclusion, 44
inverse image-, 117
Kuratowski-closure, 45
lax, 105
lifting, 93
module, 45
monadic, 77
open-set, 45
oplax, 105
$\mathcal{V}$-powerset, 180
powerset, contravariant, 45, 50
powerset, covariant, 44
preserves colimits, 63
preserves disjointness, 246
preserves isomorphisms, 70
preserves limits, 63
preserves monomorphisms, 63
pseudo-, 105
reflects isomorphisms, 54
reflects limits, 63
reflects monomorphisms, 63
representable, 72
self-adjoint, 50
set-of-objects, 141
topological, 130
$\mathbb{M}_0$-topological, 131
transportable, 132

ultrafilter, 75
underlying-order, 45
up-set, 75
whiskering, 46
$(\mathbb{T}, \mathcal{V})$-functor, 160, 217
closed, 418
open, 406
proper, 406
$\mathcal{V}$-functor, 150
V-functor, 111
2-functor, 105
functorial, 123

Galois correspondence, *see* adjunction
generating, 125
densely, 140
strongly, 126
generator
as a class, 126
as an object, 126
strong, 126
generic point, 264
graph
directed, 113
edge, 113
of a map, 20
vertices, 113
$(\mathbb{T}, \mathcal{V})$-graph, 216
greatest lower bound, 26
group, 45
topological, 137

Hausdorff
metric, 254
$(\mathbb{T}, \mathcal{V})$-space, 376
topology, 130
$\mathcal{P}$-Hausdorff, 437
morphism, 437
hereditary
closure operator, 124
weakly, 123
morphisms class, 433
sink, 245
$\mathcal{M}$-hereditary, 118
hom
-functor, contravariant, 45
-functor, covariant, 45
-functor, internal, 101
-object, 111
-set, 42
homeomorphism, 367
local, 447
homomorphism
of $\mathbb{T}$-algebras, 76
of frames, 29
of join-semilattices, 28
lax, 160
of meet-semilattices, 28
of monoidal categories, 98
of monoids, 19, 99
of quantales, 30
of quantales, lax, 30
of quantaloids, 107
of semilattices, 29
$\mathbb{T}$-homomorphism, 76
horizontal composition, 47, 106
hull, idempotent, 443

ideal, 34
base, 34
principal, 34
proper, 34
idempotent, 28
closure operator, 123
hull, 443
monad, 95
identity, 42, 111
functor, 44, 75
law, 42
lax, 162
monad, 75
$\mathcal{M}$-image, 118
image filter, 34
inclusion functor, 44
indiscrete
object, 127
order, 50
structure, 203
topology, 50
induced
monad, 74
order, 29, 204
induction, 37
inf-map, 27
infimum, 26
preservation, 27
initial
extension, 301
lifting, 130
object, 55
strict, 453
pre-, 454
source, 127
initially dense, 135
injection of coproduct, 60
$\mathcal{M}$-injective, 140
injective, locally, 439
$\mathcal{P}$-injective, locally, 451
integral, 148
interior
down-, 40
operation, 24
internal hom-functor, 101
interpolating, 304, 305
$\mathcal{M}$-intersection, 120
$\nu^{\mathbb{T}}$-invariant, 338
inverse
-image functor, 117
image, 117
image filter, 34

limit, 63
$\mathcal{M}$-inverse image, 117
irreducible, 268, 276
isomorphism, 23, 43
natural, 48
preservation, 70
reflection, 54

join, 26
join-semilattice, 28
homomorphism, 28
joint kernel pair, 82
jointly surjective sink, 122

Kan extension
left, 69
pointwise, 68, 69
right, 67
kernel pair, 59, 82
joint, 82
Kleisli
category, 84
composition, 84
convolution, 161
extension, 292, 313
monoid, *see* $\mathbb{T}$-monoid
$(\mathbb{T}, \mathcal{V})$-tower, 315
triple, 85
morphism, 85
KM-test space, 419
Kock–Zöberlein monad, 109
dual, 110
Kowalsky sum, 35
Kuratowski
-closure functor, 45
closure, 39

large class, 17
large-complete, 132
lattice, 29
cocontinuous, 345
complete, 29
completely distributive, 32
continuous, 344
diamond, 40
law
associativity, 42
distributive, 86
lax, 182
identity, 42
middle unit, 88
middle-interchange, 18, 47, 179
multiplication, 74
unit, 74
lax
algebra, *see* $(\mathbb{T}, \mathcal{V})$-category
comma category, 306
distributive law, 182
functor, 105
homomorphism, 160
compatible, 210
of quantales, 30
idempotent, 109
identity, 162
transformation, 106
lax extension
associative, 165
dual, 183
flat, 157
of functor, 155
of functor, flat, 156
initial, 301
of monad, 157
morphism, 207
symmetric, 397
lean, 148
least upper bound, 26
left
adjoint, 23, 49, 106
companion, 116
factorization class, 116
unitary, 163
left-cancellable, 115
weakly, 115
lifting, 108
Cartesian, 129
co-Cartesian, 129
of a functor, 93
initial, 130
of a monad, 87
limit
closed under, 64, 115
connected, 133
creation, 78
of a diagram, 57
direct, 63
directed, 63
inverse, 63
preservation, 63
reflection, 63
list monad, 75
locally
Cartesian closed, 227
compact, 270
$\mathcal{P}$-compact Hausdorff, 448
injective, 439
$\mathcal{P}$-injective, 451
$\mathcal{P}$-perfect, 448
small, 45
lower bound, 26
lower-Vietoris space, 274

map, 106
$(\mathbb{T}, \mathcal{V})$-continuous, 376
$\mathcal{V}$-continuous, 376
pointed, 243
mate, 51
maximal, 35, 41
maximum, 28
meet, 26

topological, 357
meet-semilattice, 28
homomorphism, 28
metacategory, 44
metric, 151
closure operation, 200
closure space, 200
finitary, 151
Hausdorff, 254
probabilistic, 186
relation, 147
separated, 151
symmetric, 151
ultra-, 199
metric space, 151
probabilistic, 186
ultra-, 199
middle
-interchange law, 18, 47, 179
unit law, 88
minimum, 28
modular, widely, 424
module, 22
functor, 45
$\mathcal{V}$-module, 153
monad, 74
associated, 74
Cartesian, 185
clique, 299
composite, 88
discrete presheaf, 323
double-powerset, 75
down-set, 110
down-set-filter, 345
extension, 90
filter, 75
finitary-up-set, 299
finite-powerset, 198
free-monoid, 75
idempotent, 95
identity, 75
induced, 74
lifting, 87
list, 75
morphism, 74
multiplication, 74
open-filter, 341
ordered-filter, 345
power-enriched, 288
powerset, 75
$\mathcal{V}$-powerset, 150, 180
terminal, 92
trivial, 92
ultraclique, 319
ultrafilter, 75
unit, 74
up-set, 75, 110
monadic
category, 77
strictly, 77
functor, 77
pre-, 95
strictly, 77
monic, 59
mono
-coreflective, 117
-coreflective, regular, 117
-fibration, 130
-source, 122
extremal, 126
monoid
in a category, 98
commutative, 19
homomorphism, 19, 99
ordered, 102, 255
separated, 343
in Set, 19
$\mathbb{T}$-monoid, 290
monoidal category, 96
closed, 101
homomorphism, 98
strict, 97
symmetric, 96
monomorphism, 59
extremal, 115
regular, 71
split, 71
strong, 138
monotone, 21
closure operator, 122
morphism, 42
bi-, 97
Cartesian, 127
couniversal, *see* couniversal arrow
diagonal, 46
epi-, 60
epic, 60
iso-, 43
of Kleisli triples, 85
of lax extensions, 207
of monads, 74
monic, 59
mono-, 59
quotient, 116, 140
universal, *see* universal arrow
multi-ordered set, 353, 387
multiple pullback, 59
stable, 115
multiplication
law, 74
of a monad, 74

natural
isomorphism, 48
number object, 37
transformation, 46
lax, 106
oplax, 106
pseudo-, 106
V-natural transformation, 112

naturality diagram, 46
nearly open, 223
neighborhood, 41
  filter, 41
non-expansive map, 151, 192, 199, 200
  probabilistically, 186
norm, H-valued, 386
normal
  $(\mathbb{T}, \mathcal{V})$-space, 398
  topology, 268

object, 42
  discrete, 127
  hom-, 111
  indiscrete, 127
  initial, 55
  terminal, 55
  trivial, 92
  zero-, 55
observable specialization system, 362
open, 24, 223
  -closed interchange, 449
  -filter monad, 341
  -set functor, 45
  embedding, 224
  $(\mathbb{T}, \mathcal{V})$-functor, 406
  nearly, 223
  neighborhood filter, 366
  neighborhood filter map, 366
  Scott-, 356
$\mathcal{P}$-open, 443
$(\mathbb{T}, \mathcal{V})$-open map, 406
2-open, 444
opfibration, 129
oplax
  functor, 105
  transformation, 106
opposite
  category, 43
  relation, 20
  $\mathcal{V}$-relation, 146
order, 21
  -compatible topology, 356
  -embedding, 28
  -preserving, *see* monotone
  antisymmetric, *see* separated
  discrete, 29, 50
  indiscrete, 50
  induced, 29, 204
  partial, 21
  pointwise, 22
  pre-, *see* order
  refinement, 285, 289
  separated, 21
  specialization, 29
  total, 21
  underlying, 29, 204
ordered
  -filter monad, 345
  category, 104
    conjugate, 105
  hull, 21
  monoid, 102, 255
ordered set, 21
  cocontinuous, 356
  complete, 25
  discrete, 29
  multi-, 353, 387
ordinal recursion, 38
orthogonal
  factorization system, *see* factorization system
  morphism, 114
  sink, 119

partial
  -map classifier, 241
    regular, 241
  order, *see* separated order
  product, 228
$\mathcal{M}$-partial
  -map classifier, 240
  map, 239
$\mathcal{P}$-perfect, 437
  locally, 448
point
  -separating source, 122
  of a frame, 366
pointed, 37
  map, 243
  set, 243
pointwise order, 22
power-enriched
  monad, 288
  morphism, 288
powerset
  functor, contravariant , 45
  functor, covariant, 44
  monad, 75
    finite-, 198
$\mathcal{V}$-powerset, 180
  functor, 180
  monad, 150, 180
pre-approach space, 220
premonadic, 95
preorder, *see* order
preservation
  of colimits, 63
  of infima, 27
  of isomorphisms, 70
  of limits, 63
  of suprema, 27
pretopological space, 219
$\mathcal{E}$-pretopology, 433
pretopology, fiberwise, 435
prime, completely, 366
principal
  filter, 34, 35
  filter monad morphism, 287
  ideal, 34
probabilistic

metric, 186
metric space, 186
probabilistically non-expansive map, 186
product
bi-, 108
Cartesian, 18
in a category, 57
closed under, 115
fibered, 59
partial, 228
projection, 57
tensor, 97
$\mathcal{V}$-profunctor, *see* $\mathcal{V}$-module
projection, 228
of product, 57
projective object, 140
proper
class, 17
factorization system, 114
filter, 34
$(\mathbb{T}, \mathcal{V})$-functor, 406
ideal, 34
quasi-, 432
super-, 423
$(\mathbb{T}, \mathcal{V})$-proper map, 406
$\mathcal{P}$-proper morphism, 437
pseudo
-approach space, 220
-functor, 105
-natural transformation, 106
pseudotopological space, 213
pullback, 58, 60
multiple, 59
stable, 115, 226
multiple, 115
pushout, 60

quantale, 30
Cartesian closed, 225
commutative, 30
homomorphism, 30
integral, 148
lax homomorphism, 30
lean, 148
non-trivial, 147
superior, 376
quantaloid, 107
homomorphism, 107
right-sided, 321
quasi-proper, 432
quasitopos, 245
quotient morphism, 116, 140

recursion, ordinal, 38
refinement order, 285, 289
reflection, 53
of colimits, 63
of isomorphisms, 54
of limits, 63
separated, 21
reflective
bi-, 117
epi-, 117
subcategory, 53
$\mathcal{E}$-reflective, 117
reflector, 53
reflexive
Kleisli morphism, 290
relation, 21, 233
numerical, 233
regular
-subobject classifier, 239
completely, Hausdorff, 450
epi-reflective, 117
epimorphism, 71
mono-coreflective, 117
monomorphism, 71
partial-map classifier, 241
$(\mathbb{T}, \mathcal{V})$-space, 394
topology, 394
relation, 19
auxiliary, 317
bi-, 147
dual, 20
metric, 147
opposite, 20
order, 21
reflexive, 21, 233
numerical, 233
transitive, 21
$(\mathbb{T}, \mathcal{V})$-relation, 161
reflexive, 158
transitive, 158
$\mathcal{V}$-relation, 145
dual, 146
finitary, 151
opposite, 146
reflexive, 150
separated, 151
symmetric, 151
transitive, 150
replete subcategory, 54
representable functor, 72
restriction of a filter, 35
retraction, 71
right
-cancellable, weakly, 119
adjoint, 23, 49, 106
companion, 116
factorization class, 116
unitary, 163
unitary $(\mathbb{T}, \mathcal{V})$-graph, 217
round filter, 364

Scott
-open, 356
topology, 347, 356
section, 71
self
-adjoint, 50

-dual, 45
semilattice, 28
homomorphism, 29
join-, 28
meet-, 28
separated, 21
metric, 151
monoid, 343
order, 21
reflection, 21
$\mathcal{V}$-relation, 151
$(\mathbb{T}, \mathcal{V})$-space, 263, 389
separator, *see* generator
sequence, 41
constant, 41
eventually constant, 42
sub-, 41
set, 17
-of-objects functor, 141
pointed, 243
Sierpiński space, 258
sink, 119
epi-, 122
hereditary, 245
jointly surjective, 122
slice
over, 56
under, 56
small
category, 42
class, 17
sober, 268, 277, 367
weakly, 268, 277
sobrification, 367
solution set, 65
source, 119
initial, 127
mono-, 122
point-separating, 122
space
Alexandroff, 136
approach, 192
bitopological, 200
closure, 25
KM-test, 419
metric, 151
Sierpiński, 258
T0-, 29
topological, 29
ultrametric, 199
$(\mathbb{T}, \mathcal{V})$-space, 158, 376
compact, 376
connected, 460
extremally disconnected, 399
Hausdorff, 376
normal, 398
R0, 391
R1, 391
regular, 394
separated, 263, 389
T0, 391
T1, 391
totally disconnected, 463
$\mathcal{V}$-space, 376
span, 169
spatialization, 366
specialization order, 29
specialize, 357
split
epimorphism, 71
fork, 79
monomorphism, 71
pair, 80
splitting morphism, 79
stable, 270, 277
under multiple pullbacks, 115
under pullbacks, 115
stably
closed, 420
compact, 270
v-stratum, 310
strict
finitely, 247
initial object, 453
monoidal category, 97
strictly monadic, 77
strong
epimorphism, 138
generator, 126
monomorphism, 138
strongly generating, 126
structure
discrete, 203
indiscrete, 203
morphism, 76
subbase, 269
subcategory, 72
coreflective, 54
$\mathcal{M}$-coreflective, 117
full, 44
reflective, 53
$\mathcal{E}$-reflective, 117
replete, 54
subframe, 29
subobject, 117
classifier, 239
regular-, 239
$\mathcal{M}$-subobject, 117
classifier, 238
subsequence, 41
sup
-dense, 304
-map, 27
sup-dense, 304
supercompact, 377
superior, 376
superproper, 423
supremum, 26
preservation, 27
symmetric

Cartesian product, 19
extension, lax, 397
metric, 151
monoidal category, 96
$\mathcal{V}$-relation, 151

taut, 225
tensor
in a monoidal category, 97
of a quantale, 30
tensored $\mathcal{V}$-category, 215
terminal
monad, 92
object, 55
weakly, 249
thin category, 105
three-chain, 30
top, 28
topological
functor, 130
group, 137
meet, 357
pre-, space, 219
pseudo-, space, 213
space, 29
space, bi-, 200
$\mathbb{M}_0$-topological
functor, 131
topology, 29
Alexandroff, 136
base, 29
on a category, 433
compact, 191
locally, 270
connected, 459
derived, 435
discrete, 50
extremally disconnected, 398
fiberwise, 435
Hausdorff, 130
normal, 268
order-compatible, 356
R0, 392
R1, 392
regular, 394
Scott, 347, 356
subbase, 269
T0, 29, 389
T1, 392
Tychonoff, 451
Zariski, 317
$\mathcal{E}$-topology, 433
topos, 245
quasi-, 245
total order, 21
totally below, 32
transfinite induction, *see* ordinal recursion
transitive
Kleisli morphism, 290
relation, 21
transportable functor, 132
triangular identity, 49
trivial
factorization system, 116
monad, 92
object, 92
two-chain, 30
Tychonoff
space, 451
theorem, 380, 442
$\mathcal{P}$-Tychonoff, 448

ultraclique, 318
monad, 319
ultrafilter, 35
functor, 75
monad, 75
ultrametric, 199
space, 199
underlying
-order functor, 45
order, 29, 204
unique diagonalization property, 114
unit
of an adjunction, 49
law, 74
middle, 88
of a monad, 74
unitary
$(\mathbb{T}, \mathcal{V})$-graph, 217
left, 163
$(\mathbb{T}, \mathcal{V})$-relation, 164
right, 163
universal, 226
arrow, 52
weakly, 66
property, 52
up
-closed, 26
-closure, 26
-directed, 34
up-set, 26
finitary, 299
monad, 299
functor, 75
monad, 75, 110
upper bound, 26

vertical composition, 46, 106
vertices, 113
Vietoris space, lower-, 274

way-below, 268
weak
pullback, 173
preservation, 184
weakly
left-cancellable, 115
right-cancellable, 119
sober, 268, 277

universal arrow, 66
wellpowered, 121
whiskering
functor, 46
left-, 409
right-, 409

Yoneda
embedding, 48
Lemma, 48

Zariski topology, 317
zero-object, 55
Zorn's Lemma, 41

www.ingramcontent.com/pod-product-compliance
Lightning Source LLC
LaVergne TN
LVHW102010210926
848483LV00010B/243
9781107063945